General Chemistry
(English-Chinese Bilingual Edition)
普通化学
（英汉双语版）

杨 娟 编著

图书在版编目（CIP）数据

普通化学：英、汉 / 杨娟编著 . —北京：北京大学出版社，2023.12
ISBN 978-7-301-34579-5

Ⅰ . ①普… Ⅱ . ①杨… Ⅲ . ①普通化学 – 教材 – 英、汉 Ⅳ . ①O6

中国国家版本馆CIP数据核字（2023）第204002号

书　　名	普通化学（英汉双语版）
	PUTONG HUAXUE（YING-HAN SHUANGYUBAN）
著作责任者	杨　娟　编著
责 任 编 辑	郑月娥
标 准 书 号	ISBN 978-7-301-34579-5
出 版 发 行	北京大学出版社
地　　址	北京市海淀区成府路205号　100871
网　　址	http://www.pup.cn　新浪微博：@北京大学出版社
电 子 邮 箱	编辑部 lk2@pup.cn　总编室 zpup@pup.cn
电　　话	邮购部 010-62752015　发行部 010-62750672　编辑部 010-62767347
印 刷 者	天津中印联印务有限公司
经 销 者	新华书店
	889毫米×1194毫米　16开本　29.75印张　1000千字
	2023年12月第1版　2023年12月第1次印刷
定　　价	180.00元

未经许可，不得以任何方式复制或抄袭本书之部分或全部内容。
版权所有，侵权必究
举报电话：010-62752024　电子邮箱：fd@pup.cn
图书如有印装质量问题，请与出版部联系，电话：010-62756370

前　　言

本书是根据笔者近年来在北京大学化学与分子工程学院讲授的普通化学课程的课件整理而成的英汉双语教材。该课程是面向高等学校化学专业一年级本科生的核心课程，采用英汉双语教学，介绍化学的基本概念、原理、方法及其发展过程。学生在学习化学专业基础知识的同时，也提升了化学专业英文水平。课程以科学方法论为纲贯穿始终，注重培养学生的科学思维，力求激发其自主探究问题的兴趣。该课程于 2020 年入选首批国家级一流本科课程。

本教材的内容可分为以下四个部分：

1. 科学方法论（第一章）。

2. 化学反应原理（第二、三、四章），包括气体、液体、固体与溶液，化学热力学、化学动力学，以及联系热力学和动力学的化学平衡原理。

3. 四大化学平衡（第五、六、九章），包括酸碱电离平衡、沉淀溶解平衡、氧化还原平衡和配位解离平衡。其中氧化还原平衡与电化学相关联，配位解离平衡与配位化学相关联。

4. 物质结构与性质（第七、八章），包括原子结构、分子结构和晶体结构。

衷心感谢在本书编写过程中给予笔者诸多指导和帮助的师长、同事和学生们。感谢 92 岁高龄的周公度老师耐心细致地阅读完本书，并提出了非常宝贵的意见和建议。感谢北大化学学院无机所的各位同事，特别是王颖霞、卞祖强、王炳武、郑捷、黄闻亮等老师，和他们的讨论令笔者受益良多。北大化学学院 2020、2021 和 2022 级的多位同学均参与了本书的阅读和纠错环节。特别感谢 2021 级赵思诚同学，他花费了大量的时间和精力，协助笔者绘制了本教材的多幅图片。非常感谢课程助教陈少闯、潘高翔、艾宇航、赵润涛，以及和沛淼、陈卓昇、张青欣、刘宇轩、林宇晗、周稚坤、秦家鹏、姜弈臣、杜予硕、朱晗宇、关涛、汤凯麟、陈硕航、苏亚、王康安、王姝睿、李宇宸等同学，他们为本书的完善付出了诸多努力。感谢北京大学出版社郑月娥编辑的审改，感谢北京大学教材建设委员会项目的经费支持。感谢笔者的家人，没有他们的大力支持，也就没有本书的付梓。

最后需要说明的是，编写本教材时笔者首先完成了英文部分，再将其翻译成中文。在翻译过程中优先考虑的是英汉双语之间的对照性，因水平有限，不可避免地牺牲了一部分中文的流畅性和简洁性，还望读者见谅。限于笔者能力，书中错误在所难免，敬请专家、同行和读者批评指正。

杨娟
2023 年 4 月于北京大学

Contents

Chapter 1 Introduction ... 2
   1.1 Science and Scientific Methods ... 4
   1.2 Chemistry and Research ... 8

Chapter 2 Gases, Liquids, Solids, and Solutions ... 10
   2.1 Simple Gas Laws and Ideal Gas Equation ... 12
   2.2 Mixtures of Gases ... 16
   2.3 Kinetic-Molecular Theory of Gases ... 20
   2.4 Distribution of Molecular Speeds ... 26
   2.5 Real Gases and van der Waals Equation ... 32
   2.6 Properties of Liquids and Solids ... 36
   2.7 Phase Transition and Phase Diagram ... 46
   2.8 Properties of Solutions ... 52
   Extended Reading Materials ... 56
   Problems ... 56

Chapter 3 Thermochemistry and Chemical Equilibria ... 60
   3.1 Some Terminologies in Thermochemistry ... 62
   3.2 The First Law of Thermodynamics ... 68
   3.3 Heats of Reaction and Enthalpy Change ... 74
   3.4 Hess's Law and Standard Enthalpies of Formation ... 84
   3.5 Spontaneity and the Concept of Entropy ... 90
   3.6 Criteria for Spontaneous Change ... 102
   3.7 Dynamic Equilibrium and the Equilibrium Constants ... 108
   3.8 The Reaction Quotient and Le Châtelier's Principle ... 112
   3.9 Gibbs Free Energy Change and Equilibrium ... 114
   Extended Reading Materials ... 118
   Problems ... 120

Chapter 4 Chemical Kinetics ... 124
   4.1 The Rate of a Chemical Reaction ... 126

目 录

第 1 章 绪论 ··· 3
  1.1 科学与科学方法 ··· 5
  1.2 化学与科研 ··· 9

第 2 章 气体、液体、固体和溶液 ··· 11
  2.1 简单气体定律与理想气体状态方程 ··· 13
  2.2 混合气体 ··· 17
  2.3 气体分子运动论 ··· 21
  2.4 气体分子的速率分布 ··· 27
  2.5 实际气体与范德华方程 ··· 33
  2.6 液体和固体的性质 ··· 37
  2.7 相变与相图 ··· 47
  2.8 溶液的性质 ··· 53
  拓展阅读材料 ··· 55
  习题 ··· 57

第 3 章 化学热力学与化学平衡 ··· 61
  3.1 化学热力学的一些术语 ··· 63
  3.2 热力学第一定律 ··· 69
  3.3 反应热与焓变 ··· 75
  3.4 盖斯定律与标准生成焓 ··· 85
  3.5 自发性与熵的概念 ··· 91
  3.6 自发变化的判据 ··· 103
  3.7 动态平衡与平衡常数 ··· 107
  3.8 反应商与勒夏特列原理 ··· 111
  3.9 吉布斯自由能变与平衡 ··· 115
  拓展阅读材料 ··· 119
  习题 ··· 119

第 4 章 化学动力学 ··· 125
  4.1 化学反应速率 ··· 127

- 4.2 Effect of Concentration on Reaction Rates: The Rate Laws ... 128
- 4.3 Order of Reaction ... 132
- 4.4 The Effect of Temperature on Reaction Rates ... 138
- 4.5 Theoretical Models of Chemical Kinetics ... 140
- 4.6 Reaction Mechanisms ... 150
- 4.7 Catalysis and Catalytic Chemistry ... 156
- Extended Reading Materials ... 160
- Problems ... 162

Chapter 5 Acid-Base Equilibria and Precipitation-Dissolution Equilibria ... 166
- 5.1 Theory of Acids and Bases ... 168
- 5.2 Self-Ionization of Water and the pH Scale ... 174
- 5.3 Ionization Equilibria of Acids and Bases ... 178
- 5.4 Common-Ion Effect and Buffer Solutions ... 186
- 5.5 Acid-Base Indicators and Titration ... 190
- 5.6 Solubility Product Constant and Its Relationship with Solubility ... 194
- 5.7 Criteria for Precipitation and Its Completeness ... 198
- 5.8 Dissolution and Transformation of Precipitates ... 200
- 5.9 Fractional Precipitation ... 204
- Extended Reading Materials ... 208
- Problems ... 208

Chapter 6 Redox Reactions and Electrochemistry ... 212
- 6.1 Some Terminologies in Redox Reactions ... 214
- 6.2 Electrode Potentials and Cell Potential ... 216
- 6.3 Standard Electrode Potentials and Standard Cell Potential ... 220
- 6.4 Relationship Between E°_{cell}, ΔG°, and Redox Equilibrium Constant K ... 224
- 6.5 Potential Diagrams ... 226
- 6.6 Nernst Equation and Concentration Cells ... 230
- 6.7 Electrolytic Cells and Overpotential ... 234
- 6.8 Battery and Its Applications ... 238
- Extended Reading Materials ... 242
- Problems ... 244

Chapter 7 Atomic Structure ... 248
- 7.1 The Nuclear Atom ... 250
- 7.2 Quantum Theory ... 256
- 7.3 Atomic Spectra and Bohr Theory ... 266
- 7.4 The Nature of Microscopic Particles ... 274
- 7.5 Quantum Mechanical Model of Hydrogen-Like Species ... 280
- 7.6 Quantum Mechanical Results of Hydrogen-Like Species ... 290

- 4.2 浓度对反应速率的影响：速率方程 ... 129
- 4.3 反应级数 ... 133
- 4.4 温度对反应速率的影响 ... 139
- 4.5 化学动力学理论模型 ... 141
- 4.6 反应机理 ... 151
- 4.7 催化作用与催化化学 ... 157
- 拓展阅读材料 ... 161
- 习题 ... 163

第 5 章 酸碱电离平衡与沉淀溶解平衡 ... 167
- 5.1 酸碱理论 ... 169
- 5.2 水的自耦电离与 pH ... 175
- 5.3 酸碱电离平衡 ... 179
- 5.4 同离子效应与缓冲溶液 ... 187
- 5.5 酸碱指示剂与酸碱滴定 ... 191
- 5.6 溶度积常数及其与溶解度的关系 ... 195
- 5.7 沉淀生成与沉淀完全的判据 ... 197
- 5.8 沉淀的溶解与转化 ... 199
- 5.9 分步沉淀 ... 203
- 拓展阅读材料 ... 207
- 习题 ... 209

第 6 章 氧化还原反应与电化学 ... 213
- 6.1 氧化还原反应的一些术语 ... 215
- 6.2 电极电势与电池电动势 ... 217
- 6.3 标准电极电势与标准电池电动势 ... 221
- 6.4 $E_{池}^{\ominus}$、$\Delta G^{\ominus}$ 与氧化还原平衡常数 K 的关系 ... 223
- 6.5 元素电势图 ... 227
- 6.6 能斯特方程与浓差电池 ... 231
- 6.7 电解池与超电势 ... 235
- 6.8 电池及其应用 ... 239
- 拓展阅读材料 ... 243
- 习题 ... 245

第 7 章 原子结构 ... 249
- 7.1 核型原子 ... 251
- 7.2 量子理论 ... 257
- 7.3 原子光谱与玻尔理论 ... 267
- 7.4 微观粒子的特性 ... 275
- 7.5 类氢物种的量子力学模型 ... 281
- 7.6 类氢物种的量子力学结论 ... 291

7.7　Multielectron Species and Electron Configurations ········· 306
7.8　The Periodic Law and the Periodic Table ········· 314
7.9　Periodic Properties of the Elements ········· 316
Extended Reading Materials ········· 326
Problems ········· 328

Chapter 8　Molecular Structure and Crystal Structure ········· 332

8.1　Lewis Theory ········· 334
8.2　Shape of Molecules ········· 338
8.3　Polarity of Molecules ········· 342
8.4　Valence-Bond Theory ········· 346
8.5　Molecular Orbital Theory ········· 358
8.6　Bonding in Metals and Band Theory ········· 368
8.7　Intermolecular Forces ········· 374
8.8　Crystal Structure ········· 378
8.9　Various Types of Crystals and Their Structures ········· 386
Extended Reading Materials ········· 394
Problems ········· 396

Chapter 9　Complex-Ion Equilibria and Coordination Compounds ········· 400

9.1　Basic Concepts in Coordination Compound ········· 402
9.2　Isomerism ········· 406
9.3　Crystal Field Theory ········· 414
9.4　Properties of Coordination Compounds ········· 420
9.5　Complex-Ion Equilibria ········· 424
9.6　Applications of Coordination Chemistry ········· 430
Extended Reading Materials ········· 434
Problems ········· 434

Appendix ········· 438

Appendix A　Names of Chemical Compounds ········· 440
Appendix B　Table of Mathematical and Physical Constants ········· 450
Appendix C　Tables of Chemical Data ········· 451

7.7	多电子物种与电子组态	307
7.8	元素周期律与元素周期表	313
7.9	元素性质的周期性	317
拓展阅读材料		327
习题		329

第8章 分子结构与晶体结构 333

8.1	路易斯理论	335
8.2	分子的形状	339
8.3	分子的极性	341
8.4	价键理论	347
8.5	分子轨道理论	359
8.6	金属键与能带理论	369
8.7	分子间作用力	375
8.8	晶体结构	379
8.9	各种晶体类型及其结构	385
拓展阅读材料		395
习题		397

第9章 配位解离平衡与配合物 401

9.1	配合物的基本概念	403
9.2	异构现象	407
9.3	晶体场理论	415
9.4	配合物的性质	421
9.5	配位解离平衡	425
9.6	配位化学的应用	433
拓展阅读材料		435
习题		437

附录 439

附录 A	化合物的命名	441
附录 B	数理常数表	450
附录 C	化学数据表	451

Chapter 1 Introduction

Welcome to the world of Chemistry! Chemistry is a charming branch of science that is sometimes called the central science. As a primary discipline, chemistry contains many subdisciplines, including but not limited to inorganic chemistry, analytical chemistry, organic chemistry, physical chemistry, materials chemistry, theoretical chemistry, and chemical biology. General chemistry is the foundation of all chemistry. In general chemistry, basic chemical concepts, principles, methods as well as their development processes are systematically introduced. By learning general chemistry, students are expected to build up basic background for further professional studies in chemistry. It is also required for the students to understand basic chemical methods and principles, and more importantly, to analyze and solve problems in chemical studies step by step.

In this introductory chapter, we will first learn science and scientific methods in general, and then focus on chemistry and research. As this textbook might be your first college-level chemistry book in English, it is recommended that you first read the brief introduction on how to name chemical compounds in **Appendix A**.

1.1 Science and Scientific Methods

1.2 Chemistry and Research

第1章 绪论

欢迎来到化学的世界！化学是一门迷人的科学分支，时常被称为中心科学。作为一级学科，化学包含许多分支学科，包括但不限于无机化学、分析化学、有机化学、物理化学、材料化学、理论化学和化学生物学。普通化学是化学的基础，系统地介绍了化学的基本概念、原理、方法及其发展过程。通过普通化学的学习，同学们将为进一步的化学专业课程学习奠定基础，不仅要求能理解基本的化学方法和原理，更为重要的是，还要求能逐步分析和解决化学研究中的问题。

在本章绪论中，我们首先将总体上学习科学与科学方法，然后重点学习化学与科研。由于这本教科书可能是你的第一本大学水平的英文化学书，建议你先阅读**附录 A** 中关于化合物命名的简要介绍。

1.1 科学与科学方法

1.2 化学与科研

1.1 Science and Scientific Methods

Science

The word **science**, which comes from the Latin word *scientia*, meaning "knowledge", is a systematic enterprise that builds and organizes knowledge in the form of testable explanations and predictions about the universe. There are two keywords in this definition. One is "a systematic enterprise", which means that science should be carried out in an orderly and planned way. The other is "testable", which suggests that science can be tested by experiments, and the results of the experiments should be reproducible. Science can be used not only to explain natural phenomena but also to predict future events.

Of all the driving forces of science, two are most essential and of great importance. The most primitive and direct driving force of science is to improve the quality of everyday life. The humankind wishes to live a better life and that is the intuitive reason why we do science. The other equally important driving force of science is curiosity, a strong desire of the humankind to understand the mysterious universe and to explore the unknown.

Deduction and **induction** are two basic and important scientific methods. Deduction starts with certain **basic assumptions**, and then follows certain logical rules to make conclusions. For example, Euclidean geometry was constructed based on five axioms, which are its basic assumptions. The ancient Greek philosopher Aristotle assumed that the universe is made of five fundamental elements: earth, water, fire, air, and aether, the first four of which make the earth and the last one of which fills in the universe. Aristotle believed that all other substances were formed by combinations of these five elements. Nowadays we know that Aristotle's conclusions are incorrect mainly because his five-element assumption is false. Therefore, assumptions are vital in deduction.

Induction, on the other hand, makes no initial assumptions but makes careful observations of natural phenomena, and then formulates a generalization (or a natural law) to describe the observed phenomena. A good example of induction is Nicolaus Copernicus's statement of the earth revolving around the sun. Copernicus did not make any assumptions about whether the earth revolves around the sun or the sun revolves around the earth. His final conclusion, which can be viewed as a natural law, was made based on the careful observations of himself and many previous astronomers. Comparing deduction with induction, we notice that the major difference in these two scientific methods is with or without the initial assumptions.

Natural laws are concise statements about natural phenomena, often in the form of mathematical expressions. For instance, the ideal gas law states

$$pV = nRT \tag{1.1}$$

where p is the pressure, V is the volume, n is the amount, T is the thermodynamic temperature of an ideal gas, and R is the molar gas constant. The first law of thermodynamics gives

$$\Delta U = q + w \tag{1.2}$$

where ΔU is the change of internal energy, q is the heat, and w is the work of a system. Both **Equations (1.1)** and **(1.2)** are natural laws formulated from natural phenomena in concise mathematical form. The success of a natural law depends on its ability to explain observations, and more importantly, to predict new phenomena. Copernicus's work was a great success because he was able to predict future positions of the planets more accurately than his contemporaries. However, we should always remember that natural laws are not absolute truth, and that each natural law has its particular scope of application. For example, **Equation (1.1)** is the ideal gas law and suits only for ideal gases. In the case of real gases, this law needs to be modified according to further experimental results, leading to van der Waals Equation. Meanwhile, Copernicus believed that the earth revolves around the sun in a circular orbit. His work was refined a half-century later by Johannes Kepler, who showed that planets travel in elliptical orbits instead of circular orbits. Therefore, in order to

1.1 科学与科学方法

科学

"科学" 一词来源于拉丁语 "scientia"，意为"知识"，它是一项以对宇宙的可验证的解释和预测的形式来构建和组织知识的系统性事业。在这个定义中有两个关键词：一个是"系统性事业"，这意味着科学应以有序且有计划的方式来进行；另一个是"可验证的"，这表明科学可以通过实验来验证，并且实验结果应该是可重复的。科学不仅可用于解释自然现象，还可用于预测未来的事件。

在科学的所有驱动力中，有两种是最基本且最重要的。科学最原始且最直接的驱动力是为了提高日常生活的质量。人类希望过上更好的生活，这是我们从事科学研究的直观原因。另一个同等重要的科学驱动力是好奇心，即人类理解神秘宇宙和探索未知世界的强烈愿望。

推演法和**归纳法**是两种基本而重要的科学方法。推演法从一些**基本假定**开始，然后遵循某些逻辑规则得出结论。例如，欧几里德几何是基于五条公理构建的，这五条公理即欧几里德几何的基本假定。古希腊哲学家亚里士多德认为，宇宙由五种基本元素组成：地、水、火、风和以太，其中前四种组成地球而最后一种充满宇宙。亚里士多德认为，所有其他物质均由这五种元素组合而成。现在我们知道亚里士多德的结论之所以不对，主要原因是他的"五元素"假定是错误的。因此，基本假定在推演法中至关重要。

归纳法不做任何初始假定，而是对自然现象进行仔细观察，然后得出一个概括规律（即自然定则）来描述观察到的现象。尼古拉斯·哥白尼关于地球围绕太阳转的陈述就是归纳法的一个好例子。哥白尼并没有对地球围绕太阳转还是太阳围绕地球转做出任何假设。他的最终结论（可视为一条自然定则）是基于他本人和之前的许多天文学家的仔细观察得出的。比较推演法和归纳法，我们注意到这两种科学方法的主要区别在于有无初始假定。

自然定则是关于自然现象的简明陈述，通常以数学表达式的形式呈现。例如，理想气体状态方程为

$$pV = nRT \tag{1.1}$$

其中 p 是理想气体的压强，V 是体积，n 是物质的量，T 是热力学温度，R 是摩尔气体常数。热力学第一定律给出

$$\Delta U = q + w \tag{1.2}$$

其中 ΔU 是体系内能的改变量，q 是热，w 是功。**式（1.1）**和**（1.2）**都是由自然现象得出的、以简明数学形式呈现的自然定则。自然定则的成功取决于其解释观测结果的能力，以及更为重要的是，其预测新现象的能力。哥白尼的工作非常成功，因为他能够比同时代人更为准确地预测行星未来的位置。然而我们应该永远记住，自然定则并非绝对真理，每条自然定则都有其特定的适用范围。例如，**式（1.1）**是理想气体状态方程，仅适用于理想气体。对于实际气体，该方程需要根据进一步的实验结果进行修正，从而得出范德华方程。此外，哥白尼认为地球以圆形轨道围绕太阳旋转。半个世纪后，约

verify a natural law, scientists design experiments to show whether the conclusions deduced from the natural laws are supported by experimental results or not. If the experimental results support the natural law, it means that this natural law is valid within the experimental conditions. If not, modifications of natural laws might be required according to the results of further experiments.

Natural laws can explain and predict natural phenomena. However, to understand why natural laws can explain and predict natural phenomena involves two concepts: hypothesis and theory. A **hypothesis** is a tentative explanation of a natural law. If a hypothesis survives testing by experiments, it is then upgraded into a theory. A **theory** is a model that can be used to explain natural laws and make further predictions about the natural phenomena. As new evidences accumulate, most theories undergo modification, and some are even completely discarded. For instance, classical mechanics are valid in studying macroscopic systems, however, in the cases of microscopic systems, classical mechanics are completely discarded and replaced by quantum mechanics.

Scientific Methods

The **scientific method** is the combination of observation, experimentation, and the formulation of natural laws, hypotheses, and theories. Starting from natural observations, it first formulates natural laws, and then postulates hypothesis for a tentative explanation. In order to verify the hypothesis, various experiments are designed by scientists. If the experimental results show that the hypothesis is inadequate, the hypothesis is iteratively modified until it survives testing by experiments and upgrades into a theory. A theory is an amplified hypothesis that can make predictions about the future phenomena. In order to verify the theory, more experiments are designed by scientists to test the predictions of the theory. If the experiments show that the theory is inadequate, the theory is iteratively modified until it survives all the performed experiments and upgrades into a well-established theory. Even a well-established theory is still not absolute truth. Over time, some well-established theories may require further modification or even be completely replaced by new theories as new observations or experiments accumulate.

The flow diagram in **Figure 1.1** summarizes the above scientific methods and illustrates the process of how science evolves. It also shows us the different stages of learning. At the beginning of science evolution, numerous natural or experimental observations are conducted. These observations give basic description of phenomena, answer the fundamental question "what is it", and belong to the first stage of learning. From these numerous observations, scientists summarize natural laws, which can not only explain the observations but also predict future phenomena. Modifications of natural laws are necessary as new experiments accumulate. The natural laws show the general patterns of phenomena, which correspond to the second stage of learning and answer the more advanced question "how is it". From "what is it" to "how is it", the scientific method induction is involved. In this process, no initial assumptions but only careful observations are made, and natural laws are formulated to describe the observed phenomena. At the final stage of science evolution, also at the last stage of learning, the ultimate question "why is it" needs to be answered. In order to understand why natural laws can apply to numerous observations, theories based on models are developed. Theories can not only explain but also predict new natural laws and new observations. Modifications of theories are necessary as new evidence accumulates. From "how is it" to "why is it", the scientific method deduction is involved. All theories start from several basic assumptions, and make conclusions following rigorous logical rules.

For example, in understanding the behavior of gases, scientists first made numerous observations and collected enormous data about the p-V, V-T, and V-n relations, which answered the first level question "what is it". Later, scientists formulated simple gas laws (Boyle's law from the p-V relation, Charlie's law from the V-T relation, and Avogadro's law from the V-n relation) and the ideal gas equation, which answered the second level question "how is it". In order to understand the "why is it" question, or why the product of pV happens to equal the product of nRT, scientists developed a so-called "kinetic-molecular theory of gases", which will be discussed in **Section 2.3**.

翰尼斯·开普勒对他的工作进行了改进，证明了行星以椭圆轨道而非圆形轨道运行。因此，为了验证自然定则，科学家们设计实验来测试从自然定则推导出的结论是否得到实验结果的支持。如果实验结果支持自然定则，就意味着该自然定则在实验条件下是有效的。否则，根据进一步实验的结果，可能需要对自然定则进行修正。

自然定则可以解释和预测自然现象。然而，要理解为什么自然定则可以解释和预测自然现象，需要用到两个概念：假说和理论。**假说**是对自然定则的试探性的解释。如果假说通过了实验的检验，它就升级成为理论。**理论**是一种可用于解释自然定则并对自然现象做出进一步预测的模型。随着新证据的积累，大多数理论都会经历修正，有的理论甚至被完全抛弃。例如，经典力学在研究宏观体系时有效，然而在研究微观体系时，经典力学被完全抛弃，取而代之的是量子力学。

科学方法

科学方法是观察、实验以及得出自然定则、假说和理论的组合。从自然观察出发，首先形成自然定则，然后提出假说进行试探性解释。为了验证假说，科学家们设计了各种实验。如果实验表明该假说不充分，则其将被反复修正，直到通过实验验证并升级为理论。理论是一种可以预测未来现象的加强版假说。为了验证理论，科学家们设计了更多的实验来测试。如果实验表明该理论不充分，则其也将被反复修正，直到通过所有实验验证，并升级为成熟理论。成熟理论同样并非绝对真理。随着时间的推移，随着新的观察或实验的积累，一些成熟理论也可能需要进一步修正，甚至被新理论完全取代。

图 1.1 总结了以上科学方法并说明了科学演化的过程，也展示了学习的不同阶段。在科学演化之初，人们进行了大量的自然或实验观察。这些观察给出了对现象的基本描述，回答了"它是什么"的基本问题，属于学习的第一阶段。从这些大量观察中，科学家们总结出自然定则，不仅可以解释观察结果，还可以预测未来的现象。随着新实验的积累，可以对自然定则进行必要的修正。自然定则展示了现象的通用规律，对应学习的第二阶段，并回答了更高级的问题"它是怎样的"。从"它是什么"到"它是怎样的"，涉及科学方法中的归纳法。在此过程中没有初始假定，只进行了仔细观察，并得出自然定则来描述观察到的现象。在科学演化的最终阶段，亦即学习的最终阶段，需要回答终极问题"它为什么会这样"。为了理解为什么自然定则可以应用于大量观察，人们发展了基于模型的理论。理论不仅可以解释，而且可以预测新的自然定则和新的观察结果。随着新证据的积累，可以对理论进行必要的修正。从"它是怎样的"到"它为什么会这样"，涉及科学方法中的推演法。所有理论都是从几条基本假定出发，并遵循严格的逻辑规则得出结论。

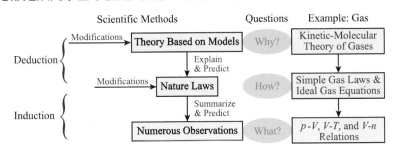

Figure 1.1 The scientific method illustrated.

图 1.1 科学方法论示意图。

1.2 Chemistry and Research

Chemistry

Chemistry is a branch of science that studies the composition, structure, properties and change of matter. Chemistry is chiefly concerned with atoms and molecules, and their interactions and transformations. You may have already been familiar with atoms and molecules, however, at current stage you might not quite understand the interactions and transformations of atoms and molecules. These are of great importance in chemistry and will be discussed in detail throughout this textbook. In short, chemistry is the science of change. In fact, this is also how the Chinese term for chemistry is coined.

Chemistry is sometimes called the central science. It lies near the heart of many matters of public concerns. Chemistry has direct relationship with many other sciences and technologies, including but not limited to energy, materials, biomedicine, and many modern technologies. Historically, chemists practiced chemistry by manipulating the existing materials in their environment. The production of metals, beer, wine, and soap all involved chemistry. With modern knowledge, chemists can even create new materials that do not exist naturally and that often exhibit unusual properties. Thus, motor fuels and thousands of chemicals used in the manufacture of plastics, synthetic fabrics, pharmaceuticals, and pesticides can all be made from petroleum. Modern chemical knowledge is also needed to understand the processes that sustain life and to understand and control processes that are detrimental to the environment, such as the formation of smog and the destruction of stratospheric ozone.

In spite of its irreplaceable role in improving the quality of everyday life, chemistry seems to be misunderstood by many people in modern society. From the shouting "we hate chemistry" in a TV commercial to the food labels stating "no chemicals added", the references to chemistry and chemicals are often misleading. In fact, all materials are made up of chemicals and chemical-free consumer products do not exist at all. Therefore, what we should keep in mind is that: chemistry is not the problem; it's the solution to many problems.

Research

Research comprises "creative work undertaken on a systematic basis in order to increase the stock of knowledge, including knowledge of humanity, culture and society, and the use of this stock of knowledge to devise new applications". The first part of this definition relates to science and the last part associates with technology, both of which are important topics of research. Specifically, research is used to (1) establish or confirm facts, (2) reaffirm the results of previous work, (3) solve new or existing problems, (4) support theorems, or (5) develop new theories. Of these five levels of content, (1) and (2) focus on confirming the already existing work and still belong to research because the results of research should be reproducible in independent laboratories, but are less superior to the last three, which show the essence of research.

The **frontiers** of research refer to the highly active fields near the boundary between known and unknown. All research frontiers aim to expand the stock of knowledge, that is, to push the boundary in the direction from known towards unknown. What we have read and learnt in textbooks nowadays are the research frontiers many years ago. Some current research frontiers might be written into the future textbooks. Research as well as the frontier of research are illustrated in **Figure 1.2**.

例如，为理解气体的行为，科学家们首先进行了大量观察，并收集了关于 p-V、V-T 和 V-n 关系的大量数据，这些数据回答了第一层问题"它是什么"。后来，科学家们得出了简单气体定律（波义耳定律来自 p-V 关系，查理定律来自 V-T 关系，阿伏伽德罗定律来自 V-n 关系）和理想气体状态方程，这回答了第二层问题"它是怎样的"。为了理解"它为什么会这样"的问题，即为什么 pV 的乘积恰好等于 nRT 的乘积，科学家们建立了称为"气体分子运动论"的理论，将在 **2.3 节**中讨论。

1.2 化学与科研

化学

化学是研究物质的组成、结构、性质及变化的科学分支，主要研究原子和分子及其相互作用和转化。你可能已经熟悉了原子和分子，但是，现阶段你可能还不太了解原子和分子的相互作用和转化。这些在化学中非常重要，将在本教材中详细讨论。简言之，化学是变化的科学。事实上，汉语中的"化学"一词也正是这样创造出来的。

化学常被称为中心科学，处于许多公众关心问题的核心。化学与许多其他科学和技术有直接关系，包括但不限于能源、材料、生物医药及许多现代技术。历史上，化学家们以环境中现有的物质来进行化学实践。金属、啤酒、葡萄酒和肥皂等的生产都涉及化学。有了现代知识，化学家们甚至可以创造出自然界中不存在的新材料，这些材料往往表现出不同寻常的特性。汽车燃料以及用于制造塑料、合成纤维、药品和杀虫剂的数千种化学品均可用石油制成。现代化学知识还可用于理解维持生命的过程，以及理解和控制对环境有害的过程，如雾霾的形成和平流层臭氧的破坏等。

尽管化学在改善日常生活质量方面发挥着不可替代的作用，但它似乎常常被现代社会的许多人误解。从曾经电视广告中大喊的"我们恨化学"到食品标签上写着的"无化学品添加"，对化学和化学品的提及往往是误导性的。事实上，所有材料均由化学品制成，根本不存在不含化学品的消费品。因此，我们应该记住：化学不是问题，而是许多问题的解决方法。

科研

科研包括"在系统基础上用以增加人类整体知识储量（包括人文、文化和社会知识储量）的原创性工作，以及将这些知识储量用于设计新应用的工作"。该定义的前半部分与科学相关，后半部分与技术相关，而两者都是重要的科研课题。具体来说，科研可用于：(1) 建立或确认事实；(2) 重新确认之前工作的结果；(3) 解决新的或现存的问题；(4) 支持定理；(5) 发展新理论。在这五个层次的内容中，(1) 和 (2) 侧重于确认已经存在的工作，虽然仍属于科研范畴，但因科研成果应该在独立实验室中可重复，故不如后三个更能体现科研的本质。

科研的**前沿**指处于已知和未知之间边界附近的高度活跃的领域。所有科研前沿的目标都是为了拓展整体知识储量，即将知识的边界由已知向未知的方向推进。现在我们在教科书中所读到和学到的知识都曾是许多年前的科研前沿，而一些当前的科研前沿也可能会被写入未来的教科书中。科研与科研前沿如**图 1.2** 所示。

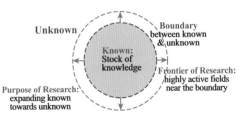

Figure 1.2 Research and the frontier of research.

图 1.2 科研与科研前沿。

Chapter 2 Gases, Liquids, Solids, and Solutions

The three common states of matter are gas, liquid, and solid. A well-recognized fourth state of matter is called plasma, which will be introduced in the extended reading materials at the end of this chapter. In this chapter, we will start from gases because historically gases are the starting point of learning and exploring thermochemistry. Meanwhile, gases are the illustrative examples to preliminarily understand the basic chemical principles. The study and development of gas laws and theories can be viewed as a miniature of how science has evolved that we have learnt in **Chapter 1**. We will study from ideal gases to real gases, from practice to theory, and from macroscopic to microscopic. After gases, we will learn liquids and solids, both of which are called the condensed phases. Finally, we will move to a type of mixture—solutions.

2.1 Simple Gas Laws and Ideal Gas Equation

2.2 Mixtures of Gases

2.3 Kinetic-Molecular Theory of Gases

2.4 Distribution of Molecular Speeds

2.5 Real Gases and van der Waals Equation

2.6 Properties of Liquids and Solids

2.7 Phase Transition and Phase Diagram

2.8 Properties of Solutions

第 2 章　气体、液体、固体和溶液

物质的三种常见状态为：气态、液态和固态。等离子态是被普遍承认的物质的第四态，详见章末拓展阅读材料。本章我们将从气体开始，因为历史上气体是学习和研究热化学的起点，同时气体也是初步理解化学基本原理的一个典型范例。气体定律和理论的研究及发展可视为我们在**第 1 章**学到的科学发展演化过程的一个缩影。我们将按照从理想气体到实际气体、从实践到理论、从宏观到微观的顺序来学习气体。接下来我们会学习统称为凝聚态的两种状态：液态和固态。最后我们会过渡到一种混合物：溶液。

2.1　简单气体定律与理想气体状态方程

2.2　混合气体

2.3　气体分子运动论

2.4　气体分子的速率分布

2.5　实际气体与范德华方程

2.6　液体和固体的性质

2.7　相变与相图

2.8　溶液的性质

2.1 Simple Gas Laws and Ideal Gas Equation

Our real world is a very complicated world containing many parameters that may simultaneously affect one another. The ability to distinguish some non-trivial parameters from a series of trivial parameters, or to put forward some simplified models from a complex system, is an important ability that is required in science and research.

Basic Assumptions of the Ideal Gas

In the case of gases, the **ideal gas** is a model simplified from the real gas that meets the following two basic assumptions:
1) Ideal gas molecules are separated by great distances and the intermolecular forces are negligible. This means that all collisions are elastic with conserved energy and momentum.
2) Ideal gas molecules are tiny and do not occupy any space. Thus, ideal gas molecules can be viewed as point masses.

At very high temperatures and very low pressures, the distance between real gas molecules is great and the intermolecular forces between them are weak enough to be neglected. Meanwhile, the volume of real gases is much greater than the occupied space of real gas molecules themselves so that the occupied molecular space is also negligible. Under these circumstances, the two basic assumptions of ideal gas are satisfied, and the properties of real gas are close to those of ideal gas. Therefore, we can regard ideal gas as an extreme case of real gas at very high temperatures and very low pressures, and it is of practical significance to study the properties of ideal gas. Notice that simplifying a real object first, constructing an ideal model from it for the convenience of study, and then applying the ideal model back to the real object after necessary modifications, is a commonly used method in research to deal with complex systems. More discussion about real gases can be found in **Section 2.5**.

Pressure of Gases

The physical quantity **pressure** is defined as an exerted force per unit area, or the ratio of a force to the area over which it is distributed. A gas is composed of a very large number of molecules in constant and random motion, frequently colliding with one another. When putting into a vessel, the gas molecules collide with the vessel walls and exert a force on the vessel walls. The pressure of a gas in a vessel equals the exerted force divided by the area of the vessel walls.

Assume that an ideal gas molecule with a mass m and a velocity v collides perpendicularly to a vessel wall of area A. According to the first assumption of ideal gas, the intermolecular forces are negligible and the collision is elastic with conserved energy and momentum, so the velocity of the molecule is $-v$ after the collision. The momentum change of the gas molecule in a total collision period of t is given by

$$\Delta p = (-mv) - mv = -2mv \tag{2.1}$$

The momentum change should equal the impulse of the exerted force of the vessel wall to the molecule F', as

$$F't = -2mv$$

So,

$$F' = -\frac{2mv}{t} \tag{2.2}$$

The exerted force of the molecule to the vessel wall F is the reaction of the force of the vessel wall to the molecule F'. According to Newton's third law of motion, action and reaction are equal and opposite. Therefore,

$$F = -F' = \frac{2mv}{t}$$

The force F is in the same direction as the motion of molecule, perpendicular to the vessel wall, and thus give a pressure to the vessel wall as

$$p = \frac{F}{A} = \frac{2mv}{At} \tag{2.3}$$

2.1 简单气体定律与理想气体状态方程

我们的世界是一个包含多种同时互相影响因素的复杂世界。能够从一系列非关键因素中区分出关键因素，或者说从复杂体系中提炼出简化模型，是科学研究中的一种非常重要的能力。

理想气体的基本假定

对于气体这一研究体系，**理想气体**是人们从实际气体中简化出、符合以下两条基本假定的一种模型：
1) 理想气体分子间距离极大，分子间作用力可忽略不计。这意味着所有碰撞都是弹性的，能量和动量均守恒。
2) 理想气体分子极小，不占据体积，故可将理想气体分子视为质点。

在高温和低压条件下，实际气体分子间距离相当大，分子间作用力极其微弱，可忽略不计。同时，实际气体的体积远大于气体分子自身占据的空间，因此气体分子自身占据的空间也可忽略。这种条件下的实际气体满足理想气体的两条基本假定，其性质很接近理想气体。因此理想气体可视为实际气体在高温低压条件下的一种极限情况，研究理想气体的性质具有实际意义。注意，这种先简化实际研究对象，从中构建出理想模型以便于研究，再经过一些必要的修正，将理想模型应用回实际研究对象的方法，是科学研究中处理复杂体系时的一种常用方法。关于实际气体的更多讨论详见 **2.5 节**。

气体的压强

物理量**压强**定义为单位面积上的作用力，或力与其作用面积的比值。气体由大量做连续不断、无规则运动且经常相互碰撞的分子组成。盛放在容器中的气体，其分子碰撞器壁并对器壁施加压力。容器内气体的压强等于气体对器壁的作用力除以器壁面积。

假设有一质量为 m、速率为 v 的理想气体分子，沿垂直方向碰撞面积为 A 的器壁。根据理想气体第一条基本假定，分子间作用力可忽略不计，该碰撞为弹性碰撞，能量和动量均守恒，故碰撞后气体分子的速率为 $-v$。在总时间为 t 的碰撞过程中，气体分子动量的改变量为

$$\Delta p = (-mv) - mv = -2mv \tag{2.1}$$

动量的改变量等于器壁对分子的作用力 F' 的冲量，即

$$F't = -2mv$$

故

$$F' = -\frac{2mv}{t} \tag{2.2}$$

分子对器壁的作用力 F 是器壁对分子的作用力 F' 的反作用力。根据牛顿第三运动定律，两者大小相等、方向相反。因此

$$F = -F' = \frac{2mv}{t}$$

作用力 F 与分子运动的方向一致，垂直于器壁，对器壁产生的压强为

$$p = \frac{F}{A} = \frac{2mv}{At} \tag{2.3}$$

For gas molecules with motions not perpendicular to the vessel wall, its component in the perpendicular direction should be considered instead.

A frequently asked question here is how to distinguish the collision period and the collision time. The former is the period of time between two successive collisions whereas the latter is the actual time of the molecule colliding with the vessel wall. It is the collision period that should be used to derive the pressure of gas in **Equation (2.3)**. Here, we need to consider the difference between macroscopic and microscopic quantities. Microscopically, the collisions of gas molecules with the vessel walls are discontinuous. However, since the gas comprises a great number of molecules and the collision period is extremely short, the pressure of gas, which is a macroscopic concept, arises from an average of a great number of collisions to the vessel walls in a period of time and is still continuous. An analogy to this is that when you standing in a heavy rain with an umbrella, although the raindrops are discontinuous, you will feel a nearly continuous and steady force on your umbrella exerted by the heavy rain due to the very large number of raindrops and very short period.

Simple Gas Laws

About the gas pressure p, volume V, thermodynamic temperature T, and amount n of gases, scientists have formulated some **simple gas laws**, as summarized in **Figure 2.1** and listed as follows:

1) **Boyle's Law**: the volume of gas is inversely proportional to the pressure of gas at constant n and T, or
$$V \propto 1/p \tag{2.4}$$
2) **Charles's Law**: the volume of gas is directly proportional to the thermodynamic temperature of gas at constant n and p, or
$$V \propto T \tag{2.5}$$
3) **Avogadro's Law**: the volume of gas is directly proportional to the amount of gas at constant p and T, or
$$V \propto n \tag{2.6}$$

The relationship between the thermodynamic (absolute or Kelvin) temperature T and the Celsius temperature t is given by
$$T(K) = t(°C) + 273.15 \tag{2.7}$$
According to Charles's Law, the gas volume becomes negative at temperatures below −273.15°C. Therefore, we define −273.15°C or 0 K as the absolute zero of temperature since the volume of gas cannot be negative.

Ideal Gas Equation

In addition to p, T, and n, the volume V of an ideal gas is not related to any other physical quantities. We can combine the above three simple gas laws, all involving V, into the following relationship:
$$V \propto \frac{nT}{p} \tag{2.8}$$
Define the **molar gas constant** R as the ratio of V to nT/p, we then have
$$R = \frac{V}{nT/p} = \frac{pV}{nT} \tag{2.9}$$
or
$$pV = nRT \tag{2.10}$$
Equation (2.10) is called the **ideal gas equation**, which includes four variables and one constant R.

Standard Conditions of Temperature and Pressure

Because the properties of gas such as density and molar volume depend on temperature and pressure, it is practically useful to define a set of **standard conditions of temperature and pressure (STP)** so that we can compare the properties of different gases at these standard conditions. STP was defined differently in the past. In this textbook, we follow the definition of STP recommended by the International Union of Pure and

对于运动方向与器壁不垂直的气体分子，可以考虑其在垂直方向的分速率。

此处的一个常见问题是如何区分碰撞过程的时间和碰撞时间。这两个概念中前者为相邻两次碰撞之间的时间间隔，而后者为分子碰撞器壁的实际时间。在**式 (2.3)** 中应使用碰撞过程的时间来推导气体的压强，这里需考虑宏观量与微观量的差异。尽管微观上气体分子对器壁的碰撞是不连续的，但由于气体中含有极其大量的分子，且碰撞过程的时间间隔极其短暂，作为宏观概念的压强来自一段时间内对器壁大量碰撞的平均效应，因此仍表现为连续的。一个近似的类比是在大雨中撑伞静立，虽然雨点打在伞上是不连续的，但由于雨点多、间隔短，感觉雨点对伞的作用是近似连续和均匀的。

简单气体定律

关于气体的压强 p、体积 V、热力学温度 T 和物质的量 n，科学家们总结出如下几个**简单气体定律**，如**图 2.1** 所示，其内容如下：

1) **波义耳定律**：当 n 和 T 恒定时，气体体积 V 与压强 p 成反比，即

$$V \propto \frac{1}{p} \tag{2.4}$$

2) **查理定律**：当 n 和 p 恒定时，气体体积 V 与热力学温度 T 成正比，即

$$V \propto T \tag{2.5}$$

3) **阿伏伽德罗定律**：当 p 和 T 恒定时，气体体积 V 与其物质的量 n 成正比，即

$$V \propto n \tag{2.6}$$

热力学温度（绝对温度或开氏温度）T 与摄氏温度 t 的换算关系为

$$T(\text{K}) = t(°\text{C}) + 273.15 \tag{2.7}$$

根据查理定律，$-273.15°\text{C}$ 以下理想气体的体积将为负值，而气体体积不可能为负值，因此将 $-273.15°\text{C}$ 或 0 K 定义为**绝对零度**。

理想气体状态方程

除 p、T 和 n 之外，理想气体体积 V 不与其他物理量相关。综合以上三个与 V 相关的简单气体定律，可得如下关系：

$$V \propto \frac{nT}{p} \tag{2.8}$$

定义**摩尔气体常数** R 为 V 与 nT/p 的比值，则有

$$R = \frac{V}{nT/p} = \frac{pV}{nT} \tag{2.9}$$

即

$$pV = nRT \tag{2.10}$$

式 (2.10) 即为**理想气体状态方程**，式中包含四个变量及一个常数 R。

标准温度和压强

由于气体的很多性质（如密度、摩尔体积等）均与温度和压强相关，因此定义一组**温度和压强的标准状况**（STP，简称标况）非常有用，这样可以在标况下比较不同气体的性质。历史上 STP 的定义各不相同，本教材采用国际纯粹与应用化学联合会（IUPAC）自

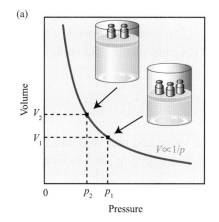

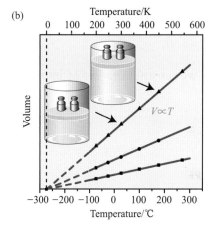

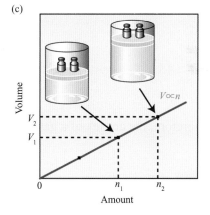

Figure 2.1 The simple gas laws. (a) Boyle's law. (b) Charles's law. (c) Avogadro's law.

图 2.1 简单气体定律：(a) 波义耳定律；(b) 查理定律；(c) 阿伏伽德罗定律。

Applied Chemistry (IUPAC) since 1982, as

$$T = 0°C = 273.15 \text{ K} \quad \text{and} \quad p = 1 \text{ bar} = 10^5 \text{ Pa}$$

The **molar volume** V_m, which is defined as the volume per mole of gas, is 22.711 L for an ideal gas at STP. Consequently, the value of the molar gas constant R can be calculated from V_m, as

$$R = \frac{pV}{nT} = \frac{10^5 \text{ Pa} \times 22.711 \times 10^{-3} \text{ m}^3}{1 \text{ mol} \times 273.15 \text{ K}}$$

$$= 8.3145 \text{ Pa m}^3 \text{ mol}^{-1} \text{ K}^{-1}$$

The value and unit of R can be either 8.3145 J mol^{-1} K^{-1}, 8.3145 kPa L mol^{-1} K^{-1}, or 0.082057 atm L mol^{-1} K^{-1}. The values of some constants involved in this textbook are tabulated in **Appendix B**.

Applications of the Ideal Gas Equation

The ideal gas equation can be recast into different forms for different applications. For example, starting from the definition of molar volume $V_m = V/n$, we can directly derive

$$V_m = \frac{RT}{p} \tag{2.11}$$

If we substitute $n = m/M$ into the ideal gas equation, we can reform it into

$$pV = \frac{m}{M}RT$$

or

$$M = \frac{m}{n} = \frac{mRT}{pV} \tag{2.12}$$

where M is the **molar mass** and m is the mass of the gas.

Another application is to calculate the density of a gas. With the definition of **density** given by $\rho = m/V$, we then have

$$p\frac{m}{\rho} = \frac{m}{M}RT$$

or

$$\rho = \frac{pM}{RT} = \frac{M}{V_m} \tag{2.13}$$

Therefore, the density of a gas is proportional to its molar mass.

2.2 Mixtures of Gases

Gases often exist in the form of mixtures. Apparently, the simple gas laws and the ideal gas equation apply to individual gases as well as to a mixture of nonreactive gases, because all these gas laws were derived initially based on the behavior of air, which is clearly a mixture.

Basic Concepts

A system composed of two or more gases is called a **mixture gas**, and each individual gas in the mixture is called a **component gas**. In the mixture gas air, N_2, O_2, CO_2, etc. are all its component gases. For a mixture gas that is composed of N different component gases, if the amount of the ith component gas is represented by n_i and the total amount of the mixture gas is represented by n_{tot}, we can easily have

$$n_{tot} = \sum_{i=1}^{N} n_i$$

The **mole fraction** of the ith component gas x_i is defined as

1982 年以来的版本，将 STP 定义为

$$T = 0°C = 273.15 \text{ K} \quad 且 \quad p = 1 \text{ bar} = 10^5 \text{ Pa}$$

气体的**摩尔体积** V_m 定义为 1 mol 气体的体积，在 STP 下理想气体的摩尔体积为 22.711 L。由 V_m 可相应计算摩尔气体常数 R 的值为

$$R = \frac{pV}{nT} = \frac{10^5 \text{ Pa} \times 22.711 \times 10^{-3} \text{ m}^3}{1 \text{ mol} \times 273.15 \text{ K}}$$

$$= 8.3145 \text{ Pa m}^3 \text{ mol}^{-1} \text{ K}^{-1}$$

R 也可取 8.3145 J mol^{-1} K^{-1}、8.3145 kPa L mol^{-1} K^{-1} 或 0.082057 atm L mol^{-1} K^{-1} 等数值和单位。本教材中涉及的一些常数的值列在**附录 B** 中。

理想气体状态方程的应用

对于不同的应用，理想气体状态方程可改写成不同的形式。例如，根据摩尔体积的定义式 $V_m = V/n$，可直接推导

$$V_m = \frac{RT}{p} \tag{2.11}$$

将 $n = m/M$ 代入理想气体状态方程，可得

$$pV = \frac{m}{M}RT$$

即

$$M = \frac{m}{n} = \frac{mRT}{pV} \tag{2.12}$$

式中 M 是气体的**摩尔质量**，m 是气体的质量。

理想气体状态方程的另一应用是计算气体密度，将**密度**的定义式 $\rho = m/V$ 代入，可得

$$p\frac{m}{\rho} = \frac{m}{M}RT$$

即

$$\rho = \frac{pM}{RT} = \frac{M}{V_m} \tag{2.13}$$

故气体的密度与其摩尔质量成正比。

2.2 混合气体

气体常以混合物的形式存在。显然，简单气体定律和理想气体状态方程对于单一气体和互不反应的混合气体均适用，因为气体定律最初就是由空气的数据推导出的，而空气显然是混合气体。

基本概念

由两种或两种以上气体组成的体系称为**混合气体**，组成混合气体的每种气体都称为**组分气体**。在混合气体空气中，N_2、O_2、CO_2 等均为其组分气体。如果混合气体由 N 种组分气体组成，其中第 i 种组分气体的物质的量用 n_i 表示，混合气体总物质的量用 $n_总$ 表示，易得

$$n_总 = \sum_{i=1}^{N} n_i$$

$$x_i = \frac{n_i}{n_{tot}} \tag{2.14}$$

Apparently,
$$1 = \sum_{i=1}^{N} x_i$$

If we apply the ideal gas equation to a mixture gas, we can add a subscript "tot" to all the variables to indicate that these are the total values. The total pressure p_{tot} and the total volume V_{tot} of the mixture gas can be determined by the total amount n_{tot}, as

$$p_{tot} = \frac{n_{tot}RT}{V_{tot}} \quad \text{and} \quad V_{tot} = \frac{n_{tot}RT}{p_{tot}}$$

respectively.

Dalton's Law of Partial Pressure

John Dalton (1766—1844) made an important contribution to the study of mixture gases. He defined that the **partial pressure** of a component gas in a mixture gas as the pressure that the component gas would exert if it were in the vessel alone, or if it were with the total volume of the vessel. According to the above definition, for a mixture gas that is composed of N different component gases, the partial pressure of the ith component gas is given by

$$p_i = \frac{n_i RT}{V_{tot}} \tag{2.15}$$

Dalton's law of partial pressure states that the total pressure of a mixture gas equals the sum of the partial pressures of its component gases, as shown in **Figure 2.2(a)**. Dalton's law of partial pressure can be expressed as

$$p_{tot} = \sum_{i=1}^{N} p_i \tag{2.16}$$

Law of Partial Volume

Similar to the definition of partial pressure, we can define that the **partial volume** of a component gas in a mixture gas as the volume that the component gas would individually occupy if it were at the total pressure. According to the above definition, for a mixture gas that is composed of N different component gases, the partial volume of the ith component gas is given by

$$V_i = \frac{n_i RT}{p_{tot}} \tag{2.17}$$

Law of partial volume states that the total volume of a mixture gas equals the sum of the partial volume of its component gases, as shown in **Figure 2.2(b)**. The law of partial volume can be expressed as

$$V_{tot} = \sum_{i=1}^{N} V_i \tag{2.18}$$

Applications of Laws of Partial Pressure and Partial Volume

The key to understanding partial pressure and partial volume is that one should always remember to use the total volume to calculate the partial pressure, and to use the total pressure to calculate the partial volume. Both **Equations (2.15)** and **(2.17)** can be remembered together by the following equation

$$p_{tot}V_i = p_i V_{tot} = n_i RT \tag{2.19}$$

Considering the ratio of the ith component gas to the total mixture gas, we have

$$\frac{p_i}{p_{tot}} = \frac{n_i RT / V_{tot}}{n_{tot} RT / V_{tot}} = \frac{n_i}{n_{tot}} = x_i$$

and

$$\frac{V_i}{V_{tot}} = \frac{n_i RT / p_{tot}}{n_{tot} RT / p_{tot}} = \frac{n_i}{n_{tot}} = x_i$$

第 i 种组分气体的**摩尔分数** x_i 定义为

$$x_i = \frac{n_i}{n_\text{总}} \tag{2.14}$$

显然有

$$1 = \sum_{i=1}^{N} x_i$$

如果将理想气体状态方程应用于混合气体，可对所有变量都加一个下标"总"来表示混合气体的总值。混合气体的总压强 $p_\text{总}$ 和总体积 $V_\text{总}$ 可由总物质的量 $n_\text{总}$ 决定，分别有

$$p_\text{总} = \frac{n_\text{总} RT}{V_\text{总}} \quad 且 \quad V_\text{总} = \frac{n_\text{总} RT}{p_\text{总}}$$

道尔顿分压定律

约翰·道尔顿（1766—1844）对混合气体的研究作出了重要贡献。他将混合气体中某一组分气体的**分压**定义为，该组分气体单独占据容器（即具有容器的总体积）时所具有的压强。根据上述定义，对于由 N 种组分气体组成的混合气体，其中第 i 种组分气体的分压可表示为

$$p_i = \frac{n_i RT}{V_\text{总}} \tag{2.15}$$

道尔顿分压定律指出，混合气体的总压等于各组分气体的分压之和，如**图 2.2 (a)** 所示。道尔顿分压定律可用下式表示：

$$p_\text{总} = \sum_{i=1}^{N} p_i \tag{2.16}$$

分体积定律

与分压的定义类似，混合气体中某一组分气体的**分体积**可定义为，该组分气体单独具有总压时所占据的体积。根据上述定义，对于由 N 种组分气体组成的混合气体，其中第 i 种组分气体的分体积可表示为

$$V_i = \frac{n_i RT}{p_\text{总}} \tag{2.17}$$

分体积定律指出，混合气体的总体积等于各组分气体的分体积之和，如**图 2.2（b）** 所示。分体积定律可用下式表示：

$$V_\text{总} = \sum_{i=1}^{N} V_i \tag{2.18}$$

分压定律和分体积定律的应用

理解分压和分体积概念的关键在于，总是采用总体积来计算分压、采用总压来计算分体积。**式（2.15）**和**（2.17）**可统一记为

$$p_\text{总} V_i = p_i V_\text{总} = n_i RT \tag{2.19}$$

考虑第 i 种组分气体与总混合气体的比值，有

$$\frac{p_i}{p_\text{总}} = \frac{n_i RT / V_\text{总}}{n_\text{总} RT / V_\text{总}} = \frac{n_i}{n_\text{总}} = x_i$$

$$\frac{V_i}{V_\text{总}} = \frac{n_i RT / p_\text{总}}{n_\text{总} RT / p_\text{总}} = \frac{n_i}{n_\text{总}} = x_i$$

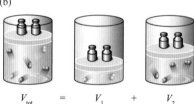

Figure 2.2 Laws of partial pressure and partial volume illustrated. (a) Dalton's law of partial pressure. (b) Law of partial volume.

图 2.2 气体分压定律和分体积定律示意图：(a) 道尔顿分压定律；(b) 分体积定律。

Therefore,

$$\frac{p_i}{p_{tot}} = \frac{V_i}{V_{tot}} = \frac{n_i}{n_{tot}} = x_i \tag{2.20}$$

Considering the ratio of the ith component gas to the jth component gas, we have

$$\frac{p_i}{p_j} = \frac{V_i}{V_j} = \frac{n_i}{n_j} = \frac{x_i}{x_j} \tag{2.21}$$

A commonly used example to understand the laws of partial pressure and partial volume involves the experiment of collecting gases over water in a pneumatic trough. These gases should be unreactive with and essentially insoluble in water, such as H_2, O_2, and N_2. The collected gas is a mixture gas of two components: the desired gas and water vapor. When measuring the pressure by a barometer, the barometric pressure equals the total pressure of the wet gas that is the sum of the partial pressures of two component gases. Therefore,

$$p_{bar} = p_{tot} = p_{gas} + p_{H_2O}$$

or

$$p_{gas} = p_{bar} - p_{H_2O}$$

where p_{H_2O} is called the vapor pressure of water. It is the partial pressure of water vapor in dynamic equilibrium with liquid water. p_{H_2O} depends only on the temperature of water. It is the highest partial pressure of water vapor that can be maintained at this temperature. If the partial pressure of water vapor in a sample is developed to be higher than p_{H_2O} at a certain temperature, some of the water vapor will condense into liquid water whereas the vapor pressure remains constant. More discussion about vapor pressure can be found in **Section 2.6**.

2.3 Kinetic-Molecular Theory of Gases

In **Section 1.1** we have introduced the three stages of how science has evolved and have used how scientists understand the behavior of gases as an example. In this chapter, we have already learnt the first stage of collecting p-V, V-T, and V-n data, and the second stage of formulating the simple gas laws as well as the ideal gas equation. Now, we move to the third and final stage to understand why the product of pV equals the product of nRT. This can be derived from a theory first developed during the mid-nineteenth century called the **kinetic-molecular theory of gases**.

Basic Assumptions

All theories start from some basic assumptions, so does the kinetic-molecular theory of gases. There are three basic assumptions in this theory, listed as follows:
1) A gas is composed of a very large number of particles (molecules or atoms, etc.) with point masses. These particles are in constant, random, straight line motion.
2) Rapid and elastic collisions happen between particles, or between particles and the vessel walls. The total energy and momentum remain constant.
3) Particles are separated by great distances. No force is present between the particles except during collisions. The volume of particles themselves is negligible.

Derivation of the Pressure-Volume Equation

All theories follow rigorous logical rules to make conclusions. Here, we introduce a frequently used simplification method in research that first decomposes a complicated three-dimensional (3D) problem into one-dimensional (1D), and then returns from 1D to 3D after some simplified derivation. This method is valid in kinetic-molecular theory of gases because in this case all three dimensions (x, y, and z) are equivalent, which means that you cannot distinguish any one dimension from the others.

Let us first consider the derivation only in the x direction (decomposed into 1D). Similar to what we

因此

$$\frac{p_i}{p_\text{总}} = \frac{V_i}{V_\text{总}} = \frac{n_i}{n_\text{总}} = x_i \tag{2.20}$$

考虑第 i 种组分气体与第 j 种组分气体的比值，有

$$\frac{p_i}{p_j} = \frac{V_i}{V_j} = \frac{n_i}{n_j} = \frac{x_i}{x_j} \tag{2.21}$$

理解分压定律和分体积定律的一个常用例子是在水槽中采用排水法收集气体的实验。这些气体应该是不与水反应且基本不溶于水的气体，如 H_2、O_2、N_2 等。收集到的气体是包含两种组分气体的混合气体：要收集的气体和水蒸气。这时使用气压计测量到的压强是混合湿气的总压，等于两种组分气体的分压之和。故

$$p_\text{测} = p_\text{总} = p_\text{气} + p_\text{水}$$

即

$$p_\text{气} = p_\text{测} - p_\text{水}$$

其中 $p_\text{水}$ 称为水的蒸气压，是水蒸气与液态水达到动态平衡时的分压。$p_\text{水}$ 只与温度有关，是在该温度下水蒸气所能达到的最高分压。如果达到某温度下的 $p_\text{水}$ 后仍进一步提高水的分压，则部分水蒸气将凝结为液态水，而蒸气压保持不变。关于蒸气压的更多讨论详见 **2.6 节**。

2.3 气体分子运动论

在 **1.1 节**中，我们以科学家们对气体行为的理解为例，介绍了科学发展演化的三个阶段。本章我们已经学习了第一阶段关于 p-V、V-T 和 V-n 数据的收集，第二阶段简单气体定律和理想气体状态方程的提出。现在我们过渡到第三即最终阶段，理解为什么 pV 的乘积能够等于 nRT 的乘积。这可以从一个最早在 19 世纪中叶发展起来的理论（即**气体分子运动论**）推导得到。

基本假定

所有理论均从基本假定开始，气体分子运动论也不例外。该理论有以下三点基本假定：
1) 气体由大量可视为质点的微粒（分子、原子等）组成，这些微粒做连续不断、无规则的直线运动。
2) 微粒之间以及微粒与器壁之间发生的碰撞是快速且弹性的，总能量和动量均守恒。
3) 微粒之间的距离很大，除碰撞之外微粒间没有作用力，微粒自身的体积可忽略不计。

压强体积方程的推导

所有的理论都是通过严密的逻辑规律来推导结论。这里我们介绍一种科研中常见的简化推导方法，即将一个三维空间的复杂问题先降至一维、经过简化的推导之后再由一维回归三维的方法。该方法在气体分子运动论中有效，是因为在这种情况下所有三个维度（即 x、y 和 z）均是等价的，即无法将三维中的任一维度与其他维度区分开。

have used in the derivation of pressure in **Section 2.1**, we assume that a particle with a mass m inside a cubic vessel with an edge l collides perpendicularly to a vessel wall with a velocity v_x. According to Newton's second law of motion, the force component in the x direction, f_x, of the particle exerting to the vessel wall during an elastic collision is given by

$$f_x = ma_x = m\frac{\Delta v_x}{\Delta t} = \frac{\Delta(mv_x)}{\Delta t} = \frac{\Delta p_x}{\Delta t}$$

where the momentum change is

$$\Delta p_x = mv_x - (-mv_x) = 2mv_x$$

and $\Delta t = 2l/v_x$ is the period of time between two successive collisions of the particle on a particular vessel wall. Note that the particle travels $2l$ before it can collide to the same wall of a cubic vessel. Therefore,

$$f_x = \frac{\Delta p_x}{\Delta t} = \frac{2mv_x}{2l/v_x} = \frac{mv_x^2}{l} \tag{2.22}$$

For a collection of gas (called an ensemble) comprises N particles in total, there is a distribution of speeds (or velocities), since particles may have a variety of speeds. If N_i denotes the number of gas particles with a speed of v_{x_i}, we then have

$$N = \sum_i N_i$$

For this ensemble of gas, the x component of the resultant force, F_x, on this vessel wall equals the sum of all the x component of the individual forces exerting by individual particles, and is given by

$$F_x = \sum_{i=1}^{N} f_{x_i} = \frac{N_1 mv_{x_1}^2}{l} + \frac{N_2 mv_{x_2}^2}{l} + \cdots + \frac{N_i mv_{x_i}^2}{l} + \cdots$$

$$= \frac{m}{l}\left(N_1 v_{x_1}^2 + N_2 v_{x_2}^2 + \cdots + N_i v_{x_i}^2 + \cdots\right) \tag{2.23}$$

The above **Equation (2.23)** is very complicated and not easy to remember or to use. A common way in research under such circumstances is to make some new definitions. In general, new definitions are usually coined when the following two requirements are met:
1) It makes the expressions much more concise;
2) It has a specific physical meaning.

Here, we coin a new definition of the **mean-square speed**, $\overline{v_x^2}$, as

$$\overline{v_x^2} = \frac{1}{N}\sum_{i=1}^{N} v_{x_i}^2 = \frac{N_1 v_{x_1}^2}{N} + \frac{N_2 v_{x_2}^2}{N} + \cdots + \frac{N_i v_{x_i}^2}{N} + \cdots$$

$$= \frac{1}{N}\left(N_1 v_{x_1}^2 + N_2 v_{x_2}^2 + \cdots + N_i v_{x_i}^2 + \cdots\right) \tag{2.24}$$

Using this definition, **Equation (2.23)** can be rewritten as

$$F_x = \frac{Nm\overline{v_x^2}}{l} \tag{2.25}$$

Obviously, **Equation (2.25)** is much more concise than **Equation (2.23)**. Meanwhile, the mean-square speed has a specific physical meaning, which is the algebraic average of the squares of the molecular speed. Later in this section, we will find that $\overline{v_x^2}$ is directly proportional to the thermodynamic temperature of the gas, and thus gives direct measurements of the temperature.

After the above simplified derivation in 1D, we now return to 3D. Based on vector algebra, we have

$$v^2 = v_x^2 + v_y^2 + v_z^2$$

Again, as x, y, and z dimensions are identical, it can be easily derived that

$$\overline{v^2} = \overline{v_x^2} + \overline{v_y^2} + \overline{v_z^2} = 3\overline{v_x^2} \tag{2.26}$$

Therefore,

我们先来考虑 x 方向（降至一维）的推导过程。与 **2.1 节**中压强的推导方法类似，假定在边长为 l 的立方体容器中有一质量为 m 的气体微粒，以 v_x 的速率垂直碰撞某一器壁。根据牛顿第二运动定律，该气体微粒弹性撞击器壁时对器壁施加的力在 x 方向的分量（f_x）为

$$f_x = ma_x = m\frac{\Delta v_x}{\Delta t} = \frac{\Delta(mv_x)}{\Delta t} = \frac{\Delta p_x}{\Delta t}$$

其中动量的改变量为

$$\Delta p_x = mv_x - (-mv_x) = 2mv_x$$

$\Delta t = 2l/v_x$ 为该微粒连续两次撞击同一器壁的时间间隔，注意气体微粒需要移动 $2l$ 的路程才能与立方体容器的同一器壁再次发生碰撞。因此

$$f_x = \frac{\Delta p_x}{\Delta t} = \frac{2mv_x}{2l/v_x} = \frac{mv_x^2}{l} \tag{2.22}$$

对于一个由 N 个微粒组成的气体集合（称为系综），大量微粒具有不同的运动速率，存在一个速率分布。如果用 N_i 表示速率为 v_{x_i} 的气体微粒的数目，则有

$$N = \sum_i N_i$$

对于此气体系综，大量微粒碰撞某一器壁的合力在 x 方向的分量（F_x），等于所有微粒碰撞的分力在 x 方向的分量之和，可表示为

$$F_x = \sum_{i=1}^{N} f_{x_i} = \frac{N_1 mv_{x_1}^2}{l} + \frac{N_2 mv_{x_2}^2}{l} + \cdots + \frac{N_i mv_{x_i}^2}{l} + \cdots$$

$$= \frac{m}{l}\left(N_1 v_{x_1}^2 + N_2 v_{x_2}^2 + \cdots + N_i v_{x_i}^2 + \cdots\right) \tag{2.23}$$

式（2.23）比较复杂，既不好记也不好用。这种情况在科研中的常规解决办法是：给出新的定义。总体而言，满足如下两个条件时通常会给出新的定义：

1) 新定义可将复杂的表达式大幅简化；
2) 新定义具有明确的物理意义。

这里就给出一个新的定义：定义**均方速率** $\overline{v_x^2}$ 为

$$\overline{v_x^2} = \frac{1}{N}\sum_{i=1}^{N} v_{x_i}^2 = \frac{N_1 v_{x_1}^2}{N} + \frac{N_2 v_{x_2}^2}{N} + \cdots + \frac{N_i v_{x_i}^2}{N} + \cdots$$

$$= \frac{1}{N}\left(N_1 v_{x_1}^2 + N_2 v_{x_2}^2 + \cdots + N_i v_{x_i}^2 + \cdots\right) \tag{2.24}$$

采用该定义，**式（2.23）**可改写为

$$F_x = \frac{Nm\overline{v_x^2}}{l} \tag{2.25}$$

式（2.25）显然远比**式（2.23）**形式简明。同时均方速率具有明确的物理意义，它是分子速率平方的算术平均值。在本节后续，我们可以看到均方速率与气体的热力学温度成正比，是温度高低的直接体现。

推导出上述一维的简化公式后，我们再回归三维。由矢量代数，可得

$$v^2 = v_x^2 + v_y^2 + v_z^2$$

同样，由于 x、y 和 z 三维均等价，经过简单推导可得

$$\overline{v^2} = \overline{v_x^2} + \overline{v_y^2} + \overline{v_z^2} = 3\overline{v_x^2} \tag{2.26}$$

$$p = \frac{F_x}{A} = \frac{Nm\overline{v_x^2}/l}{l^2} = \frac{Nm\overline{v_x^2}}{l^3} = \frac{Nm\overline{v^2}}{3V}$$

or

$$pV = \frac{1}{3}Nm\overline{v^2} \tag{2.27}$$

Equation (2.27) is called the **pressure-volume equation**.

Significance of the Pressure-Volume Equation

In **Equation (2.27)**, all the variables on the left side (p and V) are macroscopic physical quantities that can be measured experimentally for an ensemble of gas. All the variables on the right side (N, m, and $\overline{v^2}$) are microscopic properties for individual gas molecules. Therefore, the significance of the pressure-volume equation is that it shows the connections between the macro-world and the micro-world. It is an equation that relates the macroscopic physical quantities to some microscopic properties. The pressure-volume equation reveals that some macroscopic physical quantities such as p and V can be understood microscopically by the mass, mean-square speed, and total number of individual molecules. Meanwhile, it also shows that some microscopic properties of individual molecules can be directly calculated from some experimentally measurable macroscopic quantities.

Meaning of Temperature

To further derive the ideal gas equation from the pressure-volume equation, we need to explore the meaning of temperature. Due to the requirement of some complex background knowledge, we will not derive and explain the detail here but rather use directly the following conclusion: The thermodynamic temperature T of an ideal gas is proportional to the **average translational kinetic energy**, $\overline{E_k}$ or $\overline{e_k}$, of its molecules. This idea can be expressed either macroscopically in **Equation (2.28)** or microscopically in **Equation (2.29)**, as

$$\overline{E_k} = \frac{3}{2}RT \tag{2.28}$$

or

$$\overline{e_k} = \frac{3}{2}k_BT \tag{2.29}$$

where R is the molar gas constant, and k_B is the **Boltzmann constant** that is related to R and the **Avogadro constant** N_A by

$$k_B = R/N_A \tag{2.30}$$

Note that $\overline{E_k}$ is the average translational kinetic energy of 1 mol gas molecules, and $\overline{e_k}$ is the average translational kinetic energy of an individual gas molecule. It can be easily derived that

$$\overline{e_k} = \overline{E_k}/N_A \tag{2.31}$$

In physics, the kinetic energy is generally defined as

$$E_k = \frac{1}{2}mv^2$$

The average translational kinetic energy of gas molecules is related to the mean-square speed by

$$\overline{E_k} = \frac{1}{2}M\overline{v^2} \tag{2.32}$$

$$\overline{e_k} = \frac{1}{2}m\overline{v^2} \tag{2.33}$$

where M is the molar mass and m is the mass of an individual gas molecule.

Therefore, the mean-square speed is directly proportional to the temperature of the gas, as

$$\overline{v^2} = \frac{3RT}{M} = \frac{3k_BT}{m} \tag{2.34}$$

因此
$$p = \frac{F_x}{A} = \frac{Nm\overline{v_x^2}/l}{l^2} = \frac{Nm\overline{v_x^2}}{l^3} = \frac{Nm\overline{v^2}}{3V}$$

即
$$pV = \frac{1}{3}Nm\overline{v^2} \tag{2.27}$$

式（2.27）被称为**压强体积方程**。

压强体积方程的意义

在**式（2.27）**中，方程左侧的所有变量（p 和 V）均为气体系综的可通过实验测量的宏观物理量，而方程右侧的所有变量（N、m 和 $\overline{v^2}$）均为单个气体分子的微观性质。因此压强体积方程的意义在于，它显示了宏观世界和微观世界之间的联系，是一个将宏观物理量与微观性质相关联的方程。压强体积方程揭示了 p 和 V 这样的宏观物理量可以在微观上通过单个气体分子的质量、均方速率和总气体分子数目来理解，同时也表明一些分子层面的微观性质可以通过实验上可测量的宏观量来直接计算。

温度的含义

为了从压强体积方程进一步推导出理想气体状态方程，我们需要探索温度的含义。由于需要用到一些较为复杂的背景知识，这里我们不做详细推导和解读，而只直接给出如下结论：理想气体的热力学温度 T 与气体分子的**平均平动能** $\overline{E_k}$ 或 $\overline{e_k}$ 成正比。可用宏观表达**式（2.28）**或微观表达**式（2.29）**表示为

$$\overline{E_k} = \frac{3}{2}RT \tag{2.28}$$

$$\overline{e_k} = \frac{3}{2}k_B T \tag{2.29}$$

其中 R 是摩尔气体常数，k_B 称为**玻尔兹曼常数**，其与 R 和**阿伏伽德罗常数** N_A 的联系为

$$k_B = R/N_A \tag{2.30}$$

注意 $\overline{E_k}$ 是 1 mol 气体分子的平均平动能，而 $\overline{e_k}$ 是单个气体分子的平均平动能。易得

$$\overline{e_k} = \overline{E_k}/N_A \tag{2.31}$$

物理学上动能的常规定义为

$$E_k = \frac{1}{2}mv^2$$

气体分子的平均平动能与其均方速率的联系为

$$\overline{E_k} = \frac{1}{2}M\overline{v^2} \tag{2.32}$$

$$\overline{e_k} = \frac{1}{2}m\overline{v^2} \tag{2.33}$$

其中 M 为气体的摩尔质量，而 m 为单个气体分子的质量。

因此，均方速率与气体的温度成正比，有

$$\overline{v^2} = \frac{3RT}{M} = \frac{3k_B T}{m} \tag{2.34}$$

从压强体积方程可推导

From the pressure-volume equation, we can derive

$$pV = \frac{1}{3}Nm\overline{v^2} = \frac{2}{3}N\left(\frac{1}{2}m\overline{v^2}\right) = \frac{2}{3}nN_A\left(\frac{3}{2}k_BT\right)$$

or

$$pV = nRT$$

So far, all the derivation in the kinetic-molecular theory of gas has been completed. Starting from the three basic assumptions and following a series of rigorous logical derivation, we obtain the final expression of the ideal gas equation.

Relating the temperature of gas to the average translational kinetic energy or to the mean-square speed of gas molecules helps us to understand what is happening microscopically when gases with different temperature are mixed with each other. Molecules in the hotter gas, on average, have higher kinetic energies or move faster than do the molecules in the colder gas. When the two gases are mixed, molecules in the hotter gas transfer some kinetic energy through collisions with molecules in the colder gas. The transfer of energy continues until the average kinetic energies of the molecules in the mixture become equal, that is until the temperatures become equalized. Finally, **Equations (2.28)** and **(2.29)** provide a new way of understanding the absolute zero of temperature: It is the temperature at which translational molecular motion should cease.

Here, you may wonder what "translational" means in average translational kinetic energy. Translation is a movement that changes the position of an object by moving every point the same distance in a given direction, without rotation, vibration, or change in size. Ideal gas molecules can be viewed as point masses without any volume, and thus do not have rotational or vibrational motions. The kinetic energy of an ideal gas refers only to its translational kinetic energy.

2.4 Distribution of Molecular Speeds

As discussed in **Section 2.3**, not all the molecules in a gas travel at the same speed, and there is a distribution of molecular speeds in an ensemble of gas. In this section, we will discuss the distribution formula of molecular speeds, its measurement, and its applications.

Maxwell Distribution of Molecular Speeds

Due to the very large number of gas molecules, the exact speed of each molecule is unknown, but predictions of how many molecules have a particular speed can be made statistically. In 1860, James C. Maxwell (1831—1879) derived the famous **Maxwell distribution of molecular speeds**, as given by the following equation

$$F(v) = 4\pi\left(\frac{M}{2\pi RT}\right)^{3/2} v^2 \exp(-Mv^2/2RT) \tag{2.35}$$

where $F(v)$ is the probability density function that gives the probability, per unit speed, of finding a molecule with a speed near v. In a simplified understanding, $F(v)$ can be approximately considered as the probability or fraction of molecules that have a speed between $v-0.5$ and $v+0.5$. **Equation (2.35)** is a complicated equation, which we will not attempt to derive but rather accept as valid for qualitative or semi-quantitative interpretation.

Mathematically, the fraction of molecules (dN/N) with speeds in a very small range dv near v can be written as

$$\frac{dN}{N} = F(v)dv$$

Here, $F(v)$ is the length, dv is the width and dN/N gives the area of a near-rectangle, as shown in **Figure 2.3(a)**. The fraction of molecules ($\Delta N/N$) with speeds between v_1 and v_2 is then given by

$$\frac{\Delta N}{N} = \int_{v_1}^{v_2} F(v)dv \tag{2.36}$$

$$pV = \frac{1}{3}Nm\overline{v^2} = \frac{2}{3}N\left(\frac{1}{2}m\overline{v^2}\right) = \frac{2}{3}nN_A\left(\frac{3}{2}k_BT\right)$$

即

$$pV = nRT$$

至此完成了气体分子运动论的全部推导，从三条基本假定出发，经过一系列逻辑严密的推导，得到了理想气体状态方程。

将气体的温度与气体分子的平均平动能或均方速率相联系，有助于从微观上理解不同温度的气体相互混合时所发生的情况。平均而言，较热气体中的分子比较冷气体中的分子具有更高的动能或更快的运动速率。当两种气体混合时，较热气体中的分子通过与较冷气体中的分子碰撞来传递动能。能量的转移一直持续到混合气体中分子的平均动能相等为止，也就是说直到温度相等为止。最后，**式（2.28）**和**（2.29）**提供了一种理解绝对零度的新方法：绝对零度是分子平动停止时的温度。

这里你可能想知道平均平动能中的"平动"究竟是什么意思。平动是一种通过在指定方向上将每个点都移动相同距离的方式来改变物体位置的运动，其中不发生任何旋转、振动或改变大小的运动。理想气体分子可视为没有任何体积的质点，因此不存在转动和振动，理想气体的动能即为其平动能。

2.4 气体分子的速率分布

如 **2.3 节**所述，并非所有气体分子均以相同的速率运动，气体系综里存在分子的速率分布。本节我们将讨论分子速率的分布公式、测量方法及其应用。

麦克斯韦速率分布

由于气体分子数量众多，每个分子确切的运动速率是未知的，但可以从统计上预测有多少分子具有特定的速率。1860 年詹姆斯·C. 麦克斯韦（1831—1879）提出著名的**麦克斯韦速率分布**方程，形式如下：

$$F(v) = 4\pi\left(\frac{M}{2\pi RT}\right)^{3/2} v^2 \exp\left(-Mv^2/2RT\right) \quad (2.35)$$

其中 $F(v)$ 是概率密度函数，给出了找到速率处于 v 附近单位速率间隔中的分子的概率。在简化理解中，$F(v)$ 可近似认为是速率介于 $v-0.5$ 和 $v+0.5$ 之间的分子在所有分子中的概率或所占分数。**式 (2.35)** 形式较为复杂，我们不进行推导，只接受其合理性并运用其进行定性或半定量的理解。

数学上，速率处于 v 附近极小间隔 dv 中的分子在所有分子中所占的分数（dN/N）为

$$\frac{dN}{N} = F(v)dv$$

在如**图 2.3 (a)** 所示的近长方形中，$F(v)$ 和 dv 分别表示长和宽，dN/N 给出近长方形的面积。速率介于 v_1 和 v_2 之间的分子在所有分子中所占的分数（$\Delta N/N$）为

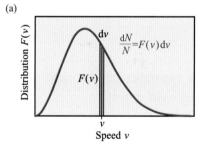

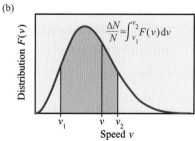

Figure 2.3 **Maxwell distribution of molecular speeds.** (a) The fraction of molecules (dN/N) with speeds in a very small range dv near v. (b) The fraction of molecules (ΔN/N) with speeds between v_1 and v_2.

图 2.3 分子速率的麦克斯韦分布图：(a) 速率处于 v 附近极小间隔 dv 中的分子在所有分子中所占的分数（dN/N）；(b) 速率介于 v_1 和 v_2 之间的分子在所有分子中所占的分数（ΔN/N）。

which can be represented by the total area under the $F(v)$ curve between $v = v_1$ and $v = v_2$, as shown in **Figure 2.3(b)**. As the fraction of molecules with all possible speeds, from 0 to $+\infty$, must equal 1, we then have

$$\int_0^{+\infty} F(v) \, dv = 1 \tag{2.37}$$

This means that the total area under the $F(v)$ curve equals 1.

It can be seen from **Figure 2.3** that $F(v)$ increases first and then decreases as v increases from 0 to $+\infty$, peaking at a certain speed. The Maxwell formula demonstrates that the shape of distribution depends on the product of two opposing factors: a factor that is proportional to v^2, increasing from a value of 0 with v, and another exponential factor $\exp(-Mv^2/2RT)$, decreasing from a value of 1 with v. The v^2 factor favors the presence of molecules with high speeds and is responsible for there being few molecules with speeds near zero. The exponential factor favors low speeds and limits the number of molecules that can have high speeds. The combined effect of these two factors makes $F(v)$ increases from a value of 0, reaches a maximum, and then decreases as v further increases. Notice that the distribution curve goes to infinity and is not symmetrical about its maximum.

Equation (2.35) shows that the distribution of molecular speeds depends on temperature T and molar mass M. **Figure 2.4** illustrates how the distribution of molecular speeds varies with respect to temperature and molar mass. When we compare the distributions for $N_2(g)$ at 300 K and 1000 K, we see that the range of speeds broadens and the distribution shifts toward high speeds as the temperature increases. The distributions for $N_2(g)$ and $He(g)$ at 300 K reveal that the lighter the gas, the broader the range of speeds. However, the total area under all three curves are still the same, equaling 1, no matter how the temperature and molar mass vary.

Three Characteristic Molecular Speeds

In **Section 2.3**, we have defined the mean-square speed $\overline{v^2}$ and learnt that it is directly proportional to the thermodynamic temperature of gas. In this section, we define three characteristic molecular speeds in Maxwell distribution, identified in **Figure 2.5**, as

1) The **root-mean-square speed** v_{rms}, which is the root of the mean-square speed, as

$$v_{rms} = \sqrt{\overline{v^2}} = \sqrt{\frac{1}{N}\sum_{i=1}^{N} v_i^2} = \sqrt{\frac{3RT}{M}} \tag{2.38}$$

We can remember the name of root-mean-square speed by taking the root of the mean of the squares of the molecular speeds.

2) The **average speed** v_{av} or $\overline{v}$, which is the algebraic average or mean of the molecular speeds, as

$$v_{av} = \overline{v} = \frac{1}{N}\sum_{i=1}^{N} v_i \tag{2.39}$$

3) The **most probable speed** or **modal speed** v_m, which is the speed with the highest probability density and can be calculated from

$$\frac{dF(v)}{dv} = 0 \tag{2.40}$$

Mathematically, it can be calculated that the three characteristic speeds in Maxwell distribution are proportional to each other by the following ratio

$$v_m : v_{av} : v_{rms} = \sqrt{2} : \sqrt{\frac{8}{\pi}} : \sqrt{3} \approx 1 : 1.13 : 1.22 \tag{2.41}$$

Therefore, we have

$$v_{av} = \sqrt{\frac{8RT}{\pi M}} \tag{2.42}$$

and

$$v_m = \sqrt{\frac{2RT}{M}} \tag{2.43}$$

$$\frac{\Delta N}{N} = \int_{v_1}^{v_2} F(v)\,\mathrm{d}v \quad (2.36)$$

可用 $F(v)$ 曲线下在 $v = v_1$ 和 $v = v_2$ 之间的总覆盖面积来表示，如**图 2.3 (b)** 所示。由于具有全部可能速率（从 0 到 $+\infty$）的分子占比必为 1，则有

$$\int_0^{+\infty} F(v)\,\mathrm{d}v = 1 \quad (2.37)$$

这表明 $F(v)$ 曲线下的总覆盖面积等于 1。

从**图 2.3** 可见，随着 v 从 0 增加到 $+\infty$，$F(v)$ 曲线先上升后下降，具有一个峰值速率。麦克斯韦方程表明，分布曲线的形状由两个具有相反效应的项的乘积决定：一项与 v^2 成正比，随 v 增加从 0 开始增加；另一项为指数项 $\exp(-Mv^2/2RT)$，随 v 增加从 1 开始下降。v^2 项有利于高速率分子的存在，并导致速率接近零的分子数极少。指数项有利于低速率分子的存在，限制了高速率分子的数量。两项的综合作用使得 $F(v)$ 从 0 开始增加，达到极大值，然后随着 v 继续增加而减小。注意分布曲线一直延伸到无穷远，且并不关于极大值对称。

式（2.35） 表明分子速率的分布与温度 T 和摩尔质量 M 有关，**图 2.4** 给出了分子速率分布随温度和摩尔质量的变化曲线。比较氮气在 300 K 和 1000 K 时的分布曲线，可以看到随温度升高速率分布范围变宽，且高速率分子的占比更多。氮气和氦气在 300 K 时的分布曲线表明，较轻气体的速率分布范围更宽。但无论温度和摩尔质量如何变化，所有三条曲线下的总覆盖面积均为 1。

三种特征分子速率

在 **2.3 节**中定义了均方速率 $\overline{v^2}$，并了解其值正比于气体的热力学温度。本节定义麦克斯韦分布中的三种特征气体分子速率（如**图 2.5** 所示），分别为：

1) **方均根速率** v_{rms}：定义为均方速率的平方根，即

$$v_{\mathrm{rms}} = \sqrt{\overline{v^2}} = \sqrt{\frac{1}{N}\sum_{i=1}^{N} v_i^2} = \sqrt{\frac{3RT}{M}} \quad (2.38)$$

所谓方均根速率，表示将分子速率先平方、再平均、再开平方根。

2) **平均速率** v_{av} 或 $\overline{v}$：定义为分子速率的算术平均值，即

$$v_{\mathrm{av}} = \overline{v} = \frac{1}{N}\sum_{i=1}^{N} v_i \quad (2.39)$$

3) **最概然速率** v_{m}：定义为具有最高概率密度的速率，可通过下式计算

$$\frac{\mathrm{d}F(v)}{\mathrm{d}v} = 0 \quad (2.40)$$

数学上可计算，麦克斯韦分布中的三种特征速率存在如下比例关系：

$$v_{\mathrm{m}} : v_{\mathrm{av}} : v_{\mathrm{rms}} = \sqrt{2} : \sqrt{\frac{8}{\pi}} : \sqrt{3} \approx 1 : 1.13 : 1.22 \quad (2.41)$$

因此有

$$v_{\mathrm{av}} = \sqrt{\frac{8RT}{\pi M}} \quad (2.42)$$

$$v_{\mathrm{m}} = \sqrt{\frac{2RT}{M}} \quad (2.43)$$

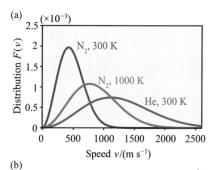
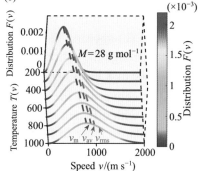
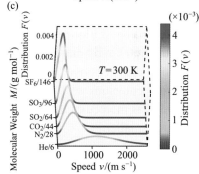

Figure 2.4 Distribution of molecular speeds depending on temperature and molar mass. (a) Molecular speed distributions for N_2 at 300 K and 1000 K, and He at 300 K. (b) Effect of temperature on molecular speed distribution. (c) Effect of molar mass on molecular speed distribution.

图 2.4 分子速率分布随温度和摩尔质量的变化图： (a) 300 K 和 1000 K 下 N_2 及 300 K 下 He 的分子速率分布图；(b) 分子速率分布随温度的变化图；(c) 分子速率分布随摩尔质量的变化图。

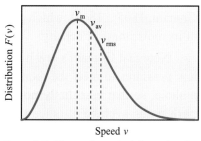

Figure 2.5 Three characteristic molecular speeds in Maxwell distribution: root-mean-square speed v_{rms}, average speed v_{av}, and modal speed v_{m}, with $v_{\mathrm{rms}} > v_{\mathrm{av}} > v_{\mathrm{m}}$.

图 2.5 麦克斯韦分布中的三种特征分子速率： 方均根速率 v_{rms}、平均速率 v_{av} 和最概然速率 v_{m}，其中 $v_{\mathrm{rms}} > v_{\mathrm{av}} > v_{\mathrm{m}}$。

Experimental Determination of Molecular Speed Distribution

The distribution of molecular speeds can be determined experimentally by an apparatus shown in **Figure 2.6**. The apparatus comprises a temperature-controllable oven, an evacuated chamber, and a molecular detector. The gas molecules heated in the oven at a constant temperature emerge through a small hole into the evacuated chamber. After passing through a series of slits called collimators, the gas molecules are herded into a molecular beam moving in the same direction. The density of molecules in the evacuated chamber is kept very low so that few collisions between them will not disturb the beam. The molecular beam then passes through a series of rotating disks rotating at the same speed, each with a slit cut that is offset from one another by a certain angle. The molecules passing through the first rotating disk will pass through the second disk only if their velocities meet the requirement that they arrive at the disk at the exact moment that the second slit appears. Thus, for a given rotating speed, only those molecules with the appropriate velocity can pass through the entire series of disks and finally reach the detector to be counted. Changing the rotating speed of the disk leads to the detection of molecules with different velocity.

After converting the rotating speed into the corresponding molecular speed, the detected number of molecules with different velocity can be plotted with respect to the molecular speed to obtain a distribution curve similar to those in **Figure 2.4**. By varying the temperature of the oven, data at different temperatures can be collected. By changing the gas in the oven, plots of different gas molecules can be made.

Diffusion and Effusion

According to **Figure 2.4**, He(g) at 300 K has a most probable speed of approximately 1100 m s^{-1}, which is about 4000 km h^{-1} and is even faster than the speed of an airplane. It seems that gas molecules could travel very long distances over a very short period of time, but this is not quite the case. Gas molecules undergo numerous collisions with one another and constantly change their directions. Therefore, although the net rate at which gas molecules move in a particular direction does depend on their average speeds, the actual values are much smaller.

Both diffusion and effusion are related to the molecular speeds of gas, and can be viewed as applications of molecular speeds. **Diffusion** is the migration of molecules as a result of random molecular motion. The diffusion of two or more gases leads to the mixing of molecules and a homogeneous mixture will soon be produced in a closed container, as shown in **Figure 2.7(a)**. A related concept, **effusion**, is the escape of gas molecules from their container through a tiny pinhole into a vacuum. The effusion of a mixture of two gases is given by **Figure 2.7(b)**. The effusion rate of gas molecules is directly proportional to their molecular speeds, meaning that molecules with high speeds effuse faster than molecules with low speeds. For two different gases at the same temperature and pressure, the effusion rate is proportional to their root-mean-square speeds as

$$\frac{R_1}{R_2} = \frac{(v_{\text{rms}})_1}{(v_{\text{rms}})_2} = \sqrt{\frac{3RT/M_1}{3RT/M_2}} = \sqrt{\frac{M_2}{M_1}} = \sqrt{\frac{\rho_2}{\rho_1}} \tag{2.44}$$

Equation (2.44) is a kinetic-theory statement of a nineteenth-century law called **Graham's law** (Thomas Graham, 1805—1869), which states that the effusion rate of a gas is inversely proportional to the square root of its molar mass or density.

Note that Graham's law has serious limitations. It can be used to describe effusion only for gases at very low pressures to meet the requirement that the mean free path of the gas is greater than the diameter of the pinhole, so that essentially no collisions occur as gas molecules pass through the pinhole. The **mean free path** is the average distance that a gas molecule travels between successive collisions, and the mean free path increases with decreasing gas pressure. Graham's law was originally proposed to describe the diffusion rate of gases, but it actually does not quantitatively apply to diffusion. Although gases of low molar mass do diffuse faster than those of higher molar mass, because numerous collisions occur between the gas molecules during diffusion, the molecular speeds of gas are changed so that quantitative predictions about the diffusion rates cannot be made based on Graham's law.

分子速率分布的实验测定

分子速率分布可采用如**图 2.6** 所示装置进行实验测定，该装置由可控温的分子炉源、真空腔和分子检测器组成。在分子炉源中恒温加热的气体分子通过小孔进入真空腔，在通过一系列称为准直器的狭缝后，聚集成一束沿同一方向运动的分子束。真空腔内的分子密度保持极低值，这样分子之间几乎不发生碰撞，不会对分子束产生干扰。接下来分子束穿过一系列以相同速率旋转的旋转盘，每个盘上都开了一道狭缝且错开一定角度。在那些通过了第一个旋转盘的分子中，只有速率满足当其到达第二个旋转盘时狭缝恰好出现的分子，才能穿过第二个旋转盘。因此，对于给定的旋转速率，只有具有合适速率的分子才能通过所有旋转盘，最终到达检测器而被计数。改变旋转盘速率，即可检测出具有不同速率的分子。

将旋转速率转换为相应的分子速率，以检测到的具有不同速率的分子数对分子速率作图，即可得到类似于**图 2.4** 中的分布曲线。改变分子炉源的温度，可收集不同温度下的数据。改变炉源中的气体种类，可绘制不同气体分子的速率分布图。

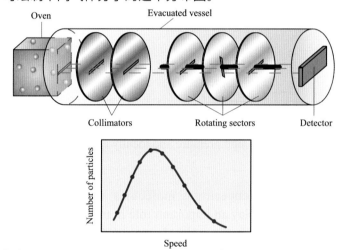

Figure 2.6 The experimental steup to determine the distribution of molecular speeds.

图 2.6 测量分子速率分布的实验装置示意图。

扩散与隙流

由**图 2.4** 可知，300 K 时氦气的最概然速率约为 1100 m s^{-1}，合 4000 km h^{-1}，比飞机的飞行速率还快。虽然看起来气体分子似乎在短时间内即可移动很远的距离，但事实并非如此。气体分子彼此会发生大量碰撞并不断改变运动方向，因此气体分子在一个特定方向上的净运动速率虽也与其平均速率相关，但实际数值远小于后者。

扩散和隙流均与气体分子运动速率相关，可视为气体分子速率的应用。**扩散**即分子的迁移，是分子随机运动的结果。两种或两种以上气体的扩散导致了分子的混合，在密闭容器中很快会形成均匀的混合物，如**图 2.7(a)** 所示。另一个相关概念**隙流**，是指气体分子从容器中通过小孔

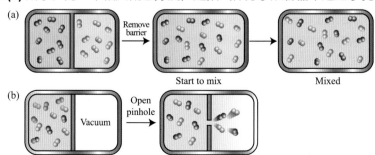

Figure 2.7 Illustrative diagrams of diffusion and effusion. (a) Diffusion is the passage of one substance through another. (b) Effusion is the passage of a substance through a pinhole into a vacuum.

图 2.7 扩散和隙流的示意图：(a) 扩散是一种物质通过另一种物质的过程；(b) 隙流是物质通过小孔进入真空的过程。

2.5 Real Gases and van der Waals Equation

We have discussed the behaviors, laws, and theories of ideal gases in the past several sections. In this section, we apply the ideal gas model back to the real case and study the nonideal or real gases. On the contrary to the two basic assumptions of the ideal gas, for **real gases:**

1) The intermolecular forces are not negligible. Attractive forces are present at far distance and repulsive forces are present only at very close distance.
2) The volume of molecules is not negligible. Molecules do occupy some space.

Gases tend to behave ideally at high temperature and low pressure, and behave nonideally at low temperature and high pressure. We can understand this by analyzing the volume and pressure terms of real gases.

Volume of Real Gases

Real gas molecules do occupy some space and cannot be pressurized infinitely under high pressure. The final result of pressurization is the volume of the molecules V_{mlc} themselves. The volume of real gases will be greater than that of ideal gases and the difference is V_{mlc}. Thus,

$$V_{real} = V_{ideal} + V_{mlc} > V_{ideal}$$

If the occupied volume of 1 mol molecules is b, then the occupied volume of n mol molecules is given by

$$V_{mlc} = nb$$

and

$$V_{ideal} = V_{real} - nb \tag{2.45}$$

At very high pressure ($p \to +\infty$), $V_{real} \to nb$ and $V_{ideal} \to 0$. Compared to V_{real}, the volume of gas molecules V_{mlc} is significant and nonnegligible. At very low pressure ($p \to 0$), $V_{real} \to +\infty$ and $V_{ideal} \to +\infty$. Compared to V_{real}, V_{mlc} is small and negligible. Thus, gases behave nonideally at very high pressure and ideally at very low pressure.

Pressure of Real Gases

The pressure of gas is the result of collisions between gas molecules and the vessel walls. The nonnegligible intermolecular forces between real gases play an important role in modifying the pressure. The intermolecular forces behave as repulsive forces only when the molecules are at very close distance, about to collide with one another. Other than during collisions, the intermolecular forces behave as attractive forces and serve as the major forces between real gas molecules. When one molecule (denoted as **A**) collides with the vessel wall, the attractive forces of other molecules to molecule **A** cause it to exert less force on the wall than if these attractions did not exist. Consequently, the pressure of real gases is smaller than that of ideal gases. If we define the difference between p_{ideal} and p_{real} as p_{inner}, then

$$p_{real} = p_{ideal} - p_{inner} < p_{ideal}$$

p_{inner} is the inner pressure that accounts for the inner attractive forces between molecule **A** and other molecules. It is proportional to the concentration of both molecule **A** and other molecules, and can be represented by

$$p_{inner} \propto \left(\frac{n_A}{V}\right)\left(\frac{n_{others}}{V}\right)$$

Since attractive forces are present between all the gas molecules and all molecules collide with the vessel wall, one cannot distinguish any one molecule from the others. Any molecule can be "molecule **A**" and simultaneously be "other molecules". The concentrations of both "molecule **A**" and "other molecules" are the same, equaling the concentration of gas molecules in the vessel. Therefore,

$$p_{inner} \propto \left(\frac{n}{V}\right)^2$$

Let $a = p_{inner} / (n/V)^2$, then

向真空的逃逸，两种气体混合物的隙流如**图 2.7(b)** 所示。气体分子的隙流速率正比于其分子速率，即高速率分子比低速率分子的隙流更快。等温等压下两种不同气体的隙流速率与其方均根速率成正比，有

$$\frac{R_1}{R_2} = \frac{(v_{\text{rms}})_1}{(v_{\text{rms}})_2} = \sqrt{\frac{3RT/M_1}{3RT/M_2}} = \sqrt{\frac{M_2}{M_1}} = \sqrt{\frac{\rho_2}{\rho_1}} \quad (2.44)$$

式（2.44）是 19 世纪**格雷厄姆定律**（托马斯·格雷厄姆，1805—1869）的运动论表述，指出气体分子的隙流速率与其摩尔质量或密度的平方根成反比。

值得一提的是，格雷厄姆定律有严格的限制。它只能用于描述极低压气体的隙流，要求气体的平均自由程大于小孔直径，这样当分子通过小孔时基本不会发生碰撞。**平均自由程**是气体分子在连续碰撞之间移动的平均距离，随压强降低而增加。格雷厄姆定律最初被用于描述气体的扩散速率，但实际上它并不能定量地应用于扩散。虽然摩尔质量小的气体确实比摩尔质量大的扩散更快，但由于扩散过程中气体分子之间会发生大量碰撞，改变了气体分子的运动速率，因此不能用格雷厄姆定律对扩散速率进行定量预测。

2.5 实际气体与范德华方程

前几节中我们讨论了理想气体的行为、规律和理论，本节将把理想气体模型应用回实际情况中，研究非理想气体即实际气体。与理想气体的两个基本假定相反，对于**实际气体**：
1) 分子间作用力不可忽略，远距离下存在吸引力，而仅在极近距离时才存在排斥力。
2) 分子的体积不可忽略，分子占据一定的空间。

在高温低压下气体行为趋向于理想状态，而在低温高压下气体行为为非理想状态。这可以通过分析实际气体的体积项和压强项来理解。

实际气体的体积

实际气体分子占据一定的空间，在高压下不能无限压缩，压缩的最终结果是占据分子自身的体积 $V_{\text{分}}$。因此实际气体的体积比理想气体更大，两者差值为 $V_{\text{分}}$，即

$$V_{\text{实}} = V_{\text{理}} + V_{\text{分}} > V_{\text{理}}$$

如果 1 mol 分子占据的体积为 b，则 n mol 分子占据的体积为

$$V_{\text{分}} = nb$$

且有

$$V_{\text{理}} = V_{\text{实}} - nb \quad (2.45)$$

在极高压 $(p \to +\infty)$ 下，$V_{\text{实}} \to nb$，$V_{\text{理}} \to 0$；与 $V_{\text{实}}$ 相比，气体分子自身的体积 $V_{\text{分}}$ 很重要且不可忽略。在极低压 $(p \to 0)$ 下，$V_{\text{实}} \to +\infty$，$V_{\text{理}} \to +\infty$；与 $V_{\text{实}}$ 相比，$V_{\text{分}}$ 很小可忽略不计。因此，气体的行为在高压下为非理想状态，低压下为理想状态。

实际气体的压强

气体的压强是气体分子与器壁碰撞的结果，实际气体中不可忽略的分子间作用力在调节压强中起到重要作用。只有当分子极其靠

$$p_{\text{inner}} = a\left(\frac{n}{V}\right)^2$$

and

$$p_{\text{ideal}} = p_{\text{real}} + \frac{an^2}{V^2} \tag{2.46}$$

In **Equation (2.46)**, V should be V_{real} and the subscript "real" is omitted.

At very low temperature, the molecular speeds are low and the average translational kinetic energy of gas molecules is small. Compared to a small kinetic energy (which tends to expand the gas), the inner attractive forces between molecules (which tends to hold the gas together) are significant and nonnegligible. At very high temperature, gas molecules are in rapid motion and the intermolecular forces of attraction is negligible. Thus, gases behave nonideally at very low temperature and ideally at very high temperature.

Van der Waals Equation

In the ideal gas equation of $pV = nRT$, p actually stands for p_{ideal} and V for V_{ideal}. The subscript "ideal" is omitted since only ideal gases are involved. For real gases, we can substitute the derived Equations (2.46) of p_{ideal} and (2.45) of V_{ideal} into the ideal gas equation as

$$p_{\text{ideal}} V_{\text{ideal}} = \left(p_{\text{real}} + \frac{an^2}{V_{\text{real}}^2}\right)(V_{\text{real}} - nb) = nRT$$

Omitting the subscript "real", we then have the famous **van der Waals equation** (Johannes van der Waals, 1837—1923) for real gases as

$$\left(p + \frac{an^2}{V^2}\right)(V - nb) = nRT \tag{2.47}$$

When $n = 1$, the volume of 1 mol molecules is the molar volume $V_{\text{m}} = V/n$. Van der Waals equation can be reformed into a more concise formula in terms of V_{m} as

$$\left(p + \frac{a}{V_{\text{m}}^2}\right)(V_{\text{m}} - b) = RT \tag{2.48}$$

a and b are called the **van der Waals constants** and should be positive by definition. The values of a and b vary with different gases, as listed in **Table 2.1**.

Table 2.1 The van der Waals Constants and Compressibility Factors for Some Gases
表 2.1 一些气体的范德华常数及压缩系数

Name (名称)	a/(bar L^2 mol^{-2})	b/(L mol^{-1})	Compressibility Factor (压缩系数) Z	Name (名称)	a/(bar L^2 mol^{-2})	b/(L mol^{-1})	Compressibility Factor (压缩系数) Z
H_2	0.2452	0.0265	1.006	C_3H_8	9.39	0.0905	
He	0.0346	0.0238	1.005	C_2H_6	5.580	0.0651	0.922
Ideal gas (理想气体)	0	0	1	CH_4	2.303	0.0431	0.983
N_2	1.370	0.0387	0.998	C_2H_4	4.612	0.0582	
CO	1.472	0.0395	0.997	F_2	1.171	0.0290	
O_2	1.382	0.0319	0.994	NO	1.46	0.0289	
O_3	3.570	0.0487		NO_2	5.36	0.0443	
Ne	0.208	0.0167		N_2O	3.852	0.0444	0.945
CO_2	3.658	0.0429	0.950	NF_3	3.58	0.0545	0.965
Cl_2	6.343	0.0542		SO_2	6.865	0.0568	
NH_3	4.225	0.0371	0.887	SF_6	7.857	0.0879	0.880
C_2H_2	4.516	0.0522		CCl_4	20.01	0.1281	
C_4H_{10}	13.89	0.1164					

Notice that the equations for real gases have many different formulas and van der Waals equation is only the most commonly used form of the formulas. It reproduces the observed behavior of real gases with

近、几乎彼此碰撞时，分子间作用力才表现为排斥力。而在碰撞过程之外，分子间作用力均表现为吸引力，是实际气体分子间的主要作用力。当一个分子（记为 A）与器壁碰撞时，其他分子对分子 A 的吸引力使其施加给器壁的力比吸引力不存在时更小。因此，实际气体的压强小于理想气体。如果将 $p_{理}$ 和 $p_{实}$ 的差值定义为 $p_{内}$，则

$$p_{实} = p_{理} - p_{内} < p_{理}$$

$p_{内}$ 是分子 A 与其他分子之间的吸引力所引起的内压，它正比于分子 A 和其他分子的浓度，可表示为

$$p_{内} \propto \left(\frac{n_A}{V}\right)\left(\frac{n_{其他}}{V}\right)$$

由于所有气体分子间均存在吸引力且所有分子都会与器壁碰撞，因此任何一个分子与其他分子均无法区分。任何分子都可以是"分子 A"，同时也可以是"其他分子"。"分子 A"和"其他分子"的浓度相同，都等于容器中气体分子的浓度，因此

$$p_{内} \propto \left(\frac{n}{V}\right)^2$$

令 $a = p_{内}/(n/V)^2$，则

$$p_{内} = a\left(\frac{n}{V}\right)^2$$

$$p_{理} = p_{实} + \frac{an^2}{V^2} \tag{2.46}$$

在**式（2.46）**中，V 应为 $V_{实}$，省略了下标"实"。

在极低温下，分子运动速率很慢，气体分子的平均平动能很小；相比于很小的动能（倾向于使气体膨胀），分子间吸引力（倾向于使气体聚集）很重要且不可忽略。在极高温下，气体分子快速运动，分子间吸引力可忽略。因此，气体的行为在低温下为非理想状态，高温下为理想状态。

范德华方程

在理想气体状态方程 $pV = nRT$ 中，p 实际上代表 $p_{理}$，V 代表 $V_{理}$，由于只涉及理想气体，所以省略了下标"理"。对于实际气体，将导出的 $p_{理}$ 表达式 (2.46) 和 $V_{理}$ 表达式 (2.45) 代入理想气体状态方程，可得

$$p_{理}V_{理} = \left(p_{实} + \frac{an^2}{V_{实}^2}\right)(V_{实} - nb) = nRT$$

省略下标"实"，即为著名的实际气体的**范德华方程**（约翰尼斯·范德华，1837—1923）：

$$\left(p + \frac{an^2}{V^2}\right)(V - nb) = nRT \tag{2.47}$$

当 $n = 1$ 时，1 mol 分子的体积即摩尔体积 $V_m = V/n$。范德华方程可改写为以 V_m 表示的更简明的形式：

$$\left(p + \frac{a}{V_m^2}\right)(V_m - b) = RT \tag{2.48}$$

a 和 b 称为**范德华常数**，根据定义应为正值。如**表 2.1** 所列，不同气

moderate accuracy. Van der Waals equation is most accurate for gases comprising approximately spherical molecules that have small dipole moments.

Compressibility Factor

When applying an ideal model back to the real object, some comparison between the ideal model and the real object is necessary. A generally applied method in research is to define a ratio between the real and ideal parameters so that the value of this ratio tells how far the real object deviates from the ideal model. For gases, in order to compare the behaviors of real gases and ideal gases, we define a **compressibility factor** Z as

$$Z = \frac{pV}{nRT} \quad (2.49)$$

For ideal gases, $pV = nRT$ and $Z = 1$ are valid under all conditions. For real gases, p and V should be the measured pressure p_{real} and volume V_{real}. According to van der Waals equation

$$nRT = \left(p_{real} + \frac{an^2}{V_{real}^2}\right)(V_{real} - nb) = p_{ideal} V_{ideal}$$

$$Z = \frac{p_{real} V_{real}}{nRT} = \frac{p_{real} V_{real}}{\left(p_{real} + \frac{an^2}{V_{real}^2}\right)(V_{real} - nb)} = \frac{p_{real} V_{real}}{p_{ideal} V_{ideal}}$$

For real gases, $Z \neq 1$ in most cases, and $Z = 1$ may accidentally hold under very special conditions. Therefore, the value of Z, or how far Z is away from 1, at a certain condition reveals how far the behaviors of this real gas at this condition deviates from the behaviors of ideal gases.

Figure 2.8 plots the compressibility factor for three real gases at 0 °C as a function of pressure. The ideal gas is represented by a horizontal dashed line at $Z = 1$. At very low pressure ($p \to 0$), all real gases tend to behave ideally with $Z \to 1$. As the pressure of a real gas increases, Z first decreases ($Z < 1$) and then increases ($Z > 1$). At very high pressure ($p \to +\infty$), all real gases tend to behave nonideally with $Z \to +\infty$.

To understand which parameter accounts for $Z < 1$ or $Z > 1$, we can mathematically split Z into two terms, as

$$Z = \frac{p}{p + \frac{an^2}{V^2}} \cdot \frac{V}{V - nb} \quad (2.50)$$

Since both a and b are positive values, both

$$\frac{p}{p + \frac{an^2}{V^2}} < 1 \quad \text{and} \quad \frac{V}{V - nb} > 1$$

are valid for all gases at all conditions. Whether $Z < 1$ or $Z > 1$ depends on which one of these two terms dominates over the other. $Z < 1$ when an^2/V^2 is nonnegligible compared to p and the intermolecular forces of attraction is significant, for example, for real gases at low temperature. $Z > 1$ when nb is nonnegligible compared to V and the molecular volume is significant, for example, for real gases at high pressure.

2.6 Properties of Liquids and Solids

As we have discussed in **Section 2.5**, the intermolecular forces of attraction play an important role in causing the behaviors of real gases to deviate from ideality. When these forces are sufficiently strong, a gas condenses to a liquid. Both liquids and solids are called the **condensed phases**, which are phases with significant intermolecular forces that is nonnegligible. The origin of intermolecular forces, which are the interactions between molecules, arises from the permanent and momentary unequal distribution of electron density within molecules. The intermolecular forces include but are not limited to van der Waals forces and hydrogen bonding. More discussion about intermolecular forces can be found in **Section 8.7**.

体具有不同的 a 和 b 值。

注意实际气体方程存在许多不同的公式，范德华方程只是这些公式中最常用的形式，它以中等精度呈现了观察到的实际气体的行为。对于由具有较小偶极矩的近球形分子构成的气体，范德华方程最为精确。

压缩系数

当把理想模型应用到实际对象时，理想模型与实际对象之间的比较是必需的。研究中普遍采用的方法是，定义一个实际参数与理想参数的比值，用该比值反映实际对象与理想模型的偏离程度。对于气体，为比较实际气体和理想气体的行为，可定义**压缩系数** Z 为

$$Z = \frac{pV}{nRT} \tag{2.49}$$

对于理想气体，$pV = nRT$ 和 $Z = 1$ 在所有条件下均成立。对于实际气体，p 和 V 应为测得的压强 $p_{实}$ 和体积 $V_{实}$。根据范德华方程

$$nRT = \left(p_{实} + \frac{an^2}{V_{实}^2}\right)(V_{实} - nb) = p_{理}V_{理}$$

$$Z = \frac{p_{实}V_{实}}{nRT} = \frac{p_{实}V_{实}}{\left(p_{实} + \frac{an^2}{V_{实}^2}\right)(V_{实} - nb)} = \frac{p_{实}V_{实}}{p_{理}V_{理}}$$

实际气体在大多数情况下 $Z \neq 1$，而只有在非常特殊的条件下，$Z = 1$ 才可能偶然成立。因此，在一定条件下的 Z 值，或者说 Z 偏离 1 的程度，反映了该条件下实际气体与理想气体行为的偏离程度。

图 2.8 给出了三种实际气体在 0℃ 下的压缩系数对压强的函数图像，理想气体用 $Z = 1$ 的水平虚线表示。极低压（$p \to 0$）下，所有实际气体行为均趋于理想状态，$Z \to 1$。随着实际气体压强的增加，Z 先减小（$Z < 1$）后增大（$Z > 1$）。极高压（$p \to +\infty$）下，所有实际气体行为均趋于非理想状态，$Z \to +\infty$。

为了理解哪些参数导致了 $Z < 1$ 或 $Z > 1$，可以从数学上将 Z 分解为两项：

$$Z = \frac{p}{p + \frac{an^2}{V^2}} \cdot \frac{V}{V - nb} \tag{2.50}$$

由于 a 和 b 均为正值，故

$$\frac{p}{p + \frac{an^2}{V^2}} < 1 \quad 且 \quad \frac{V}{V - nb} > 1$$

上式对所有气体在所有条件下均成立。$Z < 1$ 还是 $Z > 1$ 取决于这两项中的哪一项占主导。当 an^2/V^2 相比于 p 不可忽略，即分子间吸引力显著时，$Z < 1$，如低温下的实际气体。当 nb 相比于 V 不可忽略，即分子体积显著时，$Z > 1$，如高压下的实际气体。

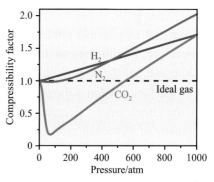

Figure 2.8 The behavior of real gases. The compressibility factor is plotted as a function of pressure for three different gases at 0℃.

图 2.8 实际气体的行为：0 ℃ 下三种不同气体的压缩系数随压强的变化图。

2.6 液体和固体的性质

正如 **2.5 节**中所讨论，分子间吸引力在导致实际气体行为偏离

In this section, we will briefly introduce some properties of liquids and solids that are related to the intermolecular forces. For gas molecules, we have learnt that the intermolecular forces of attraction are present at far distance and the repulsive forces are present only at very close distance. For liquids and solids, the attractive forces, which are even more significant than for gases, still serve as the major intermolecular forces. These intermolecular forces of attraction account for many properties of liquids and solids.

Surface Tension of Liquids

Have you ever observed the phenomena of a paper clip made of stainless-steel floating on water, as pictured in **Figure 2.9(a)**? The density of steel is much higher than that of water and steel should not float on water. It is the surface tension of water that overcomes the gravity of the paper clip, allowing it to remain suspended on the surface of water.

A liquid comprises interior molecules and surface molecules, which experience different intermolecular forces, as shown in **Figure 2.9(b)**. The interior molecules experience the intermolecular forces of attraction from all other molecules in all directions, and the net force is balanced. The molecules at the surface, on the other hand, are attracted only by other surface molecules and by molecules below the surface, neglecting the attractions from the vapor and air molecules. Thus, the surface molecules experience net attractive forces towards the bulk of liquid. These attractive forces hold the liquid together and explain why drops of a liquid are normally in spherical shape when other forces such as gravity are not taken into consideration.

We can also analyze this from a perspective of energy. Compared to surface molecules, the interior molecules have more neighbors and experience more attractive forces. These attractive forces make interior molecules more stable or in lower energy states than surface molecules. To lower the total energy, as many molecules as possible tend to enter the bulk of a liquid whereas as few as possible remain at the surface. Therefore, liquids tend to maintain a minimal surface area. To increase the surface area of a liquid requires additional energy or work to move molecules from the low-energy interior to the high-energy surface.

Surface tension is the energy, or work, required to increase the unit surface area of a liquid, often represented by the Greek letter γ (gamma). Surface tension has the SI unit of $J\ m^{-2}$. As temperature increases, so does the intensity of molecular motion, and the intermolecular forces of attraction become less significant. Less work is required to extend the surface of a liquid, meaning that surface tension decreases with increased temperature.

Surface tension not only explains the shape of a liquid drop but also determines the spreading behaviors of a liquid on a surface. If a liquid drop spreads into a film across a surface, we say that the liquid wets the surface. Whether a drop of liquid wets a surface or retains its spherical shape and stands on the surface depends on the strengths of two types of intermolecular forces: the **cohesive forces** (F_C) between the liquid molecules and the **adhesive forces** (F_A) between the liquid molecules and the surface. If cohesive forces are strong compared with adhesive forces, a drop maintains its spherical shape. Otherwise, the drop spreads into a film across the surface.

Water wets many surfaces, such as glass and certain fabrics, as shown on the left side of **Figure 2.10(a)**. In these cases, $F_A > F_C$. Changing water to mercury, the glass will not be wetted because the cohesive force in mercury, consisting of metallic bonds between Hg atoms, are strong and $F_C > F_A$. If glass is coated with a film of oil or grease, water no longer wets the surface and water droplets stand on the glass, as shown on the right side of **Figure 2.10(a)**. The intensity of F_C does not change since the properties of water retain, but the intensity of F_A decreases significantly because the intermolecular forces between water and oil are much weaker than those between water and glass, so that F_A becomes even weaker than F_C ($F_A < F_C$). You can understand this as if these two forces are "pulling" the surface water molecules in opposite directions. These water molecules will be pulled to the surface if F_A exceeds F_C, and thus spread with a large surface area and wet the surface. On the contrary, if F_C exceeds F_A water molecules will be pulled to the bulk of water and stand on the surface like droplets.

Surfactants are compounds that can significantly lower the surface tension of a liquid. For example, soap is a surfactant that can decrease the surface tension of water. If you dip a soap in the water with a paper clip floating on it, you will observe that the paper clip sinks to the bottom soon. This happens because the significantly lowered surface tension of water could no longer support the paper clip. Surfactants are usually organic compounds that are **amphiphilic**, meaning that they contain both hydrophobic groups ("tails")

理想状态时起到重要作用。当这些力足够强时，气体凝结为液体。液态和固态统称为**凝聚态**，是存在显著且不可忽略的分子间作用力的态。分子间作用力即分子之间的相互作用，源自分子内电子密度的永久和瞬时的不均匀分布。分子间作用力包括但不限于范德华力和氢键。关于分子间作用力的更多讨论详见 **8.7 节**。

本节我们将简要介绍液体和固体的一些与分子间作用力相关的性质。对于气体分子，我们已经了解分子间吸引力存在于较远距离，而排斥力仅存在于极近距离。对于液体和固体而言，吸引力仍是最主要的分子间作用力，甚至比在气体中更为显著，这些分子间吸引力可以解释液体和固体的许多性质。

液体的表面张力

你观察过如**图 2.9（a）**所示的不锈钢回形针漂浮在水面上的现象吗？钢的密度远大于水，因此不应该浮在水上。正是由于水的表面张力克服了回形针的重力，使其保持悬浮在水面上。

液体由内部分子和表面分子组成，它们受到不同的分子间作用力，如**图 2.9（b）**所示。内部分子受到来自各个方向的所有其他分子的吸引力，其合力是平衡的。忽略蒸气和空气分子的吸引力，表面分子只被其他表面分子及表面以下的分子吸引。因此，表面分子受到的净吸引力朝向液体内部，该净吸引力使液体聚集在一起，这也解释了在不考虑重力等其他力时，液滴为什么通常呈球形。

我们也可以从能量的角度来分析。与表面分子相比，内部分子有更多的近邻分子，受到了更多的吸引力。这些吸引力使内部分子比表面分子更稳定或者说处于能量更低的状态。为降低总能量，尽可能多的分子倾向于进入液体内部，而尽可能少的分子保留在表面。因此，液体趋向于保持最小的表面积。为增加液体的表面积，需要额外的能量或功才能将分子从能量较低的内部移到能量较高的表面。

表面张力是增加液体单位表面积所需的能量或功，通常用希腊字母 γ（gamma）表示，国际单位为 $J\ m^{-2}$。随着温度的升高，分子运动的强度也随之增加，分子间吸引力变得更不显著，扩展液体表面所需的功减少，这意味着表面张力随温度升高而变小。

表面张力不仅可以解释液滴的形状，还决定了液体在表面的铺展行为。如果液滴在表面铺展成膜，称液体浸润了表面。液滴是浸润表面还是保持球形立在表面上，取决于两种分子间作用力的强度：液体分子之间的**内聚力**（F_C）和液体分子与表面之间的**黏附力**（F_A）。如果内聚力强于黏附力，液滴将保持球形。反之，液滴将在表面铺展成膜。

水可以浸润玻璃和某些纤维等多种表面，如**图 2.10（a）**左图所示，此时 $F_A > F_C$。将水换成水银，玻璃将不会浸润，这是因为 Hg 原子间存在金属键，水银的内聚力很强，$F_C > F_A$。如果玻璃表面涂有一层油脂，则水不再浸润该表面，水滴会立在玻璃表面上，如**图 2.10（a）**右图所示。由于水的性质不变，F_C 强度不变，但水和油之间的分子间作用力远小于水和玻璃之间的作用力，F_A 强度显著降低，因此 F_A 变得比 F_C 更弱（$F_A < F_C$）。这可以理解为两个力朝相反方向"拉"表面的水分子，如果 F_A 超过 F_C，水分子将被拉到表面，铺展成较大的表面积并浸润表面。相反，如果 F_C 超过 F_A，水分子将被拉到内部，并像液滴一样立在表面上。

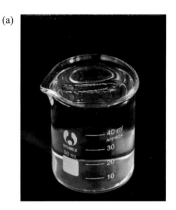

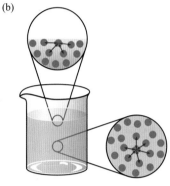

Figure 2.9 Surface tension illustrated. (a) A supported stainless-steel paper clip on the surface of water due to surface tension. (b) Difference in attractive forces experienced by interior and surface molecules.

图 2.9 表面张力示意图：(a) 不锈钢回形针由于表面张力漂浮在水面上；(b) 液体内部和表面分子所受吸引力的差别。

and hydrophilic groups ("heads"). Therefore, a surfactant contains both a water-insoluble (or oil-soluble) component and a water-soluble (or oil-insoluble) component. Surfactants will diffuse in water and adsorb at interfaces between air and water, or at the interface between oil and water in the case when water is mixed with oil. The water-insoluble hydrophobic group may extend out of the bulk water phase, into the air or into the oil phase, whereas the water-soluble head group remains in the water phase.

Surface tension or wetting behavior can also explain the meniscus formation [**Figure 2.10(b)**] or capillary phenomenon [**Figure 2.10(c)**]. Because water can wet glass, when water is put into a glass container, a **concave meniscus**, the bottom of which is below the level of water-glass contact line, is formed. When mercury is put into a glass container, since mercury does not wet glass, a **convex meniscus**, the top of which is above the level of mercury-glass contact line, is formed. Similarly, the water level inside a glass capillary tube is noticeably higher than outside, called the **capillary action**. Because of the strong adhesive forces between water and glass, a thin film of water spreads up the inside walls of the capillary and the pressure below the meniscus falls slightly. Atmospheric pressure then pushes a column of water up the tube to eliminate the pressure difference. The smaller the diameter of the capillary, the higher the column of water rises. Conversely, mercury in a glass capillary tube will have a lower level than the mercury outside the capillary.

Viscosity of Liquids

Viscosity is the resistance of a liquid to flow, often represented by the Greek letter η (eta) and with the SI unit of Pa s. When a liquid flows, one portion of the liquid moves with respect to neighboring portions. Cohesive forces within the liquid create an internal friction, which reduces the rate of flow. Therefore, the stronger the intermolecular forces of attraction, the greater the viscosity. Because intermolecular forces of attraction can be offset by higher molecular kinetic energies, viscosity generally decreases with increased temperature for liquids. One of the methods to measure the viscosity of a liquid is to time the fall of a steel ball through a certain depth of liquid. The greater the viscosity of the liquid, the longer it takes for the ball to fall.

Vaporization Enthalpy of Liquids

Vaporization or **evaporation** is the passage of molecules from the surface of a liquid into the gaseous state. To understand vaporization requires the basic understanding about Boltzmann distribution. Like the molecular speeds of gases can be represented statistically by Maxwell distribution, the energy of an ensemble of molecules can be represented statistically by **Boltzmann distribution**, as

$$f(\varepsilon) = AT^{-3/2}\varepsilon^{1/2}\exp(-\varepsilon/k_{B}T) \qquad (2.51)$$

where A is a constant given by $A = 2/\sqrt{\pi k_B^3}$, k_B is the Boltzmann constant, and $f(\varepsilon)$ is the probability density function that gives the probability, per unit energy, of finding a molecule with an energy near ε. Mathematically, the fraction of molecules ($\Delta N/N$) with energies between ε_1 and ε_2 is given by

$$\frac{\Delta N}{N} = \int_{\varepsilon_1}^{\varepsilon_2} f(\varepsilon)\,d\varepsilon \qquad (2.52)$$

which can also be represented by the total area under the $f(\varepsilon)$ curve between $\varepsilon = \varepsilon_1$ and $\varepsilon = \varepsilon_2$. As the fraction of molecules with all possible energies, from 0 to $+\infty$, must equal 1, we then have

$$\int_{0}^{+\infty} f(\varepsilon)\,d\varepsilon = 1 \qquad (2.53)$$

This means that the total area under the $f(\varepsilon)$ curve should be 1. In many cases, we are particularly interested in the fraction of molecules with energies above a certain value ε_0, which can be calculated as

$$\frac{\Delta N}{N} = \int_{\varepsilon_0}^{+\infty} f(\varepsilon)\,d\varepsilon \propto \exp(-\varepsilon_0/k_{B}T) \qquad (2.54)$$

as shown in **Figure 2.11**. More discussion about Boltzmann distribution can be found in **Section 3.5**.

If the minimum kinetic energy required for an ensemble of molecules to escape from the surface of a liquid is denoted by ε, only those molecules with energies higher than or equal to ε, the fraction of which is proportional to $\exp(-\varepsilon/k_{B}T)$, can vaporize into the gaseous state. Therefore, vaporization occurs more readily with:

表面活性剂是一类能显著降低液体表面张力的化合物，例如肥皂就是一种能降低水的表面张力的表面活性剂。如果把肥皂浸入漂浮着回形针的水中，回形针很快就会沉入水底。这是因为水的表面张力显著降低，无法再支撑回形针。表面活性剂通常都是**两亲性**的有机化合物，这意味着它们既包含疏水基团（尾基）又包含亲水基团（头基），因此同时含有非水溶（即油溶）组分和水溶（即非油溶）组分。表面活性剂可扩散进水中并吸附在空气与水的界面处，或者在油水混合时吸附在油和水的界面处。非水溶的疏水基团会从水相内部伸到空气或油相中，而水溶性的头基则保留在水相中。

表面张力或浸润行为也可解释弯液面的形成 [**图 2.10（b）**] 或毛细现象 [**图 2.10 (c)**]。当把水放入玻璃容器时，水浸润玻璃形成**凹液面**，其底部低于水和玻璃的接触线。当把水银放入玻璃容器中，水银不浸润玻璃形成**凸液面**，其顶部高于水银和玻璃的接触线。类似地，玻璃毛细管内的水位明显高于外部，称为**毛细作用**。由于水和玻璃之间存在较强的黏附力，一薄层水膜在毛细管内铺展开，弯液面下方的压强略微下降，随后大气压将水柱推上毛细管以消除压差。毛细管直径越小，水柱上升得越高。相反，玻璃毛细管中的水银液面会低于毛细管外。

液体的黏度

黏度是液体流动的阻力，通常用希腊字母 η（eta）表示，国际单位为 Pa s。当液体流动时，其中一部分相对其邻近部分移动，液体的内聚力产生内摩擦力，从而降低流速。因此分子间吸引力越强，黏度就越大。由于分子间吸引力可以被更高的分子动能所抵消，液体的黏度通常随温度升高而降低。测量液体黏度的方法之一，是使钢球在一定深度的液体中下落并计时，液体黏度越大，钢球下落所需的时间就越长。

液体的蒸发焓

蒸发是分子从液体表面变为气态的过程。为理解蒸发，需要对玻尔兹曼分布有一些基本的了解。就像气体分子的速率可用麦克斯韦分布从统计上表示一样，分子系综的能量也可以用**玻尔兹曼分布**从统计上表示如下：

$$f(\varepsilon) = AT^{-3/2}\varepsilon^{1/2}\exp(-\varepsilon/k_BT) \tag{2.51}$$

其中 $A = 2/\sqrt{\pi k_B^3}$ 为常数，k_B 是玻尔兹曼常数，$f(\varepsilon)$ 是概率密度函数，给出了找到能量处于 ε 附近单位能量间隔中的分子的概率。数学上，能量介于 ε_1 和 ε_2 之间的分子在所有分子中所占的分数（$\Delta N/N$）为

$$\frac{\Delta N}{N} = \int_{\varepsilon_1}^{\varepsilon_2} f(\varepsilon)d\varepsilon \tag{2.52}$$

它可以用 $f(\varepsilon)$ 曲线下在 $\varepsilon = \varepsilon_1$ 和 $\varepsilon = \varepsilon_2$ 之间的总覆盖面积来表示。由于具有全部可能能量（从 0 到 $+\infty$）的分子总占比必为 1，则有

$$\int_0^{+\infty} f(\varepsilon)d\varepsilon = 1 \tag{2.53}$$

这表明 $f(\varepsilon)$ 曲线下的总覆盖面积为 1。在许多情况下，我们对能量高于某一特定值 ε_0 的分子所占分数特别感兴趣，可以计算为

$$\frac{\Delta N}{N} = \int_{\varepsilon_0}^{+\infty} f(\varepsilon)d\varepsilon \propto \exp(-\varepsilon_0/k_BT) \tag{2.54}$$

如**图 2.11** 所示。关于玻尔兹曼分布的更多讨论详见 **3.5 节**。

Figure 2.10 Phenomena explained by surface tension. (a) Wetting (left) and nonwetting (right) of a surface. (b) Concave (left) and convex (right) meniscus formation. (c) Capillary action.

图 2.10 表面张力可解释的现象：(a) 表面浸润（左）和不浸润（右）行为；(b) 凹液面（左）和凸液面（右）的形成；(c) 毛细现象。
(Photos taken by Min Lyu with aid of Yanzi Ma.)

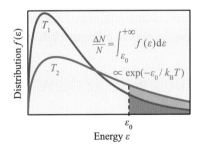

Figure 2.11 Boltzmann distribution of energy. The fraction of molecules ($\Delta N/N$) with energies above ε_0 at two different temperatures $T_1 < T_2$ are shown as shaded areas.

图 2.11 玻尔兹曼能量分布图：阴影面积表示两个不同温度（$T_1 < T_2$）下能量高于 ε_0 的分子在所有分子中所占的分数（$\Delta N/N$）。

1) increased temperature: more molecules have sufficient kinetic energy to overcome intermolecular forces of attraction in the liquid. The value of $\exp(-\varepsilon/k_B T)$ increases with increased temperature.
2) decreased strength of intermolecular forces: the kinetic energy needed to overcome intermolecular forces of attraction is less, and more molecules have enough energy to escape. The value of $\exp(-\varepsilon/k_B T)$ increases with decreased ε.
3) increased surface area of the liquid: a greater proportion of the liquid molecules are at the surface.

Because the molecules escaped through evaporation are much more energetic than average, the average kinetic energy of the remaining molecules decreases, and the temperature of the liquid falls. To vaporize a liquid at a constant temperature, the excess kinetic energy carried away by the vaporizing molecules must be compensated by adding heat to the liquid. The **enthalpy of vaporization** is the quantity of heat that must be absorbed if a certain quantity of liquid is vaporized at a constant temperature. Enthalpy of vaporization is often represented by ΔH_{vap} and has a common unit of kJ mol^{-1}. More discussion about the concept of enthalpy can be found in **Section 3.3**.

Because heat is required to add to vaporize molecules from liquid to gaseous state, vaporization is an **endothermic** process and ΔH_{vap} is always positive. The differences in ΔH_{vap} are the result of intermolecular forces. Generally, higher intermolecular forces of attraction yield higher ΔH_{vap}. The conversion of a gas or vapor to a liquid is called **condensation**. The **enthalpy of condensation** (ΔH_{cond}) is equal in magnitude but opposite in sign to ΔH_{vap} as

$$\Delta H_{cond} = -\Delta H_{vap}$$

Therefore, ΔH_{cond} is always negative and condensation is an **exothermic** process.

Vapor Pressure of Liquids

In a container with both liquid and vapor present, vaporization and condensation occur simultaneously. If sufficient liquid is present, eventually a **dynamic equilibrium** is reached in which the amount of vapor remains constant. Dynamic equilibrium always implies that two opposing processes are occurring simultaneously and at equal rates. As a result, there is no net change with time once equilibrium has been established. A symbolic representation of the liquid-vapor equilibrium is shown as

$$\text{liquid} \underset{\text{condensation}}{\overset{\text{vaporization}}{\rightleftharpoons}} \text{vapor}$$

If a dynamic equilibrium between vaporization and condensation is reached, the equilibrium pressure of vapor is called the **vapor pressure** of the liquid. Liquids with high vapor pressures at room temperature are said to be **volatile**, and those with very low vapor pressures are **nonvolatile**. Whether a liquid is volatile or not is determined primarily by the strengths of its intermolecular forces. The weaker these forces are, the more volatile the liquid is and the higher its vapor pressure is.

As an excellent approximation, the vapor pressure of a liquid depends only on the particular liquid and its temperature. Vapor pressure depends on neither the amount of liquid nor that of vapor, as long as some of each is present at equilibrium. A graph of vapor pressure as a function of temperature is known as a **vapor pressure curve**, as shown in **Figure 2.12(a)**. The vapor pressure of a liquid increases exponentially with temperature, as

$$p = C\exp(-\Delta H_{vap}/RT) \tag{2.55}$$

By taking the natural logarithm (ln) of the vapor pressure, we have

$$\ln p = -\frac{\Delta H_{vap}}{RT} + \ln C \tag{2.56}$$

If plotting $\ln p$ as a function of $1/T$, the relationship is of a straight line with a slope of $-\Delta H_{vap}/R$ and an intercept of $\ln C$, as shown in **Figure 2.12(b)**. It is customary to eliminate $\ln C$ by subtracting **Equation (2.56)** for two different temperatures, resulting in a form called **Clausius-Clapeyron equation** (Rudolf Clausius, 1822—1888; Benoit Clapeyron, 1799—1864) as

$$\ln\frac{p_2}{p_1} = -\frac{\Delta H_{vap}}{R}\left(\frac{1}{T_2}-\frac{1}{T_1}\right) \tag{2.57}$$

如果用 ε 表示某系综里分子从液体表面逃逸所需的最小动能，则只有那些能量大于或等于 ε 的分子才可以蒸发进入气态，其在所有分子中所占的分数与 $\exp(-\varepsilon/k_\text{B}T)$ 成正比。因此，蒸发在以下情况更容易发生：

1) 温度升高：更多分子具有足够的动能以克服液体中的分子间吸引力，$\exp(-\varepsilon/k_\text{B}T)$ 随温度升高而增加。
2) 分子间作用力强度降低：克服分子间吸引力所需的动能减少，更多分子具有足够的能量逃逸，$\exp(-\varepsilon/k_\text{B}T)$ 随 ε 降低而增加。
3) 液体表面积增加：位于表面的液体分子所占的分数更大。

由于蒸发过程中逃逸分子的能量远大于平均能量，剩余分子的平均动能降低，液体的温度下降。为在恒温下蒸发液体，必须通过给液体增加热量来补偿蒸发掉的分子带走的多余动能。**蒸发焓**是在恒温下蒸发一定量液体所需吸收的热量，通常用 $\Delta H_\text{蒸发}$ 表示，常用单位为 kJ mol^{-1}。关于焓的概念的更多讨论详见 **3.3 节**。

由于将分子从液态蒸发到气态需要加热，因此蒸发是一个**吸热**过程，蒸发焓始终为正值。蒸发焓的差异是分子间作用力的结果，较高的分子间吸引力通常导致较高的蒸发焓。将气体或蒸气转化为液体的过程称为**凝结**，凝结焓 ($\Delta H_\text{凝结}$) 与蒸发焓大小相等、符号相反：

$$\Delta H_\text{凝结} = -\Delta H_\text{蒸发}$$

因此，凝结焓始终为负值，凝结是一个**放热**过程。

液体的蒸气压

在同时存在液体和蒸气的容器中，蒸发和凝结总是同时发生。如果存在足够的液体，终将达到**动态平衡**，蒸气的量保持不变。动态平衡总是意味着两个相反的过程以相等的速率同时发生，因此一旦建立平衡，就不会有随时间的净变化。气液平衡可用如下符号表示：

$$\text{液体} \underset{\text{凝结}}{\overset{\text{蒸发}}{\rightleftharpoons}} \text{蒸气}$$

如果在蒸发和凝结之间达到动态平衡，蒸气的平衡压强称为液体的**蒸气压**。室温下具有较高蒸气压的液体称为**挥发性**液体，而具有极低蒸气压的液体称为**非挥发性**液体。液体是否易挥发主要取决于其分子间作用力的强度，分子间作用力越弱，液体越易挥发，蒸气压越高。

作为良好的近似，液体的蒸气压只取决于液体自身及其温度。只要平衡时既有液体又有蒸气，蒸气压就与液体的量以及蒸气的量均无关。蒸气压随温度变化的曲线称为**蒸气压曲线**，如**图 2.12（a）**所示。液体的蒸气压随温度升高呈指数级增加

$$p = C\exp(-\Delta H_\text{蒸发}/RT) \quad (2.55)$$

对蒸气压取自然对数（ln），有

$$\ln p = -\frac{\Delta H_\text{蒸发}}{RT} + \ln C \quad (2.56)$$

以 $\ln p$ 对 $1/T$ 作图，可得一条斜率为 $-\Delta H_\text{蒸发}/R$、截距为 $\ln C$ 的直线，如**图 2.12（b）**所示。通常取两种不同温度下的**式 (2.56)** 相减以消除 $\ln C$，可得如下形式的**克劳修斯-克拉贝龙方程**（鲁道夫·克劳修斯，1822—1888；伯努瓦·克拉贝龙，1799—1864）：

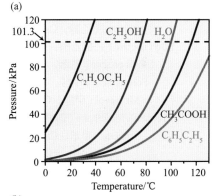

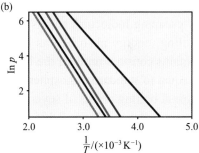

Figure 2.12 Vapor pressure curves of diethyl ether ($C_2H_5OC_2H_5$), ethanol (C_2H_5OH), water (H_2O), acetic acid (CH_3COOH), and ethylbenzene ($C_6H_5C_2H_5$). (a) The vapor pressure is plotted with respect to temperature. The normal boiling points are the corresponding temperatures at the intersection of the dashed horizontal line at p = 101.3 kPa with the vapor pressure curves. (b) ln$p$ is plotted with respect to $1/T$.

图 2.12 乙醚($C_2H_5OC_2H_5$)、乙醇(C_2H_5OH)、水(H_2O)、乙酸(CH_3COOH)和乙苯($C_6H_5C_2H_5$)的蒸气压曲线：(a) 蒸气压对温度作图。正常沸点为 p = 101.3 kPa 的水平虚线与蒸气压曲线的交点所对应的温度；(b) ln$p$ 对 $1/T$ 作图。

Note that ΔH_{vap} is actually not a constant but dependent on temperature. However, the dependence of ΔH_{vap} on temperature is not significant. If the temperature change is not considerable in a process, meaning that T_2 does not differ appreciably from T_1, ΔH_{vap} can be viewed as a constant and the above Clausius-Clapeyron equation is validly applicable. **Table 2.2** lists the enthalpies of vaporization of some common liquids at 298 K and at their normal boiling points.

The (saturated) vapor pressure of water is the highest partial pressure of water vapor that can be maintained at a certain temperature. If a sample of air at T develops a water vapor pressure higher than the saturated vapor pressure, it is expected that some of the vapor to condense to liquid water. **Relative humidity** (RH) is the ratio of the partial pressure of water vapor to the saturated vapor pressure of water at the same temperature, expressed on a percent basis. Water vapor with the same partial pressure results in higher relative humidity in cool air than in warm air.

Table 2.2 Enthalpies of Vaporization (ΔH_{vap}) of Some Common Liquids at 298 K and at the Normal Boiling Points (T_b)
表 2.2 一些常见液体在 298 K 及正常沸点（T_b）下的蒸发焓（$\Delta H_{蒸发}$）

Name (Formula)[名称（化学式）]	ΔH_{vap}(298 K)/(kJ mol^{-1})	$\Delta H_{vap}(T_b)$/(kJ mol^{-1})	T_b/K
Acetic acid (乙酸，CH_3COOH)	23.36	23.70	391.1
Acetone (丙酮，CH_3COCH_3)	30.99	29.10	329.20
Benzene (苯，C_6H_6)	33.83	30.72	353.24
Diethyl ether (乙醚，$C_2H_5OC_2H_5$)	27.10	26.52	307.7
Ethanol (乙醇，C_2H_5OH)	42.32	38.56	351.44
Ethyl acetate (乙酸乙酯，$CH_3COOC_2H_5$)	35.60	31.94	350.26
Ethylbenzene (乙苯，$C_6H_5C_2H_5$)	42.24	35.57	409.31
Formic acid (甲酸，$HCOOH$)	20.10	22.69	374
Methanol (甲醇，CH_3OH)	37.43	35.21	337.8
Tetrachloromethane (四氯化碳，CCl_4)	32.43	29.82	350.0
Toluene (甲苯，$C_6H_5CH_3$)	38.01	33.18	383.78
Water (水，H_2O)	43.98	40.65	373.12

When a liquid is heated in a container open to the atmosphere, there is a particular temperature at which vaporization occurs throughout the liquid rather than simply at the surface. If the pressure exerted by escaping molecules equals that exerted by molecules of the atmosphere, **boiling** is said to occur. During boiling, energy absorbed as heat is used only to convert molecules of liquid to vapor. The temperature remains constant until all the liquid has boiled away. The temperature at which the vapor pressure of a liquid is equal to standard atmospheric pressure (1 atm = 760 mmHg = 101.3 kPa) is the normal **boiling point**. The normal boiling points of several liquids can be determined from the intersection of the dashed horizontal line at p = 1 atm in **Figure 2.12(a)** with the vapor pressure curves for the liquids. The boiling point of a liquid varies significantly with barometric pressure. Shift the dashed horizontal line to higher or lower pressures, and the new points of intersection with the vapor pressure curves appear at different temperatures. This explains lower boiling point of water at high altitudes where the barometric pressures are below 1 atm.

Some Properties of Solids

Here, we only introduce some properties of solids that are directly related to the liquid or gaseous states of matter. Some other properties of solids such as ductility and conductivity will be discussed in **Section 8.6**.

As a crystalline solid is heated, its atoms, ions, or molecules vibrate more vigorously. Eventually a temperature is reached at which these vibrations disrupt the ordered crystalline structure. The atoms, ions, or molecules can slip past one another, and the solid loses its definite shape and is converted to a liquid. This process is called **melting**, or fusion, and the temperature at which it occurs is the **melting point**. The reverse process, the conversion of a liquid to a solid, is called **freezing**, or solidification, and the temperature at which it occurs is the **freezing point**. The melting point of a solid and the freezing point of its liquid are identical. At this temperature, solid and liquid coexist in a dynamic equilibrium. The process of melting is endothermic and the process of freezing is exothermic. The quantity of heat required to melt a certain quantity of solid at a constant temperature is the **enthalpy of fusion** (ΔH_{fus}).

$$\ln\frac{p_2}{p_1} = -\frac{\Delta H_{蒸发}}{R}\left(\frac{1}{T_2}-\frac{1}{T_1}\right) \qquad (2.57)$$

注意 $\Delta H_{蒸发}$ 实际上并不是常数，而与温度相关，但 $\Delta H_{蒸发}$ 对温度的依赖性并不显著。如果过程中温度变化不大，即 T_2 与 T_1 没有明显差异，则 $\Delta H_{蒸发}$ 可视为常数，上述克劳修斯-克拉贝龙方程可有效适用。**表 2.2** 列出了一些常见液体在 298 K 及其正常沸点下的蒸发焓。

水的（饱和）蒸气压是水蒸气在一定温度下可维持的最高分压。如果温度为 T 的空气样品中水的蒸气压高于其饱和蒸气压，则部分水蒸气将凝结为液态水。**相对湿度**（RH）是相同温度下水蒸气的分压与其饱和蒸气压的比值，以百分数表示。分压相等的水蒸气在冷空气中的相对湿度高于暖空气中。

当液体在敞口容器中加热时，某个特定温度下蒸发不仅发生在表面，还会发生在整个液体中。当逃逸分子的压强等于大气压时，液体即会**沸腾**。沸腾时吸收的热能仅用于将液体分子转化为蒸气而温度保持不变，直至液体全部沸腾。液体的蒸气压等于标准大气压（1 atm = 760 mmHg = 101.3 kPa）时的温度，即为其正常**沸点**。从**图 2.12（a）**中 $p=1$ atm 的水平虚线与几种液体蒸气压曲线的交点，可得这几种液体的正常沸点。液体的沸点随蒸气压变化很大，将水平虚线移至更高或更低压强处，与蒸气压曲线的新交点将出现在不同的温度下。这解释了在大气压低于 1 atm 的高海拔地区，水的沸点变低。

固体的性质

这里我们只介绍固体的一些与物质的液态或气态直接相关的性质。关于固体其他性质（如延展性和导电性）的更多讨论详见 **8.6 节**。

加热晶态固体时，其原子、离子或分子振动得更剧烈，最终将达到一个振动会破坏有序晶体结构的温度。此时原子、离子或分子可以彼此滑动，固体将失去其特定形状并转化为液体，这个过程称为**熔化**或融化，发生的温度称为**熔点**。其逆过程即液体转化为固体的过程称为**凝固**或固化，发生的温度即为**凝固点**。固体的熔点和液体的凝固点相同，在此温度下固态和液态以动态平衡的形式共存。熔化过程吸热，凝固过程放热。恒温下熔化一定量固体所需的热量即为**熔化焓**（$\Delta H_{熔化}$）。

将固体从熔点以下加热至转变为略高于熔点的液体，追踪此过程的温度变化，可绘制出一张温度对时间变化的**加热曲线**，熔化时的温度保持不变。反向进行此过程，从液体开始将其冷却为固体，即可得**冷却曲线**。冷却曲线的形状理论上应该是从左至右翻转的加热曲线，但实验测定的冷却曲线却并非如此。冷却时温度可能降至凝固点以下而不出现任何固体，这种情况称为**过冷**。要在凝固点时从液体开始形成晶态固体，液体中必须含有一些小微粒，晶体才能从这些小微粒上开始生长。如果液体中含有的微粒数极少，它在凝固之前可能会过冷一段时间。一旦过冷液体开始凝固，温度会在凝固完成时回升到正常凝固点。

与液体一样，固体也能释放蒸气，由于存在更强的分子间作用力，固体通常不如液体易挥发。分子从固态直接进入蒸气状态称为**升华**，发生的温度称为**升华点**。其逆过程即分子从蒸气状态进入固态的过程称为**凝华**。当升华和凝华以相同速率发生时，固体与其蒸气之间存在动态平衡，此时蒸气具有的特征压强称为**升华压**，升华压随温

If tracing the changes of temperature that occur as a solid is heated from below the melting point to produce a liquid somewhat above the melting point, a **heating curve** of temperature versus time is produced. The temperature remains constant while melting occurs. Run this process backward by starting with a liquid and cooling it to a solid, a **cooling curve** is obtained. Theoretically, the appearance of the cooling curve is that of a heating curve flipped from left to right. However, an experimentally determined cooling curve does not look like that. The temperature may drop blow the freezing point without any solid appearing. This condition is known as **supercooling**. For a crystalline solid to start forming from a liquid at the freezing point, the liquid must contain some small particles on which crystals can grow. If a liquid contains a very limited number of particles, it may supercool for a time before freezing. When a supercooled liquid does begin to freeze, however, the temperature rises back to the normal freezing point while freezing is completed.

Like liquids, solids can also give off vapors, although solids are generally not as volatile as liquids due to the stronger intermolecular forces. The direct passage of molecules from the solid to the vapor state is called **sublimation**. The temperature at which sublimation occurs is the **sublimation point**. The reverse process, the passage of molecules from the vapor to the solid state, is called **deposition**. When sublimation and deposition occur at equal rates, a dynamic equilibrium exists between a solid and its vapor, which exerts a characteristic pressure called the **sublimation pressure**. A plot of sublimation pressure as a function of temperature is called a **sublimation pressure curve**. The **enthalpy of sublimation** (ΔH_{sub}) is the quantity of heat needed to convert a certain quantity of solid to vapor at a constant temperature. At the sublimation point, sublimation (solid → vapor) is equivalent to melting (solid → liquid) followed by vaporization (liquid → vapor). This suggests the following relationship among ΔH_{sub}, ΔH_{fus}, and ΔH_{vap}:

$$\Delta H_{sub} = \Delta H_{fus} + \Delta H_{vap} \tag{2.58}$$

The value of ΔH_{sub} obtained with **Equation (2.58)** can replace ΔH_{vap} in Clausius-Clapeyron **Equation (2.57)**, so that sublimation pressures can be calculated as a function of temperature as

$$\ln p = -\frac{\Delta H_{sub}}{RT} + \ln C \tag{2.59}$$

and

$$\ln \frac{p_2}{p_1} = -\frac{\Delta H_{sub}}{R}\left(\frac{1}{T_2} - \frac{1}{T_1}\right) \tag{2.60}$$

Three familiar solids with significant sublimation pressures are ice, dry ice (solid CO_2), and iodine. In a cold climate, snow may disappear from the ground even though the temperature is below 0°C. Under these conditions, the snow does not melt but rather sublimes. The sublimation pressure of ice at 0°C is 4.58 mmHg. If the air is not already saturated with water vapor, the ice will sublime.

2.7 Phase Transition and Phase Diagram

Under certain pressure and temperature conditions, the three common states of matter (gas, liquid, and solid) can transform from and into one another. The transitions between gaseous, liquid, and solid states of matter are called **phase transition** or **phase change**. Note that although the terms "phase" and "state of matter" tend to be used synonymously, there is a small distinction between them. The states of matter are distinctively referred to solid, liquid, and gaseous states. A **phase** is a region in a system throughout which all physical and chemical properties of a material are essentially uniform. There can be several immiscible phases of the same state of matter. For example, a mixture of water and oil, although both in liquid state, will spontaneously separate into two phases. The dynamic equilibrium between two phases during a phase transition is called **phase equilibrium**. A chart showing the conditions at which thermodynamically distinct phases existor coexist at equilibrium is called a **phase diagram**. Here, we only introduce the simplest phase diagram: the pressure-temperature diagram of a simple pure substance.

度变化的曲线称为**升华压曲线**。恒温下将一定量固体转化为蒸气所需的热量称为**升华焓**（$\Delta H_{升华}$）。在升华点升华（固体→蒸气）的过程相当于先熔化（固体→液体）再蒸发（液体→蒸气）的过程。这表明 $\Delta H_{升华}$、$\Delta H_{熔化}$ 和 $\Delta H_{蒸发}$ 之间存在如下关系：

$$\Delta H_{升华} = \Delta H_{熔化} + \Delta H_{蒸发} \tag{2.58}$$

从**式（2.58）**中得到的 $\Delta H_{升华}$ 可以替代克劳修斯 - 克拉贝龙方程 [**式（2.57）**] 中的 $\Delta H_{蒸发}$，因此可通过下式计算升华压对温度的函数：

$$\ln p = -\frac{\Delta H_{升华}}{RT} + \ln C \tag{2.59}$$

$$\ln \frac{p_2}{p_1} = -\frac{\Delta H_{升华}}{R}\left(\frac{1}{T_2} - \frac{1}{T_1}\right) \tag{2.60}$$

冰、干冰（固态 CO_2）和碘是三种常见的具有显著升华压的固体。即使在温度低于 0ºC 的寒冷天气中，雪也可能从地面消失，这时雪并非融化了而是升华了。冰在 0ºC 下的升华压为 4.58 mmHg，如果空气中水蒸气的分压低于升华压，冰就会升华。

2.7 相变与相图

在一定压强和温度条件下，物质的三种常见状态（气、液和固态）可以相互转化。物质在气、液、固态之间的转变称为**相变**。注意"相"和"物质状态"这两个术语在通常语境下可互换使用，但它们之间还存在一点小区别。物质状态（简称物态）特指固、液、气等状态，而**相**指的是体系中材料的物理和化学性质完全均匀一致的部分，同一状态的物质可能存在几个不互溶的相。例如，水和油的混合物虽然都是液态，但会自发地分成两相。相变过程中两相之间的动态平衡称为**相平衡**。用来表示热力学上不同相单独存在或平衡共存条件的图称为**相图**。这里我们只介绍最简单的相图：单一纯净物的外压 - 温度图。

单相区域

图 2.13（a）为某单一纯净物的相图示意图，y 轴和 x 轴分别表示体系的外压和温度。相图中点的分布给出了单相区的轮廓，绿色、蓝色和黄色点坐标分别代表固、液和气相为稳定相时的外压和温度。如 X 表示气相区域中外压为 p_X、温度为 T_X 的一个点，这意味着当外压恒为 p_X、温度恒为 T_X 时，不论花费多长时间，该物质终将全部变成气相。也就是说，在 p_X 和 T_X 的条件下，该物质在热力学上以纯气相的形式存在。

在外压 - 温度图中，固相区通常位于左上角，液相区位于中上部，而气相区位于右下部。相图的右上角形成了一种称为超临界流体的特殊物态，在**图 2.13（a）**中用紫色点表示，超临界流体将在本节后续内容中讨论。

直线和曲线

将**图 2.13（a）**中相同颜色的点连接成区域，即可得如**图 2.13（b）**

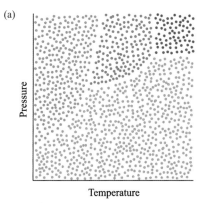

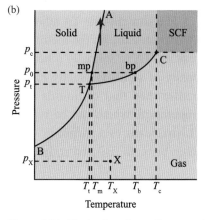

Figure 2.13 Illustrative phase diagrams of a simple pure substance. (a) The outline of a phase diagram is suggested by the distribution of points. (b) The different regions of a phase diagram correspond to single phases. Green, blue, yellow, and purple represent solid, liquid, gas, and supercritical fluid phases, respectively. Abbreviation: T-triple point; C-critical point; mp-melting point; bp-boiling point; SCF-supercritical fluid.

图 2.13 单一纯净物的相图示意图：(a) 相图的轮廓由点的分布给出；(b) 相图上的不同区域对应不同的单相区。绿色、蓝色、黄色和紫色分别代表固相、液相、气相和超临界流体相。缩写：T—三相点；C—临界点；mp—熔点；bp—沸点；SCF—超临界流体。

Single-Phase Regions

Figure 2.13(a) shows illustrative phase diagrams of a simple pure substance. The y and x axes indicate the external pressure and the temperature of the system, respectively. The distribution of points in the phase diagram suggests the outline of single-phase regions. The green, blue, and yellow points identify the external pressures and temperatures at which solid, liquid, and gas are the stable phases, respectively. For example, X indicates a point in the gas region with an external pressure of p_X and a temperature of T_X. This means that holding a constant external pressure at p_X and a constant temperature at T_X, this substance will eventually all become gas, no matter how long this process is taken. That is, under the conditions of p_X and T_X, this substance thermodynamically exists as a pure gaseous state.

In a pressure-temperature diagram, solid phase generally located in the upper-left corner, liquid in the upper-middle part, and gas in the bottom-right part. In the upper-right corner of a phase diagram, a special state of matter called supercritical fluid is formed, which is represented by purple points in **Figure 2.13(a)** and will be discussed later in this section.

Straight and Curved Lines

Connecting the points of the same color in **Figure 2.13(a)** into regions, we then have the simplified phase diagram shown in **Figure 2.13(b)**. Here, straight (TA) or curved (TB and TC) lines where single-phase regions adjoin represent equilibria between the two corresponding phases. All points on these lines denote the dynamic equilibrium states in which the two corresponding phases coexist, and also indicate the boundary conditions of the two corresponding phases.

The curved line TB adjoining gas phase with solid phase is the sublimation pressure curve, and can be described by Clausius-Clapeyron equation of sublimation with ΔH_{sub}. The curved line TC adjoining gas phase with liquid phase is the vapor pressure curve, and can be described by Clausius-Clapeyron equation of vaporization with ΔH_{vap}. There must always be a discontinuity between these two curves, since these two equations have different slopes.

The straight-line TA adjoins solid and liquid phases and is the fusion curve. It represents the effect of external pressure on the melting point of the substance. Since melting is not significantly affected by external pressure, TA is normally a nearly vertical line. The slope of TA is positive for most substances, because the density of solids is usually higher than that of liquids, and consequently, the volume of solids is generally smaller than that of liquids for most substances. Therefore, increasing external pressure at a constant temperature, which can be represented by a short vertical upward arrow across the fusion curve in **Figure 2.13(b)**, leads to a shrink in volume and a transition from liquid to solid. Some other substances, such as water, bismuth (Bi) and antimony (Sb), show unusual negative slope of TA.

Triple Point

In **Figure 2.13(b)**, the point T is called the **triple point** and defines the unique pressure (p_t) and temperature (T_t) at which the three common states of matter coexist in equilibrium. It is also the intersection of TA, TB, and TC. At the triple point, the vapor pressure of the liquid and the sublimation pressure of the solid are the same, both equaling the triple point pressure p_t.

The melting point (T_m) and boiling point (T_b) of a pure substance can also be read directly from its phase diagram. The intersection between the horizontal $p = 1$ atm line and the fusion curve defines the normal melting point of this pure substance. Although numerically T_m may not differ significantly from the triple point temperature T_t, the physical meanings are totally different. The intersection between the horizontal line at $p = 1$ atm and the vapor pressure curve defines the normal boiling point. For some substances, such as CO_2, the triple point pressure p_t is greater than 1 atm, which means that the horizontal $p = 1$ atm line do not intersect the fusion curve and the vapor pressure curve but the sublimation curve. This intersection then defines the normal sublimation point (T_s) and explains the sublimation behavior of those substances at atmospheric pressure.

Critical Point and Supercritical Fluids

Another special point in **Figure 2.13(b)** is the **critical point** C. It is the highest point on the vapor

所示的简化相图。这里连接单相区的直线（TA）或曲线（TB 和 TC）代表相应的两相恰好处于平衡。这些线上的所有点表示处于两个对应相共存的动态平衡状态，也代表两个对应相存在的边界条件。

气相与固相邻接的曲线 TB 即为升华压曲线，可用带 $\Delta H_{升华}$ 的克劳修斯-克拉贝龙升华方程描述。气相与液相邻接的曲线 TC 为蒸气压曲线，可用带 $\Delta H_{蒸发}$ 的克劳修斯-克拉贝龙蒸发方程描述。由于这两个方程具有不同的斜率，因此这两条曲线一定是不连续的。

固相与液相邻接的直线 TA 为熔化曲线，代表外压对物质熔点的影响。由于外压对熔化没有显著影响，TA 通常为一条近似竖直的直线。对于大多数物质，TA 的斜率为正，因为固体的密度通常高于液体，大多数物质固态的体积小于液态。因此恒温下增加外压会导致体积收缩，发生从液态到固态的转变，这一过程可用**图 2.13（b）**中的一条穿过熔化曲线且竖直向上的短箭头表示。水、铋（Bi）和锑（Sb）等一些物质的 TA 线具有反常的负斜率。

三相点

在**图 2.13（b）**中，T 点称为**三相点**，定义了物质的气、液、固三相处于平衡共存时独特的外压（p_t）和温度（T_t），它也是 TA、TB 和 TC 的交点。在三相点处，液体的蒸气压和固体的升华压相等，均等于三相点的外压 p_t。

纯净物的熔点（T_m）和沸点（T_b）也可直接从相图中读取。$p = 1$ atm 的水平线与熔化曲线之间的交点定义了该纯净物的正常熔点。虽然数值上 T_m 可能与三相点温度 T_t 没有显著差异，但物理意义完全不同。$p=1$ atm 的水平线与蒸气压曲线的交点定义了正常沸点。对于某些物质如 CO_2，三相点外压 p_t 大于 1 atm，这意味着 $p=1$ atm 的水平线不与熔化曲线和蒸气压曲线相交，而是与升华曲线相交，其交点定义了正常升华点（T_s），并解释了这些物质在常压下的升华行为。

临界点与超临界流体

图 2.13（b）中的另一个特殊点是**临界点** C，它是蒸气压曲线上的最高点，代表了液相可能存在的最高温度。气体只能在低于**临界温度** T_c 时发生液化；当温度在 T_c 以上，无论如何加压均不能使气体液化。在 T_c 下使气体液化所需的最低外压称为**临界压强** p_c。在 T_c 和 p_c 下气体的摩尔体积称为**临界体积** V_c。**表 2.3** 列出了一些常见物质的临界数据。

如果气体的 T_c 高于室温（25ºC），则在室温时只需施加高于 p_c 的外压即可实现液化，这类气体称为**可凝聚气体**或**非永久气体**。但如果气体的 T_c 低于室温，仅靠加压不能使其液化，只有同时将温度降至 T_c 以下并将外压增至 p_c 以上，才能使其液化，这类气体称为**永久气体**。物质的沸点 T_b 总是低于临界温度 T_c，如果 T_c 和 T_b 均高于室温，则该物质在室温和常压下即为液体或固体。

此外，"气相"和"蒸气"这两个词的差别也与临界点相关。蒸气指的是处于临界温度以下的气相物质，通过简单加压而无须降温即可凝结成液体。因此，临界温度同样也是蒸气所能存在（或者与液相或固相共存）的最高温度。当指代位于相图右下部的单相区时，应该用"气相区"而非"蒸气区"。

pressure curve and represents the highest temperature at which the liquid can exist. A gas can be liquefied only at temperatures below its **critical temperature** T_c. Above T_c, the gas cannot be liquefied no matter how high the external pressure is applied. The minimum external pressure required to liquefy a gas at T_c is called the **critical pressure** p_c. The molar volume of a gas at T_c and p_c is called the **critical volume** V_c. **Table 2.3** lists the critical data of some common substances.

Table 2.3 The Critical Constants of Some Common Substances
表 2.3 一些常见物质的临界常数

Substance (物质)	T_b/K	T_c/K	p_c/bar	V_c/(cm^3 mol^{-1})	Substance (物质)	T_b/K	T_c/K	p_c/bar	V_c/(cm^3 mol^{-1})
Permanent Gases (永久气体)									
He	4.222	5.1953	51.953	0.22746	F_2	85.04	144.41	51.724	66
H_2	20.388	33.14	12.964	65	Ar	87.302	150.687	48.63	75
Ne	27.097	44.49	26.786	42	O_2	90.188	154.581	50.43	73
N_2	77.355	126.192	33.9	90	CH_4	111.6	190.56	46.0	99
Nonpermanent Gases (非永久气体)									
CO_2	194.6	304.13	73.75	94	NH_3	239.82	405.56	113.57	69.8
C_2H_6	184.5	305.36	48.8	146	Cl_2	239.11	417.0	79.91	123
HCl	188	324.7	83.1	81	C_4H_{10}	272.6	425.2	37.9	257
C_3H_8	231.04	369.9	42.5	199					
Liquids or Solids (液体或固体)									
C_5H_{12}	309.21	469.7	33.7	310	C_6H_6	353.23	562.0	49.0	257
C_6H_{14}	341.87	507.5	30.3	366.0	H_2O	373.12	647.10	220.6	56
C_7H_{16}	371.53	540.1	27.4	428	I_2	457.6	819	117	155

If T_c of a gas is higher than room temperature (25 °C), the liquefaction of this gas can be accomplished at room temperature just by applying an external pressure higher than its p_c. This gas is classified as a **condensable gas** or a **nonpermanent gas**. If T_c of a gas is lower than room temperature, however, simply applying an external pressure cannot lead to the liquefaction of this gas. Simultaneously lowering temperature to a value below T_c and adding external pressure to a value above p_c are required for its liquefaction. This gas is classified as a **permanent gas**. The boiling point T_b of a substance is always lower than its critical temperature T_c. If both T_c and T_b are higher than room temperature, this substance is then a liquid or a solid at room temperature and atmospheric pressure.

In addition, the difference between "gas" and "vapor" is also related to the critical point. A vapor is a gaseous substance at temperatures below its critical temperature, meaning that the vapor can be condensed into a liquid by simply increasing the external pressure and without reducing the temperature. Therefore, the critical temperature is also the highest temperature at which a vapor can exist, or co-exist with a liquid or a solid. We should use "gas phase" instead of "vapor phase" to refer to the single-phase region in the bottom-right part of the phase diagram.

If a liquid is heated slowly and continuously in a sealed container, its temperature and vapor pressure will rise gradually along the vapor pressure curve in the phase diagram, and the liquid and gas phases will be in continuous dynamic equilibrium. In this heating process, the density of the liquid decreases, and that of the vapor increases. Eventually, when reaching the end of the vapor pressure curve which is at the critical point, the density of the liquid equals that of the vapor. The liquid and gas phases become identical and indistinguishable. Meanwhile, the surface tension of the liquid approaches zero. The interface between the liquid and vapor eventually disappears. This state of matter is neither liquid nor gas. It is called a **supercritical fluid** (SCF) and is located at the upper-right corner of a phase diagram when the temperature is above T_c and the external pressure is above p_c.

The characteristic properties of a SCF include:

1) It has high density like a liquid, and low viscosity like a gas.

在密封容器中缓慢且持续地加热液体，其温度和蒸气压将沿相图中的蒸气压曲线逐渐升高，液相和气相将处于连续的动态平衡状态。在此加热过程中，液体的密度降低而蒸气的密度增加。当最终到达蒸气压曲线的末端即临界点时，液体和蒸气的密度相等，液相和气相变得完全相同、无法区分。同时液体的表面张力接近于零，液相和气相之间的界面最终消失。这种状态的物质既不是液体也不是气体，称为**超临界流体**（SCF），位于相图右上角温度高于 T_c、外压大于 p_c 的区域。

超临界流体的特性包括：
1) 它具有像液体一样高的密度和像气体一样低的黏度。
2) 它可以像气体一样穿过固体，像液体一样溶解物质。
3) 接近临界点时，压强或温度的微小改变即会导致超临界流体密度的大幅变化，从而允许对其许多性质进行"微调"。

超临界二氧化碳（SC-CO_2）是一种典型的超临界流体，已成为重要的商业和工业溶剂。SC-CO_2 的临界温度为 304.2 K，因此 CO_2 是一种可凝聚气体，临界压强为 72.8 atm。SC-CO_2 具有化学稳定性、无毒、不易燃、成本低、易获得等特点。相对较低的 T_c 以及 CO_2 的稳定性，使其几乎不会导致加工材料变性。通过调节 SC-CO_2 的压强可改变加工材料在其中的溶解度，从而实现选择性萃取。在工艺完成之后，只需恢复至大气压即可轻松去除 SC-CO_2。SC-CO_2 被广泛用于食品工业、聚合物生产和加工中的分离，或用作干洗溶剂等。

相图举例

这里我们给出一些代表性的相图示例。碘的相图是最简单的相图之一，如**图 2.14（a）**所示。碘的三相点位于 386.8 K 和 0.1195 atm。$p=1$ atm 的水平线分别与熔化曲线和蒸气压曲线相交于其正常熔点（T_m = 386.9 K）和正常沸点（T_b=457.6 K）处。在 0.1195~1 atm 的范围内，熔化过程基本不受外压影响，因此 T_m 和 T_t 的温度几乎相同。碘的临界点位于 819 K 和 9.55 atm。

图 2.14（b）所示的二氧化碳的相图与碘的相图的主要区别在于，其三相点压强（p_t = 5.1118 atm）大于 1 atm，$p = 1$ atm 的水平线与升华曲线相交于其正常升华点（T_s = 194.6855 K）。所以固态 CO_2 在常压下不会熔化而会升华，因此被称为"干冰"。由于能保持较低温度且不会熔化产生液体，干冰被广泛用于冷冻和食品保鲜。

图 2.15（a）给出的水的相图示意图呈现了几个新特征。一是熔化曲线具有负斜率；也就是说，冰的熔点随外压增加而降低。另一个特征是**多型性**，即固态物质存在超过一种形式或晶体结构的现象。普通冰（称为冰Ⅰ）在常压下存在，而其他形式的冰只存在于高压下。注意其他形式的冰的熔化曲线均具有正斜率。实际上，固体存在多型性更多地体现为一种规则而非例外。当存在多型性时，相图中的三相点比常规的固-液-气三相点更多。例如，冰Ⅰ、冰Ⅲ和液态水在 251.2 K 和 209.9 MPa 下处于平衡状态。水的常规三相点为 273.16 K，与水的冰点（约 273.15 K）相差 0.01 K。三相点温度是纯水的固-液-气平衡温度，而冰点是冰与被空气饱和的水溶液在一个大气压下的平衡温度。因此，水的冰点随时间和地点而变化，三相点温度则有一个恒定值。在国际单位制中，1 K 被定义为纯水的

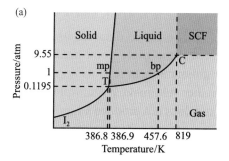

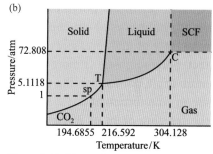

Figure 2.14 Illustrative phase diagrams. (a) The phase diagram of iodine. (b) The phase diagram of carbon dioxide. Abbreviation: sp-sublimation point.

图 2.14 相图示意图：(a) 碘的相图；(b) 二氧化碳的相图。缩写：sp—升华点。

2) It can effuse through solids like a gas, and dissolve materials like a liquid.
3) Close to the critical point, small changes in external pressure or temperature result in large changes in density of a SCF, allowing many properties to be "fine-tuned".

Supercritical carbon dioxide (SC-CO_2) is a typical SCF that becomes an important commercial and industrial solvent. The critical temperature of SC-CO_2 is 304.2 K, which makes CO_2 a condensable gas, and the critical pressure is 72.8 atm. SC-CO_2 is chemically stable, non-toxic, non-flammable, low-cost, and readily available. The relatively low T_c and the stability of CO_2 causes little denaturing of the processed materials. Selective extraction of the processed materials can be achieved by varying their solubility with the pressure of SC-CO_2. After the process, SC-CO_2 can be easily removed by simply restoring the atmospheric pressure. SC-CO_2 is widely used in separations in food industry, polymer production and processing, or as a dry-cleaning solvent, etc.

Examples of Phase Diagrams

Here, we give some representative examples of phase diagrams. One of the simplest phase diagrams is that of iodine, shown in **Figure 2.14(a)**. The triple point of iodine is at 386.8 K and 0.1195 atm. The horizontal line at $p = 1$ atm intersects the fusion and vapor pressure curves at its normal melting point (T_m = 386.9 K) and normal boiling point (T_b = 457.6 K), respectively. Melting is essentially unaffected by pressure in the limited range from 0.1195 to 1 atm, so T_m and T_t are at almost the same temperature. The critical point of iodine is at 819 K and 9.55 atm.

The phase diagram of carbon dioxide, shown in **Figure 2.14(b)**, differs from that of iodine mainly in that the triple point pressure (p_t = 5.1118 atm) is greater than 1 atm. The $p = 1$ atm line intersects the sublimation curve at its normal sublimation point (T_s = 194.6855 K). Therefore, solid CO_2 does not melt but rather sublimate at atmospheric pressure, and so is called "dry ice". Because it maintains a low temperature and does not produce a liquid by melting, dry ice is widely used in freezing and preserving foods.

The illustrative phase diagram of water given in **Figure 2.15(a)** presents several new features. One is that the fusion curve has a negative slope; that is, the melting point of ice decreases with increasing pressure. Another feature is **polymorphism**, which is the phenomena that a solid material may exist in more than one form or crystal structure. Ordinary ice, called ice I, exists under ordinary pressures. The other forms exist only at high pressures. Note that the fusion curves for the other forms of ice have positive slopes. In fact, polymorphism is more the rule than the exception among solids. Where it occurs, a phase diagram has more triple points than the usual solid-liquid-vapor triple point. For example, ice I, ice III, and liquid water are in equilibrium at 251.2 K and 209.9 MPa. The usual triple point of water is at 273.16 K. It differs from the freezing point of water (about 273.15 K) by 0.01 K. Triple point temperature is the solid-liquid-vapor equilibrium temperature for pure water. However, the freezing point is the equilibrium temperature of ice and the air-saturated water solution at atmospheric pressure. Therefore, the freezing point of water varies from time to time, and from place to place. The triple point temperature has a constant value. In the SI unit, 1 K is defined as 1/273.16 of the triple point temperature of pure water.

The red lines and curves in **Figure 2.15(b)** show the 3D phase diagram of water, in which the third dimension is the molar volume (in logarithm scale). The regular pressure-temperature phase diagram can be found as the pT projection (magenta curves) of the 3D phase diagram. At the triple point, the three common states of matter coexist but may differ in their relative composition; that is, as far as the three common states coexist, the molar volume can vary in a range. Therefore, this triple-coexist state is represented as a triple "point" line in the 3D phase diagram (red line), the pT projection of which corresponds to the regular triple point. The critical point C is still a point in the 3D phase diagram.

2.8 Properties of Solutions

Gas, liquid, solid, as well as supercritical fluid that we have discussed in the previous sections are different phases of a pure substance. Solution is a type of dispersion system that is not a pure substance but a mixture.

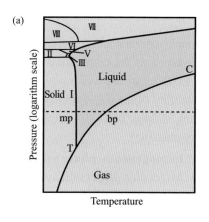 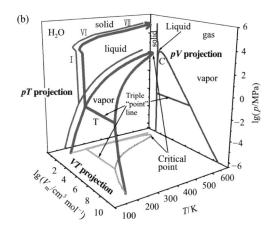

Figure 2.15 The phase diagram of water. (a) The illustrative p-T phase diagram. (b) The 3D phase diagram.

图 2.15 水的相图：(a) 外压 - 温度相图示意图；(b) 三维相图。
(Source: https://biomodel.uah.es/Jmol/plots/phase-diagrams/)

三相点温度的 1/273.16。

图 2.15（b）中红色的直线和曲线展示了水的三维相图，其中第三维是摩尔体积（对数标尺）。正常外压 - 温度相图即为三维相图在 pT 平面的投影（紫红色曲线）。三相点处物质的三种常见状态共存，但其相对组成可能不同；也就是说，只要三种常见状态共存，其摩尔体积就可在一定范围内变化。因此，该三相共存态在三维相图中显示为一条三相"点"直线（红线），其在 pT 平面的投影对应于正常三相点。在三维相图中，临界点 C 仍显示为一个点。

2.8 溶液的性质

我们在前几节中讨论的气、液、固体以及超临界流体均为纯净物的不同相态，而溶液是一种分散系，属于混合物而非纯净物。

分散系

一种或多种物质的微粒分散在另一种物质的连续相中形成的体系称为**分散系**。被分散的物质称为**分散质**，另一种物质的连续相称为**分散剂**或**分散介质**。分散系总是由两种或两种以上纯净物组成的混合物。按照分散质微粒的大小以及分散系的稳定性，分散系可分为分子分散系（< 1 nm，稳定）、胶体分散系（1~1000 nm，半稳定）和粗分散系（> 1000 nm，非常不稳定）。

只含一个相的体系称为**单相体系**或**均相体系**，含两个或两个以上相的体系称为**多相体系**。分子分散系具有均匀一致的物理和化学性质，是单相体系；胶体分散系和粗分散系是多相体系。由足以沉降的颗粒组成的分散系称为**悬浊液**，如泥浆等，属于粗分散系。

溶液

分子、原子或离子等分散在分散剂中形成的单相均匀混合物称为**溶液**，属于分子分散系。溶液中的分散质称为**溶质**，分散剂称为**溶剂**。一般来说，溶剂是溶液中量最大的组分，溶液的物态通常与溶剂相同。根据溶剂的物态，溶液可分为固溶体、液溶体和气溶体。

Dispersion System

A **dispersion system** is a system in which distributed particles of one or more materials are dispersed in a continuous phase of another material. The dispersed materials are called **dispersates** and the continuous phase of another material is called the **dispersant** or **dispersing medium**. Dispersion systems are always mixtures of two or more pure materials. According to the size of dispersate particles and the stability of the dispersion system, dispersion systems can be classified into molecular dispersion systems (< 1 nm, stable), colloidal dispersion systems (1~1000 nm, semi-stable), and coarse dispersion systems (> 1000 nm, very unstable).

A system consisting of only one phase is called a **single-phase system** or **homogeneous phase system**. A system consisting of two or more phases is called a **multiple phase system**. Molecular dispersion systems show uniform physical and chemical properties and are single-phase systems. Colloidal and coarse dispersion systems are multiple phase systems. A dispersion system of particles sufficiently large for sedimentation is called a **suspension**, such as slurry, which belongs to coarse dispersion systems.

Solution

Dispersion of molecules, atoms, or ions in a dispersant to form a homogeneous mixture of a single phase is called a **solution**, which belongs to molecular dispersion systems. In a solution, the dispersate is called the **solute** and the dispersant is called the **solvent**. In general, the solvent is the component that is present in the greatest quantity in a solution, and the solution usually takes the same state as the solvent. According to the state of solvent, solutions can be classified as solid solutions, liquid solutions, and gaseous solutions. The term **aqueous solution** is used when the solvent of a solution is water.

For instance, air is a gaseous solution in which the solutes O_2 and other gases dissolve in the gaseous solvent N_2. H_2 gas can be dissolved in some metals, especially Pd, to form gas-in-solid solutions that can be used for H_2 storage. Alloys can be viewed as solid-in-solid solutions. Taking the Cu alloys as an example, brass is an alloy of Zn in Cu (Cu-Zn alloy); white copper is Cu-Ni alloy; bronze is other Cu-metal alloys, typically Cu-Sn alloy.

Solution Concentration

The **concentration** of a solution is a measure of the quantity of solute in a given quantity of solvent or solution. Qualitatively, a **dilute solution** is a solution of relatively low concentration, and a **concentrated solution** is a solution of relatively high concentration. Quantitatively, concentration can be classified into mass concentration, molar concentration, number concentration, and volume concentration, which are defined as the mass, amount, number of entities, and volume, respectively, of a solute in a solution. Some commonly used concentrations and related quantities are listed in **Table 2.4**.

Table 2.4　Some Commonly Used Concentrations and Related Quantities
表 2.4　一些常用浓度及相关量

Physical Quantities (物理量)	Symbol (符号)	Numerator (分子)	Denominator (分母)	Multiplier (乘法因子)	Unit (单位)
Mass fraction (质量分数)	w	Mass of solute (溶质质量)	Mass of solution (溶液质量)		1
Volume fraction (体积分数)	φ	Volume of solute (溶质体积)	Volume of solution (溶液体积)		1
Mass/volume percent (质量体积百分数)		Mass of solute (溶质质量)	Volume of solution (溶液体积)	100%	%, g mL^{-1}
Mole fraction (摩尔分数)	x	Amount of solute (溶质物质的量)	Total amounts of all components (所有物质总物质的量)		1
molar concentration (物质的量浓度)	c	Amount of solute (溶质物质的量)	Volume of solution (溶液体积)		mol L^{-1}
mass concentration (质量浓度)	m	Amount of solute (溶质物质的量)	Mass of solvent (溶剂质量)		mol kg^{-1}

Parts-per notation is a set of pseudo units to describe small values of dimensionless quantities, such as mole

当溶液的溶剂为水时，称为**水溶液**。

例如，空气是一种气溶体，其中溶质 O_2 和其他气体溶解在气相溶剂 N_2 中。H_2 可以溶解在一些金属（特别是钯）中，形成可用于储氢的气-固溶体。合金可视为固-固溶体，以铜合金为例，黄铜是含锌的铜合金（铜锌合金）；白铜是铜镍合金；青铜则是含其他金属的铜合金，通常为铜锡合金。

溶液浓度

溶液的**浓度**是对一定量溶剂或溶液中所含溶质的量的量度。定性地说，**稀溶液**是浓度相对较低的溶液，而**浓溶液**是浓度相对较高的溶液。定量而言，浓度可分为质量浓度、物质的量浓度、数浓度和体积浓度，分别定义了溶液中溶质的质量、物质的量、微粒数和体积。**表 2.4** 列出了一些常用浓度及其相关量。

pp 表示法是一组用于描述某种量纲为 1 的量（如摩尔分数或质量分数）的微量成分的准单位。由于这些分数是按数量来计量的，因此是量纲为 1 的纯数字。常用的 pp 表示法有百万分之一（ppm，$1/10^6$）、十亿分之一（ppb，$1/10^9$）和万亿分之一（ppt，$1/10^{12}$）。对于极稀水溶液，其溶液密度非常接近水的密度，因此有 1 ppm ≈ 1 mg L^{-1}，1 ppb ≈ 1 μg L^{-1}，1 ppt ≈ 1 ng L^{-1}。

溶解度

除无限混溶的两种物质组成的溶液外，所有溶液均存在一个不会再溶解更多溶质的浓度，此时的溶液称为**饱和**溶液。在饱和溶液中，溶解和结晶之间达到动态平衡。一定温度和压强下饱和溶液的浓度称为溶质在指定溶剂中的**溶解度**，通常用 100 g 溶剂所能溶解溶质的最大克数表示。**溶解度曲线**给出了溶质的溶解度随温度的变化关系。

常用于预测溶解度的一条经验规律是相似相溶原理。极性溶质在极性溶剂中溶解度较大，如乙醇（CH_3CH_2OH）和水能无限互溶；非极性溶质在非极性溶剂中溶解度较大，如萘在苯中溶解度很大；极性物质和非极性物质难以互溶，如油水不互溶。

拓展阅读材料　等离子体

等离子体（Plasma）是除固、液、气三态之外被普遍承认的物质的第四种基本状态。当温度持续升高时，气态分子首先解离成单个原子，然后逐渐失去核外电子，最终成为由高度电离的原子核及脱离原子核吸引的自由电子组成的等离子体。等离子体存在于恒星和聚变反应堆中，也可通过加热中性气体或将其置于强电磁场中人工产生。

"Plasma"一词来源于古希腊语，意思是"可塑物质"或"果冻"。1928 年，欧文·朗缪尔（1881—1957）首次引入该词来描述电离的气体。朗缪尔对他观察到的等离子体进行了如下描述："除了电极附近的鞘层中含有极少电子外，电离气体中含有数量大致相等的离子和电子，因此产生的空间电荷非常小。我们将用等离子体这个名称来描述包含离子和电子电荷平衡的区域。"

根据含有等离子体的环境温度和密度，产生的等离子体可以部

fraction or mass fraction. Because these fractions are quantity-per-quantity measures, they are pure numbers with no associated units of measurement. The commonly used notations are parts-per-million (ppm, $1/10^6$), parts-per-billion (ppb, $1/10^9$), and parts-per-trillion (ppt, $1/10^{12}$). For very dilute aqueous solutions, since the density of solution is very close to that of water, we then have 1 ppm $\approx$ 1 mg L^{-1}, 1 ppb $\approx$ 1 μg L^{-1}, and 1 ppt $\approx$ 1 ng L^{-1}.

Solubility

Unless two substances are miscible, there exists a concentration at which no further solute will dissolve in a solution. At this point, the solution is said to be **saturated**. In a saturated solution, a dynamic equilibrium is reached between dissolution and crystallization. The concentration of the saturated solution at a certain temperature and pressure is called the **solubility** of the solute in the given solvent, normally shown as the maximum grams of solute dissolved in 100 g solvent. A **solubility curve** shows the solubility of solutes with respect to temperature.

A popular and useful rule to predict solubility is that "like dissolves likes". Polar solutes dissolve good in polar solvents, such as ethanol (CH_3CH_2OH) and water can be infinitely miscible. Nonpolar solutes dissolve good in nonpolar solvents, such as naphthalene in benzene. However, polar and nonpolar substances are not miscible, such as oil and water.

Extended Reading Materials Plasma

Plasma is the well-recognized fourth fundamental state of matter other than solid, liquid, and gas. When temperature increases continuously, a gas molecule first dissociates into atoms, then loses its electrons one by one, and finally becomes a plasma consisting of a highly electrified collection of nuclei and free electrons escaped from the nuclei. Plasma is present in stars and fusion reactors, and can also be generated artificially by heating a neutral gas or subjecting it to a strong electromagnetic field.

The word *plasma* comes from Ancient Greek, meaning "moldable substance" or "jelly". The term "plasma" was first introduced as a description of ionized gas by Irving Langmuir (1881—1957) in 1928. Langmuir described the plasma he observed as follows: "Except near the electrodes, where there are sheaths containing very few electrons, the ionized gas contains ions and electrons in about equal numbers so that the resultant space charge is very small. We shall use the name plasma to describe this region containing balanced charges of ions and electrons."

Based on the temperature and density of the environment that contains a plasma, partially ionized and fully ionized forms of plasma may be produced. Neon signs and lightning are examples of partially ionized plasma. The earth's ionosphere (50~1,000 km altitude in the atmosphere) is a plasma and the magnetosphere (~70,000 km away from the earth) contains plasma in the earth's surrounding space environment. The interior of the sun is an example of fully ionized plasma, along with the solar corona and stars.

Problems

2.1 A compound is 85.6% carbon by mass. The rest is hydrogen. When 10.0 g of the compound is evaporated at 50.0°C, the vapor occupies 6.30 L at 1.00 atm pressure. What is the molecular formula of the compound?

2.2 The amount of ozone in a mixture of gases can be determined by passing the mixture through a solution of excess potassium iodide. Ozone reacts with the iodide ion as follows:

$$O_3(g) + 3I^-(aq) + H_2O(l) \rightarrow O_2(g) + I_3^-(aq) + 2OH^-(aq)$$

The amount of I_3^- produced is determined by titrating with thiosulfate ion, $S_2O_3^{2-}$:

$$I_3^-(aq) + 2S_2O_3^{2-}(aq) \rightarrow 3I^-(aq) + S_4O_6^{2-}(aq)$$

A mixture of gases occupies a volume of 53.2 L at 18°C and 0.993 atm. The mixture is passed slowly through a solution containing an excess of KI to ensure that all the ozone reacts. The resulting solution requires 26.2 mL of 0.1359 mol L^{-1} $Na_2S_2O_3$ to titrate to the end point. Calculate the mole fraction of ozone in the original mixture.

分电离或完全电离。霓虹灯和闪电是部分电离的等离子体的例子。地球的电离层（大气中 50~1000 km 高度处）是一个等离子体层，而地磁层（距离地球约 70 000 km 远）还包含了地球周围空间环境中的等离子体。太阳内部、日冕以及恒星则是完全电离的等离子体的例子。

习题

2.1 某化合物质量的 85.6% 为碳，其余为氢。当 10.0 g 该化合物在 50.0ºC 下蒸发时，蒸气在 1.00 atm 的压强下占 6.30 L。试求该化合物的分子式。

2.2 使气体混合物通过过量的碘化钾溶液，可测定气体混合物中臭氧含量。臭氧与碘离子的反应如下：

$$O_3(g) + 3I^-(aq) + H_2O(l) \rightarrow O_2(g) + I_3^-(aq) + 2OH^-(aq)$$

产生的 I_3^- 的量，可通过硫代硫酸根离子（$S_2O_3^{2-}$）溶液的滴定来测定：

$$I_3^-(aq) + 2S_2O_3^{2-}(aq) \rightarrow 3I^-(aq) + S_4O_6^{2-}(aq)$$

在 18ºC 和 0.993 atm 下，气体混合物的体积为 53.2 L。使混合物缓慢通过含有过量 KI 的溶液，以确保所有臭氧发生反应。所得溶液需用 26.2 mL 0.1359 mol L^{-1} Na$_2$S$_2$O$_3$ 溶液滴定至终点。计算原始混合物中臭氧的摩尔分数。

2.3 在 30.1ºC 下，0.1052 g H$_2$O（l）样品在 8.050 L 干燥空气样品（体积恒定）中完全蒸发。空气必须冷却至以下哪个温度才能达到 80.0% 的相对湿度？水的蒸气压值：20ºC，17.54 mmHg；19ºC，16.48 mmHg；18ºC，15.48 mmHg；17ºC，14.53 mmHg；16ºC，13.63 mmHg；15ºC，12.79 mmHg。

2.4 对于 300 K 和 1.0 atm 的 N$_2$（g）样品，试计算：(a) v_{rms}；(b) v_{rms} 处分子速率分布函数的值。

2.5 探空气球是一个充满 H$_2$（g）并携带一套仪器（即其负载）的橡胶气囊。由于气囊、气体以及负载的总质量小于相应体积空气的质量，气球将会上升，并逐渐膨胀。已知气囊质量为 1200 g；负载质量为 1700 g；在 0.00ºC 和 1.00 atm 下，气球中 H$_2$（g）的量为 3.25 m^3；在最大高度处气球的直径为 7.5 m。试估算球形气球所能上升的最大高度区间。气压和温度随高度的变化如**表 P2.1** 所列。

2.6 按室温下表面张力增加的顺序排列以下物质：(a) CH$_3$OH；(b) CCl$_4$；(c) CH$_3$CH$_2$OCH$_2$CH$_3$。解释你的答案。

2.7 固体对二氯苯（C$_6$H$_4$Cl$_2$）易升华，可用作驱虫剂。根据给出的数据，估算 25ºC 时 C$_6$H$_4$Cl$_2$ 的升华压：mp=53.1ºC；54.8ºC 时蒸气压为 10.0 mmHg；$\Delta H_{熔化}$=17.88 kJ mol^{-1}；$\Delta H_{蒸发}$=72.22 kJ mol^{-1}。说明估算时所需做出的假定。

2.8 我们知道液体的蒸发焓通常是温度的函数。如果我们希望考虑温度变化，则不能使用**式（2.57）**中给出形式的克劳修斯-克拉贝龙方程。相反，我们必须回到克劳修斯-克拉贝龙方程所基于的微分方程，并将其重新整合为一个新的表达式。我们的

2.3 A 0.1052 g sample of $H_2O(l)$ evaporates completely at 30.1°C in an 8.050 L sample of dry air (constant volume). To which one of the following temperatures must the air be cooled to give a relative humidity of 80.0%? Vapor pressures of water: 20°C, 17.54 mmHg; 19°C, 16.48 mmHg; 18°C, 15.48 mmHg; 17°C, 14.53 mmHg; 16°C, 13.63 mmHg; 15°C, 12.79 mmHg.

2.4 Consider a sample of $N_2(g)$ at 300 K and 1.0 atm. Calculate (a) v_{rms} and (b) the value of the molecular speed distribution function at v_{rms}.

2.5 A sounding balloon is a rubber bag filled with $H_2(g)$ and carrying a set of instruments (the payload). Because this combination of bag, gas, and payload has a smaller mass than a corresponding volume of air, the balloon rises. As the balloon rises, it expands. Estimate the maximum height range to which a spherical balloon can rise given the mass of bag, 1200 g; payload, 1700 g; quantity of $H_2(g)$ in balloon, 3.25 m³ at 0.00°C and 1.00 atm; diameter of balloon at maximum height, 7.5 m. Air pressure and temperature as functions of altitude are listed in **Table P2.1**.

Table P2.1（表 P2.1）

Altitude (高度) /km	Pressure (压强) /mbar	Temperature (温度) /K	Altitude (高度) /km	Pressure (压强) /mbar	Temperature (温度) /K
0	1.0×10^3	288	30	1.2×10^1	230
5	5.4×10^2	256	40	2.9×10^0	250
10	2.7×10^2	223	50	8.1×10^{-1}	250
20	5.5×10^1	217	60	2.3×10^{-1}	256

2.6 Rank the following in order of increasing surface tension at room temperature: (a) CH_3OH; (b) CCl_4; (c) $CH_3CH_2OCH_2CH_3$. Explain your answer.

2.7 Because solid p-dichlorobenzene, $C_6H_4Cl_2$, sublimes rather easily, it has been used as a moth repellent. From the data given, estimate the sublimation pressure of $C_6H_4Cl_2(s)$ at 25°C: mp = 53.1°C; vapor pressure is 10.0 mmHg at 54.8°C; ΔH_{fus} = 17.88 kJ mol⁻¹; ΔH_{vap} = 72.22 kJ mol⁻¹. State the assumptions that you need to make for the estimation.

2.8 We have learned that the enthalpy of vaporization of a liquid is generally a function of temperature. If we wish to take this temperature variation into account, we cannot use Clausius-Clapeyron equation in the form given in **Equation (2.57)**. Instead, we must go back to the differential equation upon which Clausius-Clapeyron equation is based and reintegrate it into a new expression. Our starting point is the following equation describing the rate of change of vapor pressure with temperature in terms of the enthalpy of vaporization, the temperature and the difference in molar volumes of the vapor (V_g), and liquid (V_l):

$$\frac{dp}{dT} = \frac{\Delta H_{vap}}{T(V_g - V_l)}$$

Because in most cases the volume of 1 mol vapor greatly exceeds the molar volume of liquid, we can treat the V_l term as if it were zero. Also, unless the vapor pressure is unusually high, we can treat the vapor as if it was an ideal gas. Make appropriate substitutions into the above expression, and separate the p and dp terms from the T and dT terms. The appropriate substitution for ΔH_{vap} means expressing it as a function of temperature. Finally, integrate the two sides of the equation between the limits p_1 and p_2 on one side, T_1 and T_2 on the other.

(a) Derive an equation for the vapor pressure of $C_2H_4(l)$ as a function of temperature, if ΔH_{vap} = 15 971 + 14.55 T − 0.160 T^2 (in J mol⁻¹).

(b) Use the equation derived in (a), together with the fact that the vapor pressure of $C_2H_4(l)$ at 120 K is 10.16 Torr (1 Torr = 1 mmHg = 133.322 Pa), to determine the normal boiling point of ethylene.

2.9 An aqueous solution has 109.2 g KOH/L solution, and the solution density is 1.09 g mL⁻¹. In order to use 100.0 mL of this solution to prepare 0.250 mol L⁻¹ KOH, what mass of which component, KOH or H_2O, would you add to the 100.0 mL of solution?

出发点是以下方程，该方程描述了蒸气压随温度的变化率与蒸发焓、蒸气（V_g）与液体（V_l）的摩尔体积之差以及温度的关系：

$$\frac{dp}{dT} = \frac{\Delta H_{蒸发}}{T(V_g - V_l)}$$

由于大多数情况下，1 mol 蒸气的体积远大于液体的摩尔体积，我们可将 V_l 项视为零。此外，除非蒸气压异常高，否则我们可以将蒸气视为理想气体。对上述表达式进行适当的替代，并将 p 和 dp 项与 T 和 dT 项分开。$\Delta H_{蒸发}$ 的适当替代意味着将其表示为温度的函数。最后，对等式两侧进行积分：一侧从 p_1 积分到 p_2，另一侧从 T_1 积分到 T_2。

(a) 已知 $\Delta H_{蒸发} = 15\,971 + 14.55\,T - 0.160\,T^2$（单位：J mol^{-1}），试推导 $C_2H_4(l)$ 蒸气压随温度变化的方程。

(b) 采用 (a) 中推导的方程，根据 120 K 下 $C_2H_4(l)$ 的蒸气压为 10.16 Torr (1 Torr = 1 mmHg = 133.322 Pa)，确定乙烯的正常沸点。

2.9 某水溶液中每升溶液含 109.2 g KOH，溶液密度为 1.09 g mL^{-1}。为了使用 100.0 mL 该溶液制备 0.250 mol L^{-1} KOH 溶液，你会向 100.0 mL 该溶液中添加多少质量的哪种组分：KOH 还是 H_2O？

Chapter 3 Thermochemistry and Chemical Equilibria

Thermodynamics is a branch of science concerned with heat, temperature, and their relationship to energy and work. **Thermochemistry**, as a sub-branch of thermodynamics, is the study of the energy and heat exchange associated with chemical reactions and/or physical transformations. The content of thermochemistry focuses on: heat exchange during a chemical reaction; prediction of whether a reaction is spontaneous or non-spontaneous and favorable or unfavorable; the limit of a chemical reaction, which can be represented in terms of chemical equilibria.

3.1   Some Terminologies in Thermochemistry

3.2   The First Law of Thermodynamics

3.3   Heats of Reaction and Enthalpy Change

3.4   Hess's Law and Standard Enthalpies of Formation

3.5   Spontaneity and the Concept of Entropy

3.6   Criteria for Spontaneous Change

3.7   Dynamic Equilibrium and the Equilibrium Constants

3.8   The Reaction Quotient and Le Châtelier's Principle

3.9   Gibbs Free Energy Change and Equilibrium

第 3 章　化学热力学与化学平衡

热力学是研究热、温度及其与能量和功的关系的科学分支。**化学热力学**（也称**热化学**）则是研究化学反应及相应物态变化过程中的能量及热交换的热力学分支。化学热力学的内容聚焦于：化学反应中的热交换；预测反应的自发性及条件的有利性；化学反应的限度（可用化学平衡来表示）。

3.1　化学热力学的一些术语

3.2　热力学第一定律

3.3　反应热与焓变

3.4　盖斯定律与标准生成焓

3.5　自发性与熵的概念

3.6　自发变化的判据

3.7　动态平衡与平衡常数

3.8　反应商与勒夏特列原理

3.9　吉布斯自由能变与平衡

3.1 Some Terminologies in Thermochemistry

In this section, we introduce some basic terms called **terminologies**. The systematic understandings of any subject always start from some terminologies.

System, Surroundings, and the Universe

Let us think of the **universe** as being comprised of a system and its surroundings. A **system** is the part of the universe that we choose to study. The **surroundings** are the remaining part of the universe outside the system, with which the system interacts. Whatever in the universe that we are particularly interested in and take into consideration can be considered as a system. Whatever outside this system in the universe must be the surroundings of this system. Therefore, the universe consists of only a system and its surroundings. If we choose the surroundings to be a new system, then the original system will become the new surroundings of the new system.

The interactions between a system and its surroundings comprise the exchange of energy (in forms of heat and work, which will be discussed later in this section) and matter. According to the content of interactions, systems can be classified into three types:
1) **Open system**: both energy and matter can be freely exchanged with its surroundings;
2) **Closed system**: only energy but no matter can be exchanged with its surroundings;
3) **Isolated system**: neither energy nor matter can be exchanged with its surroundings.

It seems that a fourth type of system in which only matter but no energy can be exchanged with its surroundings should also be defined. However, no such system exists because no system can exchange only matter but no energy with its surroundings. This can be understood by two means: First, the exchange of matter is always accompanied by the exchange of energy; Secondly, matter can also be considered as a form of energy according to Einstein mass-energy equation $E = mc^2$.

If the entire universe is taken as a system, then its surroundings must be zero. The zero surroundings can exchange neither energy nor matter with the system. Therefore, the universe can be viewed as a very large isolated system.

Properties of a System

A **property of a system** is a characteristic that is measurable and the value of which describes the state of the system. According to the relationship with the amount of matter, the properties of a system can be classified into intensive and extensive properties. An **intensive property** is one that is independent of the amount of matter observed, such as density, pressure, molar mass, and temperature. Since intensive properties do not depend on the amount of matter but on the nature of substances, they can be used to identify substances. An **extensive property** is one that is dependent on the quantity of matter observed, such as mass, volume, and amount of substances.

If you mix two portions of the same substances at the same conditions, the properties whose values do not change should be intensive properties, and the properties whose values are additive with respect to the amounts of the two portions should be extensive properties. For example, when a glass of water is mixed with another glass of water of the same conditions, the extensive properties such as mass and volume are additive, but the intensive properties such as density and temperature remain unchanged. An intensive property can always be written as the ratio of two extensive properties. For example, $\rho = m/V$, $M = m/n$, $p = nRT/V$, etc.

State and State Functions

A **specified state** of a system means that all properties of the system are determined completely and do not vary with time, normally in thermodynamic equilibrium, which will be discussed later in **Section 3.7**. For example, if three variables out of the total four (p, V, n, and T) for an ideal gas are defined, then the state of this ideal gas is specified. A **state function** (or **function of state**) is a physical quantity that is determined completely by the state of the system. The state function has a unique value for a specified state of a system.

3.1 化学热力学的一些术语

本节我们将介绍一些**基本术语**。对任何学科的系统理解总是从术语开始。

体系、环境和宇宙

让我们将**宇宙**看作由一个体系及其周围环境组成。所谓**体系**就是我们所选择研究的那一部分宇宙；**环境**则是宇宙中除体系之外的其他部分，与体系之间存在相互作用。宇宙中任何我们特别感兴趣并加以考虑的部分，均可视为体系；宇宙中除体系之外的所有部分必定是该体系的环境。因此，宇宙只由体系及其环境组成。如果选择原环境作为一个新体系，那么原体系必将成为新体系的新环境。

体系及其环境之间的相互作用包括能量交换（以热和功的形式，将在本节后续内容中讨论）和物质交换。根据相互作用的内容，体系可分为三类：
1) **开放体系**：可与其环境自由交换能量和物质的体系；
2) **封闭体系**：与其环境只有能量交换而没有物质交换的体系；
3) **孤立体系**：不能与其环境交换能量和物质的体系。

看起来似乎还应定义第四类体系，即与其环境只有物质交换而没有能量交换的体系。但是这类体系并不存在，因为没有体系与其环境只存在物质交换而不发生能量交换。这可以从两个方面来理解：第一，物质交换时总伴随有能量交换；第二，根据爱因斯坦质能方程 $E=mc^2$，物质也可被认为是能量的一种形式。

如果把整个宇宙当作体系，那么它的环境必然为零。为零的环境既不能与体系发生能量交换，也不能发生物质交换。因此，宇宙可视为一个巨大的孤立体系。

体系的性质

体系的性质指其可测量的属性，其值描述了体系的状态。根据与物质的量的关系，体系的性质可分为强度性质和广度性质。**强度性质**与物质的量无关，如密度、压强、摩尔质量和温度等。由于强度性质与物质的量无关而只与物质的本性有关，它们可用于鉴定物质。**广度性质**与物质的量相关，如质量、体积和物质的量等。

混合相同条件下两份同样的物质，其值不发生变化的性质是强度性质，而值可以由两份相加得到的则是广度性质。例如，把一杯水与同样条件的另一杯水混合，其质量和体积等广度性质可直接加和，而密度和温度等强度性质的值则保持不变。强度性质总可以写成两个广度性质的比值的形式，如 $\rho = m/V$、$M = m/n$、$p = nRT/V$ 等。

状态和状态函数

我们说一个体系具有**确定的状态**，是指体系的各种性质完全确定且不随时间变化，通常指处于热力学平衡态（将在 **3.7 节**讨论）。例如，对于理想气体体系，如果四个变量（p、V、T 和 n）中有三个确定，则理想气体具有确定的状态。**状态函数**是完全由体系的状态

This value depends only on the exact condition or state of the system but not on how that state is reached.

For a specified state of a system, all state functions must have unique values. If the state of the system changes, the values of state functions will also change. The state before the change is called the **initial state** and the state after the change is called the **final state**. Once the initial and final states of a system are specified, the changes in the values of state functions must also have unique values, as given by

$$\Delta f = f_f - f_i \tag{3.1}$$

where f is any state function of the system and Δf is the change of this state function. The subscripts f and i stand for the final and initial states, respectively. If the final state happens to be exactly the same as the initial state, which means the system goes in a cycle, the changes in the values of all state functions must be zero, as

$$\Delta f_{cycle} = f_i - f_i = 0 \tag{3.2}$$

Process and Path

A **process** is a passage of a system from an initial state to a final state of thermodynamic equilibrium. The states of the system can be varied in a process. A **path** is the detailed pathway through which a system passes in a process. When processes are discussed, we focus on the initial and final states. When paths are discussed, not only the initial and final states but also all the detailed pathways matter.

The commonly involved processes in thermochemistry include:
1) **Isothermal process**: a process occurring at a constant temperature. $\Delta T = 0$ or $T_i = T_f$.
2) **Isobaric process**: a process occurring at a constant pressure. $\Delta p = 0$ or $p_i = p_f$.
3) **Isochoric process**: a process occurring at a constant volume. $\Delta V = 0$ or $V_i = V_f$.
4) **Adiabatic process**: a process occurring with no heat exchange between the system and its surroundings. $q = 0$.

The values of all state functions of a system must be independent of the path. A function whose value is dependent on the path of a process is termed a **path-dependent function**. Obviously, all path-dependent functions must be non-state functions.

Energy, Heat, and Work

Energy (E) is the capacity to do work; it is an extensive property of a system. Energy is transferable between the system and its surroundings in a process via fundamental interactions, and is convertible in different forms but can neither be created nor destroyed. **Heat** (q) is the energy exchanged between the system and its surroundings as a result of a temperature difference. When the temperatures of a system and its surroundings are different, energy will always be transferred from the one with a higher temperature to the other with a lower temperature in the form of heat until the same temperature is reached. **Work** (w) is the energy exchanged between the system and its surroundings other than heat. Therefore, a system and its surroundings can only exchange energy in two forms: either heat or work.

Work is usually represented by the action of a force acting from a distance. Among all the forms of work, we need to pay special attention to the pressure-volume work because it is the most commonly involved work in thermochemistry. The **pressure-volume work** (p-V work) is the work involved in the expansion and compression of gases, and can be represented by

$$w = F_{ext} d = -p_{ext} \Delta V \tag{3.3}$$

where F_{ext} is the external force, d is the distance, $p_{ext} = F_{ext}/A$ is the external pressure, $\Delta V = V_f - V_i = -Ad$ is the change in the volume of gases, and A is the area to which the external force exerts. Notice that the pressure used to calculate the p-V work in **Equation (3.3)** must be the external pressure, which may or may not equal the internal pressure of gases. Work can take many other forms in addition to p-V work, such as electric work and surface work. We only take the p-V work into consideration in this chapter. Discussion about electric work can be found in **Section 6.4**.

Notice that sometimes a terminology may not be specified as "of the system" or "of the surroundings". By default, it is always referred to "of the system" because the system is what we are particularly interested in. For example, if not specified, the energy always refers to the energy of the system, and the process always refers to that of the system, etc.

决定的物理量，对于具有确定状态的体系，状态函数具有确定的值，且该值只与体系的确切条件或状态有关，而与怎样到达该状态无关。

对于体系的某一确定状态，所有状态函数均具有确定的值。如果体系的状态发生改变，状态函数的值也会随之改变。变化前的状态称为**始态**，变化后的状态称为**终态**。一旦体系的始态和终态确定，状态函数的改变量也必为如下定值：

$$\Delta f = f_{终} - f_{始} \tag{3.1}$$

其中 f 是体系的任意状态函数，Δf 是该状态函数的改变量，下标"终"和"始"分别代表终态和始态。如果终态恰好与始态完全相同，说明体系经历了一个循环，所有状态函数的改变量必然为零，即

$$\Delta f_{循环} = f_{始} - f_{始} = 0 \tag{3.2}$$

过程与途径

过程指体系从始态到达终态（二者均为热力学平衡态）的历程，在过程中体系的状态可以发生改变。**途径**则是体系在一个过程中所经历的具体路径。在讨论过程时，我们重点关注始态和终态；而在讨论途径时，我们关注的不止始态和终态，具体的路径也很重要。

化学热力学的常见过程有：
1) **恒温过程**：温度保持不变的过程，$\Delta T = 0$ 或 $T_{始} = T_{终}$。
2) **恒压过程**：压强保持不变的过程，$\Delta p = 0$ 或 $p_{始} = p_{终}$。
3) **恒容过程**：体积保持不变的过程，$\Delta V = 0$ 或 $V_{始} = V_{终}$。
4) **绝热过程**：体系与其环境没有热交换的过程，$q = 0$。

所有状态函数的值必定与途径无关，值与途径相关的函数称为**途径相关函数**。显然，所有途径相关函数均不是状态函数。

能量、热和功

能量（E）是做功的能力，是体系的广度性质，可以在一个过程中通过基本相互作用在体系与其环境之间交换，也可以在各种形式之间转化，但既不会凭空产生，也不会凭空消失。**热**（q）是体系与其环境之间由于温度不同而导致的能量交换。当体系与其环境温度不同时，能量总是以热的形式从高温物体向低温物体转移，直至温度相等。**功**（w）是体系与其环境之间除热之外的其他所有能量交换。因此，体系与其环境之间有且只有两种能量交换形式：要么是热，要么是功。

功通常可用力作用一段距离的形式来表示。在所有形式的功中，我们要特别注意压强体积功，因为它是化学热力学中最常涉及的功。压强体积功（简称**体积功**）是气体在膨胀和压缩过程中所做的功，可表示为

$$w = F_{外}d = -p_{外}\Delta V \tag{3.3}$$

其中 $F_{外}$ 是外力，d 是距离，$p_{外} = F_{外}/A$ 是外压，$\Delta V = V_{终} - V_{始} = -Ad$ 是气体体积的改变量，A 是外力作用的面积。注意用**式 (3.3)** 计算体积功时使用的压强必须是外压，它可能等于也可能不等于气体自身的压强。除体积功外，还有许多其他形式的功，如电功和表面功等，在本章中我们只考虑体积功。关于电功的讨论详见 **6.4 节**。

注意有的术语并没有指明是"体系的"还是"环境的"，默认情况下总是指"体系的"，因为体系才是我们所特别感兴趣的。例如，如果没有指明，能量总是指体系的能量，过程总是指体系所经历的

Kinetic Energy and Potential Energy

The word *kinetic* means "motion" in Greek. **Kinetic energy** (E_k) is the energy of motion and can be given by

$$E_k = mv^2 / 2$$

where m is the mass and v is the velocity of an object. **Potential energy** (E_p or V) is the stored energy that has the potential to do work. The potential energy is associated with the attractive or repulsive forces between objects, and can result from position, condition, composition, etc. of the system.

The relationship between energy and work and the conversion between kinetic and potential energies can be explained by the behavior of a bouncing ball. To lift the ball to a starting position, we must apply a force over a distance to overcome the gravity. The work we do is "stored" in the ball as potential energy. When we release the ball, it is pulled toward Earth's center by the force of gravity. Potential energy is converted into kinetic energy in this process. The kinetic energy reaches its maximum just as the ball strikes the ground. On its rebound, the kinetic energy of the ball decreases and its potential energy increases. If the collision of the ball with the ground is not elastic, the ball reaches a lower height than the initial position.

Internal Energy and Thermal Energy

Internal energy (U) is the total energy attributed to the particles of matter and their interactions within a system. Internal energy is an extensive property of the system. It generally comprises the translational (U_T), rotational (U_R), vibrational (U_V), electronic (U_E), and nuclear (U_N) energies, and can be given by

$$U = U_T + U_R + U_V + U_E + U_N \tag{3.4}$$

Sometimes the internal energy can also be written as

$$U = U_T + U_R + U_V + U_{EE} + U_{NE} + U_{NN} \tag{3.5}$$

where U_{EE}, U_{NE}, and U_{NN} represent the energies associated with the repulsive forces between electrons, the attractive forces between nuclei and electrons, and the repulsive forces between the nuclei, respectively.

Thermal energy is the kinetic energy associated with random molecular motion. Thermal energy depends only on temperature and the amount of matter. Approximately, thermal energy can be understood as the translational energy (U_T) in **Equations (3.4)** and **(3.5)**. It is also the total translational kinetic energy (e_k) related to the average translational kinetic energy ($\overline{e_k}$) that we have previously discussed in **Section 2.3**. Heat exchange between a system and its surroundings involves only the exchange of thermal energy.

The ideal gas is the simplest system with no rotational or vibrational motions. The internal energy of an ideal gas depends only on temperature and amount of gas. For a given amount of ideal gases, the internal energy is a function of temperature and can be written as $U(T)$, that is, $\Delta U = 0$ if $\Delta T = 0$.

Summary

As a summary of the above terminologies, let us consider a process of a system from an initial state to a final state. In this very process, not only the system goes from its initial state to final state, but the surroundings and universe also change from their initial states to final states. Since the universe can be viewed as a very large isolated system, every state functions of the universe remains constant during the process. However, the state functions of both the system and its surroundings may change. The reason why these state functions change is because of the interactions between the system and its surroundings. These interactions cause the exchange of matter and energy between the system and its surroundings. The energy exchange in this process can only exist in two forms: either heat or work. Heat is the energy exchange due to temperature difference, and it involves only the exchange of thermal energy. Work is the energy exchange other than heat, and it involves the exchange of all energies other than thermal energy.

Energy (including internal energy, thermal energy, etc.) is an extensive state function of the system that has unique values in specified states. The change in energy has a unique value in a process and is also an extensive state function. Both heat and work, however, are not the properties of the system. Both heat and work are not dependent on the state but only exist in a process. Their values can be different for different paths of the same process. Therefore, both heat and work are not state functions but path-dependent functions.

过程等。

动能与势能

"Kinetic"一词在希腊语中是"运动"的意思。**动能**（E_k）即运动的能量，可用下式表示：

$$E_k = mv^2/2$$

其中 m 是物体的质量，v 是其速率。**势能**（E_p 或 V）是储存起来、可用于做功的能量。势能与物体间的吸引力或排斥力相关联，通常与体系的位置、条件、化学组成等有关。

能量和功的关系以及动能和势能之间的转化可用弹跳球的行为来说明。将一个小球举至某初始位置，需要对它施加一个力并作用一段距离以克服重力。所做的功以势能的形式"储存"在小球里。当释放小球时，它被重力拉向地心，在此过程中势能转化为动能，且动能在小球刚碰到地面的瞬间达到最大。回弹时小球的动能减小而势能增大，如果与地面的碰撞非弹性，则小球到达的高度会比初始位置低一些。

内能与热能

内能（U）指体系内物质微粒及其相互作用所包含的所有能量，是体系的广度性质。它通常包含平动能（$U_平$）、转动能（$U_转$）、振动能（$U_振$）、电子能（$U_电$）和核能（$U_核$），可用下式表示：

$$U = U_平 + U_转 + U_振 + U_电 + U_核 \qquad (3.4)$$

有时也可写作

$$U = U_平 + U_转 + U_振 + U_{电电} + U_{核电} + U_{核核} \qquad (3.5)$$

其中 $U_{电电}$、$U_{核电}$ 和 $U_{核核}$ 分别表示与电子间的排斥力、核与电子间的吸引力以及核间的排斥力相关的能量。

热能是与分子无规则运动相关的动能，只与温度和物质的量有关。热能可近似地理解为**式（3.4）**和**（3.5）**中的平动能（$U_平$），它也是与 **2.3 节**讨论的平均平动能（$\overline{e_k}$）相关的总平动能（e_k）。体系与其环境之间的热交换只包含热能的交换。

理想气体是最简单的体系，不存在转动和振动，其内能只与温度和物质的量有关。对于一定量的理想气体，内能是温度的函数，可写为 $U(T)$，即 $\Delta T = 0$ 时 $\Delta U = 0$。

小结

作为对以上术语的小结，让我们考虑体系从始态变化至终态的某个过程。在此过程中，不仅体系从始态变为终态，环境和宇宙也从始态变为终态。由于宇宙可视为一个巨大的孤立体系，宇宙的所有状态函数在此过程中均保持不变，但体系和环境的状态函数则可能发生变化。状态函数改变的原因来自体系与其环境之间的相互作用，这些相互作用导致了体系与其环境之间物质和能量的交换。在此过程中，能量交换有且只有两种形式：热或功。热是由于温度不同而导致的能量交换，只涉及热能的交换。功是除热之外的其他能量交换，涉及除热能之外的其他能量的交换。

能量（包括内能、热能等）是体系的广度状态函数，在确定状态下具有确定的值。能量的改变量在一个过程中也具有确定的值，因此也是广度状态函数。热和功均不是体系的性质，不与某个状态

3.2 The First Law of Thermodynamics

Based on the terminologies that we have learnt in the previous section, three laws of thermodynamics will be introduced in this chapter. Here, we start with the first law of thermodynamics.

Change of Internal Energy

Internal energy is the total energy of a system, and a system contains only internal energy. The internal energy changes during a process. The absolute values of internal energy cannot be determined and are not important; it is the change of internal energy in the process that matters. Since the change of internal energy in this process can only exist in two forms: either heat or work, the change of internal energy (ΔU) must equal the sum of heat (q) and work (w), given by

$$\Delta U = q + w \tag{3.6}$$

This is known as the **first law of thermodynamics**. This equation must be applied to a process, during which various paths may exist with different values of heat and work. However, the change of internal energy remains the same and equals the sum of heat and work in various paths.

Signs of Heat and Work

During a process, heat may be either absorbed or given off by the system, and work can be done either by the system or on the system. As a result, the internal energy of the system may either increase or decrease. In order for the consistency of **Equation (3.6)** under all circumstances, the signs of heat and work are defined as follows:

1) Whatever causes the *decrease* in internal energy ($\Delta U = U_f - U_i < 0$ or $U_f < U_i$) carries a *negative* sign. Thus, if heat is *given off* by the system, $q < 0$. If work is done *by* the system, $w < 0$.
2) Whatever causes the *increase* in internal energy ($\Delta U = U_f - U_i > 0$ or $U_f > U_i$) carries a *positive* sign. Thus, if heat is *absorbed* by the system, $q > 0$. If work is done *on* the system, $w > 0$.

For the p-V work in a gaseous system, when gases expand work is done by the system and $w < 0$. This is consistent with **Equation (3.3)**, as $w = -p_{ext}\Delta V < 0$ since $\Delta V = V_f - V_i > 0$ in this expansion process. Similarly, when gases compress work is done on the system and $w > 0$. This also agrees with $w = -p_{ext}\Delta V > 0$ as $\Delta V = V_f - V_i < 0$ in the compression process.

Law of Conservation of Energy

The first law of thermodynamics is also called the **law of conservation of energy**, because it can be derived that the total energy of the universe is conserved, which means that the total energy of the universe remains constant. The derivation is given as follows:

For any system in the universe in a process, we have

$$\Delta U_{sys} = \left(U_{sys}\right)_f - \left(U_{sys}\right)_i = q + w$$

If taking the surroundings as a new system and applying the first law of thermodynamics to it, we have

$$\Delta U_{surr} = \left(U_{surr}\right)_f - \left(U_{surr}\right)_i = -q - w$$

For the entire universe,

$$\Delta U_{univ} = \Delta U_{sys} + \Delta U_{surr} = 0$$

or

$$\left(U_{univ}\right)_f = \left(U_{sys}\right)_f + \left(U_{surr}\right)_f = \left(U_{sys}\right)_i + \left(U_{surr}\right)_i = \left(U_{univ}\right)_i$$

Therefore, the total energy of the universe is conserved.

Since an isolated system does not exchange energy with its surroundings, the total energy of any isolated system must always be conserved. We have discussed in **Section 3.1** that the universe can be viewed as a very large isolated system with zero surroundings to interact. From this aspect, it is also proven that the total energy of the universe is conserved.

相关，而只存在于过程中。热和功的值在同一过程的不同途径中可能不同。因此，热和功均不是状态函数而是途径相关函数。

3.2 热力学第一定律

在上节的术语基础上，本章会介绍热力学三大定律，这里从热力学第一定律开始。

内能的改变量

内能是体系各种能量的总和，一个体系也只含有内能。内能在一个过程中发生变化，其绝对数值既不可测也不重要，重要的是在过程中内能的改变量。由于此过程中内能的改变量有且只有两种形式：热或功，故内能的改变量（ΔU）一定等于热（q）和功（w）之和，即

$$\Delta U = q + w \tag{3.6}$$

上式被称为**热力学第一定律**。该式必须应用于一个过程，其中可能存在多条途径，具有不同的热和功的值，但是在各种途径中内能的改变量均相同，始终等于热和功之和。

热和功的符号

在一个过程中，体系可能吸热也可能放热，可能做功也可能被做功，其结果导致体系的内能可能增加也可能减少。为了在所有条件下保持**式 (3.6)** 的一致性，热和功的符号定义为：

1) 凡是使体系内能减少（$\Delta U = U_{终} - U_{始} < 0$ 或 $U_{终} < U_{始}$）的量，均带负号。因此，体系放热 $q < 0$，体系做功 $w < 0$。
2) 凡是使体系内能增加（$\Delta U = U_{终} - U_{始} > 0$ 或 $U_{终} > U_{始}$）的量，均带正号。因此，体系吸热 $q > 0$，体系被做功 $w > 0$。

对于气体体系的体积功，当气体膨胀时体系对外做功，$w < 0$。由于膨胀过程 $\Delta V = V_{终} - V_{始} > 0$，$w = -p_{外}\Delta V < 0$，这与**式 (3.3)** 相符。同样，当气体压缩时外界对体系做功，$w > 0$。这也与 $w = -p_{外}\Delta V > 0$ 相符，因为压缩过程 $\Delta V = V_{终} - V_{始} < 0$。

能量守恒定律

热力学第一定律也称**能量守恒定律**，因为可以推导出宇宙的总能量守恒，即宇宙的总能量保持定值。其推导如下：

对于宇宙中任意体系的某个过程，有

$$\Delta U_{体系} = \left(U_{体系}\right)_{终} - \left(U_{体系}\right)_{始} = q + w$$

如果将原环境看作一个新体系，并对其应用热力学第一定律，可得

$$\Delta U_{环境} = \left(U_{环境}\right)_{终} - \left(U_{环境}\right)_{始} = -q - w$$

对于整个宇宙而言

$$\Delta U_{宇宙} = \Delta U_{体系} + \Delta U_{环境} = 0$$

即

$$\left(U_{宇宙}\right)_{终} = \left(U_{体系}\right)_{终} + \left(U_{环境}\right)_{终} = \left(U_{体系}\right)_{始} + \left(U_{环境}\right)_{始} = \left(U_{宇宙}\right)_{始}$$

因此，宇宙的总能量守恒。

由于孤立体系不与其环境交换能量，任何孤立体系的总能量总

Reversible and Irreversible Expansion Processes

Next, we are going to discuss the reversible and irreversible processes. Let us first consider three hypothetical experiments pictured in **Figures 3.1~3.3**. In experiment **A** [**Figure 3.1(a)**], two identical weights (each with a mass of $m/2$) are loaded in a weightless apparatus (including a cylinder, a piston, a pan, some wires, etc.) with some quantities of ideal gases confined by the cylinder walls and the piston. The space above the piston is a vacuum. The cylinder is contained in a water bath to keep the temperature of the gas constant. After the thermodynamic equilibrium is reached, set this as the initial state of the system, and assume that $p_i = 4$ atm and $V_i = 3$ L. At some moment, one of the two weights is removed, leaving the total mass on the pan to be $m/2$. Wait until thermodynamic equilibrium is reached again and set this as the final state. From the ideal gas equation, we can easily calculate that $p_f = p_i/2 = 2$ atm and $V_f = 6$ L. During the entire process from the initial state to the final state, the external pressure of gas remains constant, equaling the pressure exerted by $m/2$, which is the same as p_f. Therefore, we can calculate that $w_A = -p_f(V_f - V_i) = -2$ atm $\cdot$ (6-3) L $= -6$ atm L. The absolute value of w_A can be represented in the pressure-volume plot as the area of a rectangle with a length of p_f and a width of $V_f - V_i$ [**Figure 3.1(b)**].

In experiment **B** [**Figure 3.2(a)**], everything is exactly the same as in experiment **A** except that the two weights are replaced by four identical but lighter weights (each with a mass of $m/4$). The total mass loaded on the pan is still m so that $p_i = 4$ atm and $V_i = 3$ L hold at the initial state. At some moment, one of the four weights is removed, leaving the total mass on the pan to be $3m/4$. After the equilibrium is reached, this is an intermediate state with $p_m = 3p_i/4 = 3$ atm and $V_m = 4$ L. Then another weight is removed, leaving the total mass on the pan to be $m/2$ again. After the equilibrium is reached, for the final state we also have $p_f = 2$ atm and $V_f = 6$ L. In this process, the initial and final states are exactly the same as in experiment **A**, however, the path is different with an intermediate state involved. The work can be calculated as the sum of two terms: $w_B = w_1 + w_2 = -p_m(V_m - V_i) - p_f(V_f - V_m) = -3$ atm $\cdot$ (4-3) L -2 atm $\cdot$ (6-4) L $= -7$ atm L. The work in experiment **B** is less than that in experiment **A**. The absolute value of w_B can be represented in the pressure-volume plot as the area of the sum of two rectangles, one with a length of p_m and a width of $V_m - V_i$, and the other with a length of p_f and a width of $V_f - V_m$ [**Figure 3.2(b)**].

In experiment **C** [**Figure 3.3(a)**], the weights are replaced by some sand with a total mass m. The sand is removed grain by grain, and in the final state half of the sand is removed. In this process, the initial and final states are still the same, but the removal of each grain corresponds to an intermediate state and the path involves many intermediate states. The work in experiment **C** can be calculated as

$$w_C = w_1 + w_2 + w_3 + \cdots$$
$$= -p_{m1}(V_{m1} - V_i) - p_{m2}(V_{m2} - V_{m1}) - p_{m3}(V_{m3} - V_{m2}) + \cdots$$

The absolute value of w_C can be represented in the pressure-volume plot as the total area from V_i to V_f under the p-V curve [**Figure 3.3(b)**].

Comparing the three areas representing $|w_A|$, $|w_B|$, and $|w_C|$, it is not difficult to conclude that $|w_A| < |w_B| < |w_C|$. As the work done in gas expansion is always negative, we have

$$0 > w_A > w_B > w_C \tag{3.7}$$

Since all three processes have the same initial and final states with constant temperature, and the internal energy of an ideal gas is a function of temperature, we have

$$\Delta U_A = \Delta U_B = \Delta U_C = 0 \quad \text{and} \quad 0 < q_A < q_B < q_C \tag{3.8}$$

Experiment **C** proceeds in a pseudo reversible fashion. A **reversible process** is one that can be made to reverse its direction when an infinitesimal change is made in a system variable. For example, adding a grain of sand rather than removing one would reverse the expansion in experiment **C**. However, this process is not quite reversible because grains of sand have more than an infinitesimal mass. A reversible process is an ideal process, in which the system is always in equilibrium with its surroundings. On the contrary, experiments **A** and **B** are **irreversible processes** with stepwise paths.

The absolute value of work done in a reversible process can be represented as the total area under the p-V curve [**Figure 3.3(b)**]. Mathematically, for an isothermal reversible process from (p_i, V_i, T) to (p_f, V_f, T) with only p-V work, we can take the differential form of the p-V work, as

守恒。我们在 **3.1 节**讨论了宇宙可视为一个巨大的孤立体系，没有环境与之作用。从这个角度也可以证明宇宙的总能量守恒。

可逆与不可逆膨胀过程

接下来我们要讨论可逆与不可逆过程，让我们先考虑如**图 3.1~3.3** 所示的三个假想实验。在实验 **A**[**图 3.1(a)**] 中，气缸壁与活塞间装有一定量理想气体，将两个完全相同的砝码（质量均为 $m/2$）放在质量不计的装置（包括气缸、活塞、托盘、缆绳等）上。活塞以上为真空，气缸处于水浴中，以保持气体恒温。达到热力学平衡时，将此状态设为体系的始态，假设 $p_{始} = 4$ atm，$V_{始} = 3$ L。在某一时刻移去两个砝码之一，托盘上的总质量变为 $m/2$。待重新达到热力学平衡后，将此状态设为终态。由理想气体状态方程，易得 $p_{终}=p_{始}/2=2$ atm，$V_{终} = 6$ L。在从始态至终态的整个过程中，气体的外压均保持不变，为 $m/2$ 所施加的压强，等于 $p_{终}$。因此，可以算出 $w_A = -p_{终}(V_{终} - V_{始}) = -2$ atm · $(6-3)$ L $= -6$ atm L，其绝对值在压强 - 体积图中可用一个长为 $p_{终}$、宽为 $V_{终} - V_{始}$ 的矩形的面积来表示 [**图 3.1(b)**]。

在实验 **B**[**图 3.2(a)**] 中，所有条件与实验 **A** 几乎完全相同，除了将两个砝码替换为四个较轻的砝码（质量均为 $m/4$）。托盘上的总载重仍为 m，因此始态仍为 $p_{始} = 4$ atm，$V_{始} = 3$ L。在某一时刻移去四个砝码之一，托盘上的总质量变为 $3m/4$。当达到平衡时，设其为中间态，$p_{中} = 3p_{始}/4 = 3$ atm，$V_{中} = 4$ L。然后再移去一个砝码，托盘上的总质量重新变为 $m/2$。当达到平衡时，对于终态仍有 $p_{终} = 2$ atm，$V_{终} = 6$ L。在此过程中，始态和终态与实验 **A** 完全相同，但途径不同，经历了一个中间态。功可由两项之和计算：$w_B = w_1 + w_2 = -p_{中}(V_{中} - V_{始}) - p_{终}(V_{终} - V_{中}) = -3$ atm · $(4-3)$ L $- 2$ atm · $(6-4)$ L $= -7$ atm L。实验 **B** 的功小于实验 **A**。w_B 的绝对值在压强 - 体积图中可用两个矩形的面积之和来表示，其中一个长为 $p_{中}$、宽为 $V_{中} - V_{始}$，另一个长为 $p_{终}$、宽为 $V_{终} - V_{中}$ [**图 3.2(b)**]。

在实验 **C**[**图 3.3(a)**] 中，将砝码替换为总质量为 m 的沙子。一粒一粒地移去沙子，使终态时共有一半沙子被移去。在此过程中，始态和终态仍与之前相同，但每移去一粒沙子就对应了一个中间态，这一途径包含了许多个中间态。实验 **C** 的功可由下式得出：

$$w_C = w_1 + w_2 + w_3 + \cdots$$
$$= -p_{中1}(V_{中1} - V_{始}) - p_{中2}(V_{中2} - V_{中1}) - p_{中3}(V_{中3} - V_{中2}) + \cdots$$

w_C 的绝对值在压强 - 体积图中可用 p-V 曲线下从 $V_{始}$ 到 $V_{终}$ 的总面积来表示 [**图 3.3(b)**]。

比较代表 $|w_A|$、$|w_B|$ 和 $|w_C|$ 的三块面积，不难得出 $|w_A| < |w_B| < |w_C|$。由于气体膨胀总做负功，有

$$0 > w_A > w_B > w_C \tag{3.7}$$

因为三个过程具有相同的始态和终态，且气体温度保持恒定，而理想气体的内能是温度的函数，故

$$\Delta U_A = \Delta U_B = \Delta U_C = 0 \quad 且 \quad 0 < q_A < q_B < q_C \tag{3.8}$$

实验 **C** 通过近似可逆的方式进行。**可逆过程**是可以向正逆两个方向无限小地改变体系某一变量的过程。例如，加入而非移去一粒沙子将逆转实验 **C** 的膨胀过程。该过程并非完全可逆，因为沙子的质量并非无限小。可逆过程是理想过程，其中体系总是与其环境处

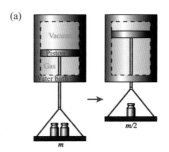

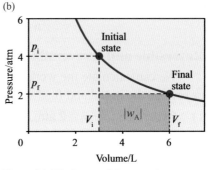

Figure 3.1 The irreversible expansion process A. (a) The schematic setup. (b) The absolute value of the work $|w_A|$ shown as an area in the p-V plot.

图 3.1 不可逆膨胀过程 A： (a) 实验装置图；(b) p-V 图中以面积表示的功的绝对值 $|w_A|$。

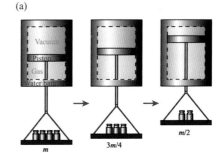

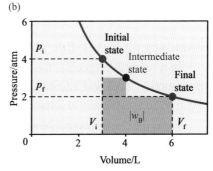

Figure 3.2 The irreversible expansion process B. (a) The schematic setup. (b) The absolute value of the work $|w_B|$ shown as an area in the p-V plot.

图 3.2 不可逆膨胀过程 B： (a) 实验装置图；(b) p-V 图中以面积表示的功的绝对值 $|w_B|$。

$$dw = -pdV$$

Therefore,

$$w_{rev} = -\int_{V_i}^{V_f} pdV = -\int_{V_i}^{V_f} \frac{nRT}{V}dV = -nRT\int_{V_i}^{V_f} \frac{1}{V}dV$$
$$= nRT\ln\frac{V_i}{V_f} = nRT\ln\frac{p_f}{p_i} \tag{3.9}$$

In experimental **C**, $w_C \approx -nRT\ln 2 = -p_f V_f \ln 2 = -8.32$ atm L.

In general, the work done by the gas in a reversible expansion is always smaller in value (with signs) than that in any irreversible expansion. Consequently, the heat absorbed in a reversible expansion is larger in value than that in any irreversible expansion. Thus, we can conclude that for an ideal gas isothermal expansion process

$$w_{rev} < w_{ir} < 0, \quad q_{rev} > q_{ir} > 0, \quad \text{and} \quad \Delta U_{rev} = \Delta U_{ir} = 0 \tag{3.10}$$

Reversible and Irreversible Compression Processes

We can calculate the work and heat in an ideal gas isothermal compression process in a similar way. Let us reverse the initial and final states in experiments **A**, **B**, and **C**.

1) Experiment **A**: $p_i = 2$ atm, $V_i = 6$ L, $p_f = 4$ atm, and $V_f = 3$ L.
   $w_A = -p_f(V_f - V_i) = -4$ atm · $(3-6)$ L $= 12$ atm L.
2) Experiment **B**: $p_i = 2$ atm, $V_i = 6$ L, $p_m = 3$ atm, $V_m = 4$ L, $p_f = 4$ atm, and $V_f = 3$ L.
   $w_B = w_1 + w_2 = -p_m(V_m - V_i) - p_f(V_f - V_m) = -3$ atm · $(4-6)$ L -4 atm · $(3-4)$ L $= 10$ atm L.
3) Experiment **C**: $p_i = 2$ atm, $V_i = 6$ L, $p_f = 4$ atm, and $V_f = 3$ L.

$$w_C \approx nRT\ln\frac{V_i}{V_f} = nRT\ln 2 = p_f V_f \ln 2 = 8.32 \text{ atm L}.$$

The absolute values of w_A, w_B, and w_C can be represented as the corresponding areas in the pressure-volume plots shown in **Figure 3.4**, and the work done in gas compression is always positive. Therefore,

$$w_A > w_B > w_C > 0, \quad q_A < q_B < q_C < 0, \quad \text{and} \quad \Delta U_A = \Delta U_B = \Delta U_C = 0 \tag{3.11}$$

Both experiments **A** and **B** are irreversible processes with stepwise paths and experiment **C** is pseudo reversible. Notice that in experiment **C** the work done in compression equals the opposite of the work done in expansion. However, in experiments **A** and **B**, the work done in compression is greater than the opposite of the work done in the corresponding expansion.

In general, the work done on the gas in a reversible compression is always smaller in value (with signs) than that in any irreversible compression. Consequently, the heat given off in a reversible compression is always larger in value than that in any irreversible compression. Thus, we can conclude that for an ideal gas isothermal compression process

$$0 < w_{rev} < w_{ir}, \quad 0 > q_{rev} > q_{ir}, \quad \text{and} \quad \Delta U_{rev} = \Delta U_{ir} = 0 \tag{3.12}$$

Summary on Reversible and Irreversible Processes

In a reversible process, the system is always in equilibrium with its surroundings. A reversible process may take countless steps and infinite time to complete. It is an ideal process that does not exist in reality, but under certain circumstances, especially at some equilibrium conditions, many real processes can be simplified as reversible processes.

From the above reversible and irreversible isothermal expansion and compression processes of ideal gases, we can conclude that the work involved in a reversible process (no matter expansion or compression) is the minimum possible work, and that the heat involved in a reversible process is the maximum possible heat, as

$$w_{rev} < w_{ir}, \quad q_{rev} > q_{ir}, \quad \text{and} \quad \Delta U_{rev} = \Delta U_{ir} \tag{3.13}$$

Meanwhile, the work in a reversible compression exactly equals the opposite of the work in a reversible expansion. However, the work in an irreversible compression is always greater than the opposite of the work in the corresponding irreversible expansion, as

于平衡。相反，实验 **A** 和 **B** 中存在逐级进行的途径，是**不可逆过程**。

可逆过程中所做功的绝对值可用 p-V 曲线下的总面积来表示 [**图 3.3(b)**]。数学上，对于一个从 $(p_{始}, V_{始}, T)$ 到 $(p_{终}, V_{终}, T)$、只有体积功的恒温可逆过程，体积功的微分形式为

$$dw = -pdV$$

因此

$$w_{可逆} = -\int_{V_{始}}^{V_{终}} pdV = -\int_{V_{始}}^{V_{终}} \frac{nRT}{V} dV = -nRT \int_{V_{始}}^{V_{终}} \frac{1}{V} dV$$
$$= nRT \ln \frac{V_{始}}{V_{终}} = nRT \ln \frac{p_{终}}{p_{始}} \quad (3.9)$$

实验 **C** 中，$w_C \approx -nRT \ln 2 = -p_{终} V_{终} \ln 2 = -8.32 \text{ atm L}$。

一般而言，可逆膨胀过程气体做的功，在数值上（包含符号）总小于任何不可逆膨胀的功。相应地，可逆膨胀过程吸收的热，在数值上总大于任何不可逆膨胀的热。因此，对于理想气体恒温膨胀过程，有

$$w_{可逆} < w_{不可逆} < 0, \quad q_{可逆} > q_{不可逆} > 0, \quad \Delta U_{可逆} = \Delta U_{不可逆} = 0 \quad (3.10)$$

可逆与不可逆压缩过程

我们可用类似的方法计算理想气体恒温压缩过程的功和热。让我们将实验 **A**、**B** 和 **C** 的始态和终态互换。

1) 实验 **A**：$p_{始} = 2 \text{ atm}, V_{始} = 6 \text{ L}, p_{终} = 4 \text{ atm}, V_{终} = 3 \text{ L}$
$w_A = -p_{终}(V_{终} - V_{始}) = -4 \text{ atm} \cdot (3 - 6) \text{ L} = 12 \text{ atm L}$

2) 实验 **B**：$p_{始} = 2 \text{ atm}, V_{始} = 6 \text{ L}, p_{中} = 3 \text{ atm}, V_{中} = 4 \text{ L}, p_{终} = 4 \text{ atm}, V_{终} = 3 \text{ L}$
$w_B = w_1 + w_2 = -p_{中}(V_{中} - V_{始}) - p_{终}(V_{终} - V_{中}) = -3 \text{ atm} \cdot (4-6) \text{ L} - 4 \text{ atm} \cdot (3-4) \text{ L} = 10 \text{ atm L}$

3) 实验 **C**：$p_{始} = 2 \text{ atm}, V_{始} = 6 \text{ L}, p_{终} = 4 \text{ atm}, V_{终} = 3 \text{ L}$
$w_C \approx nRT \ln(V_{始}/V_{终}) = nRT \ln 2 = p_{终} V_{终} \ln 2 = 8.32 \text{ atm L}$

w_A、w_B 和 w_C 的绝对值可用**图 3.4** 压强 - 体积图中的对应面积来表示，气体压缩总做正功。因此

$$w_A > w_B > w_C > 0, \quad q_A < q_B < q_C < 0, \quad \Delta U_A = \Delta U_B = \Delta U_C = 0 \quad (3.11)$$

实验 **A** 和 **B** 均为含有逐级进行途径的不可逆过程，而实验 **C** 为准可逆过程。注意，实验 **C** 中压缩过程所做的功等于膨胀过程所做功的相反数。而实验 **A** 和 **B** 中，压缩过程所做的功大于膨胀过程所做功的相反数。

一般而言，可逆压缩过程对气体所做的功，在数值上（包含符号）总小于任何不可逆压缩的功。相应地，可逆压缩过程释放的热，在数值上总大于任何不可逆压缩的热。因此，对于理想气体恒温压缩过程，有

$$0 < w_{可逆} < w_{不可逆}, \quad 0 > q_{可逆} > q_{不可逆}, \quad \Delta U_{可逆} = \Delta U_{不可逆} = 0 \quad (3.12)$$

可逆与不可逆过程小结

在可逆过程中体系与其环境总处于平衡。一个可逆过程的完成可能需要通过无数步、经历无限长时间。它是一个现实中不存在的理想过程，但在特定情况下，尤其在某些平衡条件下，许多实际过

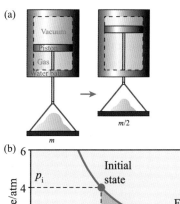

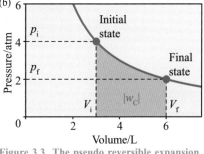

Figure 3.3 The pseudo reversible expansion process C. (a) The schematic setup. (b) The absolute value of the work $|w_C|$ shown as an area in the p-V plot.

图 3.3 准可逆膨胀过程 C： (a) 实验装置图；(b) p-V 图中以面积表示的功的绝对值 $|w_C|$。

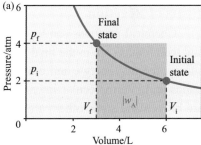

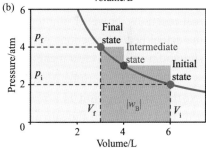

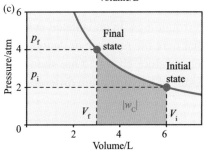

Figure 3.4 The absolute values of work are shown as areas in the p-V plots. (a) The irreversible compression process **A**. (b) The irreversible compression process **B**. (c) The pseudo reversible compression process **C**.

图 3.4 p-V 图中以面积表示的功的绝对值：(a) 不可逆压缩过程 **A**；(b) 不可逆压缩过程 **B**；(c) 准可逆压缩过程 **C**。

$$\left(w_{\text{comp}}\right)_{\text{rev}} = -\left(w_{\text{exp}}\right)_{\text{rev}} \quad \text{and} \quad \left(w_{\text{comp}}\right)_{\text{ir}} > -\left(w_{\text{exp}}\right)_{\text{ir}} \tag{3.14}$$

If we expand/compress an ideal gas from an initial state to a final state, and then compress/expand it back to the original initial state to form a cycle, the total work in this cycle is identically zero if the entire process is carried out reversibly. The total work must be positive if any part of the cycle is done irreversibly, as

$$\left(w_{\text{cycle}}\right)_{\text{rev}} = 0 \quad \text{and} \quad \left(w_{\text{cycle}}\right)_{\text{ir}} > 0 \tag{3.15}$$

Consequently,

$$\left(q_{\text{comp}}\right)_{\text{rev}} = -\left(q_{\text{exp}}\right)_{\text{rev}} \quad \text{and} \quad \left(q_{\text{comp}}\right)_{\text{ir}} < -\left(q_{\text{exp}}\right)_{\text{ir}} \tag{3.16}$$

$$\left(q_{\text{cycle}}\right)_{\text{rev}} = 0 \quad \text{and} \quad \left(q_{\text{cycle}}\right)_{\text{ir}} < 0 \tag{3.17}$$

3.3 Heats of Reaction and Enthalpy Change

In the last two sections, we have introduced some concepts as well as the first law of thermodynamics. In this section, we will apply the above concepts and natural laws to understand the heat effects that accompany real chemical reactions. In some chemical reactions such as combustion reactions, heat is violently released when the reactants burn. We would like to learn and understand how heats and energies are involved in those reactions so that we can make better use of the released heats and energies.

Heats of Reaction

Chemical reactions can be viewed as a special type of process in which the reactants are in the initial state and the products are in the final state of the system. In this reaction process, energy is converted among chemical energy, thermal energy, and other forms of energies. **Chemical energy** is the energy associated with chemical bonds and intermolecular interactions; it belongs to potential energy. During a reaction some chemical bonds are broken and some others are formed. Consequently, the chemical energy of a system will change and some may appear in the form of heat. The **heat of reaction** (q_{rxn}) is defined as the quantity of heat exchanged between a system and its surroundings when a chemical reaction occurs within the system at *constant temperature*. In real reactions, however, the temperature of the system may either increase or decrease to different extent, and the actual heat of reactions will obviously differ with respect to the temperature changes. In order to give q_{rxn} a straightforward and clear meaning, constant temperature is required in the above definition.

Chemical reactions are classified as exothermic and endothermic according to the value of q_{rxn}. An **exothermic reaction** is a reaction with negative q_{rxn} ($q_{\text{rxn}} < 0$). In a non-isolated system, the released q_{rxn} will be given off to the surroundings. In an isolated system, no heat is exchanged between the system and its surroundings and thus the excess q_{rxn} will cause a temperature increase in the system. An **endothermic reaction** is one with positive q_{rxn} ($q_{\text{rxn}} > 0$). In a non-isolated system, the absorbed q_{rxn} will be given off by the surroundings. In an isolated system, the absorbed q_{rxn} will cause a temperature decrease in the system. Notice that here the increase or decrease in temperature is a result of q_{rxn}; that is, the released or absorbed q_{rxn} causes temperature change. One should not confuse it with the definition of heat that is the energy exchange as a result of a temperature difference; that is, the temperature difference causes heat exchange.

q_V and q_p

Consider a chemical reaction as a process, in which the reactants are in the initial state and the products are in the final state, as given by

$$\text{Reactants} \rightarrow \text{Products}$$
$$\text{(initial state)} \quad \text{(final state)}$$
$$U_i \quad\quad\quad U_f$$
$$\Delta U = U_f - U_i$$

程可简化为可逆过程。

从以上理想气体的可逆与不可逆恒温膨胀和压缩过程中可得结论：可逆过程（不论膨胀还是压缩）涉及的功是所有可能的功的最小值，可逆过程涉及的热是所有可能的热的最大值，即

$$w_{可逆} < w_{不可逆}, \quad q_{可逆} > q_{不可逆}, \quad \Delta U_{可逆} = \Delta U_{不可逆} \quad (3.13)$$

同时，可逆压缩的功恰好等于可逆膨胀的功的相反数。但不可逆压缩的功总大于相应不可逆膨胀的功的相反数，即

$$(w_{压缩})_{可逆} = -(w_{膨胀})_{可逆} \quad 且 \quad (w_{压缩})_{不可逆} > -(w_{膨胀})_{不可逆} \quad (3.14)$$

我们将理想气体从始态先进行膨胀或压缩至终态，再将其压缩或膨胀回原来的始态，以形成一个循环，如果整个过程均以可逆的方式进行，则此循环的总功恒为零。如果循环有任何部分以不可逆的方式进行，那么总功必为正，即

$$(w_{循环})_{可逆} = 0 \quad 且 \quad (w_{循环})_{不可逆} > 0 \quad (3.15)$$

相应地，

$$(q_{压缩})_{可逆} = -(q_{膨胀})_{可逆} \quad 且 \quad (q_{压缩})_{不可逆} < -(q_{膨胀})_{不可逆} \quad (3.16)$$

$$(q_{循环})_{可逆} = 0 \quad 且 \quad (q_{循环})_{不可逆} < 0 \quad (3.17)$$

3.3 反应热与焓变

前两节我们已经介绍了一些概念以及热力学第一定律。本节我们会应用上述概念和定律，来理解实际化学反应中伴随的热效应。在一些化学反应如燃烧反应中，反应物发生燃烧并剧烈放热。我们希望学习并理解这些反应中涉及的热和能量，这样就可以更好地利用这些热和能量。

反应热

化学反应可视为一类特殊的过程，其中反应物是体系的始态而生成物为终态。在此反应过程中，能量在化学能、热能和其他能之间转化。**化学能**是与化学键和分子间作用力相关联的能量，属于势能。在反应过程中一些化学键断裂而另一些键形成，体系的化学能会随之变化，有的会以热的形式出现。**反应热**（$q_{反应}$）定义为体系在恒温下发生化学反应时与环境之间交换的热量。然而在实际反应中，体系的温度可能会不同程度地升高或降低，实际的反应热显然会随温度的改变而有所不同。为了使 $q_{反应}$ 具有直接而清晰的含义，上述定义中需加上恒温条件。

根据 $q_{反应}$ 的值可将化学反应分为放热反应和吸热反应。**放热反应**是 $q_{反应}$ 为负值（$q_{反应} < 0$）的反应。在非孤立体系中，释放的 $q_{反应}$ 会进入其环境。孤立体系与其环境没有热交换，因此多余的 $q_{反应}$ 会使体系温度升高。**吸热反应**是 $q_{反应}$ 为正值（$q_{反应} > 0$）的反应。在非孤立体系中，吸收的 $q_{反应}$ 由其环境给出。而在孤立体系中，吸收的 $q_{反应}$ 会使体系温度降低。注意此处所说的温度升高或降低是由于 $q_{反应}$ 所致，即释放或吸收的 $q_{反应}$ 导致了温度的变化。请不要与热的定义混淆，热是由于温度不同而导致的能量交换，即由于温度差造成了热交换。

According to the first law of thermodynamics, $\Delta U = q + w$. Since in this reaction process $q = q_{rxn}$, we have
$$\Delta U = q_{rxn} + w$$
The work involved in most chemical reactions is only p-V work that is associated with the expansion and compression of gases. That is, most chemical reactions happen in conditions without work other than p-V work. If a reaction occurs at *constant volume* and without non-p-V work, then
$$\Delta V = 0 \quad \text{and} \quad w = -p_{ext}\Delta V = 0$$
We denote the heat of a constant-volume reaction as q_V, and then
$$\Delta U = q_{rxn} + w = q_{rxn} = q_V \tag{3.18}$$
The **heat of reaction at constant volume** is equal to ΔU. Even for a reaction that does not happen at constant volume in reality, the equation $\Delta U = q_V$ is still valid because q_V then stands for the hypothetical heat of reaction if it *did occur* at constant volume.

The equation $\Delta U = q_V$ seems simple. However, the world we live in is a constant-pressure world! Most reactions happen in beakers, flasks, and other containers open to the atmosphere and under the *constant pressure* of the atmosphere. We denote the **heat of reaction under constant pressure** as q_p. It is q_p that matters and suits most reactions.

For a typical reaction happening under constant pressure and without non-p-V work, we have
$$\Delta U = q_{rxn} + w = q_p - p\Delta V = q_V$$
Here, the subscript "ext" in p_{ext} is omitted for conciseness. Then,
$$q_p = q_V + p\Delta V = \Delta U + p\Delta V$$
Applying $\Delta U = U_f - U_i$ and $\Delta V = V_f - V_i$, we have
$$q_p = (U_f - U_i) + p(V_f - V_i) = (U_f + pV_f) - (U_i + pV_i)$$
As this is a constant-pressure reaction, $p = p_i = p_f$. Therefore,
$$q_p = (U_f + p_f V_f) - (U_i + p_i V_i) = (U + pV)_f - (U + pV)_i = \Delta(U + pV) \tag{3.19}$$

Enthalpy and Enthalpy Change

In **Equation (3.19)**, q_p can be written as the difference of two terms: one is the final state of $U + pV$ and the other is the initial state of $U + pV$. We can then define a new state function, called **enthalpy** (H), as the sum of the internal energy and the pressure-volume product of a system, as
$$H = U + pV \tag{3.20}$$
Equation (3.19) can be reformed as
$$q_p = H_f - H_i = \Delta H \tag{3.21}$$
Therefore, the physical meaning of **enthalpy change** (ΔH) is the heat of reaction under constant pressure and without non-p-V work. Both H and ΔH are extensive functions of the state. Similar as internal energy, the absolute values of enthalpy cannot be determined and are not important; it is the enthalpy change in the reaction that matters.

As discussed previously in **Section 2.3**, new definitions are usually coined when the following two requirements are met:
1) It makes the expressions much more concise;
2) It has a specific physical meaning.

Here, both requirements are met, so new definitions of enthalpy and enthalpy change are coined.

It is worthy to mention that each formula has its specific scope of application. Taking ΔH as an example, in general, we have
$$\Delta H = H_f - H_i = \Delta U + \Delta(pV) \tag{3.22}$$
For reactions occurring under constant pressure, we have

q_V 和 q_p

将化学反应看作一个过程，其中反应物是始态、生成物是终态，可表示为

$$\text{反应物} \rightarrow \text{生成物}$$
$$(\text{始态}) \quad (\text{终态})$$
$$U_\text{始} \qquad U_\text{终}$$
$$\Delta U = U_\text{终} - U_\text{始}$$

根据热力学第一定律，$\Delta U = q + w$。由于此反应过程中 $q = q_\text{反应}$，有

$$\Delta U = q_\text{反应} + w$$

大多数化学反应涉及的功只有与气体膨胀和压缩相关的体积功，即通常都发生在没有非体积功的条件下。如果一个反应发生在恒容且没有非体积功的条件下，则

$$\Delta V = 0 \quad \text{且} \quad w = -p_\text{外} \Delta V = 0$$

将恒容条件下的反应热记作 q_V，则有

$$\Delta U = q_\text{反应} + w = q_\text{反应} = q_V \tag{3.18}$$

恒容反应热与 ΔU 相等。对于实际上并没有在恒容条件下发生的反应，等式 $\Delta U = q_V$ 仍成立，因为此时 q_V 代表反应如果发生在恒容条件下的假想反应热。

等式 $\Delta U = q_V$ 看起来简单，但我们实际生活在一个恒压的世界里！大多数反应在烧杯、烧瓶以及其他敞口容器中发生，处在恒定的大气压条件下。我们将**恒压反应热**记作 q_p，这才是对于大多数反应而言真正重要且适宜的反应热。

对于一个发生在恒压且没有非体积功时的典型反应，有

$$\Delta U = q_\text{反应} + w = q_p - p\Delta V = q_V$$

这里为简洁见，略去了 $p_\text{外}$ 中的下标"外"。有

$$q_p = q_V + p\Delta V = \Delta U + p\Delta V$$

将 $\Delta U = U_\text{终} - U_\text{始}$ 和 $\Delta V = V_\text{终} - V_\text{始}$ 代入，有

$$q_p = (U_\text{终} - U_\text{始}) + p(V_\text{终} - V_\text{始}) = (U_\text{终} + pV_\text{终}) - (U_\text{始} + pV_\text{始})$$

对于恒压反应，$p = p_\text{始} = p_\text{终}$，故

$$q_p = (U_\text{终} + p_\text{终}V_\text{终}) - (U_\text{始} + p_\text{始}V_\text{始}) = (U + pV)_\text{终} - (U + pV)_\text{始} = \Delta(U + pV) \tag{3.19}$$

焓与焓变

在**式（3.19）**中，q_p 可写作两项之差：一项是终态的 $U + pV$，另一项是始态的 $U + pV$。我们可以定义一个新的状态函数，称为**焓**（H），作为体系的内能和压强与体积之积的和，即

$$H = U + pV \tag{3.20}$$

式（3.19）可改写为

$$q_p = H_\text{终} - H_\text{始} = \Delta H \tag{3.21}$$

因此，**焓变**（ΔH）的物理意义即为恒压且没有非体积功条件下的反应热。H 和 ΔH 均为广度状态函数。与内能类似，焓的绝对数值既不能确定也不重要，重要的是反应过程中的焓变。

正如前述 **2.3 节**所讨论的，满足如下两个条件时通常会给出新的定义：

$$\Delta H = H_f - H_i = \Delta U + p\Delta V \tag{3.23}$$

For reactions occurring at constant pressure and volume, we have

$$\Delta H = H_f - H_i = \Delta U \tag{3.24}$$

Measuring Heats of Reaction Experimentally

Previously, we have defined q_{rxn} as the quantity of heat exchanged between a system and its surroundings when a chemical reaction occurs within the system at *constant temperature*. However, real reactions may not always occur at constant temperature and the temperature of the system may either increase or decrease after the reaction. Then q_{rxn} can be viewed as the quantity of heat exchanged between a system and its surroundings as the system is restored to its initial temperature. In practical cases, we do not physically restore the system to its initial temperature. Instead, we calculate the quantity of heat that *would be* exchanged in this restoration. To do this, a probe (thermometer) is placed within the system to record the temperature change produced by the reaction. Then, we use the temperature change and other data of the system to calculate q_{rxn} that would have occurred at constant temperature. Experimentally, q_{rxn} is determined on a **calorimeter**, a device for measuring quantities of heat.

A bomb calorimeter [**Figure 3.5(a)**] is a device designed ideally for measuring q_V evolved in a combustion reaction. The system is everything within the double-walled outer jacket of the calorimeter that holds constant volume. This includes the bomb and its content, the water in which the bomb is immersed, the thermometer, the stirrer, and so on. The system is isolated from its surroundings. When the combustion reaction occurs, chemical energy is converted to thermal energy, and the temperature of the system rises. q_{rxn} is the quantity of heat that the system would have to lose to its surroundings to be restored to its initial temperature. This quantity of heat, in turn, is just the negative of the thermal energy gained by the calorimeter and its contents ($q_{calorim}$), as

$$q_V = -q_{calorim} = -(q_{bomb} + q_{water} + \cdots)$$

If the calorimeter is assembled in exactly the same way each time we use it—that is, use the same bomb, add the same quantity of water, and so on—we can define a heat capacity of the calorimeter. The **heat capacity** (C) of a system is the quantity of heat required to raise the temperature of the system by one degree Celsius. Sometimes, the **specific heat** (c) is used instead, which is the quantity of heat required to raise the temperature by one degree Celsius per unit mass. Therefore,

$$q = C\Delta T = cm\Delta T \tag{3.25}$$

The specific heats of substances are somewhat temperature dependent. At 25°C, the specific heat of water is 4.18 J g^{-1} °C^{-1}. However, the specific heats of substances do not vary significantly with respect to temperature. When temperature changes only in a small range, the specific heat can be assumed to be a constant. For example, the specific heat of water can be assumed as a constant between 0~100 °C.

Another type of calorimeter called the "coffee-cup" calorimeter [**Figure 3.5(b)**] is generally used to measure q_p of an aqueous reaction. We mix the reactants in a styrofoam cup under constant atmospheric pressure and measure the temperature change. Styrofoam is a good heat insulator, so there is very little heat transfer between the cup and the surrounding air. We treat the system (the cup and its contents) as an isolated system. Similarly, q_p is also the negative of the quantity of heat producing the temperature change in the calorimeter: $q_p = -q_{calorim}$.

Extent of Reaction

To calculate the actual q_{rxn} for a specific reaction, one must consider how much extent the reaction proceeds or how many moles of reactants are consumed. However, since the stoichiometric coefficients are generally different for different reactants, the value of q_{rxn} varies when different reactants are used, which is very inconvenient. In order to overcome this inconvenience, we use the **extent of reaction** ξ (Greek, xi) to measure the extent to which the reaction has proceeded, that is, how many *moles of reaction* have occurred.

Consider a general reaction written as

$$a\mathrm{A} + b\mathrm{B} + \cdots \rightleftharpoons g\mathrm{G} + h\mathrm{H} + \cdots$$

We regard it as an equation (replacing $\rightleftharpoons$ with =) and reformulate it as

$$0 = -a\mathrm{A} - b\mathrm{B} + g\mathrm{G} + h\mathrm{H} + \cdots$$

1) 新定义可将复杂的表达式大幅简化；
2) 新定义具有明确的物理意义。

这里两个条件均满足，因此我们给出了焓与焓变的定义。

值得一提的是，每个公式都有其特定的适用范围。以 ΔH 为例，一般情况下，有

$$\Delta H = H_\text{终} - H_\text{始} = \Delta U + \Delta(pV) \tag{3.22}$$

对于发生在恒压条件下的反应，有

$$\Delta H = H_\text{终} - H_\text{始} = \Delta U + p\Delta V \tag{3.23}$$

对于发生在恒压恒容条件下的反应，有

$$\Delta H = H_\text{终} - H_\text{始} = \Delta U \tag{3.24}$$

反应热的实验测定

此前我们定义了 $q_\text{反应}$ 为恒温条件下发生化学反应时体系与其环境交换的热量。但实际反应可能不总在恒温条件下发生，体系的温度在反应后可能升高或降低。那么 $q_\text{反应}$ 可视为使体系重新回到初始温度时与其环境所交换的热量。在实例中，我们不用将体系物理地复原至初始温度，而是计算在此复原过程中所交换的热量。为达到这一目的，可在体系中放入一个探针（即温度计）来记录反应引起的温度变化。然后用温度变化及其他数据来计算反应假如发生在恒温条件下的 $q_\text{反应}$。实验中，$q_\text{反应}$ 由测量热量的**量热计**测得。

弹式量热计 [**图 3.5(a)**] 是用于测量燃烧反应的 q_V 的装置。其体系为保持恒容的双层量热计外壳内的所有部件，包括氧弹及其内容物、将氧弹浸入的水、温度计、搅拌器等，该体系可视为孤立体系。当燃烧反应发生时，化学能转化为热能，体系温度升高。$q_\text{反应}$ 为体系若要复原至初始温度所需对环境释放的热量。该热量即为量热计及其内容物所获热量（$q_\text{量热计}$）的相反数，即

$$q_V = -q_\text{量热计} = -(q_\text{氧弹} + q_\text{水} + \cdots)$$

如果我们每次使用量热计时均以完全相同的方式组装，即使用同一个氧弹、加入同等量的水等，我们即可确定量热计的热容。一个体系的**热容**（C）是使该体系升高 1℃ 所需的热量。有时也用**比热**（c）替代，它是使单位质量的物质升高 1℃ 所需的热量。故

$$q = C\Delta T = cm\Delta T \tag{3.25}$$

物质的比热一定程度上与温度相关。在 25℃ 下，水的比热为 4.18 J g^{-1} ℃$^{-1}$。但物质的比热随温度变化并不显著，当温度只在一个小范围内变化时，比热可视为常数。例如，水的比热在 0~100 ℃ 之间均可视为常数。

另一种量热计称为"咖啡杯式"量热计 [**图 3.5(b)**]，一般用于测量水溶液反应的 q_p。在恒定大气压下，将聚苯乙烯泡沫塑料杯里的反应物混合并测量其温度变化。泡沫塑料是良好的热绝缘体，因此杯子与周围空气几乎没有热交换。将体系（咖啡杯及其内容物）视为孤立体系，q_p 同样等于使量热计温度发生改变的热量的相反数：$q_p = -q_\text{量热计}$。

反应进度

为计算某特定反应的实际 $q_\text{反应}$，我们必须考虑反应进行到了何种程度，或者说消耗了多少摩尔的反应物。但由于不同反应物一般具有不同的化学计量数，$q_\text{反应}$ 的值也会随使用的反应物不同而变化，

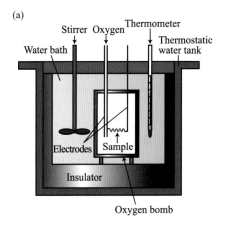

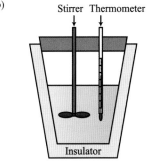

Figure 3.5 The calorimeters to measure the heat of reaction experimentally. (a) A bomb calorimeter to measure q_V. (b) A "coffee-cup" calorimeter to measure q_p.

图 3.5 实验中测量反应热的量热计：(a) 测量 q_V 的弹式量热计; (b) 测量 q_p 的"咖啡杯式"量热计。

or simply as
$$0 = \sum_X v_X X \tag{3.26}$$

where the symbol Σ (Greek, sigma) means "the sum of"; X stands for various substances involved in this reaction; and v_X is the stoichiometric coefficient for X, which is defined to be negative for reactants and positive for products. Suppose an infinitesimal amount of the reaction $d\xi$ has occurred, and the change of the amount of X can be represented by

$$dn_X = v_X d\xi$$

The extent of reaction is then defined as

$$d\xi = \frac{dn_X}{v_X} \tag{3.27}$$

Considering finite changes instead of infinitesimal changes, the extent of reaction is

$$\Delta\xi = \xi_f - \xi_i = \frac{\Delta n_X}{v_X} = \frac{(n_f)_X - (n_i)_X}{v_X}$$

At the beginning of a reaction, there is no product so $\xi_i = 0$ and $\xi_f = \Delta\xi$. For simplification purpose, sometimes ξ is used instead of ξ_f or $\Delta\xi$.

The advantage of using ξ to measure the extent in which the reaction has proceeded is that the values of ξ do not vary with respect to the substances used, no matter reactants or products. For example, let us consider the exothermic ammonia synthesis reaction

$$N_2(g) + 3H_2(g) \rightarrow 2NH_3(g) + q_{rxn}$$

If 10 mol N_2 are consumed in a period of time, 30 mol H_2 must also be consumed and 20 mol NH_3 must be produced according to the above equation. We can calculate ξ with respect to N_2, H_2 and NH_3 as

$$\xi_{N_2} = \frac{-10 \text{ mol}}{-1} = 10 \text{ mol}, \; \xi_{H_2} = \frac{-30 \text{ mol}}{-3} = 10 \text{ mol}, \text{ and } \xi_{NH_3} = \frac{20 \text{ mol}}{2} = 10 \text{ mol}$$

Therefore,

$$\xi_{N_2} = \xi_{H_2} = \xi_{NH_3}$$

We can think $\xi = 1$ mol as that 1 mole of the above reaction has proceeded. This process consumes 1 mol N_2 and 3 mol H_2, produces 2 mol NH_3, and gives off a total heat of q_{rxn}.

If the exothermic ammonia synthesis reaction is written as

$$\frac{1}{2}N_2(g) + \frac{3}{2}H_2(g) \rightarrow NH_3(g) + \frac{1}{2}q_{rxn}$$

it means that when 1 mole of the above reaction has proceeded ($\xi = 1$ mol), it consumes ½ mol N_2 and ³⁄₂ mol H_2, produces 1 mol NH_3, and gives off a total heat of ½ q_{rxn}.

Enthalpy Change in Chemical Reactions

Chemical reactions happen typically under constant pressure and without non-p-V work. For those reactions

$$q_p = \Delta H = \Delta U + p\Delta V = \Delta U - w$$

For example, consider the following reaction

$$2CO(g) + O_2(g) \rightarrow 2CO_2(g)$$

The heat of this reaction is measured as -566.0 kJ at a constant temperature of 298.15 K, that is, $\Delta H = q_p = -566.0$ kJ. At $\xi = 1$ mol, 2 mol CO and 1 mol O_2 to produce 2 mol CO_2. The value of the p-V work is given by

$$w = -p\Delta V = -p(V_f - V_i) = -RT(n_f - n_i) = -0.0083145 \times 298.15 \times (2 - 2 - 1) \text{ kJ} = 2.5 \text{ kJ}$$

The change in internal energy is

$$\Delta U = \Delta H + w = (-566.0 + 2.5) \text{ kJ} = -563.5 \text{ kJ}$$

这非常不方便。为克服这一不便，我们用**反应进度** ξ（希腊字母，xi）来衡量反应进行的程度，即发生了多少摩尔的反应。

考虑如下反应通式：

$$a\mathrm{A} + b\mathrm{B} + \cdots \rightleftharpoons g\mathrm{G} + h\mathrm{H} + \cdots$$

将其看作等式（用 = 替代 $\rightleftharpoons$），可变形为

$$0 = -a\mathrm{A} - b\mathrm{B} + g\mathrm{G} + h\mathrm{H} + \cdots$$

或简写为

$$0 = \sum_{\mathrm{X}} \nu_{\mathrm{X}} \mathrm{X} \qquad (3.26)$$

其中符号 $\sum$（希腊字母，sigma）代表"之和"；X 代表参与反应的各种物质；ν_{X} 为 X 对应的化学计量数，定义其值对反应物为负、对生成物为正。假设发生了无限小量的反应 $\mathrm{d}\xi$，则 X 的物质的量的变化可表示为

$$\mathrm{d}n_{\mathrm{X}} = \nu_{\mathrm{X}} \mathrm{d}\xi$$

故反应进度可定义为

$$\mathrm{d}\xi = \frac{\mathrm{d}n_{\mathrm{X}}}{\nu_{\mathrm{X}}} \qquad (3.27)$$

考虑一定量的变化而非无限小的变化，则反应进度为

$$\Delta\xi = \xi_{终} - \xi_{始} = \frac{\Delta n_{\mathrm{X}}}{\nu_{\mathrm{X}}} = \frac{(n_{终})_{\mathrm{X}} - (n_{始})_{\mathrm{X}}}{\nu_{\mathrm{X}}}$$

反应刚开始没有生成物，故 $\xi_{始} = 0$ 且 $\xi_{终} = \Delta\xi$。为简便起见，有时直接用 ξ 替代 $\xi_{终}$ 或 $\Delta\xi$。

使用 ξ 来衡量反应进行程度的优势在于，ξ 不随计算时使用的物质（不论反应物还是生成物）而改变。例如，考察合成氨这一放热反应

$$\mathrm{N_2(g)} + 3\mathrm{H_2(g)} \rightarrow 2\mathrm{NH_3(g)} + q_{反应}$$

若在某段时间内消耗了 10 mol $\mathrm{N_2}$，那么根据以上方程式必然同时消耗 30 mol $\mathrm{H_2}$ 并生成 20 mol $\mathrm{NH_3}$。用 $\mathrm{N_2}$、$\mathrm{H_2}$ 和 $\mathrm{NH_3}$ 来分别计算 ξ，有

$$\xi_{\mathrm{N_2}} = \frac{-10\ \mathrm{mol}}{-1} = 10\ \mathrm{mol}, \qquad \xi_{\mathrm{H_2}} = \frac{-30\ \mathrm{mol}}{-3} = 10\ \mathrm{mol}$$

$$\xi_{\mathrm{NH_3}} = \frac{20\ \mathrm{mol}}{2} = 10\ \mathrm{mol}$$

因此，

$$\xi_{\mathrm{N_2}} = \xi_{\mathrm{H_2}} = \xi_{\mathrm{NH_3}}$$

可以认为 $\xi = 1$ mol 即为发生了 1 mol 上述反应，此过程消耗了 1 mol $\mathrm{N_2}$ 和 3 mol $\mathrm{H_2}$，生成了 2 mol $\mathrm{NH_3}$，并放出了 $q_{反应}$ 的总热量。

若把合成氨的放热反应写作

$$\frac{1}{2}\mathrm{N_2(g)} + \frac{3}{2}\mathrm{H_2(g)} \rightarrow \mathrm{NH_3(g)} + \frac{1}{2}q_{反应}$$

则表示当发生 1 mol 上述反应时（$\xi = 1$ mol），消耗了 ½ mol $\mathrm{N_2}$ 和 ³⁄₂ mol $\mathrm{H_2}$，生成了 1 mol $\mathrm{NH_3}$，并放出了 ½ $q_{反应}$ 的总热量。

化学反应中的焓变

典型的化学反应通常发生在恒压且没有非体积功的条件下。对于这些反应

$$q_p = \Delta H = \Delta U + p\Delta V = \Delta U - w$$

In this textbook, all heats of reactions are treated as ΔH unless otherwise stated.

Some chemical reactions may involve changes in the state of matter. The enthalpy changes of the same reaction will be different if the states of some reactants or products are different. When a liquid vaporizes, if the temperature of the liquid remains constant, the liquid must absorb heat from its surroundings to compensate the energy carried off by the vaporizing molecules. This heat required to vaporize a fixed quantity of liquids at a constant temperature is the **enthalpy of vaporization** (ΔH_{vap}) that we have learnt in **Section 2.6**. For the vaporization of 1 mol water

$$H_2O(l) \rightarrow H_2O(g) \quad \Delta H_{vap} = 44.0 \text{ kJ} \, (298.15 \text{ K})$$

The heat required to melt a fixed quantity of solids at a constant temperature is called the **enthalpy of fusion** (ΔH_{fus}). For the melting of 1 mol ice

$$H_2O(s) \rightarrow H_2O(l) \quad \Delta H_{fus} = 6.01 \text{ kJ} \, (273.15 \text{ K})$$

Standard States and Standard Enthalpy Changes

The measured enthalpy change for a reaction has a unique value only if the reactants (initial state) and products (final state) are precisely described. A typical way that scientists do under such circumstances is to define a particular state as the standard state for both reactants and products. Therefore, the **standard enthalpy change** ($\Delta H°$) is defined as the enthalpy change in a reaction in which all reactants and products are in their standard states.

The **standard state** of a solid or a liquid substance is the pure substance at $p° = 1$ bar and at the temperature of interest. The standard state of a gas is the pure ideal gas at $p° = 1$ bar and the temperature of interest. The standard state of a solution is the ideal solution with a concentration of $m° = 1$ mol kg^{-1} or $c° = 1$ mol L^{-1} and the temperature of interest. Although temperature is not involved in the definition of a standard state, it still must be specified in tabulated values of $\Delta H°$, because $\Delta H°$ depends on temperature. The values given in this textbook are all for 298.15 K (25.0°C) unless otherwise stated.

Thermochemical Equation

A **thermochemical equation** is a balanced stoichiometric chemical equation that includes the standard molar enthalpy of reaction $\Delta_r H_m°(T)$. In $\Delta_r H_m°(T)$, the subscript "r" stands for "reaction" and "m" for "molar"; the degree symbol (°) denotes that all substances involved are at their standard states. Therefore, **standard molar enthalpy** of a reaction is the standard heat of the reaction when 1 mol the reaction has proceeded ($\xi = 1$ mol) at T. $\Delta_r H_m°(T)$ can be abbreviated as $\Delta H°$, and if not specified, the temperature is set to be 298.15 K or 25.0°C by default. For example,

$$H_2(g) + \frac{1}{2}O_2(g) \rightarrow H_2O(l) \quad \Delta H° = -285.8 \text{ kJ mol}^{-1}$$

$$2H_2(g) + O_2(g) \rightarrow 2H_2O(l) \quad \Delta H° = -571.6 \text{ kJ mol}^{-1}$$

$$\frac{1}{2}N_2(g) + O_2(g) \rightarrow NO_2(g) \quad \Delta H° = 33.2 \text{ kJ mol}^{-1}$$

When writing a thermochemical equation, one must pay special attention in that:

1) The equation must be balanced with proper stoichiometric coefficients, which can be either integers or fractions.
2) The states of substances must be included in the thermochemical equation, as changes in the state of matter always involve enthalpy change. If substances with different allotropes or crystalline structures are involved, it also needs to be specified in the thermochemical equation. For example,

$$H_2(g) + \frac{1}{2}O_2(g) \rightarrow H_2O(g) \quad \Delta H° = -241.8 \text{ kJ mol}^{-1}$$

$$C(graphite, s) + O_2(g) \rightarrow CO_2(g) \quad \Delta H° = -393.5 \text{ kJ mol}^{-1}$$

$$C(diamond, s) + O_2(g) \rightarrow CO_2(g) \quad \Delta H° = -395.4 \text{ kJ mol}^{-1}$$

3) The value of $\Delta_r H_m°(T)$ associated with the corresponding stoichiometric coefficients and states of substances must be included in the thermochemical equation.

例如，考虑如下反应

$$2CO(g) + O_2(g) \to 2CO_2(g)$$

测得该反应在 298.15 K 下的反应热为 -566.0 kJ，即 $\Delta H = q_p = -566.0$ kJ。当 $\xi = 1$ mol 时，2 mol CO 与 1 mol O_2 生成 2 mol CO_2。体积功的值为

$$w = -p\Delta V = -p(V_{终} - V_{始}) = -RT(n_{终} - n_{始})$$
$$= -0.0083145 \times 298.15 \times (2 - 2 - 1) \text{ kJ} = 2.5 \text{ kJ}$$

内能的改变量为

$$\Delta U = \Delta H + w = (-566.0 + 2.5) \text{ kJ} = -563.5 \text{ kJ}$$

在本教材中除非特别说明，所有反应热均指 ΔH。

有些化学反应可能涉及物态变化，如果某些反应物或生成物的物态不同，即使同一反应的焓变也会不同。当液体蒸发时，若其温度保持不变，则必须从环境吸热来抵消蒸发分子所带走的能量。恒温下蒸发一定量液体所需的热，即为我们在 **2.6 节**已经学过的**蒸发焓**（$\Delta H_{蒸发}$）。对于 1 mol 水的蒸发，有

$$H_2O(l) \to H_2O(g) \quad \Delta H_{蒸发} = 44.0 \text{ kJ} \; (298.15 \text{ K})$$

恒温下熔化一定量固体所需的热，称为**熔化焓**（$\Delta H_{熔化}$）。对于 1 mol 冰的熔化，有

$$H_2O(s) \to H_2O(l) \quad \Delta H_{熔化} = 6.01 \text{ kJ} \; (273.15 \text{ K})$$

标准状态与标准焓变

只有当反应物（始态）和生成物（终态）均严格确定时，一个反应测得的焓变才具有定值。这种情况下科学家们的常规做法是，定义某一特殊状态作为反应物和生成物的标准状态，而**标准焓变**（$\Delta H^{\ominus}$）则定义为所有反应物和生成物均处于标准状态的反应焓变。

固态和液态物质的**标准状态**（简称标态）是在 $p^{\ominus} = 1$ bar 和指定温度下的纯物质，气体的标态是 $p^{\ominus} = 1$ bar 和指定温度下的纯理想气体，溶液的标态是指定温度下浓度为 $m^{\ominus} = 1$ mol kg^{-1} 或 $c^{\ominus} = 1$ mol L^{-1} 的理想溶液。虽然标态的定义中并不涉及温度，但在 $\Delta H^{\ominus}$ 的列表值中仍需指定温度，因为 $\Delta H^{\ominus}$ 与温度相关。如果没有明确指出，在本教材中给出的均为 298.15 K（25.0℃）下的值。

热化学方程式

热化学方程式是包含反应标准摩尔焓变 $\Delta_r H_m^{\ominus}(T)$ 的配平的化学计量方程式。在 $\Delta_r H_m^{\ominus}(T)$ 中，下标 "r" 代表 "反应"，"m" 代表 "摩尔"，符号 "$\ominus$" 表示参与反应的所有物质均处于标态。因此，反应的**标准摩尔焓变**是在温度 T 下发生 1 mol 反应（$\xi = 1$ mol）时的标准反应热。$\Delta_r H_m^{\ominus}(T)$ 可简写为 $\Delta H^{\ominus}$，如果没有特别说明，温度默认为 298.15 K 或 25.0℃。例如，

$$H_2(g) + \frac{1}{2}O_2(g) \to H_2O(l) \quad \Delta H^{\ominus} = -285.8 \text{ kJ mol}^{-1}$$

$$2H_2(g) + O_2(g) \to 2H_2O(l) \quad \Delta H^{\ominus} = -571.6 \text{ kJ mol}^{-1}$$

$$\frac{1}{2}N_2(g) + O_2(g) \to NO_2(g) \quad \Delta H^{\ominus} = 33.2 \text{ kJ mol}^{-1}$$

在书写热化学方程式时，需要特别注意以下三点：

1) 热化学方程式必须用合适的化学计量数配平，计量数可以是整

Dependence of Enthalpy Change on Temperature

The enthalpy change for a reaction is temperature-dependent, and thus, the temperature needs to be specified when indicating the enthalpy change. In general, ΔH does not vary significantly with respect to temperature, so that sometimes ΔH can be assumed to be a constant when the temperature changes only in a small range. This is generally the case when Clausius-Clapeyron equation is applicable in **Section 2.6**.

The difference in ΔH for a reaction at two different temperatures is determined by the quantity of heat involved in changing the reactants and products from one temperature to the other under constant pressure. An **enthalpy diagram** is a diagrammatic representation of enthalpy changes in a process, as shown in **Figure 3.6**. The enthalpy change can be calculated using $q=cm\Delta T$ by considering the specific heats of all substances involved in the reaction. If the state of any substance has changed during the temperature change process, the corresponding enthalpy change in the state change also needs to be considered. Therefore, one can calculate ΔH for a reaction at any temperature of interest from ΔH at 298.15 K.

3.4 Hess's Law and Standard Enthalpies of Formation

One of the advantages in using the concept of enthalpy lies in that a great number of heats of reactions can be calculated from a small number of experimental measurements. This is done based on Hess's law and the concept of standard enthalpy of formation.

Hess's Law

As discussed previously in the previous section, ΔH is an extensive state function and is directly proportional to the amounts of substances in a system. Given that

$$N_2(g) + 3H_2(g) \rightarrow 2NH_3(g) \quad \Delta H_1^\circ = -91.8 \text{ kJ mol}^{-1}$$

we then have

$$\frac{1}{2}N_2(g) + \frac{3}{2}H_2(g) \rightarrow NH_3(g) \quad \Delta H_2^\circ = \frac{1}{2}\Delta H_1^\circ = -45.9 \text{ kJ mol}^{-1}$$

Meanwhile, ΔH changes sign when a process is reversed, such as

$$2NH_3(g) \rightarrow N_2(g) + 3H_2(g) \quad \Delta H_3^\circ = -\Delta H_1^\circ = 91.8 \text{ kJ mol}^{-1}$$

In the mid-nineteenth century, the Russian chemist Germain H. Hess (1802—1850) first proposed **Hess's law** based on numerous thermochemistry data. It states that the overall heat of a reaction is always the same no matter a reaction happens in one step or multiple steps, in real or in hypothesis. For example

$$\frac{1}{2}N_2(g) + \frac{1}{2}O_2(g) \rightarrow NO(g) \quad \Delta H_4^\circ = 91.3 \text{ kJ mol}^{-1}$$

$$NO(g) + \frac{1}{2}O_2(g) \rightarrow NO_2(g) \quad \Delta H_5^\circ = -58.1 \text{ kJ mol}^{-1}$$

we then have

$$\frac{1}{2}N_2(g) + O_2(g) \rightarrow NO_2(g) \quad \Delta H_6^\circ = \Delta H_4^\circ + \Delta H_5^\circ = 33.2 \text{ kJ mol}^{-1}$$

Hess's law was proposed when the first law of thermodynamics and many other thermodynamic laws were not discovered yet. From the prospective of modern science, Hess's law can be derived directly from the nature of ΔH as an extensive state function; that is, ΔH always has the same value, as long as the reaction happens from the same initial state to the same final state. However, historically Hess's law has provided many important experimental evidences leading to the discovery of the first law of thermodynamics. We should objectively evaluate the contribution of a scientific work to its specific historical background.

Hess's law can be applied to calculate the ΔH values that cannot be measured precisely or directly, which is where Hess's law is of the greatest significance. For example, to obtain the standard enthalpy change for the reaction

数也可以是分数。

2) 热化学方程式中必须包含物质状态，因为物态变化总牵涉焓变。如果还涉及了具有不同同素异形体或晶体结构的物质，也需要在热化学方程式中标明。例如

$$H_2(g) + \frac{1}{2}O_2(g) \rightarrow H_2O(g) \quad \Delta H^\ominus = -241.8 \text{ kJ mol}^{-1}$$

$$C(石墨, s) + O_2(g) \rightarrow CO_2(g) \quad \Delta H^\ominus = -393.5 \text{ kJ mol}^{-1}$$

$$C(金刚石, s) + O_2(g) \rightarrow CO_2(g) \quad \Delta H^\ominus = -395.4 \text{ kJ mol}^{-1}$$

3) 热化学方程式中必须包含与相应化学计量数和物态对应的 $\Delta_r H_m^\ominus(T)$ 值。

焓变与温度的关系

反应的焓变与温度相关，因此给出焓变时必须指定温度。一般而言，ΔH 不随温度显著变化，故当温度只在小范围变化时，ΔH 有时可视为常数。**2.6 节**的克劳修斯 - 克拉贝龙方程适用时，就是这种情况。

反应在两个不同温度下 ΔH 的差值，由恒压下将反应物和生成物从一个温度变化到另一个温度时所涉及的热量决定。**焓图**是某个过程中焓变的示意图，如**图 3.6** 所示。焓变可由 $q=cm\Delta T$、通过参与反应的所有物质的比热计算出。如果在温度变化过程中涉及了物态变化，则该物态变化中的相应焓变也要纳入考虑。因此，可由 298.15 K 下的 ΔH 计算出反应在任意指定温度下的 ΔH。

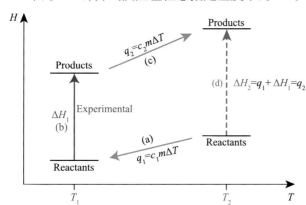

Figure 3.6　The enthalpy diagram to conceptualize ΔH as a function of temperature.

图 3.6　表示 ΔH 是温度函数的焓图。

3.4　盖斯定律与标准生成焓

使用焓的概念优势之一在于，可以从较少量的实验测量值计算得到大量反应热数据，而这是基于盖斯定律和标准生成焓的概念实现的。

盖斯定律

如上节所述，ΔH 是广度状态函数，与体系的物质的量成正比。已知

$$N_2(g) + 3H_2(g) \rightarrow 2NH_3(g) \quad \Delta H_1^\ominus = -91.8 \text{ kJ mol}^{-1}$$

有

$$\frac{1}{2}N_2(g) + \frac{3}{2}H_2(g) \rightarrow NH_3(g) \quad \Delta H_2^\ominus = \frac{1}{2}\Delta H_1^\ominus = -45.9 \text{ kJ mol}^{-1}$$

此外，当过程反转时 ΔH 改变符号，如

$$3C(\text{graphite},s) + 4H_2(g) \rightarrow C_3H_8(g) \quad \Delta H° = ?$$

If we try to get graphite and hydrogen to react, a slight reaction will occur, but it will not go to completion. Furthermore, the product will not be limited to propane (C_3H_8); several other hydrocarbons will form as well. The fact is that we cannot directly measure $\Delta H°$ for this reaction. Instead, we may resort to an indirect calculation from the $\Delta H°$ values that can be established by experiments as follows:

$$(a) \; C_3H_8(g) + 5O_2(g) \rightarrow 3CO_2(g) + 4H_2O(l) \quad \Delta H_a° = -2219.9 \text{ kJ mol}^{-1}$$
$$(b) \; C(\text{graphite},s) + O_2(g) \rightarrow CO_2(g) \quad \Delta H_b° = -393.5 \text{ kJ mol}^{-1}$$
$$(c) \; H_2(g) + \frac{1}{2}O_2(g) \rightarrow H_2O(l) \quad \Delta H_c° = -285.8 \text{ kJ mol}^{-1}$$

The above three reactions are the combustion reactions of C_3H_8, C, and H_2, respectively, the $\Delta H°$ values of which can all be measured experimentally. By applying Hess's law, we can combine these three equations by $3(b) + 4(c) - (a)$ to obtain

$$3C(\text{graphite},s) + 4H_2(g) \rightarrow C_3H_8(g) \quad \Delta H° = 3\Delta H_b° + 4\Delta H_c° - \Delta H_a° = -103.8 \text{ kJ mol}^{-1}$$

Standard Enthalpies of Formation

Previously, we have discussed that the absolute values of enthalpy cannot be determined and are not important; it is the enthalpy change in the reaction that matters. What scientists usually do under such circumstances is to choose an arbitrary zero for enthalpy. A similar case is what scientists have done in defining the height of a mountain. The absolute value of the height of a mountain also cannot be determined because there is no absolute zero for the height. By agreement, scientists define the elevation of a mountain as the vertical distance between the mountaintop and the mean sea level. The mean sea level, then, is the arbitrarily chosen zero for elevation. All other points on the earth are relative to this zero elevation. For example, the elevation of Mt. Everest is 8848 m.

Although this zero point is chosen arbitrarily, it still meets some basic requirements:
1) It should be a constant value that does not vary (or at least does not vary significantly) with conditions;
2) It should be convenient to use;
3) It should be a well-known basis.

The mean sea level meets all the above requirements, and so should the arbitrary zero for enthalpy. Since all substances are made of elements, scientists choose the enthalpies of certain forms of the elements to be the arbitrary zero for enthalpy. Then, the enthalpies of other substances can be determined relative to this zero.

The **standard enthalpy of formation** [$\Delta_f H_m°(T)$, or $\Delta H_f°$ for short at 298.15 K] of a substance is the enthalpy change that occurs in the formation of 1 mol substance from the reference forms of the elements, all in their standard states. The **reference forms** of the elements are the most stable forms (with a few exceptions) of the elements at 1 bar and the given temperature. In $\Delta H_f°$, the degree symbol denotes that the enthalpy change is a standard enthalpy change, and the subscript "f" signifies that this is a formation reaction of the substance from its elements. Because the formation of the reference form of an element from itself is no change at all, $\Delta H_f°$ of a pure element in its reference form is 0.

Examples of the reference forms of several elements include:

$$\text{Na}(s), H_2(g), N_2(g), O_2(g), C(\text{graphite}), Br_2(l)$$

For carbon, there is a measurable enthalpy difference between two naturally existing allotropes: graphite and diamond. Since

$$C(\text{graphite},s) \rightarrow C(\text{diamond},s) \quad \Delta H° = 1.9 \text{ kJ mol}^{-1}$$

graphite, the more stable form with lower enthalpy, is chosen as the reference form. Thus, we assign that $\Delta H_f°(\text{graphite}) = 0$, and $\Delta H_f°(\text{diamond}) = 1.9 \text{ kJ mol}^{-1}$. Although bromine can be obtained in either the gaseous or liquid state at 298.15 K, $Br_2(l)$ is the most stable form since

$$Br_2(l) \rightarrow Br_2(g) \quad \Delta H° = 30.9 \text{ kJ mol}^{-1}$$

The standard enthalpies of formation are $\Delta H_f°(Br_2(l)) = 0$ and $\Delta H_f°(Br_2(g)) = 30.9 \text{ kJ mol}^{-1}$.

$$2NH_3(g) \rightarrow N_2(g) + 3H_2(g) \quad \Delta H_3^\ominus = -\Delta H_1^\ominus = 91.8 \text{ kJ mol}^{-1}$$

19 世纪中叶，俄国化学家杰曼·H. 盖斯（1802—1850）基于大量热化学数据首次提出**盖斯定律**：不管化学反应是一步完成还是分步完成，是实际发生还是假想发生，其总和热效应均相同。如

$$\frac{1}{2}N_2(g) + \frac{1}{2}O_2(g) \rightarrow NO(g) \quad \Delta H_4^\ominus = 91.3 \text{ kJ mol}^{-1}$$

$$NO(g) + \frac{1}{2}O_2(g) \rightarrow NO_2(g) \quad \Delta H_5^\ominus = -58.1 \text{ kJ mol}^{-1}$$

有

$$\frac{1}{2}N_2(g) + O_2(g) \rightarrow NO_2(g) \quad \Delta H_6^\ominus = \Delta H_4^\ominus + \Delta H_5^\ominus = 33.2 \text{ kJ mol}^{-1}$$

盖斯定律提出时，热力学第一定律和许多其他热力学定律都还没有被发现。从现代科学的角度看，盖斯定律可由 ΔH 作为广度状态函数的本质直接导出，即从同一始态到同一终态，化学反应的 ΔH 总相等。然而，历史上盖斯定律为热力学第一定律的发现提供了许多重要的实验证据。我们应该在特定历史背景下客观地评价科研工作的贡献。

盖斯定律可用于计算一些难以准确测定或根本不能直接测定的 ΔH，这也正是盖斯定律最为重要的意义所在。例如，为得到如下反应的标准焓变：

$$3C(\text{石墨}, s) + 4H_2(g) \rightarrow C_3H_8(g) \quad \Delta H^\ominus = ?$$

如果我们尝试将石墨和氢气直接作用，反应非常轻微，不会进行完全。此外，生成物不限于丙烷（C_3H_8），其他几种碳氢化合物也会形成。事实上，我们无法直接测得该反应的 ΔH。相反，可以利用实验可测的如下 ΔH 来进行间接计算：

(a) $C_3H_8(g) + 5O_2(g) \rightarrow 3CO_2(g) + 4H_2O(l) \quad \Delta H_a^\ominus = -2219.9 \text{ kJ mol}^{-1}$

(b) $C(\text{石墨}, s) + O_2(g) \rightarrow CO_2(g) \quad \Delta H_b^\ominus = -393.5 \text{ kJ mol}^{-1}$

(c) $H_2(g) + \frac{1}{2}O_2(g) \rightarrow H_2O(l) \quad \Delta H_c^\ominus = -285.8 \text{ kJ mol}^{-1}$

上述三个反应分别是 C_3H_8、C 和 H_2 的燃烧反应，其 ΔH 均可通过实验测量。应用盖斯定律，通过 3(b) + 4(c) − (a) 组合这三个反应，可得

$$3C(\text{石墨}, s) + 4H_2(g) \rightarrow C_3H_8(g)$$

$$\Delta H^\ominus = 3\Delta H_b^\ominus + 4\Delta H_c^\ominus - \Delta H_a^\ominus = -103.8 \text{ kJ mol}^{-1}$$

标准生成焓

此前我们讨论过，焓的绝对数值既不能确定也不重要，重要的是反应过程中的焓变。在这种情况下，科学家们通常会给焓选择一个任意零点。这与科学家们确定山的高度类似，山的高度的绝对数值也不能确定，因为并不存在高度的绝对零点。科学家们约定，将平均海平面至山顶的垂直距离定义为海拔，而平均海平面即是给海拔选择的任意零点。地球上其他点的海拔都是相对于该零点而言，如珠穆朗玛峰的海拔为 8848 m。

尽管零点的选择是任意的，仍需满足一些基本要求：
1) 它应是一个不随条件变化（或至少变化不显著）的常数；
2) 它应使用方便；
3) 它应是一个众所周知的基准。

An exception in which the reference form is not the most stable form is the element phosphorus. Although solid white phosphorus has a higher enthalpy than solid red phosphorus, it is still chosen as the reference form due to historical reasons.

$$P(s, white) \rightarrow P(s, red) \quad \Delta H° = -17.6 \text{ kJ mol}^{-1}$$

The standard enthalpies of formation are $\Delta H_f°(P(s, white)) = 0$ and $\Delta H_f°(P(s, red)) = -17.6$ kJ mol^{-1}.

Standard enthalpies of formation for a number of substances at 298.15 K are tabulated in **Table 3.1** and **Appendix C.1**. Both positive and negative standard enthalpies of formation are possible. The standard enthalpies of formation are closely related to molecular structure. For example, among various hydrocarbons, alkanes such as CH_4, C_2H_6, etc. have negative $\Delta H_f°$ values but alkenes and alkynes such as C_2H_4 and C_2H_2 have positive $\Delta H_f°$ values. This indicates that alkanes are more stable or less reactive than alkenes and alkynes.

Table 3.1 Standard Molar Enthalpies of Formation ($\Delta H_f°$) of Some Common Substances at 298.15 K
表 3.1 298.15 K 时一些常见物质的标准摩尔生成焓 ($\Delta H_f°$)

Substance (物质)	$\Delta H_f°$/(kJ mol^{-1})	Substance (物质)	$\Delta H_f°$/(kJ mol^{-1})
AgF(s)	−204.6	$CH_3OH(l)$	−239.2
AgCl(s)	−127.0	$C_2H_5OH(l)$	−277.6
AgBr(s)	−100.4	HF(g)	−273.3
AgI(s)	−61.8	HCl(g)	−92.3
$AgNO_3$(s)	−124.4	HBr(g)	−36.3
$BaCl_2$(s)	−855.0	HI(g)	26.5
$BaSO_4$(s)	−1473.2	H_2O(l)	−285.8
$Ca(OH)_2$(s)	−985.2	H_2O(g)	−241.8
CCl_4(l)	−128.2	H_2O_2(l)	−187.8
CO(g)	−110.5	H_2S(g)	−20.6
CO_2(g)	−393.5	H_2SO_4(l)	−814.0
CH_4(g)	−74.6	NH_3(g)	−45.9
C_2H_2(g)	227.4	NO(g)	91.3
C_2H_4(g)	52.4	NO_2(g)	33.2
C_2H_6(g)	−84.0	N_2O_4(g)	11.1
C_3H_8(g)	−103.8	N_2H_4(l)	50.6
n-C_4H_{10}(g)	−125.7	SO_2(g)	−296.8
C_6H_6(l)	49.1	SO_3(g)	−395.7
C_6H_6(g)	82.9		

Standard Enthalpies of Reaction

The enthalpy changes of reactions are very important and commonly used data. However, no chemistry handbooks can list the standard enthalpies of reaction ($\Delta H°_{rxn}$, or simply $\Delta H°$) of all chemical reactions since there are numerous chemical reactions to be listed one by one. Instead, chemistry handbooks have listed the $\Delta H_f°$ data of thousands of commonly used chemical compounds, from which the $\Delta H°$ values of numerous chemical reactions can be calculated. The significance of standard enthalpy of formation is that it allows us to obtain a great number of $\Delta H°$ values from much smaller number of $\Delta H_f°$ data.

Consider a general reaction written as

$$\text{Reactants} \rightarrow \text{Products} \quad \Delta H° = ?$$

We can construct two hypothetical reactions as

$$\text{Elements in reference forms} \rightarrow \text{Reactants} \quad \Delta H° = \Delta H_f° (\text{reactants})$$

$$\text{Elements in reference forms} \rightarrow \text{Products} \quad \Delta H° = \Delta H_f° (\text{products})$$

The elements to form reactants and those to form products must be the same according to the law of conservation of matter. By applying Hess's law, we then have

$$\Delta H° = \sum_X \nu_X \Delta H_f°(X) \tag{3.28}$$

where X stands for various substances involved in this reaction, and ν_X is the stoichiometric coefficient for X.

平均海平面满足以上所有要求，焓的任意零点也应满足。由于所有物质均由元素组成，科学家们选择元素的指定单质的焓值作为焓的任意零点。这样，其他物质的焓即可相对此零点确定下来。

物质的**标准生成焓** [$\Delta_f H_m^\ominus(T)$，或在 298.15 K 下简写为 $\Delta H_f^\ominus$] 是在标态下由元素的指定单质生成 1 mol 该物质的焓变。元素的**指定单质**通常为 1 bar 和指定温度下该元素最稳定的单质（存在少数例外）。在 $\Delta H_f^\ominus$ 中，符号"$\ominus$"表示该焓变为标准焓变，下标"f"代表这是一个由单质生成该物质的反应。由于从指定单质生成其自身的过程并未发生任何改变，指定单质自身的生成焓为零。

几种元素的指定单质举例如下：

$$\text{Na(s), H}_2\text{(g), N}_2\text{(g), O}_2\text{(g), C(石墨), Br}_2\text{(l)}$$

碳的两种天然存在的同素异形体（石墨和金刚石）之间存在可测量的焓值差异。由于

$$\text{C(石墨,s)} \rightarrow \text{C(金刚石,s)} \quad \Delta H^\ominus = 1.9 \text{ kJ mol}^{-1}$$

焓值更低、更为稳定的石墨被选为指定单质。令 $\Delta H_f^\ominus$(石墨) = 0，则 $\Delta H_f^\ominus$(金刚石) = 1.9 kJ mol^{-1}。尽管溴在 298.15 K 下能以气体或液体的形式存在，由于

$$\text{Br}_2\text{(l)} \rightarrow \text{Br}_2\text{(g)} \quad \Delta H^\ominus = 30.9 \text{ kJ mol}^{-1}$$

Br$_2$(l) 更为稳定，因此标准生成焓为 $\Delta H_f^\ominus$(Br$_2$(l)) = 0，$\Delta H_f^\ominus$(Br$_2$(g)) = 30.9 kJ mol^{-1}。

磷是一个例外，其指定单质并非最稳定单质，尽管固体白磷的焓高于固体红磷，由于历史原因白磷仍被选为指定单质。

$$\text{P(s,白磷)} \rightarrow \text{P(s,红磷)} \quad \Delta H^\ominus = -17.6 \text{ kJ mol}^{-1}$$

其标准生成焓为 $\Delta H_f^\ominus$(P(s, 白磷))=0，$\Delta H_f^\ominus$(P(s, 红磷))=−17.6 kJ mol^{-1}。

一些物质在 298.15 K 下的标准生成焓列于**表 3.1** 及**附录 C.1** 中，其值可能为正也可能为负。标准生成焓与分子结构密切相关，例如，在各种碳氢化合物中，甲烷、乙烷等烷烃的 $\Delta H_f^\ominus$ 均为负值，而乙烯、乙炔等烯烃和炔烃的 $\Delta H_f^\ominus$ 均为正值。这表明，烷烃比烯烃和炔烃更稳定、反应性更低。

反应的标准焓变

反应的焓变是十分重要且普遍应用的数据。但没有一本化学手册能列出所有化学反应的标准焓变（$\Delta H_{反应}^\ominus$，或简写为 $\Delta H^\ominus$），因为化学反应的数量过于庞大，不能一一列出。相反，化学手册中列出了上千种常用化合物的 $\Delta H_f^\ominus$ 数据，从这些数据即可计算出大量化学反应的 $\Delta H^\ominus$。标准生成焓的重要性就在于，可以从少量的 $\Delta H_f^\ominus$ 数据得到大量的 $\Delta H^\ominus$。

对于如下反应通式：

$$\text{反应物} \rightarrow \text{生成物} \quad \Delta H^\ominus = ?$$

我们可以构造如下两个假想反应：

$$\text{元素的指定单质} \rightarrow \text{反应物} \quad \Delta H^\ominus = \Delta H_f^\ominus(\text{反应物})$$

$$\text{元素的指定单质} \rightarrow \text{生成物} \quad \Delta H^\ominus = \Delta H_f^\ominus(\text{生成物})$$

根据物质守恒定律，组成反应物的元素和组成生成物的元素一定相同。应用盖斯定律，有

Standard Enthalpies of Ionic Reactions in Solutions

Many chemical reactions in aqueous solutions can be considered as reactions between ions and are represented by net ionic equations. When the ionic reactions in solutions are concerned, the zero point of the reference forms of the elements for enthalpy is not applicable any more. Another zero point needs to be defined in these cases. Since $H^+(aq)$ is the simplest ion in aqueous solutions, we arbitrarily choose it as the zero point for enthalpy of ions. For the neutralization reaction between a strong acid and a strong base, we have

$$H^+(aq) + OH^-(aq) \rightarrow H_2O(l) \quad \Delta H° = -55.8 \text{ kJ mol}^{-1}$$

Given that $\Delta H_f°(H^+) = 0$ and $\Delta H_f°(H_2O) = -285.8 \text{ kJ mol}^{-1}$, we can obtain

$$\Delta H_f°(OH^-) = \Delta H_f°(H_2O) - \Delta H_f°(H^+) - \Delta H°$$
$$= [-285.8 - 0 - (-55.8)] \text{ kJ mol}^{-1} = -230.0 \text{ kJ mol}^{-1}$$

The $\Delta H_f°$ of other ions in solutions can be calculated in a similar way. The $\Delta H_f°$ values in solutions depend on the solute concentration, and data listed in **Table 3.2** and **Appendix C.1** are for ideal solutions or dilute aqueous solutions (about 1 mol L^{-1}) at 298.15 K.

Table 3.2 Standard Molar Enthalpies of Formation ($\Delta H_f°$) of Some Common Ions in Aqueous Solutions at 298.15 K

表 3.2 298.15 K 时一些常见离子在水溶液中的标准摩尔生成焓（$\Delta H_f^\ominus$）

Ion（离子）	$\Delta H_f°$/(kJ mol^{-1})	Ion（离子）	$\Delta H_f°$/(kJ mol^{-1})
Cations（阳离子）			
H^+	0	Cu^{2+}	64.8
Li^+	−278.5	Fe^{2+}	−89.1
Na^+	−240.1	Zn^{2+}	−153.9
K^+	−252.4	Cd^{2+}	−75.9
NH_4^+	−132.5	Pb^{2+}	−1.7
Ag^+	105.6	Mn^{2+}	−220.8
Mg^{2+}	−466.9	Fe^{3+}	−48.5
Ca^{2+}	−542.8	Al^{3+}	−531.0
Ba^{2+}	−537.6		
Anions（阴离子）			
OH^-	−230.0	MnO_4^-	−541.4
F^-	−332.6	CO_3^{2-}	−677.1
Cl^-	−167.2	SO_4^{2-}	−909.3
Br^-	−121.6	$S_2O_3^{2-}$	−652.3
I^-	−55.2	PO_4^{3-}	−1277.4
NO_3^-	−207.4		

3.5 Spontaneity and the Concept of Entropy

In previous sections of this chapter, we have focused on the heat exchange during a chemical reaction. From this section, we move to another important part of thermochemistry: the prediction of whether a reaction is spontaneous or nonspontaneous, favorable or unfavorable.

Spontaneous and Nonspontaneous Processes

In real life, we all have the experience that water always flows from high to low on its own, and that the melting of ice occurs spontaneously at temperatures above 0°C. If we want to move water from low to high, or freeze water to ice at temperatures above 0°C, some external actions are required. More examples of spontaneous processes include the unwinding of a spring-wound toy, diffusion of gases, and rusting of iron, etc.

A **spontaneous process** is a process that occurs in a system where no external action is necessary to make it continue. Conversely, a **nonspontaneous process** is a process that will not occur unless any external action is continuously applied. A chemical reaction can also be considered as a process. Consider the

$$\Delta H^{\ominus} = \sum_X \nu_X \Delta H_f^{\ominus}(X) \tag{3.28}$$

其中 X 代表参与反应的各种物质，ν_X 为 X 对应的化学计量数。

溶液中离子反应的标准焓变

许多水溶液中的化学反应可视为离子之间的反应，并用离子方程式来表示。当考虑溶液中的离子反应时，将元素指定单质的焓值作为零点就不再适用了，这时需要定义另一个零点。由于 $H^+(aq)$ 是水溶液中最简单的离子，我们选择其作为离子焓的任意零点。对于强酸和强碱之间的中和反应，有

$$H^+(aq) + OH^-(aq) \rightarrow H_2O(l) \quad \Delta H^{\ominus} = -55.8 \text{ kJ mol}^{-1}$$

已知 $\Delta H_f^{\ominus}(H^+) = 0$ 以及 $\Delta H_f^{\ominus}(H_2O) = -285.8 \text{ kJ mol}^{-1}$，可得

$$\Delta H_f^{\ominus}(OH^-) = \Delta H_f^{\ominus}(H_2O) - \Delta H_f^{\ominus}(H^+) - \Delta H^{\ominus}$$
$$= [-285.8 - 0 - (-55.8)] \text{ kJ mol}^{-1} = -230.0 \text{ kJ mol}^{-1}$$

溶液中其他离子的 $\Delta H_f^{\ominus}$ 可用类似方法计算。溶液的 $\Delta H_f^{\ominus}$ 与浓度相关，**表 3.2** 及**附录 C.1** 中所列数据为 298.15 K 下理想溶液或稀水溶液（浓度约 1 mol L^{-1}）的值。

3.5 自发性与熵的概念

在本章前几节中，我们聚焦于化学反应的热效应。从本节开始，我们转向热化学的另一重要部分：预测一个反应自发还是非自发、条件有利还是不利。

自发与非自发过程

在实际生活中，水总是自发地从高处向低处流，在 0°C 以上冰总会自发地融化。如果我们想把水从低处移到高处，或者在 0°C 以上将水凝结成冰，就必须提供一些外加作用。更多自发过程的例子包括松开紧好发条的玩具、气体的扩散以及铁的生锈等。

自发过程是无须持续外加作用即可自动发生的过程。相反，**非自发过程**是没有持续的外加作用就不会自动发生的过程。将化学反应视为一个过程，考虑 HCl(aq) 和 NaOH(aq) 的中和反应，其离子方程式为

$$H^+(aq) + OH^-(aq) \rightarrow H_2O(l)$$

$H^+(aq)$ 和 $OH^-(aq)$ 的浓度会持续下降直至终态，此时基本上所有 $H^+(aq)$ 和 $OH^-(aq)$ 均转化为 $H_2O(l)$，称这个反应是自发的。再来考虑其逆反应：$H_2O(l)$ 电离成 $H^+(aq)$ 和 $OH^-(aq)$。我们不能说逆反应完全不可能发生，但如果没有任何干预（如电解），逆反应不会自动发生，我们称逆反应是非自发的。

关于自发与非自发过程的结论小结如下：
1) 如果一个过程自发，那么其逆过程必然非自发。
2) 自发与非自发过程都可能发生，但只有自发过程能在没有外加作用干预的情况下自动发生，非自发过程的发生要求存在一个持续的外加作用。
3) 自发与非自发过程均为热力学概念，与过程发生的速率（动力

neutralization of NaOH (aq) with HCl (aq), the net ionic equation of which is
$$H^+(aq) + OH^-(aq) \rightarrow H_2O(l)$$
The concentrations of H^+ (aq) and OH^-(aq) decrease until a final state is reached in which essentially all the H^+ (aq) and OH^-(aq) have been converted to H_2O (l). We say that this reaction is spontaneous. Now consider the reverse reaction: the decomposition of H_2O (l) into H^+ (aq) and OH^-(aq). We should not say that the reverse reaction is impossible, but without any intervention (such as by electrolysis) the reverse reaction will not occur on its own. We say that the reverse reaction is nonspontaneous.

Conclusions on spontaneous and nonspontaneous processes are summarized as follows:
1) If a process is spontaneous, its reverse process is then nonspontaneous.
2) Both spontaneous and nonspontaneous processes are possible, but only spontaneous processes will occur without intervention. Nonspontaneous processes require the system to be acted on continuously by an external agent.
3) Both spontaneous and nonspontaneous processes are thermodynamic concepts, and have no inevitable relationship with the rate of a process, which is a kinetic concept. Spontaneous processes can be either fast or slow.
4) Spontaneity depends on conditions. In certain conditions a process can be spontaneous, but in other conditions it may be nonspontaneous.

The Concept of Entropy

The prediction of the spontaneity of a process is very important and useful, because we can then take advantage of this spontaneity. For example, we can build hydropower plants to make electricity by making use of the spontaneity of water flowing from high to low. Here, the scientific question we need to answer is: what are the general criteria for spontaneous changes?

As objects always fall from high to low spontaneously, it seems that the potential energy can serve as the criterion for this spontaneity. However, in other cases, such as **free expansion**, which is the spontaneous expansion of an ideal gas into a vacuum, potential energy could not explain the spontaneity. Therefore, potential energy cannot serve as the general criterion for all spontaneous changes. Historically, potential energy, internal energy, heat, and enthalpy have been used as the criteria for spontaneous changes but all failed. The general criteria for all spontaneous changes are still needed.

Later, scientists have concluded that the spontaneity of a process is affected by two key factors simultaneously. One is the change in energy, and the system tends to reach the lowest energy. The other is the change in the degree of disorder, and the system tends to reach the highest degree of disorder. The concept **entropy** (S) is then coined as the measurement of the degree of disorder in a system. The greater the entropy, the higher the degree of disorder in a system.

Let us consider free expansion again. As shown in **Figure 3.7**, a gas container with a central barrier is kept in an isothermal environment. Initially, the left half contains an ideal gas of 1.00 bar pressure, and the right half is a vacuum. When a pinhole is opened in the barrier, the gas spontaneously expands from the left into the right. After this expansion, the molecules are dispersed throughout the apparatus, with essentially equal numbers of molecules in both parts and a pressure of 0.50 bar. In this isothermal expansion process, $\Delta U = 0$ since the internal energy depends only on temperature. $\Delta H = 0$ because $\Delta(pV) = \Delta(nRT) = 0$ for a fixed amount of an ideal gas. This means that the spontaneity in this expansion is not caused by the energy factor. All gas molecules are in the left half initially, and are equally distributed in the container finally. The degree of disorder in the system increases and so does the entropy of the system. The spontaneity in this expansion is caused by the entropy factor.

Boltzmann Equation for Entropy

The concept of entropy should be understood from a microscopic view. Particles (atoms, ions, molecules, etc.) can be distributed among many available microscopic energy levels called **states**, such as the electronic, vibrational, rotational, and translational energy levels. The particular way that a number of particles are distributed among these states and still give rise to the same total energy is called a **microstate**. The Austrian physicist Ludwig Boltzmann (1844—1906) first made the conceptual breakthrough to associate the number of states of a system with the number of microstates. Entropy is the thermodynamic property of a system that can be directly related to

学概念）无必然联系。自发过程可快可慢。
4) 自发性与条件相关，在一定条件下自发的过程，当条件改变时可能变为非自发。

熵的概念

预测一个过程的自发性十分重要且用途广泛，因为这样就可以利用这种自发性。例如，我们可以利用水从高向低流的自发性来建造水电站发电。那么需要回答的科学问题是：自发变化的通用判据是什么？

由于物体总是自发地从高向低落，势能似乎可以作为这种自发性的判据。然而在如**自由膨胀**（即理想气体自发向真空膨胀）等其他情况下，势能并不能解释其自发性。因此，势能不能作为所有自发变化的通用判据。历史上，势能、内能、热和焓均曾被用作自发变化的判据，但都失败了。仍需要一个适用于所有自发变化的通用判据。

后来，科学家们得出结论：一个过程的自发性同时受两个关键因素影响。一个是能量变化，体系趋向于达到最低能量。另一个是混乱度变化，体系趋向于达到最高混乱度。其后，**熵**（S）作为衡量体系混乱度的概念被提出。熵越大，体系的混乱度就越高。

让我们再次考虑自由膨胀的例子。如**图 3.7** 所示，将一个中间设有隔板的气体容器置于恒温环境中，起初左侧装有 1.00 bar 理想气体，右侧为真空。在隔板上开一个小孔，气体会自发从左侧膨胀至右侧。膨胀后气体分子在整个装置中均匀分散，两侧的分子数基本相等，压强均为 0.50 bar。在此恒温膨胀过程中，由于内能只与温度有关，故 $\Delta U = 0$。而对于一定量理想气体，$\Delta(pV) = \Delta(nRT) = 0$，故 $\Delta H = 0$。这意味着该膨胀过程的自发性不是由能量因素导致的。所有气体分子起初都在左侧，而最终在两侧平均分布。体系的混乱度增加，熵也增加了。该膨胀过程的自发性是由熵的因素导致的。

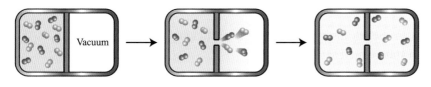

Figure 3.7 Illustrative diagram for the isothermal free expansion of an ideal gas into a vacuum.

图 3.7 理想气体向真空恒温自由膨胀示意图。

熵的玻尔兹曼方程

熵的概念应该从微观角度来理解。粒子（原子、离子、分子等）可以分布在许多被称为**态**的可用微观能级上，如电子能级、振动能级、转动能级和平动能级等。大量粒子分布在这些态上且具有相同总能量的特殊方式，称为一个**微观状态**。奥地利物理学家路德维希·玻尔兹曼（1844—1906）首次提出概念上的突破，将体系的态数与微观状态数联系起来。熵是体系的热力学性质，它可以通过如下**玻尔兹曼方程**与体系的可用微观状态数直接关联：

$$S = k_B \ln W \quad (3.29)$$

其中 S 是熵；W 是微观状态数；k_B 为玻尔兹曼常数，满足 $k_B = R/N_A$，其值为 1.38×10^{-23} J K^{-1}。

给定数量的粒子所能占据的态越多，体系的微观状态就越多，熵也就越大。考虑一个只有两个态的简单体系，这两个态的能量分别为 $\varepsilon = 0$ 和 2 个能量单位。将标记为 **A** 和 **B** 的两个可分辨粒子分布在这两

the number of available microstates of the system by **Boltzmann equation**, as

$$S = k_B \ln W \tag{3.29}$$

where S is the entropy, W is the number of microstates, and k_B is the Boltzmann constant which fits $k_B = R/N_A$ and has a value of 1.38×10^{-23} J K^{-1}.

For a given number of particles, the more states the particles can occupy, the more microstates the system has, and the greater the entropy is. Consider a simple system that has only two states, with energy $\varepsilon = 0$ and 2 energy units, respectively. Two distinguishable particles labeled **A** and **B** are to be distributed in these two states. If the total energy of the system is 2 energy units, then there are only 2 microstates available, as shown in **Figure 3.8(a)**, and the entropy of this system is $k_B \ln 2$. Consider another system that has one more state than the first system, with energy $\varepsilon = 0$, 1, and 2 energy units, respectively. Again, two distinguishable particles labeled **A** and **B** are to be distributed in these states. Given the same total energy of 2 energy units, the available number of microstates now becomes 3 [**Figure 3.8(b)**], larger than the number of microstates available in the first system. The entropy of the second system is $k_B \ln 3$, greater than that of the first system.

For a larger system which consists of 4 different states, with energy $\varepsilon = 0$, 1, 2, and 3 energy units, 3 distinguishable particles labeled **A, B**, and **C** are distributed among these states. If the total energy of the system is 6 energy units, how many available microstates can be generated? In this case, we first need to consider how many different distributions of particles can give rise to the same total energy. If all three particles are positioned in the $\varepsilon = 2$ state, the number of microstate $W_1 = 3!/3! = 1$. If two particles are distributed in the $\varepsilon = 3$ state, it means that the third particle must be positioned in the $\varepsilon = 0$ state to satisfy a total energy of 6. For this particular distribution the number of microstate $W_2 = 3!/(2!1!) = 3$. If one particle is distributed in the $\varepsilon = 3$ state, in order for a total energy of 6, the other two particles must be positioned in the $\varepsilon = 1$ and 2 states, respectively, and $W_2 = 3!/(1!1!1!) = 6$. There is no distribution other than the above three that can meet the total energy of 6 requirement. Therefore, the total number of available microstates is given by $W = W_1 + W_2 + W_3 = 10$. The entropy of this system is then $k_B \ln 10$ (**Figure 3.9**).

Statistically, for a system consisting of m different states each with an energy of ε_i ($i = 1, 2, \ldots, m$), N distinguishable particles are distributed among these states. If altogether there are k different distributions that can give rise to a total energy of E, then the total number of microstates W is given by

$$W = \sum_{j=1}^{k} W_j = W_1 + W_2 + \cdots + W_j + \cdots + W_k \tag{3.30}$$

where W_j is the number of microstates in the jth distribution. In this particular jth distribution, if n_i is the number of particles distributed in the ith state which satisfies

$$N = \sum_{i=1}^{m} n_i = n_1 + n_2 + \cdots + n_i + \cdots + n_m \tag{3.31}$$

and

$$E = \sum_{i=1}^{m} n_i \varepsilon_i = n_1 \varepsilon_1 + n_2 \varepsilon_2 + \cdots + n_i \varepsilon_i + \cdots + n_m \varepsilon_m \tag{3.32}$$

then W_j can be calculated by

$$W_j = \frac{N!}{\prod_{i=1}^{m} n_i!} = \frac{N!}{n_1! n_2! \cdots n_i! \cdots n_m!} \tag{3.33}$$

where the symbol $\prod$ (Greek, pi) means "the product of", and $N!$ (called the factorial of N) is the product of all positive integers less than or equal to N.

To understand why the spontaneity in free expansion is caused by the entropy factor, let us consider the particle-in-a-1D-box model. This model is rather complicated, and here we will not attempt to derive or discuss the detail but rather accept the following conclusion: the quantized energy of a matter wave in a 1D box is given by

$$E_n = \frac{n^2 h^2}{8mL^2} \tag{3.34}$$

个态上。如果体系的总能量为 2 个能量单位，则一共只存在 2 种可用的微观状态，如**图 3.8(a)** 所示，体系的熵为 $k_B \ln 2$。考虑另一体系，它比前一体系多了一个态，能量分别为 $\varepsilon = 0$、1 和 2 个能量单位。同样将标记为 **A** 和 **B** 的两个可分辨粒子分布在这些态上，总能量依然为 2 个能量单位。现在可用的微观状态数变为 3 [**图 3.8(b)**]，大于前一体系可用的微观状态数。第二个体系的熵为 $k_B \ln 3$，大于前一体系的熵。

对于一个由 4 个不同态组成的更大体系，态的能量分别为 $\varepsilon = 0$、1、2 和 3 个能量单位，将标记为 **A**、**B** 和 **C** 的三个可分辨粒子分布在这些态上。如果体系的总能量为 6 个能量单位，一共有多少种可用的微观状态？这种情况下我们首先需要考虑，为得到相同的总能量，粒子一共有多少种不同的分布情况。如果所有三个粒子均处于 $\varepsilon = 2$ 态，微观状态数为 $W_1 = 3!/3! = 1$。如果有两个粒子分布在 $\varepsilon = 3$ 态，这意味着第三个粒子必定位于 $\varepsilon = 0$ 态，才能满足总能量为 6 个能量单位。对于这种特定的分布，微观状态数为 $W_2 = 3!/(2!1!) = 3$。如果有一个粒子分布在 $\varepsilon = 3$ 态，为使总能量为 6 个能量单位，其余两个粒子必须分别处于 $\varepsilon = 1$ 和 2 态，$W_3 = 3!/(1!1!1!) = 6$。除以上三种分布外，再没有其他任何分布方式能满足总能量为 6 个能量单位的要求。因此，全部可用微观状态的总数为 $W = W_1 + W_2 + W_3 = 10$，体系的熵为 $k_B \ln 10$（**图 3.9**）。

统计上，对于一个由 m 个不同的态组成的体系，每个态对应的能量分别为 ε_i ($i = 1, 2, \cdots, m$)，将 N 个可分辨粒子分布在这些态上，如果一共有 k 种不同的分布方式使总能量均为 E，那么微观状态的总数 W 为

$$W = \sum_{j=1}^{k} W_j = W_1 + W_2 + \cdots + W_j + \cdots + W_k \quad (3.30)$$

其中 W_j 是第 j 种分布方式所对应的微观状态数。在这第 j 种分布方式中，如果 n_i 是分布在第 i 个态上的粒子数，应满足

$$N = \sum_{i=1}^{m} n_i = n_1 + n_2 + \cdots + n_i + \cdots + n_m \quad (3.31)$$

$$E = \sum_{i=1}^{m} n_i \varepsilon_i = n_1 \varepsilon_1 + n_2 \varepsilon_2 + \cdots + n_i \varepsilon_i + \cdots + n_m \varepsilon_m \quad (3.32)$$

那么 W_j 可通过下式计算：

$$W_j = \frac{N!}{\prod_{i=1}^{m} n_i!} = \frac{N!}{n_1! n_2! \cdots n_i! \cdots n_m!} \quad (3.33)$$

其中符号 $\prod$（希腊字母，pi）代表"之积"，$N!$（称为 N 的阶乘）是所有小于或等于 N 的正整数的乘积。

为了理解为什么自由膨胀的自发性是由熵的因素导致的，让我们考虑一维势箱模型。该模型较为复杂，这里我们不做推导和详细讨论，只接受如下结论：一维势箱中物质波的量子化能量为

$$E_n = \frac{n^2 h^2}{8mL^2} \quad (3.34)$$

其中 m 是粒子的质量；L 是一维势箱的长度；h 为普朗克常数，其值为 6.62607×10^{-34} J·s；$n = 1, 2, 3, \cdots$ 是量子数。因此，对于长度为 L 的一维势箱中粒子，其量子化的能量为

$$E_1 = \frac{h^2}{8mL^2}, \quad E_2 = \frac{4h^2}{8mL^2} = 4E_1, \quad E_3 = \frac{9h^2}{8mL^2} = 9E_1, \quad \cdots$$

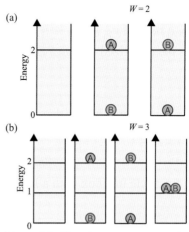

Figure 3.8 The microstates and entropies of the two systems consist of two distinguishable particles labeled A and B, with a total energy of 2 energy units. (a) Particles are distributed in two states, with energy $\varepsilon = 0$ and 2 energy units, respectively. (b) Particles are distributed in three states, with energy $\varepsilon = 0$, 1, and 2 energy units, respectively.

图 3.8 由标记为 A 和 B 的两个可分辨粒子组成、总能量为 2 个能量单位的两个体系的微观状态数和熵：(a) 粒子分布在能量分别为 $\varepsilon = 0$ 和 2 个能量单位的两个能级上；(b) 粒子分布在能量分别为 $\varepsilon = 0$、1 和 2 个能量单位的三个能级上。

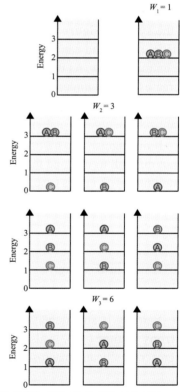

Figure 3.9 The microstate and entropy of a system in which three distinguishable particles labeled A, B, and C, with a total energy of 6 energy units, are distributed in four states, with energy $\varepsilon = 0$, 1, 2, and 3 energy units, respectively.

图 3.9 由标记为 A、B 和 C 的三个可分辨粒子组成的体系的微观状态数和熵，该体系的总能量为 6 个能量单位，粒子分布在能量分别为 $\varepsilon = 0$、1、2 和 3 个能量单位的四个能级上。

where m is the mass of the particle, L is the length of the 1D box, h is the Planck's constant with a value of 6.62607×10^{-34} J s, and $n = 1,2,3,\cdots$ is a quantum number. Therefore, for a particle in a 1D box with a length of L, the quantized energies are

$$E_1 = \frac{h^2}{8mL^2}, \quad E_2 = \frac{4h^2}{8mL^2} = 4E_1, \quad E_3 = \frac{9h^2}{8mL^2} = 9E_1, \cdots$$

For a particle in a box with a length of $2L$, the quantized energies are

$$E_1' = \frac{h^2}{8m(2L)^2} = \frac{1}{4}E_1, \quad E_2' = \frac{4h^2}{8m(2L)^2} = E_1, \quad E_3' = \frac{9h^2}{8m(2L)^2} = \frac{9}{4}E_1, \cdots$$

For a particle in a box with a length of $3L$, the quantized energies are

$$E_1'' = \frac{h^2}{8m(3L)^2} = \frac{1}{9}E_1, \quad E_1'' = \frac{4h^2}{8m(3L)^2} = \frac{4}{9}E_1, \quad E_3'' = \frac{9h^2}{8m(3L)^2} = E_1, \cdots$$

Figure 3.10(a) shows the representative energy level diagrams for a particle in 1D boxes of length L, $2L$, and $3L$. As the length of the box increases, the separation between the energy levels decreases, resulting in an increase in accessible energy levels. For example, there are 2, 5, and 8 thermally accessible energy levels, indicated by the rainbow band, for particles in the boxes of length L, $2L$, and $3L$, respectively. Extending this model to three-dimensional space and a great number of gas molecules, as the length or volume of the system increases, the number of accessible states increases. Therefore, the total number of available microstates increases significantly, and so does the entropy of the system. In the case of free expansion shown in **Figure 3.7**, there are more available translational energy levels among which the gas molecules can be distributed in the expanded volume. The consequently increased W and S explain the spontaneity in the free expansion process.

The particle-in-a-1D-box model can also be used to understand the effect of temperature on the entropy of the system [**Figure 3.10(b)**]. At low temperatures, the gas molecules have a low average translational kinetic energy and can occupy only a few of the relatively low energy levels. The value of W is small, and the entropy is low. As the temperature increases, the average translational kinetic energy of the gas molecules increases and the molecules have access to a larger number of energy levels. Thus, the number of W increases and the entropy rises. The above conclusions are equally valid for liquids and solids. More discussion about quantized energy and matter waves can be found in **Chapter 7**.

If distinguishable particles in the order of N_A are distributed among different states in the order of N_A, it is not difficult to imagine that the total number of available microstates will be astronomical. To estimate the magnitude of W in this case, we can apply $k_B = 1.38 \times 10^{-23}$ J K^{-1} and the entropy in the order of 10^2 J K^{-1} for 1 mol substance into **Equation (3.29)**, as

$$\ln W = \frac{S}{k_B} = \frac{\sim 10^2 \text{ J K}^{-1}}{1.38 \times 10^{-23} \text{ J K}^{-1}} \approx 10^{25}$$

This results in $W = \exp(10^{25})$, which is an extremely large number.

Entropy Change

Entropy is an extensive function of the state. Like other state functions, its change in a process, the **entropy change** (ΔS), is also an extensive function of the state. The entropy change can be understood both microscopically and macroscopically. In a microscopic view,

$$\Delta S = S_f - S_i = k_B \left(\ln W_f - \ln W_i \right) \tag{3.35}$$

Macroscopically, the entropy change equals the ratio of the heat of reaction and the thermodynamic temperature in an isothermal reversible process, given by

$$\Delta S = \frac{q_{rev}}{T} \tag{3.36}$$

Even if the actual process is not an isothermal reversible process, **Equation (3.36)** is still valid because q_{rev} then stands for the corresponding hypothetical heat of reaction if the process *did occur* reversibly at a constant temperature T.

对于长度为 $2L$ 的势箱中粒子，其量子化的能量为

$$E'_1 = \frac{h^2}{8m(2L)^2} = \frac{1}{4}E_1, \ E'_2 = \frac{4h^2}{8m(2L)^2} = E_1, \ E'_3 = \frac{9h^2}{8m(2L)^2} = \frac{9}{4}E_1, \ \cdots$$

对于长度为 $3L$ 的势箱中粒子，其量子化的能量为

$$E''_1 = \frac{h^2}{8m(3L)^2} = \frac{1}{9}E_1, \ E''_2 = \frac{4h^2}{8m(3L)^2} = \frac{4}{9}E_1, \ E''_3 = \frac{9h^2}{8m(3L)^2} = E_1, \ \cdots$$

图 3.10(a) 给出了长度分别为 L、$2L$ 和 $3L$ 的一维势箱中粒子的能级示意图。随着势箱长度的增加，能级之间的间隔减小，从而导致可达能级数增加。例如，对于长度为 L、$2L$ 和 $3L$ 的势箱中粒子，分别有 2、5 和 8 个热可达能级（用彩虹色带表示）。将该模型扩展到三维空间和大量气体分子，随着体系长度或体积的增大，可达态的数量增加。因此，可用的微观状态总数显著增加，体系的熵也显著增加。在**图 3.7** 所示的自由膨胀情况下，膨胀后体积增大，存在更多的可用平动能级供气体分子分布，由此导致了 W 和 S 的增加，也解释了自由膨胀过程的自发性。

一维势箱模型还可用于理解温度对体系熵的影响 [**图 3.10(b)**]。低温下气体分子的平均平动能较低，只能占据相对较低的几个能级，W 值较小，熵也较小。随着温度的升高，气体分子的平均平动能增加，分子可以达到更多的能级，因此 W 值增加，熵也增加。上述结论对液体和固体同样适用。关于量子化能量和物质波的更多讨论详见**第 7 章**。

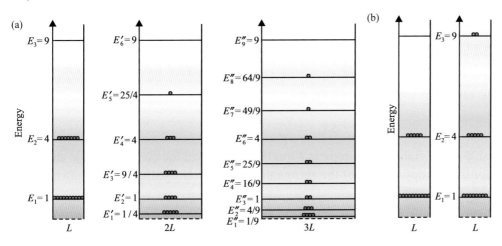

Figure 3.10 Energy levels for particles in a 1D box. (a) The energy levels become more numerous and closer together as the length of the box increases. (b) More energy levels become accessible in a box of fixed length as the temperature increases.

图 3.10 一维势箱中粒子的能级图：(a) 随着势箱长度的增加，能级变得更多、间隔更密；(b) 随着温度的升高，长度固定的势箱中粒子可达的能级变得更多。

如果将 N_A 量级个可分辨粒子分布在 N_A 量级个不同的态上，不难想象可用的微观状态总数将会是一个天文数字。为估算该情况下 W 的量级，可将 $k_B = 1.38 \times 10^{-23}$ J K^{-1} 以及 1 mol 物质的熵通常在 10^2 J K^{-1} 量级代入**式（3.29）**，有

$$\ln W = \frac{S}{k_B} = \frac{\sim 10^2 \text{ J K}^{-1}}{1.38 \times 10^{-23} \text{ J K}^{-1}} \approx 10^{25}$$

可得 $W = \exp(10^{25})$，这是一个极其庞大的数值。

熵变

熵是一个广度状态函数，与其他状态函数一样，在一个过程中熵的改变量即**熵变**（ΔS），也是广度状态函数。熵变既可以从微观上理解，也可以从宏观上理解。从微观角度看

The relationship between **Equations (3.35)** and **(3.36)** can be derived rigorously but is rather complicated and will not be presented here. However, a simplified derivation method, which considers just two from the numerous energy levels representing the surroundings, can be shown. **Figure 3.11** shows the two particular states, **A** and **B**, where state **A** is the most probable initial state and state **B** is the most probable final state upon the addition of an infinitesimal amount of heat (q_{rev}). The amount of heat used is just the energy difference ($\Delta\varepsilon$) between the two levels i and j. Before the process, there are n_i and n_j particles in the levels i and j, respectively. The corresponding number of microstates is given by

$$W_A = \frac{N!}{n_1! n_2! \cdots n_i! n_j! \cdots}$$

During the process, a single particle absorbs the heat and jumps from level i to level j. After the process, there are $n_i - 1$ and $n_j + 1$ particles in the levels i and j, respectively. The corresponding number of microstates is given by

$$W_B = \frac{N!}{n_1! n_2! \cdots (n_i - 1)!(n_j + 1)! \cdots}$$

Therefore, the ratio can be written as

$$\frac{W_B}{W_A} = \frac{n_i! n_j!}{(n_i - 1)!(n_j + 1)!} = \frac{n_i}{n_j + 1} \approx \frac{n_i}{n_j}$$

The last approximation is valid because both n_i and n_j are very large numbers. According to Boltzmann distribution law,

$$\frac{n_j}{n_i} = \exp\left(-\frac{\Delta\varepsilon}{k_B T}\right) \tag{3.37}$$

which relates the population of level i to that of level j. Again, the temperature of **B** approximately equals that of **A** because both n_i and n_j are very large numbers.

The entropy change of the surroundings in this process is given by

$$\Delta S_{surr} = k_B \ln\left(\frac{W_B}{W_A}\right) = k_B \ln\left[\exp\left(\frac{\Delta\varepsilon}{k_B T}\right)\right] = \frac{\Delta\varepsilon}{T}$$

The heat of reaction in this reversible process equals $\Delta\varepsilon$, so

$$\Delta S_{surr} = \frac{q_{rev}}{T}$$

It shows that the relationship of the statistical entropy in **Equation (3.29)** and the thermodynamic entropy in **Equation (3.36)** is true for the entropy change of the surroundings. This relationship can be shown to hold for the system as well.

The following situations generally produce an increase in entropy:

1) Pure liquids are formed from solids. In the melting process, a crystalline solid is replaced by a less structured liquid. Molecules that were relatively fixed in position in the solid, being limited to vibrational motion, are now free to move about a bit. The molecules have gained some translational and rotational motion. The number of accessible microscopic energy levels has increased, and so has the entropy. $S_{liquid} > S_{solid}$.

2) Gases are formed from either solids or liquids. Molecules in the gaseous state, because they can move within a large free volume, have many more accessible energy levels than do those in the liquid or solid state. In the gas, energy can be spread over a much greater number of microscopic energy levels than in the liquid or solid. The entropy of the gaseous state is much higher than that of the liquid or solid state. $S_{gas} \gg S_{liquid} > S_{solid}$.

3) Liquid solutions are formed from solids and liquids. In the dissolving process, a crystalline solid and a pure liquid are replaced by a mixture of ions and solvent molecules in the solution state. Although there might be some decrease in entropy associated with the clustering of solvent molecules, such as water, around the ions because of ion-dipole forces, the increase in entropy that accompanies the destruction of the crystalline lattice of the solid predominates. For the overall dissolving process, the entropy increases. $S_{soln} > (S_{solvent} + S_{solute})$.

4) The temperature of the substance increases. Increased temperature means an increased number of

$$\Delta S = S_{终} - S_{始} = k_B \left(\ln W_{终} - \ln W_{始} \right) \quad (3.35)$$

而宏观上，熵变等于恒温可逆过程的反应热与热力学温度之比，即

$$\Delta S = \frac{q_{可逆}}{T} \quad (3.36)$$

即使实际过程并非恒温可逆，**式（3.36）**仍然有效，因为此时$q_{可逆}$代表该过程如果在恒温T下可逆地发生时所对应的假想反应热。

式（3.35）和**（3.36）**之间的关系可以严格推导，但相当复杂，此处不予介绍。下面仅给出一种简化版推导方法，只考虑了环境的众多能级中的两个。**图3.11**给出了两个特殊的态**A**和**B**，其中**A**是最可能的始态，**B**是吸收一个无限小的热（$q_{可逆}$）之后最可能的终态，此热量恰好等于能级i和j之间的能量差（$\Delta\varepsilon$）。在此过程之前，能级i和j上分别存在n_i和n_j个粒子，对应的微观状态数为

$$W_A = \frac{N!}{n_1! n_2! \cdots n_i! n_j! \cdots}$$

在此过程中，一个粒子吸热后从能级i跃迁至能级j。在此过程之后，能级i和j上分别存在$n_i - 1$和$n_j + 1$个粒子，对应的微观状态数为

$$W_B = \frac{N!}{n_1! n_2! \cdots (n_i - 1)! (n_j + 1)! \cdots}$$

因此，两个微观状态数之比可写成

$$\frac{W_B}{W_A} = \frac{n_i! n_j!}{(n_i - 1)!(n_j + 1)!} = \frac{n_i}{n_j + 1} \approx \frac{n_i}{n_j}$$

最后一步的近似是合理的，因为n_i和n_j都是非常大的数。根据玻尔兹曼分布定律

$$\frac{n_j}{n_i} = \exp\left(-\frac{\Delta\varepsilon}{k_B T} \right) \quad (3.37)$$

这将能级i上的粒子数与能级j上的粒子数联系起来。同样，由于n_i和n_j都是非常大的数，**B**和**A**的温度近似相等。

在此过程中环境的熵变为

$$\Delta S_{环境} = k_B \ln\left(\frac{W_B}{W_A} \right) = k_B \ln\left[\exp\left(\frac{\Delta\varepsilon}{k_B T} \right) \right] = \frac{\Delta\varepsilon}{T}$$

该可逆过程的反应热等于$\Delta\varepsilon$，故

$$\Delta S_{环境} = \frac{q_{可逆}}{T}$$

这表明对于环境的熵变，**式（3.29）**表示的统计熵与**式（3.36）**表示的热力学熵之间的关系是正确的，可以证明此关系对于体系同样成立。

下述情况通常会导致熵增：
1) 由固体生成纯液体。在熔化过程中，晶态固体被结构更不规整的液体取代。固体中分子受振动的限制被相对固定在某个位置附近，熔化后可以更为自由地移动，分子获得了一些平动和转动的自由。可达的微观能级数增加，熵也增加。$S_{液} > S_{固}$。
2) 由固体或液体生成气体。气态分子可以在很大的自由体积内运动，因此比液态或固态分子具有更多的可达能级。气体的能量可以分散在比液体或固体更多的微观能级上，气态的熵远高于固态或液态。$S_{气} \gg S_{液} > S_{固}$。

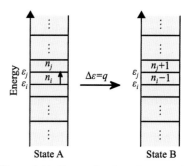

Figure 3.11 Energy levels and populations representing the most probable initial (A) and final (B) states of the surroundings upon the addition of an infinitesimal amount of heat (q_{rev}).

图3.11 环境最可能的始态（A）以及在吸收了一个无限小的热（$q_{可逆}$）之后最可能的终态（B）的能级及粒子数示意图。

accessible energy levels for the increased molecular motion, whether it be the vibrational motion of atoms or ions in a solid, or the translational and rotational motion of molecules in a liquid or gas.

The Third Law of Thermodynamics

Unlike the cases with internal energy and enthalpy, there is an absolute zero for entropy, and it is possible to determine the absolute entropy values. For a system consisting of m different states, with energy $\varepsilon = 0$, 1, 2, … energy units, N distinguishable particles are distributed among these states. Both m and N are very large numbers, in the order of N_A. If the total energy of the system is 0, meaning the ground state, how many microstate can be generated and how much is the entropy? In order to give a total energy of 0, all particles must be distributed in the $\varepsilon = 0$ state, and no particle should be positioned in the higher energy states. Therefore, the total number of available microstates must be $W = N!/(N!0!0!...) = 1$, and the entropy is then $k_B \ln 1 = 0$.

The above case is only an illustrative example. In fact, the lowest possible energy of a system cannot be zero. It corresponds to $n = 1$ in the particle-in-a-box model, as

$$E_1 = \frac{h^2}{8mL^2} \tag{3.38}$$

This energy is called **zero-point-energy**. In this lowest possible energy state, all particles are distributed at the lowest energy level ($n = 1$), and the temperature of the system must be 0 K. The total number of available microstates is still 1 and the entropy is still 0. This corresponds to the absolute zero for entropy. The **third law of thermodynamics** can be stated as follows: the entropy of a pure perfect crystal at 0 K is zero. The absolute entropy of 1 mol substances in its standard state is called the **standard molar entropy** [$S_m^\circ(T)$, or S° for short at 298.15 K]. Standard molar entropies of a number of substances at 298.15 K are tabulated in **Table 3.3** and **Appendix C.1**.

Table 3.3 Standard Molar Entropies (S°) of Some Common Substances at 298.15 K
表3.3 298.15 K 时一些常见物质的标准摩尔熵($S^\ominus$)

Substance (物质)	S°/(J mol^{-1} K^{-1})	Substance (物质)	S°/(J mol^{-1} K^{-1})
AgCl(s)	96.3	$C_2H_5OH(l)$	160.7
AgBr(s)	107.1	HF(g)	173.8
AgI(s)	115.5	HCl(g)	186.9
$AgNO_3$(s)	140.9	HBr(g)	198.7
$BaCl_2$(s)	123.7	HI(g)	206.6
$BaSO_4$(s)	132.2	$H_2O(l)$	70.0
$Ca(OH)_2$(s)	83.4	$H_2O(g)$	188.8
CO(g)	197.7	$H_2O_2(l)$	109.6
CO_2(g)	213.8	H_2S(g)	205.8
CH_4(g)	186.3	$H_2SO_4(l)$	156.9
C_2H_2(g)	200.9	NH_3(g)	192.8
C_2H_4(g)	219.3	NO(g)	210.8
C_2H_6(g)	229.2	NO_2(g)	240.1
C_3H_8(g)	270.3	N_2O_4(g)	304.4
$C_6H_6(l)$	173.4	$N_2H_4(l)$	121.2
C_6H_6(g)	269.2	SO_2(g)	248.2
$CH_3OH(l)$	126.8	SO_3(g)	256.8

In the equilibrium between two phases, the exchange of heat can be carried out reversibly, and the quantity of exchanged heat is equal to the enthalpy change for the transition, ΔH_{tr}. Therefore,

$$\Delta S_{tr} = \frac{\Delta H_{tr}}{T_{tr}} \tag{3.39}$$

The general symbol "tr" can be replaced by the more specific phase transition term, such as "vap" for the vaporization of a liquid and "fus" for the melting of a solid. If the transitions involve standard state conditions, the degree sign (°) is also used.

A useful generalization of **Equation (3.39)**, known as **Trouton's rule** (Frederick T. Trouton, 1863—1922), states that for many liquids at their normal boiling points, the standard molar entropy

3) 由固体和液体生成溶液。在溶解过程中，晶态固体和纯液体被处于溶液状态的离子和溶剂分子的混合物所取代。尽管由于离子-偶极作用，溶剂分子（如水）在离子附近的聚集可能导致熵的降低，但固体晶格破坏带来的熵增占主导地位，整个溶解过程的熵增加。$S_{溶液} > (S_{溶剂} + S_{溶质})$。
4) 物质的温度升高。温度升高意味着分子运动的可达能级数增加，分子运动可以是固体中原子或离子的振动，也可以是液体或气体中分子的平动和转运。

热力学第三定律

与内能和焓的情况不同，熵存在绝对零点，绝对熵值是可以确定的。对于由能量分别为 $\varepsilon = 0, 1, 2, \cdots$ 个能量单位的 m 个不同态组成的体系，将 N 个可分辨粒子分布在这些态上。N 和 m 均为非常大的数（在 N_A 量级）。如果体系的总能量为 0（即基态），一共有多少种微观状态？熵又是多少呢？为使总能量为 0，所有粒子都必须分布在 $\varepsilon = 0$ 的态上，不能有任何粒子处在更高能级上。因此，可用的微观状态总数必为 $W = N!/(N!0!0!\cdots) = 1$，熵为 $k_B \ln 1 = 0$。

上述情况只是一个说明性的示例。事实上体系的最低能量不可能为 0，对应的是势箱模型中 $n = 1$ 的情况，有

$$E_1 = \frac{h^2}{8mL^2} \tag{3.38}$$

此能量称为**零点能**。在这种可能的最低能量状态下，所有粒子均处于最低能级（$n = 1$），体系的温度必须为 0 K。可用的微观状态总数仍为 1，熵仍为 0，这对应于熵的绝对零点。**热力学第三定律**可表述为：完美纯晶体在 0 K 时的熵为零。1 mol 物质在标态下的绝对熵值称为**标准摩尔熵** [$S_m^\ominus(T)$，或在 298.15 K 下简写为 $S^\ominus$]，一些物质在 298.15 K 下的标准摩尔熵列于**表 3.3** 及**附录 C.1** 中。

两相平衡时的热交换可以可逆地进行，交换的热量等于相变过程的焓变 $\Delta H_{相变}$。故

$$\Delta S_{相变} = \frac{\Delta H_{相变}}{T_{相变}} \tag{3.39}$$

下标"相变"可替换为更具体的相变方式，如"蒸发"对应于液体蒸发过程，"熔化"对应于固体熔化过程。如果该相变涉及标态条件，则应使用符号（$^\ominus$）。

作为**式（3.39）**的有用推广，**乔顿规则**（弗雷德里克·T. 乔顿，1863—1922）指出：许多处于正常沸点的液体，其标准摩尔蒸发熵的值均约为 87 J mol^{-1} K^{-1}，即

$$\Delta S_{蒸发}^\ominus = \frac{\Delta H_{蒸发}^\ominus}{T_{沸点}} \approx 87 \text{ J mol}^{-1} \text{ K}^{-1} \tag{3.40}$$

在 1 bar 下将 1 mol 分子从液态转化为蒸气，所增加的可达微观能级对于不同液体均大致相当，因此 $\Delta S_{蒸发}^\ominus$ 也大致相当。乔顿规则不适用的例子也很好理解，例如水或乙醇中，分子间氢键导致其熵比常规液体更低，因此蒸发过程的熵增大于正常值，$\Delta S_{蒸发}^\ominus > 87$ J mol^{-1} K^{-1}。

从熵的绝对零点出发，我们可以确定物质在其他温度和压强条件下的熵变。将这些熵变加起来，即可得绝对熵的数值。**图 3.12** 给出了确定绝对熵与温度的函数关系的方法。当发生相变时，可用**式**

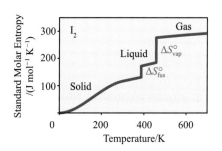

Figure 3.12 Standard molar entropy as a function of temperature.

图 3.12 标准摩尔熵与温度的函数关系图。

change of vaporization has a value of about 87 J mol^{-1} K^{-1}, as

$$\Delta S°_{vap} = \frac{\Delta H°_{vap}}{T_{bp}} \approx 87 \text{ J mol}^{-1} \text{ K}^{-1} \tag{3.40}$$

The increased accessibility of microscopic energy levels produced in transferring 1 mol molecules from liquid to vapor at 1 bar is roughly comparable for different liquids, so similar values of $\Delta S°_{vap}$ are expected. Instances in which Trouton's rule fails are also understandable. For example, in water and in ethanol, hydrogen bonding among molecules produces a lower entropy than would otherwise be expected in the liquid state. Consequently, the entropy increase in the vaporization process is greater than normal, and so $\Delta S°_{vap} > 87$ J mol^{-1} K^{-1}.

From the absolute zero for entropy, we can evaluate entropy changes as the substance is brought to other conditions of temperature and pressure. Adding together these entropy changes, we can then obtain a numerical value of the absolute entropy. **Figure 3.12** illustrates the method for determining absolute entropy as a function of temperature. Where phase transition occurs, **Equation (3.39)** is used to evaluate the corresponding entropy changes. Over temperature ranges in which there are no transitions, $\Delta S°$ values are obtained from measurements of specific heats as a function of temperature.

The standard molar entropy change [$\Delta_r S°_m(T)$, or $\Delta S°$ for short at 298.15 K] for a general reaction can be calculated with a similar form to **Equation (3.28)**, as

$$\Delta S° = \sum_X v_X S°(X) \tag{3.41}$$

where X stands for various substances involved in this reaction, and v_X is the stoichiometric coefficient for X.

3.6 Criteria for Spontaneous Change

In the previous section, we have discussed the concepts of entropy and entropy change, and how they are related to the number of microstates in the system. However, the scientific question still remains: what are the general criteria for spontaneous changes? In this section, we will answer this question.

Entropy Change for an Isolated System

Recall from **Section 3.2** that the work done in a reversible process is always less than that in an irreversible process, and that the heat involved in a reversible process is always more than that in an irreversible process. In expressions,

$$w_{rev} < w_{ir} \quad \text{and} \quad q_{rev} > q_{ir}$$

Combining with **Equation (3.36)**, we have

$$\Delta S = \frac{q_{rev}}{T} > \frac{q_{ir}}{T} \tag{3.42}$$

If a process is not reversible, it must be irreversible, and vice versa. **Equation (3.42)** can be rewritten by splitting situations as

$$\begin{cases} \Delta S = \frac{q_{rev}}{T} > \frac{q}{T}, & q = q_{ir} \text{ for all irreversible processes} \\ \Delta S = \frac{q_{rev}}{T} = \frac{q}{T}, & q = q_{rev} \text{ for all reversible processes} \end{cases}$$

For any given process which can be either reversible or irreversible, the general expression is

$$\Delta S \geqslant \frac{q}{T} \tag{3.43}$$

where q is the heat involved in the process. $q = q_{ir}$ for irreversible processes and $q = q_{rev}$ for reversible processes.

If the system is an isolated system, it means that the system does not interact with its surroundings and $q = 0$. Therefore, the entropy change for an isolated system must be greater than or equal to 0, or

（3.39）来确定对应的熵变。在没有发生相变的温度范围内，$\Delta S^{\ominus}$ 可通过测量比热与温度的函数关系来获得。

一般反应的标准摩尔熵变 [$\Delta_r S_m^{\ominus}(T)$，或在 298.15 K 下简写为 $\Delta S^{\ominus}$] 可用与式（3.28）类似的式子计算，有

$$\Delta S^{\ominus} = \sum_X \nu_X S^{\ominus}(X) \tag{3.41}$$

其中 X 代表参与反应的各种物质，ν_X 为 X 对应的化学计量数。

3.6 自发变化的判据

上节我们讨论了熵和熵变的概念，以及它们与体系微观状态数的关系。然而，科学问题仍然存在：自发变化的通用判据是什么？本节我们会回答这一问题。

孤立体系的熵变

回顾 **3.2 节**，可逆过程的功总小于不可逆过程，可逆过程的热总大于不可逆过程，可表达为

$$w_{可逆} < w_{不可逆} \quad 且 \quad q_{可逆} > q_{不可逆}$$

结合式（3.36），可得

$$\Delta S = \frac{q_{可逆}}{T} > \frac{q_{不可逆}}{T} \tag{3.42}$$

如果一个过程不是可逆的，则必定是不可逆的；反之亦然。式（3.42）可分情况重写为

$$\begin{cases} \Delta S = \dfrac{q_{可逆}}{T} > \dfrac{q}{T}, & \text{对于所有不可逆过程 } q = q_{不可逆} \\ \Delta S = \dfrac{q_{可逆}}{T} = \dfrac{q}{T}, & \text{对于所有可逆过程 } q = q_{可逆} \end{cases}$$

对于任意给定过程，不论可逆还是不可逆，通式均为

$$\Delta S \geqslant \frac{q}{T} \tag{3.43}$$

其中 q 是该过程所涉及的热，对于不可逆过程 $q = q_{不可逆}$，对于可逆过程 $q = q_{可逆}$。

如果体系为孤立体系，这意味着该体系不与环境相互作用，$q=0$。因此孤立体系的熵变必定大于或等于 0，即

$$\Delta S \geqslant 0 \tag{3.44}$$

其中对于不可逆过程 $\Delta S > 0$，对于可逆过程 $\Delta S = 0$。

回顾 **3.5 节**，由于非自发过程需要持续的外加作用，而这些外加作用一定会导致体系与环境之间的相互作用，因此，孤立体系所能发生的不可逆过程必定是自发的。式（3.44）可以重新解释为：在孤立体系中，自发过程 $\Delta S > 0$，处于平衡的过程 $\Delta S = 0$。因此，孤立体系自发变化的判据为 $\Delta S > 0$。

热力学第二定律

由于宇宙可视为一个巨大的孤立体系，式（3.44）也适用于整个宇宙。然而，当考虑整个宇宙时，由于以下两个原因，式（3.44）

$$\Delta S \geqslant 0 \tag{3.44}$$

where $\Delta S > 0$ for irreversible processes and $\Delta S = 0$ for reversible processes.

Recall from **Section 3.5** that nonspontaneous processes require some external action to be applied continuously and that such external action must always cause some interactions between the system and its surroundings. Therefore, the irreversible processes that an isolated system can undergo must be spontaneous. **Equation (3.44)** can be reinterpreted as that in an isolated system, $\Delta S > 0$ for spontaneous processes and $\Delta S = 0$ for processes at equilibrium. Consequently, the criteria for spontaneous changes in an isolated system are $\Delta S > 0$.

The Second Law of Thermodynamics

As the universe can be viewed as a very large isolated system, **Equation (3.44)** is also applicable to the entire universe. However, when considering the entire universe, the equal sign in **Equation (3.44)** should be removed due to the following two reasons:
1) Among the countless processes happened in the entire universe, there must always be some irreversible processes to which the equal sign does not apply.
2) Reversible processes are ideal processes that do not exist in reality.

Therefore, we have

$$\Delta S_{univ} = \Delta S_{sys} + \Delta S_{surr} > 0 \tag{3.45}$$

Equation (3.45) is one way of stating the **second law of thermodynamics**. It states that *all spontaneous processes produce an increase in the entropy of the universe*. The general criteria for any spontaneous changes happening in the universe are $\Delta S_{univ} > 0$.

Gibbs Free Energy and Gibbs Free Energy Change

Although **Equation (3.45)** can be used as the general criteria for spontaneity, it would be very inconvenient to apply. It requires to calculate all the entropy changes in the entire universe to evaluate whether a process is spontaneous or not. It would be surely preferable to have an alternative criterion that can be applied only to the system, which is the part that we are interested in and the properties of which are easy to measure. Then there is no need to worry about any changes happening in the surroundings, which are difficult to calculate.

Let us consider a hypothetical process involving only *p-V* work at constant temperature and pressure. This process is accompanied by a heat effect, q_p, which is equal to ΔH_{sys}. The heat effect experienced by the surroundings (q_{surr}) is given by

$$q_{surr} = -q_p = -\Delta H_{sys} \tag{3.46}$$

If the surroundings are large enough, the path by which heat enters or leaves the surroundings can be made *reversible*, leading to nearly no change in the temperature of the surroundings. Then the entropy change in the surroundings is given by

$$\Delta S_{surr} = \frac{q_{rev}}{T} = \frac{q_{surr}}{T} = -\frac{\Delta H_{sys}}{T} \tag{3.47}$$

Therefore,

$$-T\Delta S_{univ} = -T\left(\Delta S_{sys} + \Delta S_{surr}\right) = \Delta H_{sys} - T\Delta S_{sys} \tag{3.48}$$

All the variables on the right side of **Equation (3.48)** involve only the system, and the term ΔS_{univ} on the left side serves as the criteria for spontaneous change. If we define a new state function, called the **Gibbs free energy** (*G*) (Josiah W. Gibbs, 1839—1903), as the difference of the enthalpy and the temperature-entropy product of a system, as

$$G = H - TS \tag{3.49}$$

the **Gibbs free energy change** (ΔG) for a process at constant temperature is

$$\Delta G = \Delta H - T\Delta S \tag{3.50}$$

Criteria for Spontaneous Change

With the definitions of the Gibbs free energy and the Gibbs free energy change, **Equation (3.48)** can be

中的等号应该去掉：
1) 在整个宇宙所发生的无数个过程中，一定总会有一些不可逆过程，使得等号不适用。
2) 可逆过程是实际不存在的理想过程。

因此，我们有

$$\Delta S_{宇宙} = \Delta S_{体系} + \Delta S_{环境} > 0 \tag{3.45}$$

式（3.45）是**热力学第二定律**的一种表述形式，它指出所有自发过程均会使宇宙的熵增加。因此，宇宙中发生任何自发变化的通用判据是：$\Delta S_{宇宙} > 0$。

吉布斯自由能与吉布斯自由能变

尽管**式（3.45）**可用作自发性的通用判据，但使用起来很不方便，它需要计算整个宇宙的所有熵变来评估一个过程是否自发。当然最好有一个替代判据，仅需应用于我们所感兴趣且性质易于测定的体系，这样就不必担心环境会发生任何难以计算的变化。

让我们考虑一个恒温恒压下只涉及体积功的假想过程，该过程的热效应为 q_p，其值等于 $\Delta H_{体系}$。则环境所经历的热效应为

$$q_{环境} = -q_p = -\Delta H_{体系} \tag{3.46}$$

如果环境足够大，热进出环境的途径可视为可逆，使环境温度几乎不变。则环境的熵变为

$$\Delta S_{环境} = \frac{q_{可逆}}{T} = \frac{q_{环境}}{T} = -\frac{\Delta H_{体系}}{T} \tag{3.47}$$

故

$$-T\Delta S_{宇宙} = -T\left(\Delta S_{体系} + \Delta S_{环境}\right) = \Delta H_{体系} - T\Delta S_{体系} \tag{3.48}$$

式（3.48）右侧的所有变量均只涉及体系，左侧的 $\Delta S_{宇宙}$ 可用作自发变化的判据。如果将体系的焓和温度与熵之积的差值定义为一个新的状态函数，称为**吉布斯自由能**（G）（约西亚·W. 吉布斯，1839—1903），即

$$G = H - TS \tag{3.49}$$

则恒温过程的**吉布斯自由能变**（ΔG）为

$$\Delta G = \Delta H - T\Delta S \tag{3.50}$$

自发变化的判据

有了吉布斯自由能和吉布斯自由能变的定义，**式（3.48）**可改写为

$$\Delta G_{体系} = \Delta H_{体系} - T\Delta S_{体系} = -T\Delta S_{宇宙} \tag{3.51}$$

因此，用一个只涉及体系的变量 $\Delta G_{体系}$ 即可评估过程的自发性，我们得到了恒温恒压下自发变化的最终判据。为简明起见，通常省略 $\Delta G_{体系}$ 的下标"体系"。

对恒温恒压下只有体积功的过程而言，如下结论成立：
1) 若 ΔG 为负值（$\Delta G < 0$），过程自发。
2) 若 ΔG 等于零（$\Delta G = 0$），过程处于平衡。
3) 若 ΔG 为正值（$\Delta G > 0$），过程非自发。

根据 ΔH 和 ΔS 的符号，ΔG 与自发性共有以下四种可能情况：
1) 若 ΔH 为负值且 ΔS 为正值，则 ΔG 必为负值，该过程在所有温度下均自发。

reformed as

$$\Delta G_{sys} = \Delta H_{sys} - T\Delta S_{sys} = -T\Delta S_{univ} \quad (3.51)$$

Therefore, ΔG_{sys}, which is a variable involving only the system, can be used to evaluate the spontaneity in a process. We then have our final criteria for spontaneous change at constant T and p. The subscript "sys" in ΔG_{sys} is often omitted for conciseness.

For a process occurring with only p-V work at constant T and p, the following statements are valid:
1) If ΔG is negative ($\Delta G<0$), the process is spontaneous.
2) If ΔG is zero ($\Delta G = 0$), the process is at equilibrium.
3) If ΔG is positive ($\Delta G>0$), the process is nonspontaneous.

Based on the signs of ΔH and ΔS, there are altogether four possibilities for ΔG and spontaneity:
1) If ΔH is negative and ΔS is positive, ΔG must be negative and the process is spontaneous at all temperatures.
2) If ΔH is positive and ΔS is negative, ΔG must be positive and the process is nonspontaneous at all temperatures.
3) If both ΔH and ΔS are positive, whether a process is spontaneous or not depends on temperature. In general, it is spontaneous at higher temperatures and nonspontaneous at lower temperatures.
4) If both ΔH and ΔS are negative, whether a process is spontaneous or not also depends on temperature. In general, it is spontaneous at lower temperatures and nonspontaneous at higher temperatures.

Meanwhile, if a process occurs with works other than p-V work (w_{non}) at constant T and p, then the following statements are valid:
1) If $\Delta G<w_{non}$, the process is spontaneous.
2) If $\Delta G = w_{non}$, the process is at equilibrium.
3) If $\Delta G>w_{non}$, the process is nonspontaneous.

Standard Gibbs Free Energy Change

Because the Gibbs free energy is related to enthalpy, the absolute values of the Gibbs free energy cannot be determined either. It is the Gibbs free energy change ΔG in the reaction that matters. The Gibbs free energy change in the standard states is called the **standard Gibbs free energy change** ($\Delta G°$), which is given by

$$\Delta G° = \Delta H° - T\Delta S° \quad (3.52)$$

The standard state conventions are the same as those introduced to enthalpy change in **Section 3.3**.

Table 3.4 The Standard Molar Gibbs Free Energies of Formation ($\Delta G_f°$) of Some Common Substances at 298.15 K
表 3.4 298.15 K 时一些常见物质的标准摩尔生成吉布斯自由能（$\Delta G_f^\ominus$）

Substance（物质）	$\Delta G_f°/(kJ\ mol^{-1})$	Substance（物质）	$\Delta G_f°/(kJ\ mol^{-1})$
AgCl(s)	−109.8	$C_2H_5OH(l)$	−174.8
AgBr(s)	−96.9	HF(g)	−275.4
AgI(s)	−66.2	HCl(g)	−95.3
$AgNO_3$(s)	−33.4	HBr(g)	−53.4
$BaCl_2$(s)	−806.7	HI(g)	1.7
$BaSO_4$(s)	−1362.2	H_2O(l)	−237.1
$Ca(OH)_2$(s)	−897.5	H_2O(g)	−228.6
CO(g)	−137.2	H_2O_2(l)	−120.4
CO_2(g)	−394.4	H_2S(g)	−33.4
CH_4(g)	−50.5	H_2SO_4(l)	−690.0
C_2H_2(g)	209.9	NH_3(g)	−16.4
C_2H_4(g)	68.4	NO(g)	87.6
C_2H_6(g)	−32.0	NO_2(g)	51.3
C_3H_8(g)	−23.4	N_2O_4(g)	99.8
C_6H_6(l)	124.5	N_2H_4(l)	149.3
C_6H_6(g)	129.7	SO_2(g)	−300.1
CH_3OH(l)	−166.6	SO_3(g)	−371.1

2) 若 ΔH 为正值且 ΔS 为负值，则 ΔG 必为正值，该过程在所有温度下均非自发。
3) 若 ΔH 和 ΔS 均为正值，则过程是否自发取决于温度。一般而言，高温下自发、低温下非自发。
4) 若 ΔH 和 ΔS 均为负值，则过程是否自发也取决于温度。一般而言，低温下自发、高温下非自发。

同时，如果一个过程发生在恒温恒压下且存在非体积功（$w_\text{非}$），则如下结论成立：
1) 若 $\Delta G < w_\text{非}$，过程自发。
2) 若 $\Delta G = w_\text{非}$，过程处于平衡。
3) 若 $\Delta G > w_\text{非}$，过程非自发。

标准吉布斯自由能变

由于吉布斯自由能与焓相关，其绝对值也无法确定，因此重要的是反应过程中的吉布斯自由能变 ΔG。标态下的吉布斯自由能变称为**标准吉布斯自由能变**（$\Delta G^\ominus$），由下式给出：

$$\Delta G^\ominus = \Delta H^\ominus - T\Delta S^\ominus \tag{3.52}$$

标态的规定与 **3.3 节**介绍焓变时的规定相同。

物质的**标准生成吉布斯自由能** [$\Delta_\text{f} G_\text{m}^\ominus(T)$，或在 298.15 K 下简写为 $\Delta G_\text{f}^\ominus$] 是在标态下由元素的指定单质生成 1 mol 该物质的吉布斯自由能变。该定义使得标态下元素指定单质的 $\Delta G_\text{f}^\ominus$ 为 0。一些物质在 298.15 K 下的标准生成吉布斯自由能列在**表 3.4** 及**附录 C.1** 中。某物质的 $\Delta G_\text{f}^\ominus$ 可由下式计算：

$$\Delta G_\text{f}^\ominus = \Delta H_\text{f}^\ominus - T\Delta S^\ominus \tag{3.53}$$

一般反应的标准吉布斯自由能变 [$\Delta_\text{r} G_\text{m}^\ominus(T)$，或在 298.15 K 下简写为 $\Delta G^\ominus$] 可由下式计算：

$$\Delta G^\ominus = \sum_\text{X} \nu_\text{X} \Delta G_\text{f}^\ominus(\text{X}) \tag{3.54}$$

其中 X 代表参与反应的各种物质，ν_X 为 X 对应的化学计量数。

3.7 动态平衡与平衡常数

基于此前介绍的热力学状态函数，我们将在本节和接下来的两节讨论关于化学反应的一个非常重要的问题：**化学平衡**。所有化学反应都是可逆的，这意味着它们可以同时正向和逆向地进行。然而反应进行的程度有限，不能简单地由化学计量数计算。不同反应的反应限度可能存在显著差异，可以用化学平衡和平衡常数来表示。

动态平衡

在化学反应中，当反应物和生成物的浓度不再随时间变化，且不再观测到体系性质的变化时，就达到了平衡条件。化学平衡一般可分为四大类：酸碱电离平衡、沉淀溶解平衡、氧化还原平衡和配位解离平衡。这些平衡的细节将在**第 5、6 和 9 章**中讨论。

化学平衡具有如下特点：
1) 化学平衡始终是**动态平衡**，其中两个相反的反应以同等速率进

The **standard Gibbs free energy of formation** [$\Delta_f G_m^\circ(T)$, or ΔG_f° for short at 298.15 K] of a substance is the Gibbs free energy change in the formation of 1 mol substance from the reference forms of the elements, all in their standard states. This definition leads to values of zero for ΔG_f° of the elements in reference forms in their standard states. The standard Gibbs free energies of formation for a number of substances at 298.15 K are tabulated in **Table 3.4** and **Appendix C.1**. The ΔG_f° value of a substance can be calculated by

$$\Delta G_f^\circ = \Delta H_f^\circ - T\Delta S^\circ \tag{3.53}$$

The standard Gibbs free energy change [$\Delta_r G_m^\circ(T)$, or ΔG° for short at 298.15 K] for a general reaction can be calculated from

$$\Delta G^\circ = \sum_X \nu_X \Delta G_f^\circ(X) \tag{3.54}$$

where X stands for various substances involved in this reaction, and ν_X is the stoichiometric coefficient for X.

3.7 Dynamic Equilibrium and the Equilibrium Constants

Based on the previously introduced thermodynamic state functions, we are going to discuss in this and the next two sections a very important question about chemical reactions: the **chemical equilibrium**. All chemical reactions are reversible, meaning that they can process both forwardly and reversely at the same time. However, the limit to which the reactions proceed is limited, and cannot be calculated simply from the stoichiometric coefficients. The extent of reactions may differ from one another significantly, and can be represented by chemical equilibrium and the equilibrium constant.

Dynamic Equilibrium

In a chemical reaction, when the concentrations of both reactants and products have no further tendency to change with time and no change in the properties of the system is observed, an equilibrium condition is reached. Chemical equilibria can be classified into four general categories: acid-base equilibria, precipitation-dissolution equilibria, oxidation-reduction equilibria, and complex-ion equilibria. The details of these equilibria will be discussed in **Chapters 5, 6,** and **9**.

The characteristics of chemical equilibrium are listed as follows:
1) The chemical equilibrium is always a **dynamic equilibrium**, in which two opposing reactions take place at equal rates; that is, the forward reaction proceeds at the same rate as the reverse reaction. More discussion about reaction rates can be found in **Chapter 4**.
2) The states of the system are specified at equilibrium, and are independent of how the equilibrium condition is reached.
3) A chemical reaction tends to reach its equilibrium spontaneously.
4) The chemical equilibrium depends on conditions. When the conditions change, the original equilibrium is destroyed and a new equilibrium will be established under the new conditions.

Equilibrium Constant Expression

For a general reversible reaction represented by

$$a\text{A} + b\text{B} + \cdots \rightleftharpoons g\text{G} + h\text{H} + \cdots$$

When a dynamic equilibrium is reached at temperature T, the equilibrium constant expression can be written as

$$K = \frac{a_G^g a_H^h \cdots}{a_A^a a_B^b \cdots} = \prod_X (a_X)^{\nu_X} \tag{3.55}$$

where K is the **thermodynamic equilibrium constant**, X stands for various substances involved in this reaction, ν_X is the stoichiometric coefficient for X, and a_X is the **activity** for X. The equilibrium constant K is a function of temperature.

行，即正向反应与逆向反应发生的速率相等。关于反应速率的更多讨论详见**第 4 章**。
2) 达到平衡时体系具有确定的状态，并与如何达到平衡的过程无关。
3) 化学反应有自发达到平衡的趋势。
4) 化学平衡与条件相关，当条件改变时，原有平衡被破坏，在新的条件下会建立新的平衡。

平衡常数表达式

对于具有如下通式的可逆反应：
$$aA + bB + \cdots \rightleftharpoons gG + hH + \cdots$$
当在温度 T 下达到动态平衡时，平衡常数表达式可写为

$$K = \frac{a_G^g a_H^h \cdots}{a_A^a a_B^b \cdots} = \prod_X (a_X)^{v_X} \tag{3.55}$$

其中 K 是**热力学平衡常数**，X 代表参与反应的各种物质，v_X 为 X 对应的化学计量数，a_X 是 X 的**活度**。平衡常数 K 是温度的函数。

如果 X 为溶液，则其活度 a_X 与浓度 [X] 之间的关系为

$$a_X = \frac{\gamma_X [X]}{c^{\ominus}} \tag{3.56}$$

其中 γ_X 是 X 的**活度系数**，量纲为 1；$c^{\ominus}$ 代表 1 mol kg^{-1} 或 1 mol L^{-1} 的标准浓度，其单位与 [X] 相同。因此，a_X 的量纲也为 1。对于理想溶液或稀溶液，$\gamma_X = 1$。**式（3.55）**可改写为

$$K = \left(\frac{1}{c^{\ominus}}\right)^{\Delta n} \prod_X [X]^{v_X} = \left(\frac{1}{c^{\ominus}}\right)^{\Delta n} K_c \tag{3.57}$$

其中

$$\Delta n = \sum_X v_X \quad \text{且} \quad K_c = \prod_X [X]^{v_X} \tag{3.58}$$

K_c 是以浓度形式表示的平衡常数，其单位与 $(c^{\ominus})^{\Delta n}$ 相同。而 K 始终是一个量纲为 1 的数。

如果 X 为气体，则其活度 a_X 与分压 p_X 之间的关系为

$$a_X = \frac{\gamma_X p_X}{p^{\ominus}} \tag{3.59}$$

其中 γ_X 是 X 的活度系数，量纲为 1；$p^{\ominus}$ 代表 1 bar 的标准压强。对于理想气体或低压气体，$\gamma_X = 1$。气体浓度 [X] 与其分压 p_X 之间的关系为

$$[X] = \frac{n}{V} = \frac{p_X}{RT} \tag{3.60}$$

因此，**式（3.55）**可改写为

$$K = \left(\frac{1}{p^{\ominus}}\right)^{\Delta n_{\text{气}}} \prod_X (p_X)^{v_X} = \left(\frac{RT}{p^{\ominus}}\right)^{\Delta n_{\text{气}}} \prod_X [X]^{v_X}$$

或

$$K = \left(\frac{1}{p^{\ominus}}\right)^{\Delta n_{\text{气}}} K_p = \left(\frac{RT}{p^{\ominus}}\right)^{\Delta n_{\text{气}}} K_c \tag{3.61}$$

其中 $\Delta n_{\text{气}}$ 是参与反应的气态物质的化学计量数之差，K_p 是以分压形式表示的平衡常数。

如果参与反应的一种或多种物质是纯固体或纯液体，由于它们的浓度或分压在化学反应过程中不发生改变，其活度可设为 1。因此，

If X is a solution, the activity a_X is related to its concentration [X] by

$$a_X = \frac{\gamma_X [X]}{c^\circ} \tag{3.56}$$

where γ_X is the dimensionless **activity coefficient** of X; and c° stands for the standard concentration of 1 mol kg^{-1} or 1 mol L^{-1}, which takes the same unit as [X]. Therefore, a_X is also dimensionless. For ideal solutions or solutions with dilute concentrations, $\gamma_X = 1$. **Equation (3.55)** can be rewritten as

$$K = \left(\frac{1}{c^\circ}\right)^{\Delta n} \prod_X [X]^{\nu_X} = \left(\frac{1}{c^\circ}\right)^{\Delta n} K_c \tag{3.57}$$

where

$$\Delta n = \sum_X \nu_X \quad \text{and} \quad K_c = \prod_X [X]^{\nu_X} \tag{3.58}$$

and K_c is the equilibrium constant in terms of concentration, which takes the same unit as $(c^\circ)^{\Delta n}$. K, on the other hand, is always a dimensionless quantity.

If X is a gas, the activity a_X is related to its partial pressure p_X by

$$a_X = \frac{\gamma_X p_X}{p^\circ} \tag{3.59}$$

where γ_X is the dimensionless activity coefficient of X, and p° stands for the standard pressure of 1 bar. For ideal gases or gases with low pressure, $\gamma_X = 1$. The gas concentration [X] is related to its partial pressure p_X by

$$[X] = \frac{n}{V} = \frac{p_X}{RT} \tag{3.60}$$

Therefore, **Equation (3.55)** can be rewritten as

$$K = \left(\frac{1}{p^\circ}\right)^{\Delta n_{gas}} \prod_X (p_X)^{\nu_X} = \left(\frac{RT}{p^\circ}\right)^{\Delta n_{gas}} \prod_X [X]^{\nu_X}$$

or

$$K = \left(\frac{1}{p^\circ}\right)^{\Delta n_{gas}} K_p = \left(\frac{RT}{p^\circ}\right)^{\Delta n_{gas}} K_c \tag{3.61}$$

where Δn_{gas} is the difference in the stoichiometric coefficients of gaseous substances involved in the reaction, and K_p is the equilibrium constant in terms of partial pressure.

If one or more substances involved in the reaction are pure solids or liquids, since their concentrations or partial pressures do not change during a chemical reaction, their activities are set equal to 1. Consequently, the equilibrium constant expression does not contain terms for pure solids or liquids. For example, for the decomposition reaction of calcium carbonate into calcium oxide and carbon dioxide:

$$CaCO_3(s) \rightleftharpoons CaO(s) + CO_2(g)$$

both calcium carbonate and calcium oxide are pure solids, and thus only the gas phase carbon dioxide is included in the equilibrium constant expressions, as

$$K_c = [CO_2] \quad \text{and} \quad K_p = p_{CO_2}$$
$$K = K_p / p^\circ = K_c RT / p^\circ$$

Relationships Involving Equilibrium Constants

Since the stoichiometric coefficients of the reaction substances are involved in the equilibrium constant expression, we must always make sure that the expression for K matches the corresponding balanced equation. For a general reversible reaction represented by

$$aA + bB + \cdots \rightleftharpoons gG + hH + \cdots \quad K = \prod_X (a_X)^{\nu_X}$$

If the above equation is reversed, the value of K should be inverted

平衡常数表达式中无须包含纯固体或纯液体的项。例如，对于碳酸钙分解为氧化钙和二氧化碳的反应：

$$CaCO_3(s) \rightleftharpoons CaO(s) + CO_2(g)$$

碳酸钙和氧化钙均为纯固体，因此平衡常数表达式中只包含气相二氧化碳，即

$$K_c = [CO_2] \quad 且 \quad K_p = p_{CO_2}$$

$$K = K_p / p^\ominus = K_c RT / p^\ominus$$

平衡常数之间的关系

由于平衡常数表达式涉及反应物质的化学计量数，必须始终确保 K 的表达式与相应的平衡方程式相匹配。对于具有如下通式的可逆反应：

$$a\mathrm{A} + b\mathrm{B} + \cdots \rightleftharpoons g\mathrm{G} + h\mathrm{H} + \cdots \quad K = \prod_\mathrm{X}(a_\mathrm{X})^{\nu_\mathrm{X}}$$

如果反转上述方程式，K 也应取倒数：

$$g\mathrm{G} + h\mathrm{H} + \cdots \rightleftharpoons a\mathrm{A} + b\mathrm{B} + \cdots \quad K_{反} = \prod_\mathrm{X}(a_\mathrm{X})^{-\nu_\mathrm{X}} = \frac{1}{K}$$

如果平衡常数为 K_1 和 K_2 的两个方程式相加可得第三个方程式，则第三个方程式的平衡常数 K_3 应为

$$K_3 = K_1 \cdot K_2$$

如果方程式的系数乘以同一因子，则平衡常数应提高到相应的次幂。同样，如果方程式的系数除以同一因子，则平衡常数应取相应的方根。

平衡常数的量级

原则上每个化学反应都存在一个平衡常数，但它们的值可能有显著差异。在许多情况下，平衡常数的量级而非其准确的值，即可充分表明反应限度。

如果一个反应进行完全，这意味着逆反应可忽略不计，净反应只正向进行。然后一个或多个反应物将被耗尽，**式（3.55）** 中分母的某一项接近零，使 K 成为一个非常大的数。反过来，一个非常大的 K 也表示正反应进行完全，且逆反应几乎不发生。同样，一个非常小的 K 表示正反应几乎不发生，且逆反应进行完全。

通常我们采用 10^{10} 和 10^{-10} 作为 K 的分界值来区分反应限度：
1) 当 $K > 10^{10}$ 时，正反应进行完全，逆反应几乎不发生。
2) 当 $K < 10^{-10}$ 时，正反应几乎不发生，逆反应进行完全。
3) 当 $10^{-10} \leqslant K \leqslant 10^{10}$ 时，反应限度可通过改变温度、浓度和分压等条件进行调节。

3.8 反应商与勒夏特列原理

当反应处于平衡条件时，参与反应的所有物质浓度均保持不变，反应不存在净变化。在最终达到平衡条件之前，我们希望能够预测反应净变化的方向，即反应将向哪个方向建立平衡。在这些情况下，

$$gG + hH + \cdots \rightleftharpoons aA + bB + \cdots \quad K_{\text{rev}} = \prod_X (a_X)^{-v_X} = \frac{1}{K}$$

If two equations with equilibrium constants K_1 and K_2 add up to form a third equation, the equilibrium constant of the third equation K_3 should be

$$K_3 = K_1 \cdot K_2$$

If the coefficients in an equation is multiplied by a common factor, the equilibrium constant should be raised to the corresponding power. Similarly, if the coefficients in an equation is divided by a common factor, the equilibrium constant should take the corresponding root.

Magnitude of Equilibrium Constants

In principle, every chemical reaction has an equilibrium constant, but their values may differ from one another significantly. In many cases, the magnitude of equilibrium constants, rather than the accurate values, can sufficiently demonstrate the limit of reactions.

If a reaction goes to completion, it means that the reverse reaction is negligible and the net reaction proceeds only in the forward direction. Then one or more of the reactants is used up. A term in the denominator in **Equation (3.55)** approaches zero and makes K a very large number. Conversely, a very large value of K also signifies that the forward reaction goes to completion and the reverse reaction does not occur to any significant extent. Similarly, a very small value of K signifies that the forward reaction does not occur to any significant extent and the reverse reaction goes to completion.

In general, we use the magnitudes of 10^{10} and 10^{-10} as the demarcation points in K to distinguish the limit of reaction:

1) If $K > 10^{10}$, the forward reaction goes to completion and the reverse reaction does not occur to any significant extent;
2) If $K < 10^{-10}$, the forward reaction does not occur to any significant extent and the reverse reaction goes to completion;
3) If $10^{-10} \leqslant K \leqslant 10^{10}$, the limit of reaction is adjustable by varying the conditions such as temperature, concentration, and partial pressure.

3.8 The Reaction Quotient and Le Châtelier's Principle

When a reaction is at its equilibrium condition, the concentrations of all substances involved in the reaction remain constant and there is no net change of the reaction. Before the equilibrium condition is finally reached, we would like to have the ability to predict the direction of net change of the reaction, i.e., the direction to which the equilibrium will be established. In these cases, we need to use a concept called the reaction quotient.

Reaction Quotient

The **reaction quotient** is designated Q and defined as the ratio of activities of products to the activities of reactants at the initial state of reaction given by

$$Q = \prod_X (a_X)^{v_X}_{\text{init}} \tag{3.62}$$

Note that Q has the exactly same form as K except that the initial activities are used in the expression of Q but the equilibrium activities are used in that of K. Q_p and Q_c can be defined in a similar way to K_p and K_c, respectively.

Predicting the Direction of Reaction

If there is only reactants and no products at all in the initial state, $Q = 0 < K$. A net reaction must occur to the right and start to produce some products. As it does, the activities of products increase, the activities of reactants decrease, and the value of Q increases until eventually $Q = K$ when an equilibrium is established.

需要使用反应商的概念。

反应商

反应商用 Q 表示，定义为在反应的初始状态下生成物与反应物活度之比：

$$Q = \prod_X (a_X)^{v_X}_{初始} \tag{3.62}$$

注意 Q 与 K 的表达式形式完全相同，只是 Q 的表达式使用初始活度，而 K 的表达式使用平衡活度。Q_p 和 Q_c 的定义方式与 K_p 和 K_c 类似。

预测反应的方向

如果初始状态下只有反应物而完全没有生成物，则 $Q = 0 < K$。净反应必定向右发生，并开始产生一些生成物。这样，生成物活度增加而反应物活度降低，Q 增大，直至最终建立平衡时达到 $Q = K$。如果初始状态下只有生成物而完全没有反应物，$Q = \infty > K$。净反应必定向左发生，并开始产生一些反应物。这样，生成物活度降低而反应物活度增加，Q 减小，直至最终建立平衡时达到 $Q = K$。

因此，Q 和 K 的比较可用于预测反应方向：

1) 若 $Q = K$，则反应处于平衡，且反应没有净变化。
2) 若 $Q < K$，随着反应的进行，生成物活度增加而反应物活度降低，将导致从左至右（即正向）的净变化，净反应将自发正向进行以达到平衡。
3) 若 $Q > K$，随着反应的进行，生成物活度降低且反应物活度增加，将导致从右至左（即逆向）的净变化，净反应将自发逆向进行以达到平衡。

化学平衡的移动

由于化学平衡取决于反应条件（如压强、浓度、温度等），当条件改变时，原有平衡被破坏，在新的条件下会建立新的平衡，这称为**平衡的移动**。我们希望不仅能够预测平衡将向哪个方向移动，还能够利用条件变化，将平衡向我们所希望的特定方向移动。

当在一定条件下建立平衡时，满足 $Q = K$。如果参与反应的一些物质的浓度或分压改变，Q 将会改变。如果反应温度改变，K 也会改变。在这两种情况下，无论 Q 或 K 的改变均会使 $Q \neq K$，原有平衡被破坏。如果 $Q < K$ 或 $Q/K < 1$，平衡将向右移动；而如果 $Q > K$ 或 $Q/K > 1$，平衡将向左移动，直到在新条件下建立新的平衡。无论 Q 或 K 怎样变化，Q 和 K 的比值总决定了平衡移动的方向。此外，Q/K 的值还决定了原平衡与新平衡之间移动的程度。Q/K 的值越远离 1，原平衡与新平衡之间的差异就越大。

勒夏特列原理

1884 年，法国化学家亨利·L. 勒夏特列（1850—1936）总结了一条预测平衡移动方向的基本规则：当平衡体系受到温度、压强或反应物种浓度变化的影响时，体系会通过达到一个新的平衡来应对，新的平衡将部分抵消该变化所造成的影响。这称为**勒夏特列原理**。

If there is only products and no reactants at all in the initial state, $Q = \infty > K$. A net reaction must occur to the left and start to produce some reactants. As it does, the activities of products decrease, the activities of reactants increase, and the value of Q decreases until eventually $Q = K$ when an equilibrium is established.

Therefore, the comparison of Q and K can be used to predict the direction of reaction:
1) If $Q = K$, the reaction is at its equilibrium and there is no net change of the reaction.
2) If $Q<K$, the activities of products should increase and the activities of reactants should decrease, as the reaction proceeds. It will lead to a net change occurring from left to right, i.e., in the forward direction. The net reaction will spontaneously occur in the forward direction to reach the equilibrium.
3) If $Q>K$, the activities of products should decrease and the activities of reactants should increase, as the reaction proceeds. It will lead to a net change occuring from right to left, i.e., in the reverse direction. The net reaction will spontaneously occur in the reverse direction to reach the equilibrium.

Shift of a Chemical Equilibrium

As chemical equilibrium depends on conditions (such as pressure, concentration, temperature, etc.), when the conditions change, the original equilibrium is destroyed and a new equilibrium will be established under the new conditions, which is called the **shift of an equilibrium**. We would like to have the abilities not only to predict to which direction the equilibrium will shift, but also to take advantage of the condition change to shift the equilibrium to the particular direction that we prefer.

When an equilibrium is established at certain conditions, $Q = K$ is satisfied. If the concentration or the pressure of some substances involved in the reaction change, the value of Q will change. If the temperature of reaction changes, the value of K will also change. In both cases, the change in either Q or K makes $Q \neq K$ and the original equilibrium is destroyed. If $Q<K$ or $Q/K< 1$, the equilibrium will shift to the right, and if $Q>K$ or $Q/K> 1$, the equilibrium will shift to the left, until a new equilibrium is established under the new conditions. No matter how Q or K changes, the ratio of Q and K always determines the direction in which the equilibrium will shift. Moreover, the value of Q/K also determines the magnitude of shift between the original and new equilibria. The further the value of Q/K is away from 1, the larger the difference between the original and new equilibria is.

Le Châtelier's Principle

In 1884, the French chemist Henry L. Le Châtelier (1850—1936) summarized an essential rule for predicting the direction in which an equilibrium shifts: When an equilibrium system is subjected to a change in temperature, pressure, or concentration of a reacting species, the system responds by attaining a new equilibrium that partially offsets the impact of the change. This is called **Le Châtelier's principle**.

3.9 Gibbs Free Energy Change and Equilibrium

In **Section 3.6**, we have introduced a new state function called the Gibbs free energy (G), and discussed how the Gibbs free energy change (ΔG) can serve as the criteria for a spontaneous process occurring at constant T and p. We have also showed how to calculate the standard Gibbs free energy change ($\Delta G°$) from standard enthalpy change ($\Delta H°$) and standard entropy change ($\Delta S°$). However, most reactions happen at nonstandard states. Under such circumstances, we need to work with the nonstandard Gibbs free energy change.

Nonstandard Gibbs Free Energy Change

To obtain the relationship between ΔG and $\Delta G°$, let us consider a reaction of ideal gas molecules, for which $\Delta H = \Delta H°$ applies because the enthalpy of an ideal gas is a function of temperature only and is independent of pressure. Therefore,

$$\Delta G = \Delta H - T\Delta S = \Delta H° - T\Delta S$$

Next, we need to obtain a relationship between ΔS and $\Delta S°$. We then consider the isothermal expansion process of an ideal gas for which $\Delta U = 0$ and $q = -w$. If the expansion occurs reversibly, the work is given by

3.9 吉布斯自由能变与平衡

3.6 节我们引入了一个称为吉布斯自由能（G）的新状态函数，并讨论了吉布斯自由能变（ΔG）如何作为恒温恒压下自发过程发生的判据。我们还展示了怎样从标准焓变（$\Delta H^\ominus$）和标准熵变（$\Delta S^\ominus$）来计算标准吉布斯自由能变（$\Delta G^\ominus$）。然而，大多数反应发生在非标态下，这时我们需要使用非标准吉布斯自由能变。

非标准吉布斯自由能变

为获得 ΔG 和 $\Delta G^\ominus$ 之间的关系，让我们考虑一个理想气体分子的反应，由于理想气体的焓仅为温度的函数而与压强无关，$\Delta H = \Delta H^\ominus$ 成立。因此

$$\Delta G = \Delta H - T\Delta S = \Delta H^\ominus - T\Delta S$$

接下来，我们需要得到 ΔS 和 $\Delta S^\ominus$ 之间的关系。考虑理想气体恒温膨胀过程，$\Delta U = 0$ 且 $q = -w$。如果此膨胀以可逆的方式进行，其功为

$$w_{可逆} = nRT \ln \frac{p_{终}}{p_{始}}$$

恒温可逆膨胀的热为

$$q_{可逆} = -w_{可逆} = -nRT \ln \frac{p_{终}}{p_{始}}$$

因此，恒温膨胀的摩尔熵变可写为

$$\Delta S = \frac{q_{可逆}}{nT} = -R \ln \frac{p_{终}}{p_{始}} = S_{终} - S_{始}$$

若令始态为标态，这意味着 $p_{始} = p^\ominus = 1 \text{ bar}$ 且 $S_{始} = S^\ominus$，则

$$S_{终} = S^\ominus - R \ln \frac{p_{终}}{p^\ominus} = S^\ominus - R \ln p_{终}$$

即

$$S = S^\ominus - R \ln p$$

对于具有如下通式的气相可逆反应：

$$a\text{A}(g) + b\text{B}(g) + \cdots \rightleftharpoons g\text{G}(g) + h\text{H}(g) + \cdots$$

非标准熵变 ΔS 可写为

$$\Delta S = \Delta S^\ominus - R \ln Q \tag{3.63}$$

其中 Q 为该气相反应的反应商，由下式给出：

$$Q = \prod_X (p_X)_{初始}^{\nu_X}$$

因此，非标准吉布斯自由能变与反应商的关系为

$$\Delta G = \Delta H^\ominus - T\Delta S = \Delta H^\ominus - T\left(\Delta S^\ominus - R \ln Q\right)$$

即

$$\Delta G = \Delta G^\ominus + RT \ln Q \tag{3.64}$$

虽然**式（3.64）**从气相反应中推导出，但普遍适用于所有反应。

标准吉布斯自由能变

如果一个反应已处于平衡态，$\Delta G = 0$ 且 $Q = K$。将其应用于**式（3.64）**，即可将标准吉布斯自由能变与平衡常数联系起来，有

$$\Delta G^\ominus = -RT \ln K \tag{3.65}$$

$$w_{rev} = nRT \ln \frac{p_f}{p_i}$$

The reversible isothermal heat of expansion is

$$q_{rev} = -w_{rev} = -nRT \ln \frac{p_f}{p_i}$$

Therefore, the molar entropy change for the isothermal expansion can be written as

$$\Delta S = \frac{q_{rev}}{nT} = -R \ln \frac{p_f}{p_i} = S_f - S_i$$

Let us set the initial state to be the standard state, which means $p_i = p° = 1$ bar and $S_i = S°$, then

$$S_f = S° - R \ln \frac{p_f}{p°} = S° - R \ln p_f$$

or

$$S = S° - R \ln p$$

For a general reversible gaseous reaction represented by

$$a\text{A}(g) + b\text{B}(g) + \cdots \rightleftharpoons g\text{G}(g) + h\text{H}(g) + \cdots$$

nonstandard entropy change ΔS can be written as

$$\Delta S = \Delta S° - R \ln Q \tag{3.63}$$

where Q is the reaction quotient for this gaseous reaction given by

$$Q = \prod_X (p_X)_{init}^{v_X}$$

Therefore, the nonstandard Gibbs free energy change can be related to the reaction quotient as

$$\Delta G = \Delta H° - T\Delta S = \Delta H° - T(\Delta S° - R \ln Q)$$

or

$$\Delta G = \Delta G° + RT \ln Q \tag{3.64}$$

Although derived from a gaseous reaction, **Equation (3.64)** is generally applied to all reactions.

Standard Gibbs Free Energy Change

If a reaction is already at its equilibrium, $\Delta G = 0$ and $Q = K$. Applying those into **Equation (3.64)**, we can then relate the standard Gibbs free energy change to the equilibrium constant as

$$\Delta G° = -RT \ln K \tag{3.65}$$

This means that the tabulation of thermodynamic data in **Appendix C.1** can serve as a direct source of countless equilibrium constant values at 298.15 K. In **Section 3.7**, the magnitudes of 10^{10} and 10^{-10} have been used as the demarcation points in K to distinguish the limit of reaction. The corresponding demarcation values in $\Delta G°$ can be calculated as ± 57 kJ mol^{-1}. Therefore,

1) If $\Delta G° < -57$ kJ mol^{-1}, equilibrium favors products;
2) If $\Delta G° > 57$ kJ mol^{-1}, equilibrium favors reactants;
3) If -57 kJ mol$^{-1} \leqslant \Delta G° \leqslant 57$ kJ mol^{-1}, equilibrium calculation is necessary when the conditions vary.

van't Hoff Equation

Since temperature is not defined in the standard state, $\Delta G°$ is dependent on temperature and can be written as

$$-RT \ln K = \Delta G° = \Delta H° - T\Delta S°$$

Dividing by $-RT$, we have

$$\ln K = -\frac{\Delta H°}{RT} + \frac{\Delta S°}{R} \tag{3.66}$$

If both $\Delta H°$ and $\Delta S°$ do not vary significantly with temperature, we can assume their values to be constant. Plotting $\ln K$ with respect to $1/T$, we will obtain a straight line with a slope of $-\Delta H°/R$ and an intercept of $\Delta S°/R$. Applying two different temperatures into **Equation (3.66)** and subtracting one from the other, we then have

这意味着**附录 C.1** 的热力学数据可作为 298.15 K 下无数平衡常数的直接来源。**3.7 节**中，10^{10} 和 10^{-10} 被用作平衡常数 K 的分界值来区分反应限度。可计算相应的 $\Delta G^\ominus$ 分界值为 ± 57 kJ mol^{-1}，故：

1) 当 $\Delta G^\ominus < -57$ kJ mol^{-1} 时，平衡有利于生成物；
2) 当 $\Delta G^\ominus > 57$ kJ mol^{-1} 时，平衡有利于反应物；
3) 当 -57 kJ mol$^{-1} \leqslant \Delta G^\ominus \leqslant 57$ kJ mol^{-1} 时，条件改变时需进行必要的平衡计算。

范特霍夫方程

由于标态并未定义温度，$\Delta G^\ominus$ 与温度相关，可写为

$$-RT \ln K = \Delta G^\ominus = \Delta H^\ominus - T\Delta S^\ominus$$

方程两边除以 $-RT$，可得

$$\ln K = -\frac{\Delta H^\ominus}{RT} + \frac{\Delta S^\ominus}{R} \qquad (3.66)$$

如果 $\Delta H^\ominus$ 和 $\Delta S^\ominus$ 均不随温度显著变化，我们可假定其值为常数。将 $\ln K$ 对 $1/T$ 作图，可得一条斜率为 $-\Delta H^\ominus/R$、截距为 $\Delta S^\ominus/R$ 的直线。将两个不同温度代入**式（3.66）**并相减，可得

$$\ln \frac{K_2}{K_1} = -\frac{\Delta H^\ominus}{R}\left(\frac{1}{T_2} - \frac{1}{T_1}\right) \qquad (3.67)$$

式（3.67） 最早由雅各布·H. 范特霍夫（1852—1911）推导出，因此被称为**范特霍夫方程**。

我们在 **2.6 节**学过的克劳修斯-克拉贝龙方程 [**式（2.57）**] 可视为**式（3.67）**的特殊情况，对应于气-液相平衡，其平衡常数为饱和蒸气压，且有 $\Delta H^\ominus = \Delta H_{蒸发}$。

吉布斯自由能变、平衡与自发变化的方向

这里我们来总结自发变化的结论性判据，及其与吉布斯自由能变和平衡的关系。如**图 3.13** 所示，对于某个假想反应，将吉布斯自由能对反应进度作图，反应物位于最左侧，生成物位于最右侧，均处于标态。$\Delta G^\ominus$ 是生成物与反应物之间的标准吉布斯自由能变，平衡点位于纯反应物和纯生成物之间的某处。由于所有化学反应均包含正反应和逆反应，且正反应和逆反应自发进行的方向均为吉布斯自由能降低（$\Delta G < 0$）的方向，这意味着平衡点必须是吉布斯自由能最小的点。换言之，吉布斯自由能的减少是任何恒温恒压下发生的化学反应自发达到平衡的驱动力。在平衡状态下，吉布斯自由能已经达到了最小值，因此没有反应的驱动力，也就没有进一步的净反应。

在**图 3.13(a)** 中，$\Delta G^\ominus$ 很小（不论为正值或负值），平衡混合物中同时含有大量反应物和生成物，且位于纯反应物和纯生成物的中点附近。$Q > K$ 的混合物位于平衡点右侧，将沿吉布斯自由能降低的方向向左发生自发变化，最终达到平衡。同样，$Q < K$ 的混合物位于平衡点左侧，在达到平衡前将自发产生更多的生成物。

在**图 3.13(b)** 中，$\Delta G^\ominus$ 很大且为负值，平衡混合物包含大量生成物，反应物的量可忽略不计。平衡点接近纯生成物的极限。因此，在达到平衡前反应基本进行完全。在**图 3.13(c)** 中，$\Delta G^\ominus$ 很大且为正值，平衡混合物包含大量反应物，生成物的量可忽略不计。平衡点接近纯反应物的极限，在达到平衡前反应几乎不发生。

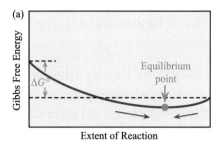
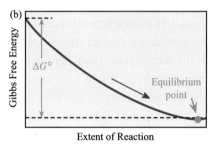
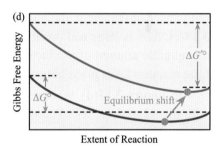

Figure 3.13 Gibbs free energy change, equilibrium, and the direction of spontaneous change. (a) ΔG° is small, so the equilibrium mixture lies about midway between reactants and products. (b) ΔG° is large and negative, so the equilibrium point lies close to the products. (c) ΔG° is large and positive, so the equilibrium point lies close to the reactants. (d) ΔG° varies with temperature, and the shape of curve and position of equilibrium point also change, leading to the shift of the equilibrium.

图 3.13 吉布斯自由能变、平衡及自发变化的方向：(a) $\Delta G^\ominus$ 很小，平衡混合物位于反应物和生成物的中点附近；(b) $\Delta G^\ominus$ 很大且为负值，平衡点靠近生成物；(c) $\Delta G^\ominus$ 很大且为正值，平衡点靠近反应物；(d) $\Delta G^\ominus$ 随温度改变，曲线的形状及平衡点的位置也发生改变，导致了平衡点的移动。

$$\ln\frac{K_2}{K_1} = -\frac{\Delta H^\circ}{R}\left(\frac{1}{T_2} - \frac{1}{T_1}\right) \tag{3.67}$$

Equation (3.67) was first derived by Jacobus H. van't Hoff (1852—1911) and thus is often referred to as **van't Hoff equation**.

Clausius-Clapeyron **Equation (2.57)** that we have learnt in **Section 2.6** can be considered as a special case of **Equation (3.67)**. The circumstance is a liquid-vapor phase equilibrium, in which the equilibrium constants are the saturated vapor pressures and $\Delta H^\circ = \Delta H_{vap}$.

Gibbs Free Energy Change, Equilibrium, and the Direction of Spontaneous Change

Here, we summarize the conclusive criteria for spontaneous change and the relationship between the Gibbs free energy change and equilibrium. As shown in **Figure 3.13**, the Gibbs free energy is plotted against the extent of reaction for a hypothetical reaction. The reactants are on the left and the products are on the right, both in their standard states. ΔG° is the standard Gibbs free energy change between the products and reactants. The equilibrium point lies somewhere between pure reactants and pure products. Since every chemical reaction consists of both a forward and a reverse reaction, and the direction of spontaneous change in both the forward and reverse reactions is the direction in which the Gibbs free energy decreases ($\Delta G < 0$), it means that the equilibrium point must be the minimum in the Gibbs free energy. In other words, the decrease in the Gibbs free energy serves as the driving force for the spontaneity of any chemical reaction to reach its equilibrium under constant temperature and pressure. At equilibrium the Gibbs free energy is already at its minimum, so there is no driving force for the reaction and there is no further net reaction.

In **Figure 3.13(a)**, ΔG° is small, no matter positive or negative, so the equilibrium mixture contains significant amounts of both reactants and products and lies about midway between pure reactants and pure products. Mixtures with $Q > K$ are to the right of the equilibrium point and undergo spontaneous change to the left in the direction of the lower Gibbs free energy, eventually reaching equilibrium. Similarly, mixtures with $Q < K$ are to the left of the equilibrium point and spontaneously yield more products before reaching equilibrium.

In **Figure 3.13(b)**, ΔG° is large and negative, so the equilibrium mixture contains a significant amount of products and a negligible amount of reactants. The equilibrium point lies close to the extreme of pure products. Consequently, the reaction goes essentially to completion before equilibrium is reached. In **Figure 3.13(c)**, ΔG° is large and positive, so the equilibrium mixture contains a significant amount of reactants and a negligible amount of products. The equilibrium point lies close to the extreme of pure reactants, and very little reaction takes place before equilibrium is reached.

In **Figure 3.13(d)**, when temperature varies, the Gibbs free energy of both reactants and products varies with it, and so does ΔG°. The shape of the curve between reactants and products changes, and the position of the equilibrium point also changes, leading to the shift of the equilibrium.

Extended Reading Materials Nanothermodynamics

The study of thermodynamics of small systems was first investigated theoretically based on statistical mechanical models by Terrell L. Hill (1917—2014) in the 1960s. Over the next 25 years, his interests shifted drastically and he almost dropped this research field except for very occasional applications. With the explosion of interest in experimental and theoretical nanoscience in the 1990s, Hill's book *Thermodynamics of Small Systems* was republished in 1994, and he returned to this field of research. In 2000, Professor Ralph V. Chamberlin of Arizona State University published a paper in *Nature* and used the term "**nanothermodynamics**" for the first time, by applying Hill's theory to solve problems of ferromagnetism. Soon, nanothermodynamics was accepted by scientists in the research field. Hill's main contribution was to extend the thermodynamics of Josiah W. Gibbs to ensembles of small systems, particularly in the fields of biochemistry and molecular biology.

Nanothermodynamics is applicable to microscopic systems that do not reach the thermodynamic limit.

在**图 3.13(d)** 中，当温度改变时，反应物和生成物的吉布斯自由能均发生变化，$\Delta G^{\ominus}$ 也随之变化。在反应物和生成物之间的曲线形状会发生改变，平衡点的位置也会改变，从而导致平衡的移动。

拓展阅读材料　纳米热力学

20 世纪 60 年代，泰瑞尔·L. 希尔（1917—2014）基于统计力学模型，首次从理论上研究了小体系的热力学。在接下来的 25 年里，他的兴趣发生了巨大的转变，除极偶尔的应用外，他几乎放弃了这个研究领域。20 世纪 90 年代，随着人们对实验和理论纳米科学兴趣的爆发，希尔的《小体系热力学》一书于 1994 年再版，他也重新回到了这个研究领域。2000 年，亚利桑那州立大学的拉尔夫·V. 钱伯林教授在《自然》期刊上发表了一篇论文，在应用希尔的理论来解决与铁磁性相关的一些问题时，首次使用了"**纳米热力学**"这一术语。不久，纳米热力学就被该研究领域的科学家们所接受。希尔的主要贡献在于，将约西亚·W. 吉布斯的热力学扩展到小体系系综，尤其是生物化学和分子生物学领域。

纳米热力学适用于未达到热力学极限的微观体系。对于由大量（N 个）粒子组成的系综，**热力学极限**指的是体积随粒子数呈正比增加的极限，即

$$N \to \infty, \quad V \to \infty \quad \text{且} \quad \frac{N}{V} = \text{常数}$$

在热力学极限以上，普通宏观热力学有效。与普通宏观热力学相比，纳米热力学具有以下主要特性：
1) 粒子数更少，比表面积更大；
2) 内能不再是一种广度性质，而与纳米粒子的形状有关；
3) 热和功的概念变得模糊；
4) 勒夏特列原理可能不再适用。

习题

3.1　判断在**表 P3.1** 给出的相应过程中，下列热化学量是大于（>）、小于（<）或等于（=）零。

3.2　根据融化的冰的量测量反应放热值的量热器，称为冰量热器。现在考虑 25.0℃ 和 744 mmHg 下，在空气中恒压燃烧 0.100 L 甲烷气体 $CH_4(g)$，释放的热量用于融化 9.53 g 0℃ 的冰。
(a) 写出 CH_4 完全燃烧的方程式，并确定在上述情况下该燃烧是否完全，已知冰的 $\Delta H_{熔化}$ = 6.01 kJ mol^{-1}。
(b) 在 CH_4 不完全燃烧时，假定产生 CO(g)，且 $H_2O(l)$ 是燃烧的另一产物。通过一个具有小整数系数的单一方程，尽可能地表示该燃烧反应。

3.3　在某一参考来源的 H_2SO_4 条目下，列出了许多标准生成焓。例如，纯 $H_2SO_4(l)$ 的 $\Delta H_f^{\ominus}$ = −814.0 kJ mol^{-1}（以下 $\Delta H_f^{\ominus}$ 单位同此）。每摩尔 H_2SO_4 中：含 1 mol H_2O 的溶液，−841.8；含 10 mol H_2O 的溶液，−880.5；含 50 mol H_2O 的溶液，−886.8；含 100 mol

For an ensemble composed of a very large number (N) of particles, the **thermodynamic limit** is the limit where the volume increases in proportion with the number of particles, as

$$N \to \infty, \quad V \to \infty, \quad \text{and} \quad \frac{N}{V} = \text{constant}$$

Above thermodynamic limit, ordinary macroscopic thermodynamics is valid. Compared to ordinary macroscopic thermodynamics, nanothermodynamics has the following major characteristics:

1) There are less particles and larger surface area;
2) The internal energy is no longer an extensive property, and depends on the shape of the nanoparticle;
3) The concept of heat and work become blurred;
4) Le Châtelier's principle might no longer apply.

Problems

3.1 Indicate the following thermochemical quantities are greater than (>), less than (<), or equal to (=) zero in the corresponding processes given in **Table P3.1**.

Table P3.1 (表 P3.1)

Process (过程)	q	w	ΔU	ΔH	ΔS	ΔG
Ideal gas isothermal free expansion (理想气体恒温自由膨胀)						
$H_2O(l, p^\ominus, 298\ K) \to H_2O(s, p^\ominus, 298\ K)$						
Benzene (l, $p^\ominus$, 10 K above its boiling T) → Benzene (g, $p^\ominus$, 10 K above its boiling T) 苯 (l, $p^\ominus$, 沸点以上 10 K) → 苯 (g, $p^\ominus$, 沸点以上 10 K)						

3.2 A calorimeter that measures an exothermic heat of reaction by the quantity of ice that can be melted is called an ice calorimeter. Now consider that 0.100 L of methane gas, $CH_4(g)$, at 25.0°C and 744 mmHg is burned at constant pressure in air. The heat liberated is used to melt 9.53 g ice at 0°C.

(a) Write an equation for the complete combustion of CH_4 and determine in the above case whether the combustion is complete or not, given that $\Delta H_{fus}(ice) = 6.01\ kJ\ mol^{-1}$.

(b) In the case of incomplete combustion of CH_4, assume that $CO(g)$ is produced and $H_2O(l)$ is another product of the combustion. Represent the combustion as best you can through a single equation with small whole numbers as coefficients.

3.3 Under the entry H_2SO_4, a reference source lists many values for the standard enthalpy of formation. For example, for pure $H_2SO_4(l)$, $\Delta H_f^\circ = -814.0\ kJ\ mol^{-1}$; for a solution with 1 mol H_2O per mole of H_2SO_4, -841.8; with 10 mol H_2O, -880.5; with 50 mol H_2O, -886.8; with 100 mol H_2O, -887.7; with 500 mol H_2O, -890.5; with 1000 mol H_2O, -892.3; with 10,000 mol H_2O, -900.8; and with 100,000 mol H_2O, -907.3.

(a) Explain why these above values are not all the same.

(b) The value of $\Delta H_f^\circ(H_2SO_4(aq))$ in an infinitely dilute solution is $-909.3\ kJ\ mol^{-1}$. What data can you cite to confirm this value? Explain.

(c) If 500.0 mL of 1.00 mol L^{-1} $H_2SO_4(aq)$ is prepared from pure $H_2SO_4(l)$, what is the approximate change in temperature that should be observed? Assume that the $H_2SO_4(l)$ and $H_2O(l)$ are at the same temperature initially and that the specific heat of the $H_2SO_4(aq)$ is about 4.2 J g^{-1} $°C^{-1}$.

3.4 Suppose that you have a setup similar to the one depicted in **Figure 3.1(a)** except that there are two different weights rather than two equal weights. One weight is a steel cylinder 10.00 cm in diameter and 25 cm long, the other weight produces a pressure of 745 Torr. The temperature of the gas in the cylinder in which the expansion takes place is 25.0°C. The piston restraining the gas has a diameter of 12.00 cm, and the height of the piston above the base of the gas cylinder is 8.10 cm. The density of the steel is 7.75 g cm^{-3}. How much work is done when the steel cylinder is suddenly removed from the piston?

3.5 How enthalpy changes with temperature for a constant-pressure process is given by

H₂O 的溶液，−887.7；含 500 mol H₂O 的溶液，−890.5；含 1000 mol H₂O 的溶液，−892.3；含 10 000 mol H₂O 的溶液，−900.8；含 100 000 mol H₂O 的溶液，−907.3。

(a) 解释为什么上述 $\Delta H_f^\ominus$ 不完全相同。

(b) 在无限稀溶液中，$\Delta H_f^\ominus(H_2SO_4(aq))$ 为 $-909.3 \text{ kJ mol}^{-1}$。你可以引用哪些数据来确认此值？请解释。

(c) 如果用纯 $H_2SO_4(l)$ 制备 500.0 mL 1.00 mol L⁻¹ $H_2SO_4(aq)$，应观察到的温度变化大致为多少？假定 $H_2SO_4(l)$ 和 $H_2O(l)$ 的初始温度相同，$H_2SO_4(aq)$ 的比热约为 $4.2 \text{ J g}^{-1} \text{ °C}^{-1}$。

3.4 假设有一个类似于**图 3.1(a)** 所示的装置，但有两个不同砝码而非两个等重砝码。一个砝码是直径 10.00 cm、长 25 cm 的钢制圆柱体，另一个砝码产生的压强为 745 Torr。发生膨胀的气缸内气体的温度为 25.0°C。限制气体的活塞直径为 12.00 cm，活塞距离气缸底座的高度为 8.10 cm。钢的密度为 7.75 g cm^{-3}。突然将圆柱形砝码从活塞上取下，需要做多少功？

3.5 恒压过程的焓随温度的变化由下式给出：

$$q_p = C_p \Delta T$$

其中 C_p 是物质的热容。严格地说，恒压下物质的热容是焓随温度变化的切线斜率，即

$$C_p = \frac{dH}{dT}$$

热容是一个广度量，通常用摩尔热容 $C_{p,m}$（即 1 mol 物质的热容，为强度性质）来表示。恒压热容可用于估算由温度变化而引起的焓变。对于温度的微小变化，

$$dH = C_p dT \quad \text{（恒压）}$$

为评估特定温度变化（从 T_1 到 T_2）所对应的焓变，有

$$\int_{H(T_1)}^{H(T_2)} dH = H(T_2) - H(T_1) = \int_{T_1}^{T_2} C_p dT$$

若假定 C_p 与温度无关，则可得第一个等式，即

$$\Delta H = q_p = C_p \Delta T$$

另一方面，热容是温度的函数，一个便捷的经验表达式为

$$C_{p,m} = a + bT + \frac{c}{T^2}$$

计算将 N_2 从 25.0°C 加热到 100.0°C 的摩尔焓变。N_2 的摩尔热容为

$$C_{p,m} = \left[28.58 + 3.77 \times 10^{-3} T - \frac{0.5 \times 10^5}{T^2}\right] \text{ J mol}^{-1} \text{ K}^{-1}$$

3.6 在 25°C 和 1.00 atm 下，1.100 L 烧瓶内含有与 100.0 mL 饱和水溶液相接触的 $CO_2(g)$，其中 $[CO_2(aq)] = 3.29 \times 10^{-2} \text{ mol L}^{-1}$。

(a) 计算 25°C 时平衡 $CO_2(g) \rightleftharpoons CO_2(aq)$ 的 K_{co}。

(b) 如果向烧瓶中加入 0.01000 mol 放射性 ¹⁴CO_2，当重新建立平衡时，在气相和水溶液中各有多少摩尔 ¹⁴CO_2？

3.7 **表 P3.2** 给出了 298.15 K 时两种固态 HgI_2 的数据。假定在 HgI_2(红)→HgI_2(黄) 的相变过程中，25 °C 时的 $\Delta H^\ominus$ 和 $\Delta S^\ominus$ 与平衡温度 127 °C 时的值相同，试估算表中两个缺失值。

3.8 某体系具有四种态，其能量 ε 分别为 0、1、2 和 3 个能量单位，

$$q_p = C_p \Delta T$$

where C_p is the heat capacity of the substance. Strictly speaking, the heat capacity of a substance at constant pressure is the slope of the tangent line representing the variation of enthalpy with temperature, that is

$$C_p = \frac{dH}{dT}$$

Heat capacity is an extensive quantity and is usually quoted as molar heat capacity $C_{p,m}$, the heat capacity of 1 mol substances; an intensive property. The heat capacity at constant pressure is used to estimate the change in enthalpy due to a change in temperature. For infinitesimal changes in temperature,

$$dH = C_p dT \quad \text{(at constant pressure)}$$

To evaluate the change in enthalpy for a particular temperature change, from T_1 to T_2, we write

$$\int_{H(T_1)}^{H(T_2)} dH = H(T_2) - H(T_1) = \int_{T_1}^{T_2} C_p dT$$

If we assume that C_p is independent of temperature, then we recover the first equation, as

$$\Delta H = q_p = C_p \Delta T$$

On the other hand, we often find that the heat capacity is a function of temperature; a convenient empirical expression is

$$C_{p,m} = a + bT + \frac{c}{T^2}$$

Calculate the change in molar enthalpy of N_2 when it is heated from 25.0°C to 100.0°C. The molar heat capacity of N_2 is given by

$$C_{p,m} = \left[28.58 + 3.77 \times 10^{-3} T - \frac{0.5 \times 10^5}{T^2} \right] \text{J mol}^{-1} \text{K}^{-1}$$

3.6 A 1.100 L flask at 25°C and 1.00 atm pressure contains $CO_2(g)$ in contact with 100.0 mL of a saturated aqueous solution in which $[CO_2(aq)] = 3.29 \times 10^{-2}$ mol L^{-1}.

(a) Calculate the value of K_c at 25°C for the equilibrium $CO_2(g) \rightleftharpoons CO_2(aq)$.

(b) If 0.01000 mol of radioactive $^{14}CO_2$ is added to the flask, how many moles of the $^{14}CO_2$ will be found in the gas phase and in the aqueous solution when equilibrium is re-established?

3.7 Data are given in **Table P3.2** for the two solid forms of HgI_2 at 298.15 K. Estimate values for the two missing entries by assuming that for the transition $HgI_2(red) \rightarrow HgI_2(yellow)$, the values of $\Delta H°$ and $\Delta S°$ at 25°C have the same values as they do at the equilibrium temperature of 127°C.

Table P3.2 (表 P3.2)

Substances (物质)	$\Delta H_f°$/(kJ mol^{-1})	$S°$/(J mol^{-1} K^{-1})	$\Delta G_f°$/(kJ mol^{-1})
HgI_2 (red, 红)	−105.4	180.0	−101.7
HgI_2 (yellow, 黄)	−102.9	?	?

3.8 For a system that has four states, with energy $\varepsilon = 0, 1, 2,$ and 3 energy units, respectively. Five distinguishable particles are to be distributed in these states. If the total energy of the system is 10 energy units, what is the number of microstates of the system?

3.9 The normal boiling point of cyclohexane, C_6H_{12}, is 80.7°C. Estimate the temperature at which the vapor pressure of cyclohexane is 100.0 mmHg by using Trouton's rule.

3.10 The standard molar entropy of solid hydrazine, N_2H_4, at its melting point of 1.53°C is 67.15 J mol^{-1} K^{-1}. The enthalpy of fusion is 12.66 kJ mol^{-1}. For $N_2H_4(l)$ in the interval from 1.53°C to 298.15 K, the molar heat capacity at constant pressure is given by the expression $C_p = [97.78 + 0.0586(T - 280)]$ J mol^{-1} K^{-1}. Calculate the standard molar entropy of $N_2H_4(l)$ at 298.15 K.

将五个可分辨粒子分布在这些态上。如果体系的总能量为 10 个能量单位，体系的微观状态数是多少？

3.9 环己烷 C_6H_{12} 的常规沸点为 80.7℃。采用乔顿规则，估算环己烷的蒸气压为 100.0 mmHg 时的温度。

3.10 固体肼 N_2H_4 在其熔点 1.53 ℃ 时的标准摩尔熵为 67.15 J mol^{-1} K^{-1}，熔化焓为 12.66 kJ mol^{-1}。$N_2H_4(l)$ 在 1.53 ℃ 至 298.15 K 范围内的恒压摩尔热容，可由下式给出：
$$C_p = [97.78 + 0.0586(T - 280)] \text{ J mol}^{-1} \text{ K}^{-1}$$
试计算 298.15 K 时 $N_2H_4(l)$ 的标准摩尔熵。

Chapter 4 Chemical Kinetics

In thermochemistry, we focus on the direction and extent of a chemical reaction. However, thermochemistry cannot tell us how long will it take for a chemical reaction to reach its equilibrium. This belongs to the scope of chemical kinetics and is vital in the practical applications of chemical reactions. For example, although the conversion of the ammonia synthesis reaction is quite high according to its thermodynamical data: $\Delta G°$ (298 K) = −32.8 kJ mol^{-1} and K (298 K) = 5.6×10^5, there is no practical value of this reaction in ambient conditions due to the extremely slow reaction rate. Chemical kinetics concerns on how rates of chemical reactions are measured, or can be predicted, and how reaction-rate data are used to deduce the probable reaction mechanisms.

4.1   The Rate of a Chemical Reaction

4.2   Effect of Concentration on Reaction Rates: The Rate Laws

4.3   Order of Reaction

4.4   The Effect of Temperature on Reaction Rates

4.5   Theoretical Models of Chemical Kinetics

4.6   Reaction Mechanisms

4.7   Catalysis and Catalytic Chemistry

第 4 章 化学动力学

在热化学中,我们关注化学反应的方向和限度。但热化学并不能告诉我们一个化学反应需要多长时间才能达到平衡,这属于化学动力学的范畴,在化学反应的实际应用中至关重要。例如,虽然根据合成氨反应的热力学数据:$\Delta G^{\ominus}(298\text{ K}) = -32.8 \text{ kJ mol}^{-1}$ 和 $K(298\text{ K}) = 5.6 \times 10^5$,其转化率相当高,但由于反应速率极慢,在常温常压下该反应没有实际价值。化学动力学关注如何测量和预测化学反应的速率,以及如何使用反应速率的数据来推导可能的反应机理。

4.1 化学反应速率

4.2 浓度对反应速率的影响:速率方程

4.3 反应级数

4.4 温度对反应速率的影响

4.5 化学动力学理论模型

4.6 反应机理

4.7 催化作用与催化化学

4.1 The Rate of a Chemical Reaction

The rate of a chemical reaction tells us how fast a reaction can happen. It is represented by the change in the concentration of a reacting species, either a reactant or a product, over time.

Average Rate of Reaction and Its Measurement

The average rate of reaction is the average change in concentration over a period of time. Consider a general reaction written as

$$a\,\text{A} + b\,\text{B} + \cdots \rightarrow g\,\text{G} + h\,\text{H} + \cdots$$

the **average rate of reaction** is defined as

$$\bar{R}_{\text{rxn}} = -\frac{1}{a}\frac{\Delta[\text{A}]}{\Delta t} = -\frac{1}{b}\frac{\Delta[\text{B}]}{\Delta t} = \frac{1}{g}\frac{\Delta[\text{G}]}{\Delta t} = \frac{1}{h}\frac{\Delta[\text{H}]}{\Delta t} \tag{4.1}$$

Because the changes in concentration of reactants are negative and those of products are positive, a negative sign is incorporated into the above definition for reactants. To obtain a single, positive quantity, it is necessary to divide all rates by the appropriate stoichiometric coefficients. The rate of reaction is a quantity with a dimension of concentration·(time)$^{-1}$. The commonly used unit is mol L^{-1} s^{-1}.

In **Section 3.3**, we have learnt to use the extent of reaction ξ to measure the extent to which the reaction has proceeded. The average rate of reaction can also be defined in terms of the extent of reaction. Recall that for a general reaction represented by

$$0 = \sum_{\text{X}} v_{\text{X}} \text{X}$$

where X stands for various substances involved in the reaction, and v_{X} is the stoichiometric coefficient for X, the extent of reaction is given by

$$\Delta\xi = \xi_{\text{f}} - \xi_{\text{i}} = \frac{\Delta n_{\text{X}}}{v_{\text{X}}} = \frac{V\Delta[\text{X}]}{v_{\text{X}}}$$

The average rate of reaction is given by

$$\bar{R}_{\text{rxn}} = \frac{1}{v_{\text{X}}}\frac{\Delta[\text{X}]}{\Delta t} = \frac{1}{v_{\text{X}}}\frac{\Delta n_{\text{X}}/V}{\Delta t} = \frac{1}{V}\frac{\Delta\xi}{\Delta t} \tag{4.2}$$

To measure the average rate of a chemical reaction, we need to measure both the change in concentration of a reacting species and the change in time. For example, for the hydrogen peroxide decomposition reaction

$$\text{H}_2\text{O}_2(\text{aq}) \rightarrow \text{H}_2\text{O}(\text{l}) + \frac{1}{2}\text{O}_2(\text{g})$$

the change in time can be measured with a stopwatch or other timing devices. The change in concentration can be calculated either by measuring the volume of the released O_2 or by taking small samples of the reaction mixture from time to time and titrating H_2O_2 by $KMnO_4$ in an acidic solution. **Table 4.1** lists data for this decomposition reaction at every 500 s intervals. The average rates of reaction over each 500 s are calculated in the last column. As can be seen, the average rate of reaction is not constant. It decreases continuously with the remaining concentration of H_2O_2, as the reaction proceeds.

Instantaneous Rate of Reaction and Its Measurement

Figure 4.1 plots the concentration of H_2O_2 with respect to the reaction time for the data listed in **Table 4.1**. A curve can be traced from a number of data points, and the remaining [H_2O_2] can be read from this curve at any point of interest corresponding to any instant of time. We can also draw a tangent line at this point of interest. The slope of this tangent line determines the **instantaneous rate of reaction** at this particular time of interest and can be written in a general form as

4.1 化学反应速率

化学反应速率告诉我们反应发生的快慢，可以用反应物种（反应物或生成物）的浓度随时间的改变量来表示。

平均反应速率及其测量方法

平均反应速率是一段时间内浓度的平均改变量。考虑具有如下通式的反应：

$$a\,A + b\,B + \cdots \rightarrow g\,G + h\,H + \cdots$$

平均反应速率定义为

$$\bar{R}_{反应} = -\frac{1}{a}\frac{\Delta[A]}{\Delta t} = -\frac{1}{b}\frac{\Delta[B]}{\Delta t} = \frac{1}{g}\frac{\Delta[G]}{\Delta t} = \frac{1}{h}\frac{\Delta[H]}{\Delta t} \tag{4.1}$$

由于反应物的浓度变化为负值而生成物的浓度变化为正值，因此在上述定义中对反应物添加了一个负号。为得到单一的正值，需将所有速率均除以适当的化学计量数。反应速率的单位为浓度单位·(时间单位)$^{-1}$，常用单位为 mol L^{-1} s^{-1}。

3.3 节我们已经学过使用反应进度 ξ 来衡量反应进行的程度。平均反应速率也可以用反应进度来定义。回顾一下，对于具有如下通式的反应：

$$0 = \sum_{X} \nu_X X$$

其中 X 是参与反应的各种物质，ν_X 为 X 的化学计量数，反应进度为

$$\Delta\xi = \xi_{终} - \xi_{始} = \frac{\Delta n_X}{\nu_X} = \frac{V\Delta[X]}{\nu_X}$$

则平均反应速率可表示为

$$\bar{R}_{反应} = \frac{1}{\nu_X}\frac{\Delta[X]}{\Delta t} = \frac{1}{\nu_X}\frac{\Delta n_X / V}{\Delta t} = \frac{1}{V}\frac{\Delta\xi}{\Delta t} \tag{4.2}$$

为测量化学反应的平均速率，我们需要测量反应物种的浓度变化以及时间的变化。例如，对于过氧化氢的分解反应：

$$H_2O_2(aq) \rightarrow H_2O(l) + \frac{1}{2}O_2(g)$$

时间的变化可用秒表或其他计时装置来测量。浓度的变化可以通过测量释放 O_2 的体积来计算，或者通过不时地采集少许反应混合物样品，并在酸性溶液中用 KMnO$_4$ 溶液滴定 H$_2$O$_2$ 来计算。**表 4.1** 列出了该分解反应每隔 500 s 的数据，最后一列计算了每 500 s 的平均反应速率。可以看到，平均反应速率并非常数，随着反应的进行，其值会随剩余 H$_2$O$_2$ 浓度的减少而持续降低。

瞬时反应速率及其测量方法

图 4.1 是根据**表 4.1** 所列数据绘制的 H$_2$O$_2$ 浓度随反应时间的变化图。对多个数据点进行拟合，可以得到一条曲线，从该曲线可以读取任意时刻、任意感兴趣的点所对应的剩余 [H$_2$O$_2$]。我们也可以在感兴趣的某点上作一条切线，该切线的斜率决定了这个特定时刻的**瞬时反应速率**，其通式可写为

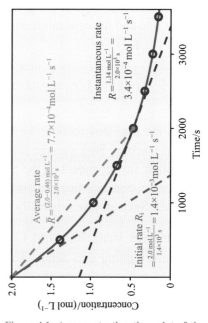

Figure 4.1 A concentration-time plot of the kinetic data for the reaction

$$H_2O_2(aq) \rightarrow H_2O(l) + \frac{1}{2}O_2(g).$$

图 4.1 由反应 $H_2O_2(aq) \rightarrow H_2O(l) + \frac{1}{2}O_2(g)$ 的动力学数据绘制的浓度 - 时间图。

$$R_{rxn} = \frac{1}{\nu_X} \frac{d[X]}{dt} = \frac{1}{V} \frac{d\xi}{dt} \tag{4.3}$$

where the differential forms are given by

$$\frac{d[X]}{dt} = \lim_{\Delta t \to 0} \frac{\Delta[X]}{\Delta t} \quad \text{and} \quad \frac{d\xi}{dt} = \lim_{\Delta t \to 0} \frac{\Delta \xi}{\Delta t}$$

The instantaneous rate of reaction signifies the rate of reaction at any instant of time within an infinitesimal time interval. Its value cannot be measured directly by experiments, but can be calculated from the slope of a tangent line to a concentration-time curve. Unless otherwise specified, the rate of reaction always refers to the instantaneous rate of reaction.

Table 4.1 Kinetic Data for the Reaction $H_2O_2(aq) \to H_2O(l) + \frac{1}{2}O_2(g)$

表 4.1 反应 $H_2O_2(aq) \to H_2O(l) + \frac{1}{2}O_2(g)$ 的动力学数据

t/(s)	Δt/(s)	$[H_2O_2]$ /(mol L^{-1})	$\Delta[H_2O_2]$ /(mol L^{-1})	$\bar{R}_{反应} = -\frac{\Delta[H_2O_2]}{\Delta t}$ / (mol L^{-1} s^{-1})
0	500	2.00	−0.61	1.2×10^{-3}
500	500	1.39	−0.43	8.6×10^{-4}
1000	500	0.96	−0.29	5.8×10^{-4}
1500	500	0.67	−0.21	4.2×10^{-4}
2000	500	0.46	−0.14	2.8×10^{-4}
2500	500	0.32	−0.10	2.0×10^{-4}
3000		0.22		

Initial Rate of Reaction

The rate when the reactants are just brought together to start the reaction is called the **initial rate of reaction**. It is the instantaneous rate of reaction at $t = 0$ and can be obtained from the tangent line to the concentration-time curve at the initial point. An alternative way is to measure the concentration of the chosen reactant as soon as possible after mixing, or obtaining $\Delta[X]$ for a very short time interval $\Delta t \to 0$ at essentially $t = 0$. These two approaches give the same result if the time interval used is limited to that in which the tangent line and the concentration-time curve practically coincide.

4.2 Effect of Concentration on Reaction Rates: The Rate Laws

In the previous section, we have observed that in the hydrogen peroxide decomposition reaction the rates of reaction, both average and instantaneous, decrease continuously with $[H_2O_2]$. It means that the rates of reaction are dependent on the concentrations of reactants. It is one of the major goals in chemical kinetics to derive an equation that can predict the relationship between the rates of reaction and the concentrations of reactants. Such an experimentally determined equation is called a **rate law**.

Elementary Reactions and Complex Reactions

Before going into detail about the rate laws, let us first classify chemical reactions into elementary reactions and complex reactions according to the reaction processes. An **elementary reaction** is a chemical reaction in which one or more of the chemical species react directly to form products in a single reaction step. For example, the reaction

$$NO_2 + CO \to NO + CO_2$$

forms the products in a single collision between the reactant molecules NO_2 and CO at high temperature, and thus it is an elementary reaction at high temperature. On the other hand, the reaction

$$H_2 + I_2 \to 2HI$$

does not form the products in a single collision between the reactant molecules H_2 and I_2. It is not an elementary

$$R_{反应} = \frac{1}{\nu_X}\frac{d[X]}{dt} = \frac{1}{V}\frac{d\xi}{dt} \tag{4.3}$$

其中的微分形式可由下式给出：

$$\frac{d[X]}{dt} = \lim_{\Delta t \to 0}\frac{\Delta[X]}{\Delta t} \quad 且 \quad \frac{d\xi}{dt} = \lim_{\Delta t \to 0}\frac{\Delta\xi}{\Delta t}$$

瞬时反应速率表示在无限小的时间间隔内任意时刻的反应速率。其值不能通过实验直接测量，但可以通过浓度 - 时间曲线的切线斜率来计算。如果没有特别说明，反应速率均指瞬时反应速率。

初始反应速率

当反应物刚聚在一起、即将开始反应时的速率称为**初始反应速率**，它是 $t=0$ 时刻的瞬时反应速率，可以由浓度 - 时间曲线的初始点的切线斜率得到。另一种方法是在混合后尽快测量所选反应物的浓度，即在 $t=0$ 附近的极短时间间隔 $\Delta t \to 0$ 内获得 $\Delta[X]$。如果所用时间间隔短于切线与浓度 - 时间曲线实际重合的间隔，则这两种方法会给出相同结果。

4.2 浓度对反应速率的影响：速率方程

上节我们观察到在 H_2O_2 的分解反应中，平均和瞬时反应速率均随 $[H_2O_2]$ 的减少而持续降低，这意味着反应速率与反应物浓度相关。推导出一个能够预测反应速率与反应物浓度之间关系的方程，是化学动力学的主要目标之一。这样的由实验确定的方程称为**速率方程**。

基元反应与复杂反应

在详细讨论速率方程之前，让我们先根据反应过程将化学反应分为基元反应和复杂反应。**基元反应**是一种或多种化学物种在单一反应步骤中直接反应形成生成物的化学反应。例如

$$NO_2 + CO \rightarrow NO + CO_2$$

在高温下，反应物 NO_2 和 CO 分子经过一次碰撞即可形成生成物，因此高温下该反应为基元反应。另一方面

$$H_2 + I_2 \rightarrow 2HI$$

不是由反应物 H_2 和 I_2 分子经过一次碰撞形成生成物，因此不是基元反应，称为**复杂反应**。其实际反应过程包含以下三个基元反应：

(1) $I_2 \rightarrow 2I$

(2) $I + H_2 \rightarrow HI + H$

(3) $I + H \rightarrow HI$

这三个基元反应称为复杂反应的**基元过程**。有时我们并不区分基元反应和复杂反应的基元过程，认为两者均是基元反应。

基元反应的质量作用定律

对于一个具有如下通式的恒温基元反应：

$$a\,A + b\,B \rightarrow g\,G + h\,H$$

其速率方程可写为

$$R_{反应} = -\frac{1}{a}\frac{d[A]}{dt} = k[A]^a[B]^b \tag{4.4}$$

reaction but a **complex reaction**. The actual reaction process consists of three elementary reactions as follows:

(1) $I_2 \rightarrow 2I$

(2) $I + H_2 \rightarrow HI + H$

(3) $I + H \rightarrow HI$

These three elementary reactions are called the **elementary processes** of the complex reaction. Sometimes we do not distinguish elementary reaction and elementary process of a complex reaction. Both are considered as elementary reactions.

Law of Mass Action of an Elementary Reaction

For a general isothermal elementary reaction represented by

$$a A + b B \rightarrow g G + h H$$

the rate law can be written as

$$R_{rxn} = -\frac{1}{a}\frac{d[A]}{dt} = k[A]^a[B]^b \tag{4.4}$$

where [A] and [B] represent the concentrations of reactants, a and b stand for the coefficients in the balanced elementary reaction equation that can only be positive whole numbers, and k is the **rate constant** that does not vary with the concentrations of reactants and is a function of temperature for a given reaction. **Equation (4.4)** is called the **law of mass action** of an elementary reaction. Because there is a differential form in the equation, it is also referred to as the differential rate law.

As far as an elementary reaction is concerned, the relationship between the rate of reaction and the concentrations of reactants always follows the law of mass action. That is, if we know a reaction is elementary and the balanced chemical equation is provided, we can directly write down its differential rate law according to the law of mass action. The exponents a and b are called the **orders** in A and B, respectively. If $a = 1$, we say that the reaction is first order in A. If $b = 2$, the reaction is second order in B, and so on. The **overall order of reaction** is the sum of all the exponents. The overall order of an elementary reaction is generally less than or equal to 2, because that elementary reactions are either unimolecular or bimolecular and the probability for a termolecular elementary reaction is negligible.

Differential Rate Law of a Complex Reaction

For a general isothermal complex reaction represented by

$$a A + b B + \cdots \rightarrow g G + h H + \cdots$$

the differential rate law can be written as

$$R_{rxn} = -\frac{1}{a}\frac{d[A]}{dt} = k[A]^m[B]^n \cdots \tag{4.5}$$

where [A], [B], ... represent the concentrations of reactants, m, n, ... are the orders in the corresponding reactants, and k is the rate constant that is a function of temperature.

The major differences between the differential rate law of a complex reaction and the law of mass action of an elementary reaction include the following:

1) For elementary reactions, $m = a$ and $n = b$ must apply. Consequently, if $m \neq a$ or $n \neq b$, it is definitely not an elementary reaction. For complex reactions, it is not necessary that $m = a$ and $n = b$. This means that we cannot directly write down the differential rate law of a complex reaction according to its balanced chemical equation.

2) The orders in elementary reactions must be positive whole numbers. The orders in complex reactions can be either positive, negative, zero, or even fractional.

3) Elementary reactions can be considered as reversible processes. Complex reactions are not reversible.

4) Both **Equations (4.4)** and **(4.5)** are called the differential rate law. Only **Equation (4.4)** that suits an elementary reaction can be called the law of mass action.

其中 [A] 和 [B] 表示反应物浓度；a 和 b 代表配平基元反应方程式的系数且只能为正整数；k 是不随反应物浓度变化的**速率常数**，且对于给定反应 k 只是温度的函数。**式 (4.4)** 称为基元反应的**质量作用定律**，由于方程中存在微分形式，因此也称微分速率方程。

就基元反应而言，反应速率与反应物浓度之间的关系始终遵循质量作用定律。也就是说，如果我们知道一个反应是基元反应，且配平的化学方程式已知，我们即可根据质量作用定律直接写出其微分速率方程。幂指数 a 和 b 分别称为 A 和 B 的**级数**。如果 $a=1$，我们称该反应是 A 的一级反应；如果 $b=2$，则反应是 B 的二级反应；以此类推。**总反应级数**是所有幂指数之和。基元反应的总级数一般小于或等于 2，因为基元反应通常是单分子反应或双分子反应，而三分子基元反应发生的概率可忽略不计。

复杂反应的微分速率方程

对于一个具有如下通式的恒温复杂反应：
$$a\text{A} + b\text{B} + \cdots \rightarrow g\text{G} + h\text{H} + \cdots$$
其微分速率方程可写为
$$R_{反应} = -\frac{1}{a}\frac{d[\text{A}]}{dt} = k[\text{A}]^m[\text{B}]^n \cdots \quad (4.5)$$
其中 [A]、[B]…表示反应物浓度；m、n…是对应反应物的级数；k 为速率常数，是温度的函数。

复杂反应的微分速率方程与基元反应的质量作用定律之间存在如下主要区别：
1) 对于基元反应，必然有 $m = a$ 且 $n = b$。因此，若 $m \neq a$ 或 $n \neq b$，则必然不是基元反应。对于复杂反应，不要求 $m = a$ 且 $n = b$。这意味着我们不能根据配平的化学方程式，直接写出复杂反应的微分速率方程。
2) 基元反应的级数必须是正整数，而复杂反应的级数可以为正整数、负整数、零、甚至为分数。
3) 基元反应可视为可逆过程，而复杂反应是不可逆过程。
4) **式（4.4）**和**（4.5）**均可称为微分速率方程，但只有适用于基元反应的**式（4.4）**才能称为质量作用定律。

初始速率法

比较化学动力学与化学热力学，可以看到平衡常数表达式与配平的化学方程式中的化学计量数始终一致，但速率方程表达式并不总与化学计量数相符合。虽然基元反应的速率方程表达式可根据配平的化学方程式直接写出，复杂反应的速率方程表达式却必须由实验数据来确定，通常可采用初始速率法。

例如，对于如下反应：
$$2\text{HgCl}_2(aq) + \text{C}_2\text{O}_4^{2-}(aq) \rightarrow \text{Hg}_2\text{Cl}_2(s) + 2\text{Cl}^-(aq) + 2\text{CO}_2(g)$$
其速率方程可尝试写为
$$R_{反应} = k[\text{HgCl}_2]^m[\text{C}_2\text{O}_4^{2-}]^n$$
我们可用上节所述方法来计算其初始反应速率。**表 4.2** 列出了具有不同初始反应物浓度的三个实验所得到的初始反应速率。比较实验 1 与实验 2，当 $\text{C}_2\text{O}_4^{2-}$ 的初始浓度翻倍而 HgCl_2 的初始浓度保持不变时，

Method of Initial Rates

Comparing chemical kinetics with thermochemistry, we can see that the equilibrium constant expression is always consistent with the stoichiometric coefficients in the balanced chemical equation. However, the rate law expression does not always accord with those stoichiometric coefficients. While the rate law expression of an elementary reaction can be written directly based on its balanced chemical equation, that of a complex reaction must be determined by experimental data, for which the method of initial rates can be used.

As an example, for the following reaction

$$2HgCl_2(aq) + C_2O_4^{2-}(aq) \rightarrow Hg_2Cl_2(s) + 2Cl^-(aq) + 2CO_2(g)$$

the tentative rate law can be written as

$$R_{rxn} = k[HgCl_2]^m [C_2O_4^{2-}]^n$$

We can calculate the initial rate of reaction using the method described in the previous section. The initial rates of reaction for three experiments with different initial concentrations of reactants are listed in **Table 4.2**. Comparing experiments 1 to 2, the initial rates of reaction quadruples when the initial concentration of $C_2O_4^{2-}$ is doubled while the initial concentration of $HgCl_2$ remains constant. It means that the reaction is second order in $C_2O_4^{2-}$. Similarly, it is not difficult to conclude that the reaction is first order in $HgCl_2$ by comparing the data in experiments 2 and 3. After determining $m = 1$ and $n = 2$, the value of rate constant can be calculated accordingly, as

$$k = \frac{R_1}{[HgCl_2]_1 [C_2O_4^{2-}]_1^2} = \frac{1.8 \times 10^{-5} \text{ mol L}^{-1} \text{ min}^{-1}}{0.105 \text{ mol L}^{-1} \times (0.15 \text{ mol L}^{-1})^2} = 7.6 \times 10^{-3} \text{ (mol L}^{-1})^{-2} \text{ min}^{-1}$$

Note that the unit of rate constant varies with the overall order of reaction. Therefore, the final rate law of this reaction is determined by experimental data as

$$R_{rxn} = [7.6 \times 10^{-3} \text{ (mol L}^{-1})^{-2} \text{ min}^{-1}][HgCl_2][C_2O_4^{2-}]^2$$

Table 4.2　Kinetic Data for the Reaction $2HgCl_2(aq) + C_2O_4^{2-}(aq) \rightarrow 2Cl^-(aq) + 2CO_2(g) + Hg_2Cl_2(s)$

表 4.2　反应 $2HgCl_2(aq) + C_2O_4^{2-}(aq) \rightarrow 2Cl^-(aq) + 2CO_2(g) + Hg_2Cl_2(s)$ 的动力学数据

Experiment（实验）	[HgCl$_2$]/(mol L^{-1})	[C$_2$O$_4^{2-}$]/(mol L^{-1})	Initial Rate（初始速率）/(mol L^{-1} min^{-1})
1	0.105	0.15	1.8×10^{-5}
2	0.105	0.30	7.1×10^{-5}
3	0.052	0.30	3.5×10^{-5}

4.3　Order of Reaction

Chemical reactions can be classified as zero-, first-, second-, third-order, etc., according to their overall orders of reaction. We will discuss those reactions, especially the relationship between concentrations of reactants and time, one by one in this section.

Zero-Order Reactions

An overall **zero-order reaction** has a rate law in which the sum of all exponents is equal to 0. Here, we take the following reaction in which a single reactant A decomposes to products as an illustrative example:

$$a\text{A} \rightarrow \text{products}$$

Since the reaction is zero order, its rate law must have the form

$$R_{rxn} = -\frac{1}{a}\frac{d[A]}{dt} = k[A]^0 = k \tag{4.6}$$

If we move all terms of concentration to the left and all other terms to the right, then

$$d[A] = -ak\,dt$$

初始反应速率增至 4 倍，说明该反应是 $C_2O_4^{2-}$ 的二级反应。类似地，通过比较实验 2 和实验 3 的数据，不难得出结论：该反应是 $HgCl_2$ 的一级反应。确定 $m=1$ 和 $n=2$ 后，该反应速率常数的值可相应计算为

$$k = \frac{R_1}{[HgCl_2]_1[C_2O_4^{2-}]_1^2} = \frac{1.8 \times 10^{-5}\ mol\ L^{-1}\ min^{-1}}{0.105\ mol\ L^{-1} \times (0.15\ mol\ L^{-1})^2}$$
$$= 7.6 \times 10^{-3}\ (mol\ L^{-1})^{-2}\ min^{-1}$$

注意速率常数的单位随总反应级数而变化。因此，该反应由实验数据确定的最终速率方程为

$$R_{反应} = [7.6 \times 10^{-3}\ (mol\ L^{-1})^{-2}\ min^{-1}][HgCl_2][C_2O_4^{2-}]^2$$

4.3 反应级数

根据总反应级数，化学反应可分为零级、一级、二级、三级等。我们将在本节逐一讨论这些反应，特别是反应物浓度与时间的关系。

零级反应

一个总**零级反应**的速率方程中所有幂指数之和等于 0。这里我们以如下由单个反应物 A 分解为生成物的反应为例：

$$aA \rightarrow 生成物$$

由于反应为零级，其速率方程必有如下形式：

$$R_{反应} = -\frac{1}{a}\frac{d[A]}{dt} = k[A]^0 = k \quad (4.6)$$

将所有包含浓度的项移至左侧，所有其他项移至右侧，则有

$$d[A] = -ak dt$$

始态 $t=0$ 且 $[A]=[A]_0$。用 $[A]_t$ 表示任意给定时刻 t 时 A 的浓度。将上述方程两边从 0 到 t 积分，有

$$\int_{[A]_0}^{[A]_t} d[A] = \int_0^t (-ak dt)$$

可得

$$[A]_t - [A]_0 = -akt$$

或

$$[A]_t = [A]_0 - akt \quad (4.7)$$

由于推导过程涉及积分，**式（4.7）**称为零级反应的**积分速率方程**。

反应的**半衰期** $t_{1/2}$（或**半寿期**）是消耗一半反应物所需的时间，即反应物浓度下降至初始值的一半所用的时间。这意味着在 $t=t_{1/2}$ 时刻，$[A]_t=[A]_0/2$。对于零级反应，不难计算

$$t_{1/2} = \frac{[A]_0}{2ak} \quad (4.8)$$

零级反应的特征包括：
1) 其浓度 - 时间图是一条具有负斜率 $-ak$ 和正截距 $[A]_0$ 的直线。
2) k 的单位与反应速率的单位相同，均为浓度单位·(时间单位)$^{-1}$。
3) 半衰期与反应物的初始浓度成正比，与反应速率常数成反比。

At the initial state, $t = 0$ and $[A] = [A]_0$. At any given time t, the concentration of A is represented by $[A]_t$. We can integrate the above equation from 0 to t on both sides, as

$$\int_{[A]_0}^{[A]_t} d[A] = \int_0^t (-akdt)$$

We have

$$[A]_t - [A]_0 = -akt$$

or

$$[A]_t = [A]_0 - akt \tag{4.7}$$

This is called the **integrated rate law** of a zero-order reaction since integration is involved in deriving this equation.

The **half-life** ($t_{1/2}$) of a reaction is the time required for one-half of a reactant to be consumed. It is the time during which the concentration of a reactant decreases to one-half of its initial value. It means that at $t = t_{1/2}$, $[A]_t = [A]_0/2$. It is not difficult to calculate that for a zero-order reaction:

$$t_{1/2} = \frac{[A]_0}{2ak} \tag{4.8}$$

The characteristics of a zero-order reaction include the following:
1) The concentration-time graph is a straight line with a negative slope of $-ak$ and a positive intercept of $[A]_0$.
2) The unit of k is the same as the unit of the rate of reaction, with a dimension of concentration·(time)$^{-1}$.
3) The half-life is proportional to the initial reactant concentration, and inversely proportional to the reaction constant. As the reaction proceeds, the half-life decreases continuously with decreasing [A].
4) The rate of reaction is a constant that does not vary with the reactant concentration. This normally happens for heterogeneous reactions on a surface where the active sites on the surface are limited, so that the rate of reaction is dependent on the concentration of the active sites which is a constant.

First-Order Reactions

An overall **first-order reaction** has a rate law with the sum of all exponents equal to 1. We still take the following reaction in which a single reactant A decomposes to products as an illustrative example:

$$aA \rightarrow \text{products}$$

Given that the reaction is first order, its rate law must have the form

$$R_{rxn} = -\frac{1}{a}\frac{d[A]}{dt} = k[A] \tag{4.9}$$

Similarly, we move all terms of concentration to the left and all other terms to the right as

$$\frac{d[A]}{[A]} = -akdt$$

and integrate it from 0 to t on both sides, as

$$\int_{[A]_0}^{[A]_t} \frac{d[A]}{[A]} = \int_0^t (-akdt)$$

$$\ln[A]_t - \ln[A]_0 = -akt$$

We then have

$$\ln[A]_t = \ln[A]_0 - akt$$

or

$$\ln\frac{[A]_t}{[A]_0} = -akt \tag{4.10}$$

According to the above integrated rate law, the half-life of a first-order reaction can be calculated as

随着反应的进行，[A] 逐渐下降，半衰期持续变短。
4) 反应速率是不随反应物浓度而变化的常数。零级反应通常为发生在某个表面的异相反应，由于表面活性位点数有限，因此反应速率取决于活性位点的浓度，而活性位点的浓度是常数。

一级反应

一个总**一级反应**的速率方程中所有幂指数之和等于 1。我们仍以如下由单个反应物 A 分解为生成物的反应为例：

$$a\text{A} \rightarrow 生成物$$

由于该反应为一级，其速率方程必有如下形式：

$$R_{反应} = -\frac{1}{a}\frac{d[A]}{dt} = k[A] \tag{4.9}$$

类似地，将所有包含浓度的项移至左侧，所有其他项移至右侧，有

$$\frac{d[A]}{[A]} = -ak\,dt$$

两边从 0 到 t 积分，有

$$\int_{[A]_0}^{[A]_t} \frac{d[A]}{[A]} = \int_0^t (-ak\,dt)$$

$$\ln[A]_t - \ln[A]_0 = -akt$$

可得

$$\ln[A]_t = \ln[A]_0 - akt$$

或

$$\ln\frac{[A]_t}{[A]_0} = -akt \tag{4.10}$$

根据上述积分速率方程，一级反应的半衰期可计算为

$$t_{1/2} = \frac{\ln 2}{ak} \tag{4.11}$$

一级反应的特征包括：
1) 将反应物浓度的自然对数对时间作图，可得到一条具有负斜率 $-ak$ 和正截距 $\ln[A]_0$ 的直线。
2) 由于 $\ln([A]_t/[A]_0)$ 量纲为 1，因此 k 的单位为（时间单位）$^{-1}$。
3) 半衰期是不随反应物浓度变化的常数。这可用于判定一个反应是否为一级反应。
4) 反应速率随时间持续降低。一级反应最熟悉的例子之一是放射性物质的衰变。无论给定时刻样品中放射性原子数有多少，$t_{1/2}$ 时将变为原来的一半；$2t_{1/2}$ 时为原来的四分之一；以此类推。

二级反应

一个总**二级反应**的速率方程中所有幂指数之和等于 2。以如下由单个反应物 A 分解为生成物的二级反应为例：

$$a\text{A} \rightarrow 生成物$$

其速率方程为

$$R_{反应} = -\frac{1}{a}\frac{d[A]}{dt} = k[A]^2 \tag{4.12}$$

将所有包含浓度的项移至左侧，所有其他项移至右侧，有

$$t_{1/2} = \frac{\ln 2}{ak} \tag{4.11}$$

The characteristics of a first-order reaction include the following:
1) The plot of the natural logarithm of reactant concentration versus time is a straight line with a negative slope of $-ak$ and a positive intercept of $\ln[A]_0$.
2) Because $\ln([A]_t / [A]_0)$ is dimensionless, the unit of k is with a dimension of $(\text{time})^{-1}$.
3) The half-life is a constant and does not vary with the reactant concentration. This can be used to determine whether a reaction is first-order or not.
4) The rate of reaction decreases continuously with time. One of the most familiar examples of a first-order reaction is radioactive decay. Whatever number of radioactive atoms are in a sample at a given moment, there will be half that number in $t_{1/2}$; a quarter of that number in $2t_{1/2}$; and so on.

Second-Order Reactions

An overall **second-order reaction** has a rate law in which the sum of all exponents is equal to 2. For an illustrative second-order reaction involving the decomposition of a single reactant A to products, as

$$aA \rightarrow \text{products}$$

its rate law is given by

$$R_{rxn} = -\frac{1}{a}\frac{d[A]}{dt} = k[A]^2 \tag{4.12}$$

Moving all terms of concentration to the left and all other terms to the right, we have

$$\frac{d[A]}{[A]^2} = -ak\,dt$$

Integrating it from 0 to t on both sides, as

$$\int_{[A]_0}^{[A]_t} \frac{d[A]}{[A]^2} = \int_0^t (-ak\,dt)$$

$$-\left(\frac{1}{[A]_t} - \frac{1}{[A]_0}\right) = -akt$$

We then have

$$\frac{1}{[A]_t} = \frac{1}{[A]_0} + akt \tag{4.13}$$

According to the above integrated rate law, the half-life of a second-order reaction can be calculated as

$$t_{1/2} = \frac{1}{ak[A]_0} \tag{4.14}$$

The characteristics of a second-order reaction include the following:
1) The plot of the reciprocal of reactant concentration versus time is a straight line with a positive slope of ak and a positive intercept of $1/[A]_0$.
2) The unit of k is with a dimension of $(\text{concentration})^{-1} \cdot (\text{time})^{-1}$.
3) The half-life is inversely proportional to both the rate constant and the initial concentration. Because the starting concentration is always one-half that of the previous half-life, each successive half-life is twice as long as the one before it.

Pseudo-Nth-Order Reactions

Consider the hydrolysis of sucrose which is an overall second-order reaction:

$$C_{12}H_{22}O_{11}(aq) + H_2O(l) \rightarrow C_6H_{12}O_6(\text{glucose, aq}) + C_6H_{12}O_6(\text{fructose, aq})$$

If we measure the half-life, we will find a constant half-life of about 8.4 h at 15 °C, which is a characteristic for a

$$\frac{d[A]}{[A]^2} = -akdt$$

两边从 0 到 t 积分，有

$$\int_{[A]_0}^{[A]_t} \frac{d[A]}{[A]^2} = \int_0^t (-akdt)$$

$$-\left(\frac{1}{[A]_t} - \frac{1}{[A]_0}\right) = -akt$$

可得

$$\frac{1}{[A]_t} = \frac{1}{[A]_0} + akt \quad (4.13)$$

根据上述积分速率方程，二级反应的半衰期可计算为

$$t_{1/2} = \frac{1}{ak[A]_0} \quad (4.14)$$

二级反应的特征包括：
1) 将反应物浓度的倒数对时间作图，可得一条具有正斜率 ak 和正截距 $1/[A]_0$ 的直线。
2) k 的单位为（浓度单位）$^{-1}\cdot$（时间单位）$^{-1}$。
3) 半衰期与速率常数和初始浓度均成反比。由于后一个半衰期的初始浓度总是前一个半衰期初始浓度的一半，所以后一个半衰期总是前一个半衰期的两倍长。

准 N 级反应

考虑蔗糖水解这个总二级反应：

$C_{12}H_{22}O_{11}(aq) + H_2O(l) \rightarrow C_6H_{12}O_6($葡萄糖$, aq) + C_6H_{12}O_6($果糖$, aq)$

若测量其半衰期，会发现在 15°C 时约恒为 8.4 h，而这是一级反应的特征。如何解释这两个事实之间的矛盾？这是因为水的摩尔浓度约为 55.5 mol L^{-1}，远高于蔗糖的浓度，故在整个反应过程中基本保持不变。虽然蔗糖水解是总二级反应，但基本保持不变的 $[H_2O]$ 可以与原速率常数 k 结合形成另一个速率常数 k'，即

$$R_{反应} = k[C_{12}H_{22}O_{11}][H_2O] \approx k'[C_{12}H_{22}O_{11}]$$

则反应速率表现得并不依赖于 $[H_2O]$。因此，该反应表现为水的零级反应和 $C_{12}H_{22}O_{11}$ 的一级反应，所以表现为总一级反应，存在恒定的半衰期。

通过保持一个或多个反应物浓度不变，而使更高级数的反应表现得类似一级反应，这样的反应称为**准一级反应**，可用一级反应动力学的方法来处理。类似地，在某些条件下，其他更高级数的反应也可以表现得类似更低级数的反应。如果某反应表现得类似 N 级反应，则称为**准 N 级反应**，可用 N 级反应动力学的方法来处理。

反应动力学小结

到目前为止，我们已经学习了反应速率、速率常数和反应级数，这些都总结在**表 4.3** 和**图 4.2** 中。绘制反应速率的各种数据（对数、倒数等）与时间的关系图，找到其中的线性关系，是确定反应级数的常用方法。我们还了解了微分速率方程和积分速率方程。我们应该记住，微分速率方程描述了反应速率与反应物浓度之间的关系，而积分

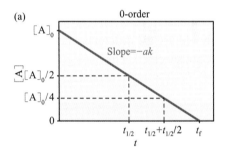

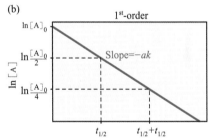

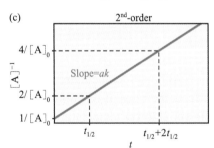

Figure 4.2 The straight-line plots for the reaction $aA \rightarrow$ products. (a) [A] vs. time plot of the zero-order reaction. (b) ln[A] vs. time plot of the first-order reaction. (c) 1/[A] vs. time plot of the second-order reaction.

图 4.2 反应 $aA \rightarrow$ 生成物 的直线图：(a) 零级反应的 [A] 对时间作图；(b) 一级反应的 ln[A] 对时间作图；(c) 二级反应的 1/[A] 对时间作图。

first-order reaction. How to explain the contradiction between these two facts? This is because the concentration of water, which is about 55.5 mol L^{-1} in molarity and much more excessive than that of sucrose, remains essentially constant throughout the reaction. Although the hydrolysis of sucrose is an overall second-order reaction, then early constant [H$_2$O] can be combined with the original rate constant k to form another rate constant k', as

$$R_{rxn} = k[C_{12}H_{22}O_{11}][H_2O] \approx k'[C_{12}H_{22}O_{11}]$$

Then the rate of reaction does not appear to depend on [H$_2$O]. So, the reaction appears to be zero-order in H$_2$O, first-order in C$_{12}$H$_{22}$O$_{11}$, and first-order overall with a constant half-life.

A higher-order reaction that behaves like a first-order reaction by holding one or more reactant concentration constant is called a **pseudo-first-order reaction**. These reactions can be treated with the methods of first-order reaction kinetics. Similarly, other reactions of higher order can be made to behave like reactions of lower order under certain conditions. If the reactions behave like a Nth-order reaction, they are called the **pseudo-Nth-order reactions** and can be treated with the methods of Nth-order reaction kinetics.

Summary on Orders of Reaction

So far, we have learnt about rates of reaction, rate constants, and orders of reaction, which are summarized in **Table 4.3** and **Figure 4.2**. Plotting various rates of reaction data (logarithm, reciprocal, etc.) versus time to find one that yields a straight line is a commonly used method to determine the order of reaction. We have also learnt about the differential rate laws and the integrated rate laws. We should remember that the differential rate laws describe the relationship between rate of reaction and concentrations of reactants, whereas the integrated rate laws tell the relationship between reactant concentrations and time. Therefore, in order to calculate the rate of reaction, we should use the differential rate laws. If time is involved in the calculation, the integrated rate laws should be used. If not specified, a rate law generally refers to the differential rate law.

Table 4.3　A Summary of Kinetics for the Hypothetical Reaction aA → Products

表 4.3　反应 aA→ 生成物的动力学小结

Order (级数)	Differential Rate Law (微分速率方程)	Integrated Rate Law (积分速率方程)	Straight Line (直线)	$k =$	Unit of k (k的单位)	Half-Life (半衰期)
0	$R_{rxn} = -\dfrac{1}{a}\dfrac{d[A]}{dt} = k$	$[A]_t = [A]_0 - akt$	[A] vs. time	$-$slope	mol L^{-1} s^{-1}	$t_{1/2} = \dfrac{[A]_0}{2ak}$
1	$R_{rxn} = -\dfrac{1}{a}\dfrac{d[A]}{dt} = k[A]$	$\ln[A]_t = \ln[A]_0 - akt$	ln[A] vs. time	$-$slope	s^{-1}	$t_{1/2} = \dfrac{\ln 2}{ak}$
2	$R_{rxn} = -\dfrac{1}{a}\dfrac{d[A]}{dt} = k[A]^2$	$\dfrac{1}{[A]_t} = \dfrac{1}{[A]_0} + akt$	1/[A] vs. time	slope	L mol^{-1} s^{-1}	$t_{1/2} = \dfrac{1}{ak[A]_0}$

4.4　The Effect of Temperature on Reaction Rates

Although temperature itself does not appear in the rate law, we all have the experience that temperature affects the rates of reaction significantly. In general, the rate of reaction is faster at higher temperatures and slower at lower temperatures. For example, food is apt to deteriorate in summer but can be stored for a longer time in the refrigerator.

Arrhenius Equation

The effect of temperature on the rates of reaction is shown in its effect on the rate constant k. After systematically studying the experimental data of the reaction rate on temperature, chemists found that the plot of the natural logarithm of rate constant versus the reciprocal of temperature yields approximately a straight line, or

$$\ln k = A + \dfrac{B}{T}$$

速率方程给出了反应物浓度与反应时间之间的关系。因此，为计算反应速率，我们应使用微分速率方程；若计算反应时间，则应采用积分速率方程。如果没有特别说明，速率方程通常指微分速率方程。

4.4 温度对反应速率的影响

虽然温度本身并没有出现在速率方程中，但我们都有温度会显著影响反应速率的经验。一般来说，较高温度下反应速率更快，较低温度下更慢。例如，食物在夏天容易腐坏，但在冰箱里可以储存更长时间。

阿伦尼乌斯方程

温度对反应速率的影响表现为对速率常数 k 的影响。在系统地研究了反应速率随温度变化的实验数据后，化学家们发现将速率常数的自然对数对温度的倒数作图，可近似得到一条直线，即

$$\ln k = A + \frac{B}{T}$$

其中 A 和 B 为常数。

1889 年，斯文特·阿伦尼乌斯（1859—1927）证明，许多化学反应的速率常数随温度的变化均符合如下经验公式：

$$\ln k = -\frac{E_a}{RT} + C$$

或

$$k = A\exp\left(-\frac{E_a}{RT}\right) \tag{4.15}$$

其中 A 称为**指前因子**，E_a 称为反应的**活化能**。可通过 $\ln k$ 对 $1/T$ 线性拟合的斜率来计算 E_a。对**式（4.15）**应用两个不同温度的情况并相减，可得

$$\ln\frac{k_2}{k_1} = -\frac{E_a}{R}\left(\frac{1}{T_2} - \frac{1}{T_1}\right) \tag{4.16}$$

式（4.15）和**（4.16）**通常都称为**阿伦尼乌斯方程**。

两个方程之间的关系

如果将化学动力学中的阿伦尼乌斯方程与热化学中的范特霍夫方程[**式（3.67）**]进行比较，会发现这两个方程在形式上非常相似：

阿伦尼乌斯方程：$\ln\dfrac{k_2}{k_1} = -\dfrac{E_a}{R}\left(\dfrac{1}{T_2} - \dfrac{1}{T_1}\right)$　或　$\ln k = -\dfrac{E_a}{RT} + C$

范特霍夫方程：$\ln\dfrac{K_2}{K_1} = -\dfrac{\Delta H}{R}\left(\dfrac{1}{T_2} - \dfrac{1}{T_1}\right)$　或　$\ln K = -\dfrac{\Delta H}{RT} + C$

历史上阿伦尼乌斯就是通过应用范特霍夫方程导出其方程的。他提出反应物 R 必须经过一个中间活化态 R* 才能转变为生成物 P，并假定 R 与 R* 处于动态平衡，即

$$R \rightleftharpoons R^* \rightarrow P$$

阿伦尼乌斯认为，反应的活化能 E_a 是 R 与 R* 之间的焓变，将其当

where A and B are constants.

In 1889, Svante Arrhenius (1859—1927) demonstrated that the rate constants of many chemical reactions vary with temperature in accordance with the empirical formula:

$$\ln k = -\frac{E_a}{RT} + C$$

or

$$k = A\exp\left(-\frac{E_a}{RT}\right) \tag{4.15}$$

where A is called the **preexponential factor** and E_a is called the **activation energy** of the reaction. E_a can be calculated from the slope of the linear fitting of $\ln k$ with respect to $1/T$. Applying two different temperatures into **Equation (4.15)** and subtracting one from the other, we then have

$$\ln \frac{k_2}{k_1} = -\frac{E_a}{R}\left(\frac{1}{T_2} - \frac{1}{T_1}\right) \tag{4.16}$$

Both **Equations (4.15)** and **(4.16)** are often referred to as **Arrhenius equation**.

Relationship between Two Equations

If we compare Arrhenius equation in chemical kinetics to van't Hoff **Equation (3.67)** in thermochemistry, we will find that these two equations are very similar in their forms as

$$\text{Arrhenius equation}: \ln\frac{k_2}{k_1} = -\frac{E_a}{R}\left(\frac{1}{T_2} - \frac{1}{T_1}\right) \quad \text{or} \quad \ln k = -\frac{E_a}{RT} + C$$

$$\text{van't Hoff equation}: \ln\frac{K_2}{K_1} = -\frac{\Delta H}{R}\left(\frac{1}{T_2} - \frac{1}{T_1}\right) \quad \text{or} \quad \ln K = -\frac{\Delta H}{RT} + C$$

In fact, historically Arrhenius derived his equation by applying van't Hoff equation. He proposed that the reactant R must go through an intermediate activated state R* to become the product P, and assumed that R is in dynamic equilibrium with R*, as

$$\text{R} \rightleftharpoons \text{R}^* \rightarrow \text{P}$$

Arrhenius believed that the activation energy E_a of the reaction is the enthalpy change between R and R*, and treated it as an equilibrium. By applying van't Hoff equation, he then derived Arrhenius equation. Because Arrhenius equation was applicable to many chemical reactions, the assumptions of activated molecule R* and activation energy E_a were accepted. Using Arrhenius equation, the experimental activation energy of a reaction can be determined by measuring the rate constants of the reaction at different temperatures.

4.5 Theoretical Models of Chemical Kinetics

In previous sections, we have discussed contents from the observations of various chemical kinetics data to the natural laws of differential and integrated rate laws as well as Arrhenius equation. In this section, we will learn two theories of chemical kinetics that can bring some insight into questions like why these natural laws apply and what the microscopic meaning of activation energy is, etc.

Collision Theory

Collision theory is the earliest reaction rate theory, which was founded in the early 20th century based on kinetic-molecular theory of gases that we have learnt in **Section 2.3**. It is mainly applicable to gas-phase bimolecular reactions, in which two gas molecules collide with each other to form the products. The main assumptions and simplified derivations of collision theory include that:

1) Molecules can be viewed as rigid hard balls. Collisions of molecules are necessary for reactions to happen

做平衡来处理，套用范特霍夫方程，即可得到阿伦尼乌斯方程。由于阿伦尼乌斯方程确实适用于不少化学反应，因此活化分子 R* 以及活化能 E_a 的设想即被接受。测定不同温度下反应的速率常数，利用阿伦尼乌斯方程可以求算出反应的实验活化能。

4.5 化学动力学理论模型

前几节我们已经讨论了从各种化学动力学数据的观测到微分和积分速率方程以及阿伦尼乌斯方程等自然定则的内容。本节我们将学习两种化学动力学理论，它们可以使我们深入理解为什么这些自然定则适用，以及活化能的微观意义是什么等问题。

碰撞理论

碰撞理论是最早的反应速率理论，它建立于 20 世纪初，基于我们在 **2.3 节**学过的气体分子运动论。它主要适用于气相双分子反应，即两个气体分子相互碰撞形成生成物的反应。碰撞理论的主要假定和简化推导过程包括：

1) 分子可视为刚性硬球，分子碰撞是发生反应的必要条件，因此反应速率与**碰撞频率** Z（即单位时间、单位体积内的碰撞次数）成正比。碰撞频率与分子浓度成正比，与温度正相关，因为随着温度升高，分子的运动速率更快。假设分子 A 与分子 B 碰撞后形成生成物，碰撞频率由下式给出：

$$Z_{AB} = Z_0 [A][B] \quad (4.17)$$

其中 Z_0 是单位浓度的 A 与 B 的碰撞频率。根据分子运动论，Z_0 可计算为

$$Z_0 = 10^6 N_A^2 \sigma_{AB} \sqrt{\frac{8 k_B T}{\pi \mu_{AB}}} \quad (4.18)$$

其中 N_A 是阿伏伽德罗常数；k_B 是玻尔兹曼常数；T 为热力学温度；σ_{AB} 是由 $\sigma_{AB} = \pi (r_A + r_B)^2$ 给出的**反应截面**，其中 r_A 和 r_B 分别是 A 和 B 的半径；μ_{AB} 是由 $1/\mu_{AB}=1/m_A + 1/m_B$ 给出的折合质量，其中 m_A 和 m_B 分别是 A 和 B 的质量。因此，Z_0 取决于 A 和 B 的大小、摩尔质量以及温度。**式 (4.18)** 中的 10^6 项是浓度的 SI 单位与常用单位 mol L^{-1} 之间的转换系数。STP 下典型气相反应的碰撞频率可计算为 10^{35} m^{-3} s^{-1} 量级。如果每次碰撞均可发生反应，则反应速率约为 10^6 mol L^{-1} s^{-1}。而气相反应的实际速率一般为 10^{-4} mol L^{-1} s^{-1} 量级，显著低于计算值。这意味着只有极少数碰撞（称为**有效碰撞**）能导致反应发生。

2) 当反应物分子发生碰撞时，必须克服它们之间的排斥力，这要求分子具有足够的运动速率或动能。我们将**阈值能量** ε_c 定义为分子碰撞后能发生反应所需的最小动能，其值与温度无关。基于分子运动论的气体分子动能的分布，可用 **2.6 节**讨论过的玻尔兹曼分布来描述。能量高于 ε_c 的分子在所有分子中所占的分数称为**能量分数** f，由下式给出：

and the rate of reaction is proportional to the **collision frequency** Z, which is the number of collisions per unit volume per unit time. Collision frequency is proportional to the concentration of molecules and positively related to temperature since molecules tend to move faster with increasing temperature. Assuming that molecules A and B collide to form the products, the collision frequency is given by

$$Z_{AB} = Z_0 [A][B] \tag{4.17}$$

where Z_0 is the collision frequency with unit concentrations of A and B. Although not derived here, Z_0 can be calculated based on the kinetic-molecular theory as

$$Z_0 = 10^6 N_A^2 \sigma_{AB} \sqrt{\frac{8k_B T}{\pi \mu_{AB}}} \tag{4.18}$$

where N_A is the Avogadro constant, k_B is the Boltzmann constant, T is the thermodynamic temperature, σ_{AB} is the **reaction cross section** given by $\sigma_{AB} = \pi (r_A + r_B)^2$, where r_A and r_B are the radii of A and B, respectively, and μ_{AB} is reduced mass given by $1/\mu_{AB} = 1/m_A + 1/m_B$, where m_A and m_B are the masses of A and B, respectively. Therefore, Z_0 is dependent on the sizes and molar masses of A and B, and temperature. The 10^6 term in **Equation (4.18)** is the resultant conversion factor between the SI unit of concentration and the commonly used unit mol L^{-1}. In a typical gas-phase reaction at STP, the calculated collision frequency is of the order of 10^{35} m^{-3} s^{-1}. If each collision results in a reaction, the rate of reaction would be about 10^6 mol L^{-1} s^{-1}. However, the actual rates are generally of the order of 10^{-4} mol L^{-1} s^{-1} for gas-phase reaction, significantly less than the calculated value. This means that only a very small fraction of collisions, called the **effective collisions**, can result in a reaction.

2) When the reactant molecules are colliding, it is necessary to overcome the repulsive forces between them, which requires sufficient molecular speed or kinetic energy. We define the **threshold energy** ε_c as the minimum kinetic energy that is required for a reaction to occur following a collision between the molecules. ε_c is independent of temperature. The distribution of the kinetic energy of gas molecules based on the kinetic-molecular theory is described by Boltzmann distribution that we have discussed in **Section 2.6**. The fraction of molecules with energy higher than ε_c is called the **energy fraction** f, given by

$$f \propto \exp\left(-\frac{\varepsilon_c}{k_B T}\right) \tag{4.19}$$

The energy fraction at a high temperature is considerably larger than that at a low temperature. Meanwhile, the higher the ε_c of a reaction, the smaller is the fraction of molecules with $\varepsilon_k > \varepsilon_c$.

3) In order for a collision to become effective, energy fraction is not the only limiting factor. All those additional factors that may further reduce the probability of effective collisions other than energy fraction is summarized into a **probability factor** P. The probability factor includes but is not limited to the following: the orientation of molecules during collisions unfavorable to result in a reaction; steric effect of complex molecules, leading to decreased probability of effective collisions; the collision time shorter than the required effective energy transfer time, so that even if the energy is enough, it is not transferred in time to the proper bond to be broken; the energy relaxation by other means before a reaction can finish.

Only effective collisions can lead to a chemical reaction, and the rate of reaction is the product of the collision frequency Z, the energy fraction f, and the probability factor P, as

$$R_{rxn} = ZfP = Z_0 P \exp\left(-\frac{\varepsilon_c}{k_B T}\right)[A][B] \tag{4.20}$$

Comparing **Equation (4.20)** with the formula of the law of mass action for a bimolecular elementary reaction, $R_{rxn} = k[A][B]$, we then have

$$k(T) = Z_0 P \exp\left(-\frac{\varepsilon_c}{k_B T}\right) = 10^6 N_A^2 \sigma_{AB} \sqrt{\frac{8k_B T}{\pi \mu_{AB}}} P \exp\left(-\frac{\varepsilon_c}{k_B T}\right) \tag{4.21}$$

Therefore, the rate constant k is dependent on the size, molar mass, temperature, threshold energy, orientation of the colliding reactant molecules, etc. Both Z_0 and f can be derived precisely based on the kinetic-molecular

$$f \propto \exp\left(-\frac{\varepsilon_c}{k_B T}\right) \quad (4.19)$$

高温下的能量分数远大于低温下的能量分数。同时，反应的 ε_c 越高，满足 $\varepsilon_k > \varepsilon_c$ 的分子所占分数就越小。

3) 为使碰撞有效，能量分数并非唯一的限制因素。除能量分数外，将所有可能进一步降低有效碰撞概率的额外因素统称为**概率因子** P。概率因子包括但不限于以下内容：碰撞时分子的空间取向不利于发生反应；复杂分子的空间位阻效应导致有效碰撞的概率降低；碰撞时间短于能量有效传递所需的时间，导致即使能量足够也不能及时传递到需要断裂的键上；在反应完成之前，能量通过其他方式发生了弛豫。

只有有效碰撞才能导致化学反应发生，因此反应速率是碰撞频率 Z、能量分数 f 和概率因子 P 的乘积，即

$$R_{反应} = ZfP = Z_0 P \exp\left(-\frac{\varepsilon_c}{k_B T}\right)[A][B] \quad (4.20)$$

将**式（4.20）**与双分子基元反应的质量作用定律公式 $R_{反应} = k[A][B]$ 相比较，有

$$k(T) = Z_0 P \exp\left(-\frac{\varepsilon_c}{k_B T}\right) = 10^6 N_A^2 \sigma_{AB} \sqrt{\frac{8 k_B T}{\pi \mu_{AB}}} P \exp\left(-\frac{\varepsilon_c}{k_B T}\right) \quad (4.21)$$

因此，速率常数 k 取决于发生碰撞反应物分子的大小、摩尔质量、温度、阈值能量及取向等。Z_0 和 f 均可根据分子运动论精确推导，但概率因子 P 不能。P 仍然是碰撞理论中的经验参数，通常作为理论和实验速率常数之间的修正因子。

如果令 $C = 10^6 N_A^2 \sigma_{AB} \sqrt{\frac{8 k_B}{\pi \mu_{AB}}} P$，**式（4.21）** 的指前因子中所有与温度无关的参数均包含在 C 中，则此式可简化为

$$k(T) = C\sqrt{T} \exp\left(-\frac{\varepsilon_c}{k_B T}\right) \quad (4.22)$$

与 **2.3 节**学过的 $\overline{e_k}$ 与 $\overline{E_k}$ 之间的关系类似，阈值能量 ε_c 也可与相应的 E_c（即 1 mol 气体分子的阈值能量）通过 $E_c = N_A \varepsilon_c$ 关联起来。故

$$k(T) = C\sqrt{T} \exp\left(-\frac{E_c}{RT}\right) \quad (4.23)$$

其中 R 为摩尔气体常数，可通过 $R = N_A k_B$ 与 k_B 相关联。

将**式（4.23）**与阿伦尼乌斯方程 [**式（4.15）**] 相比较，可推得活化能 E_a 与阈值能量 E_c 之间的关系。从阿伦尼乌斯方程

$$\ln k = -\frac{E_a}{RT} + \ln A$$

有

$$\frac{d \ln k}{dT} = -\frac{E_a}{R} \frac{d\left(\frac{1}{T}\right)}{dT} = \frac{E_a}{RT^2}$$

由碰撞理论可推导出

$$\ln k = -\frac{E_c}{RT} + \frac{1}{2}\ln T + \ln C$$

theory, but the probability factor P cannot. P is still an empirical parameter in collision theory, and usually serves as a correction factor between the theoretical and experimental rate constants.

If let $C = 10^6 N_A^2 \sigma_{AB} \sqrt{\dfrac{8k_B}{\pi \mu_{AB}}} P$, which contains all the temperature-independent parameters in the preexponential term in **Equation (4.21)**, we can then simplify the above equation as

$$k(T) = C\sqrt{T} \exp\left(-\dfrac{\varepsilon_c}{k_B T}\right) \tag{4.22}$$

In a similar manner to the relationship between $\overline{e_k}$ and $\overline{E_k}$ that we have learnt in **Section 2.3**, the threshold energy ε_c can also be related to the corresponding E_c, which is the threshold energy of 1 mol gas molecules, as $E_c = N_A \varepsilon_c$. Therefore,

$$k(T) = C\sqrt{T} \exp\left(-\dfrac{E_c}{RT}\right) \tag{4.23}$$

where R is the molar gas constant that is related to k_B by $R = N_A k_B$.

Comparing **Equation (4.23)** with Arrhenius **Equation (4.15)**, you may wonder what the relationship is between the activation energy E_a and the threshold energy E_c. From Arrhenius equation

$$\ln k = -\dfrac{E_a}{RT} + \ln A$$

we have

$$\dfrac{d \ln k}{dT} = -\dfrac{E_a}{R} \dfrac{d\left(\dfrac{1}{T}\right)}{dT} = \dfrac{E_a}{RT^2}$$

From the collision theory, we can derive

$$\ln k = -\dfrac{E_c}{RT} + \dfrac{1}{2}\ln T + \ln C$$

$$\dfrac{d \ln k}{dT} = \dfrac{E_c}{RT^2} + \dfrac{1}{2T}$$

Therefore,

$$\dfrac{d \ln k}{dT} = \dfrac{E_c}{RT^2} + \dfrac{1}{2T} = \dfrac{E_a}{RT^2}$$

$$E_a = E_c + \dfrac{1}{2}RT \tag{4.24}$$

If $E_c \gg RT/2$, it is valid to conclude that $E_a \approx E_c$. The relationship between A and C can be derived accordingly to give that $A = C\sqrt{eT}$.

Transition State Theory

The **transition state theory** (TST) was developed simultaneously by Henry Eyring (1901—1981), Meredith G. Evans (1904—1952) and Michael Polanyi (1891—1976) in 1935, based on statistical mechanics and quantum mechanics. TST is sometimes referred to as "activated-complex theory" because it involves a hypothetical species called the **activated complex** in the reaction progress. TST is also referred to as "absolute-rate theory" because it can be used to quantitatively calculate the absolute reaction rate of an elementary reaction. Here, we will only learn TST in a qualitative manner. The main assumptions of TST are listed as follows:

1) The rate of an elementary reaction can be calculated in a process where the reactants are converted into products through the activated complexes on a reaction **potential energy surface** (PES);
2) The activated complexes are in quasi-equilibrium with the reactants, and the overall reaction rate is determined by the rate of activated complexes converting into products.

$$\frac{\mathrm{d}\ln k}{\mathrm{d}T} = \frac{E_c}{RT^2} + \frac{1}{2T}$$

因此

$$\frac{\mathrm{d}\ln k}{\mathrm{d}T} = \frac{E_c}{RT^2} + \frac{1}{2T} = \frac{E_a}{RT^2}$$

$$E_a = E_c + \frac{1}{2}RT \qquad (4.24)$$

如果 $E_c \gg RT/2$，可得结论 $E_a \approx E_c$。A 和 C 之间的关系可相应推导为 $A = C\sqrt{eT}$。

过渡态理论

过渡态理论（TST）是由亨利·艾林（1901—1981）以及梅雷迪斯·G. 埃文斯（1904—1952）和迈克尔·波兰尼（1891—1976）在 1935 年同时提出的，它以统计力学和量子力学为基础。过渡态理论又称"活化配合物理论"，因为在反应进程中涉及了一种被称为**活化配合物**的假想物种。过渡态理论也称"绝对速率理论"，因为可用来定量计算基元反应的绝对反应速率。这里我们将仅以定性的方式来学习过渡态理论。该理论的主要假定有：

1) 基元反应的速率可通过反应**势能面**（PES）上由反应物经历活化配合物转化为生成物的过程计算出；
2) 活化配合物与反应物处于准平衡状态，总反应速率由活化配合物转化为生成物的速率决定。

作为说明性示例，让我们考虑一个由反应物（单原子分子 A 和双原子分子 B—C）经历活化配合物（$[A\cdots B\cdots C]^{\neq}$）转化为生成物（双原子分子 A—B 和单原子分子 C）的如下反应：

$$A + B - C \rightleftharpoons [A\cdots B\cdots C]^{\neq} \to A - B + C$$

在活化配合物中，**部分键**（…）表示 B 原子被从 B—C 分子中部分移除且与 A 分子部分连接。平衡符号表示活化配合物与反应物处于准平衡状态，箭头符号表示由活化配合物直接形成生成物。该反应可用如**图 4.3** 所示的反应曲线图来表示。**反应曲线图**是体系的势能对反应进程作图，而反应进程从左边的反应物开始，经过活化配合物的**过渡态**，最后以右边的生成物结束。反应进程可以基于一些反应坐标、沿反应途径来表示。

这里我们遇到了许多较难区分的概念，如反应曲线图、反应进程、反应途径和反应坐标等。为了理解这些概念之间的细微差别，我们需要用到反应势能面。在介绍反应势能面之前，让我们先来了解分子势能面的概念。

分子势能面是分子在特定坐标下的势能，这些坐标通常为键参数，如键长、键角、键二面角等。如果只涉及一个坐标，通常称为势能曲线。如果涉及两个或两个以上坐标，则称为势能面。虽然从图形上我们只能可视化三维曲面，但更高维的曲面可通过绘制其在三维空间的投影来进行可视化。让我们从一些简单的例子开始。

1) O_2。O_2 的键伸缩振动的势能曲线 [**图 4.4(a)**] 可用莫尔斯势能函数表示为

$$V(r) = D_e \left\{1 - \exp\left[-a(r - r_e)\right]\right\}^2$$

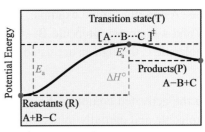

Figure 4.3 The reaction profile for the reaction $A + B - C \rightleftharpoons [A\cdots B\cdots C]^{\neq} \to A - B + C$.

图 4.3 反应 $A + B - C \rightleftharpoons [A\cdots B\cdots C]^{\neq} \to A - B + C$ 的反应曲线图。

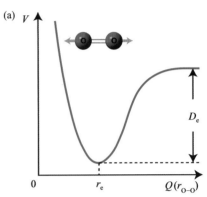

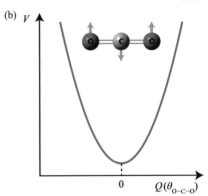

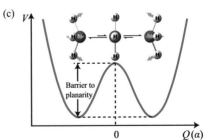

Figure 4.4 The potential energy surfaces of some molecules in terms of certain coordinates. (a) The potential energy curve for the bond stretching vibration of O_2. (b) The potential energy curve for the non-dissociate angle bending vibration of CO_2. (c) The potential energy curve for the non-dissociate inversion vibration of NH_3.

图 4.4 分子关于某些坐标的势能面：(a) O_2 的键伸缩振动的势能曲线；(b) CO_2 的非解离角弯曲振动的势能曲线；(c) NH_3 的非解离翻转振动的势能曲线。

As an illustrative example, let us consider the reaction in which the reactants of a monatomic molecule A and a diatomic molecule B—C convert into the products of a diatomic molecule A—B and a monatomic molecule C through an activated complex $[A\cdots B\cdots C]^{\neq}$, as

$$A + B - C \rightleftharpoons [A\cdots B\cdots C]^{\neq} \rightarrow A - B + C$$

In the activated complex, the **partial bonds** (⋯) indicate that atom B is partially removed from the B—C molecule and partially joined to the A molecule. The equilibrium symbol implies that the activated complex is in quasi-equilibrium with the reactants, and the arrow symbol indicates that the products are formed directly from the activated complex. This reaction can be graphically represented by a reaction profile, as shown in **Figure 4.3**. In the **reaction profile**, the potential energy of the system is plotted against the reaction progress, which starts with reactants on the left, progresses through a **transition state** of the activated complex, and ends with products on the right. The reaction progress can be represented along the reaction pathway in terms of some reaction coordinates.

Here, we encounter many concepts that are quite difficult to distinguish, such as reaction profile, reaction progress, reaction pathway, and reaction coordinate, etc. To understand the nuances between those concepts, we need to use a reaction PES. But before introducing the reaction PES, let us first learn the concept of the PES of a molecule.

The PES of a molecule is the potential energy of a molecule in terms of certain coordinates, which are generally the bond parameters such as bond length, bond angle, bond dihedral angle, etc. If only one coordinate is involved, it is often called a potential energy curve. If two or more coordinates are involved, it is then referred to as a potential energy surface. Although graphically we can only visualize 3D surfaces, higher dimensional surfaces can be plotted as projections into 3D spaces for visualization. Let us start with some simple examples.

1) O_2. The potential energy curve for the bond stretching vibration of O_2 [**Figure 4.4(a)**] can be represented by the Morse potential function, as

$$V(r) = D_e \{1 - \exp[-a(r - r_e)]\}^2$$

where the coordinate Q in **Figure 4.4(a)** is the internuclear distance, or the O—O bond length r, r_e is the equilibrium bond length of the O—O bond, D_e is the dissociation energy of the bond, and a is a structure-related parameter. When the distance between the two O atoms is infinity, there is no interaction between them and the potential energy of the system is 0. As these two O atoms approach each other, attractive forces dominate repulsive forces and the potential energy of the system keeps decreasing until a minimal potential energy is reached at r_e and the O—O bond is formed. As r becomes smaller than r_e, repulsive forces increase dramatically and the potential energy goes to infinity at $r = 0$. More discussion about the potential energy curve of a diatomic molecule can be found in **Section 8.4**.

2) CO_2. The potential energy curve for the non-dissociate angle bending vibration of CO_2 is given in **Figure 4.4(b)**. The coordinate Q is the bend angle, or the O—C—O bond angle θ. Since CO_2 is a linear molecule in its equilibrium configuration, the minimum in the curve corresponds to the linear configuration at $\theta = 0$. At very large θ, increased potential energy also leads to the dissociation of CO_2 but this is not shown in **Figure 4.4(b)**. Both curves in **Figures 4.4(a)** and **4.4(b)** have only a single minimum.

3) NH_3. **Figure 4.4(c)** shows the potential energy curve for the non-dissociate inversion vibration of NH_3. The coordinate Q is the inversion angle, or the bond dihedral angle α between N—H_1 and the H_2—N—H_3 plane. At $\alpha = 0$, all atoms in NH_3 are coplanar. However, the equilibrium structure of NH_3 is a pyramid configuration with $\alpha \neq 0$. For large amplitude of the inversion motion, the NH_3 molecule may go through the planar configuration to the reverse pyramidal configuration. These two pyramidal configurations correspond to the two equivalent minima in the potential curve. The maximum in the curve corresponds to the planar configuration at $\alpha = 0$. The energy difference between the maximum and the minima is called the energy barrier or barrier to planarity.

4) Cyclobutane C_4H_8. The potential energy curve for the ring-puckering vibration of the four-membered ring molecule cyclobutane is shown in **Figure 4.5(a)**. The equilibrium structure of cyclobutane is a non-

其中**图 4.4(a)** 的坐标 Q 是核间距，即 O—O 键键长 r；r_e 是 O—O 键的平衡键长；D_e 是该键的解离能；a 是与结构相关的参数。当两个 O 原子距离无穷远时，它们之间没有相互作用，体系的势能为 0。当两个 O 原子逐渐接近时，吸引力大于排斥力，体系的势能不断降低，直至在 r_e 处达到势能的极小值，O—O 键形成。当 r 小于 r_e 时，排斥力急剧增加，势能在 $r = 0$ 时趋于无穷大。关于双原子分子势能曲线的更多讨论详见 **8.4 节**。

2) CO_2。**图 4.4(b)** 给出了 CO_2 的非解离角弯曲振动的势能曲线。坐标 Q 是弯曲角，即 O—C—O 键键角 θ。由于 CO_2 的平衡构型为直线形分子，曲线的极小值对应于 $\theta = 0$ 的直线构型。当 θ 很大时，升高的势能也会导致 CO_2 解离 [**图 4.4(b)** 并没有展示]。**图 4.4(a)** 和 **4.4(b)** 的两条曲线均只有一个极小值。

3) NH_3。**图 4.4(c)** 给出了 NH_3 的非解离翻转振动的势能曲线。坐标 Q 是翻转角，即 N—H_1 和 H_2—N—H_3 平面的键二面角 α。当 $\alpha=0$ 时，NH_3 的所有原子均共面。然而，NH_3 的平衡结构是 $\alpha \neq 0$ 的角锥构型。对于大振幅的翻转振动，NH_3 分子可通过平面构型转变为反方向的角锥构型。这两个角锥构型对应于势能曲线上的两个等同的极小值。曲线的极大值则对应于 $\alpha = 0$ 的平面构型。极大值与极小值之间的能量差称为能垒或平面化势垒。

4) 环丁烷 C_4H_8。**图 4.5(a)** 给出了四元环分子环丁烷的环折叠振动的势能曲线。由于 CH_2—CH_2 扭转相互作用，环丁烷的平衡结构为非平面的折叠构型。尽管折叠角可作为坐标 Q，但使用折叠位移 x 更容易也更常见，其中 $2x$ 是环丁烷两条对角线之间的距离。两个折叠构型（$x\neq 0$）对应于两个等同的极小值点，而平面构型（$x=0$）为极大值点。类似的例子还有环戊烯 C_5H_8 和 1,4-环己二烯 C_6H_8 [**图 4.5(a)**]。由于双键在环折叠振动过程中表现为一个整体，这两个分子可视为"准四元环"，其势能曲线也可用折叠位移 x 来表示。

5) 环戊烷 C_5H_{10}。要描述上述分子的势能，只需要一个与分子主振动相关的坐标来绘制势能曲线就足够了。尽管这些分子具有更多的振动模式，但这一主振动是与分子的构象变化相关的最为重要的一个。然而，为描述五元环分子环戊烷的势能，需要使用与两个主要骨架振动相关的两个坐标来绘制势能面。一个是与环弯曲振动相关的弯曲位移 x，另一个是与环扭转振动相关的扭转角 τ [**图 4.5(b)**]。相应地，该分子存在一个平面构型（$x=0$ 且 $\tau=0$，对应于极大值点）、两个扭转构型（$x=0$ 且 $\tau\neq 0$，对应于两个等同的极小值点）和两个弯曲构型（$x\neq 0$ 且 $\tau=0$，对应于两个等同的鞍点）。**图 4.5(b)** 有助于我们可视化三维表面上的鞍点，它是一个方向（沿扭转角 τ）上的极大值点和另一个方向（沿弯曲位移 x）上的极小值点。类似的例子还有"准五元环"分子环己烯 C_6H_{10} [**图 4.5(b)**]。

在简要介绍了一些分子的势能曲线或势能面之后，让我们回到反应势能面。化学反应总涉及旧键的断裂和新键的形成，相应的键参数在化学反应中起着关键作用，体系的势能受键参数变化的影响。因此，反应势能面可用几个关键的键参数来表示。

例如，下述反应的势能面

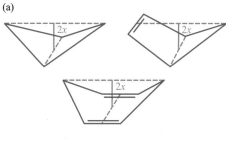

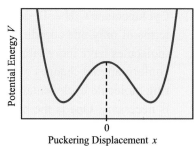

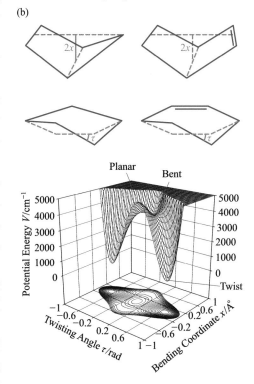

Figure 4.5 The potential energy surfaces of some molecules in terms of certain coordinates. (a) The potential energy curve for the ring-puckering vibration of four-membered ring molecule cyclobutane and pseudo-four-membered ring molecule cyclopentene and 1,4-cyclohexadiene. (b) The potential energy surface for the ring-bending and ring-twisting vibrations of five-membered ring molecule cyclopentane and pseudo-five-membered ring molecule cyclohexene.

图 4.5 分子关于某些坐标的势能面：(a) 四元环分子环丁烷与准四元环分子环戊烯和 1,4-环己二烯的环折叠振动的势能曲线；(b) 五元环分子环戊烷与准五元环分子环己烯的环弯曲和环扭转振动的势能曲面。
(Source: Rivera-Gaines, V. E. *et al.*, *J. Am. Chem. Soc.* **1991**, *113*, 9735–9742.)

planar puckered configuration due to the CH_2—CH_2 torsional interactions. Although the puckering angle can serve as the coordinate Q, however, it is easier and more common to use the puckering displacement x instead, where $2x$ is the distance between the two diagonals of the cyclobutane. The two puckered configurations ($x \neq 0$) correspond to the two equivalent minima and the planar configuration ($x = 0$) is the maximum. Similar examples are cyclopentene C_5H_8 and 1,4-cyclohexadiene C_6H_8 [**Figure 4.5(a)**]. Since double bonds behave as single units during the ring-puckering vibration, these two molecules can be considered as "pseudo-four-membered rings", whose potential energy curves can also be represented in terms of the puckering displacement x.

5) Cyclopentane C_5H_{10}. To describe the potential energy of the above molecules, a potential energy curve in terms of only one coordinate relating to one major vibration of the molecule is enough. Although these molecules have more vibrational modes, this major vibration is the most important one associated with conformational changes of the molecule. However, to describe the potential energy of the five-membered ring molecule cyclopentane, a PES in terms of two coordinates relating to two major skeleton vibrations is necessary. One is the bending displacement x relating to the ring-bending vibration, and the other is the twisting angle τ relating to the ring-twisting vibration [**Figure 4.5(b)**]. Consequently, the molecule has one planar configuration ($x = 0$ and $\tau = 0$, corresponding to the maximum), two twist configurations ($x=0$ and $\tau \neq 0$, corresponding to the two equivalent minima), and two bent configurations ($x \neq 0$ and $\tau = 0$, corresponding to the two equivalent saddle points). **Figure 4.5(b)** helps us to visualize the saddle points on a 3D surface, which is the maximum in one direction (along the twisting angle τ) and the minimum in another direction (along the bending displacement x). A similar example is the "pseudo-five-membered ring" molecule cyclohexene C_6H_{10} [**Figure 4.5(b)**].

After the brief introduction of the potential energy curves or surfaces of some molecules, let us return to the reaction PES. A chemical reaction always involves the breaking of some old bonds and the formation of some new bonds. The corresponding bond parameters play key roles in the chemical reaction and the potential energy of the system is subject to the bond parameter changes. Therefore, the reaction PES can be represented in terms of several key bond parameters.

For example, the PES of the reaction

$$A + B - C \rightleftharpoons [A \cdots B \cdots C]^{\neq} \to A - B + C$$

can be represented in terms of r_{AB} and r_{BC}, which are the internuclear distances or bond lengths of A—B and B—C, respectively. The PES of this reaction is schematically illustrated in **Figure 4.6**, in which r_{AB} and r_{BC} serve as the x- and y-coordinates, respectively, and the z-coordinate is the potential energy of the system. In this 3D surface, the reactants locate at the left valley since r_{AB} is large and r_{BC} is small at the beginning of the reaction. The products locate at the right valley where r_{AB} is small and r_{BC} is large. The transition state of the activated complex, on the other hand, corresponds to the saddle point on this 3D PES with modest lengths of both A—B and B—C, consistent with the partial bonds. The **reaction pathway** refers to the minimum energy path, highlighted by the gray curve on the PES, which leads from reactants to products through the transition state of the activated complex. The projection of the reaction pathway onto the xy-plane is called the **reaction progress**, indicated by the black curve on the contour map. Sometimes, to simplify the situation, the potential energy of the system is directly plotted against the reaction progress, and this is the reaction profile that we have discussed briefly in **Figure 4.3**. The reaction profile can also be viewed as a special projection of the reaction pathway in a direction perpendicular to the reaction progress. Although not specified in a reaction profile, we should now understand that both the reaction pathway and reaction progress are in terms of the reaction coordinates that are some key bond parameters in a chemical reaction.

Meanwhile, we should pay special attention to distinguish the two concepts of activated complex and transition state. The activated complex is the hypothetical species believed to exist with some partial bonds, which are located between the reactants and products in the reaction profile. The activated complex can have various different configurations and their potential energies lie near the saddle point on the reaction PES. The transition state, however, is a particular configuration of the activated complex that corresponds to the exact saddle point, with maximum potential energy along the reaction pathway but minimum potential energy in

$$A+B-C \rightleftharpoons [A\cdots B\cdots C]^{\neq} \rightarrow A-B+C$$

可用 r_{AB} 和 r_{BC} 来表示，它们分别是 A—B 和 B—C 的核间距（或键长）。该反应的势能面示意图如**图 4.6** 所示，其中 x 和 y 坐标分别为 r_{AB} 和 r_{BC}，z 坐标是体系的势能。在此三维曲面上，反应物位于左边的山谷处，因为反应开始时 r_{AB} 很大而 r_{BC} 很小。生成物位于右边的山谷处，那里 r_{AB} 很小而 r_{BC} 很大。另一方面，活化配合物的过渡态对应于该三维势能曲面上的鞍点，其 A—B 和 B—C 长度均适中，与部分键一致。**反应途径**指能量最小途径（标记为势能面上的灰色曲线），该途径从反应物经过活化配合物的过渡态到达生成物。反应途径在 xy 平面上的投影称为**反应进程**，用等高线图上的黑色曲线表示。有时为简化情况，常用体系的势能直接对反应进程作图，这就是我们在**图 4.3** 中简要讨论过的反应曲线图。反应曲线图也可看作反应途径在垂直于反应进程方向上的特殊投影。尽管在反应曲线图中没有具体说明，但我们现在应该明白，反应途径和反应进程都基于反应坐标，而反应坐标是化学反应中的一些关键的键参数。

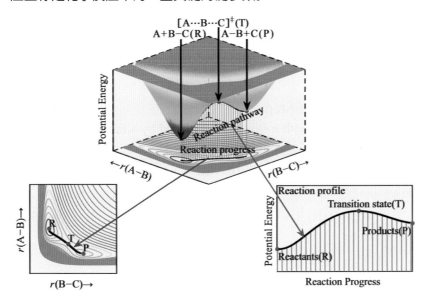

Figure 4.6 Schematic diagram of the reaction pathway and reaction progress on the 3D potential energy surface for the reaction $A+B-C \rightleftharpoons [A\cdots B\cdots C]^{\neq} \rightarrow A-B+C$.

图 4.6 反应 $A+B-C \rightleftharpoons [A\cdots B\cdots C]^{\neq} \rightarrow A-B+C$ 的反应途径与反应进程在三维势能曲面上的示意图。

同时，我们应特别注意区分活化配合物和过渡态这两个概念。活化配合物是假想存在的具有部分键的物种，位于反应曲线图的反应物和生成物之间。活化配合物可以有各种不同构型，其势能均位于反应势能面的鞍点附近。而过渡态是活化配合物的一种特殊构型，对应于精确的鞍点，沿反应途径具有极大的势能，但在垂直方向上具有极小的势能。

反应物和生成物之间的能量差就是该反应的焓变 ΔH。例如，**图 4.3** 所示的反应为吸热反应，反应物的能量低于生成物的能量。活化配合物与反应物之间的能量差就是正反应的活化能 E_a，而活化配合物与生成物之间的能量差就是逆反应的活化能 E_a'。因此，反应的焓变等于正反应和逆反应的活化能之差，即 $\Delta H = E_a - E_a'$。

the perpendicular direction.

The difference in energies between the reactants and products is ΔH of the reaction. For example, the reaction shown in **Figure 4.3** is an endothermic reaction with lower energy in reactants than in products. The difference in energies between the activated complex and the reactants is the activation energy of the forward reaction E_a. The difference in energies between the activated complex and the products is the activation energy of the reverse reaction E_a'. Therefore, the enthalpy change of a reaction equals the difference in activation energies of the forward and reverse reactions, or $\Delta H = E_a - E_a'$.

4.6 Reaction Mechanisms

In addition to the rates of reaction, orders of reaction, rate constant, and activation energy, chemical kinetics also concerns chiefly determining the reaction mechanisms and relating them to the rate laws of chemical reactions. A **reaction mechanism** is a step-by-step detailed description of the reaction pathway of a chemical reaction, each step of which is an elementary reaction. That is, a reaction mechanism is a process that decomposes a complex reaction into a series of elementary reactions. A plausible reaction mechanism must be consistent with the stoichiometry of the overall reaction as well as with the experimentally determined rate law. Reaction mechanism can help us achieve better understanding of the reaction pathways and the fundamental basis of chemical reactions, etc. In this section, we will show two typical kinds of reaction mechanisms.

A Mechanism with a Slow Step Followed by a Fast Step

The first typical kind of reaction mechanism is one with a slow step followed by a fast step, such as the gaseous reaction between H_2 and ICl producing I_2 and HCl, given by

$$H_2(g) + 2\,ICl(g) \rightarrow I_2(g) + 2\,HCl(g)$$

The experimentally determined rate law for this reaction is

$$R_{rxn} = k[H_2][ICl] \tag{4.25}$$

A plausible mechanism of this reaction can be

(1) Slow: $H_2(g) + ICl(g) \rightarrow HI(g) + HCl(g)$ $R_1 = k_1[H_2][ICl]$
(2) Fast: $HI(g) + ICl(g) \rightarrow I_2(g) + HCl(g)$ $R_2 = k_2[HI][ICl]$
Overall: $H_2(g) + 2ICl(g) \rightarrow I_2(g) + 2HCl(g)$

The species HI is produced in step (1) and consumed in step (2). It does not appear in the experimental overall rate law, and is called a **reaction intermediate**. Since step (1) is slow and step (2) is fast, it suggests that the HI consuming rate in step (2) must be just as fast as its formation rate in step (1). Therefore, the overall reaction rate is governed only by the rate at which HI is formed in step (1), and step (1) is then the **rate-determining step** in this mechanism. Consequently,

$$R_{rxn} = R_1 = k[H_2][ICl] \quad \text{and} \quad k = k_1$$

The reaction mechanism is consistent with the experimentally determined rate law.

The reaction profile for a complex reaction is more complicated than that for an elementary reaction. The reaction profile for the two steps in the above proposed mechanism is shown in **Figure 4.7**. There are two transition states of two activated complexes and one reaction intermediate. The activation energy for step (1) is greater than that for step (2), which is expected since step (1) is slower than step (2). Note the difference between a reaction intermediate and an activated complex. A reaction intermediate is a real species that has fully formed bonds and can sometimes be isolated in a complex reaction. However, an activated complex is only a hypothetical species that has partially formed bonds. The activated complex exists only momentarily and cannot be isolated in a chemical reaction.

4.6 反应机理

除了研究反应速率、反应级数、速率常数和活化能之外，化学动力学还主要关注确定反应机理，并将其与化学反应的速率方程相联系。**反应机理**是对化学反应途径的逐步详细描述，其中每一步均为一个基元反应。也就是说，反应机理是将复杂反应分解为一系列基元反应的过程。合理的反应机理必须与整个反应的化学计量数一致，并与实验确定的速率方程一致。反应机理可以帮助我们更好地理解反应途径，掌握化学反应的基础。本节我们将展示两种典型的反应机理。

先慢反应后快反应的机理

第一种典型的反应机理是先慢反应后快反应的机理，如 H_2 和 ICl 生成 I_2 和 HCl 的气相反应，其方程式为

$$H_2(g) + 2\,ICl(g) \rightarrow I_2(g) + 2\,HCl(g)$$

该反应由实验确定的速率方程为

$$R_{反应} = k[H_2][ICl] \tag{4.25}$$

该反应的一种合理的机理可能是

(1) 慢：$H_2(g) + ICl(g) \rightarrow HI(g) + HCl(g)$ $R_1 = k_1[H_2][ICl]$

(2) 快：$HI(g) + ICl(g) \rightarrow I_2(g) + HCl(g)$ $R_2 = k_2[HI][ICl]$

总反应：$H_2(g) + 2ICl(g) \rightarrow I_2(g) + 2HCl(g)$

物种 HI 在步骤（1）中生成而在步骤（2）中消耗。它并不出现在总实验速率方程中，称为**反应中间体**。由于步骤（1）较慢而步骤（2）较快，这表明步骤（2）中 HI 的消耗速率只能与步骤（1）中的生成速率一样快。因此，总反应速率由步骤（1）中 HI 的生成速率决定，步骤（1）即为该机理的**决速步**。相应地

$$R_{反应} = R_1 = k[H_2][ICl] \quad 且 \quad k = k_1$$

该反应机理与实验确定的速率方程一致。

复杂反应的反应曲线图比基元反应的更复杂。上述机理存在两个步骤，其反应曲线图如**图 4.7** 所示，图中有两个活化配合物的过渡态以及一个反应中间体。步骤（1）的活化能大于步骤（2），这是符合预期的，因为步骤（1）比步骤（2）更慢。注意反应中间体与活化配合物的区别。反应中间体是一种真实存在的物种，具有完全形成的键，在复杂反应中有时可以分离出来。而活化配合物只是具有部分键的假想物种，只在瞬间存在，无法从化学反应中分离出来。

先快平衡后慢反应的机理

第二种典型的反应机理是先快平衡后慢反应的机理，如 NO 和 O_2 生成 NO_2 的气相反应，其方程式为

$$2\,NO(g) + O_2(g) \rightarrow 2\,NO_2(g)$$

该反应由实验确定的速率方程为

$$R_{反应} = k[NO]^2[O_2] \tag{4.26}$$

尽管速率方程与化学计量数一致，但该反应仍然不是基元反应，因为单步三分子反应的机理是极不可能的。

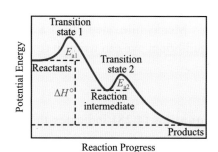

Figure 4.7 A reaction profile for a two-step mechanism.

图 4.7 两步机理的反应曲线图。

A Mechanism with a Fast Equilibrium Step Followed by a Slow Step

The second typical kind of reaction mechanism is one with a fast equilibrium step followed by a slow step, such as the gaseous reaction of NO and O_2 producing NO_2, given by

$$2\ NO\ (g) + O_2\ (g) \rightarrow 2\ NO_2\ (g)$$

The experimentally determined rate law for this reaction is

$$R_{rxn} = k[NO]^2[O_2] \qquad (4.26)$$

Even though the rate law is consistent with the stoichiometry, this reaction is still not elementary since a one-step termolecular mechanism is highly improbable.

A plausible mechanism of this reaction can be

(1) Fast: $NO(g) + NO(g) \underset{k_{-1}}{\overset{k_1}{\rightleftharpoons}} N_2O_2(g) \quad R_1 = k_1[NO]^2 = R_{-1} = k_{-1}[N_2O_2]$

(2) Slow: $N_2O_2(g) + O_2(g) \xrightarrow{k_2} 2NO_2(g) \quad R_2 = k_2[N_2O_2][O_2]$

Overall: $2NO(g) + O_2(g) \rightarrow 2NO_2(g)$

Since step (1) is fast and step (2) is slow, step (2) is the rate-determining step and the overall reaction rate is governed by R_2. Consequently

$$R_{rxn} = R_2 = k_2[N_2O_2][O_2]$$

Because N_2O_2 is an intermediate, its concentration must be eliminated from the rate law. As step (1) is fast and reaches equilibrium, the rates of both forward and reverse reactions are equal, so

$$R_1 = k_1[NO]^2 = R_{-1} = k_{-1}[N_2O_2]$$

Rearranging the above equation, we have

$$[N_2O_2] = \frac{k_1}{k_{-1}}[NO]^2$$

Therefore,

$$R_{rxn} = R_2 = k_2[N_2O_2][O_2] = \frac{k_2 k_1}{k_{-1}}[NO]^2[O_2] \qquad (4.27)$$

where the experimentally observed rate constant k is related to the other rate constants in the proposed mechanism as

$$k = \frac{k_2 k_1}{k_{-1}}$$

Meanwhile, step (1) is a fast equilibrium step, and by applying what we have learnt in chemical equilibrium, we can write an equilibrium constant expression as

$$K = \frac{[N_2O_2]}{[NO]^2} = \frac{k_1}{k_{-1}} \qquad (4.28)$$

Equation (4.28) indicates the relationship between rate constants and equilibrium constant. It also demonstrates the connection between chemical kinetics and chemical equilibrium.

Although the above mechanism is consistent with both the stoichiometry of the overall reaction and the experimentally determined rate law, it is not the only plausible mechanism. Another plausible mechanism of this reaction can be

(1) Fast: $NO(g) + O_2(g) \underset{k_{-1}}{\overset{k_1}{\rightleftharpoons}} NO_3(g) \quad R_1 = k_1[NO][O_2] = R_{-1} = k_{-1}[NO_3]$

(2) Slow: $NO_3(g) + NO(g) \xrightarrow{k_2} 2NO_2(g) \quad R_2 = k_2[NO_3][NO]$

Overall: $2NO(g) + O_2(g) \rightarrow 2NO_2(g)$

Similarly,

$$[NO_3] = \frac{k_1}{k_{-1}}[NO][O_2]$$

该反应的一种合理的机理可能是

(1) 快: $NO(g) + NO(g) \underset{k_{-1}}{\overset{k_1}{\rightleftharpoons}} N_2O_2(g)$ $R_1 = k_1[NO]^2 = R_{-1} = k_{-1}[N_2O_2]$

(2) 慢: $N_2O_2(g) + O_2(g) \xrightarrow{k_2} 2NO_2(g)$ $R_2 = k_2[N_2O_2][O_2]$

总反应: $2NO(g) + O_2(g) \rightarrow 2NO_2(g)$

由于步骤（1）较快而步骤（2）较慢，因此步骤（2）是决速步，总反应速率由 R_2 控制。相应地

$$R_{反应} = R_2 = k_2[N_2O_2][O_2]$$

由于 N_2O_2 是中间体，其浓度必须从速率方程中剔除。而步骤（1）是快平衡，正反应和逆反应的速率相等，即

$$R_1 = k_1[NO]^2 = R_{-1} = k_{-1}[N_2O_2]$$

将上述方程重排，有

$$[N_2O_2] = \frac{k_1}{k_{-1}}[NO]^2$$

故

$$R_{反应} = R_2 = k_2[N_2O_2][O_2] = \frac{k_2 k_1}{k_{-1}}[NO]^2[O_2] \quad (4.27)$$

其中实验观察到的速率常数 k 与所拟定机理中其他速率常数的关系为

$$k = \frac{k_2 k_1}{k_{-1}}$$

同时，步骤（1）为快平衡，应用我们在化学平衡中所学，可将平衡常数的表达式写为

$$K = \frac{[N_2O_2]}{[NO]^2} = \frac{k_1}{k_{-1}} \quad (4.28)$$

式（4.28）给出了速率常数与平衡常数之间的关系，也表明了化学动力学与化学平衡之间的联系。

尽管上述机理与整个反应的化学计量数以及实验确定的速率方程均一致，但它并非唯一合理的机理。该反应的另一种合理的机理可能是

(1) 快: $NO(g) + O_2(g) \underset{k_{-1}}{\overset{k_1}{\rightleftharpoons}} NO_3(g)$
$R_1 = k_1[NO][O_2] = R_{-1} = k_{-1}[NO_3]$

(2) 慢: $NO_3(g) + NO(g) \xrightarrow{k_2} 2NO_2(g)$ $R_2 = k_2[NO_3][NO]$

总反应: $2NO(g) + O_2(g) \rightarrow 2NO_2(g)$

类似地

$$[NO_3] = \frac{k_1}{k_{-1}}[NO][O_2]$$

$$R_{反应} = R_2 = k_2[NO_3][NO] = \frac{k_2 k_1}{k_{-1}}[NO]^2[O_2]$$

这一机理也符合化学计量数和实验确定的速率方程。在这个阶段，我们无法判断上述两种合理的机理中究竟哪一个才是实际的反应途径，一个看似合理的机理只意味着它没有被动力学所排除。为了确定实际的反应机理，需要采用超快反应动力学实验捕获真实反应中

$$R_{rxn} = R_2 = k_2[NO_3][NO] = \frac{k_2 k_1}{k_{-1}}[NO]^2[O_2]$$

This mechanism is also consistent with both the stoichiometry and the rate law. At this stage, we cannot tell which one of the above two plausible mechanisms is the actual reaction path. A plausible mechanism only means that it is not ruled out by kinetics. To determine the actual reaction mechanism, ultrafast reaction dynamic experiments to capture the signal of real reaction intermediates are necessary. Those experiments have captured N_2O_2 as the reaction intermediate. Therefore, the first plausible mechanism is proved by experiments to be the actual mechanism.

The Steady-State Approximation

In complex multi-step reaction mechanisms, more than one rate-determining step may exist and more than one reaction intermediates may be formed. Those reaction intermediates appear neither in the overall reaction equation nor in the experimental overall rate law. They must reach a steady-state condition in which their consuming rate equals their formation rate. This is called the **steady-state approximation** (SSA). SSA can be used in chemical kinetics to derive the overall rate law from a proposed mechanism, especially when no rate-determining step can be identified.

For example, let us reconsider the first plausible mechanism for the reaction of NO and O_2, but this time we do not know which step is fast and which one is slow. The proposed mechanism is

$$(1)\ NO(g) + NO(g) \xrightarrow{k_1} N_2O_2(g) \quad R_1 = k_1[NO]^2$$
$$(-1)\ N_2O_2(g) \xrightarrow{k_{-1}} NO(g) + NO(g) \quad R_{-1} = k_{-1}[N_2O_2]$$
$$(2)\ N_2O_2(g) + O_2(g) \xrightarrow{k_2} 2NO_2(g) \quad R_2 = k_2[N_2O_2][O_2]$$

For clarity, the equilibrium step is written as two separate steps.

The intermediate N_2O_2 is produced in step (1) and consumed in both step (−1) and step (2). Its formation rate is given by

$$R_f(N_2O_2) = R_1 = k_1[NO]^2$$

and its consumption rate is given by

$$R_c(N_2O_2) = R_{-1} + R_2 = k_{-1}[N_2O_2] + k_2[N_2O_2][O_2]$$

Using SSA, $R_f(N_2O_2) = R_c(N_2O_2)$ and

$$k_1[NO]^2 = k_{-1}[N_2O_2] + k_2[N_2O_2][O_2] = [N_2O_2](k_{-1} + k_2[O_2])$$

Rearranging to solve for $[N_2O_2]$, we have

$$[N_2O_2] = \frac{k_1[NO]^2}{k_{-1} + k_2[O_2]}$$

Both the overall reaction and step (2) involve the consumption of O_2, therefore,

$$R_{rxn} = -\frac{d[O_2]}{dt} = R_2 = k_2[N_2O_2][O_2] = \frac{k_1 k_2[NO]^2[O_2]}{k_{-1} + k_2[O_2]} \qquad (4.29)$$

This is the rate law for the proposed mechanism based on SSA. This rate law is more complicated in its form than the experimentally determined rate law given in **Equation (4.26)**, because no assumptions about the relative rates of the steps are made. If we now assume that the rate of step (−1) is much faster than that of step (2), then

$$R_{-1} = k_{-1}[N_2O_2] \gg R_2 = k_2[N_2O_2][O_2]$$

which means

$$k_{-1} \gg k_2[O_2] \quad \text{or} \quad k_{-1} + k_2[O_2] \approx k_{-1}$$

Therefore,

间体的信号。该实验已经捕获了 N_2O_2 作为反应中间体，因此，第一种合理的机理被实验证明是实际机理。

稳态近似

在复杂的多步反应机理中，可能存在不止一个决速步，也可能形成不止一个反应中间体。这些反应中间体既不出现在总反应方程式中，也不出现在总实验速率方程中。它们必须达到一个稳态条件，即其消耗速率等于生成速率，这就是所谓的**稳态近似**（SSA）。在化学动力学中，稳态近似法可用于从拟定机理推导出总速率方程，尤其是在无法确定决速步的情况下。

例如，让我们重新考虑 NO 和 O_2 反应的第一种合理的机理，但这次我们并不知道哪一步快、哪一步慢。拟定机理为

$$(1)\ NO(g) + NO(g) \xrightarrow{k_1} N_2O_2(g) \quad R_1 = k_1[NO]^2$$

$$(-1)\ N_2O_2(g) \xrightarrow{k_{-1}} NO(g) + NO(g) \quad R_{-1} = k_{-1}[N_2O_2]$$

$$(2)\ N_2O_2(g) + O_2(g) \xrightarrow{k_2} 2NO_2(g) \quad R_2 = k_2[N_2O_2][O_2]$$

为清楚起见，将平衡反应改写成两个独立的步骤（1）和（−1）。

中间体 N_2O_2 在步骤（1）中生成，在步骤（−1）和步骤（2）中消耗。其生成速率为

$$R_{生成}(N_2O_2) = R_1 = k_1[NO]^2$$

其消耗速率为

$$R_{消耗}(N_2O_2) = R_{-1} + R_2 = k_{-1}[N_2O_2] + k_2[N_2O_2][O_2]$$

使用稳态近似，$R_{生成}(N_2O_2) = R_{消耗}(N_2O_2)$，有

$$k_1[NO]^2 = k_{-1}[N_2O_2] + k_2[N_2O_2][O_2] = [N_2O_2](k_{-1} + k_2[O_2])$$

重排以解出 $[N_2O_2]$，有

$$[N_2O_2] = \frac{k_1[NO]^2}{k_{-1} + k_2[O_2]}$$

总反应和步骤（2）均涉及 O_2 的消耗，故

$$R_{反应} = -\frac{d[O_2]}{dt} = R_2 = k_2[N_2O_2][O_2] = \frac{k_1 k_2 [NO]^2 [O_2]}{k_{-1} + k_2[O_2]} \quad (4.29)$$

这就是基于稳态近似、由拟定机理推导出的速率方程。该方程的形式比**式（4.26）**给出的实验确定的速率方程更为复杂，因为并没有对各步骤的相对速率作出假定。如果我们现在假定步骤（−1）的速率比步骤（2）快得多，则

$$R_{-1} = k_{-1}[N_2O_2] \gg R_2 = k_2[N_2O_2][O_2]$$

这意味着

$$k_{-1} \gg k_2[O_2] \quad 即 \quad k_{-1} + k_2[O_2] \approx k_{-1}$$

因此

$$R_{反应} = \frac{k_1 k_2 [NO]^2 [O_2]}{k_{-1} + k_2[O_2]} \approx \frac{k_1 k_2}{k_{-1}}[NO]^2[O_2] = k[NO]^2[O_2]$$

该结论与**式（4.26）**和**（4.27）**相同。我们将在 **4.7 节**看到另一个稳态近似的例子。

$$R_{rxn} = \frac{k_1 k_2 [NO]^2 [O_2]}{k_{-1} + k_2 [O_2]} \approx \frac{k_1 k_2}{k_{-1}} [NO]^2 [O_2] = k[NO]^2 [O_2]$$

This conclusion is the same as those in **Equations (4.26)** and **(4.27)**. We will see another example of SSA in the next section.

4.7 Catalysis and Catalytic Chemistry

According to the rate law, the rate of reaction is determined by the concentrations of reactants, the orders of reactants, and the rate constant that is a function of temperature. For a given reaction, if both the concentrations of reactants and temperature are fixed, will the rate of reaction remain fixed for sure? No, the reaction rate can still be altered by using a catalyst. A **catalyst** is a substance that participates in a chemical reaction and changes its reaction rate but does not itself undergo a permanent change. The catalyst changes the reaction rate by providing an alternative reaction pathway of different activation energy. The effect of changing the reaction rate of a catalyst is called **catalysis**, and a reaction that involves a catalyst is called a **catalytical reaction**.

Catalysts can be classified into three types according to their effects on the reaction rate:
1) Positive catalysts that can speed up a reaction. If not specified, catalysts often refer to positive catalysts.
2) Negative catalysts that can slow down a reaction, also called anticatalysts.
3) Co-catalysts that do not change the reaction rate by themselves, but can improve the catalytic performance of the catalysts.

Characteristics of Catalysts

The amount, composition, and chemical properties of catalysts do not change after the reaction, but their physical shapes often vary since catalysts do participate in the reaction. Catalysts shorten the time required to reach an equilibrium by taking an alternative reaction pathway with lower activation energies for both forward and reverse reactions to increase their reaction rates to the same extent. With lower activation energies, catalysts can also make the reaction happen at milder conditions such as lower temperatures or pressures. Catalysts do not alter the equilibrium conditions of a reaction, nor can they realize reactions that are ruled out by thermochemistry.

Catalysts are highly selective in two main aspects: one is that certain reactions can only be catalyzed by certain catalysts, and the other is that different products may form from the same reactants when different catalysts are used. The performance of catalysts can be greatly reduced by adding a small amount of impurities that occupy the active sites, and this is called the poisoning of catalysts. The poisoned catalysts can be regenerated to recover their performance by removing the impurities from the active sites.

Homogeneous Catalysis

Homogeneous catalysis is a type of catalysis that involves a catalyst in the same phase (usually liquid phase or gaseous phase) as the reactants. Recall in **Section 2.7** that a phase is a region in a system throughout which all physical and chemical properties of a material are essentially uniform. Here, phase refers not only to solid, liquid, and gaseous states, but also to immiscible liquids, such as oil and water. There is no interface in a homogeneous catalytical reaction, so the efficiency and selectivity of catalysts are usually high.

The acid-catalyzed decomposition of ester is a typical homogeneous catalytical reaction, in which the reactants, products, and catalysts are all in the liquid or solution phase. Homogeneous catalysts increase the reaction rate by taking an alternative reaction pathway with a lower activation energy.

The activation energy for the decomposition of acetaldehyde (CH_3CHO) at 518°C is about 190 kJ mol^{-1}, given as

$$CH_3CHO(g) \rightarrow CH_4(g) + CO(g) \quad E_a = 190 \text{ kJ mol}^{-1}$$

4.7 催化作用与催化化学

根据速率方程，反应速率由反应物浓度、反应物级数以及速率常数决定，其中速率常数是温度的函数。对于一个指定反应，如果反应物浓度和温度均固定，那么反应速率一定会保持不变吗？答案是不一定，反应速率仍然可以通过使用催化剂来改变。**催化剂**是一种参与化学反应并改变其反应速率的物质，但其本身不会发生永久性变化。催化剂通过提供具有不同活化能的替代反应途径来改变反应速率。通过催化剂改变反应速率的效果称为**催化作用**，涉及催化剂的反应称为**催化反应**。

根据对反应速率的影响，催化剂可分为三类：
1) 正催化剂：可加快反应速率。如果没有特别说明，催化剂通常指正催化剂。
2) 负催化剂：减慢反应速率，又称阻化剂。
3) 助催化剂：自身不改变反应速率，但可以提高催化剂的催化性能。

催化剂的特性

催化剂的物质的量、组成和化学性质在反应后并不改变，但由于催化剂确实参与了反应，其物理形状往往会发生变化。催化剂对正、逆反应均采取具有较低活化能的替代反应途径，同等程度地提高其反应速率，由此来缩短达到平衡所需的时间。由于活化能较低，催化剂还可使反应在较为温和的条件（如较低的温度或压强）下发生。催化剂不会改变反应的平衡条件，也不能实现热力学所排除的反应。

催化剂具有高选择性，主要体现在两个方面：一是特定反应具有特定的催化剂；二是使用不同催化剂时，相同的反应物可能形成不同的生成物。当加入少量杂质占据活性位点后，会极大地降低催化剂的性能，这称为催化剂中毒。中毒的催化剂可通过去除活性位点上的杂质进行再生，以恢复其性能。

均相催化

均相催化是催化剂与反应物处于同一相（通常为液相或气相）的一种催化类型。回顾 **2.7 节**，相是体系中物理和化学性质完全均匀一致的部分。这里相不仅指固态、液态和气态，还指不互溶的液体，如油和水。均相催化反应没有相界面，因此催化剂的效率和选择性通常很高。

酸催化的酯分解是一类典型的均相催化反应，其中反应物、生成物和催化剂均处于液相（或溶液相）。均相催化剂通过采取具有较低活化能的替代反应途径来提高反应速率。

在 518°C 下乙醛（CH_3CHO）分解的活化能约为 190 kJ mol^{-1}，即

$$CH_3CHO(g) \to CH_4(g) + CO(g) \quad E_a = 190 \text{ kJ mol}^{-1}$$

如果使用 I_2 作为均相催化剂，活化能可降低至 136 kJ mol^{-1}，这是因为存在中间体为 CH_3I 的如下替代反应途径：

(1) $CH_3CHO(g) + I_2(g) \to CH_3I(g) + HI(g) + CO$

(2) $CH_3I(g) + HI(g) \to CH_4(g) + I_2(g)$

If I_2 is used as the homogeneous catalyst, the activation energy decreases to 136 kJ mol^{-1} because an alternative reaction pathway with an intermediate CH_3I is involved, as

(1) $CH_3CHO(g) + I_2(g) \rightarrow CH_3I(g) + HI(g) + CO(g)$

(2) $CH_3I(g) + HI(g) \rightarrow CH_4(g) + I_2(g)$

Overall: $CH_3CHO(g) \xrightarrow{I_2} CH_4(g) + CO(g) \quad E_a' = 136 \text{ kJ mol}^{-1}$

Using Arrhenius equation and assuming that the preexponential factors are the same, we can calculate that

$$\ln\frac{R'}{R} = \ln\frac{k'}{k} = \frac{E_a' - E_a}{RT} = 8.21$$

$$\therefore R' : R = 3.7 \times 10^3$$

The reaction rate with the I_2 catalyst can be about 3700 times the uncatalyzed rate.

Heterogeneous Catalysis

Heterogeneous catalysis is a type of catalysis in which the phase of the catalyst differs from that of the reactants or products. Many heterogeneous reactions are catalyzed by allowing the reactant molecules, usually in gaseous or solution phase, to adsorb on the appropriate solid surface of the catalyst, and essential reaction intermediates are found on the surface. Not all surface atoms are equally effective for catalysis. Those catalytically effective surface atoms are called **active sites**. Many transition elements and their compounds show excellent catalytic activities. The precise mechanism of heterogeneous catalysis is not totally understood yet, but in many cases the availability of electrons in d orbitals in active sites may play an important role. One of the advantages of heterogeneous catalysis is that it is easy to separate the catalysts from the products since they are in different phases.

Basically, heterogeneous catalysis involves the following four steps:
1) Adsorption of the reactants on the surface;
2) Diffusion of the reactants along the surface;
3) Reaction at an active site to form adsorbed products;
4) Desorption of the products from the surface.

Enzyme Catalysis

In addition to homogeneous and heterogeneous catalyses, **enzyme catalysis** is usually regarded as a third, separate category of catalysis because enzymes possess partial properties of both homogeneous and heterogeneous catalysts. **Enzymes** are large biological molecules responsible for the metabolic processes that sustain life. **Figure 4.8** shows the ribbon diagram of the enzyme human glyoxalase I with two purple spheres representing its two catalytic zinc ions. The characteristics of enzyme catalysis include high selectivity, high efficiency, and operating in human body conditions (neutral mild conditions).

A so-called **"lock-and-key" model** is used to describe the enzyme activity. It was first proposed by Emil Fischer (1852—1919) in 1894. The molecules upon which enzymes may act are called **substrates** (S). In this model, the substrate attaches to the enzyme (E) at an active site to form the enzyme-substrate complex (ES). The complex then decomposes to form the products (P) and the enzyme is regenerated.

$$E + S \underset{k_{-1}}{\overset{k_1}{\rightleftharpoons}} ES$$

$$ES \xrightarrow{k_2} E + P$$

The rate of the overall reaction is given by

$$R_{rxn} = \frac{d[P]}{dt} = k_2[ES]$$

Applying the SSA to the intermediate ES, we have

$$R_f(ES) = k_1[E][S] = R_c(ES) = (k_{-1} + k_2)[ES] \tag{4.30}$$

总反应：$CH_3CHO(g) \xrightarrow{I_2} CH_4(g) + CO(g)$ $E_a' = 136$ kJ mol^{-1}

使用阿伦尼乌斯方程，并假定指前因子相同，可计算得

$$\ln \frac{R'}{R} = \ln \frac{k'}{k} = \frac{E_a' - E_a}{RT} = 8.21$$

$$\therefore R' : R = 3.7 \times 10^3$$

在 I_2 催化剂的作用下，其反应速率约为无催化剂时速率的 3700 倍。

异相催化

异相催化也称非均相催化，是催化剂与反应物或生成物不处于同一相的一种催化类型。许多异相反应都是将反应物分子（通常处于气相或溶液相）吸附在固体催化剂适当的表面上来进行催化，而关键的反应中间体也在该表面上发现。并非所有的表面原子对催化作用都同等有效，那些具有有效催化作用的表面原子称为**活性位点**。许多过渡元素及其化合物均表现出优异的催化活性。异相催化的确切机理目前尚不完全清楚，但在许多情况下，活性位点上存在可用的 d 轨道电子可能具有重要作用。异相催化的优势之一在于，由于催化剂和生成物处于不同相，因此很容易将其从生成物中分离出来。

异相催化基本包括以下四个步骤：
1) 反应物在表面吸附；
2) 反应物沿表面扩散；
3) 在活性位点发生反应，形成吸附的生成物；
4) 生成物从表面脱附。

酶催化

在均相和异相催化之外，**酶催化**通常可视为第三类、独立的催化类型，因为酶同时具有均相和异相催化剂的部分特性。**酶**是一类维持生命代谢过程的生物大分子。**图 4.8** 是人类乙二醛酶 I 的缎带图，其中两个紫色小球代表两个具有催化活性的锌离子。酶催化的特性包括高选择性、高效性以及在人体条件（中性温和条件）下进行催化。

酶的活性常用"**锁钥**"模型来描述，该模型由埃米尔·菲舍尔（1852—1919）在 1894 年首次提出。酶所作用的分子称为**底物**（S）。在"锁钥"模型中，底物附着在酶（E）的活性位点上，形成酶-底物复合物（ES），然后该复合物分解形成生成物（P），而酶则被再生。

$$E + S \underset{k_{-1}}{\overset{k_1}{\rightleftharpoons}} ES$$

$$ES \xrightarrow{k_2} E + P$$

总反应速率为

$$R_{反应} = \frac{d[P]}{dt} = k_2 [ES]$$

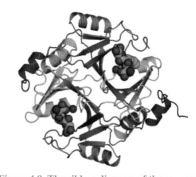

Figure 4.8 The ribbon diagram of the enzyme human glyoxalase I with two purple spheres representing its two catalytic zinc ions.

图 4.8 人类乙二醛酶 I 的缎带图，其中两个紫色小球代表两个具有催化活性的锌离子。

对中间体 ES 应用稳态近似，有

$$R_{生成}(ES) = k_1 [E][S] = R_{消耗}(ES) = (k_{-1} + k_2)[ES] \quad (4.30)$$

式（4.30）的 [ES] 可解得

$$[ES] = \frac{k_1 [E][S]}{k_{-1} + k_2}$$

尽管游离的酶的浓度 [E] 未知，但我们知道酶的总浓度 $[E]_0$，由物料守

We can solve **Equation (4.30)** for [ES]

$$[ES] = \frac{k_1[E][S]}{k_{-1} + k_2}$$

Although the concentration of free enzyme [E] is unknown, we do know the total concentration of enzyme $[E]_0$, which equals the sum of [E] and [ES] by material balance. Applying $[E] = [E]_0 - [ES]$ into the above equation, we can derive

$$[ES] = \frac{k_1[E]_0[S]}{(k_{-1} + k_2) + k_1[S]}$$

Therefore,

$$R_{rxn} = \frac{k_2 k_1 [E]_0 [S]}{(k_{-1} + k_2) + k_1[S]}$$

To make a more concise form, we can divide the numerator and denominator by k_1, as

$$R_{rxn} = \frac{k_2 [E]_0 [S]}{\left(\dfrac{k_{-1} + k_2}{k_1}\right) + [S]}$$

If we define $K_M = (k_{-1} + k_2)/k_1$, then

$$R_{rxn} = \frac{k_2 [E]_0 [S]}{K_M + [S]} \tag{4.31}$$

Experimentally, the typical effect of substrate concentration [S] on the rate of an enzyme catalytical reaction is shown in **Figure 4.9**. There is a limited concentration of active sites on the enzyme. At [S] lower than the concentration of active sites, the rate of reaction is proportional to [S], or in first-order. At [S] higher than the concentration of active sites, all active sites are saturated and the rate is independent of [S], or in zero-order. This behavior agrees well with the above derived rate law. If [S] is sufficiently low, we have

$$K_M \gg [S] \quad \text{or} \quad K_M + [S] \approx K_M$$

$$\therefore R_{rxn} = \frac{k_2}{K_M}[E]_0[S] \propto [S]$$

This rate law is first order in [S]. If [S] is sufficiently high, we get

$$K_M \ll [S] \quad \text{or} \quad K_M + [S] \approx [S]$$

$$\therefore R_{rxn} = k_2[E]_0 \propto [S]^0$$

The reaction rate is constant and is the maximum rate attainable for the particular enzyme due to the saturation of active sites. This is an example of the scientific method that we have learnt in **Section 1.1**, in which the postulated mechanisms are continually tested by subsequent experiments and modified when necessary.

Extended Reading Materials — Reaction Dynamics and Femtochemistry

Reaction dynamics is a branch of chemical kinetics that studies the process of individual chemical events on atomic scales and over very brief time periods. It considers state-to-state kinetics between reactant and product molecules in specific quantum states, and how energy is distributed among various states. The 1986 Nobel Prize in chemistry was awarded to Dudley R. Herschbach (1932—), Yuan T. Lee (1936—), and John C. Polanyi (1929—) "for their contributions concerning the dynamics of chemical elementary processes". Herschbach and Lee designed a crossed molecular beam (**Figure 4.10**) experiment, in which narrow beams of reactant molecules in selected quantum states are allowed to react in order to determine the reaction probability as a function of various quantum states of the reactant molecules.

衡可知，其等于 [E] 和 [ES] 之和。将 [E] = [E]$_0$ − [ES] 应用于上式，可得

$$[ES] = \frac{k_1[E]_0[S]}{(k_{-1}+k_2)+k_1[S]}$$

故

$$R_{反应} = \frac{k_2 k_1[E]_0[S]}{(k_{-1}+k_2)+k_1[S]}$$

为使形式更简明，可将分子和分母均除以 k_1，有

$$R_{反应} = \frac{k_2[E]_0[S]}{\left(\dfrac{k_{-1}+k_2}{k_1}\right)+[S]}$$

如果定义 $K_M = (k_{-1}+k_2)/k_1$，则

$$R_{反应} = \frac{k_2[E]_0[S]}{K_M+[S]} \tag{4.31}$$

实验上底物浓度 [S] 对酶催化反应速率的典型影响如**图 4.9** 所示。酶上活性位点的浓度有限，当 [S] 低于活性位点浓度时，反应速率与 [S] 成正比，即为一级反应。当 [S] 高于活性位点浓度时，所有活性位点均饱和，则速率与 [S] 无关，即为零级反应。这种行为与以上推导出的速率方程非常吻合。如果 [S] 足够小，有

$$K_M \gg [S] \quad 或 \quad K_M+[S] \approx K_M$$
$$\therefore R_{反应} = \frac{k_2}{K_M}[E]_0[S] \propto [S]$$

该速率方程对 [S] 而言是一级反应。如果 [S] 足够大，可得

$$K_M \ll [S] \quad 或 \quad K_M+[S] \approx [S]$$
$$\therefore R_{反应} = k_2[E]_0 \propto [S]^0$$

该反应速率为定值，是特定的酶由于活性位点饱和所能达到的最大速率。这是我们在 **1.1 节**学到的科学方法的一个示例，其中假定的机理不断地被后续实验所检验，并在必要时进行修正。

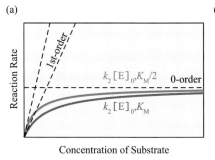
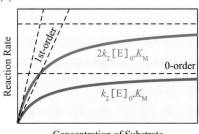
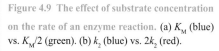

Figure 4.9 The effect of substrate concentration on the rate of an enzyme reaction. (a) K_M (blue) vs. $K_M/2$ (green). (b) k_2 (blue) vs. $2k_2$ (red).

图 4.9 底物浓度对酶反应速率的影响：(a) K_M（蓝线）与 $K_M/2$（绿线）的对比；(b) k_2（蓝线）与 $2k_2$（红线）的对比。

拓展阅读材料　反应微观动力学与飞秒化学

反应微观动力学是化学动力学的一个分支，研究的是原子尺度上以及极短时间内单个化学事件的过程。它考虑的是处于特定量子态下的反应物和生成物分子之间的态间动力学，以及能量在不同态之间的分布。1986 年诺贝尔化学奖授予了杜德利·R. 赫斯巴赫（1932— ）、李远哲（1936— ）和约翰·C. 波兰尼（1929— ），

Spectroscopic observation of reaction dynamics on extremely short timescales (usually in fs scale, 1 fs= 10^{-15} s) is known as **femtochemistry**. The steps in some reactions occur in the fs to as (1 as = 10^{-18} s) timescales, and will sometimes form reaction intermediates. Femtochemistry allows exploration of which chemical reactions take place, and investigates why some reactions occur but not others. The 1999 Nobel Prize in chemistry was awarded to Ahmed H. Zewail (1946—   ), who was recognized as the "father of femtochemistry". In addition, theoretical studies of reaction dynamics involve calculating the reaction PES as a function of reaction coordinates, and then calculating the trajectory of a point on this surface representing the state of the system, etc.

Problems

4.1   Data for the reaction A → products are listed in **Table P4.1**. Use these data to determine (a) the order of the reaction; (b) the rate constant, k; (c) the rate of the reaction at $t = 3.5$ min, from the rate law; (d) the rate of the reaction at $t = 3.5$ min, from the slope of the tangent line in the $[A]_t$ vs. t plot; (e) the initial rate of the reaction.

Table P4.1 （表 P4.1）

t/min	$[A]_t$/(mol L^{-1})	t/(min)	$[A]_t$/(mol L^{-1})
0	1.00	3	0.41
1	0.74	4	0.30
2	0.55	5	0.22

4.2   Hydroxide ion is involved in the mechanism of the following reaction but is not consumed in the overall reaction.

$$OCl^- + I^- \xrightarrow{OH^-} OI^- + Cl^-$$

(a) From the data given in **Table P4.2**, determine the order of the reaction with respect to OCl^-, I^-, and OH^-.

(b) What is the overall reaction order?

(c) Write the rate law, and determine the value of the rate constant, k.

Table P4.2 （表 P4.2）

$[OCl^-]$/(mol L^{-1})	$[I^-]$/(mol L^{-1})	$[OH^-]$/(mol L^{-1})	Formation Rate of OI^- （OI^- 的生成速率） /(mol L^{-1} s^{-1})
0.0040	0.0020	1.00	4.8×10^{-4}
0.0020	0.0040	1.00	5.0×10^{-4}
0.0020	0.0020	1.00	2.4×10^{-4}
0.0020	0.0020	0.50	4.6×10^{-4}
0.0020	0.0020	0.25	9.4×10^{-4}

4.3   The half-life for the first-order decomposition of nitramide, $NH_2NO_2(aq) \rightarrow N_2O(g) + H_2O(l)$, is 123 min at 15°C. If 165 mL of a 0.105 mol L^{-1} NH_2NO_2 solution is allowed to decompose, how long must the reaction proceed to yield 50.0 mL of $N_2O(g)$ collected over water at 15°C? Given that the barometric pressure is 756 mmHg, and the vapor pressure of water is 12.8 mmHg at 15°C.

4.4   Data in **Table P4.3** are given for the reaction 2A + B → products. Establish the order of this reaction with respect to A and to B.

Table P4.3 （表 P4.3）

Expt. 1 （实验 1），[B] = 1.00 mol L^{-1}		Expt.2 （实验 2），[B] = 0.50 mol L^{-1}	
Time （时间）/min	[A]/(mol L^{-1})	Time （时间）/min	[A]/(mol L^{-1})
0	1.000×10^{-3}	0	1.000×10^{-3}
1	0.951×10^{-3}	1	0.975×10^{-3}
5	0.779×10^{-3}	5	0.883×10^{-3}
10	0.607×10^{-3}	10	0.779×10^{-3}
20	0.368×10^{-3}	20	0.607×10^{-3}

以"表彰他们在化学基元过程的微观动力学方面的贡献"。赫斯巴赫和李远哲设计了一个交叉分子束（**图 4.10**）实验，在该实验中，反应发生在处于选定量子态的反应物分子狭束之间，用来确定反应概率与反应物分子的各种量子态之间的函数关系。

在极短时间尺度上（通常为飞秒 fs 尺度，1 fs=10^{-15} s）对反应微观动力学的光谱观测称为**飞秒化学**。某些反应步骤发生在从飞秒到阿秒（1 as = 10^{-18} s）的时间尺度上，有时会形成反应中间体。飞秒化学可以探索哪些化学反应会发生，并研究为什么某些反应发生而其他反应不发生。1999 年诺贝尔化学奖授予了艾哈迈德·H. 泽维尔（1946— ），他被公认为"飞秒化学之父"。此外，反应微观动力学的理论研究涉及计算反应势能面与反应坐标之间的函数关系，以及计算势能面上代表体系状态的点的运行轨迹等。

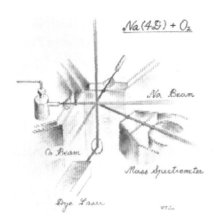

Figure 4.10 The crossed molecular beam apparatus designed by Herschbach and Lee.

图 4.10 赫斯巴赫与李远哲设计的交叉分子束装置。

习题

4.1 **表 P4.1** 列出了反应 A→ 生成物的数据。根据这些数据，试确定：(a) 反应的级数；(b) 速率常数 k；(c) 由速率方程得到的 t = 3.5 min 时的反应速率；(d) 由 $[A]_t$ 对 t 作图的切线斜率得到的 t = 3.5 min 时的反应速率；(e) 反应的初始速率。

4.2 氢氧根离子参与如下反应机理，但不会在整个反应中消耗。

$$OCl^- + I^- \xrightarrow{OH^-} OI^- + Cl^-$$

(a) 根据**表 P4.2**给出的数据，确定反应对 OCl^-、I^- 和 OH^- 的级数。
(b) 总反应级数是多少？
(c) 写出速率方程，并确定速率常数 k 的值。

4.3 硝酰胺分解为一级反应：

$$NH_2NO_2(aq) \rightarrow N_2O(g) + H_2O(l)$$

15°C 时半衰期为 123 min。如果 165 mL 0.105 mol L^{-1} NH_2NO_2 溶液发生分解，在 15°C 反应要持续多久，才能用排水集气法收集到 50.0 mL $N_2O(g)$？已知大气压为 756 mmHg，15°C 时水的蒸气压为 12.8 mmHg。

4.4 反应 2A+B→ 生成物的数据见**表 P4.3**，确定该反应对 A 和 B 的级数。

4.5 证明以下机理符合 **4.2 题**得到的碘化物-次氯酸盐反应的速率方程。

快：$OCl^- + H_2O \underset{k_{-1}}{\overset{k_1}{\rightleftharpoons}} HOCl + OH^-$

慢：$I^- + HOCl \xrightarrow{k_2} HOI + Cl^-$

快：$HOI + OH^- \underset{k_{-3}}{\overset{k_3}{\rightleftharpoons}} H_2O + OI^-$

4.6 推导水溶液中以下反应合理的机理：

$$Hg_2^{2+} + Tl^{3+} \rightarrow 2Hg^{2+} + Tl^+$$

实验观察到的速率方程为

$$R_{反应} = k\,[Hg_2^{2+}]\,[Tl^{3+}] / [Hg^{2+}]$$

4.7 对于氯气和氯仿的反应，提出了以下三步反应机理：

(1) $Cl_2(g) \underset{k_{-1}}{\overset{k_1}{\rightleftharpoons}} 2Cl(g)$

4.5 Show that the following mechanism is consistent with the rate law established for the iodide-hypochlorite reaction in **Problem 4.2**.

$$\text{Fast:} \quad OCl^- + H_2O \underset{k_{-1}}{\overset{k_1}{\rightleftharpoons}} HOCl + OH^-$$

$$\text{Slow:} \quad I^- + HOCl \xrightarrow{k_2} HOI + Cl^-$$

$$\text{Fast:} \quad HOI + OH^- \underset{k_{-3}}{\overset{k_3}{\rightleftharpoons}} H_2O + OI^-$$

4.6 Derive a plausible mechanism for the following reaction in aqueous solution, $Hg_2^{2+} + Tl^{3+} \to 2Hg^{2+} + Tl^+$, for which the observed rate law is: $R_{rxn} = k\,[Hg_2^{2+}][Tl^{3+}]/[Hg^{2+}]$.

4.7 The following three-step mechanism has been proposed for the reaction of chlorine and chloroform.

$$(1) \quad Cl_2(g) \underset{k_{-1}}{\overset{k_1}{\rightleftharpoons}} 2Cl(g)$$

$$(2) \quad Cl(g) + CHCl_3(g) \xrightarrow{k_2} HCl(g) + CCl_3(g)$$

$$(3) \quad CCl_3(g) + Cl(g) \xrightarrow{k_3} CCl_4(g)$$

The numerical values of the rate constants for these steps are $k_1 = 4.8\times10^3$, $k_{-1} = 3.6\times10^3$; $k_2 = 1.3\times10^{-2}$; $k_3 = 2.7\times10^2$. Derive the rate law and the magnitude of k for the overall reaction.

*4.8 The reaction A + B → products is first order in A, first order in B, and second order overall. Given that the starting concentrations of the reactants are $[A]_0$ and $[B]_0$, and the concentrations of the reactants at the time t are $[A]_t$ and $[B]_t$, derive the integrated rate law for this reaction in terms of $[A]_0$, $[B]_0$, $[A]_t$, $[B]_t$, t, and k, where k is the rate constant for this reaction.

*4.9 The rate of an enzyme-catalyzed reaction can be slowed down by the presence of an inhibitor (I) that reacts with the enzyme in a rapid equilibrium process. By adding this step to the mechanism for enzyme catalysis in **Section 4.7**, determine the rate of the enzyme-catalyzed reaction with inhibitor in terms of $[E]_0$, $[S]$, $[I]$, and various k as listed below:

$$E + S \underset{k_{-1}}{\overset{k_1}{\rightleftharpoons}} ES \xrightarrow{k_2} E + P$$

$$E + I \underset{k_{-3}}{\overset{k_3}{\rightleftharpoons}} EI$$

(2) $Cl(g) + CHCl_3(g) \xrightarrow{k_2} HCl(g) + CCl_3(g)$

(3) $CCl_3(g) + Cl(g) \xrightarrow{k_3} CCl_4(g)$

这些步骤的速率常数为 $k_1 = 4.8 \times 10^3$, $k_{-1} = 3.6 \times 10^3$; $k_2 = 1.3 \times 10^{-2}$; $k_3 = 2.7 \times 10^2$。试推导总反应的速率方程，并计算 k 的值。

*4.8 反应 A+B→ 生成物是 A 的一级反应、B 的一级反应，且是一个总二级反应。已知反应物的初始浓度为 $[A]_0$ 和 $[B]_0$，在时间 t 时反应物的浓度为 $[A]_t$ 和 $[B]_t$，试推导该反应的积分速率方程，以 $[A]_0$、$[B]_0$、$[A]_t$、$[B]_t$、t 及 k 的形式表示，其中 k 为该反应的速率常数。

*4.9 酶催化反应的速率可通过抑制剂（I）的存在来减慢，该抑制剂可与酶发生快平衡反应。将这一步骤添加到 **4.7 节** 的酶催化机理中，试确定抑制剂存在时酶催化反应的速率方程，以 $[E]_0$、$[S]$、$[I]$ 及各种 k 的形式表示，其中各种 k 的形式如下所示：

$$E + S \underset{k_{-1}}{\overset{k_1}{\rightleftharpoons}} ES \xrightarrow{k_2} E + P$$

$$E + I \underset{k_{-3}}{\overset{k_3}{\rightleftharpoons}} EI$$

Chapter 5 Acid-Base Equilibria and Precipitation-Dissolution Equilibria

In **Chapter 3**, we have learnt thermochemistry and chemical equilibrium. There are four general kinds of chemical equilibria: acid-base equilibrium, precipitation-dissolution equilibrium, oxidation-reduction equilibrium, and complex-ion equilibrium. We will discuss the first two kinds of chemical equilibria in this chapter. Both acid-base equilibrium and precipitation-dissolution equilibrium usually take place in solutions, the rates of reaction for which are rapid due to the relatively low activation energies (< 40 kJ mol^{-1}). Moreover, the effects of pressure and temperature on the equilibrium constants are normally negligible. Therefore, we will focus on the effect of concentrations on these two equilibria in this chapter.

5.1 Theory of Acids and Bases

5.2 Self-Ionization of Water and the pH Scale

5.3 Ionization Equilibria of Acids and Bases

5.4 Common-Ion Effect and Buffer Solutions

5.5 Acid-Base Indicators and Titration

5.6 Solubility Product Constant and Its Relationship with Solubility

5.7 Criteria for Precipitation and Its Completeness

5.8 Dissolution and Transformation of Precipitates

5.9 Fractional Precipitation

第 5 章　酸碱电离平衡与沉淀溶解平衡

第 3 章里我们学习了化学热力学和化学平衡。我们通常说的四大化学平衡包括酸碱电离平衡、沉淀溶解平衡、氧化还原平衡和配位解离平衡，本章我们将讨论前两类化学平衡。酸碱电离平衡与沉淀溶解平衡通常发生在溶液中，由于具有较低的活化能（< 40 kJ mol^{-1}），其反应速率较快，且压强和温度对这两类平衡常数的影响通常可忽略不计。因此，本章我们将重点讨论浓度对这两类平衡的影响。

5.1　酸碱理论

5.2　水的自耦电离与 pH

5.3　酸碱电离平衡

5.4　同离子效应与缓冲溶液

5.5　酸碱指示剂与酸碱滴定

5.6　溶度积常数及其与溶解度的关系

5.7　沉淀生成与沉淀完全的判据

5.8　沉淀的溶解与转化

5.9　分步沉淀

5.1 Theory of Acids and Bases

The understanding of acids and bases has developed gradually in the history of science. Several theories of acids and bases, including the definitions of acids and bases, the nature of acid-base reaction, etc., will be discussed in this section.

Arrhenius Theory of Acids and Bases

The first modern definition of acids and bases was proposed by Svante Arrhenius (1859—1927) in 1887. Arrhenius defined **acid** as a substance that dissociates in water to form hydrogen ion (H^+ or proton) and **base** as a substance that dissociates in water to form hydroxide ion (OH^-). Therefore, an acid increases $[H^+]$ and a base increases $[OH^-]$ in an aqueous solution. In this definition, the acid-base reaction can be considered as a **neutralization reaction**, written as

$$\text{Acid}(aq) + \text{Base}(aq) \rightleftharpoons \text{Salt}(aq) + H_2O(l) \tag{5.1}$$

where aq stands for aqueous. Sometimes (aq) are omitted for simplification purpose.

The limitations of Arrhenius theory of acids and bases are listed as follows:
1) It is restricted only to aqueous solutions and is not applicable to non-aqueous solutions.
2) It cannot handle non-H^+ acids such as BF_3 and non-OH^- bases such as NH_3.
3) It does not recognize the key role of solvent in ionization. For example, H^+ does not exist as a free species in an aqueous solution and should be written as hydronium ion H_3O^+.
4) The only **amphiprotic** substance, which means that it can act either as an acid or a base, is H_2O according to Arrhenius theory. It cannot explain the acid-base properties of certain salts, such as HCO_3^-.

Brønsted-Lowry Theory of Acids and Bases

In 1923, Johannes N. Brønsted (1879—1947) and Thomas M. Lowry (1874—1936) proposed independently a proton theory of acids and bases, also called Brønsted-Lowry theory of acids and bases. In this theory, acid is defined as a substance that can donate a proton (H^+) and base is a substance that can accept a proton. When an acid donates a proton, a **conjugate base** of this acid is formed, which can accept a proton to become the acid again, represented by

$$\text{Acid} \rightleftharpoons \text{Conjugate Base} + H^+$$

Similarly, when a base accepts a proton, its **conjugate acid** that can donate a proton to become this base is formed, as

$$\text{Base} + H^+ \rightleftharpoons \text{Conjugate Acid}$$

These conjugate acids and bases are called **conjugate pairs**. The acid-base reaction is a proton transfer reaction in ionization equilibrium between the conjugate pairs, shown as

$$\text{Acid 1} + \text{Base 2} \rightleftharpoons \text{Conjugate Acid 2} + \text{Conjugate Base 1}$$

or

$$H^+A + B \rightleftharpoons H^+B + A \tag{5.2}$$

where A is the conjugate base of acid H^+A and H^+B is the conjugate acid of base B. In this acid-base reaction, a proton is transferred from acid H^+A to base B. Therefore, the acid is a **proton donor** which is deprotonated and the base is a **proton acceptor** which is protonated in an acid-base reaction. Note that an acid cannot release a proton by itself; for an acid to donate a proton to become its conjugate base, there must always be a base that can accept this proton. Similarly, for a base to accept a proton, there must always be an acid that can release this proton.

The scope of acids and bases in Brønsted-Lowry theory is much wider than that in Arrhenius theory. They are extended to both aqueous and non-aqueous solutions. All Arrhenius acids are also Brønsted-Lowry acids since they can produce protons in aqueous solution. All Arrhenius bases are also Brønsted-Lowry bases since the OH^- they produce in aqueous solution can react with protons. Certain species, such as OCl^-, even though they do not form

5.1 酸碱理论

在科学史上，人类对酸碱的理解是逐渐发展起来的。本节我们将讨论几个酸碱理论，内容包括酸碱的定义以及酸碱反应的本质等。

阿伦尼乌斯酸碱理论

1887 年，斯文特·阿伦尼乌斯（1859—1927）首先提出酸碱的现代定义。阿伦尼乌斯将**酸**定义为溶解在水中能形成氢离子（H^+ 或质子）的物质，将**碱**定义为溶解在水中能形成氢氧根离子（OH^-）的物质。因此，酸和碱分别能增加水溶液中的 $[H^+]$ 和 $[OH^-]$。在该定义中，酸碱反应即为**中和反应**，可写为

$$酸(aq) + 碱(aq) \rightleftharpoons 盐(aq) + 水(l) \tag{5.1}$$

其中 aq 代表水溶液，有时为简化起见也经常省略。

阿伦尼乌斯酸碱理论具有如下局限性：
1) 仅限于水溶液，不包含非水溶液。
2) 不能解释非 H^+ 酸（如 BF_3）和非 OH^- 碱（如 NH_3）。
3) 没有揭示溶剂在电离中的关键作用，如 H^+ 在水溶液中不能单独存在，应写成水合氢离子 H_3O^+。
4) 既可作酸又可作碱的物质称为**两性物质**。阿伦尼乌斯酸碱理论中，水是唯一的两性物质，因此阿伦尼乌斯酸碱理论不能解释某些盐（如 HCO_3^-）的酸碱性。

布朗斯特 - 劳莱酸碱理论

1923 年，约翰尼斯·N. 布朗斯特（1879—1947）和托马斯·M. 劳莱（1874—1936）分别独立提出酸碱质子理论，又称布朗斯特 - 劳莱酸碱理论。在该理论中，凡是能给出质子（H^+）的物质称为酸，凡是能接受质子的物质称为碱。酸给出质子后形成其**共轭碱**，共轭碱接受质子后又变回酸，可表示为

$$酸 \rightleftharpoons 共轭碱 + H^+$$

类似地，碱接受质子后形成其**共轭酸**，共轭酸给出质子后又变回碱，可表示为

$$碱 + H^+ \rightleftharpoons 共轭酸$$

这些共轭的酸和碱称为**共轭酸碱对**。酸碱反应即为处于电离平衡的共轭酸碱对之间发生的质子转移反应，可表示为

$$酸1 + 碱2 \rightleftharpoons 共轭酸2 + 共轭碱1$$

或

$$H^+A + B \rightleftharpoons H^+B + A \tag{5.2}$$

其中 A 是酸 H^+A 的共轭碱，H^+B 是碱 B 的共轭酸。在此酸碱反应中，质子从酸 H^+A 转移到碱 B，因此酸为**质子给体**，碱为**质子受体**，在酸碱反应中酸被去质子化，而碱被质子化。注意酸自身并不能给出质子，必须有碱接受质子，酸才能给出质子变为其共轭碱。同样，必须有酸给出质子，碱才能接受质子。

布朗斯特 - 劳莱酸碱理论的酸碱范围比阿伦尼乌斯酸碱理论更

OH⁻, can accept protons in a reaction and thus are considered as Brønsted-Lowry bases but not Arrhenius bases.

The scope of amphiprotic substances in Brønsted-Lowry theory also extends from the only amphiprotic substance water in Arrhenius theory. For example, HCO_3^- can donate a proton to water to form CO_3^{2-} and H_3O^+ as

$$HCO_3^- + H_2O \rightleftharpoons H_3O^+ + CO_3^{2-}$$

Here, HCO_3^- serves as an acid and water is a base. On the other hand, HCO_3^- can also accept a proton from water to form H_2CO_3 and OH⁻ as

$$HCO_3^- + H_2O \rightleftharpoons OH^- + H_2CO_3$$

Here, HCO_3^- serves as a base and water is an acid. Therefore, HCO_3^- is amphiprotic according to Brønsted-Lowry theory. Whether the HCO_3^- solution is acidic or alkaline depends on the tendencies of the above two reactions, which will be discussed later in **Section 5.3**.

Acids and bases can be classified into strong and weak according to the ability that they can donate or accept the protons, or the degree that they can react with the solvent and ionize. Strong acids and bases ionize nearly completely, with a very large ionization constant that is usually greater than 10^6. Together with the fact that the ionization rate is quite rapid, strong acids and bases can be treated as though they were completely ions initially. For example, HCl is a strong acid with an ionization equation as

$$HCl + H_2O \rightarrow H_3O^+ + Cl^- \quad K_a = \frac{[H_3O^+][Cl^-]}{[HCl]} > 10^6 \tag{5.3}$$

where K_a is the **acid ionization constant**. An arrow is used in the above chemical equation instead of an equilibrium symbol, because the forward reaction proceeds almost to completion and the reverse reaction does not occur to any significant extent. Cl⁻, the conjugate base of the strong acid HCl, is then a very weak base that almost does not react with water to hydrolysis (the reverse of the above reaction). The stronger the acid, the weaker its conjugate base, and vice versa.

Weak acids and bases only ionize partially, forming an ionization equilibrium. For example, NH_3 is a weak base with an ionization equilibrium as

$$NH_3 + H_2O \rightleftharpoons NH_4^+ + OH^- \quad K_b = \frac{[NH_4^+][OH^-]}{[NH_3]} = 1.75 \times 10^{-5} \tag{5.4}$$

where K_b is the **base ionization constant**. The values of K_a or K_b can be used to help determine the relative strengths of weak acids and bases. Only a small percentage of weak acid and base molecules will ionize, leading to a degree of ionization much less than 1. The calculations between acid/base ionization constants and degree of ionization can be found in **Section 5.3**.

The relative strength of acids and bases depends not only on the acids and bases themselves but also on the solvents. A substance may behave differently in different solvents. For example, acetic acid HAc behaves as a weak acid in aqueous solution. It behaves as a relatively strong acid in liquid ammonia NH_3 solution since NH_3 is a stronger base that can accept a proton more easily than water. In HF solution, however, HAc behaves as a base because HF ($K_a = 6.3 \times 10^{-4}$) is a stronger acid than HAc ($K_a = 1.75 \times 10^{-5}$). HAc will accept a proton to form H_2Ac^+ in HF solution.

Water is the most commonly used solvent, and if not specified, it is the default solvent throughout the entire textbook. H_3O^+ and OH⁻ are the strongest acid and base existed in water, respectively, because stronger acids and bases will react immediately with water to form H_3O^+ and OH⁻. This is termed as the **leveling effect** of solvent. For example, both $HClO_4$ and HCl ionize completely in aqueous solution, so their acidity is all leveled down to H_3O^+. In order to differentiate that $HClO_4$ is a stronger acid than HCl, we should use a weaker base than water such as diethyl ether $C_2H_5OC_2H_5$ as the solvent, in which $HClO_4$ is completely ionized but HCl is only partially ionized. This solvent is then called the **differentiating solvent**.

Table 5.1 and **Appendix C.2** list the relative strengths and K_a values of some conjugate pairs in aqueous solution. The relative acidity of the acids on the left decreases from top to bottom and the alkalinity of the corresponding conjugate bases on the right increases consequently. The acids above H_3O^+ are all very strong acids that ionize nearly completely and cannot exist in their molecular form in water. Their conjugate bases are extremely weak. Those between H_3O^+ and H_2O are weak acids that can exist in water, descending in their acidity.

为宽泛，同时包含了水溶液和非水溶液。所有阿伦尼乌斯酸都是布朗斯特-劳莱酸，因为它们均能在水溶液中给出质子。所有阿伦尼乌斯碱都是布朗斯特-劳莱碱，因为它们在水溶液中形成的 OH^- 能与质子结合。一些物种（如 OCl^-）虽然不能形成 OH^-，但在反应中能接受质子，因此是布朗斯特-劳莱碱，却不是阿伦尼乌斯碱。

布朗斯特-劳莱酸碱理论的两性物质范围也比阿伦尼乌斯酸碱理论（水是唯一的两性物质）更为宽泛。例如，在水中 HCO_3^- 能给出质子生成 CO_3^{2-} 和 H_3O^+，有

$$HCO_3^- + H_2O \rightleftharpoons H_3O^+ + CO_3^{2-}$$

这里 HCO_3^- 是酸而水是碱。另一方面，HCO_3^- 也能从水中接受质子，生成 H_2CO_3 和 OH^-，有

$$HCO_3^- + H_2O \rightleftharpoons OH^- + H_2CO_3$$

这里 HCO_3^- 是碱而水是酸。因此，根据布朗斯特-劳莱酸碱理论，HCO_3^- 是两性物质。至于 HCO_3^- 水溶液究竟显酸性还是碱性，取决于上述两个反应向右进行的倾向性的大小，更多讨论详见 **5.3 节**。

根据给出或接受质子的能力，或者与溶剂反应后电离的程度，可将酸碱分为强酸、强碱和弱酸、弱碱。强酸、强碱基本完全电离，其电离常数通常大于 10^6，而且它们电离的速率也非常快，所以强酸、强碱可认为一开始就全部以离子的状态存在。例如，HCl 是强酸，其电离方程式为

$$HCl + H_2O \rightarrow H_3O^+ + Cl^- \quad K_a = \frac{[H_3O^+][Cl^-]}{[HCl]} > 10^6 \quad (5.3)$$

其中 K_a 称为**酸式电离常数**。由于正反应基本进行到底而逆反应几乎不发生，上述电离方程式中用箭头代替平衡符号。Cl^- 是强酸 HCl 的共轭碱，其水解反应（即上述逆反应）几乎不发生，因此它是一个极弱碱。酸的酸性越强，其共轭碱的碱性就越弱；反之亦然。

弱酸、弱碱只能部分电离，形成电离平衡。例如，NH_3 是弱碱，其电离平衡式为

$$NH_3 + H_2O \rightleftharpoons NH_4^+ + OH^- \quad K_b = \frac{[NH_4^+][OH^-]}{[NH_3]} = 1.75 \times 10^{-5} \quad (5.4)$$

其中 K_b 称为**碱式电离常数**。K_a 和 K_b 的大小可用于区分弱酸、弱碱的相对强度。只有很少的弱酸、弱碱分子会发生电离，其电离度远小于 1。酸碱电离常数与电离度的计算详见 **5.3 节**。

酸碱的相对强弱不仅与其自身相关，也取决于溶剂，同一物质在不同溶剂中可能有不同表现。例如，醋酸 HAc 在水溶液中表现为弱酸，在液氨 NH_3 溶液中表现为中强酸，因为 NH_3 是比水更强的碱，比水更容易接受质子。而 HAc 在 HF 溶液中则表现为碱，因为 HF（K_a=6.3×10^{-4}）是比 HAc（K_a=1.75×10^{-5}）更强的酸，HAc 在 HF 溶液中可以接受质子形成 H_2Ac^+。

水是最为常用的溶剂，若不加说明，本教材中默认的溶剂即为水。H_3O^+ 和 OH^- 分别是水中所能存在的最强酸和最强碱，因为更强的酸碱将立即与水反应形成 H_3O^+ 和 OH^-，这种效应称为溶剂的**拉平效应**。例如，$HClO_4$ 和 HCl 在水溶液中均完全电离，故其酸性均被拉平至 H_3O^+ 的酸性。为了区分 $HClO_4$ 是比 HCl 更强的酸，需要使用一个比水更弱的碱（如乙醚 $C_2H_5OC_2H_5$）作为溶剂。在乙醚中 $HClO_4$ 完全

Their conjugate bases, between H_2O and OH^- consequently, are weak bases in water, ascending in their alkalinity. We can compare the relative strengths of those weak acids and bases with the same concentration according to their K_a values. Those below H_2O are extremely weak acids that almost do not ionize in water. Their conjugate bases are very strong bases that ionize nearly completely and cannot exist in their molecular form in water.

Table 5.1 Relative Strengths of Some Conjugate Pairs in Aqueous Solution and K_a Values of the Acid at 298.15 K
表 5.1 水溶液中一些共轭酸碱对的相对强度及 298.15 K 时酸的 K_a

Strength (强度)	Acid (酸) Name (名称)	Formula (化学式)	K_a	Conjugate Base (共轭碱) Name (名称)	Formula (化学式)	Strength (强度)
Very Strong (极强)	Perchloric acid (高氯酸)	$HClO_4$	$\gg 1$	Perchlorate ion (高氯酸根离子)	ClO_4^-	Very Weak (极弱)
	Hydriodic acid (氢碘酸)	HI		Iodide ion (碘离子)	I^-	
	Hydrobromic acid (氢溴酸)	HBr		Bromide ion (溴离子)	Br^-	
	Hydrochloric acid (盐酸)	HCl		Chloride ion (氯离子)	Cl^-	
	Sulfuric acid (硫酸)	H_2SO_4		Hydrogen sulfate ion (硫酸氢根离子)	HSO_4^-	
	Nitric acid (硝酸)	HNO_3		Nitrate ion (硝酸根离子)	NO_3^-	
Strong (强)	Hydronium ion (水合质子)	H_3O^+	1	Water (水)	H_2O	
Weak (弱)	Sulfurous acid (亚硫酸)	H_2SO_3	1.4×10^{-2} (K_{a1})	Hydrogen sulfurous ion (亚硫酸氢根离子)	HSO_3^-	Weak (弱)
	Hydrogen sulfate ion (硫酸氢根离子)	HSO_4^-	1.0×10^{-2} (K_{a2})	Sulfate ion (硫酸根离子)	SO_4^{2-}	
	Phosphoric acid (磷酸)	H_3PO_4	6.9×10^{-3} (K_{a1})	Dihydrogen phosphate ion (磷酸二氢根离子)	$H_2PO_4^-$	
	Hydrofluoric acid (氢氟酸)	HF	6.3×10^{-4}	Fluoride ion (氟离子)	F^-	
	Nitrous acid (亚硝酸)	HNO_2	5.6×10^{-4}	Nitrite ion (亚硝酸根离子)	NO_2^-	
	Acetic acid (醋酸)	CH_3COOH	1.75×10^{-5}	Acetate ion (醋酸根离子)	CH_3COO^-	
	Carbonic acid (碳酸)	H_2CO_3	4.5×10^{-7} (K_{a1})	Hydrogen carbonate ion (碳酸氢根离子)	HCO_3^-	
	Hydrocyanic acid (氢氰酸)	HCN	6.2×10^{-10} (K_{a1})	Hydrocyanic ion (氢氰酸根离子)	CN^-	
	Ammonium ion (铵离子)	NH_4^+	5.6×10^{-10}	Ammonia (氨)	NH_3	
	Hydrogen carbonate ion (碳酸氢根离子)	HCO_3^-	4.7×10^{-11} (K_{a2})	Carbonate ion (碳酸根离子)	CO_3^{2-}	
Very Weak (极弱)	Water (水)	H_2O	1.0×10^{-14}	Hydroxide ion (氢氧根离子)	OH^-	Strong (强)
	Methanol (甲醇)	CH_3OH	3.2×10^{-16}	Methoxide ion (甲氧基离子)	CH_3O^-	Very Strong (极强)
	Ammonia (氨)	NH_3	1.8×10^{-19}	Amide ion (氨基负离子)	NH_2^-	

The favored direction of an acid-base reaction is always from the stronger to the weaker member of a conjugate pair. A reaction of a strong acid and a strong base to produce a weak acid and a weak base proceeds almost to completion, whereas a reaction of a weak acid and a weak base to produce a strong acid and a strong base forms an equilibrium with a small equilibrium constant. For example

$$HCl + OH^- \rightarrow Cl^- + H_2O \quad \text{completely}$$
very strong acid   strong base   very weak base   weak acid

$$H_2SO_4 + H_2O \rightarrow HSO_4^- + H_3O^+ \quad \text{almost completely}$$
very strong acid   weak base   very weak base   strong acid

$$HSO_4^- + H_2O \rightleftharpoons SO_4^{2-} + H_3O^+ \quad \text{in equilibrium}$$
weak acid   weak base   weak base   strong acid

Brønsted-Lowry theory is the widely used acid-base theory in solution, and most of the contents discussed in this chapter is based on Brønsted-Lowry theory of acids and bases. However, there are still some limitations to this theory. Similar to Arrhenius theory, it also cannot handle non-proton acids such as SO_3. Moreover, Brønsted-Lowry theory is limited only to proton-transfer reactions and does not elucidate the fundamental nature of acid-base reactions.

Lewis Theory of Acids and Bases

Lewis theory of acids and bases was first proposed in 1923 and later elaborated in 1938 by Gilbert N. Lewis (1875—1946). Lewis defines a base as a substance that can donate an electron pair and an acid as a substance that can receive this electron pair. Therefore, the acid-base reaction is a reaction in which a base provides an electron pair to an acid to form a **coordinate covalent bond** and to produce an **adduct**, as given by

电离而 HCl 仅部分电离，这种溶剂称为**区分溶剂**。

表 5.1 及**附录 C.2** 中列出了水溶液中一些共轭酸碱对的相对强度及 K_a。表的左侧为酸，从上至下其相对酸性逐渐减弱；右侧为对应共轭碱，其相对碱性相应地逐渐增强。H_3O^+ 以上的酸全部为极强酸，在水中基本完全电离，不能以分子形式存在，其共轭碱的碱性极弱。位于 H_3O^+ 和 H_2O 之间的酸是在水中能够存在的弱酸，酸性依次减弱。其共轭碱相应地位于 H_2O 和 OH^- 之间，是在水中能够存在的弱碱，碱性依次增强。可以通过 K_a 的大小来比较具有相同浓度的弱酸碱的相对强度。位于 H_2O 以下的酸是在水中几乎不发生电离的极弱酸，它们的共轭碱是基本完全电离的极强碱，在水中不能以分子形式存在。

酸碱反应的有利方向，总是从较强的共轭酸碱对生成较弱的共轭酸碱对。由强酸和强碱生成弱酸和弱碱的反应基本进行完全，而由弱酸和弱碱生成强酸和强碱的反应会形成平衡且平衡常数较小。例如

$$HCl + OH^- \rightarrow Cl^- + H_2O \quad \text{反应完全}$$
极强酸　强碱　极弱碱　弱酸

$$H_2SO_4 + H_2O \rightarrow HSO_4^- + H_3O^+ \quad \text{反应基本完全}$$
极强酸　弱碱　极弱碱　强酸

$$HSO_4^- + H_2O \rightleftharpoons SO_4^{2-} + H_3O^+ \quad \text{反应处于平衡}$$
弱酸　弱碱　弱碱　强酸

布朗斯特-劳莱酸碱理论是溶液中广泛应用的酸碱理论，本章的大部分内容均基于该理论。但该理论仍具有一定的局限性，与阿伦尼乌斯酸碱理论类似，它也不能解释非质子酸（如 SO_3）。此外，布朗斯特-劳莱酸碱理论仅局限于质子转移反应，且不能阐明酸碱反应的本质。

路易斯酸碱理论

路易斯酸碱理论由吉尔伯特·N. 路易斯（1875—1946）最早于 1923 年提出，后来于 1938 年阐明。路易斯将碱定义为凡是能给出电子对的物质，将酸定义为所有能接受电子对的物质。因此，酸碱反应是碱提供电子对给酸、形成**配位共价键**并生成酸碱**加合物**的反应，可表示为

$$A+B: \rightleftharpoons A \leftarrow B \quad (5.5)$$

其中 B: 是路易斯碱，为**电子对给体**，具有一对用比号（:）表示的电子对；A 是路易斯酸，为**电子对受体**；A←B 是酸碱加合物，具有一个用箭头（←）表示的配位共价键。如果其中的酸为金属离子，生成的酸碱加合物又称**配位化合物**。关于配位共价键和配位化合物的更多讨论分别详见 **8.1 节**和**第 9 章**。

路易斯碱通常为具有可共用的孤电子对的物种，如 :OH^- 和 :NH_3，其孤电子对分别位于 O 原子和 N 原子上。路易斯酸通常为具有可容纳电子对的空轨道的物种。例如，HCl 分子并不是路易斯酸，因为它不能接受电子对，但 HCl 电离出的 H^+ 具有空的 $1s$ 轨道，能与电子对形成配位共价键，因此 H^+ 是路易斯酸。酸 BF_3 与碱 NH_3 的反应，是路易斯酸碱理论的一个典型示例，可表示为

$$A+B: \rightleftharpoons A \leftarrow B \tag{5.5}$$

where B: is the Lewis base that is an **electron-pair donor**, with an electron pair represented by the ratio sign (:), A is the Lewis acid that is an **electron-pair acceptor**, and A←B is the adduct with a coordinate covalent bond represented by the arrow (←). If metal ions are involved as acids, the adducts are also referred to as **coordination compounds**. More discussion about coordinated covalent bonds and coordination compounds can be found in **Sections 8.1** and **Chapter 9**, respectively.

Lewis bases are species with lone-pair electrons available for sharing, such as :OH⁻ and :NH₃, with lone-pair electrons on the O and N atoms, respectively. Lewis acids are species that normally have vacant orbitals to accommodate electron pairs. For instance, HCl molecule is not a Lewis acid because it cannot accept an electron-pair. However, the H^+ ionized from HCl is a Lewis acid since it has a vacant $1s$ orbital that can form a coordinate covalent bond with an electron pair. A typical example of an acid-base reaction in Lewis theory is the reaction between acid BF_3 and base NH_3, shown as

Hard and Soft Acids and Bases (HSAB) Theory

Lewis acids and bases are further classified into "hard" and "soft" acids and bases by Ralph G. Pearson (1919—2022) in 1963. This theory is known as **hard and soft acids and bases** (HSAB) theory. In HSAB, "hard" refers to species with high charge density (small size and high charge states) and low polarizability; "soft" applies to species with low charge density (large size and low charge states) and high polarizability. Between hard and soft species, there are borderline species with intermediate charge density and intermediate polarizability. **Polarizability** is the relative tendency for the electron cloud of a species to be distorted from its normal shape by an external electric field. More discussion about polarizability can be found in **Section 8.7**. The commonly used hard, borderline, and soft acids and bases are summarized in **Table 5.2**.

Table 5.2 Some Common Hard, Borderline, and Soft Acids and Bases
表 5.2 一些常见的硬酸碱、交界酸碱和软酸碱

	Hard（硬）	Borderline（交界）[a]	Soft（软）[a]
Acids（酸）	H^+, Li^+, Na^+, K^+, Be^{2+}, Mg^{2+}, Ca^{2+}, Cr^{2+}, Cr^{3+}, Al^{3+}, Fe^{3+}, SO_3, BF_3	Fe^{2+}, Co^{2+}, Ni^{2+}, Cu^{2+}, Zn^{2+}, Pb^{2+}, SO_2, BBr_3	Cu^+, Au^+, Ag^+, Tl^+, Hg_2^{2+}, Pd^{2+}, Cd^{2+}, Pt^{2+}, Hg^{2+}, BH_3
Bases（碱）	F^-, OH^-, H_2O, NH_3, CO_3^{2-}, NO_3^-, Cl^-, O^{2-}, SO_4^{2-}, PO_4^{3-}, ClO_4^-	NO_2^-, SO_3^{2-}, Br^-, N_3^-, N_2, $C_6H_5\underline{N}$, $SC\underline{N}^-$	H^-, R^-, $\underline{C}N^-$, $\underline{C}O$, I^-, $S\underline{C}N^-$, R_3P, $C_6H_5^-$, R_2S

[a]The underlined element is the site to which the classification refers.（带下划线的元素是该分类所指的位置）

In HSAB, an empirical chemical bonding rule is that hard acids bind strongly to hard bases and soft acids bind strongly to soft bases. Generally speaking, the more stable interactions between acids and bases are hard-to-hard (ionic character predominantly) and soft-to-soft (covalent character predominantly). More about ionic and covalent bonding can be found in **Chapter 8**. HSAB theory is useful in qualitatively explaining the stability of coordination compounds, choosing preparative conditions, predicting the directions and products of reactions, etc.

In 1983, Pearson together with Robert G. Parr (1921—2017) extended the qualitative HSAB theory into a quantitative method by calculating the values of "absolute hardness". Later, the concept of "absolute hardness" was connected with the concept of electronegativity, which can be found in **Section 7.9**.

5.2 Self-Ionization of Water and the pH Scale

According to Brønsted-Lowry theory, the acid-base reactions are proton transfer reactions. A proton transfer reaction that happens between the solutes is called a neutralization reaction, and that between the

软硬酸碱理论（HSAB）

1963 年，拉尔夫·G. 皮尔森（1919—2022）将路易斯酸碱进一步归类为硬酸、硬碱和软酸、软碱，这一理论称为**软硬酸碱理论**（HSAB）。在该理论中，"硬"指的是具有较高电荷密度（体积小、电荷高）和较低可极化性的物种，"软"指的是具有较低电荷密度（体积大、电荷低）和较高可极化性的物种。在"硬"和"软"的物种之间，还存在具有中等电荷密度和中等可极化性的交界物种。**可极化性**是一个物种的电子云在外加电场下从其常规形状发生扭曲的相对趋势。关于可极化性的更多讨论详见 **8.7 节**。**表 5.2** 列出了常见的硬、交界和软的酸碱。

在软硬酸碱理论中，形成化学键的一个经验规则是：硬亲硬、软亲软。一般而言，酸碱之间较为稳定的相互作用存在于硬酸 - 硬碱（主要为离子性）和软酸 - 软碱（主要为共价性）之间。关于离子键和共价键的更多讨论详见**第 8 章**。软硬酸碱理论可用于定性地解释配位化合物的稳定性、选择制备条件、预测反应方向和反应产物等方面。

1983 年，皮尔森和罗伯特·G. 帕尔（1921—2017）通过计算"绝对硬度"，将通常定性的软硬酸碱理论扩展为定量的方法。后来，"绝对硬度"的概念被与电负性（详见 **7.9 节**）的概念关联在一起。

5.2 水的自耦电离与 pH

根据布朗斯特 - 劳莱酸碱理论，酸碱反应是质子转移反应。发生在溶质之间的质子转移反应称为中和反应，发生在溶质和溶剂之间的质子转移反应称为电离反应，而发生在溶剂之间的质子转移反应则称为**自耦电离反应**。

水的自耦电离

水是最重要也最常用的溶剂。在阿伦尼乌斯、布朗斯特 - 劳莱和路易斯酸碱理论中，水均为两性物质，既是共轭碱为 OH^- 的弱酸，也是共轭酸为 H_3O^+ 的弱碱。纯水中水分子之间会发生转移质子的自耦电离反应，其方程式和 $\Delta G_f^\ominus$ 为

$$H_2O(l) + H_2O(l) \rightleftharpoons H_3O^+(aq) + OH^-(aq) \quad (5.6)$$

$\Delta G_f^\ominus (\text{kJ mol}^{-1})$ -237.1 -237.1 $-237.1+0$ -157.2

$\Delta G^\ominus = [-237.1 - 157.2 - (-237.1) - (-237.1)] \text{ kJ mol}^{-1} = 79.9 \text{ kJ mol}^{-1}$

$\therefore K = \exp(-\Delta G^\ominus / RT) = 1.0 \times 10^{-14}$

该反应的 $\Delta G^\ominus$ 可计算为 79.9 kJ mol^{-1}，室温（25°C）下对应的平衡常数为 1.0×10^{-14}。水的自耦电离反应的平衡常数称为**水的离子积**（K_w），25°C 时

solute and solvent is called an ionization reaction. If the proton transfer happens between the solvents, it is then called a **self-ionization reaction**.

Self-Ionization of Water

Water is the most important and commonly used solvent. It is amphiprotic in Arrhenius, Brønsted-Lowry, and Lewis theories of acids and bases. It can be either a weak acid whose conjugate base is OH⁻ or a weak base whose conjugate acid is H_3O^+. In pure water, a self-ionization reaction happens when protons are transferred between the water molecules, the equation and ΔG_f° values of which are shown as

$$H_2O(l) + H_2O(l) \rightleftharpoons H_3O^+(aq) + OH^-(aq) \tag{5.6}$$

ΔG_f° (kJ mol⁻¹)   −237.1   −237.1   −237.1+0   −157.2

$$\Delta G^\circ = [-237.1 - 157.2 - (-237.1) - (-237.1)] \text{ kJ mol}^{-1} = 79.9 \text{ kJ mol}^{-1}$$

$$\therefore K = \exp(-\Delta G^\circ / RT) = 1.0 \times 10^{-14}$$

The ΔG° value of the reaction is calculated to be 79.9 kJ mol⁻¹, which corresponds to an equilibrium constant of 1.0×10^{-14} at room temperature (25°C). The equilibrium constant for the self-ionization of water is called the **ion product of water** (K_w). At 25°C,

$$K_w = a(H_3O^+) a(OH^-) = 1.0 \times 10^{-14} \tag{5.7}$$

Since the concentrations of H_3O^+ and OH⁻ are very dilute in pure water, it can be written as

$$K_w = [H_3O^+][OH^-] = 1.0 \times 10^{-14} \tag{5.8}$$

$$\therefore [H_3O^+] = [OH^-] = 1.0 \times 10^{-7} \text{ mol L}^{-1}$$

K_w is an equilibrium constant that depends only on temperature but not on concentrations. Because the self-ionization reaction of water is endothermic, the value of K_w increases with temperature. K_w connects the concentrations of H_3O^+ and OH⁻, and applies to pure water as well as to all aqueous solutions. Using **Equation (5.8)**, we can also calculate the concentrations of H_3O^+ and OH⁻ in acidic or alkaline solutions. For example, adding some acid into pure water so that $[H_3O^+] = 0.1$ mol L⁻¹, the concentration of OH⁻ is then

$$[OH^-] = \frac{K_w}{[H_3O^+]} = \frac{1.0 \times 10^{-14}}{0.1} \text{ mol L}^{-1} = 1.0 \times 10^{-13} \text{ mol L}^{-1}$$

All the OH⁻ in this acidic solution are ionized from water. However, the H_3O^+ have two sources: only negligible amount ionized from water (1.0×10^{-13} mol L⁻¹) and the majority from the acid (0.1 mol L⁻¹ − 1.0×10^{-13} mol L⁻¹ ≈ 0.1 mol L⁻¹).

pH and pOH

Many aqueous reactions happen at very dilute H_3O^+ concentrations, typically much lower than 1 mol L⁻¹. It is then convenient to use the pH notation to indicate the acidity or alkalinity of a solution. The term **pH** was first proposed by the Danish biochemist Søren Sørensen (1868—1939) in 1909, defined as the negative of the logarithm of [H⁺] (later restated as [H_3O^+]). Therefore,

$$pH = -\lg[H_3O^+] \tag{5.9}$$

When the pH value of a solution changes by 1 unit, the concentration of H_3O^+ changes by 10 times.

In a similar manner, the quantities **pOH** and **pK_w** can be defined as

$$pOH = -\lg[OH^-] \quad \text{and} \quad pK_w = -\lg K_w$$

respectively. Because $K_w = [H_3O^+][OH^-] = 1.0 \times 10^{-14}$, it can be easily derived that

$$pK_w = pH + pOH = 14.00 \tag{5.10}$$

For example, in a solution with $[H_3O^+] = 0.0020$ mol L⁻¹, pH = $-\lg(2.0 \times 10^{-3})$ = 2.70 and pOH = 14.00 − 2.70 = 11.30. In a solution with pH = 5.50, pOH = 8.50, $[H_3O^+] = 10^{-5.50}$ mol L⁻¹ = 3.2×10^{-6} mol L⁻¹ and [OH⁻] = $10^{-8.50}$ mol L⁻¹ = 3.2×10^{-9} mol L⁻¹.

A **neutral** aqueous solution has equal concentrations of H_3O^+ and OH⁻. Since $[H_3O^+] = [OH^-] = 1.0 \times 10^{-7}$

$$K_w = a(H_3O^+)a(OH^-) = 1.0\times10^{-14} \tag{5.7}$$

由于纯水中 H_3O^+ 和 OH^- 的浓度非常稀，上式可写为

$$K_w = [H_3O^+][OH^-] = 1.0\times10^{-14} \tag{5.8}$$

$$\therefore [H_3O^+] = [OH^-] = 1.0\times10^{-7}\ \text{mol L}^{-1}$$

平衡常数 K_w 与温度相关，不随浓度变化。由于水的自耦电离反应吸热，K_w 随温度升高而增加。K_w 将 H_3O^+ 和 OH^- 的浓度关联在一起，适用于纯水及所有水溶液。利用**式（5.8）**，还可以计算酸性或碱性溶液中 H_3O^+ 和 OH^- 的浓度。例如，在纯水中加酸使 $[H_3O^+] = 0.1\ \text{mol L}^{-1}$，则 OH^- 的浓度为

$$[OH^-] = \frac{K_w}{[H_3O^+]} = \frac{1.0\times10^{-14}}{0.1}\ \text{mol L}^{-1} = 1.0\times10^{-13}\ \text{mol L}^{-1}$$

此酸性溶液的所有 OH^- 均来自水，但 H_3O^+ 有两个来源：极少量来自水 ($1.0\times10^{-13}\ \text{mol L}^{-1}$)，绝大多数来自酸 ($0.1\ \text{mol L}^{-1} - 1.0\times10^{-13}\ \text{mol L}^{-1} \approx 0.1\ \text{mol L}^{-1}$)。

pH 和 pOH

很多水溶液反应发生在极稀 H_3O^+ 浓度时，通常远低于 $1\ \text{mol L}^{-1}$，这时采用 pH 符号来表示溶液的酸碱度较为方便。**pH** 这个术语由丹麦生物化学家索伦·索伦森（1868—1939）在 1909 年首次提出，定义为 $[H^+]$（后来修改为 $[H_3O^+]$）的负对数。故

$$\text{pH} = -\lg[H_3O^+] \tag{5.9}$$

当溶液的 pH 变化 1 个单位时，H_3O^+ 的浓度改变 10 倍。

类似地，pOH 和 pK_w 分别定义为

$$\text{pOH} = -\lg[OH^-] \quad \text{且} \quad \text{p}K_w = -\lg K_w$$

由于 $K_w = [H_3O^+][OH^-] = 1.0\times10^{-14}$，易推导得

$$\text{p}K_w = \text{pH} + \text{pOH} = 14.00 \tag{5.10}$$

例如，$[H_3O^+] = 0.0020\ \text{mol L}^{-1}$ 的水溶液中，pH $= -\lg(2.0\times10^{-3}) = 2.70$，pOH $= 14.00 - 2.70 = 11.30$。pH $= 5.50$ 的水溶液中，pOH $= 8.50$，$[H_3O^+] = 10^{-5.50}\ \text{mol L}^{-1} = 3.2\times10^{-6}\ \text{mol L}^{-1}$，$[OH^-] = 10^{-8.50}\ \text{mol L}^{-1} = 3.2\times10^{-9}\ \text{mol L}^{-1}$。

中性水溶液中 H_3O^+ 和 OH^- 的浓度相等。由于 25ºC 时纯水中 $[H_3O^+] = [OH^-] = 1.0\times10^{-7}\ \text{mol L}^{-1}$ 且 pH $= 7.00$，所有 pH $= 7.00$ 的水溶液均为中性。pH < 7.00 的溶液为**酸性**，pH > 7.00 的溶液为**碱性**。pH 和 pOH 在 2~12 范围内使用较为方便。在此范围之外的浓酸浓碱中，H_3O^+ 和 OH^- 的摩尔浓度与其真实活度的差别较大，以 mol L^{-1} 表示的实际摩尔浓度更为有用。

表 5.3 总结了一些常用材料的 pH。测量水溶液的 pH 常使用酸碱指示剂和 pH 计，更多相关讨论分别详见 **5.5 节**和 **6.6 节**。

非水溶剂的自耦电离

与水类似，许多非水液体均能充当溶剂形成溶液，也存在自耦电离平衡和对应溶剂的离子积。例如，液氨中存在

$$NH_3(l) + NH_3(l) \rightleftharpoons NH_4^+(\text{sol}) + NH_2^-(\text{sol})$$

mol L^{-1} and pH = 7.00 in pure water at 25°C, all aqueous solutions with pH = 7.00 are neutral. If pH < 7.00, the solution is **acidic**; if pH > 7.00, the solution is **basic** or **alkaline**. The pH and pOH scales are useful and convenient in the range of 2~12. Otherwise the molarities of H$_3$O$^+$ and OH$^-$ in concentrated acids and bases may differ significantly from their true activities, and the actual molarity values in mol L^{-1} would be more useful.

The pH values of some commonly used materials are summarized in **Table 5.3**. To measure the pH of an aqueous solution, both an acid-base indicator and a pH meters can be used, which will be further discussed in **Sections 5.5** and **6.6**, respectively.

Table 5.3 The pH Values of Some Common Materials
表 5.3 一些常见材料的 pH

Material (材料)	pH Value (pH)	Material (材料)	pH Value (pH)
Gastric juices (胃酸)	1.0~2.0	Blood (血液)	7.4
Vinegar (食醋)	2.4~3.4	Sea water (海水)	7.0~8.5
Carbonated water (碳酸饮料)	3.9	Baking soda (小苏打)	8.4
Beer (啤酒)	4.0~4.5	Toothpaste (牙膏)	8~9
Rainwater (雨水)	5.6	Soap (肥皂)	8~10
Urine (尿)	5~7	Milk of magnesia (镁乳)	10.5
Milk (牛奶)	6.4	Household ammonia (家用氨水)	11.9

Self-Ionization of Non-aqueous Solvents

Similar to water, many non-aqueous liquids can also serve as solvents to form solutions. There are self-ionization equilibria and the corresponding ion products for those solvents. For example, in liquid ammonia

$$NH_3(l) + NH_3(l) \rightleftharpoons NH_4^+(sol) + NH_2^-(sol)$$

where sol stands for solution. The ion product of ammonia is represented by K_{NH_3}. At 25°C,

$$K_{NH_3} = [NH_4^+][NH_2^-] = 1.0 \times 10^{-28} \quad \text{and} \quad pK_{NH_3} = 28.00$$

The self-ionization reaction of non-aqueous solvents can be either protic (proton transfer) or non-protic, with some examples given below:

1) Protic self-ionization:

$$2HAc(l) \rightleftharpoons H_2Ac^+(sol) + Ac^-(sol)$$
$$3HF(l) \rightleftharpoons H_2F^+(sol) + HF_2^-(sol)$$

In the self-ionization of HF, the F$^-$ produced in the proton transfer between two HF combines with a third HF to form HF$_2^-$.

2) Non-protic self-ionization:

$$2PF_5(l) \rightleftharpoons PF_4^+(sol) + PF_6^-(sol)$$
$$N_2O_4(l) \rightleftharpoons NO^+(sol) + NO_3^-(sol)$$

5.3 Ionization Equilibria of Acids and Bases

The proton transfer reaction that happens between the solute and solvent is called an ionization reaction. In an aqueous solution, acids and bases can transfer protons with water and form ionization equilibria.

Ionization of Strong Acids and Bases

As discussed previously in **Section 5.1**, strong acids and bases ionize nearly completely, and can be treated as though they were completely ions initially. Their ionization reactions go essentially to completion, or they can be considered to form ionization equilibria with very large acid or base ionization constants (>10^6). Meanwhile, the self-ionization of water is greatly suppressed and only occurs to a very slight extent. As a result, strong acids and bases are the only significant sources in calculation of [H$_3$O$^+$] or [OH$^-$]. The contribution due to the self-ionization of water can generally be ignored unless the solution is extremely

其中 sol 代表溶液。液氨的离子积用 K_{NH_3} 表示，25°C 时

$$K_{NH_3} = \left[NH_4^+\right]\left[NH_2^-\right] = 1.0 \times 10^{-28} \quad 且 \quad pK_{NH_3} = 28.00$$

非水溶剂的自耦电离反应可以是质子转移反应，也可以是非质子转移反应，分别举例如下：

1) 质子转移的自耦电离：

$$2HAc(l) \rightleftharpoons H_2Ac^+(sol) + Ac^-(sol)$$

$$3HF(l) \rightleftharpoons H_2F^+(sol) + HF_2^-(sol)$$

在 HF 的自耦电离中，两个 HF 之间质子转移所产生的 F^-，可与第三个 HF 结合形成 HF_2^-。

2) 非质子转移的自耦电离：

$$2PF_5(l) \rightleftharpoons PF_4^+(sol) + PF_6^-(sol)$$

$$N_2O_4(l) \rightleftharpoons NO^+(sol) + NO_3^-(sol)$$

5.3 酸碱电离平衡

溶质和溶剂之间发生的质子转移反应称为电离反应。水溶液中，酸和碱能与水发生质子传递，并形成电离平衡。

强酸强碱的电离

如 **5.1 节**所述，强酸、强碱基本完全电离，可视为初始时即全部以离子的形式存在。它们的电离基本反应完全，或者可以认为它们形成了酸碱电离常数非常大（$>10^6$）的电离平衡。同时，水的自耦电离受到极大的抑制，几乎不发生。其结果为，强酸、强碱是计算 $[H_3O^+]$ 和 $[OH^-]$ 的唯一重要来源。除了极稀溶液之外，水的自耦电离所产生的影响通常可忽略不计。例如，在 0.01 mol L^{-1} NaOH 溶液中，$[OH^-]$ = 0.01 mol L^{-1}，pOH = 2，因此 pH = 14 – 2 = 12，$[H_3O^+] = 10^{-12}$ mol L^{-1}。

对于确实极稀的强酸、强碱溶液，则需同时考虑两个电离平衡。例如，在 10^{-8} mol L^{-1} HCl 溶液中

$$HCl + H_2O \rightarrow H_3O^+ + Cl^-$$
$$\qquad\qquad\qquad 10^{-8} \quad\ 10^{-8} \quad \text{mol L}^{-1}$$

$$H_2O + H_2O \rightleftharpoons H_3O^+ + OH^-$$
$$\qquad\qquad\qquad\quad x \qquad x \quad \text{mol L}^{-1}$$

$$K_w = \left[H_3O^+\right]\left[OH^-\right] = (10^{-8} + x)x = 1.0 \times 10^{-14}$$

$$\therefore x = \left[OH^-\right] = 9.5 \times 10^{-8} \text{ mol L}^{-1}$$

$$\left[H_3O^+\right] = (10^{-8} + x) \text{ mol L}^{-1} = 1.05 \times 10^{-7} \text{ mol L}^{-1}$$

一元弱酸弱碱的电离平衡

弱酸、弱碱只发生部分电离，形成具有相应酸碱电离常数的电离平衡。酸和碱的电离是分步进行的，每一步只转移一个质子。根据分步电离的次数，酸和碱可分为一元酸碱（一步电离）和多元酸碱（多步电离）。一些一元和多元弱酸碱的电离常数详见**附录 C.2**。

dilute. For example, in 0.01 mol L^{-1} NaOH solution, [OH$^-$] = 0.01 mol L^{-1} and pOH = 2. Therefore, pH = 14 − 2 = 12 and [H$_3$O$^+$] = 10^{-12} mol L^{-1}.

If strong acid or base solutions are indeed extremely dilute, then we need to consider both ionization reactions simultaneously. For example, in the 10^{-8} mol L^{-1} HCl solution

$$HCl + H_2O \rightarrow H_3O^+ + Cl^-$$
$$10^{-8} \quad 10^{-8} \quad \text{mol L}^{-1}$$
$$H_2O + H_2O \rightleftharpoons H_3O^+ + OH^-$$
$$x \quad x \quad \text{mol L}^{-1}$$
$$K_w = [H_3O^+][OH^-] = (10^{-8} + x)x = 1.0 \times 10^{-14}$$
$$\therefore x = [OH^-] = 9.5 \times 10^{-8} \text{ mol L}^{-1}$$
$$[H_3O^+] = (10^{-8} + x) \text{ mol L}^{-1} = 1.05 \times 10^{-7} \text{ mol L}^{-1}$$

Ionization Equilibria of Monoprotic Weak Acids and Bases

Weak acids and bases only ionize partially, forming ionization equilibria with the corresponding acid and base ionization constants. The ionization of acids and bases is stepwise, with one proton transfer in each step. According to the number of stepwise ionization, acids and bases can be either monoprotic (one-step ionization) or polyprotic (multi-step ionization). The ionization constants for some monoprotic and polyprotic weak acids and bases can be found in **Appendix C.2**.

For a typical monoprotic weak acid HA with an acid ionization constant K_a, if its molarity is represented by c and α stands for the **degree of ionization**, which is defined as the ratio of ionized molecules to the overall molecules, we can write its ionization equation as

$$HA + H_2O \rightleftharpoons H_3O^+ + A^-$$

Initial:	c	0	0
Changes:	$-c\alpha$	$c\alpha$	$c\alpha$
Equilibrium:	$c(1-\alpha)$	$c\alpha$	$c\alpha$

Ignoring the self-ionization of water, we can calculate

$$K_a = \frac{[H_3O^+][A^-]}{[HA]} = \frac{c\alpha^2}{1-\alpha} \tag{5.11}$$

In some approximate or estimate calculations, 0.05 is often used as a boundary value. If $x/y < 0.05$, we consider that $x \ll y$ and $y-x \approx y$, which means that x is negligible compared to y. In this case, if the degree of ionization $\alpha < 0.05$, which in return requires $c/K_a > 380$, then $1 - \alpha \approx 1$, meaning that the ionized HA molecules are negligible compared to the overall HA molecules. Under this simplifying condition, we have

$$K_a = c\alpha^2, \quad \alpha = \frac{[H_3O^+]}{c} = \sqrt{\frac{K_a}{c}} \quad \text{and} \quad [H_3O^+] = \sqrt{K_a c}$$

If $\alpha > 0.05$ or $c/K_a < 380$, it means that the simplifying condition is not applicable and solving **Equation (5.11)** is then required. For example, to calculate the pH of 0.10 mol L^{-1} acetic acid (HAc, $K_a = 1.75 \times 10^{-5}$) and dichloroacetic acid (CHCl$_2$COOH, $K_a = 4.4 \times 10^{-2}$), we have

For HAc

$$\frac{c}{K_a} = \frac{0.10}{1.75 \times 10^{-5}} = 5.7 \times 10^3 > 380$$

$$[H_3O^+] = \sqrt{K_a c} = 1.3 \times 10^{-3} \text{ mol L}^{-1} \quad \text{and} \quad \text{pH} = 2.89$$

$$\alpha = \frac{[H_3O^+]}{c} = 0.013 < 0.05$$

For CHCl$_2$COOH

对于典型的一元弱酸 HA，其酸式电离常数为 K_a，摩尔浓度为 c，**电离度**（即已电离分子与总分子的比率）为 α，电离方程式可写为

$$HA + H_2O \rightleftharpoons H_3O^+ + A^-$$

初始浓度： c 0   0
反应浓度： $-c\alpha$ $c\alpha$ $c\alpha$
平衡浓度： $c(1-\alpha)$ $c\alpha$ $c\alpha$

忽略水的自耦电离，有

$$K_a = \frac{[H_3O^+][A^-]}{[HA]} = \frac{c\alpha^2}{1-\alpha} \tag{5.11}$$

在一些近似计算或估算中，通常使用 0.05 作为边界值，如果 $x/y < 0.05$，则认为 $x \ll y$ 或 $y - x \approx y$，这意味着相对于 y，x 可忽略不计。在这种情况下，如果电离度 $\alpha < 0.05$，反过来可计算得 $c/K_a > 380$，这时 $1-\alpha \approx 1$，即意味着相对于所有 HA 分子，已电离的 HA 分子可忽略不计。在该简化条件下，有

$$K_a = c\alpha^2, \quad \alpha = \frac{[H_3O^+]}{c} = \sqrt{\frac{K_a}{c}} \quad 且 \quad [H_3O^+] = \sqrt{K_a c}$$

如果 $\alpha > 0.05$ 或 $c/K_a < 380$，即意味着简化条件不适用，这时需求**解方程（5.11）**。例如，分别计算 0.10 mol L^{-1} 的醋酸（HAc，$K_a = 1.75 \times 10^{-5}$）和二氯乙酸（CHCl$_2$COOH，$K_a = 4.4 \times 10^{-2}$）的 pH，有：

HAc：

$$\frac{c}{K_a} = \frac{0.10}{1.75 \times 10^{-5}} = 5.7 \times 10^3 > 380$$

$$[H_3O^+] = \sqrt{K_a c} = 1.3 \times 10^{-3} \text{ mol L}^{-1} \quad 且 \quad pH = 2.89$$

$$\alpha = \frac{[H_3O^+]}{c} = 0.013 < 0.05$$

CHCl$_2$COOH：

$$\frac{c}{K_a(\text{CHCl}_2\text{COOH})} = \frac{0.10}{4.4 \times 10^{-2}} = 2.3 < 380$$

$$\text{CHCl}_2\text{COOH} + H_2O \rightleftharpoons H_3O^+ + \text{CHCl}_2\text{COO}^-$$

平衡浓度： $0.10 - x$ x x

$$K_a = \frac{[H_3O^+][\text{CHCl}_2\text{COO}^-]}{[\text{CHCl}_2\text{COOH}]} = \frac{x^2}{0.10 - x} = 4.4 \times 10^{-2}$$

$$x = [H_3O^+] = 0.048 \text{ mol L}^{-1} \quad 且 \quad pH = 1.32$$

$$\alpha = \frac{[H_3O^+]}{c} = 0.48 > 0.05$$

如果仍然使用简化条件计算二氯乙酸，可得 $[H_3O^+] = 0.067$ mol L^{-1}，相对误差约为 40%。另一方面，对 HAc 求解**方程（5.11）**可得 $[H_3O^+] = 1.3 \times 10^{-3}$ mol L^{-1}，误差仅为 1%，可忽略不计。

如果持续稀释某弱酸 HA，其电离平衡常数可写为

$$K_a = \frac{[H_3O^+][A^-]}{[HA]} = \frac{\{n(H_3O^+)/V\}\{n(A^-)/V\}}{n(HA)/V} = \frac{n(H_3O^+)n(A^-)}{n(HA)} \cdot \frac{1}{V}$$

随着 V 增加，$1/V$ 减小，为保持恒定的 K_a，比率 $n(H_3O^+) n(A^-) /n(HA)$

$$\frac{c}{K_a(\text{CHCl}_2\text{COOH})} = \frac{0.10}{4.4 \times 10^{-2}} = 2.3 < 380$$

$$\text{CHCl}_2\text{COOH} + \text{H}_2\text{O} \rightleftharpoons \text{H}_3\text{O}^+ + \text{CHCl}_2\text{COO}^-$$

Equilibrium: $0.10 - x$ x x

$$K_a = \frac{[\text{H}_3\text{O}^+][\text{CHCl}_2\text{COO}^-]}{[\text{CHCl}_2\text{COOH}]} = \frac{x^2}{0.10-x} = 4.4 \times 10^{-2}$$

$$x = [\text{H}_3\text{O}^+] = 0.048 \text{ mol L}^{-1} \quad \text{and} \quad \text{pH} = 1.32$$

$$\alpha = \frac{[\text{H}_3\text{O}^+]}{c} = 0.48 > 0.05$$

If the simplifying condition is still applied for CHCl_2COOH, then $[\text{H}_3\text{O}^+] = 0.067 \text{ mol L}^{-1}$ can be calculated, leading to a relative error of about 40%. On the other hand, solving **Equation (5.11)** for HAc still gives $[\text{H}_3\text{O}^+] = 1.3 \times 10^{-3} \text{ mol L}^{-1}$ with a small relative error of 1% that is negligible.

If a weak acid HA is diluted continuously, we can analyze the ionization equilibrium by

$$K_a = \frac{[\text{H}_3\text{O}^+][\text{A}^-]}{[\text{HA}]} = \frac{\{n(\text{H}_3\text{O}^+)/V\}\{n(\text{A}^-)/V\}}{n(\text{HA})/V} = \frac{n(\text{H}_3\text{O}^+)n(\text{A}^-)}{n(\text{HA})} \cdot \frac{1}{V}$$

As V increases, $1/V$ decreases, and the ratio $n(\text{H}_3\text{O}^+) n(\text{A}^-) / n(\text{HA})$ must increase to maintain the constant value of K_a. In turn, $n(\text{H}_3\text{O}^+)$ and $n(\text{A}^-)$ must increase and $n(\text{HA})$ must decrease, leading to a continuous increase in the degree of ionization. On the other hand, strong acids remain essentially a 100% degree of ionization while diluting. **Figure 5.1** plots percent ionization as a function of solution molarity for a weak acid (HAc) and a strong acid (HCl), with the corresponding pH labelled.

The conjugate base of HA is A^-, which acts as a monoprotic weak base with a base ionization constant K_b. The reaction between an ion (or a salt that can produce ions) and water is often called a **hydrolysis reaction**. The hydrolysis reaction of A^- can be written as

$$\text{A}^- + \text{H}_2\text{O} \rightleftharpoons \text{HA} + \text{OH}^-$$

$$K_b = \frac{[\text{HA}][\text{OH}^-]}{[\text{A}^-]} = \frac{[\text{HA}][\text{OH}^-]}{[\text{A}^-]} \cdot \frac{[\text{H}_3\text{O}^+]}{[\text{H}_3\text{O}^+]}$$

$$= \frac{[\text{HA}]}{[\text{H}_3\text{O}^+][\text{A}^-]} \cdot [\text{H}_3\text{O}^+][\text{OH}^-] = \frac{K_w}{K_a}$$

Therefore, the product of the ionization constants of either an acid and its conjugate base or a base and its conjugate acid equals the ion product of water, given by

$$K_a(\text{acid}) \cdot K_b(\text{its conjugate base}) = K_w \tag{5.12}$$

$$K_b(\text{base}) \cdot K_a(\text{its conjugate acid}) = K_w \tag{5.13}$$

Ionization Equilibria of Polyprotic Weak Acids and Bases

H_2S is a diprotic weak acid that has two ionizable H atoms to be ionized in two steps. For each step, an ionization equation with a distinctive acid ionization constant can be written. For a 0.10 mol L^{-1} H_2S solution, let us start by considering the first step only,

$$\text{H}_2\text{S} + \text{H}_2\text{O} \rightleftharpoons \text{H}_3\text{O}^+ + \text{HS}^-$$

Initial: 0.10 0 0
Changes: $-x$ x x
After first step: $0.10 - x$ x x

$$K_{a1} = \frac{[\text{H}_3\text{O}^+][\text{HS}^-]}{[\text{H}_2\text{S}]} = \frac{x^2}{0.10-x} = 8.9 \times 10^{-8}$$

必须增加，这要求 $n(\text{H}_3\text{O}^+)$ 和 $n(\text{A}^-)$ 增加而 $n(\text{HA})$ 减小，使得电离度持续上升。另一方面，强酸稀释时基本保持 100% 电离度。**图 5.1** 绘制了弱酸 (HAc) 和强酸 (HCl) 的电离百分数与溶液摩尔浓度的函数关系，并标注了相应的 pH。

HA 的共轭碱 A^- 是一元弱碱，其碱式电离常数为 K_b。离子（或能产生离子的盐）与水的反应通常称为**水解反应**。A^- 的水解反应可写为

$$\text{A}^- + \text{H}_2\text{O} \rightleftharpoons \text{HA} + \text{OH}^-$$

$$K_b = \frac{[\text{HA}][\text{OH}^-]}{[\text{A}^-]} = \frac{[\text{HA}][\text{OH}^-]}{[\text{A}^-]} \cdot \frac{[\text{H}_3\text{O}^+]}{[\text{H}_3\text{O}^+]}$$

$$= \frac{[\text{HA}]}{[\text{H}_3\text{O}^+][\text{A}^-]} \cdot [\text{H}_3\text{O}^+][\text{OH}^-] = \frac{K_w}{K_a}$$

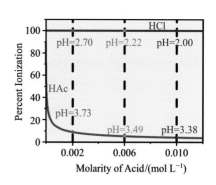

Figure 5.1 Percent ionization and the corresponding pH as a function of solution molarity for acetic acid (red) and hydrochloric acid (blue).

图 5.1 醋酸（红线）和盐酸（蓝线）的电离百分数和相应 pH 与溶液摩尔浓度的函数关系图。

因此，酸与其共轭碱的电离常数的乘积，或者碱与其共轭酸的电离常数的乘积，均等于水的离子积，可表示为

$$K_a(\text{酸}) \cdot K_b(\text{共轭碱}) = K_w \tag{5.12}$$

$$K_b(\text{碱}) \cdot K_a(\text{共轭酸}) = K_w \tag{5.13}$$

多元弱酸弱碱的电离平衡

H_2S 是二元弱酸，有两个 H 原子，可分两步电离，每一步均可写出其电离方程式，具有不同的酸式电离常数。对 $0.10\ \text{mol L}^{-1}\ \text{H}_2\text{S}$ 溶液，先只考虑第一步电离：

	$\text{H}_2\text{S} + \text{H}_2\text{O} \rightleftharpoons \text{H}_3\text{O}^+ + \text{HS}^-$

初始浓度：　　　0.10　　　　　　　0　　　0
反应浓度：　　　$-x$　　　　　　　x　　　x
第一步电离后：　$0.10 - x$　　　　x　　　x

$$K_{a1} = \frac{[\text{H}_3\text{O}^+][\text{HS}^-]}{[\text{H}_2\text{S}]} = \frac{x^2}{0.10 - x} = 8.9 \times 10^{-8}$$

$$\frac{c}{K_{a1}} \gg 380, \quad \therefore x = \sqrt{K_{a1}c} = 9.4 \times 10^{-5}\ \text{mol L}^{-1}$$

再考虑第二步电离：

$$\text{HS}^- + \text{H}_2\text{O} \rightleftharpoons \text{H}_3\text{O}^+ + \text{S}^{2-}$$

第一步电离后：　x　　　　　　　x　　　0
反应浓度：　　　$-y$　　　　　　　y　　　y
第二步电离后：　$x - y$　　　　　$x + y$　y

$$K_{a2} = \frac{[\text{H}_3\text{O}^+][\text{S}^{2-}]}{[\text{HS}^-]} = \frac{(x+y)y}{x-y} = 1 \times 10^{-19}$$

$$x \gg y \quad 且 \quad x + y \approx x - y$$

$$\therefore y = K_{a2} = 1 \times 10^{-19}\ \text{mol L}^{-1}$$

考虑两步电离，$0.10\ \text{mol L}^{-1}\ \text{H}_2\text{S}$ 溶液中各物种浓度为

$$[\text{H}_2\text{S}] = 0.10 - x \approx 0.10\ \text{mol L}^{-1}$$

$$[\text{H}_3\text{O}^+] = x + y \approx x = 9.4 \times 10^{-5}\ \text{mol L}^{-1} \quad 且 \quad \text{pH} = 4.03$$

$$[\text{HS}^-] = x - y \approx x = 9.4 \times 10^{-5}\ \text{mol L}^{-1}$$

$$[\text{S}^{2-}] = y = K_{a2} = 1 \times 10^{-19}\ \text{mol L}^{-1}$$

$$\frac{c}{K_{a1}} \gg 380, \quad \therefore x = \sqrt{K_{a1}c} = 9.4 \times 10^{-5} \text{ mol L}^{-1}$$

Then consider the second step,

$$HS^- + H_2O \rightleftharpoons H_3O^+ + S^{2-}$$

From first step:	x	x	0
Changes:	$-y$	y	y
After second step:	$x-y$	$x+y$	y

$$K_{a2} = \frac{[H_3O^+][S^{2-}]}{[HS^-]} = \frac{(x+y)y}{x-y} = 1 \times 10^{-19}$$

$$x \gg y \quad \text{and} \quad x+y \approx x-y$$

$$\therefore y = K_{a2} = 1 \times 10^{-19} \text{ mol L}^{-1}$$

Considering both ionization steps, the concentrations in 0.10 mol L^{-1} H$_2$S solution are

$$[H_2S] = 0.10 - x \approx 0.10 \text{ mol L}^{-1}$$

$$[H_3O^+] = x + y \approx x = 9.4 \times 10^{-5} \text{ mol L}^{-1} \quad \text{and} \quad pH = 4.03$$

$$[HS^-] = x - y \approx x = 9.4 \times 10^{-5} \text{ mol L}^{-1}$$

$$[S^{2-}] = y = K_{a2} = 1 \times 10^{-19} \text{ mol L}^{-1}$$

$$[OH^-] = \frac{K_w}{[H_3O^+]} = 1.1 \times 10^{-10} \text{ mol L}^{-1}$$

Therefore, for polyprotic weak acids with K_{a1} much larger than the rest, the following key statements can be made:

1) The pH value is mainly determined by the first ionization step. If $c/K_{a1} \gg 380$, $[H_3O^+] = \sqrt{K_{a1}c}$.
2) The concentration of the conjugate base produced in the first ionization step, such as [HS$^-$] for H$_2$S and [H$_2$PO$_4^-$] for H$_3$PO$_4$, approximately equals [H$_3$O$^+$].
3) The concentration of the conjugate base produced in the second ionization step, such as [S^{2-}] for H$_2$S and [HPO$_4^{2-}$] for H$_3$PO$_4$, approximately equals K_{a2} regardless of the molarity of the polyprotic acid.

Mixed weak acids can be treated in a similar way as a polyprotic acid. For example, for a solution mixed by HAc and HCN, since K_a(HAc) = 1.75×10^{-5} is much larger than K_a(HCN) = 6.2×10^{-10}, essentially all the H$_3$O$^+$ is produced by HAc. Water can be considered as a weak acid with K_a(H$_2$O) = K_w or a weak base with K_b(H$_2$O) = K_w. The self-ionization of water is negligible for any acids with $K_a \gg K_w$ or bases with $K_b \gg K_w$.

The ions produced by the ionization of polyprotic weak acids and bases can hydrolyze in water, and thus can also be treated as polyprotic weak acids and bases. For example, PO$_4^{3-}$ is a triprotic base that can hydrolyze in three steps, giving three distinctive base ionization constants as follows:

$$PO_4^{3-} + H_2O \rightleftharpoons HPO_4^{2-} + OH^- \quad K_{b1}$$
$$HPO_4^{2-} + H_2O \rightleftharpoons H_2PO_4^- + OH^- \quad K_{b2}$$
$$H_2PO_4^- + H_2O \rightleftharpoons H_3PO_4 + OH^- \quad K_{b3}$$

It is not difficult to derive the relationship between the three base ionization constants of PO$_4^{3-}$ and the three acid ionization constants of H$_3$PO$_4$, as

$$K_{a1} \cdot K_{b3} = K_{a2} \cdot K_{b2} = K_{a3} \cdot K_{b1} = K_w \tag{5.14}$$

Ionization Equilibria of Amphiprotic Substances

Amphiprotic substances such as HCO$_3^-$, HPO$_4^{2-}$, H$_2$PO$_4^-$, NH$_4$Ac can act either as acids or bases. Hydrolysis reactions happen between them and water. In order to determine the pH of their aqueous solutions, multiple ionization equilibria are needed for analysis.

Taking HCO$_3^-$ as an illustrative example, we have

$$[\text{OH}^-] = \frac{K_w}{[\text{H}_3\text{O}^+]} = 1.1\times 10^{-10}\ \text{mol L}^{-1}$$

因此，对于 K_{a1} 远大于其余酸式电离常数的多元弱酸，可得出如下主要结论：

1) pH 主要由第一步电离决定，若 $c/K_{a1} \gg 380$，则 $[\text{H}_3\text{O}^+] = \sqrt{K_{a1}c}$。
2) 第一步电离产生的共轭碱的浓度，如 H_2S 溶液中的 $[\text{HS}^-]$ 以及 H_3PO_4 溶液中的 $[\text{H}_2\text{PO}_4^-]$，约等于 $[\text{H}_3\text{O}^+]$。
3) 第二步电离产生的共轭碱的浓度，如 H_2S 溶液中的 $[\text{S}^{2-}]$ 以及 H_3PO_4 溶液中的 $[\text{HPO}_4^{2-}]$，约等于 K_{a2}，与多元弱酸的摩尔浓度无关。

混合弱酸的计算方法与多元弱酸类似。例如，对于 HAc 和 HCN 的混合溶液，由于 $K_a(\text{HAc}) = 1.75\times 10^{-5}$ 远大于 $K_a(\text{HCN}) = 6.2\times 10^{-10}$，基本上所有 H_3O^+ 均由 HAc 产生。水可视为 $K_a(\text{H}_2\text{O}) = K_w$ 的弱酸或 $K_b(\text{H}_2\text{O}) = K_w$ 的弱碱，对于 $K_a \gg K_w$ 的酸或 $K_b \gg K_w$ 的碱，水的自耦电离可忽略不计。

多元弱酸或弱碱电离所产生的离子可以水解，因此这些离子也可视为多元弱酸或弱碱。例如，PO_4^{3-} 是三元碱，可分三步水解，有如下三个不同的碱式电离常数：

$$\text{PO}_4^{3-} + \text{H}_2\text{O} \rightleftharpoons \text{HPO}_4^{2-} + \text{OH}^- \quad K_{b1}$$

$$\text{HPO}_4^{2-} + \text{H}_2\text{O} \rightleftharpoons \text{H}_2\text{PO}_4^- + \text{OH}^- \quad K_{b2}$$

$$\text{H}_2\text{PO}_4^- + \text{H}_2\text{O} \rightleftharpoons \text{H}_3\text{PO}_4 + \text{OH}^- \quad K_{b3}$$

不难推导得出，PO_4^{3-} 的三个碱式电离常数与 H_3PO_4 的三个酸式电离常数具有如下关系：

$$K_{a1}\cdot K_{b3} = K_{a2}\cdot K_{b2} = K_{a3}\cdot K_{b1} = K_w \tag{5.14}$$

两性物质的电离平衡

两性物质如 HCO_3^-、HPO_4^{2-}、H_2PO_4^- 和 NH_4Ac 等既可作酸又可作碱，它们与水之间发生水解反应。为了确定其水溶液的 pH，需要进行多重电离平衡分析。

以 HCO_3^- 为例，有

$$\text{HCO}_3^- + \text{H}_2\text{O} \rightleftharpoons \text{CO}_3^{2-} + \text{H}_3\text{O}^+ \quad K_{a2} = 4.7\times 10^{-11}$$

$$\text{HCO}_3^- + \text{H}_2\text{O} \rightleftharpoons \text{H}_2\text{CO}_3 + \text{OH}^- \quad K_{b2} = K_w/K_{a1} = 2.2\times 10^{-8}$$

由于 $K_{b2} > K_{a2}$，可以得出 HCO_3^- 水溶液呈碱性、pH>7 的定性结论。若要定量计算，需考虑 HCO_3^- 水溶液中同时存在的以下三种电离平衡：

(1) $\text{HCO}_3^- + \text{H}_2\text{O} \rightleftharpoons \text{CO}_3^{2-} + \text{H}_3\text{O}^+ \quad K_1 = K_{a2}$

(2) $\text{HCO}_3^- + \text{H}_2\text{O} \rightleftharpoons \text{H}_2\text{CO}_3 + \text{OH}^- \quad K_2 = K_w/K_{a1}$

(3) $\text{H}_3\text{O}^+ + \text{OH}^- \rightleftharpoons 2\text{H}_2\text{O} \quad K_3 = 1/K_w$

(1)+(2)+(3)，可得

$$2\text{HCO}_3^- \rightleftharpoons \text{CO}_3^{2-} + \text{H}_2\text{CO}_3 \quad K = K_1K_2K_3 = K_{a2}/K_{a1}$$

由于 HCO_3^- 的酸式电离常数和碱式电离常数相差不大，可认为生成的 CO_3^{2-} 和 H_2CO_3 浓度接近，即 $[\text{CO}_3^{2-}] \approx [\text{H}_2\text{CO}_3]$。

$$K = \frac{K_{a2}}{K_{a1}} = \frac{[\text{CO}_3^{2-}][\text{H}_2\text{CO}_3]}{[\text{HCO}_3^-]^2} \approx \frac{[\text{H}_2\text{CO}_3]^2}{[\text{HCO}_3^-]^2}\cdot\frac{[\text{H}_3\text{O}^+]^2}{[\text{H}_3\text{O}^+]^2} = \frac{[\text{H}_3\text{O}^+]^2}{K_{a1}^2}$$

$$HCO_3^- + H_2O \rightleftharpoons CO_3^{2-} + H_3O^+ \quad K_{a2} = 4.7 \times 10^{-11}$$

$$HCO_3^- + H_2O \rightleftharpoons H_2CO_3 + OH^- \quad K_{b2} = K_w / K_{a1} = 2.2 \times 10^{-8}$$

Because $K_{b2} > K_{a2}$, we can qualitatively conclude that HCO_3^- aqueous solution is alkaline and pH > 7. Quantitatively, the following three ionization equilibria exist simultaneously in HCO_3^- aqueous solution

(1) $HCO_3^- + H_2O \rightleftharpoons CO_3^{2-} + H_3O^+ \quad K_1 = K_{a2}$

(2) $HCO_3^- + H_2O \rightleftharpoons H_2CO_3 + OH^- \quad K_2 = K_w / K_{a1}$

(3) $H_3O^+ + OH^- \rightleftharpoons 2H_2O \quad K_3 = 1/K_w$

Applying (1)+(2)+(3), we have

$$2HCO_3^- \rightleftharpoons CO_3^{2-} + H_2CO_3 \quad K = K_1 K_2 K_3 = K_{a2}/K_{a1}$$

Because the acidic ionization and basic ionization constants of HCO_3^- do not differ much, the concentrations of the produced CO_3^{2-} and H_2CO_3 are close, or $[CO_3^{2-}] \approx [H_2CO_3]$.

$$K = \frac{K_{a2}}{K_{a1}} = \frac{[CO_3^{2-}][H_2CO_3]}{[HCO_3^-]^2} \approx \frac{[H_2CO_3]^2}{[HCO_3^-]^2} \cdot \frac{[H_3O^+]^2}{[H_3O^+]^2} = \frac{[H_3O^+]^2}{K_{a1}^2}$$

$$\therefore [H_3O^+] = \sqrt{K_{a1} K_{a2}} = 4.6 \times 10^{-9} \text{ mol L}^{-1} \quad \text{and} \quad pH = 8.34$$

5.4 Common-Ion Effect and Buffer Solutions

We have discussed in **Section 5.3** the ionization equilibria in an aqueous solution of a single electrolyte (either an acid, a base or an amphiprotic substance), and have learnt how to calculate its pH. In this section, we extend to aqueous solutions of both a weak electrolyte and a second source of the **common ions**, which are one of the ions produced in the ionization of the weak electrolyte.

Common-Ion Effect

The ions produced in the ionization of weak acid HA are H_3O^+ and A^-, both of which are the common ions of HA. Therefore, adding either another acid that can produce H_3O^+ or a salt that can produce A^- results in common-ion effect.

For an aqueous solution of 0.10 mol L^{-1} HAc, we have calculated in **Section 5.3** that $\alpha = 0.013$ and pH = 2.89. Now let us add into the HAc solution some NaAc, which can produce the common ion Ac^-, to make an aqueous solution that is initially 0.10 mol L^{-1} in both HAc and Ac^-. We can write the ionization equation as

$$HAc + H_2O \rightleftharpoons H_3O^+ + Ac^-$$

Initial:	0.10	0	0.10
Changes:	$-x$	x	x
Equilibrium:	$0.10 - x$	x	$0.10 + x$

$$K_a = \frac{[H_3O^+][Ac^-]}{[HAc]} = \frac{x(0.10+x)}{0.10-x} \approx x = 1.75 \times 10^{-5}$$

$$\therefore [H_3O^+] = K_a = 1.75 \times 10^{-5} \text{ mol L}^{-1} \quad \text{and} \quad pH = pK_a = 4.74$$

$$\alpha = \frac{[H_3O^+]}{c} = 0.00018 \ll 0.013$$

As can be seen, the addition of a common ion Ac^- greatly suppresses the ionization of the weak acid HAc. This can be easily understood by equilibrium shifting. The suppression of the ionization of a weak electrolyte caused by adding an ion that is a product of this ionization is called the **common-ion effect**.

For the mixture of a typical weak acid HA and its common ion A^-, we have

$$\therefore [\text{H}_3\text{O}^+] = \sqrt{K_{a1}K_{a2}} = 4.6 \times 10^{-9} \text{ mol L}^{-1} \quad 且 \quad \text{pH} = 8.34$$

5.4 同离子效应与缓冲溶液

5.3 节我们讨论了单一电解质（酸、碱或两性物质）水溶液中的电离平衡，并学习了如何计算其 pH。本节我们拓展到同时含有弱电解质及其**同离子**（即该弱电解质电离产生的某种离子）的水溶液。

同离子效应

弱酸 HA 电离产生的离子有 H_3O^+ 和 A^-，两者都是 HA 的同离子。因此，加入能产生 H_3O^+ 的另一种酸或能产生 A^- 的盐，均可导致同离子效应。

5.3 节我们计算过，0.10 mol L^{-1} HAc 水溶液的 $\alpha = 0.013$，pH = 2.89。现在我们在该 HAc 溶液中加入一些能产生同离子 Ac^- 的 NaAc，使溶液的初始 [HAc] 和 [Ac^-] 均为 0.10 mol L^{-1}，其电离方程式为

$$\text{HAc} + \text{H}_2\text{O} \rightleftharpoons \text{H}_3\text{O}^+ + \text{Ac}^-$$

初始浓度：　　　0.10　　　　　0　　　0.10
反应浓度：　　　$-x$　　　　　x　　　x
平衡浓度：　　　$0.10 - x$　　x　　　$0.10 + x$

$$K_a = \frac{[\text{H}_3\text{O}^+][\text{Ac}^-]}{[\text{HAc}]} = \frac{x(0.10 + x)}{0.10 - x} \approx x = 1.75 \times 10^{-5}$$

$$\therefore [\text{H}_3\text{O}^+] = K_a = 1.75 \times 10^{-5} \text{ mol L}^{-1} \quad 且 \quad \text{pH} = \text{p}K_a = 4.74$$

$$\alpha = \frac{[\text{H}_3\text{O}^+]}{c} = 0.00018 \ll 0.013$$

可以看到，加入同离子 Ac^- 极大地抑制了弱酸 HAc 的电离，这可以通过平衡移动来理解。由于加入一种与弱电解质电离产物相同的离子，而引起的对弱电解质电离的抑制，称为**同离子效应**。

对典型弱酸 HA 及其同离子 A^- 的混合液，有

$$K_a = \frac{[\text{H}_3\text{O}^+][\text{A}^-]}{[\text{HA}]} \quad 且 \quad [\text{H}_3\text{O}^+] = K_a \cdot \frac{[\text{HA}]}{[\text{A}^-]}$$

$$\therefore \text{pH} = \text{p}K_a - \lg\frac{[\text{HA}]}{[\text{A}^-]} = \text{p}K_a + \lg\frac{[\text{A}^-]}{[\text{HA}]}$$

A^- 是 HA 的共轭碱，我们可写出一个更为宽泛的通式：

$$\text{pH} = \text{p}K_a + \lg\frac{[共轭碱]}{[弱酸]} \tag{5.15}$$

类似地，

$$\text{pOH} = \text{p}K_b + \lg\frac{[共轭酸]}{[弱碱]} \tag{5.16}$$

上式称为**亨德森 - 哈塞巴奇方程**（劳伦斯·J. 亨德森，1878—1942；卡尔·A. 哈塞巴奇，1874—1962），可用于计算弱电解质溶液中不同物种浓度随 pH 的变化。

$$K_a = \frac{[H_3O^+][A^-]}{[HA]} \text{ and } [H_3O^+] = K_a \cdot \frac{[HA]}{[A^-]}$$

$$\therefore \text{pH} = pK_a - \lg\frac{[HA]}{[A^-]} = pK_a + \lg\frac{[A^-]}{[HA]}$$

As A^- is the conjugate base of HA, we can write a more general equation as

$$\text{pH} = pK_a + \lg\frac{[\text{conjugate base}]}{[\text{acid}]} \tag{5.15}$$

Similarly,

$$\text{pOH} = pK_b + \lg\frac{[\text{conjugate acid}]}{[\text{base}]} \tag{5.16}$$

The above equations are called **Henderson-Hasselbalch equations** (Lawrence J. Henderson, 1878—1942; Karl A. Hasselbalch, 1874—1962). They can be used to calculate the concentrations of different species with respect to pH in a weak electrolyte solution.

In the solution of HAc which is a monoprotic weak acid, the relationship between the concentrations of HAc and Ac^- can be classified into three categories according to the values of pH and pK_a as

1) If pH $< pK_a$, $[H_3O^+] > K_a$ and $[HAc] > [Ac^-]$, which means that the main species is HAc;
2) If pH $= pK_a$, $[H_3O^+] = K_a$ and $[HAc] = [Ac^-]$;
3) If pH $> pK_a$, $[H_3O^+] < K_a$ and $[HAc] < [Ac^-]$, which means that the main species is Ac^-.

For the diprotic weak acid H_2S, the species that contain S elements are H_2S, HS^-, and S^{2-}. All their concentrations vary with pH and the relationship can be classified into five categories according to the values of pH, pK_{a1}, and pK_{a2} as

1) If pH $< pK_{a1}$, $[H_3O^+] > K_{a1}$ and $[H_2S] > [HS^-] > [S^{2-}]$, which means that the main species is H_2S;
2) If pH $= pK_{a1}$, $[H_3O^+] = K_{a1}$ and $[H_2S] = [HS^-] \gg [S^{2-}]$;
3) If $pK_{a1} <$ pH $< pK_{a2}$, $K_{a1} > [H_3O^+] > K_{a2}$ and $[H_2S] < [HS^-] > [S^{2-}]$, which means that the main species is HS^-;
4) If pH $= pK_{a2}$, $[H_3O^+] = K_{a2}$ and $[HS^-] = [S^{2-}] \gg [H_2S]$;
5) If pH $> pK_{a2}$, $[H_3O^+] < K_{a2}$ and $[S^{2-}] > [HS^-] > [H_2S]$, which means that the main species is S^{2-}.

For a triprotic weak acid represented by H_3A, the concentrations of H_3A, H_2A^-, HA^{2-}, and A^{3-} vary with pH. Their relationship can be classified into seven categories according to the values of pH, pK_{a1}, pK_{a2}, and pK_{a3} in a similar manner, and can be schematically illustrated in **Figure 5.2**.

Buffer Solutions

The concentration ratio of the conjugate pairs can be controlled by adjusting the pH of the solution, and conversely the pH of the solution can also be controlled by adjusting the concentration ratio of the conjugate pairs. A practical example is a buffer solution. A **buffer solution** is a solution whose pH value changes only slightly on the addition of small amounts of either an acid or a base. Therefore, a buffer solution requires two components that do not react with each other: one can neutralize acids and the other can neutralize bases. Conjugate pairs are the common choice to make buffer solutions.

When adding the same amount of acid or base, the pH of a better buffer solution should change less. Pure water is not a good buffer. Adding 1.00 mL 1.0 mol L^{-1} either HCl or NaOH solution into 1.00 L water at pH = 7 results in a pH change by 4 units. However, add 1.00 mL 1.0 mol L^{-1} either HCl or NaOH solution into a 1.00 L HAc-NaAc solution with $[HAc] = [Ac^-] = 0.010$ mol L^{-1}, the pH change can be calculated as

Initial: pH $= pK_a = 4.74$

Adding HCl: $[HAc] = \dfrac{0.010 \times 1000 + 1.0 \times 1.00}{1000 + 1}$ mol L^{-1} = 0.011 mol L^{-1}

$[Ac^-] = \dfrac{0.010 \times 1000 - 1.0 \times 1.00}{1000 + 1}$ mol L^{-1} = 0.0090 mol L^{-1}

在一元弱酸 HAc 溶液中，根据 pH 和 pK_a 的大小关系，HAc 和 Ac$^-$ 的浓度关系可归为三类：
1) 当 pH < pK_a 时，[H$_3$O$^+$] > K_a，[HAc] > [Ac$^-$]，此时溶液中的主要物种为 HAc；
2) 当 pH = pK_a 时，[H$_3$O$^+$] = K_a，[HAc] = [Ac$^-$]；
3) 当 pH > pK_a 时，[H$_3$O$^+$] < K_a，[HAc] < [Ac$^-$]，此时溶液中的主要物种为 Ac$^-$。

对于二元弱酸 H$_2$S，溶液中含 S 元素的物种有 H$_2$S、HS$^-$ 和 S^{2-}。这些物种的浓度均随 pH 变化，根据 pH、pK_{a1} 和 pK_{a2} 的大小关系，其物种的浓度关系可归为五类：
1) 当 pH < pK_{a1} 时，[H$_3$O$^+$] > K_{a1}，[H$_2$S] > [HS$^-$] > [S^{2-}]，此时溶液中的主要物种为 H$_2$S；
2) 当 pH = pK_{a1} 时，[H$_3$O$^+$] = K_{a1}，[H$_2$S] = [HS$^-$] ≫ [S^{2-}]；
3) 当 pK_{a1} < pH < pK_{a2} 时，K_{a1} > [H$_3$O$^+$] > K_{a2}，[H$_2$S] < [HS$^-$] > [S^{2-}]，此时溶液中的主要物种为 HS$^-$；
4) 当 pH = pK_{a2} 时，[H$_3$O$^+$] = K_{a2}，[HS$^-$] = [S^{2-}] ≫ [H$_2$S]；
5) 当 pH > pK_{a2} 时，[H$_3$O$^+$] < K_{a2}，[S^{2-}] > [HS$^-$] > [H$_2$S]，此时溶液中的主要物种为 S^{2-}。

对于以 H$_3$A 表示的三元弱酸，H$_3$A、H$_2$A$^-$、HA^{2-} 和 A^{3-} 的浓度均随 pH 变化，它们的浓度关系可根据 pH、pK_{a1}、pK_{a2} 和 pK_{a3} 的大小，采用类似的方法归为七类，如**图 5.2** 所示。

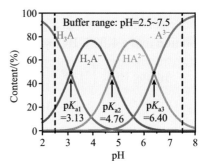

Figure 5.2 The concentrations of various species in a triprotic weak acid H$_3$A as a function of pH. The pK_a values and buffer range shown in the figure are for citric acid.

图 5.2 三元弱酸 H$_3$A 中各物种的浓度与 pH 的函数关系。图中给出了柠檬酸的 pK_a 及其缓冲范围。

缓冲溶液

共轭酸碱对的浓度比可以通过调节溶液的 pH 来控制，反之溶液的 pH 也可以通过调节共轭酸碱对的浓度比来控制，缓冲溶液就是实例之一。**缓冲溶液**是当外加少量酸或碱时能保持 pH 基本不变的溶液。因此缓冲溶液通常要求包含两种互不反应的组分：一种能中和酸，另一种能中和碱。共轭酸碱对是制备缓冲溶液的常规选择。

当加入等量酸或碱时，好的缓冲溶液 pH 变化应较小。纯水不是好的缓冲溶液，在 1.00 L pH=7 的水中加入 1.00 mL 1.0 mol L^{-1} HCl 或 NaOH 溶液，可使 pH 改变 4 个单位。但在 1.00 L [HAc] = [Ac$^-$] = 0.010 mol L^{-1} 的 HAc-NaAc 溶液中加入 1.00 mL 1.0 mol L^{-1} HCl 或 NaOH 溶液，其 pH 改变量可计算为：

初始：pH = pK_a = 4.74

加入 HCl：$[\text{HAc}] = \dfrac{0.010 \times 1000 + 1.0 \times 1.00}{1000 + 1}$ mol L^{-1} = 0.011 mol L^{-1}

$[\text{Ac}^-] = \dfrac{0.010 \times 1000 - 1.0 \times 1.00}{1000 + 1}$ mol L^{-1} = 0.0090 mol L^{-1}

∴ pH = pK_a + lg$\left(\dfrac{0.011}{0.0090}\right)$ = 4.65 且 ΔpH = −0.09

加入 NaOH：$[\text{HAc}] = \dfrac{0.010 \times 1000 - 1.0 \times 1.00}{1000 + 1}$ mol L^{-1} = 0.0090 mol L^{-1}

$[\text{Ac}^-] = \dfrac{0.010 \times 1000 + 1.0 \times 1.00}{1000 + 1}$ mol L^{-1} = 0.011 mol L^{-1}

∴ pH = pK_a + lg$\left(\dfrac{0.0090}{0.011}\right)$ = 4.83 且 ΔpH = 0.09

$$\therefore \text{pH} = \text{p}K_a + \lg\left(\frac{0.011}{0.0090}\right) = 4.65 \quad \text{and} \quad \Delta\text{pH} = -0.09$$

Adding NaOH: $[\text{HAc}] = \dfrac{0.010 \times 1000 - 1.0 \times 1.00}{1000 + 1} \text{ mol L}^{-1} = 0.0090 \text{ mol L}^{-1}$

$$[\text{Ac}^-] = \dfrac{0.010 \times 1000 + 1.0 \times 1.00}{1000 + 1} \text{ mol L}^{-1} = 0.011 \text{ mol L}^{-1}$$

$$\therefore \text{pH} = \text{p}K_a + \lg\left(\frac{0.0090}{0.011}\right) = 4.83 \quad \text{and} \quad \Delta\text{pH} = 0.09$$

Adding a small amount of a strong acid or base only slightly changes the pH of this solution, so this HAc-NaAc solution serves as a good buffer. As a comparison, adding 1.00 mL 1.0 mol L^{-1} either HCl or NaOH solution into a 1.00 L HCl solution with pH = 4.74, the corresponding pH changes are

Adding HCl: $[\text{H}_3\text{O}^+] = \dfrac{10^{-4.74} \times 1000 + 1.0 \times 1.00}{1000 + 1} \text{ mol L}^{-1} = 0.0010 \text{ mol L}^{-1}$

$$\therefore \text{pH} = 3.0 \quad \text{and} \quad \Delta\text{pH} = -1.74$$

Adding NaOH: $[\text{OH}^-] = \dfrac{1.0 \times 1.00 - 10^{-4.74} \times 1000}{1000 + 1} \text{ mol L}^{-1} = 0.0010 \text{ mol L}^{-1}$

$$\therefore \text{pOH} = 3.0, \quad \text{pH} = 11.00 \quad \text{and} \quad \Delta\text{pH} = 6.26$$

Adding a small amount of a strong acid or base changes greatly the pH of this solution. Therefore, strong acids and bases are not good buffer solutions.

For the same pH change, a good buffer solution can also accommodate more amount of acid or base. The **buffer range** is the pH range over which a buffer can effectively neutralize the added acids and bases and maintain a fairly constant pH. In practical, a range of 2 pH units, varying from pK_a− 1 to pK_a+ 1, is the maximum pH range to which a buffer solution should be exposed. For example, the HAc-NaAc buffer solution (pK_a= 4.74) has a buffer range of pH = 3.7~5.7, and the NH$_4$OH-NH$_4$Cl buffer solution (pK_a= 9.26) has a buffer range of pH = 8.3~10.3. The **buffer capacity**, which can be used to measure the ability of a buffer, is the amount of acid or base that a buffer can neutralize within its buffer range. According to Henderson-Hasselbalch equation, the ratio [conjugate base] / [acid] = 0.10 at pH = pK_a− 1, and [conjugate base] / [acid]= 10 at pH = pK_a+ 1. Therefore, within a buffer range from pK_a− 1 to pK_a+ 1, the ratio [conjugate base] / [acid] vary from 0.10 to 10.

For a buffer solution composed of the HA-A$^-$ conjugate pair to be effective, the following three conditions are necessarily required:
1) The ratio is within the limits of 0.1 < [A$^-$] : [HA] < 10.
2) The molarity of each buffer component exceeds the value of K_a by a factor of at least 100, i.e. [HA] > 100K_a and [A$^-$] > 100K_a.
3) The amounts of both buffer components are at least 10 times as great as the amount of acid or base to be neutralized, i.e. n(HA) + n(A$^-$) > 10n(acid or base).

To prepare an effective buffer solution at a certain pH, the following considerations should be taken into account:
1) Choose a conjugate pair whose pK_a is close to the required pH to be buffered, normally pK_a−1<pH< pK_a+1.
2) Adjust the ratio of the conjugate pairs to obtain the required pH. The buffer capacity is the highest with a 1:1 ratio.
3) Increase the total concentration of the buffer solution to meet the required buffer capacity.

5.5 Acid-Base Indicators and Titration

Acid-Base Indicators

An **acid-base indicator** is a substance added in small amounts to a solution so that the color of the solution

加入少量强酸或强碱只会略微改变溶液的 pH，所以这种 HAc-NaAc 溶液是好的缓冲溶液。作为比较，在 1.00 L pH = 4.74 的 HCl 溶液中加入 1.00 mL 1.0 mol L^{-1} HCl 或 NaOH 溶液，相应的 pH 改变量为

加入 HCl：$[H_3O^+] = \dfrac{10^{-4.74} \times 1000 + 1.0 \times 1.00}{1000 + 1}$ mol L^{-1} = 0.0010 mol L^{-1}

∴ pH = 3.0　且　ΔpH = −1.74

加入 NaOH：$[OH^-] = \dfrac{1.0 \times 1.00 - 10^{-4.74} \times 1000}{1000 + 1}$ mol L^{-1} = 0.0010 mol L^{-1}

∴ pOH = 3.0，　pH = 11.00　且　ΔpH = 6.26

加入少量强酸或强碱会显著改变溶液的 pH，因此，强酸、强碱不是好的缓冲溶液。

当 pH 改变量相同时，好的缓冲溶液应能缓冲更大量的外加酸碱。缓冲溶液能有效中和外加酸碱而保持 pH 几乎不变的范围称为**缓冲范围**，通常为从 pK_a− 1 到 pK_a+ 1、共计 2 个 pH 单位的范围，这也是该缓冲溶液适用的最大 pH 范围。例如，HAc-NaAc 缓冲溶液（pK_a= 4.74）的缓冲范围是 pH = 3.7 ~ 5.7，NH$_4$OH-NH$_4$Cl 缓冲溶液（pK_a= 9.26）的缓冲范围是 pH = 8.3 ~ 10.3。缓冲溶液在其缓冲范围内所能中和外来酸或碱的量称为**缓冲容量**，可用来衡量缓冲溶液的缓冲能力。根据亨德森 - 哈塞巴奇方程，当 pH = pK_a− 1 时，[共轭碱] / [弱酸] = 0.10；当 pH = pK_a+ 1 时，[共轭碱] / [弱酸] = 10。因此，在从 pK_a− 1 到 pK_a+ 1 的缓冲范围内，[共轭碱] / [弱酸] 比率在 0.10 ~ 10 之间变化。

为使由共轭酸碱对 HA-A$^-$ 组成的缓冲溶液更为有效，以下三个条件必须满足：

1) 共轭酸碱对浓度比满足 0.1 < [A$^-$] : [HA] < 10。
2) 每种组分的浓度超过 K_a 至少 100 倍，即 [HA]>100K_a 及 [A$^-$]> 100K_a。
3) 总组分物质的量超过被缓冲酸碱物质的量的 10 倍，即 n(HA) + n(A$^-$) > 10n(酸或碱)。

要配制在一定 pH 下有效的缓冲溶液，应考虑以下因素：

1) 选择 pK_a 接近所需缓冲 pH 的共轭酸碱对，通常满足 pK_a− 1 < pH <pK_a+ 1。
2) 调节共轭酸碱对浓度比以获得所需的 pH，1:1 浓度比时缓冲容量最高。
3) 增加缓冲溶液的总浓度，以满足所需的缓冲容量。

5.5 酸碱指示剂与酸碱滴定

酸碱指示剂

酸碱指示剂是少量添加至溶液中、使溶液颜色随 pH 变化的物质，通常为有机弱酸碱，其酸式（用 HIn 表示）和碱式（用 In$^-$ 表示）具有不同颜色，可表示为

$$HIn + H_2O \rightleftharpoons H_3O^+ + In^- \quad (5.17)$$

酸式颜色　　　　　　碱式颜色

changes depending on the pH. Acid-base indicators are frequently weak organic acids or bases that have different colors in the acid and base forms, represented symbolically as HIn and In^-, respectively, as given by

$$HIn + H_2O \rightleftharpoons H_3O^+ + In^- \qquad (5.17)$$
$$\text{Acid color} \qquad\qquad \text{Base color}$$

When just a small amount of indicator is added to a solution, the change in pH of the solution is negligible. Instead, the ionization equilibrium of the indicator itself is affected by the prevailing $[H_3O^+]$ in solution. In general, if more than 90% of an indicator is in the acid form, the solution takes on the acid color. If more than 90% is in the base form, the solution takes on the base color. If $[HIn] \approx [In^-]$, the solution shows an intermediate color. The complete change in color usually occurs over a range of about 2 pH units, from $pK_{HIn} - 1$ to $pK_{HIn} + 1$. The colors and pH ranges of several commonly used acid-base indicators are shown in **Table 5.4**, and summarized as follows:

1) If pH $< pK_{HIn} - 1$, $[In^-]/[HIn] < 0.10$, and the solution takes on the acid color.
2) If $pK_{HIn} - 1 \leq$ pH $\leq pK_{HIn} + 1$, $0.10 \leq [In^-]/[HIn] \leq 10$, and the solution shows an intermediate color.
3) If pH $> pK_{HIn} + 1$, $[In^-]/[HIn] > 10$, and the solution takes on the base color.

Both acid-base indicator and buffer solutions are conjugate pairs. However, buffer solutions are used in very large amounts so that the pH of the solution does not change significantly, normally within the buffer range of $pK_a - 1 \sim pK_a + 1$. The acid-base indicators, on the other hand, are used only in very small amounts in order to indicate significant pH changes by its color change.

Table 5.4 Color Changes and pH Ranges of Several Commonly Used Acid-Base Indicators
表 5.4 一些常用酸碱指示剂的颜色变化及变色 pH 范围

Acid-Base Indicator（酸碱指示剂）	Acid Color（酸式颜色）	Color Change pH（变色 pH）	Base Color（碱式颜色）
Methyl violet（甲基紫）	Yellow（黄色）	0~3.0	Violet（紫色）
Thymol blue（百里酚蓝）	Red（红色）	1.2~2.8	Yellow（黄色）
Bromphenol blue（溴酚蓝）	Yellow（黄色）	3.0~4.6	Blue-violet（蓝紫色）
Methyl orange（甲基橙）	Red（红色）	3.1~4.4	Yellow-orange（橙黄色）
Bromcresol green（溴甲酚绿）	Yellow（黄色）	3.8~5.4	Blue（蓝色）
Methyl red（甲基红）	Red（红色）	4.4~6.2	Yellow（黄色）
Chlorphenol red（氯酚红）	Yellow（黄色）	5.2~7.0	Red（红色）
Bromthymol blue（溴百里酚蓝）	Yellow（黄色）	6.0~7.6	Blue（蓝色）
Phenol red（酚红）	Yellow（黄色）	6.8~8.0	Red（红色）
Thymol blue（百里酚蓝）	Yellow（黄色）	8.0~9.6	Blue（蓝色）
Phenolphthalein（酚酞）	Colorless（无色）	8.2~10.0	Red（红色）
Thymolphthalein（百里酚酞）	Colorless（无色）	9.4~10.6	Blue（蓝色）
Alizarin yellow-R（茜素黄 R）	Yellow（黄色）	10.2~12.0	Violet（紫色）

Titration and Titration Curve

An acid-base indicator is usually prepared as a solution and a few drops of this indicator solution are added to a solution that is to be titrated to indicate its end by the color change in an acid-base titration. **Titration** is a process where a reaction is carried out by the carefully controlled addition of one solution to another. In a titration, an **analyte** solution is placed in an Erlenmeyer flask, together with a few drops of indicator. Another solution that is carefully added from a burette is called the **titrant**. The **stoichiometric point** is the point where both reactants have reacted completely and the titration should be stopped. The **end point** is the point at which the indicator changes color and indicates the titration is ended. The end point must match the stoichiometric point so that the color change marked by the end point will signal the attainment of the stoichiometric point. This match can be achieved by using an indicator whose color change occurs over a pH range that includes the pH of the stoichiometric point. In this section, we will only introduce acid-base titrations although there are also other types of titrations such as redox titration, complexometric titration, etc.

During an acid-base titration, we can always measure the pH of the titrated solution with a pH meter and plot the pH versus volume of the added titrant. Such a graph is called a **titration curve**. The titrant is normally a strong acid or base with a known concentration. The concentration of the analyte, which can be

在溶液中加入少量指示剂，溶液的 pH 变化可忽略不计，相反，指示剂自身的电离平衡会受到溶液中已存在的 $[H_3O^+]$ 的影响。一般来说，如果超过 90% 的指示剂为酸式，溶液呈酸式颜色；若超过 90% 的指示剂为碱式，溶液呈碱式颜色；如果 $[HIn] \approx [In^-]$，溶液则呈中间色。颜色的完全变化通常发生在从 $pK_{HIn} - 1$ 到 $pK_{HIn} + 1$ 的大约 2 个 pH 单位范围内。几种常用酸碱指示剂的颜色及变色 pH 范围如**表 5.4** 所示，并总结如下：

1) 当 $pH < pK_{HIn} - 1$ 时，$[In^-] / [HIn] < 0.10$，溶液呈酸式颜色。
2) 当 $pK_{HIn} - 1 \leq pH \leq pK_{HIn} + 1$ 时，$0.10 \leq [In^-] / [HIn] \leq 10$，溶液呈中间色。
3) 当 $pH > pK_{HIn} - 1$ 时，$[In^-] / [HIn] > 10$，溶液呈碱式颜色。

酸碱指示剂和缓冲溶液都是共轭酸碱对，但缓冲溶液用量很大，因此溶液的 pH 不发生显著变化，通常在 $pK_a - 1 \sim pK_a + 1$ 的缓冲范围内。而酸碱指示剂用量极少，正是通过其颜色变化来指示溶液 pH 的显著变化。

酸碱滴定与滴定曲线

通常将酸碱指示剂配制成溶液，并将数滴该指示剂溶液加入待滴定的溶液中，通过酸碱滴定过程中的颜色变化来指示滴定终点。**滴定**是将一种溶液小心地滴加到另一种溶液中来进行某种反应的过程。滴定过程中，**待测物**溶液与数滴指示剂一起放入锥形瓶中，另一种从滴定管中小心加入的溶液称为**滴定剂**。两种反应物均已完全反应、滴定应停止的点称为**化学计量点**，指示剂变色指示滴定终了的点称为**终点**。终点必须与化学计量点匹配，这种匹配可通过使用变色范围包含化学计量点所对应 pH 的指示剂来实现，这样由终点标记的颜色变化将指示达到了化学计量点。本节我们只介绍酸碱滴定，而不涉及诸如氧化还原滴定、配位解离滴定等其他类型的滴定。

在酸碱滴定过程中，我们随时可以用 pH 计测量滴定溶液的 pH，并绘制 pH 对已加入滴定剂体积的曲线图，这样得到的曲线称为**滴定曲线**。滴定剂通常是已知浓度的强酸或强碱，待测物可以是强酸碱也可以是弱酸碱，其浓度未知，可以通过分析滴定曲线来确定。下面给出几个酸碱滴定的示例：

1) 用强碱滴定强酸

用 0.100 mol L^{-1} NaOH 溶液滴定 25.00 mL 0.100 mol L^{-1} HCl 溶液的滴定曲线如**图 5.3(a)** 所示。滴定开始时，pH = $-\lg(0.100) = 1.00$，随着滴定的进行，一直到化学计量点之前，pH 均缓慢上升。在化学计量点（pH = 7）处应加入 25.00 mL NaOH 溶液，此时 pH 急剧上升。超过化学计量点后，pH 再次缓慢上升。该滴定反应的最佳指示剂是溴百里酚蓝，其变色 pH 范围为 6.0~7.6，但诸如甲基红、酚酞等在 4~10 的 pH 范围内变色的其他指示剂也适用。

2) 用强碱滴定弱酸

用 0.100 mol L^{-1} NaOH 溶液滴定 25.00 mL 0.100 mol L^{-1} HAc 溶液的滴定曲线如**图 5.3(b)** 所示。滴定开始时，pH = $-\lg(\sqrt{K_a c})$ = 2.89。因为 HAc 仅部分电离，此 pH 高于 HCl 的对应值。由于中和反应生成的 Ac$^-$ 的同离子效应，滴定初期 pH 上升较为急剧。随后直到化学计量点前的相当长一段曲线上，形成了缓冲溶液，pH 缓慢上

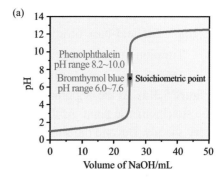

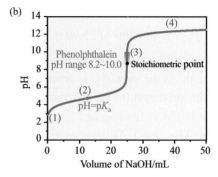

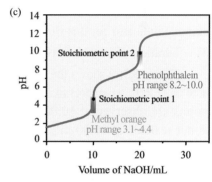

Figure 5.3 Various titration curves. (a) Titration curve for the titration of 25.00 mL 0.100 mol L^{-1} HCl with 0.100 mol L^{-1} NaOH. (b) Titration curve for the titration of 25.00 mL 0.100 mol L^{-1} HAc with 0.100 mol L^{-1} NaOH. This curve can be divided into four stages: (1) pH of a pure weak acid initially; (2) pH of a buffer solution over a broad range around pH = pK_a; (3) pH of a salt solution undergoing hydrolysis around the stoichiometric point; (4) pH of a strong base solution over a broad range beyond the stoichiometric point. (c) Titration curve for the titration of 10.00 mL 0.100 mol L^{-1} H$_3$PO$_4$ with 0.100 mol L^{-1} NaOH.

图 5.3 各种滴定曲线：(a) 用 0.100 mol L^{-1} NaOH 溶液滴定 25.00 mL 0.100 mol L^{-1} HCl 溶液的滴定曲线；(b) 用 0.100 mol L^{-1} NaOH 溶液滴定 25.00 mL 0.100 mol L^{-1} HAc 溶液的滴定曲线，可分为四个阶段：(1) 初始阶段为弱酸的 pH；(2) pH = pK_a 附近的较宽范围内为缓冲溶液的 pH；(3) 化学计量点附近为盐溶液水解的 pH；(4) 化学计量点之后的较宽范围内为强碱溶液的 pH；(c) 用 0.100 mol L^{-1} NaOH 溶液滴定 10.00 mL 0.100 mol L^{-1} H$_3$PO$_4$ 溶液的滴定曲线。

either strong or weak, is unknown and can be determined by analyzing the titration curve. Some illustrative examples of acid-base titrations are listed below.

1) Titration of a strong acid with a strong base.

The titration curve for the titration of 25.00 mL 0.100 mol L^{-1} HCl with 0.100 mol L^{-1} NaOH is shown in **Figure 5.3(a)**. At the beginning of the titration, pH = $-$lg (0.100) = 1.00. The pH increases slowly until just before the stoichiometric point. At the stoichiometric point (pH = 7), 25.00 mL NaOH should be added and the pH rises very sharply. Beyond the stoichiometric point, the pH increases slowly again. The most suitable indicator for this titration is bromthymol blue with a color change range of pH 6.0~7.6. However, other indicators whose color changes in the pH range from 4~10 such as methyl red and phenolphthalein are also suitable.

2) Titration of a weak acid with a strong base.

The titration of 25.00 mL 0.100 mol L^{-1} HAc with 0.100 mol L^{-1} NaOH is pictured in **Figure 5.3(b)**. At the beginning of the titration, pH = $-$lg ($\sqrt{K_a c}$) = 2.89, which is higher than that of HCl due to partial ionization of HAc. At the start of the titration, there is an initial rather sharp increase in pH because of the common ion effect of Ac$^-$ produced by the neutralization. Over a long section of the curve preceding the stoichiometric point, a buffer solution is formed and the pH increases only gradually. At the half-neutralization point, pH = pK_a because [HA] = [Ac$^-$]. At the stoichiometric point, 25.00 mL NaOH should be added and pH > 7 due to the hydrolysis of Ac$^-$ produced. Beyond the stoichiometric point, the pH is established entirely by the concentration of the unreacted OH$^-$ and the titration curve is identical to that of a strong acid with a strong base. The steep portion of the titration curve at the stoichiometric point occurs over a relatively short pH range (7~10). The most suitable indicator for this titration is phenolphthalein with a color change range of pH 8.2 ~ 10.0.

3) Titration of a weak polyprotic acid with a strong base.

Figure 5.3(c) shows the titration of 10.00 mL 0.100 mol L^{-1} H$_3$PO$_4$ with 0.100 mol L^{-1} NaOH. For a triprotic acid as H$_3$PO$_4$, we expect to see three separate stoichiometric points for each proton. The first two stoichiometric points are observed at 10.00 mL and 20.00 mL on the volume axis. However, a third stoichiometric point is not realized at 30.00 mL in this titration. pK_{a3} = $-$lg (4.8×10^{-13}) = 12.32 of H$_3$PO$_4$ indicates that the pH of the strongly hydrolyzed PO$_4^{3-}$ solution is beyond what can be reached by adding 0.100 mol L^{-1} NaOH to the original solution considering the diluting effect. Methyl orange (color change at pH = 3.1~4.4) and phenolphthalein are appropriate indicators for the first and second equivalence points, respectively.

5.6 Solubility Product Constant and Its Relationship with Solubility

Starting from this section, we will discuss precipitation-dissolution equilibrium (also called solubility equilibrium), which is a dynamic equilibrium existed between a chemical compound in the solid state and a solution of that compound. In solubility equilibrium, individual molecules migrate between the solid and solution phases so that the rate of dissolution equals the rate of precipitation. When this equilibrium is established, a saturated solution is formed. As mentioned earlier in **Section 2.8**, the concentration of the solute in a saturated solution is called the **solubility**, normally given as the maximum grams of solute dissolved in 100 g solvent.

Solubility Product Constant, K_{sp}

In general, when an equilibrium is established in a saturated solution between a solid solute A$_m$B$_n$ and its aqueous ions A^{n+} and B^{m-}, as

$$A_mB_n(s) \rightleftharpoons mA^{n+}(aq) + nB^{m-}(aq)$$

the equilibrium constant for this solubility equilibrium is called the **solubility product constant** (K_{sp}) and is given by

$$K_{sp} = \frac{[a(A^{n+})]^m [a(B^{m-})]^n}{a(A_mB_n)} \tag{5.18}$$

升。在化学计量点处应加入 25.00 mL NaOH 溶液，由于生成的 Ac^- 的水解，此时 pH> 7。超过化学计量点后，pH 由未反应的 OH^- 浓度决定，滴定曲线与前述强碱滴强酸的曲线一致。化学计量点附近滴定曲线急剧上升的部分，所对应的 pH 范围（7~10）相对较窄。该滴定反应的最佳指示剂是酚酞，变色 pH 范围为 8.2~10.0。

3) 用强碱滴定多元弱酸

图 5.3(c) 给出了用 0.100 mol L^{-1} NaOH 溶液滴定 10.00 mL 0.100 mol L^{-1} H_3PO_4 溶液的滴定曲线。对于 H_3PO_4 这样的三元酸，我们预期有对应于三个质子的三个分立的化学计量点。在体积轴的 10.00 mL 和 20.00 mL 处，可观测到前两个化学计量点，而该滴定的第三个化学计量点在 30.00 mL 处并未看到。磷酸的 $pK_{a3} = -\lg(4.8 \times 10^{-13}) = 12.32$，这表明考虑到稀释效果，强水解的 PO_4^{3-} 溶液的 pH 已经超出了向初始溶液中添加 0.100 mol L^{-1} NaOH 溶液所能达到的范围。甲基橙（变色 pH 范围为 3.1 ~ 4.4）和酚酞分别是第一和第二个化学计量点的合适指示剂。

5.6 溶度积常数及其与溶解度的关系

从本节开始，我们将讨论沉淀溶解平衡（也称溶解度平衡），即固态化合物与其溶液之间所存在的动态平衡。在溶解度平衡中，单个分子在固相和液相之间迁移，其溶解速率等于沉淀速率。平衡建立时，就形成了饱和溶液。如 **2.8 节**所述溶质在饱和溶液中的浓度称为**溶解度**，通常用溶解在 100 g 溶剂中的最大溶质克数表示。

溶度积常数 K_{sp}

一般来说，当固态溶质 A_mB_n 与其水合离子 A^{n+} 和 B^{m-} 在饱和溶液中建立如下平衡时

$$A_mB_n(s) \rightleftharpoons mA^{n+}(aq) + nB^{m-}(aq)$$

该溶解度平衡的平衡常数称为**溶度积常数**（K_{sp}），由下式给出：

$$K_{sp} = \frac{\left[a(A^{n+})\right]^m \left[a(B^{m-})\right]^n}{a(A_mB_n)} \tag{5.18}$$

对于纯固体，$a(A_mB_n) = 1$。对于理想溶液或稀溶液，离子的活度系数也为 1。令 $c^\ominus = 1$ mol L^{-1} 以使活度量纲为 1，**式 (5.18)** 可化简为

$$K_{sp} = \left[A^{n+}\right]^m \left[B^{m-}\right]^n \tag{5.19}$$

化合物的溶解度与温度相关，K_{sp} 也是如此。**表 5.5** 和**附录 C.2** 列出了各种化合物在 25°C 时的 K_{sp}。

K_{sp} 与溶解度的关系

K_{sp} 和溶解度均与饱和溶液中溶质的浓度有关，因此 K_{sp} 与溶质的摩尔溶解度（即摩尔浓度）之间存在确定的关系。对于理想溶液或稀溶液，K_{sp} 可直接由其溶解度计算。例如，已知 $SrSO_4$ 在 25°C 的溶解度为 0.0135 g/100 g H_2O，根据 $SrSO_4$ 的摩尔质量 $M = 183.7$ g mol^{-1}，以及 $SrSO_4$ 溶液的密度约等于水的密度 $\rho \approx 1.0$ g cm^{-3}，可计算 $SrSO_4$

For a pure solid, $a(A_mB_n)=1$. For ideal solutions or solutions with dilute concentrations, the activity coefficients of ions are also 1. By setting $c^\circ =1$ mol L^{-1} to make sure the activities are dimensionless, **Equation (5.18)** can be simplified as

$$K_{sp} = \left[A^{n+}\right]^m \left[B^{m-}\right]^n \tag{5.19}$$

The solubility of a chemical compound is temperature dependent, and so is the K_{sp}. The K_{sp} values of various chemical compounds at 25°C are listed in **Table 5.5** and **Appendix C.2**.

Table 5.5 The K_{sp} Values of Various Chemical Compounds at 298.15 K

表 5.5 298.15 K 时各种化合物的 K_{sp}

Chemical Compound（化合物）	Solubility Equilibrium（溶解度平衡）	K_{sp}
Aluminum hydroxide（氢氧化铝）	$Al(OH)_3(s) \rightleftharpoons Al^{3+}(aq) + 3OH^-(aq)$	1.3×10^{-33}
Barium carbonate（碳酸钡）	$BaCO_3(s) \rightleftharpoons Ba^{2+}(aq) + CO_3^{2-}(aq)$	2.58×10^{-9}
Barium chromate（铬酸钡）	$BaCrO_4(s) \rightleftharpoons Ba^{2+}(aq) + CrO_4^{2-}(aq)$	1.17×10^{-10}
Barium sulfate（硫酸钡）	$BaSO_4(s) \rightleftharpoons Ba^{2+}(aq) + SO_4^{2-}(aq)$	1.08×10^{-10}
Calcium carbonate（碳酸钙）	$CaCO_3(s) \rightleftharpoons Ca^{2+}(aq) + CO_3^{2-}(aq)$	3.36×10^{-9}
Calcium fluoride（氟化钙）	$CaF_2(s) \rightleftharpoons Ca^{2+}(aq) + 2F^-(aq)$	5.3×10^{-9}
Calcium sulfate（硫酸钙）	$CaSO_4(s) \rightleftharpoons Ca^{2+}(aq) + SO_4^{2-}(aq)$	4.93×10^{-5}
Chromium(III) hydroxide（氢氧化铬(III)）	$Cr(OH)_3(s) \rightleftharpoons Cr^{3+}(aq) + 3OH^-(aq)$	6.3×10^{-31}
Iron(III) hydroxide（氢氧化铁(III)）	$Fe(OH)_3(s) \rightleftharpoons Fe^{3+}(aq) + 3OH^-(aq)$	2.79×10^{-39}
Lead(II) chloride（氯化铅(II)）	$PbCl_2(s) \rightleftharpoons Pb^{2+}(aq) + 2Cl^-(aq)$	1.70×10^{-5}
Lead(II) chromate（铬酸铅(II)）	$PbCrO_4(s) \rightleftharpoons Pb^{2+}(aq) + CrO_4^{2-}(aq)$	2.8×10^{-13}
Lead(II) iodide（碘化铅(II)）	$PbI_2(s) \rightleftharpoons Pb^{2+}(aq) + 2I^-(aq)$	9.8×10^{-9}
Magnesium carbonate（碳酸镁）	$MgCO_3(s) \rightleftharpoons Mg^{2+}(aq) + CO_3^{2-}(aq)$	6.82×10^{-6}
Magnesium fluoride（氟化镁）	$MgF_2(s) \rightleftharpoons Mg^{2+}(aq) + 2F^-(aq)$	5.16×10^{-11}
Magnesium hydroxide（氢氧化镁）	$Mg(OH)_2(s) \rightleftharpoons Mg^{2+}(aq) + 2OH^-(aq)$	5.61×10^{-12}
Mercury(I) chloride（氯化亚汞）	$Hg_2Cl_2(s) \rightleftharpoons Hg_2^{2+}(aq) + 2Cl^-(aq)$	1.43×10^{-18}
Silver bromide（溴化银）	$AgBr(s) \rightleftharpoons Ag^+(aq) + Br^-(aq)$	5.35×10^{-13}
Silver carbonate（碳酸银）	$Ag_2CO_3(s) \rightleftharpoons 2Ag^+(aq) + CO_3^{2-}(aq)$	8.46×10^{-12}
Silver chloride（氯化银）	$AgCl(s) \rightleftharpoons Ag^+(aq) + Cl^-(aq)$	1.77×10^{-10}
Silver chromate（铬酸银）	$Ag_2CrO_4(s) \rightleftharpoons 2Ag^+(aq) + CrO_4^{2-}(aq)$	1.12×10^{-12}
Silver iodide（碘化银）	$AgI(s) \rightleftharpoons Ag^+(aq) + I^-(aq)$	8.52×10^{-17}
Strontium carbonate（碳酸锶）	$SrCO_3(s) \rightleftharpoons Sr^{2+}(aq) + CO_3^{2-}(aq)$	5.60×10^{-10}
Strontium sulfate（硫酸锶）	$SrSO_4(s) \rightleftharpoons Sr^{2+}(aq) + SO_4^{2-}(aq)$	3.44×10^{-7}

Relationship Between K_{sp} and Solubility

Both K_{sp} and solubility relate to the concentration of solute in a saturated solution. There is a definite relationship between K_{sp} and the molar solubility, or molarity, of the solute. For ideal solutions or solutions with dilute concentrations, K_{sp} can be directly calculated from its solubility. For example, given that the solubility of SrSO$_4$ at 25°C is 0.0135 g/100 g H$_2$O, together with the information of the molar mass of SrSO$_4$ $M = 183.7$ g mol^{-1} and the density of SrSO$_4$ solution approximately equal to that of water $\rho \approx 1.0$ g cm^{-3}, we can calculate the molarity of a saturated SrSO$_4$ solution as

$$s = \frac{n}{V} = \frac{m(SrSO_4)/M}{m_{all}/\rho} = \frac{0.0135 \text{ g}/(183.7 \text{ g mol}^{-1})}{(100+0.0135)\text{g}/(1.0 \text{ g cm}^{-3} \times 10^{-3} \text{ dm}^3/\text{cm}^3)} = 7.35 \times 10^{-4} \text{ mol dm}^{-3}$$

$$K_{sp} = \left[Sr^{2+}\right]\left[SO_4^{2-}\right] = s^2 = 5.40 \times 10^{-7}$$

Conversely, the solubility of a solute can also be calculated from its K_{sp} value. Although K_{sp} is a dimensionless

饱和溶液的摩尔浓度为

$$s = \frac{n}{V} = \frac{m(\mathrm{SrSO_4})/M}{m_{\text{总}}/\rho} = \frac{0.0135\,\mathrm{g}/(183.7\,\mathrm{g\,mol^{-1}})}{(100+0.0135)\,\mathrm{g}/(1.0\,\mathrm{g\,cm^{-3}} \times 10^{-3}\,\mathrm{dm^3/cm^3})}$$
$$= 7.35 \times 10^{-4}\,\mathrm{mol\,dm^{-3}}$$

$$K_{sp} = \left[\mathrm{Sr}^{2+}\right]\left[\mathrm{SO_4^{2-}}\right] = s^2 = 5.40 \times 10^{-7}$$

相反，溶质的溶解度也可从其 K_{sp} 计算。虽然 K_{sp} 量纲为 1，由其计算的溶解度默认为摩尔浓度。

由于多种原因，上述 $\mathrm{SrSO_4}$ 的 K_{sp} 计算值略大于**表 5.5** 中列出的相应值。其中主要原因是 $\mathrm{SrSO_4}$ 饱和溶液浓度不够稀，使离子的活度系数小于 1。其他原因包括溶质不完全分解成离子、同时存在多重平衡等。并非所有溶解的溶质均以分离的阴、阳离子形式存在于溶液中。一些离子可能以分子形式存在，溶液中的一些离子还可能结合在一起形成**离子对**（即两个带相反电荷的离子通过静电引力结合在一起），离子对形成的程度随离子电荷的增加而增加。所有不以分离的离子形式存在的额外溶质成分，均没有在 K_{sp} 中体现。

同离子效应与盐效应

同离子效应同样适用于溶解度平衡。根据勒夏特列原理，加入同离子使溶解度平衡向难溶化合物的方向移动，从而导致产生更多沉淀。因此，存在能提供同离子的第二种溶质时，难溶离子化合物的溶解度会显著降低。

如果在饱和溶液中加入非同离子的其他离子，这些其他离子往往会增加化合物的溶解度。其他离子的加入并不直接影响溶解度平衡，但它增加了溶液中的总离子浓度，使离子间的吸引力变得更为重要且不可忽略。因此，离子的活度系数小于 1，活度小于化学计量浓度，溶液中需要更高的离子浓度才能达到恒定的 K_{sp}。这种化合物的溶解度随其他离子的加入而增加的效应，称为**盐效应**。

同离子的加入也会引起总离子浓度的增加和盐效应，但盐效应导致的溶解度增加远不如同离子效应引起的溶解度降低显著。一般来说，当存在同离子效应时，盐效应可忽略不计。**图 5.4** 比较了同离子效应和盐效应。

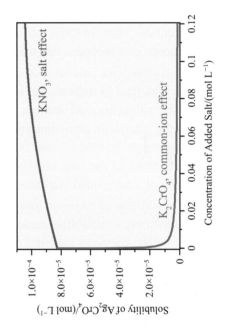

Figure 5.4 Comparison of the common-ion effect (red) and salt effect (blue) on the molar solubility of $\mathrm{Ag_2CrO_4}$.

图 5.4 同离子效应（红线）和盐效应（蓝线）对 $\mathrm{Ag_2CrO_4}$ 摩尔溶解度影响的比较。

5.7 沉淀生成与沉淀完全的判据

如上节所述，涉及 K_{sp} 的计算通常比涉及其他平衡常数的计算更容易出错。虽然 K_{sp} 可能不是很精确，但它仍能使我们做出一些正确的预测，并存在多种应用。例如，它可以用来判断溶液中的离子是否会结合形成沉淀，以及判断沉淀是否完全。

沉淀生成的判据

回顾 **3.8 节**介绍的反应商（Q），它与平衡常数 K 的形式相同，只是在表达式中使用初始活度或浓度，而不是平衡活度或浓度。沉淀溶解的反应商称为**离子积**（Q_{sp}），对下述反应

$$\mathrm{A}_m\mathrm{B}_n(s) \rightleftharpoons m\mathrm{A}^{n+}(aq) + n\mathrm{B}^{m-}(aq)$$

quantity, the calculated solubility from K_{sp} is always the molarity by default.

The above calculated K_{sp} value for $SrSO_4$ is slightly larger than the corresponding value listed in **Table 5.5** due to many reasons. One major aspect is that the concentration of the $SrSO_4$ saturated solution is not dilute enough, so that the activity coefficients of ions are less than 1. Other reasons include the incomplete dissociation of solute into ions, simultaneous equilibria, etc. Not all the dissolved solute appears in the solution as separated cations and anions. Some might remain in molecular form, and some ions in solution might join together into **ion pairs**, which are two oppositely charged ions that are held together by the electrostatic attraction. The degree of ion-pair formation increases with the ion charges. All these additional contents of solute that do not appear as separated ions are not taken into account in K_{sp}.

Common-Ion Effect and Salt Effect

The common-ion effect also applies to solubility equilibrium. According to Le Châtelier's principle, the addition of a common ion shifts the solubility equilibrium toward the undissolved compound, causing more to precipitate. Therefore, the solubility of a slightly soluble ionic compound will be significantly lowered in the presence of a second solute that furnishes a common ion.

If diverse ions that are not common ions to the compound are added to a saturated solution, those diverse ions tend to increase the solubility of the compound. The addition of diverse ions does not directly affect the solubility equilibrium. However, it increases the total ionic concentration of the solution and makes the interionic attraction more important and non-negligible. Consequently, the activity coefficients of ions become less than 1 and the activities become smaller than the stoichiometric concentration. Higher concentrations of the ions must appear in the solution to reach a constant K_{sp} value. This effect that the solubility of a compound increases with the addition of diverse ions is called the **salt effect**.

The addition of common ions also causes an increase in the total ionic concentration and the salt effect, but the increase in solubility caused by the salt effect is much less significant than the decrease in solubility caused by the common-ion effect. In general, the salt effect can be neglected when common-ion effect is present. **Figure 5.4** compares the common-ion effect and the salt effect.

5.7 Criteria for Precipitation and Its Completeness

As discussed in the previous section, calculations involving K_{sp} are generally more subject to error than those involving other equilibrium constants. Although the value of K_{sp} may not be highly accurate, it still allows us to make some correct predictions and can be applied in many useful ways. For example, it can be used to determine whether ions in a solution will combine to form a precipitate or not. It can also help to determine the completeness of the precipitation.

Criteria for Precipitation

Recall the reaction quotient (Q) that was introduced in **Section 3.8**. It has the same form as equilibrium constant K except it uses initial activities/concentrations instead of equilibrium activities/concentrations. The quotient in a precipitation-dissolution reaction is called the **ion product** (Q_{sp}). For a reaction represented by

$$A_mB_n(s) \rightleftharpoons mA^{n+}(aq) + nB^{m-}(aq)$$

$$K_{sp} = \left[A^{n+}\right]_{eq}^m \left[B^{m-}\right]_{eq}^n$$

$$Q_{sp} = \left[A^{n+}\right]_{init}^m \left[B^{m-}\right]_{init}^n \tag{5.20}$$

The value of K_{sp} is constant at a given temperature for a given reaction, but the value of Q_{sp} may change with different initial conditions. If $Q_{sp} > K_{sp}$, the ion concentrations are already higher than they would be in a saturated solution and a net change should occur from right to left. The solution is supersaturated and excess solute should precipitate from the solution. If $Q_{sp} < K_{sp}$, the solution would have been unsaturated and no precipitate would have formed from such a solution. Therefore, the criteria for determining whether ions in a

$$K_{sp} = \left[A^{n+}\right]^m_{平衡} \left[B^{m-}\right]^n_{平衡}$$

$$Q_{sp} = \left[A^{n+}\right]^m_{初始} \left[B^{m-}\right]^n_{初始} \tag{5.20}$$

对于给定反应，K_{sp} 在给定温度下恒定，但 Q_{sp} 随初始条件不同而变化。若 $Q_{sp}>K_{sp}$，则离子浓度已经高于饱和溶液的浓度，净变化将从右至左发生，溶液过饱和，多余的溶质会从溶液中沉淀出。若 $Q_{sp}<K_{sp}$，溶液未饱和，不会形成沉淀。因此，确定溶液中离子是否沉淀的判据是比较 Q_{sp} 和 K_{sp} 的大小，列出如下：

1) 当 $Q_{sp}>K_{sp}$ 时，沉淀将会生成；
2) 当 $Q_{sp}=K_{sp}$ 时，溶液刚好饱和，沉淀和溶解形成动态平衡；
3) 当 $Q_{sp}<K_{sp}$ 时，沉淀不能生成。

值得注意的是，在应用上述沉淀生成判据时，必须先考虑混合溶液所造成的稀释效应。

沉淀完全的判据

含有溶质同离子的化合物可用作沉淀剂，在溶液中加入更多沉淀剂时，会形成更多的沉淀，而溶液中残余的溶质则会减少。只有当溶液中残余溶质的量极少时，才认为溶质已沉淀完全。然而，严格来说，没有任何一个沉淀溶解平衡是沉淀完全的，根据平衡的特性，不论加入的沉淀剂如何过量，溶液中总会残留极少量溶质。

在实际应用中，由于分析天平的称量精度为 10^{-4} g，水溶液中物质的量的精度通常在 10^{-6} mol 量级。因此，沉淀完全的经验判据标准为：如果目标离子的浓度低于 10^{-6} mol L^{-1}，就认为已经沉淀完全。在其他教材中，也有将相对浓度选定为判据标准的：如果 99.9% 或更多的目标离子已经沉淀，就认为沉淀完全。

5.8 沉淀的溶解与转化

沉淀的溶解

$Q_{sp}>K_{sp}$ 时，会形成沉淀。为使沉淀溶解，需令 $Q_{sp}<K_{sp}$，这要求增加 K_{sp} 或减少 Q_{sp}。K_{sp} 是平衡常数，对于给定反应只取决于温度，因此，可通过改变温度提高 K_{sp} 来实现沉淀的溶解。大多数情况下，沉淀的溶解是通过降低离子活度或浓度，从而降低 Q_{sp} 来实现的。这可以通过添加更多溶剂、改用溶解度更低的其他溶剂、生成更难解离的弱电解质、进行氧化还原反应、生成配离子等方法来实现。氧化还原平衡和配位解离平衡将分别在**第 6 章**和**第 9 章**讨论。本节我们只讨论用酸来溶解一些沉淀。

MnS 和 CuS 在强酸中的溶解同时涉及了沉淀溶解平衡和酸碱电离平衡，其方程式为

(1) $MnS \rightleftharpoons Mn^{2+} + S^{2-}$ $K_1 = K_{sp}$

(2) $S^{2-} + H_3O^+ \rightleftharpoons HS^- + H_2O$ $K_2 = 1/K_{a2}$

(3) $HS^- + H_3O^+ \rightleftharpoons H_2S + H_2O$ $K_3 = 1/K_{a1}$

(1) + (2) + (3)，可得

solution will precipitate or not require us to compare Q_{sp} with K_{sp} are listed as follows:
1) If $Q_{sp}>K_{sp}$, precipitation should occur;
2) If $Q_{sp} = K_{sp}$, a solution is just saturated and a dynamic equilibrium is formed between precipitation and dissolution;
3) If $Q_{sp}<K_{sp}$, precipitation cannot occur.

It is worth noting that any possible dilutions caused by mixing solutions must be considered before the above criteria for precipitation are applied.

Criteria for Completeness of Precipitation

Compounds containing the common ions of a solute serve as its precipitants. As more precipitant is added to a solution, more precipitate will be formed and the remaining solute in the solution will become less. Precipitation of a solute is considered to be complete only if the amount of solute remaining in the solution is very small. However, strictly speaking, there is no "completeness" of precipitation in any precipitation-dissolution reaction. No matter how much excess of precipitant is added, there will always be some solute remaining in the solution due to the nature of an equilibrium.

In practical use, because the accuracy of an analytical balance is 10^{-4} g, the accuracy of the amount of substances in aqueous solutions is normally in the order of 10^{-6} mol. Therefore, an arbitrary rule of thumb is selected as the criteria for completeness of precipitation: precipitation is complete if the concentration of a target ion is less than 10^{-6} mol L^{-1}. In some other textbooks, a relative concentration standard is selected as the criteria: precipitation is complete if 99.9% or more of a target ion has precipitated.

5.8 Dissolution and Transformation of Precipitates

Dissolution of Precipitates

When $Q_{sp}>K_{sp}$, a precipitate is formed. In order for the precipitate to dissolve, $Q_{sp}<K_{sp}$ is necessary, which requires either an increase in K_{sp} or a decrease in Q_{sp}. K_{sp} is an equilibrium constant that depends only on temperature for a given reaction. So, dissolution of precipitates can be achieved by varying the temperature to increase K_{sp}. In most cases, the dissolution of precipitates is realized by reducing the ion activities/concentrations to decrease Q_{sp}. This can be accomplished by adding more solvents, changing to other solvents with lower solubility, forming weak electrolytes that are more difficult to dissociate, conducting oxidation-reduction reactions, forming complex ions, etc. The oxidation-reduction equilibrium and complex-ion equilibrium will be discussed in **Chapters 6** and **9**, respectively. In this section, we will only discuss the dissolution of some precipitates using acids.

The dissolution of MnS and CuS in a strong acid involves both precipitation-dissolution equilibrium and acid-base equilibrium as

$$(1)\ MnS \rightleftharpoons Mn^{2+} + S^{2-} \quad K_1 = K_{sp}$$
$$(2)\ S^{2-} + H_3O^+ \rightleftharpoons HS^- + H_2O \quad K_2 = 1/K_{a2}$$
$$(3)\ HS^- + H_3O^+ \rightleftharpoons H_2S + H_2O \quad K_3 = 1/K_{a1}$$

Applying (1)+(2)+(3), we have

$$MnS + 2H_3O^+ \rightleftharpoons Mn^{2+} + H_2S + 2H_2O$$

$$K = K_1 K_2 K_3 = \frac{K_{sp}(MnS)}{K_{a1} K_{a2}} = \frac{2.5 \times 10^{-13}}{8.9 \times 10^{-8} \times 1.0 \times 10^{-19}} = 2.8 \times 10^{13}$$

Similarly,

$$CuS + 2H_3O^+ \rightleftharpoons Cu^{2+} + H_2S + 2H_2O$$

$$\text{MnS} + 2\text{H}_3\text{O}^+ \rightleftharpoons \text{Mn}^{2+} + \text{H}_2\text{S} + 2\text{H}_2\text{O}$$

$$K = K_1 K_2 K_3 = \frac{K_{sp}(\text{MnS})}{K_{a1} K_{a2}} = \frac{2.5 \times 10^{-13}}{8.9 \times 10^{-8} \times 1.0 \times 10^{-19}} = 2.8 \times 10^{13}$$

类似地，

$$\text{CuS} + 2\text{H}_3\text{O}^+ \rightleftharpoons \text{Cu}^{2+} + \text{H}_2\text{S} + 2\text{H}_2\text{O}$$

$$K = \frac{K_{sp}(\text{CuS})}{K_{a1} K_{a2}} = \frac{6.3 \times 10^{-36}}{8.9 \times 10^{-8} \times 1.0 \times 10^{-19}} = 7.1 \times 10^{-10}$$

MnS 的总平衡常数很大，说明 MnS 的溶解是完全的；同时，CuS 的总平衡常数很小，说明 CuS 没有发生明显的溶解。因此，MnS 可以在强酸中溶解而 CuS 不能。在上述两个总平衡常数表达式中，K_{a1} 和 K_{a2} 相同而 K_{sp} 不同。显然，K_{sp} 较大的沉淀更容易在酸中溶解。

另一个例子是 $CaCO_3$ 和 CaC_2O_4 在 HAc 溶液中的溶解，总反应常数可用类似的方法计算为

$$\text{CaCO}_3 + 2\text{HAc} \rightleftharpoons \text{Ca}^{2+} + 2\text{Ac}^- + \text{CO}_2 + \text{H}_2\text{O}$$

$$K = \frac{K_{sp} K_a^2(\text{HAc})}{K_{a1} K_{a2}} = \frac{2.8 \times 10^{-9} \times (1.75 \times 10^{-5})^2}{4.5 \times 10^{-7} \times 4.7 \times 10^{-11}} = 0.041$$

$$\text{CaC}_2\text{O}_4 + \text{HAc} \rightleftharpoons \text{Ca}^{2+} + \text{Ac}^- + \text{HC}_2\text{O}_4^-$$

$$K = \frac{K_{sp} K_a(\text{HAc})}{K_{a2}} = \frac{2.32 \times 10^{-9} \times 1.75 \times 10^{-5}}{1.5 \times 10^{-4}} = 2.7 \times 10^{-10}$$

因此 $CaCO_3$ 可以溶解在 HAc 溶液中，而 CaC_2O_4 不能。在上述两个总平衡常数表达式中，两种沉淀的 K_{sp} 相近但生成的酸的强弱不同。显然，若生成的酸具有较小的 K_a，则沉淀更容易溶解。值得注意的是，不断释放的 CO_2 导致生成的碳酸浓度降低，这也有助于 $CaCO_3$ 在 HAc 溶液中的溶解。

沉淀的转化

在含有白色沉淀 $BaCO_3$ 的溶液中加入 Na_2CrO_4 溶液，可以观察到白色沉淀转化为黄色沉淀，即 $BaCrO_4$。一种沉淀转化为另一种沉淀的过程称为沉淀的转化。该反应涉及的平衡如下：

(1) $\text{BaCO}_3 \rightleftharpoons \text{Ba}^{2+} + \text{CO}_3^{2-}$ \quad $K_{sp}(\text{BaCO}_3)$

(2) $\text{Ba}^{2+} + \text{CrO}_4^{2-} \rightleftharpoons \text{BaCrO}_4$ \quad $1/K_{sp}(\text{BaCrO}_4)$

(1) + (2):

$$\text{BaCO}_3 + \text{CrO}_4^{2-} \rightleftharpoons \text{BaCrO}_4 + \text{CO}_3^{2-}$$

$$K = \frac{[\text{CO}_3^{2-}]}{[\text{CrO}_4^{2-}]} = \frac{K_{sp}(\text{BaCO}_3)}{K_{sp}(\text{BaCrO}_4)} = \frac{2.58 \times 10^{-9}}{1.17 \times 10^{-10}} = 22.1$$

由于 $BaCrO_4$ 的 K_{sp} 小于 $BaCO_3$ 的，上述总平衡常数大于 1。这说明当 $[CO_3^{2-}]/[CrO_4^{2-}] < 22.1$ 或 $[CrO_4^{2-}]/[CO_3^{2-}] > 0.045$ 时，总平衡向右移动，已沉淀的 $BaCO_3$ 将转化为 $BaCrO_4$。

再用类似的方法来考虑上述反应的逆反应：

$$\text{BaCrO}_4 + \text{CO}_3^{2-} \rightleftharpoons \text{BaCO}_3 + \text{CrO}_4^{2-}$$

$$K' = \frac{[\text{CrO}_4^{2-}]}{[\text{CO}_3^{2-}]} = \frac{K_{sp}(\text{BaCrO}_4)}{K_{sp}(\text{BaCO}_3)} = \frac{1}{22.1} = 0.045$$

$$K = \frac{K_{sp}(\text{CuS})}{K_{a1}K_{a2}} = \frac{6.3 \times 10^{-36}}{8.9 \times 10^{-8} \times 1.0 \times 10^{-19}} = 7.1 \times 10^{-10}$$

The overall equilibrium constant for MnS is quite large, which means that the dissolution of MnS goes to completion. Meanwhile, the very small overall equilibrium constant for CuS indicates that the dissolution of CuS does not occur to any significant extent. Therefore, MnS but not CuS can be dissolved in strong acids. In the above two overall equilibrium constant expressions, K_{a1} and K_{a2} are the same but K_{sp} is different. Apparently, it is easier for the precipitate with a larger K_{sp} value to dissolve in acids.

Another example is the dissolution of $CaCO_3$ and CaC_2O_4 in HAc. The overall equilibrium constants can be calculated in a similar way as

$$CaCO_3 + 2HAc \rightleftharpoons Ca^{2+} + 2Ac^- + CO_2 + H_2O$$

$$K = \frac{K_{sp}K_a^2(\text{HAc})}{K_{a1}K_{a2}} = \frac{2.8 \times 10^{-9} \times (1.75 \times 10^{-5})^2}{4.5 \times 10^{-7} \times 4.7 \times 10^{-11}} = 0.041$$

$$CaC_2O_4 + HAc \rightleftharpoons Ca^{2+} + Ac^- + HC_2O_4^-$$

$$K = \frac{K_{sp}K_a(\text{HAc})}{K_{a2}} = \frac{2.32 \times 10^{-9} \times 1.75 \times 10^{-5}}{1.5 \times 10^{-4}} = 2.7 \times 10^{-10}$$

Therefore, $CaCO_3$ but not CaC_2O_4 can be dissolved in HAc. In the above two overall equilibrium constant expressions, the K_{sp} values of two precipitates are similar but the strength of the produced acids is different. Apparently, it is easier for a precipitate to dissolve if another acid with a smaller K_a is produced. It is worth noting that the constantly released CO_2 causes the decrease in the concentration of the produced carbonate acid, which also helps the dissolution of $CaCO_3$ in HAc.

Transformation of Precipitates

Adding Na_2CrO_4 solution into a solution with some white precipitate $BaCO_3$, we can observe that the white precipitate changes into a yellow precipitate, which is $BaCrO_4$. The process where one precipitate transforms into another is called the transformation of precipitates. The chemical equilibria involved are shown as

(1) $BaCO_3 \rightleftharpoons Ba^{2+} + CO_3^{2-}$ $K_{sp}(BaCO_3)$

(2) $Ba^{2+} + CrO_4^{2-} \rightleftharpoons BaCrO_4$ $1/K_{sp}(BaCrO_4)$

(1) + (2):

$$BaCO_3 + CrO_4^{2-} \rightleftharpoons BaCrO_4 + CO_3^{2-}$$

$$K = \frac{[CO_3^{2-}]}{[CrO_4^{2-}]} = \frac{K_{sp}(BaCO_3)}{K_{sp}(BaCrO_4)} = \frac{2.58 \times 10^{-9}}{1.17 \times 10^{-10}} = 22.1$$

Because the K_{sp} of $BaCrO_4$ is smaller than that of $BaCO_3$, the above overall equilibrium constant is greater than 1. This suggests that if $[CO_3^{2-}]/[CrO_4^{2-}] < 22.1$, or $[CrO_4^{2-}]/[CO_3^{2-}] > 0.045$, the overall equilibrium will shift to the right and the precipitate $BaCO_3$ will transform into $BaCrO_4$.

Now let us consider the reverse reaction in a similar manner, as

$$BaCrO_4 + CO_3^{2-} \rightleftharpoons BaCO_3 + CrO_4^{2-}$$

$$K' = \frac{[CrO_4^{2-}]}{[CO_3^{2-}]} = \frac{K_{sp}(BaCrO_4)}{K_{sp}(BaCO_3)} = \frac{1}{22.1} = 0.045$$

This indicates that even the K_{sp} of $BaCrO_4$ is smaller than that of $BaCO_3$, if $[CrO_4^{2-}]/[CO_3^{2-}] < 0.045$ or $[CO_3^{2-}]/[CrO_4^{2-}] > 22.1$ is satisfied, $BaCrO_4$ can also transform into $BaCO_3$ under this condition.

For two precipitates with similar K_{sp} formulas such as $BaCO_3$ and $BaCrO_4$, the tendency of transformation is from the precipitate with a larger K_{sp} value into one with a smaller K_{sp} value. If the difference in K_{sp} is not significant, however, the transformation of a smaller K_{sp} precipitate into a larger K_{sp} precipitate can also be realized under certain conditions. For precipitates with different K_{sp} formulas such as AgCl and Ag_2CrO_4, the general

这表明，即使 $BaCrO_4$ 的 K_{sp} 小于 $BaCO_3$ 的，在满足 $[CrO_4^{2-}]/[CO_3^{2-}]<0.045$ 或 $[CO_3^{2-}]/[CrO_4^{2-}]>22.1$ 的条件下，$BaCrO_4$ 也可以转化为 $BaCO_3$。

对于 K_{sp} 表达式形式类似的两种沉淀，如 $BaCO_3$ 和 $BaCrO_4$，其转化趋势是从 K_{sp} 较大的沉淀转化为 K_{sp} 较小的沉淀。如果 K_{sp} 的差别不显著，在一定条件下也可以实现从 K_{sp} 较小的沉淀向 K_{sp} 较大沉淀的转化。对于 K_{sp} 表达式形式不同的两种沉淀，如 $AgCl$ 和 Ag_2CrO_4，总的转化趋势是从溶解度较大的沉淀转化为溶解度较小的沉淀。

5.9 分步沉淀

前几节的大部分讨论只涉及溶液中的一种离子或一种沉淀。在实际情况中，许多离子和沉淀可能共存于溶液中。有时只选择性地沉淀一种离子而将所有其他离子保留在溶液中，从而使该离子与其他离子分离，是一种实用的方法。这可以通过适当使用沉淀剂来实现，这一过程称为**分步沉淀**或**选择性沉淀**。实现分步沉淀的首要条件在于，被分离物质的溶解度存在显著差异。

为了实现分步沉淀，需要仔细控制条件，使得一个目标离子沉淀完全（意味着其浓度低于 10^{-6} mol L^{-1}），而其他离子尚未沉淀（意味着这些离子的 $Q_{sp}<K_{sp}$）。本节我们将给出一些采用分步沉淀分离金属硫化物和氢氧化物的示例。

二价金属硫化物的分步沉淀

在初始浓度 $[Zn^{2+}] = [Mn^{2+}] = 0.10$ mol L^{-1} 的溶液中持续充入 H_2S 气体至完全饱和，即 $[H_2S] = 0.10$ mol L^{-1}，哪种离子先沉淀？我们可以通过分步沉淀将这两种离子完全分离吗？如果可以，溶液的 pH 应控制在什么范围内？

为了回答上述问题，我们首先需要比较 ZnS 和 MnS 的 K_{sp}：$K_{sp}(ZnS) = 1.6\times 10^{-24}$ 远小于 $K_{sp}(MnS) = 2.5\times 10^{-13}$，因此 Zn^{2+} 应该先沉淀。为了完全分离这两种离子，我们需要通过控制 pH 来调控溶液中 S^{2-} 的浓度，使得 Zn^{2+} 沉淀完全而 Mn^{2+} 尚未沉淀。$[S^{2-}]$ 和 $[H_3O^+]$ 之间的关系为

$$[H_3O^+] = \sqrt{\frac{K_{a1}K_{a2}[H_2S]}{[S^{2-}]}}$$

为使 Zn^{2+} 沉淀完全，$[Zn^{2+}] < 10^{-6}$ mol L^{-1}，则

$$[S^{2-}] = \frac{K_{sp}}{[Zn^{2+}]} > 1.6\times 10^{-18} \text{ mol L}^{-1}$$

$$[H_3O^+] = \sqrt{\frac{K_{a1}K_{a2}[H_2S]}{[S^{2-}]}} < 2.4\times 10^{-5} \text{ mol L}^{-1} \quad 且 \quad pH > 4.6$$

当 Mn^{2+} 开始沉淀时，$[Mn^{2+}] = 0.10$ mol L^{-1}，则

$$[S^{2-}] = \frac{K_{sp}}{[Mn^{2+}]} = 2.5\times 10^{-12} \text{ mol L}^{-1}$$

tendency of transformation is from the precipitate with larger solubility into one with smaller solubility.

5.9 Fractional Precipitation

Most of the discussion in previous sections involves only one ion or one precipitate in a solution. In real cases, many ions and precipitates may coexist in a solution. Sometimes it is of practical use to selectively precipitate only one ion and keep all other ions in solution so that this ion can be separated from the others. This can be realized by the proper use of a precipitating reagent, and this process is called **fractional precipitation**, or selective precipitation. The primary condition for a successful fractional precipitation is that there is a significant difference in the solubilities of the substances being separated.

In order to achieve a successful fractional precipitation, it requires careful control of the conditions so that one target ion precipitates completely, meaning that its concentration is less than 10^{-6} mol L^{-1}, and the other ions have not precipitated yet, meaning that $Q_{sp} < K_{sp}$ for those ions. In this section, we will show some illustrative examples of using fractional precipitation for the separation of metal sulfides and hydroxides.

Fractional Precipitation of Divalent Metal Sulfides

For a solution initially with $[Zn^{2+}] = [Mn^{2+}] = 0.10$ mol L^{-1}, if H$_2$S gas is continuously introduced into the solution to full saturation, for which $[H_2S] = 0.10$ mol L^{-1}, which ion should precipitate first? Can we completely separate these two ions by fractional precipitation? If so, in what range should the pH of the solution be controlled?

To answer the above questions, we first need to compare the K_{sp} values of ZnS and MnS: K_{sp}(ZnS) = 1.6×10^{-24} is significantly smaller than K_{sp}(MnS) = 2.5×10^{-13}. Therefore, Zn^{2+} should precipitate first. In order for complete separation of these two ions, we need to adjust the concentration of S^{2-} in the solution by controlling the pH in a range so that Zn^{2+} precipitates completely and Mn^{2+} has not precipitated yet. The relationship between [S^{2-}] and [H$_3$O$^+$] is given by

$$[H_3O^+] = \sqrt{\frac{K_{a1}K_{a2}[H_2S]}{[S^{2-}]}}$$

For Zn^{2+} to precipitate completely, $[Zn^{2+}] < 10^{-6}$ mol L^{-1}

$$[S^{2-}] = \frac{K_{sp}}{[Zn^{2+}]} > 1.6 \times 10^{-18} \text{ mol L}^{-1}$$

$$[H_3O^+] = \sqrt{\frac{K_{a1}K_{a2}[H_2S]}{[S^{2-}]}} < 2.4 \times 10^{-5} \text{ mol L}^{-1} \quad \text{and} \quad \text{pH} > 4.6$$

For Mn^{2+} to start to precipitate, $[Mn^{2+}] = 0.10$ mol L^{-1}

$$[S^{2-}] = \frac{K_{sp}}{[Mn^{2+}]} = 2.5 \times 10^{-12} \text{ mol L}^{-1}$$

$$[H_3O^+] = \sqrt{\frac{K_{a1}K_{a2}[H_2S]}{[S^{2-}]}} = 1.9 \times 10^{-8} \text{ mol L}^{-1} \quad \text{and} \quad \text{pH} = 7.7$$

If pH < 7.7, Mn^{2+} will not precipitate. 7.7 is the highest pH for Mn^{2+} to remain unprecipitated and 4.6 is the lowest pH for Zn^{2+} to precipitate completely. Therefore, we can completely separate Zn^{2+} and Mn^{2+} using fractional precipitation by controlling pH in the range of 4.6~7.7.

For simplification and practical purpose, we only consider divalent metal ions M^{2+} here. In general, for M^{2+} to precipitate in a saturated H$_2$S solution, we have

$$[M^{2+}] = \frac{K_{sp}}{[S^{2-}]} \quad \text{and} \quad [S^{2-}] = \frac{K_{a1}K_{a2}[H_2S]}{[H_3O^+]^2}$$

$$[H_3O^+] = \sqrt{\frac{K_{a1}K_{a2}[H_2S]}{[S^{2-}]}} = 1.9 \times 10^{-8} \text{ mol L}^{-1} \quad 且 \quad pH = 7.7$$

若 pH< 7.7，Mn^{2+} 不会发生沉淀。7.7 是 Mn^{2+} 保持不发生沉淀的最高 pH，而 4.6 是 Zn^{2+} 沉淀完全所需的最低 pH。因此，通过控制 pH 在 4.6～7.7 之间，采用分步沉淀可以完全分离 Zn^{2+} 和 Mn^{2+}。

为了简化和实用的目的，我们这里只考虑二价金属离子 M^{2+}。一般来说，对于 M^{2+} 在饱和 H_2S 溶液中的沉淀，有

$$[M^{2+}] = \frac{K_{sp}}{[S^{2-}]} \quad 且 \quad [S^{2-}] = \frac{K_{a1}K_{a2}[H_2S]}{[H_3O^+]^2}$$

$$\therefore [M^{2+}] = \frac{K_{sp}}{K_{a1}K_{a2}[H_2S]} \cdot [H_3O^+]^2$$

$$\lg[M^{2+}] = \lg\left(\frac{K_{sp}}{K_{a1}K_{a2}[H_2S]}\right) - 2pH$$

由于 $[H_2S]$ 为常数，令

$$C = \lg\left(\frac{K_{sp}}{K_{a1}K_{a2}[H_2S]}\right)$$

可得
$$\lg[M^{2+}] = C - 2pH$$

上述等式表明，$\lg[M^{2+}]$ 与溶液的 pH 之间存在线性关系，斜率恒为 -2，截距 C 取决于 K_{sp}(MS)。该关系可绘制成**图 5.5(a)** 中的溶解度直线，直线上的任意点均代表 MS 沉淀与其饱和溶液之间的溶解度平衡，给出了当 M^{2+} 刚开始沉淀时对应的 $[M^{2+}]$ 和 pH。直线右侧的区域为"沉淀区"，在相应的 $[M^{2+}]$ 和 pH 时将生成 MS 沉淀。直线左侧的区域为"溶解区"，在相应的 $[M^{2+}]$ 和 pH 时不会生成 MS 沉淀。

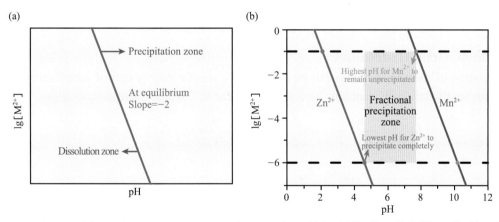

Figure 5.5 Solubility of divalent metal sulfates as a function of pH. (a) A solubility line of lg[M^{2+}] vs. pH with a slope of −2, the regions to the left and right of which are the "dissolution zone" and "precipitation zone", respectively, of M^{2+}. (b) Two solubility lines of lg[Zn^{2+}] (red) and lg[Mn^{2+}] (blue) vs. pH, where the "fractional precipitation zone" is indicated by a pale red rectangle.

图 5.5 二价金属硫化物的溶解度与 pH 的函数关系图：(a) lg[M^{2+}] 对 pH 作图呈一条斜率为 -2 的溶解度直线，直线左侧和右侧的区域分别代表 M^{2+} 的"溶解区"和"沉淀区"；(b) lg[Zn^{2+}]（红线）和 lg[Mn^{2+}]（蓝线）对 pH 作图的两条溶解度直线，二者的"分步沉淀区"用浅红色矩形表示。

如果有多个二价金属离子共存于溶液中，可绘制出一系列斜率恒为 -2、截距不同的平行线，ZnS 和 MnS 的示例如**图 5.5(b)** 所示。由于 K_{sp}(ZnS) <K_{sp}(MnS)，C(ZnS) <C(MnS)，代表 ZnS 溶解度平衡的直线位于代表 MnS 的直线左侧。在 MnS 平衡线的右侧区域，两种

$$\therefore [M^{2+}] = \frac{K_{sp}}{K_{a1}K_{a2}[H_2S]} \cdot [H_3O^+]^2$$

$$\lg[M^{2+}] = \lg\left(\frac{K_{sp}}{K_{a1}K_{a2}[H_2S]}\right) - 2\text{pH}$$

Since [H₂S] is a constant, we can set

$$C = \lg\left(\frac{K_{sp}}{K_{a1}K_{a2}[H_2S]}\right)$$

Therefore,

$$\lg[M^{2+}] = C - 2\text{pH}$$

The above equation suggests a linear relationship between lg[M²⁺] and the pH of the solution, with a constant slope of −2 and an intercept of C depending on K_{sp}(MS). This relationship can be plotted as a solubility line in **Figure 5.5(a)**. Any point on this line represents a solubility equilibrium between the MS precipitate and its saturated solution, showing the corresponding [M²⁺] and pH when M²⁺ just starts to precipitate. The region to the right of the line is the "precipitation zone", where the MS precipitate is formed at the corresponding [M²⁺] and pH. The region to the left of the line is the "dissolution zone", where no MS precipitate is formed at the corresponding [M²⁺] and pH.

If multiple divalent metal ions coexist in the solution, a series of parallel lines with a constant slope of −2 and different intercepts can be plotted. An illustrative example of ZnS and MnS is shown in **Figure 5.5(b)**. Because K_{sp}(ZnS) < K_{sp}(MnS), C(ZnS) < C(MnS) and the line representing the solubility equilibrium of ZnS is to the left of the line representing that of MnS. In the region to the right of the MnS line, both ions will precipitate. In the region to the left of the ZnS line, both ions will dissolve. In the region between these two lines, ZnS will precipitate and MnS will dissolve. In **Figure 5.5(b)**, a horizontal dashed line at lg[M²⁺] = −6 intersects the solubility lines at pH = 4.6 and 10.2, indicating the lowest pH for Zn²⁺ and Mn²⁺, respectively, to precipitate completely. Another horizontal dashed line at lg[M²⁺] = −1 intersects the solubility lines at pH = 2.1 and 7.7, indicating the highest pH for Zn²⁺ and Mn²⁺, respectively, to remain unprecipitated. The region between two solubility lines and between pH = 4.6~7.7, represented as a pale red rectangle, is the "fractional precipitation zone" where fractional precipitation can be applied to completely separate Zn²⁺ and Mn²⁺.

Fractional Precipitation of Metal Hydroxides

The fractional precipitation of metal hydroxides can be derived in a similar way to that of metal sulfides, except that the metal ions are not limited to divalent ions here. In general, for M^{n+} to precipitate as hydroxides

$$[M^{n+}] = \frac{K_{sp}}{[OH^-]^n} \quad \text{and} \quad [OH^-] = \frac{K_w}{[H_3O^+]}$$

$$\therefore [M^{n+}] = \frac{K_{sp}}{K_w^n} \cdot [H_3O^+]^n$$

$$\lg[M^{n+}] = \lg\left(\frac{K_{sp}}{K_w^n}\right) - n\text{pH}$$

Let $C = \lg\left(\frac{K_{sp}}{K_w^n}\right)$, we have

$$\lg[M^{n+}] = C - n\text{pH}$$

A linear relationship between lg[M^{n+}] and the pH of the solution is also suggested by the above equation, with a slope of −n and an intercept of C depending on the K_{sp} and n values of the hydroxides. Similar solubility lines and the "fractional precipitation zone" can be derived accordingly, and are shown in **Figure 5.6**.

离子都会沉淀。在 ZnS 平衡线的左侧区域，两种离子均会溶解。在两条平衡线之间的区域，ZnS 会沉淀而 MnS 会溶解。在**图 5.5(b)** 中，一条位于 $\lg[M^{2+}] = -6$ 的水平虚线与两条溶解度平衡线相交于 pH = 4.6 和 10.2 处，分别对应于 Zn^{2+} 和 Mn^{2+} 沉淀完全的最低 pH。另一条位于 $\lg[M^{2+}] = -1$ 的水平虚线与两条溶解度平衡线相交于 pH = 2.1 和 7.7 处，分别对应于 Zn^{2+} 和 Mn^{2+} 保持不发生沉淀的最高 pH。在两条溶解度平衡线和 pH = 4.6 ~ 7.7 之间的区域（用浅红色矩形表示），即为可以用来完全分离 Zn^{2+} 和 Mn^{2+} 的"分步沉淀区"。

金属氢氧化物的分步沉淀

金属氢氧化物的分步沉淀可以用与金属硫化物类似的方法得到，只是这里的金属离子不限于二价离子。一般来说，对于以氢氧化物沉淀的 M^{n+}，有

$$\left[M^{n+}\right] = \frac{K_{sp}}{\left[OH^-\right]^n} \quad \text{且} \quad \left[OH^-\right] = \frac{K_w}{\left[H_3O^+\right]}$$

$$\therefore \left[M^{n+}\right] = \frac{K_{sp}}{K_w^n} \cdot \left[H_3O^+\right]^n$$

$$\lg\left[M^{n+}\right] = \lg\left(\frac{K_{sp}}{K_w^n}\right) - n\text{pH}$$

令 $C = \lg\left(\dfrac{K_{sp}}{K_w^n}\right)$，有

$$\lg\left[M^{n+}\right] = C - n\text{pH}$$

上述等式也表明，$\lg[M^{n+}]$ 与溶液的 pH 之间存在线性关系，斜率为 $-n$，截距 C 取决于对应氢氧化物的 K_{sp} 和 n 值。也可相应地推导出类似的溶解度平衡线和"分步沉淀区"，如**图 5.6** 所示。

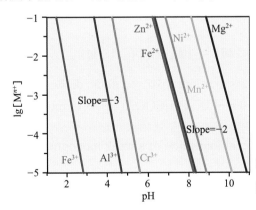

Figure 5.6 Solubility of various metal hydroxides as a function of pH.

图 5.6 各种金属氢氧化物的溶解度与 pH 的函数关系图。

拓展阅读材料　超酸与魔酸

超酸一词最早由詹姆斯·B. 科南特（1893—1978）在 1927 年提出，用于描述比传统矿物酸更强的酸。1971 年，罗纳德·J. 吉莱斯皮（1924—2021）将该定义改进为，任何比 100% 纯 H_2SO_4 酸度更大的酸。例如，三氟甲磺酸（CF_3SO_3H）和氟硫酸（HSO_3F）都是比 H_2SO_4 强约 1000 倍的超酸。大多数强超酸都是由强路易斯酸和强布朗斯特 - 劳莱酸的组合制备。例如，已知的最强超酸氟锑酸（H_2FSbF_6）是通

Extended Reading Materials — Superacids and Magic Acids

The term **superacid** was originally coined by James B. Conant (1893—1978) in 1927 to describe acids that were stronger than conventional mineral acids. This definition was refined by Ronald J. Gillespie (1924—2021) in 1971, as any acid with an acidity greater than that of 100% pure H_2SO_4. For example, both trifluoromethane sulfonic acid (CF_3SO_3H, also known as triflic acid) and fluorosulfuric acid (HSO_3F) are superacids that are about 1000 times stronger than H_2SO_4. Most strong superacids are prepared by the combination of a strong Lewis acid and a strong Brønsted-Lowry acid. As an example, the strongest known superacid fluoroantimonic acid (H_2FSbF_6) is made by dissolving antimony pentafluoride (SbF_5) in anhydrous hydrogen fluoride (HF), in which

$$SbF_5 + 2HF \rightleftharpoons SbF_6^- + H_2F^+$$

A common application of superacids is to provide an environment to create, maintain, and characterize carbocations, which are useful intermediates in many industrial reactions.

A special kind of superacid called **magic acid** was developed in the 1960s by George A. Olah (1927—2017), who was awarded the Nobel Prize in chemistry in 1994 "for his contribution to carboncation chemistry" via superacids. The magic acid was prepared by mixing SbF_5 and FSO_3H commonly at a 1:1 molar ratio. It was so-named due to the discovery of a candle dissolving in it after a Christmas party. The ability of magic acid to protonate alkanes, which under normal acidic conditions do not protonate to any extent, leads to stabilized carbocations that can be studied in greater depth and used as catalysts in organic synthesis reactions.

Problems

5.1 It is possible to write simple equations to relate pH, pK, and molarities c (mol L^{-1}) of various solutions. Three such equations are shown here.

Weak monoprotic acid: $pH = \dfrac{1}{2}pK_a - \dfrac{1}{2}\lg c$

Weak monoprotic base: $pH = pK_w - \dfrac{1}{2}pK_b + \dfrac{1}{2}\lg c$

Salt of weak monoprotic acid (pK_a) and strong base: $pH = \dfrac{1}{2}pK_w + \dfrac{1}{2}pK_a + \dfrac{1}{2}\lg c$

(a) Derive these three equations, and point out the assumptions involved in the derivations.
(b) Use these equations to determine the pH of 0.10 mol L^{-1} CH_3COOH(aq), 0.10 mol L^{-1} NH_3(aq), and 0.10 mol L^{-1} CH_3COONa(aq).

5.2 A handbook lists the following formula for the percent ionization of a weak monoprotic acid:

$$\text{Ionization}(\%) = \dfrac{100}{1 + 10^{(pK_a - pH)}}$$

(a) Derive this formula. What assumptions must you make in this derivation?
(b) Use the formula to determine the percent ionization of a formic acid solution, HCOOH, with a pH of 2.50.
(c) A 0.150 mol L^{-1} solution of propionic acid, CH_3CH_2COOH, has a pH of 2.85. What is K_a for propionic acid?

5.3 You are given 250.0 mL of 0.100 mol L^{-1} propionic acid, CH_3CH_2COOH ($K_a = 1.35 \times 10^{-5}$). You want to adjust its pH by adding an appropriate solution. What volume would you add of (a) 1.00 mol L^{-1} HCl to lower the pH to 1.00; (b) 1.00 mol L^{-1} CH_3CH_2COONa to raise the pH to 4.00; (c) water to raise the pH by 0.15 unit?

5.4 Rather than calculating the pH for different volumes of titrant, a titration curve can be established by calculating the volume of titrants required to reach certain pH values. Determine the volumes of 0.100 mol L^{-1} NaOH required to reach the following pH values at the titration of 20.00 mL of 0.150 mol L^{-1} HCl: pH = (a) 2.00; (b) 3.50; (c) 5.00; (d) 10.50; (e) 12.00. Then plot the titration curve.

过将五氟化锑（SbF_5）溶解在无水氟化氢（HF）中制备的，其离子方程式为

$$SbF_5 + 2HF \rightleftharpoons SbF_6^- + H_2F^+$$

超酸的常见应用是提供环境以产生、维持和表征碳正离子，而碳正离子是许多工业反应的有用中间体。

20世纪60年代，乔治·A. 奥拉（1927—2017）发展了一种被称为**魔酸**的特殊超酸，他凭借超酸"对碳正离子化学的贡献"被授予1994年诺贝尔化学奖。魔酸可由 SbF_5 和 FSO_3H 按1:1摩尔比混合来制备，它之所以如此命名，是因为在一次圣诞派对后发现其竟然可以溶解蜡烛。在常规酸性条件下，烷烃一般不会质子化，而魔酸使烷烃质子化的能力，导致稳定的碳正离子可以被更深入地研究，并用作有机合成反应的催化剂。

习题

5.1 可以用简单的等式来关联各种溶液的 pH、pK 以及摩尔浓度 c（mol L^{-1}）。这里给出了三个这样的等式：

一元弱酸：$\text{pH} = \frac{1}{2}\text{p}K_a - \frac{1}{2}\lg c$

一元弱碱：$\text{pH} = \text{p}K_w - \frac{1}{2}\text{p}K_b + \frac{1}{2}\lg c$

一元弱酸（pK_a）与强碱的盐：$\text{pH} = \frac{1}{2}\text{p}K_w + \frac{1}{2}\text{p}K_a + \frac{1}{2}\lg c$

(a) 推导这三个等式，并指出推导过程中涉及的假定。
(b) 使用这些等式来确定 0.10 mol L^{-1} $CH_3COOH(aq)$、0.10 mol L^{-1} $NH_3(aq)$ 和 0.10 mol L^{-1} $CH_3COONa(aq)$ 的 pH。

5.2 某手册列出了如下一元弱酸的电离百分数公式：

$$\text{电离百分数} = \frac{100}{1+10^{(\text{p}K_a - \text{pH})}}\%$$

(a) 推导此公式。在推导过程中必须做出哪些假定？
(b) 使用此公式来确定 pH 为 2.50 的甲酸 HCOOH 溶液的电离百分数。
(c) 0.150 mol L^{-1} 丙酸 CH_3CH_2COOH 溶液的 pH 为 2.85，丙酸的 K_a 是多少？

5.3 给你 250.0 mL 0.100 mol L^{-1} 丙酸 CH_3CH_2COOH（$K_a = 1.35\times10^{-5}$），要通过添加适当的溶液来调节其 pH。你将添加多少体积的 (a) 1.00 mol L^{-1} HCl 溶液以使 pH 降至 1.00；(b) 1.00 mol L^{-1} CH_3CH_2COONa 溶液以使 pH 升至 4.00；(c) 水以使 pH 提高 0.15 个单位？

5.4 可以通过计算达到特定 pH 所需滴定剂的体积来绘制一条滴定曲线，而不是计算加入不同体积滴定剂时的 pH。确定在滴定 20.00 mL 0.150 mol L^{-1} HCl 溶液时，达到以下 pH 所需的 0.100 mol L^{-1} NaOH 溶液的体积：pH= (a) 2.00；(b) 3.50；(c) 5.00；(d) 10.50；(e) 12.00。绘制其滴定曲线。

5.5 在某些情况下，两种混合酸的滴定曲线与单一酸的滴定曲线外观相同；而在其他情况下，外观不同。

5.5 In some cases, the titration curve for a mixture of two acids has the same appearance as that for a single acid; in other cases it does not.

(a) Sketch the titration curve for the titration with 0.200 mol L^{-1} NaOH of 25.00 mL of a solution that is 0.100 mol L^{-1} in HCl and 0.100 mol L^{-1} in HNO$_3$. Does this curve differ in any way from what would be obtained in the titration of 25.00 mL of 0.200 mol L^{-1} HCl with 0.200 mol L^{-1} NaOH? Explain.

(b) The titration curve shown in **Figure P5.1** was obtained when 10.00 mL of a solution containing both HCl and H$_3$PO$_4$ was titrated with 0.216 mol L^{-1} NaOH. From this curve, determine the stoichiometric molarities of both the HCl and the H$_3$PO$_4$.

(c) A 10.00 mL solution that is 0.0400 mol L^{-1} H$_3$PO$_4$ and 0.0150 mol L^{-1} NaH$_2$PO$_4$ is titrated with 0.0200 mol L^{-1} NaOH. Sketch the titration curve.

5.6 A 2.50 g sample of Ag$_2$SO$_4$(s) is added to a beaker containing 0.150 L of 0.025 mol L^{-1} BaCl$_2$.

(a) Write an equation for any reaction that occurs by proper calculations.

(b) Describe the final contents of the beaker, that is, the masses of any precipitates present and the concentrations of the ions in the solution.

5.7 A series of manipulations are conducted with saturated Mg(OH)$_2$(aq). Calculate [Mg^{2+}(aq)] at each of the following stages.

(a) 0.500 L of saturated Mg(OH)$_2$(aq) is in contact with Mg(OH)$_2$(s).

(b) 0.500 L of H$_2$O is added to the 0.500 L of solution in part (a), and the solution is vigorously stirred. Undissolved Mg(OH)$_2$(s) remains.

(c) 100.0 mL of the clear solution in part (b) is removed and added to 0.500 L of 0.100 mol L^{-1} HCl(aq).

(d) 25.00 mL of the clear solution in part (b) is removed and added to 250.0 mL of 0.065 mol L^{-1} MgCl$_2$(aq).

(e) 50.00 mL of the clear solution in part (b) is removed and added to 150.0 mL of 0.150 mol L^{-1} KOH(aq).

(a) 绘制用 0.200 mol L^{-1} NaOH 溶液滴定 25.00 mL 某溶液（其中 HCl 和 HNO$_3$ 的浓度均为 0.100 mol L^{-1}）的滴定曲线。该曲线与用 0.200 mol L^{-1} NaOH 溶液滴定 25.00 mL 0.200 mol L^{-1} HCl 溶液的结果有何不同？解释之。

(b) 当用 0.216 mol L^{-1} NaOH 溶液滴定 10.00 mL 含有 HCl 和 H$_3$PO$_4$ 的溶液时，获得如**图 P5.1** 所示滴定曲线。根据该曲线，确定 HCl 和 H$_3$PO$_4$ 的化学计量点的摩尔浓度。

(c) 用 0.0200 mol L^{-1} NaOH 溶液滴定 10.00 mL H$_3$PO$_4$ 浓度为 0.0400 mol L^{-1}、NaH$_2$PO$_4$ 浓度为 0.0150 mol L^{-1} 的溶液。绘制滴定曲线。

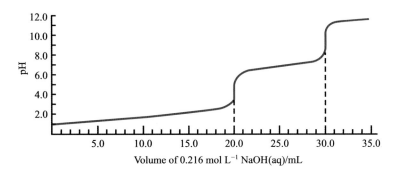

Figure P5.1（图 P5.1）

5.6 将 2.50 g Ag$_2$SO$_4$ 样品加入含有 0.150 L 0.025 mol L^{-1} BaCl$_2$ 溶液的烧杯中。

(a) 通过适当计算，写出发生的任何反应的方程式。

(b) 描述烧杯内最终所含的物质，即存在的任何沉淀的质量、溶液中离子的浓度等。

5.7 对饱和 Mg(OH)$_2$(aq) 进行一系列操作。计算下列每个阶段的 [Mg^{2+}(aq)]：

(a) 与 Mg(OH)$_2$(s) 相接触的 0.500 L 饱和 Mg(OH)$_2$(aq)。

(b) 在 0.500 L (a) 溶液中加入 0.500 L H$_2$O 并剧烈搅拌溶液。存在未溶解的 Mg(OH)$_2$(s) 残留。

(c) 取出 100.0 mL (b) 的澄清溶液，加入 0.500 L 0.100 mol L^{-1} HCl(aq) 中。

(d) 取出 25.00 mL (b) 的澄清溶液，加入 250.0 mL 0.065 mol L^{-1} MgCl$_2$(aq) 中。

(e) 取出 50.00 mL (b) 的澄清溶液，加入 150.0 mL 0.150 mol L^{-1} KOH(aq) 中。

Chapter 6 Redox Reactions and Electrochemistry

In this chapter, we will first discuss oxidation-reduction reactions, also called **redox reactions**, which simultaneously involve both oxidation and reduction half-reactions. Then we will focus on electrochemistry, from electrodes and their potentials to electrochemical cells and their potentials. The concepts such as oxidation states, batteries, and electrolysis will be introduced. The relationship between electrochemistry, equilibrium, and thermodynamics will also be discussed.

6.1 Some Terminologies in Redox Reactions

6.2 Electrode Potentials and Cell Potential

6.3 Standard Electrode Potentials and Standard Cell Potential

6.4 Relationship Between E°_{cell}, ΔG°, and Redox Equilibrium Constant K

6.5 Potential Diagrams

6.6 Nernst Equation and Concentration Cells

6.7 Electrolytic Cells and Overpotential

6.8 Battery and Its Applications

第 6 章　氧化还原反应与电化学

本章我们将首先讨论**氧化还原反应**，它同时包含了氧化和还原两个半反应。然后我们将重点介绍电化学，从电极及其电势到电化学电池及其电动势。我们还将介绍诸如氧化态、电池和电解等概念，并讨论电化学、平衡与热力学之间的关系。

6.1　氧化还原反应的一些术语

6.2　电极电势与电池电动势

6.3　标准电极电势与标准电池电动势

6.4　$E_{池}^{\ominus}$、$\Delta G^{\ominus}$ 与氧化还原平衡常数 K 的关系

6.5　元素电势图

6.6　能斯特方程与浓差电池

6.7　电解池与超电势

6.8　电池及其应用

6.1 Some Terminologies in Redox Reactions

In this section, we will introduce some basic terminologies in redox reactions, such as oxidation states, oxidation/reduction, oxidizing/reducing agents, oxidizing/reducing products, etc. We will also learn how to balance redox equations.

Oxidation States

The **oxidation state** (also called **oxidation number**) is the hypothetical charge that an atom would have if all bonds to atoms of different elements were 100% ionic, meaning that 100% transfer of electrons between the bonded atoms. The oxidation state is an indicator of the degree of oxidation of an atom in a chemical species. For example, given that the bond between Na and Cl in NaCl is ionic, Na atom loses one electron and shows a charge of +1 whereas the Cl atom gains one electron and has a charge of −1. Consequently, the oxidation states of Na and Cl in NaCl are +1 and −1, respectively. Although the bond between H and Cl in HCl is covalent, the hypothetical charges of H and Cl would be +1 and −1, respectively, if the bond is assumed to be 100% ionic. Therefore, the oxidation states of H and Cl in HCl are also +1 and −1, respectively. Further discussion about ionic and covalent bondings can be found in **Chapter 8**.

When assigning oxidation states for chemical species, some conventional rules are generally applied. The oxidation state of the atom in a free element (uncombined element) is 0. The algebraic sum of the oxidation states of all atoms in a neutral molecule must be 0, while that of an ion equals the net charge of the ion. In general, in their compounds, the oxidation states for group 1 and 2 metals are +1 and +2, respectively. In binary compounds with metals, the oxidation states for group 17, 16, and 15 elements are −1, −2 and −3, respectively. In most cases H has an oxidation state of +1, but its oxidation state becomes −1 in metal hydrides. The oxidation state of O is −2 in most cases, but becomes −1 in peroxides.

The concept of the oxidation state should not be confused with the concept of **valence**, which is a measure of the combining power of other atoms when an atom forms a chemical species. Valence is usually used in terms such as valence electrons, valence bond, etc., and there is always a non-negative integer with no sign. Oxidation state, however, can be fractional, and can be either positive, negative, or zero. Examples of the oxidation state and valency of atoms in some chemical species are shown in **Table 6.1**.

Table 6.1　Examples of the Oxidation State and Valence for Atoms in Some Chemical Species
表 6.1　一些化学物质中原子的氧化态和价的示例

Species（物质）	Atom（原子）	Oxidation State（氧化态）	Valence（价）
Cl_2	Cl	0	1
P_4	P	0	3
H_2O	H	+1	1
	O	−2	2
H_2O_2	H	+1	1
	O	−1	2
Fe_3O_4	Fe	+8/3	2 or 3（2 或 3）
	O	−2	2
CH_3CH_2OH	C	−2	4
	H	+1	1
	O	−2	2

Oxidation and Reduction Half-Reactions

A redox reaction must involve both an oxidation half-reaction and a reduction half-reaction that occur simultaneously. The overall reaction is the sum of the two half-reactions. The **oxidation** half-reaction is a process in which the oxidation state of some element increases as electrons are lost, and the **reduction** half-reaction is a process in which the oxidation state of some element decreases as electrons are gained. Oxidation

6.1 氧化还原反应的一些术语

本节我们将介绍氧化还原反应的一些基本术语，如氧化态、氧化/还原、氧化剂/还原剂、氧化产物/还原产物等。我们还将学习如何配平氧化还原方程式。

氧化态

氧化态（也称**氧化数**）是假定所有不同元素的原子之间的化学键均为 100% 离子性时，某个原子所带的假想电荷数，而 100% 离子性意味着成键原子之间存在 100% 电子转移。氧化态是物质中原子氧化程度的量度。例如，假定 NaCl 中 Na 和 Cl 之间的键是离子键，Na 原子失去一个电子，电荷数为 +1，而 Cl 原子得到一个电子，电荷数为 −1。因此，NaCl 中 Na 和 Cl 的氧化态分别为 +1 和 −1。尽管 HCl 中 H 和 Cl 之间的键是共价键，但如果假定这根键为 100% 离子性，则 H 和 Cl 所带的假想电荷数分别为 +1 和 −1。因此，HCl 中 H 和 Cl 的氧化态也分别为 +1 和 −1。关于离子键和共价键的更多讨论详见**第 8 章**。

在给物质分配氧化态时，通常会应用一些常规规则。单质（即未化合的元素）中原子的氧化态为 0。中性分子中所有原子氧化态的代数和必须为 0，而离子中所有原子氧化态的代数和等于离子的净电荷数。通常，第 1、2 族金属在其化合物中的氧化态分别为 +1 和 +2。第 17、16 和 15 族元素在二元金属化合物中的氧化态分别为 −1、−2 和 −3。大多数情况下 H 的氧化态为 +1，但在金属氢化物中为 −1。大多数情况下 O 的氧化态为 −2，但在过氧化物中为 −1。

氧化态的概念不应与**价**的概念混淆。价是衡量一个原子形成化学物质时与其他原子化合能力的量度，通常用于价电子、价键等术语中。价总是自然数，没有正负符号；而氧化态可以是分数，可以为正、为负或等于 0。**表 6.1** 给出了一些化学物质中原子的氧化态和价的示例。

氧化还原半反应

氧化还原反应必须包含同时发生的氧化半反应和还原半反应，总反应是两个半反应的加和。**氧化**半反应是某些元素失去电子、氧化态升高的过程，而**还原**半反应是某些元素得到电子、氧化态降低的过程。氧化和还原半反应必须总是同时发生，与氧化相关的电子总数必须与还原相关的电子总数相等。

在氧化还原反应中，含有氧化态降低的元素并能使另一种物质被氧化的物质，称为**氧化剂**。类似地，含有氧化态升高的元素并能使另一种物质被还原的物质，称为**还原剂**。在氧化还原反应中，氧化剂总是得到电子，自身被还原形成**还原产物**；而还原剂总是失去电子，自身被氧化形成**氧化产物**。

例如，如下氧化还原反应

$$Zn(s) + Cu^{2+}(aq) \rightarrow Zn^{2+}(aq) + Cu(s)$$

同时涉及氧化半反应 $Zn(s) \rightarrow Zn^{2+}(aq) + 2e^-$ 和还原半反应 $Cu^{2+}(aq) + 2e^- \rightarrow Cu(s)$。其中 Zn(s) 是氧化态由 0 升高到 +2 的还原剂，Cu^{2+}(aq) 是氧化态由 +2 降低至 0 的氧化剂。同时，Zn(s) 失去两个电子，被

and reduction half-reactions must always occur together, and the total number of electrons associated with the oxidation must equal that of the reduction.

In a redox reaction, the substance that contains an element whose oxidation state decreases and causes another substance to be oxidized is called the **oxidizing agent**, or **oxidant**. Similarly, the substance that contains an element whose oxidation state increases and causes another substance to be reduced is called the **reducing agent**, or **reductant**. In a redox reaction, the oxidizing agent always gains electrons and is itself reduced to form the **reducing product**, whereas the reducing agent always loses electrons and is itself oxidized to form the **oxidizing product**.

For example, the following redox reaction

$$Zn(s) + Cu^{2+}(aq) \rightarrow Zn^{2+}(aq) + Cu(s)$$

involves simultaneously an oxidation half-reaction $Zn(s) \rightarrow Zn^{2+}(aq) + 2e^-$, in which $Zn(s)$ is the reducing agent whose oxidation state increases from 0 to +2, and a reduction half-reaction $Cu^{2+}(aq) + 2e^- \rightarrow Cu(s)$, in which $Cu^{2+}(aq)$ is the oxidizing agent whose oxidation state decreases from +2 to 0. Meanwhile, $Zn(s)$ loses two electrons and is oxidized into the oxidizing product $Zn^{2+}(aq)$, whereas $Cu^{2+}(aq)$ gains these two electrons and is reduced into the reducing product $Cu(s)$.

Balancing Redox Equations

In a redox reaction, electrons are transferred from the reducing agent to the oxidizing agent. Therefore, the changes in oxidation states can be used to keep track of the total number of transferred electrons in order to balance the redox equations. In doing so, we should:

1) Determine the product according to the experimental phenomena and indicate the reaction conditions. Pay special attention to the acidic or basic medium in which the reaction is carried out. Species that can react with protons or that are unstable in the presence of protons should not appear in an acidic environment.
2) Calculate the increase and decrease, respectively, in oxidation states. If the oxidation states change for more than one element, add up all the changes.
3) Balance the redox equations are based on both electron conservation and atom conservation. Electrons and atoms are neither created nor destroyed in a chemical reaction. The total number of electrons and atoms must be conserved before and after the reaction.

6.2 Electrode Potentials and Cell Potential

Starting from this section, we will learn **electrochemistry**, which is the branch of physical chemistry concerned with the relationship between electrical potential and identifiable chemical change.

Electrode Potentials

In electrochemistry, an **electrode** is an electrical conductor that is used to make contact with a conductive electrolyte. A strip of metal, M, is a typical electrode. When this metal electrode is immersed in a solution containing ions of the same metal, M^{n+}, a **half-cell** is prepared. This half-cell can be denoted by $M(s)|M^{n+}(aq)$, where | stands for the boundary between different phases.

Two kinds of interactions exist at the interface between $M(s)$ and $M^{n+}(aq)$ in this half-cell:
1) $M(s)$ loses n electrons and is oxidized into $M^{n+}(aq)$;
2) $M^{n+}(aq)$ gains n electrons and is reduced into $M(s)$.

A redox equilibrium is quickly established between $M(s)$ and $M^{n+}(aq)$, represented by

$$M(s) \underset{\text{reduction}}{\overset{\text{oxidation}}{\rightleftharpoons}} M^{n+}(aq) + ne^- \qquad (6.1)$$

In this redox equilibrium, M^{n+} is called the **oxidizing species** and M is the **reducing species**. Both M^{n+} and M are termed a **redox couple**, usually represented by M^{n+}/M, with the oxidizing species before a slash sign (/)

氧化为氧化产物 Zn^{2+}(aq)；而 Cu^{2+}(aq) 得到两个电子，被还原为还原产物 Cu(s)。

氧化还原方程式的配平

在氧化还原反应中，电子从还原剂转移到氧化剂。因此，为配平氧化还原方程，可用氧化态的变化来追踪电子转移总数。为此，我们应该：
1) 根据实验现象确定生成物，并标明反应条件。特别注意反应是在酸性还是碱性介质中进行。能与质子反应或在质子存在下不稳定的物种，不应出现在酸性环境中。
2) 分别计算氧化态的升高和降低值。如果不止一种元素的氧化态发生了变化，则将所有变化值相加。
3) 按照电子守恒和原子守恒的原则来配平氧化还原方程式。电子和原子在化学反应中既不会产生也不会破坏。反应前后的电子和原子总数必须守恒。

6.2 电极电势与电池电动势

从本节开始我们将学习**电化学**，这是涉及电势与可识别化学变化之间关系的物理化学分支。

电极电势

在电化学中，**电极**是与导电电解质相接触的电导体。一根金属条 M 就是一个典型的电极。将这根金属电极浸泡在含有同种金属离子 M^{n+} 的溶液中，即形成一个**半电池**，可用 $M(s)|M^{n+}(aq)$ 表示，其中 | 代表不同相之间的界面。

在此半电池的 M(s) 和 M^{n+}(aq) 之间的界面上，存在两种相互作用：
1) M(s) 失去 n 个电子，被氧化成 M^{n+}(aq)；
2) M^{n+}(aq) 得到 n 个电子，被还原成 M(s)。

M(s) 和 M^{n+}(aq) 之间快速地建立了一个氧化还原平衡，可表示为

$$M(s) \underset{\text{还原}}{\overset{\text{氧化}}{\rightleftharpoons}} M^{n+}(aq) + ne^- \tag{6.1}$$

在此氧化还原平衡中，M^{n+} 称为**氧化物种**，M 称为**还原物种**。M^{n+} 和 M 统称为**氧化还原电对**，通常用 M^{n+}/M 表示，其中斜线符号（/）之前为氧化物种，之后为还原物种。

每个电极都存在一个**电极电势**，源自电极与电解质之间的相界面上所形成的电势差，氧化还原电对 M^{n+}/M 的电极电势可以用 $E(M^{n+}/M)$ 表示。然而，由于上述平衡而在电极或电解质中产生的任何变化均非常小，无法测量。因此，不能精确地测量单个电极的电势。相反，必须将两个不同的半电池组合起来，形成一个**电化学电池**，才能进行间接的测量。具体来说，必须测量电子从一个半电池的电极流向另一个半电池的电极的趋势。根据发生氧化还是还原半反应，可将电极进行分类。发生氧化半反应的电极称为**阳极**，发生还原半反应的电极称为**阴极**。阴离子通常会向阳极迁移，而阳离子会向阴极迁移。

and the reducing species after it.

Each electrode has an **electrode potential**, denoted as $E(M^{n+}/M)$ for the redox couple M^{n+}/M, that originates in the potential difference developed at the interface between the electrode and the electrolyte. However, any changes produced at the electrode or in the electrolyte as a consequence of the above equilibrium are too slight to measure. Consequently, the potential of individual electrodes cannot be measured precisely. Instead, measurements must be made indirectly based on a combination of two different half-cells to form an **electrochemical cell**. Specifically, we must measure the tendency of electrons to flow from the electrode of one half-cell to the electrode of the other. Electrodes are classified according to whether oxidation or reduction takes place there. If oxidation occurs, the electrode is called the **anode**. If reduction takes place, the electrode is called the **cathode**. Anions will generally migrate toward the anode and cations toward the cathode.

Cell Potential and Cell Diagram

Figure 6.1 shows an electrochemical cell made by the combination of two half-cells, one with a Zn electrode in contact with $Zn^{2+}(aq)$, and the other with a Cu electrode in contact with $Cu^{2+}(aq)$. The two electrodes are joined by wires to a voltmeter, and the two solutions are joined by a third electrolyte solution (called a **salt bridge**, such as KNO_3) in a U-tube. The ends of the salt bridge are plugged with a porous material that allows ions to migrate but prevents the bulk flow of liquid. As the arrows suggested in **Figure 6.1**, Zn(s) loses 2 electrons to become $Zn^{2+}(aq)$ and serves as the anode. These 2 electrons pass through the wires and the voltmeter to the Cu cathode, where they are gained by $Cu^{2+}(aq)$ to produce a deposit of metallic Cu(s). Simultaneously, anions (NO_3^-) from the salt bridge migrate into the Zn half-cell and neutralize the positive charge of the excess Cu^{2+}; cations (K^+) from the salt bridge migrate into the Cu half-cell and neutralize the negative charge of the excess anions. The overall reaction that occurs as the electrochemical cell spontaneously produces electric current is

$$\text{Oxidation}: Zn(s) \rightarrow Zn^{2+}(aq) + 2e^-$$
$$\text{Reduction}: Cu^{2+}(aq) + 2e^- \rightarrow Cu(s)$$
$$\text{Overall}: Zn(s) + Cu^{2+}(aq) \rightarrow Zn^{2+}(aq) + Cu(s) \tag{6.2}$$

The reading on the voltmeter (1.104 V) is called the **cell potential**, or **cell voltage**, or **electromotive force** (emf), and is represented by E_{cell}. Cell potential is the potential difference between a cathode and anode in an electrochemical cell, i.e.

$$E_{cell} = E_{cathode} - E_{anode} \tag{6.3}$$

For simplicity, a **cell diagram** is used to show the components of an electrochemical cell in a symbolic way. The following conventions are generally accepted in writing cell diagrams:
1) The anode where oxidation occurs is placed on the left side of the diagram;
2) The cathode where reduction occurs is placed on the right side of the diagram;
3) Each boundary between different phases is represented by a single vertical line (|);
4) The boundary between half-cell components, commonly a salt bridge, is represented by a double vertical line (||). Species in aqueous solution are placed on either side of the double vertical line. Different species within the same solution are separated from each other by a comma.

The cell diagram corresponding to both **Figure 6.1** and **Reaction (6.2)** is customarily written as

$$\text{(anode)} \quad Zn(s)|Zn^{2+}(aq)||Cu^{2+}(aq)|Cu(s) \quad \text{(cathode)} \quad E_{cell} = 1.104 \text{ V}$$

Half-cell   Salt   Half-cell   Cell potential
(oxidation)   bridge   (reduction)   (cell voltage or emf)

Classification of Electrochemical Cells

According to the direction of energy conversion, electrochemical cells can be classified into voltaic cells and electrolytic cells. **Voltaic cells**, also called **Galvanic cells**, are electrochemical cells that

电池电动势及电池符号图

图 6.1 给出了由两个半电池组合而成的一个电化学电池，在其中一个半电池中 Zn 电极与 $Zn^{2+}(aq)$ 相接触，在另一个半电池中 Cu 电极与 $Cu^{2+}(aq)$ 相接触。两个电极通过导线与电压表相连，两种溶液通过 U 形管中的第三种电解质溶液（称为**盐桥**，如 KNO_3）连接在一起。盐桥的两端用多孔材料堵塞，这种材料仅允许离子迁移通过，但阻止了液体的大量流动。如**图 6.1** 中的箭头所示，作为阳极的 Zn(s) 失去 2 个电子变成 $Zn^{2+}(aq)$，这两个电子通过导线和电压表到达 Cu 阴极，在那里被 $Cu^{2+}(aq)$ 得到，并产生金属 Cu(s) 沉积。同时，阴离子（NO_3^-）从盐桥迁移到 Zn 半电池，中和了过量 Zn^{2+} 的正电荷；而阳离子（K^+）从盐桥迁移到 Cu 半电池，中和了过量阴离子的负电荷。当电化学电池自发产生电流时，发生的总反应是

氧化反应：$Zn(s) \rightarrow Zn^{2+}(aq) + 2e^-$

还原反应：$Cu^{2+}(aq) + 2e^- \rightarrow Cu(s)$

总反应：$Zn(s) + Cu^{2+}(aq) \rightarrow Zn^{2+}(aq) + Cu(s)$ （6.2）

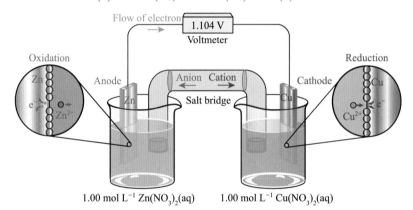

Figure 6.1 The reaction $Zn(s) + Cu^{2+}(aq) \rightarrow Zn^{2+}(aq) + Cu(s)$ in an electrochemical cell.

图 6.1 电化学电池中发生的反应 $Zn(s) + Cu^{2+}(aq) \rightarrow Zn^{2+}(aq) + Cu(s)$。

电压表的读数（1.104 V）称为**电池电势**、**电池电压**或**电池电动势**（emf），用 $E_\text{池}$ 表示。电池电动势是电化学电池中阴极和阳极之间的电势差，即

$$E_\text{池} = E_\text{阴极} - E_\text{阳极} \quad (6.3)$$

为简单起见，可用**电池符号图**以符号的形式显示电化学电池的组件。在书写电池符号图时，通常采用以下惯例：
1) 发生氧化反应的阳极位于图的左侧；
2) 发生还原反应的阴极位于图的右侧；
3) 不同相之间的界面用一条竖线 | 表示；
4) 半电池组分之间的界面（通常为盐桥）用一条双竖线 ‖ 表示，水溶液中的物种位于双竖线两侧，同一溶液中的不同物种之间用逗号分隔。

与**图 6.1** 和**反应（6.2）**对应的电池符号图通常可记为

（阳极）　$Zn(s)|Zn^{2+}(aq)\|Cu^{2+}(aq)|Cu(s)$　（阴极）　$E_\text{池} = 1.104\ V$
　　　　　半电池　　盐桥　　半电池　　　　　　电池电势
　　　　　（氧化）　　　　　（还原）　（电池电压或电池电动势）

电化学电池的分类

根据能量转换的方向，电化学电池可分为原电池和电解池。**原**

produce electricity as a result of spontaneous reactions. **Electrolytic cells** are electrochemical cells in which nonspontaneous changes are driven by electricity. Therefore, voltaic cells generate electric energy from chemical energy, whereas electrolytic cells convert electric energy into chemical energy. The electrochemical cell shown in **Figure 6.1** is a typical voltaic cell. More discussion about electrolytic cells can be found in **Section 6.7**.

6.3 Standard Electrode Potentials and Standard Cell Potential

As discussed earlier, the potential of individual electrodes cannot be measured precisely in a direct way, but can be established indirectly by measuring the cell potential, which is one of the most precise scientific measurements of all. Therefore, by choosing a particular half-cell as the arbitrary zero for electrode potential, the potentials of other half-cells can be compared with this reference. The commonly accepted reference is the standard hydrogen electrode.

Standard Hydrogen Electrode

Figure 6.2 shows a **standard hydrogen electrode** (SHE), which is a hydrogen electrode in standard conditions. The SHE involves equilibrium established on the surface of an inert metal electrode (such as platinum) between H_3O^+ ions from a solution at unit activity ($a_{H_3O^+} = 1$) and H_2 gas molecules at unit pressure ($p_{H_2} = 1$ bar). For simplicity, H^+ is usually written instead of H_3O^+. This equilibrium reaction produces a particular potential of about 4.4±0.2 V, suggested by IUPAC in 1986 from thermochemistry data, but this potential is arbitrarily declared to be exact zero at all temperatures, as

$$2H^+(a=1) + 2e^- \xrightleftharpoons[]{\text{on Pt}} H_2(g, 1\text{ bar}) \quad E^\circ(H^+/H_2) = 0 \text{ V}$$

The diagram for SHE is

$$\text{Pt}(s) | H_2(g, 1\text{ bar}) | H^+(a=1) \quad E^\circ(H^+/H_2) = 0 \text{ V}$$

which consists of three phases (solid platinum, gaseous H_2, and aqueous H^+) separated by two boundaries. In some cases, $[H^+] = 1$ mol L^{-1} and $p_{H_2} = 1$ atm are used instead, and this modified SHE is represented by

$$\text{Pt}(s) | H_2(g, 1\text{ atm}) | H^+(1\text{ mol L}^{-1}) \quad E^\circ(H^+/H_2) \approx 0 \text{ V}$$

Standard Electrode Potentials

A **standard electrode potential** is the electrode potential when all substances involved in the electrode reaction are at their standard conditions. The standard electrode potential is usually represented by $E^\circ(M^{n+}/M)$. The E° value is normally determined by comparing it with an SHE, for which $E^\circ(H^+/H_2) = 0$. The standard electrode potential measures the tendency for a reduction reaction to occur at an electrode at standard conditions. Standard conditions mean that ionic species are present in aqueous solution at $a = 1$ (approximately 1 mol L^{-1}), and gases are at $p = 1$ bar (approximately 1 atm). If no metallic substance is indicated, an inert metallic electrode, such as platinum, is used as default.

As E° measures the tendency for a reduction reaction to occur, a higher E° value indicates a greater tendency for the reduction process to occur, and thus the oxidizing species M^{n+} shows a stronger oxidation ability. Consequently, a lower E° value suggests a lesser tendency for the reduction process to occur, and the reducing species M exhibits a stronger reducing ability. **Table 6.2** lists the standard electrode potentials at 25°C in a descending order for some common reduction half-reactions. Notice that the E° values for the same redox couple might be different in acidic or basic solutions, demonstrating different oxidizing abilities in different environments. For example, $E^\circ(O_3/O_2) = 2.076$ V in acidic solution is greater than $E^\circ(O_3/O_2) = 1.24$ V in basic solution, indicating a greater oxidizing ability for O_3 in an acidic environment.

电池也称**伏特电池**，是一种通过自发反应产生电能的电化学电池。**电解池**是由电能驱动发生非自发变化的电化学电池。因此，原电池将化学能转化为电能，而电解池则将电能转化为化学能。**图 6.1** 所示的电化学电池是一个典型的原电池。关于电解池的更多讨论详见 **6.7 节**。

6.3 标准电极电势与标准电池电动势

如前所述，单个电极的电势不能直接精确地测量，但可以通过测量电池电动势来间接确定，这是最为精确的科学测量方法之一。因此，通过选择一个特定的半电池作为电极电势的任意零点，就可以将其他半电池的电势与此参比零点进行比较来确定。广为接受的参比电极为标准氢电极。

标准氢电极

图 6.2 给出了**标准氢电极**（SHE，即标态下的氢电极）示意图。标准氢电极涉及在惰性金属电极（如铂）表面上、单位活度（$a_{H_3O^+} = 1$）的 H_3O^+ 离子溶液与单位压强（$p_{H_2} = 1$ bar）的气态 H_2 分子之间所建立的平衡。为简单起见，通常用 H^+ 替代 H_3O^+。该平衡反应产生了约为 4.4±0.2 V 的特定电势，此数值由 IUPAC 于 1986 年基于热化学数据提出。但在任意温度下，该电势均被设定为精确的零点，即

$$2H^+(a=1) + 2e^- \underset{Pt}{\rightleftharpoons} H_2(g, 1\,bar) \quad E^{\ominus}(H^+/H_2) = 0\,V$$

标准氢电极的电极符号图为

$$Pt(s)\big|H_2(g, 1\,bar)\big|H^+(a=1) \quad E^{\ominus}(H^+/H_2) = 0\,V$$

它包含被两个界面分隔开的三个相（固态铂、气态 H_2 以及 H^+ 水溶液）。在某些情况下，可用 $[H^+]=1$ mol L^{-1} 和 $p_{H_2} = 1$ atm 替代。修改后的标准氢电极可表示为

$$Pt(s)\big|H_2(g, 1\,atm)\big|H^+(1\,mol\,L^{-1}) \quad E^{\ominus}(H^+/H_2) \approx 0\,V$$

标准电极电势

标准电极电势是指当参与电极反应的所有物质均处于标态时的电极电势，通常用 $E^{\ominus}(M^{n+}/M)$ 表示，其值可通过与标准氢电极进行比较来确定，而 $E^{\ominus}(H^+/H_2) = 0$。标准电极电势反映了标态下该电极发生还原反应的趋势。标态意味着，水溶液中离子物种的 $a=1$（约 1 mol L^{-1}），气体的 $p=1$ bar（约 1 atm）。如果未标明金属物质，则默认采用惰性金属电极（如铂）。

由于 $E^{\ominus}$ 反映发生还原反应的趋势，$E^{\ominus}$ 越大，表明该还原过程发生的趋势越大，因此对应的氧化物种 M^{n+} 具有更强的氧化能力。相应地，$E^{\ominus}$ 越小，表明该还原过程发生的趋势越小，因此对应的还原物种 M 具有更强的还原能力。**表 6.2** 列出了 25°C 时一些常见的还原半反应的标准电极电势，按降序排列。注意，同一氧化还原电对在酸性或碱性溶液中的 $E^{\ominus}$ 值可能不同，表明其在不同环境下的氧化能力不同。例如，酸性溶液中的 $E^{\ominus}(O_3/O_2)=2.076$ V，大于碱性

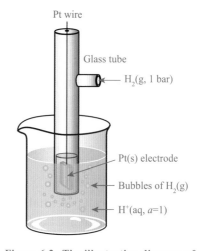

Figure 6.2 The illustrative diagram of a standard hydrogen electrode (SHE).

图 6.2 标准氢电极（SHE）示意图。

Table 6.2 Standard Electrode Potentials for Some Common Reduction Half-Reactions at 25°C
表 6.2 25°C 时一些常见的还原半反应的标准电极电势

Reduction Half-Reaction（还原半反应）	$E°$/V	Reduction Half-Reaction（还原半反应）	$E°$/V
Standard acidic solution (标准酸性溶液 pH = 0)			
$F_2(g) + 2H^+(aq) + 2e^- \rightarrow 2HF(aq)$	+3.053	$I_2(s) + 2e^- \rightarrow 2I^-(aq)$	+0.5355
$O_3(g) + 2H^+(aq) + 2e^- \rightarrow O_2(g) + H_2O(l)$	+2.076	$Cu^{2+}(aq) + 2e^- \rightarrow Cu(s)$	+0.3419
$S_2O_8^{2-}(aq) + 2e^- \rightarrow 2SO_4^{2-}(aq)$	+2.010	$SO_4^{2-}(aq) + 4H^+(aq) + 2e^- \rightarrow SO_2(g) + 2H_2O(l)$	+0.172
$H_2O_2(aq) + 2H^+(aq) + 2e^- \rightarrow 2H_2O(l)$	+1.776	$Sn^{4+}(aq) + 2e^- \rightarrow Sn^{2+}(aq)$	+0.151
$MnO_4^-(aq) + 8H^+(aq) + 5e^- \rightarrow Mn^{2+}(aq) + 4H_2O(l)$	+1.507	$S(s) + 2H^+(aq) + 2e^- \rightarrow H_2S(g)$	+0.142
$PbO_2(s) + 4H^+(aq) + 2e^- \rightarrow Pb^{2+}(aq) + 2H_2O(l)$	+1.455	$2H^+(aq) + 2e^- \rightarrow H_2(g)$	0
$Cr_2O_7^{2-}(aq) + 14H^+(aq) + 6e^- \rightarrow 2Cr^{3+}(aq) + 7H_2O(l)$	+1.36	$Pb^{2+}(aq) + 2e^- \rightarrow Pb(s)$	−0.1262
$Cl_2(g) + 2e^- \rightarrow 2Cl^-(aq)$	+1.35827	$Sn^{2+}(aq) + 2e^- \rightarrow Sn(s)$	−0.1375
$O_2(g) + 4H^+(aq) + 4e^- \rightarrow 2H_2O(l)$	+1.229	$Fe^{2+}(aq) + 2e^- \rightarrow Fe(s)$	−0.447
$MnO_2(s) + 4H^+(aq) + 2e^- \rightarrow Mn^{2+}(aq) + 2H_2O(l)$	+1.224	$Zn^{2+}(aq) + 2e^- \rightarrow Zn(s)$	−0.7618
$2IO_3^-(aq) + 12H^+(aq) + 10e^- \rightarrow I_2(s) + 6H_2O(l)$	+1.195	$Al^{3+}(aq) + 3e^- \rightarrow Al(s)$	−1.676
$Br_2(l) + 2e^- \rightarrow 2Br^-(aq)$	+1.066	$Mg^{2+}(aq) + 2e^- \rightarrow Mg(s)$	−2.372
$NO_3^-(aq) + 4H^+(aq) + 3e^- \rightarrow NO(g) + 2H_2O(l)$	+0.957	$Na^+(aq) + e^- \rightarrow Na(s)$	−2.71
$Ag^+(aq) + e^- \rightarrow Ag(s)$	+0.7996	$Ca^{2+}(aq) + 2e^- \rightarrow Ca(s)$	−2.868
$Fe^{3+}(aq) + e^- \rightarrow Fe^{2+}(aq)$	+0.771	$K^+(aq) + e^- \rightarrow K(s)$	−2.931
$O_2(g) + 2H^+(aq) + 2e^- \rightarrow H_2O_2(aq)$	+0.695	$Li^+(aq) + e^- \rightarrow Li(s)$	−3.0401
Standard basic solution (标准碱性溶液 pH = 14)			
$O_3(g) + H_2O(l) + 2e^- \rightarrow O_2(g) + 2OH^-(aq)$	+1.24	$O_2(g) + 2H_2O(l) + 4e^- \rightarrow 4OH^-(aq)$	+0.401
$OCl^-(aq) + H_2O(l) + 2e^- \rightarrow Cl^-(aq) + 2OH^-(aq)$	+0.81	$2H_2O(l) + 2e^- \rightarrow H_2(g) + 2OH^-(aq)$	−0.8277

Standard Cell Potential

A **standard cell potential**, $E°_{cell}$, is the potential difference of a cell formed from two standard electrodes. It is always the difference between the standard cathode potential and the standard anode potential, as

$$E°_{cell} = E°_{cathode} - E°_{anode} \tag{6.4}$$

If we form an electrochemical cell by using SHE as the anode and a standard electrode M^{n+}(1 mol L^{-1})|M(s) as the cathode, since $E°_{anode} = E°(H^+/H_2) = 0$, we then have $E°_{cell} = E°_{cathode}$. In this case, the measured cell potential equals exactly the standard electrode potential of the cathode, or

$$E°_{cell}\left(Pt(s)\middle|H_2(g, 1\,atm)\middle|H^+(1\,mol\,L^{-1})\middle\|M^{n+}(1\,mol\,L^{-1})\middle|M(s)\right) = E°(M^{n+}/M)$$

For example, from **Table 6.2** we can find that $E°(Cu^{2+}/Cu) = 0.342$ V and $E°(Zn^{2+}/Zn) = -0.762$ V. The following three standard cell potentials can be calculated as:

1) $Pt(s)\middle|H_2(g,1\,atm)\middle|H^+(1\,mol\,L^{-1})\middle\|Cu^{2+}(1\,mol\,L^{-1})\middle|Cu(s)$

$$E°_{cell} = E°(Cu^{2+}/Cu) - E°(H^+/H_2) = 0.342\,V$$

$$H_2(g) + Cu^{2+}(aq) \rightarrow 2H^+(aq) + Cu(s)$$

2) $Pt(s)\middle|H_2(g,1\,atm)\middle|H^+(1\,mol\,L^{-1})\middle\|Zn^{2+}(1\,mol\,L^{-1})\middle|Zn(s)$

$$E°_{cell} = E°(Zn^{2+}/Zn) - E°(H^+/H_2) = -0.762\,V$$

$$H_2(g) + Zn^{2+}(aq) \rightarrow 2H^+(aq) + Zn(s)$$

3) $Zn(s)\middle|Zn^{2+}(1\,mol\,L^{-1})\middle\|Cu^{2+}(1\,mol\,L^{-1})\middle|Cu(s)$

$$E°_{cell} = E°(Cu^{2+}/Cu) - E°(Zn^{2+}/Zn) = [0.342 - (-0.762)]\,V = 1.104\,V$$

溶液中的 $E^{\ominus}(O_3/O_2) = 1.24$ V，表明在酸性环境中 O_3 的氧化能力更强。

标准电池电动势

标准电池电动势（$E^{\ominus}_{池}$）是将两个标准电极连成电池时的电势差。它总是标准阴极电势与标准阳极电势之间的差值，即

$$E^{\ominus}_{池} = E^{\ominus}_{阴极} - E^{\ominus}_{阳极} \tag{6.4}$$

如果用标准氢电极作为阳极，用某标准电极 M^{n+}(1 mol L^{-1})|M(s) 作为阴极，来组建一个电化学电池，由于 $E^{\ominus}_{阳极} = E^{\ominus}(H^+/H_2) = 0$，则有 $E^{\ominus}_{池} = E^{\ominus}_{阴极}$。在这种情况下，测得的标准电池电动势恰好等于阴极的标准电极电势，即

$$E^{\ominus}_{池}\left(Pt(s)\big|H_2(g, 1\,atm)\big|H^+\left(1\,mol\,L^{-1}\right)\|M^{n+}\left(1\,mol\,L^{-1}\right)\big|M(s)\right)$$
$$= E^{\ominus}\left(M^{n+}/M\right)$$

例如，从表 6.2 中可查到，$E^{\ominus}(Cu^{2+}/Cu) = 0.342$ V，$E^{\ominus}(Zn^{2+}/Zn) = -0.762$ V，则以下三个标准电池电动势可计算为：

1) $Pt(s)\big|H_2(g, 1\,atm)\big|H^+\left(1\,mol\,L^{-1}\right)\|Cu^{2+}\left(1\,mol\,L^{-1}\right)\big|Cu(s)$

$$E^{\ominus}_{池} = E^{\ominus}\left(Cu^{2+}/Cu\right) - E^{\ominus}\left(H^+/H_2\right) = 0.342 \text{ V}$$
$$H_2(g) + Cu^{2+}(aq) \rightarrow 2H^+(aq) + Cu(s)$$

2) $Pt(s)\big|H_2(g, 1\,atm)\big|H^+\left(1\,mol\,L^{-1}\right)\|Zn^{2+}\left(1\,mol\,L^{-1}\right)\big|Zn(s)$

$$E^{\ominus}_{池} = E^{\ominus}\left(Zn^{2+}/Zn\right) - E^{\ominus}\left(H^+/H_2\right) = -0.762 \text{ V}$$
$$H_2(g) + Zn^{2+}(aq) \rightarrow 2H^+(aq) + Zn(s)$$

3) $Zn(s)\big|Zn^{2+}\left(1\,mol\,L^{-1}\right)\|Cu^{2+}\left(1\,mol\,L^{-1}\right)\big|Cu(s)$

$$E^{\ominus}_{池} = E^{\ominus}\left(Cu^{2+}/Cu\right) - E^{\ominus}\left(Zn^{2+}/Zn\right) = [0.342 - (-0.762)]\text{ V} = 1.104 \text{ V}$$
$$Zn(s) + Cu^{2+}(aq) \rightarrow Zn^{2+}(aq) + Cu(s)$$

6.4 $E^{\ominus}_{池}$、$\Delta G^{\ominus}$ 与氧化还原平衡常数 K 的关系

6.3 节最后的三个电化学电池中，第一个和第三个的氧化还原反应可以自发进行，相应的 $E^{\ominus}_{池}$ 为正值。而第二个的氧化还原反应不能自发进行，$E^{\ominus}_{池}$ 为负值。因此，$E^{\ominus}_{池}$ 也可用来确定氧化还原反应的自发性。在 $E^{\ominus}_{池}$、$\Delta G^{\ominus}$ 以及氧化还原平衡常数 K 之间，必定存在某种关系。

$E^{\ominus}_{池}$ 与 $\Delta G^{\ominus}$ 的关系

回顾 **3.6 节**，对于恒温恒压下只有体积功的过程，其吉布斯自由能变小于或等于 0，即

$$\Delta G \leqslant 0 \tag{6.5}$$

其中不可逆过程 $\Delta G < 0$，可逆过程 $\Delta G = 0$。如果将所有功分为体积功（w）和非体积功（$w_{非}$），则意味着 $w_{非} = 0$，或 $\Delta G \leqslant w_{非}$。原电池的氧化还原反应会自发进行，电池通过做电功（$w_{电}$）将化学能转化为电能。这种情况下 $w_{非} = w_{电} \neq 0$，且有

$$Zn(s) + Cu^{2+}(aq) \rightarrow Zn^{2+}(aq) + Cu(s)$$

6.4 Relationship Between E°_{cell}, ΔG°, and Redox Equilibrium Constant K

In the three electrochemical cells at the end of **Section 6.3**, the first and third redox reactions can happen spontaneously, and the corresponding E°_{cell} are positive. The second redox reaction, however, could not happen spontaneously, and E°_{cell} is negative. Therefore, E°_{cell} can also be used to determine the spontaneity of redox reactions. There must be some relationship between E°_{cell}, ΔG° and the redox equilibrium constant K.

Relationship Between E°_{cell} and ΔG°

Recall from **Section 3.6** that for a process happening with only p-V work at constant T and p, the Gibbs free energy change is less than or equal to 0, or

$$\Delta G \leqslant 0 \tag{6.5}$$

where $\Delta G < 0$ for irreversible processes and $\Delta G = 0$ for reversible processes. If we classify all works into p-V work (w) and non-p-V work (w_{non}), it then means that $w_{non} = 0$, or $\Delta G \leqslant w_{non}$. In a voltaic cell, a redox reaction happens spontaneously, and electric work (w_{elec}) are done by the cell to convert chemical energy into electric energy. In this case, $w_{non} = w_{elec} \neq 0$, and

$$\Delta G \leqslant w_{elec} \tag{6.6}$$

where $\Delta G < w_{elec}$ for irreversible processes and $\Delta G = w_{elec}$ for reversible processes.

For a typical electrode reaction

$$M^{n+}(aq) + ne^- \rightarrow M(s)$$

when 1 mol the above electrode reaction happens, the quantity of electricity passed through the electrode is given by

$$q = nF$$

where F is the electric charge for 1 mol electrons, called the **Faraday constant**, and

$$F = eN_A = 96485 \text{ C mol}^{-1} \tag{6.7}$$

The electric work that a voltaic cell can do can be calculated as

$$w_{elec} = -qU = -nFE_{cell}$$

For a reversible electrochemical cell with only electric work, we then have

$$\Delta G = w_{elec} = -nFE_{cell}$$

If all substances involved in this electrochemical cell are in their standard states, then

$$\Delta G^\circ = -nFE^\circ_{cell} \tag{6.8}$$

where n is the number of electrons (also called charge number) involved in an electrochemical reaction.

Criteria for Spontaneous Change in Redox Reactions

Again, recall from **Section 3.6** that the main criterion for spontaneous changes occurring with only p-V work at constant T and p is that $\Delta G < 0$. If $\Delta G = -nFE_{cell} < 0$, it requires that $E_{cell} > 0$. This suggests that E_{cell} can also be used to predict the direction of spontaneous change in a redox reaction as follows:

1) If $E_{cell} > 0$, $\Delta G < 0$ and the reaction occurs spontaneously in the forward direction;
2) If $E_{cell} = 0$, $\Delta G = 0$ and the reaction is at equilibrium;
3) If $E_{cell} < 0$, $\Delta G > 0$ and the reaction occurs spontaneously in the reverse direction.

Indirect Calculation of Standard Electrode Potentials

Not all electrodes can be combined with SHE to form voltaic cells with measurable cell potentials, which equal the difference between the electrode potential and SHE potential. The standard potentials of

$$\Delta G \leq w_{电} \tag{6.6}$$

其中不可逆过程 $\Delta G < w_{电}$，可逆过程 $\Delta G = w_{电}$。

对于一个典型的电极反应

$$M^{n+}(aq) + ne^- \to M(s)$$

当发生 1 mol 上述电极反应时，通过电极的电量为

$$q = nF$$

其中 F 是 1 mol 电子的电量，称为**法拉第常数**，有

$$F = eN_A = 96485 \text{ C mol}^{-1} \tag{6.7}$$

原电池可做的电功为

$$w_{电} = -qU = -nFE_{池}$$

对于一个只做电功的可逆电化学电池，有

$$\Delta G = w_{电} = -nFE_{池}$$

如果此电化学电池涉及的所有物质均处于标态，有

$$\Delta G^{\ominus} = -nFE_{池}^{\ominus} \tag{6.8}$$

其中 n 为电化学反应涉及的电子数（也称电荷数）。

氧化还原反应中自发过程的判据

同样，回顾 **3.6 节**，恒温恒压下只有体积功的自发变化，其主要判据为 $\Delta G < 0$。如果 $\Delta G = -nFE_{池} < 0$，则要求 $E_{池} > 0$。这表明 $E_{池}$ 也可用于预测氧化还原反应中自发变化的方向：

1) 若 $E_{池} > 0$，则 $\Delta G < 0$，反应自发正向进行；
2) 若 $E_{池} = 0$，则 $\Delta G = 0$，反应处于平衡状态；
3) 若 $E_{池} < 0$，则 $\Delta G > 0$，反应自发反向进行。

标准电极电势的间接计算

并非所有电极都能与标准氢电极组成具有可测电池电动势的原电池，使其电池电动势等于该电极电势与标准氢电极电势之差。这些电极的标准电势可以从相应的 $\Delta G^{\ominus}$ 直接计算，或者从其他标准电极电势间接计算。

例如，已知 $E^{\ominus}(Fe^{2+}/Fe) = -0.447 \text{ V}$，$E^{\ominus}(Fe^{3+}/Fe^{2+}) = +0.771 \text{ V}$，我们可以假定由这些电极与标准氢电极组成标准原电池，来间接计算 $E^{\ominus}(Fe^{3+}/Fe)$，有

(1) $Fe^{2+}(aq) + 2e^- \to Fe(s)$

$$E_{池1}^{\ominus} = E^{\ominus}(Fe^{2+}/Fe) = -0.447 \text{ V} \quad \Delta G_1^{\ominus} = -n_1 F E_{池1}^{\ominus}$$

(2) $Fe^{3+}(aq) + e^- \to Fe^{2+}(aq)$

$$E_{池2}^{\ominus} = E^{\ominus}(Fe^{3+}/Fe^{2+}) = 0.771 \text{ V} \quad \Delta G_2^{\ominus} = -n_2 F E_{池2}^{\ominus}$$

(3) $Fe^{3+}(aq) + 3e^- \to Fe(s)$

$$\Delta G_3^{\ominus} = -n_3 F E_{池3}^{\ominus} = \Delta G_1^{\ominus} + \Delta G_2^{\ominus} = -n_1 F E_{池1}^{\ominus} - n_2 F E_{池2}^{\ominus}$$

$$E_{池3}^{\ominus} = \frac{n_1 F E_{池1}^{\ominus} + n_2 F E_{池2}^{\ominus}}{n_3 F} = \frac{[2 \times (-0.447) + 1 \times 0.771]}{3} \text{ V} = -0.041 \text{ V}$$

$$\therefore E^{\ominus}(Fe^{3+}/Fe) = E_{池3}^{\ominus} = -0.041 \text{ V}$$

如 **3.6 节**所述，ΔG 是一个具有加和性的广度性质。而电池电动势 $E_{池}$ 是一个不具有加和性的强度性质，因此不能通过直接相加来计算 $E_{池}$。

those electrodes can be calculated either directly from the corresponding $\Delta G°$, or indirectly from some other standard electrode potentials.

For example, given that $E°(Fe^{2+}/Fe) = -0.440$ V and $E°(Fe^{3+}/Fe^{2+}) = +0.771$ V, we can calculate indirectly the value of $E°(Fe^{3+}/Fe)$ by hypothetically forming standard voltaic cells from these electrodes and SHE, as

(1) $Fe^{2+}(aq) + 2e^- \rightarrow Fe(s)$

$E°_{cell1} = E°(Fe^{2+}/Fe) = -0.447$ V $\quad \Delta G°_1 = -n_1 F E°_{cell1}$

(2) $Fe^{3+}(aq) + e^- \rightarrow Fe^{2+}(aq)$

$E°_{cell2} = E°(Fe^{3+}/Fe^{2+}) = 0.771$ V $\quad \Delta G°_2 = -n_2 F E°_{cell2}$

(3) $Fe^{3+}(aq) + 3e^- \rightarrow Fe(s)$

$\Delta G°_3 = -n_3 F E°_{cell3} = \Delta G°_1 + \Delta G°_2 = -n_1 F E°_{cell1} - n_2 F E°_{cell2}$

$E°_{cell3} = \dfrac{n_1 F E°_{cell1} + n_2 F E°_{cell2}}{n_3 F} = \dfrac{[2 \times (-0.447) + 1 \times 0.771]}{3}$ V $= -0.041$ V

$\therefore E°(Fe^{3+}/Fe) = E°_{cell3} = -0.041$ V

As previously discussed in **Section 3.6**, ΔG is an extensive property with additivity. Cell potential E_{cell}, however, is an intensive property that is not additive. Therefore, E_{cell} cannot be calculated by direct addition. Instead, we should calculate E_{cell} using a weighted method, where the electron number n serves as the corresponding weight.

Relationship Between $E°_{cell}$ and Redox Equilibrium Constant K

In **Section 3.9**, we have derived the relationship between the standard Gibbs free energy change $\Delta G°$ and the equilibrium constant K as

$$\Delta G° = -RT \ln K$$

Combining the above equation with **Equation (6.8)**, we then have

$$\Delta G° = -nFE°_{cell} = -RT \ln K$$

$$\therefore E°_{cell} = \frac{RT}{nF} \ln K \tag{6.9}$$

At 25°C, by applying $R = 8.314$ J mol^{-1} K^{-1}, $T = 298.15$ K, and $F = 96485$ C mol^{-1}, we can obtain

$$E°_{cell} = \frac{0.02569 \text{ V}}{n} \ln K = \frac{0.0592 \text{ V}}{n} \lg K \tag{6.10}$$

This is the relationship between standard cell potential $E°_{cell}$ and the redox equilibrium constant K. **Figure 6.3** summarizes several important relationships among thermodynamics, equilibrium, and electrochemistry.

6.5 Potential Diagrams

Potential diagrams are a series of schematic diagrams that show the standard electrode potentials between some species of the same element but with different oxidation states in aqueous solution. Here, we will only introduce Latimer diagram (Wendell M. Latimer, 1893—1955) and Frost diagram (Arthur A. Frost, 1909—2002).

Latimer Diagram

Latimer diagram is also called the reduction potential diagram. In a Latimer diagram, species of the same element are listed in decreasing oxidation states from left to right. The corresponding numerical values of the standard electrode potential $E°$ in volts are written over a horizontal line joining the two neighboring species involved in a redox couple.

相反，我们应该使用加权法计算 $E_{池}$，其中电子数 n 可作为相应的权重。

$E_{池}^{\ominus}$ 与氧化还原平衡常数 K 的关系

3.9 节我们推导了标准吉布斯自由能变 $\Delta G^{\ominus}$ 与平衡常数 K 之间的关系为

$$\Delta G^{\ominus} = -RT \ln K$$

将上述方程与**式 (6.8)** 结合，可得

$$\Delta G^{\ominus} = -nFE_{池}^{\ominus} = -RT \ln K$$

$$\therefore E_{池}^{\ominus} = \frac{RT}{nF} \ln K \tag{6.9}$$

25°C 时，将 $R = 8.314 \text{ J mol}^{-1} \text{ K}^{-1}$、$T = 298.15 \text{ K}$、$F = 96485 \text{ C mol}^{-1}$ 代入，可得

$$E_{池}^{\ominus} = \frac{0.02569 \text{ V}}{n} \ln K = \frac{0.0592 \text{ V}}{n} \lg K \tag{6.10}$$

这就是标准电池电动势 $E_{池}^{\ominus}$ 与氧化还原平衡常数 K 之间的关系。**图 6.3** 总结了热力学、平衡以及电化学的几个重要关系。

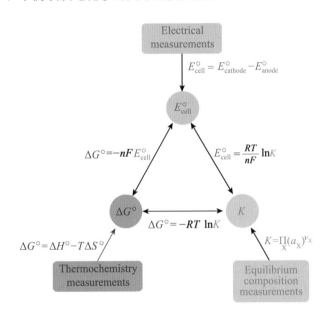

Figure 6.3 A summary of several important relationships among thermodynamics, equilibrium, and electrochemistry.

图 6.3 热力学、平衡以及电化学之间的几个重要关系的总结。

6.5 元素电势图

元素电势图是表示水溶液中同一元素具有不同氧化态物种之间的标准电极电势的一系列示意图。这里我们只介绍拉蒂默电势图（温德尔·M. 拉蒂默，1893—1955）和弗罗斯特电势图（亚瑟·A. 弗罗斯特，1909—2002）。

拉蒂默电势图

拉蒂默电势图也称还原电势图。在拉蒂默电势图中，同一元素的物种按氧化态递减的顺序从左到右排列，将对应的标准电极电势 $E^{\ominus}$ 值写在连接氧化还原电对的两个相邻物种的水平线上，单位为伏特。

例如，在 $[H^+]=1 \text{ mol L}^{-1}$ 或 pH=0 的标准酸性水溶液中，Cl 的拉蒂默电势图为

For example, the Latimer diagram for Cl in a standard acidic aqueous solution with $[H^+] = 1$ mol L^{-1} or pH = 0 is given by

$$\overset{+7}{ClO_4^-} \xrightarrow{1.189} \overset{+5}{ClO_3^-} \xrightarrow{1.214} \overset{+3}{HClO_2} \xrightarrow{1.645} \overset{+1}{HClO} \xrightarrow{1.611} \overset{0}{Cl_2} \xrightarrow{1.358} \overset{-1}{Cl^-}$$

The oxidation states for Cl in each species are labelled over the chemical formulas. A series of half-reactions with the corresponding $E°$ can be derived from this Latimer diagram by balancing them with the predominant species that existed in an acidic aqueous solution (usually H^+ and H_2O), such as

$$ClO_4^-(aq) + 2H^+(aq) + 2e^- \rightarrow ClO_3^-(aq) + H_2O(l) \quad E°(ClO_4^-/ClO_3^-) = +1.189 \text{ V}$$

$$ClO_3^-(aq) + 3H^+(aq) + 2e^- \rightarrow HClO_2(aq) + H_2O(l) \quad E°(ClO_3^-/HClO_2) = +1.214 \text{ V}$$

$$2HClO(aq) + 2H^+(aq) + 2e^- \rightarrow Cl_2(g) + 2H_2O(l) \quad E°(HClO/Cl_2) = +1.611 \text{ V}$$

Since the above three half-reactions all involve H^+, the pH of the solution will affect these three potentials.

In a standard basic aqueous solution in which pOH = 0 and pH = 14, the Latimer diagram for Cl is

$$\overset{+7}{ClO_4^-} \xrightarrow{0.36} \overset{+5}{ClO_3^-} \xrightarrow{0.33} \overset{+3}{ClO_2^-} \xrightarrow{0.66} \overset{+1}{ClO^-} \xrightarrow{0.26} \overset{0}{Cl_2} \xrightarrow{1.358} \overset{-1}{Cl^-}$$

Note that only $E°$ for the Cl_2/Cl^- couple is the same in both acidic and basic solutions because H^+ is not involved in its half-reaction.

Latimer diagram shows $E°$ for two adjacent species directly. The $E°$ for any two nonadjacent species can be calculated indirectly by combining the standard Gibbs free energy changes of the corresponding half-reactions. For example

$$E°(ClO_3^-/Cl_2) = \frac{[2 \times 1.214 + 2 \times 1.645 + 1 \times 1.611]}{5} \text{ V} = 1.466 \text{ V (acidic)}$$

Latimer diagram can also show the tendency of **disproportionation** and **comproportionation** of a species. A species tends to disproportionate into its two neighbors if the potential on the right of the species in a Latimer diagram is higher than that on the left. In the following schematic Latimer diagram

$$\overset{+n}{M^{n+}} \xrightarrow{E°_M(L)} \overset{0}{M} \xrightarrow{E°_M(R)} \overset{-n}{M^{n-}}$$

(1) $M^{n+}(aq) + ne^- \rightarrow M(s) \quad E°(M^{n+}/M) = E°_M(L)$

(2) $M(s) + ne^- \rightarrow M^{n-}(aq) \quad E°(M/M^{n-}) = E°_M(R)$

(2) − (1):

$$2M(s) \rightarrow M^{n+}(aq) + M^{n-}(aq) \quad E° = E°_M(R) - E°_M(L)$$

If $E°_M(R) > E°_M(L)$, then $E° > 0$. It means that the above disproportionation reaction will happen spontaneously, and that the species M has a tendency to disproportionate into its two neighbors M^{n+} and M^{n-}, which indicates that M is inherently unstable. Similarly, a species tends to comproportionate from its two neighbors if the potential on the right of the species in a Latimer diagram is lower than that on the left.

Frost Diagram

Frost diagram is also called the oxidation state diagram. In a Frost diagram, the numerical value of $NE°$ is plotted against N for a specific element X, where N is the oxidation state of the element X, and $E°$ is the standard electrode potential for the redox couple $X(N)/X(0)$. Because $NE°$ is proportional to the standard Gibbs free energy change in the following half-reaction, as

$$X(N) + Ne^- \rightarrow X(0) \quad NE°(X(N)/X(0)) = -\Delta G°/F \quad (6.11)$$

the species lying lowest in Frost diagram corresponds to that with the lowest Gibbs free energy, and consequently, to the most stable state of the element in an aqueous solution. Therefore, Frost diagram provides an important guide to identify the oxidation states of an element that are inherently stable or unstable. The steeper the line joining any two points (left to right) in a Frost diagram, the higher the standard potential of the corresponding couple. Meanwhile, a species is liable to undergo disproportionation if its

$$\overset{+7}{ClO_4^-} \xrightarrow{1.189} \overset{+5}{ClO_3^-} \xrightarrow{1.214} \overset{+3}{HClO_2} \xrightarrow{1.645} \overset{+1}{HClO} \xrightarrow{1.611} \overset{0}{Cl_2} \xrightarrow{1.358} \overset{-1}{Cl^-}$$

每个物种中 Cl 的氧化态都标记在化学式上。通过将其与酸性水溶液中存在的主要物种（通常为 H^+ 和 H_2O）进行配平，可从拉蒂默电势图中导出具有对应 $E^\ominus$ 值的一系列半反应，如

$$ClO_4^-(aq) + 2H^+(aq) + 2e^- \rightarrow ClO_3^-(aq) + H_2O(l) \quad E^\ominus(ClO_4^-/ClO_3^-) = +1.189 \text{ V}$$

$$ClO_3^-(aq) + 3H^+(aq) + 2e^- \rightarrow HClO_2(aq) + H_2O(l) \quad E^\ominus(ClO_3^-/HClO_2) = +1.214 \text{ V}$$

$$2HClO(aq) + 2H^+(aq) + 2e^- \rightarrow Cl_2(g) + 2H_2O(l) \quad E^\ominus(HClO/Cl_2) = +1.611 \text{ V}$$

由于上述三个半反应均涉及 H^+，因此溶液的 pH 会影响这三个电势。

在 pOH = 0、pH = 14 的标准碱性水溶液中，Cl 的拉蒂默电势图为

$$\overset{+7}{ClO_4^-} \xrightarrow{0.36} \overset{+5}{ClO_3^-} \xrightarrow{0.33} \overset{+3}{ClO_2^-} \xrightarrow{0.66} \overset{+1}{ClO^-} \xrightarrow{0.26} \overset{0}{Cl_2} \xrightarrow{1.358} \overset{-1}{Cl^-}$$

注意，只有 Cl_2/Cl^- 电对的 $E^\ominus$ 值在酸性和碱性溶液中均相同，这是因为 H^+ 不参与其半反应。

拉蒂默电势图直接给出了两个相邻物种的 $E^\ominus$ 值。任何两个非相邻物种的 $E^\ominus$ 值均可通过相应半反应的标准吉布斯自由能变来间接计算。例如

$$E^\ominus(ClO_3^-/Cl_2) = \frac{[2 \times 1.214 + 2 \times 1.645 + 1 \times 1.611]}{5} \text{ V} = 1.466 \text{ V}（酸性）$$

拉蒂默电势图还可显示物种的**歧化**和**归中**趋势。如果拉蒂默电势图中某一物种右侧的电势值高于左侧的电势值，说明该物种具有歧化为其左右两个物种的趋势。在如下拉蒂默电势示意图中

$$\overset{+n}{M^{n+}} \xrightarrow{E_M^\ominus(L)} \overset{0}{M} \xrightarrow{E_M^\ominus(R)} \overset{-n}{M^{n-}}$$

(1) $M^{n+}(aq) + ne^- \rightarrow M(s) \quad E^\ominus(M^{n+}/M) = E_M^\ominus(L)$

(2) $M(s) + ne^- \rightarrow M^{n-}(aq) \quad E^\ominus(M/M^{n-}) = E_M^\ominus(R)$

(2)−(1)：

$$2M(s) \rightarrow M^{n+}(aq) + M^{n-}(aq) \quad E^\ominus = E_M^\ominus(R) - E_M^\ominus(L)$$

如果 $E_M^\ominus(R) > E_M^\ominus(L)$，则 $E^\ominus > 0$。这意味着上述歧化反应将自发进行，物种 M 具有歧化为其左右两个物种 M^{n+} 和 M^{n-} 的趋势，表明 M 本质上不稳定。类似地，如果拉蒂默电势图中某一物种右侧的电势低于左侧的电势，则其左右两个物种有归中为该物种的趋势。

弗罗斯特电势图

弗罗斯特电势图也称氧化态电势图。弗罗斯特电势图是某特定元素 X 的 $NE^\ominus$ 对 N 的作图，其中 N 为元素 X 的氧化态，$E^\ominus$ 是氧化还原电对 X(N)/X(0) 的标准电极电势。由于 $NE^\ominus$ 与半反应的标准吉布斯自由能变成正比：

$$X(N) + Ne^- \rightarrow X(0) \quad NE^\ominus(X(N)/X(0)) = -\Delta G^\ominus/F \quad (6.11)$$

弗罗斯特电势图中位置最低的物种对应的吉布斯自由能最低，它也对应于水溶液中该元素的最稳定存在状态。因此，弗罗斯特电势图为确定某元素固有的稳定或不稳定的氧化态提供了重要的指导。弗罗斯特电势图中连接任意两点（从左到右）的直线越陡，对应电对的标准电势就越高。同时，如果某个物种对应的点位于连接其相邻两个物种直线的上方，则该物种易发生歧化反应。如果某个中间物种位于连接两

point lies above the straight line connecting its two neighbors. Two species tend to comproportionate into an intermediate species that lies blow the straight line joining the terminal species.

Figure 6.4 shows the Frost diagram for N in both standard acidic (pH = 0) and basic (pH = 14) aqueous solutions. From this Frost diagram, we can conclude that:

1) NH_4^+ is the most stable species in an acidic solution and N_2 is the most stable species in a basic solution.
2) The higher oxides and oxoacids of N are unstable in an acidic solution but relatively stable in a basic solution.
3) Hydroxylamine (NH_2OH) is particularly unstable and tends to disproportionate regardless of pH.

6.6 Nernst Equation and Concentration Cells

Data listed in **Table 6.2** as well as in Latimer diagrams are standard electrode potentials at 298.15 K and standard conditions, i.e., at $a = 1$ or $p = 1$ bar. The standard cell potential can be obtained by calculating the difference between two standard electrode potentials. However, what we need in most cases are cell potentials in nonstandard conditions. Those nonstandard cell potentials can be calculated with respect to the standard cell potentials by using Nernst equation.

Nernst Equation

Recall from **Section 3.9** that the nonstandard Gibbs free energy change can be related to the reaction quotient as

$$\Delta G = \Delta G^\circ + RT \ln Q$$

Substituting ΔG and ΔG° by $-nFE_{cell}$ and $-nFE^\circ_{cell}$, respectively, we have

$$-nFE_{cell} = -nFE^\circ_{cell} + RT \ln Q$$

$$\therefore E_{cell} = E^\circ_{cell} - \frac{RT}{nF} \ln Q \tag{6.12}$$

The above equation was first proposed by Walther Nernst (1864—1941) in 1889 and thus was named after him as **Nernst equation**. At 298.15 K

$$E_{cell} = E^\circ_{cell} - \frac{0.02569 \text{ V}}{n} \ln Q = E^\circ_{cell} - \frac{0.0592 \text{ V}}{n} \lg Q \tag{6.13}$$

If the oxidizing and reducing species are represented by ox and red, respectively, a redox equilibrium can be generally written as

$$m\text{ox} + ne^- \rightleftharpoons q\text{red}$$

Nernst equation for this electrode potential can be calculated as

$$E = E^\circ - \frac{RT}{nF} \ln \frac{[\text{red}]^q}{[\text{ox}]^m} \tag{6.14}$$

This electrode potential increases with the concentration of the oxidizing species, and decreases as the concentration of the reducing species increases.

Concentration Cells

Concentration cells are voltaic cells consisting of two half-cells with identical electrodes but different ion concentrations. Because the two electrodes are identical, the standard electrode potentials are numerically equal, as

$$E^\circ_{cathode} = E^\circ_{anode}$$

Consequently,

$$E^\circ_{cell} = E^\circ_{cathode} - E^\circ_{anode} = 0$$

个终端物种直线的下方，则这两个物种倾向于归中为中间物种。

图 6.4 给出了酸性（pH=0）和碱性（pH=14）标准水溶液中 N 的弗罗斯特电势图。从该图中，我们可得出以下结论：
1) NH_4^+ 是酸性溶液中最稳定的物种，N_2 是碱性溶液中最稳定的物种。
2) N 的高价氧化物和含氧酸在酸性溶液中不稳定，但在碱性溶液中相对稳定。
3) 羟胺（NH_2OH）特别不稳定，无论 pH 如何，均会发生歧化反应。

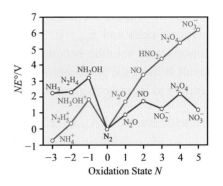

Figure 6.4 Frost diagram for nitrogen in both standard acidic (pH = 0, red) and basic (pH = 14, blue) aqueous solutions. The steeper the slope of a line, the higher the standard potential for the corresponding redox couple.

图 6.4 酸性（pH = 0，红线）和碱性（pH = 14，蓝线）标准水溶液中氮的弗罗斯特电势图：直线的斜率越大，对应氧化还原电对的标准电势越高。

6.6 能斯特方程与浓差电池

表 6.2 和拉蒂默电势图中列出的数据是 298.15 K 和标态下（即 $a = 1$ 或 $p = 1$ bar）的标准电极电势。标准电池电动势可通过计算两个标准电极电势之间的差值来获得。但在大多数情况下，我们需要的是非标态下的电池电动势。这些非标态电池电动势可用能斯特方程相对于标准电池电动势来进行计算。

能斯特方程

回顾 **3.9 节**，非标准吉布斯自由能变与反应商相关，为

$$\Delta G = \Delta G^\ominus + RT \ln Q$$

用 $-nFE_{池}$ 和 $-nFE_{池}^\ominus$ 分别替换 ΔG 和 $\Delta G^\ominus$，有

$$-nFE_{池} = -nFE_{池}^\ominus + RT \ln Q$$

$$\therefore E_{池} = E_{池}^\ominus - \frac{RT}{nF} \ln Q \quad (6.12)$$

上述方程是沃尔特·能斯特（1864—1941）于 1889 年首次提出的，因此以他的名字命名为**能斯特方程**。298.15 K 时

$$E_{池} = E_{池}^\ominus - \frac{0.02569 \text{ V}}{n} \ln Q = E_{池}^\ominus - \frac{0.0592 \text{ V}}{n} \lg Q \quad (6.13)$$

如果用 ox 和 red 分别表示氧化和还原物种，氧化还原平衡通式可写为

$$m\text{ox} + ne^- \rightleftharpoons q\text{red}$$

该电极电势的能斯特方程为

$$E = E^\ominus - \frac{RT}{nF} \ln \frac{[\text{red}]^q}{[\text{ox}]^m} \quad (6.14)$$

电极电势随氧化物种浓度的增大而增大，随还原物种浓度的增大而减小。

浓差电池

浓差电池是由两个电极完全相同但离子浓度不同的半电池组成的原电池。由于两个电极相同，标准电极电势的数值相等，即

$$E_{阴极}^\ominus = E_{阳极}^\ominus$$

因此

$$E_{池}^\ominus = E_{阴极}^\ominus - E_{阳极}^\ominus = 0$$

但由于离子浓度不同，$Q \neq 1$，两个半电池之间仍然存在电势差，$E_{池} \neq 0$。浓差电池总是发生自发变化，使得浓溶液变稀，而稀释溶液变浓。最后的结果就好像只是简单地把溶液混合在一起。但在浓

However, because the ion concentrations differ and $Q \neq 1$, there is still a potential difference between the two half-cells and $E_{cell} \neq 0$. The spontaneous change in a concentration cell always occurs such that the concentrated solution becomes more dilute, and the diluted solution becomes more concentrated. The final result is as if the solutions were simply mixed. In a concentration cell, however, the natural tendency for entropy to increase in a mixing process is used as a means of producing electricity.

Measurement of K_{sp} by Electrode Potentials

As ion concentrations affect the electrode potential, we can design some concentration cells to determine the K_{sp} values for some insoluble compounds from the electrode potentials by directly measuring the cell potentials. For example, it is not easy to accurately measure the concentrations of Ag^+ and Cl^- for the insoluble compound AgCl by using some conventional chemical analysis methods. However, by measuring the cell potential of the following concentration cell

$$Ag(s)|Ag^+(\text{satd AgCl}) \| Ag^+(0.100 \text{ mol L}^{-1})|Ag(s) \quad E_{cell} = 0.229 \text{ V}$$

in which the anode is a silver electrode placed in a saturated aqueous solution of silver chloride and the cathode is another silver electrode placed in a solution with $[Ag^+] = 0.100 \text{ mol L}^{-1}$, we can calculate that

$$E_{cell} = E°_{cell} - \frac{0.0592}{n} \lg Q = 0 - \frac{0.0592 \text{ V}}{1} \lg \frac{[Ag^+]}{0.100} = 0.229 \text{ V}$$

$$[Ag^+] = [Cl^-] = 1.35 \times 10^{-5} \text{ mol L}^{-1}$$

$$\therefore K_{sp} = [Ag^+][Cl^-] = 1.8 \times 10^{-10}$$

Alternative Standard Electrodes

The SHE is not the most convenient to use because it requires highly flammable hydrogen gas to be bubbled over the platinum electrode. Other electrodes, such as the silver-silver chloride electrode and the calomel electrode, can be used as secondary standard electrodes or reference electrodes. In an Ag-AgCl electrode, an AgCl-coated silver wire is immersed in a 1.0 mol L^{-1} KCl solution, giving the electrode

$$Ag(s)|AgCl(s)|Cl^-(1.0 \text{ mol L}^{-1})$$

with a half-cell reaction of

$$AgCl(s) + e^- \rightarrow Ag(s) + Cl^-(aq)$$

Its electrode potential has been measured against SHE to be 0.22233 V at 25°C. Since all components of this electrode are in their standard states, the standard electrode potential of the Ag-AgCl electrode is 0.22233 V at 25°C.

In a calomel electrode, mercurous chloride (also called calomel, Hg_2Cl_2) is mixed with mercury to form a paste, which is in contact with liquid mercury, and the whole setup is immersed in either a 1.0 mol L^{-1} solution of KCl (called the **normal calomel electrode**, NCE) or a saturated solution of KCl (called the **saturated calomel electrode**, SCE). The NCE is

$$Hg(l)|Hg_2Cl_2(s)|Cl^-(1.0 \text{ mol L}^{-1}) \quad E° = 0.2681 \text{ V}$$

and the SCE is

$$Hg(l)|Hg_2Cl_2(s)|Cl^-(\text{satd KCl}) \quad E° = 0.2412 \text{ V}$$

The half-cell reaction of both is

$$Hg_2Cl_2(s) + 2e^- \rightarrow 2Hg(l) + 2Cl^-(aq)$$

Glass Electrode and pH Electrode

In electrochemistry, the concentration of an unknown solution can be determined by measuring the potential difference between a **reference electrode** and an **indicator electrode** that consists of the unknown solution. One of the most commonly used indicator electrodes is the **ion-selective electrode**, which converts the concentration of a specific ion dissolved in an unknown solution into an electrical potential. A **glass electrode** is a type of ion-

差电池的混合过程中，熵增的自然趋势被用于产生电能。

由电极电势计算 K_{sp}

由于离子浓度影响电极电势，我们可以设计一些浓差电池，通过直接测量电池电动势，从电极电势中确定一些难溶化合物的 K_{sp}。例如，采用常规化学分析方法准确测定难溶化合物 AgCl 中 Ag^+ 和 Cl^- 的浓度并不容易。而通过测量以下浓差电池的电池电动势

$$Ag(s)\mid Ag^+(饱和AgCl)\parallel Ag^+(0.100\ mol\ L^{-1})\mid Ag(s) \quad E_{池}=0.229\ V$$

其中阳极是置于饱和氯化银水溶液中的银电极，阴极是置于 $[Ag^+]=0.100\ mol\ L^{-1}$ 溶液中的另一个银电极，我们可计算得

$$E_{池}=E_{池}^{\ominus}-\frac{0.0592}{n}\lg Q=0-\frac{0.0592\ V}{1}\lg\frac{[Ag^+]}{0.100}=0.229\ V$$

$$[Ag^+]=[Cl^-]=1.35\times10^{-5}\ mol\ L^{-1}$$

$$\therefore K_{sp}=[Ag^+][Cl^-]=1.8\times10^{-10}$$

二级标准电极

标准氢电极使用并不方便，因为它要求在铂电极上将高度易燃的氢气鼓泡。其他电极（如银-氯化银电极和甘汞电极）可用作二级标准电极或参比电极。在 Ag-AgCl 电极中，将涂有 AgCl 的银线浸入 $1.0\ mol\ L^{-1}$ KCl 溶液中，可得电极

$$Ag(s)\mid AgCl(s)\mid Cl^-(1.0\ mol\ L^{-1})$$

其半电池反应为

$$AgCl(s)+e^-\rightarrow Ag(s)+Cl^-(aq)$$

在 25ºC 下相对于标准氢电极，其电极电势可测量为 0.22233 V。由于该电极的所有组分均处于标态，因此 Ag-AgCl 电极在 25ºC 时的标准电极电势为 0.22233 V。

在甘汞电极中，氯化亚汞（也称甘汞，Hg_2Cl_2）与汞混合形成糊状物，并与液态汞相接触，整个装置浸泡在 $1.0\ mol\ L^{-1}$ 的 KCl 溶液（称为**标准甘汞电极**，NCE）或 KCl 饱和溶液（称为**饱和甘汞电极**，SCE）中。标准甘汞电极为

$$Hg(l)\mid Hg_2Cl_2(s)\mid Cl^-(1.0\ mol\ L^{-1}) \quad E^{\ominus}=0.2681\ V$$

饱和甘汞电极为

$$Hg(l)\mid Hg_2Cl_2(s)\mid Cl^-(饱和KCl) \quad E^{\ominus}=0.2412\ V$$

两者的半电池反应均为

$$Hg_2Cl_2(s)+2e^-\rightarrow 2Hg(l)+2Cl^-(aq)$$

玻璃电极与 pH 电极

在电化学中，未知溶液的浓度可通过测量**参比电极**和由未知溶液组成的**指示电极**之间的电势差来确定。**离子选择性电极**是最常用的指示电极之一，它将溶解在未知溶液中的特定离子浓度转换为电势信号。**玻璃电极**是一类由掺杂玻璃膜制成、对特定离子（如 H^+、K^+、NH_4^+、Cl^- 等）敏感的离子选择性电极。玻璃电极在化学分析仪器中起着重要作用。

pH 电极是一种可测量未知溶液中 H^+ 浓度的电极。pH 电极是组

selective electrode made of a doped glass membrane that is sensitive to a particular ion, such as H^+, K^+, NH_4^+, Cl^-, etc. Glass electrodes play an important part in the instrumentation for chemical analysis.

The **pH electrode** is an electrode that can measure the H^+ concentration in an unknown solution. The pH electrode is a combination electrode that consists of a H^+-sensitive glass electrode and an Ag-AgCl electrode, which serves as an internal reference electrode. The glass electrode contains a H^+-sensitive glass membrane that separates the unknown solution from a HCl solution with $[H^+] = 1.0$ mol L^{-1} and another Ag-AgCl electrode. Although called an electrode, the pH electrode is actually a concentration cell, which can be represented as

$$Ag(s)|AgCl(s)|Cl^-(1.0 \text{ mol L}^{-1}), H^+(1.0 \text{ mol L}^{-1})|\text{glass membrane}|H^+(\text{unknown})$$
$$\|Cl^-(1.0 \text{ mol L}^{-1})\|AgCl(s)|Ag(s)$$

As $E^\circ_{cell} = 0$ for concentration cells, the Gibbs free energy change corresponding to the dilution of H^+ from a known concentration of 1.0 mol L^{-1} to the concentration of the unknown solution is the source of the potential difference across the glass membrane.

$$\Delta G = -nFE_{cell} = G(\text{unknown}) - G(\text{known})$$
$$= G^\circ + RT \ln[H^+] - G^\circ - RT \ln 1.0 = RT \ln[H^+]$$
$$\therefore E_{cell} = -\frac{RT}{nF} \ln[H^+]$$

At 298.15 K

$$E_{cell} = 0.0592 \text{pH} \tag{6.15}$$

Therefore, the voltage measured by the pH electrode is directly proportional to the pH of the unknown solution. In practical cases, a buffer solution is used instead of the 1.0 mol L^{-1} HCl solution to maintain a relatively constant pH. Consequently,

$$E_{cell} = 0.0592 \text{pH} + C$$

where C is a constant that can be calibrated by using a standard buffer solution with known pH. A device that contains a pH electrode and can directly display the pH of an unknown solution in pH units is called a **pH meter**.

6.7 Electrolytic Cells and Overpotential

In **Section 6.2**, we have classified electrochemical cells into voltaic cells and electrolytic cells according to the direction of energy conversion as well as to the spontaneity of corresponding reactions. The previous sections in this chapter mainly focus on voltaic cells, and we will discuss electrolytic cells in this section.

Electrolysis and Electrolytic Cells

The cell shown in **Figure 6.1** is a voltaic cell in which electrons flow from the zinc electrode (the anode) to the copper electrode (the cathode). The following reaction happens spontaneously and converts chemical energy into electric energy:

$$Zn(s) + Cu^{2+}(aq) \rightarrow Zn^{2+}(aq) + Cu(s) \quad E^\circ_{cell} = 1.104 \text{ V}$$

Now suppose that the same cell is connected to an external electric source with a voltage greater than 1.104 V, as shown in **Figure 6.5**. This connection is made so that electrons are forced into the zinc electrode (now the cathode) and removed from the copper electrode (now the anode). The overall reaction in this case is driven by the external electric source continuously and thus is nonspontaneous. It is the reverse of the above voltaic reaction, and E°_{cell} is negative.

$$\text{Oxidation}: Cu(s) \rightarrow Cu^{2+}(aq) + 2e^-$$

合电极,由一个对 H^+ 敏感的玻璃电极和一个 Ag-AgCl 内参比电极组成。玻璃电极包含一个对 H^+ 敏感的玻璃膜,可将未知溶液与 $[H^+]$=1.0 mol L^{-1} 的 HCl 溶液以及另一个 Ag-AgCl 电极分隔开。虽然称为电极,但 pH 电极实际上是一个浓差电池,可表示为

$$Ag(s)|AgCl(s)|Cl^-(1.0\ mol\ L^{-1}), H^+(1.0\ mol\ L^{-1})|玻璃膜|H^+(未知)$$
$$\|Cl^-(1.0\ mol\ L^{-1})|AgCl(s)|Ag(s)$$

由于浓差电池的 $E^\ominus_{池}= 0$,H^+ 从已知浓度 1.0 mol L^{-1} 稀释到未知溶液所对应的吉布斯自由能变,即为玻璃膜电势差的来源。

$$\Delta G = -nFE_{池} = G(未知溶液) - G(已知溶液)$$
$$= G^\ominus + RT\ln[H^+] - G^\ominus - RT\ln 1.0 = RT\ln[H^+]$$
$$\therefore E_{池} = -\frac{RT}{nF}\ln[H^+]$$

298.15 K 时

$$E_{池} = 0.0592\text{pH} \tag{6.15}$$

因此,pH 电极测得的电动势(或电压)与未知溶液的 pH 成正比。在实际情况下,常使用缓冲溶液替代 1.0 mol L^{-1} HCl 溶液,以保持相对恒定的 pH。故

$$E_{池} = 0.0592\text{pH} + C$$

其中 C 为常数,可用已知 pH 的标准缓冲溶液进行校准。含有 pH 电极并能以 pH 单位直接显示未知溶液 pH 的装置,称为 **pH 计**。

6.7 电解池与超电势

在 **6.2 节** 中,根据能量转换的方向以及相应反应的自发性,我们将电化学电池分为原电池和电解池。本章的前述章节主要关注原电池,本节将讨论电解池。

电解与电解池

图 6.1 所示电池为原电池,其中电子从锌电极(阳极)流向铜电极(阴极)。以下反应自发进行,并将化学能转化为电能:

$$Zn(s) + Cu^{2+}(aq) \to Zn^{2+}(aq) + Cu(s) \quad E^\ominus_{池} = 1.104\ V$$

现在假定将同一电池连接到一个电压大于 1.104 V 的外加电源,如**图 6.5** 所示。此连接迫使电子进入锌电极(现在的阴极)且从铜电极(现

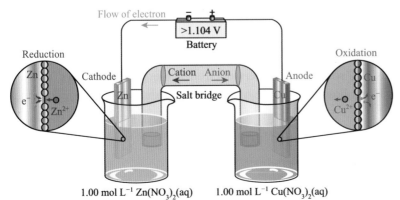

Figure 6.5 The non-spontaneous reaction $Cu(s) + Zn^{2+}(aq) \to Cu^{2+}(aq) + Zn(s)$ driven by an external electric source in an electrolytic cell.

图 6.5 电解池在外加电源的驱动下发生的非自发反应 $Cu(s) + Zn^{2+}(aq) \to Cu^{2+}(aq) + Zn(s)$。

$$\text{Reduction}: Zn^{2+}(aq) + 2e^- \rightarrow Zn(s)$$

$$\text{Overall}: Cu(s) + Zn^{2+}(aq) \rightarrow Cu^{2+}(aq) + Zn(s) \tag{6.16}$$

$$E^\circ_{cell} = E^\circ(Zn^{2+}/Zn) - E^\circ(Cu^{2+}/Cu) = (-0.762 - 0.342)\,V = -1.104\,V$$

This cell is an electrolytic cell which converts electric energy into chemical energy. Reversing the direction of the electron flow changes the voltaic cell into an electrolytic cell. The process in which nonspontaneous reactions are driven by external electric sources is called **electrolysis**.

Overpotential

Although theoretically the electrolytic reaction in **Equation (6.16)** can happen continuously at an external electric voltage of 1.104 V, the actual voltage must exceed 1.104 V in order for this redox event to be experimentally observed. The potential difference between the thermodynamically determined theoretical cell voltage (E_{cell}) and the actual decomposition voltage (E_{expt}) for an electrolytic cell is called **overpotential**. The overpotential of a cell consists of the anode overpotential (η_{anode}) and cathode overpotential ($\eta_{cathode}$), given by

$$\eta = E_{expt} - E_{cell} = \eta_{anode} + \eta_{cathode} \tag{6.17}$$

Overpotentials can be classified into the following three main categories:

1) Activation overpotential (also called electron transfer overpotential): arises from the activation energy necessary to transfer an electron from an electrode to an analyte. The activation overpotential is particularly common when gases are involved, and its values may differ significantly for different gas on different electrode. For example, the activation overpotential of H_2 on a platinum electrode is negligible, but that on a mercury electrode is about 1.5 V.

2) Concentration overpotential: the potential difference caused by differences in the concentration of charge-carriers between bulk solution and the electrode surface. It occurs when electrochemical reaction is sufficiently rapid so that the concentration of the charge-carriers is lower at the electrode surface than in the bulk solution. The rate of reaction is then dependent on the ability of charge-carriers to reach the electrode surface. The concentration overpotential can be reduced by vigorous stirring and increasing temperature.

3) Resistance overpotential: the potential difference due to resistance of wires and electrolytes, as well as junctions formed at electrode surfaces and interfaces.

Predicting Electrolysis Products

In practice, many factors will affect an electrolytic reaction. To predict the electrolysis products of an electrolytic cell, we need to consider all those complicating factors including the following:

1) Overpotential. The actual decomposition voltage for an electrolytic event to be experimentally observed may differ from the theoretical value due to the existence of overpotential. This may cause different electrolysis products at different electrodes, especially when gases are involved.

2) Competing electrode reactions. Different ions and molecules in solution may compete with each other to produce different electrolysis products. For example, in the electrolysis of molten KCl with inert electrodes such as platinum, only K^+ and Cl^- ions are available to conduct the following electrolytic half-reactions:

Oxidation: $2Cl^-(molten) \rightarrow Cl_2(g) + 2e^-$

Reduction: $K^+(molten) + e^- \rightarrow K(l)$

However, in the electrolysis of aqueous KCl solution with inert electrodes, there are $H_2O(l)$, $H^+(aq)$, and $OH^-(aq)$ in addition to $K^+(aq)$ and $Cl^-(aq)$. The following possible oxidation and reduction half-reactions may compete.

$$\text{Oxidation}: 2Cl^-(aq) \rightarrow Cl_2(g) + 2e^- \qquad -E^\circ(Cl_2/Cl^-) = -1.36\,V$$

$$2H_2O(l) \rightarrow O_2(g) + 4H^+(aq) + 4e^- \qquad -E^\circ(O_2/H_2O) = -1.23\,V$$

在的阳极）上离开。在这种情况下，总反应由外加电源持续驱动，因此是非自发的。它与上述原电池反应相反，$E_{池}^{\ominus}$ 为负值。

氧化反应：$Cu(s) \rightarrow Cu^{2+}(aq) + 2e^-$

还原反应：$Zn^{2+}(aq) + 2e^- \rightarrow Zn(s)$

总反应：$Cu(s) + Zn^{2+}(aq) \rightarrow Cu^{2+}(aq) + Zn(s)$ (6.16)

$$E_{池}^{\ominus} = E^{\ominus}(Zn^{2+}/Zn) - E^{\ominus}(Cu^{2+}/Cu) = (-0.762 - 0.342) \text{ V} = -1.104 \text{ V}$$

这种电池是一种将电能转化为化学能的电解池。翻转电子流动的方向，即可将原电池变为电解池。由外加电源驱动发生非自发反应的过程，称为**电解**。

超电势

虽然理论上讲，**式（6.16）**中的电解反应在 1.104 V 的外加电压下即可持续发生，但实际电压必须超过 1.104 V，才能从实验上观察到该氧化还原事件。热力学确定的理论电解电压（$E_{池}$）与电解池实际的分解电压（$E_{实验}$）之间的电势差，称为**超电势**（也称过电势）。电池的超电势由阳极超电势（$\eta_{阳极}$）和阴极超电势（$\eta_{阴极}$）组成，有

$$\eta = E_{实验} - E_{池} = \eta_{阳极} + \eta_{阴极} \quad (6.17)$$

超电势可分为以下三大类：

1) 活化超电势（也称电子转移超电势）：源自将电子从电极转移到待电解物质所需的活化能。当涉及气体时，活化超电势特别常见，不同电极上不同气体的活化超电势可能会显著不同。例如，H_2 在铂电极上的活化超电势可忽略不计，但在汞电极上的超电势约为 1.5 V。

2) 浓差超电势：指由溶液本体与电极表面之间的载流子浓度差引起的电势差。当电化学反应足够快，使得电极表面的载流子浓度低于溶液本体时，就会发生这种情况。这时的反应速率取决于载流子到达电极表面的能力。剧烈搅拌或升高温度均可降低浓差超电势。

3) 电阻超电势：指由于导线和电解质以及在电极表面和界面处形成的结产生的电阻而导致的超电势。

电解产物的预测

实践中影响电解反应的因素有很多。为了预测电解池的电解产物，我们需要考虑所有以下复杂因素：

1) 超电势。由于存在超电势，实验观察到的电解事件的实际分解电压与理论值可能不同。这可能会在不同的电极上产生不同的电解产物，尤其当涉及气体时。

2) 竞争性电极反应。溶液中不同的离子和分子可能相互竞争，产生不同的电解产物。例如，使用惰性电极（如铂）电解熔融 KCl 时，只存在 K^+ 和 Cl^- 离子，可进行如下电解半反应：

氧化半反应：$2Cl^-(熔融) \rightarrow Cl_2(g) + 2e^-$

还原半反应：$K^+(熔融) + e^- \rightarrow K(l)$

然而，使用惰性电极电解 KCl 水溶液时，除 $K^+(aq)$ 和 $Cl^-(aq)$ 之外，还存在 $H_2O(l)$、$H^+(aq)$ 和 $OH^-(aq)$。如下可能的氧化半反应和还原半反应会相互竞争。

氧化半反应：$2Cl^-(aq) \rightarrow Cl_2(g) + 2e^-$ $\quad\quad -E^{\ominus}(Cl_2/Cl^-) = -1.36 \text{ V}$

Reduction: $K^+(aq) + e^- \rightarrow K(s)$ $E°(K^+/K) = -2.93$ V

$2H_2O(l) + 2e^- \rightarrow H_2(g) + 2OH^-(aq)$ $E°(H_2O/H_2) = -0.83$ V

The reduction of $H^+(aq)$ into $H_2(g)$ is not considered because the concentration of $H^+(aq)$ is negligible in neutral KCl solution. Because $E°(K^+/K) \ll E°(H_2O/H_2)$ and $\eta(H_2$ on Pt$) \approx 0$, we can conclude that the cathode product is $H_2(g)$. As $E°(Cl_2/Cl^-)$ and $E°(O_2/H_2O)$ do not differ significantly but $\eta(O_2$ on Pt$) \gg \eta(Cl_2$ on Pt$)$, the oxidation reaction of Cl^- predominates and the anode product is $Cl_2(g)$.

3) States of the reacting species. The reactants and products of an electrolytic reaction may be in nonstandard states so that $E \neq E°$. This may favor different redox half-reactions and result in different electrolysis products.

4) Nature of the electrodes. An inert electrode only provides a surface for an electrolysis half-reaction to occur, but itself does not actually involve in the reaction. However, an active electrode can itself participate in the redox half-reactions. We must take into consideration the active electrode in order to predict the electrolysis products.

6.8 Battery and Its Applications

A **battery** is a device that stores chemical energy and supplies direct current by electrochemical means. Batteries may consist of one voltaic cell, or more joined in a series fashion to increase the total voltage. A good battery should have high voltage, large capacity, low self-discharge, long lifetime, and should be safe and portable, etc. In this section, we will introduce three types of batteries: primary cells, secondary cells, and fuel cells. A special type of fuel cell, called flow battery, will be discussed in the extended reading materials at the end of this chapter.

Primary Cells

A **primary cell** is a non-rechargeable electrochemical cell in which electricity is generated by an irreversible chemical reaction. In primary cells, the cathode is the **positive electrode** and the anode is the **negative electrode**. Dry cell (also called Leclanché cell), alkaline cell, and silver-zinc cell (also called button battery) are examples of primary cells.

Dry cell was invented by the French chemist Georges Leclanché (1839—1882) in the 1860s. This cell contains a zinc anode, an inert carbon (graphite) cathode with MnO_2, and a moist paste electrolyte of NH_4Cl, $ZnCl_2$, and carbon black. The cell voltage is about 1.5 V, with the following redox half-reactions:

Positive electrode (+, cathode): $2MnO_2 + 2H^+ + 2e^- \rightarrow 2MnO(OH)$

Negative electrode (−, anode): $Zn + 2NH_4Cl \rightarrow [Zn(NH_3)_2]Cl_2 + 2H^+ + 2e^-$

Dry cell is cheap to make, but it has some drawbacks such as low energy density, quick voltage drop, and short shelf-life.

A superior form of dry cell is alkaline cell, which uses the alkaline electrolyte NaOH or KOH instead of NH_4Cl. The cell voltage is still about 1.5 V, and the redox half-reactions are:

Positive electrode: $2MnO_2 + 2H^+ + 2e^- \rightarrow 2MnO(OH)$

Negative electrode: $Zn + 2OH^- \rightarrow Zn(OH)_2 + 2e^-$

Since zinc does not dissolve as readily in an alkaline medium as in an acidic medium, the alkaline cell does a better job of maintaining its voltage and has a longer shelf-life.

Although the silver-zinc cell has been developed to be rechargeable, its cyclicity is not good and it is mainly considered as a primary cell. The cell diagram of a silver-zinc cell is

$$Zn(s), ZnO(s) | KOH(satd) | Ag_2O(s), Ag(s)$$

The cell voltage is 1.8 V, and the redox half-reactions on discharging are:

Positive electrode: $Ag_2O + H_2O + 2e^- \rightarrow 2Ag + 2OH^-$

$$2H_2O(l) \rightarrow O_2(g) + 4H^+(aq) + 4e^- \quad -E^\ominus(O_2/H_2O) = -1.23 \text{ V}$$

还原半反应：$K^+(aq) + e^- \rightarrow K(s) \quad E^\ominus(K^+/K) = -2.93 \text{ V}$

$$2H_2O(l) + 2e^- \rightarrow H_2(g) + 2OH^-(aq) \quad E^\ominus(H_2O/H_2) = -0.83 \text{ V}$$

不必考虑将 $H^+(aq)$ 还原为 $H_2(g)$，因为中性 KCl 溶液中的 $H^+(aq)$ 浓度可忽略不计。由于 $E^\ominus(K^+/K) \ll E^\ominus(H_2O/H_2)$ 且 $\eta(H_2$ 在 Pt 上$) \approx 0$，可以得出结论：阴极产物为 $H_2(g)$。由于 $E^\ominus(Cl_2/Cl^-)$ 和 $E^\ominus(O_2/H_2O)$ 无显著差异，但 $\eta(O_2$ 在 Pt 上$) \gg \eta(Cl_2$ 在 Pt 上$)$，Cl^- 的氧化半反应占主导，阳极产物为 $Cl_2(g)$。

3) 反应物种的状态。电解反应的反应物和生成物可能处于非标态，因此 $E \neq E^\ominus$。这可能有利于不同的氧化还原半反应，并导致不同的电解产物。

4) 电极的性质。惰性电极仅为电解半反应的发生提供表面，但其自身实际上并不参与反应。而活性电极自身可以参与氧化还原半反应。为了预测电解产物，我们必须考虑活性电极。

6.8 电池及其应用

电池是一种储存化学能并通过电化学方式提供直流电的装置。电池可以由一个原电池组成，也可以将多个原电池以串联的方式相连接，以增加总电压。好的电池应该具有高电压、大容量、低自放电率、长寿命、安全便携等特点。本节我们将介绍三类电池：一级电池、二级电池以及燃料电池。章末拓展阅读材料中会介绍一类称为液流电池的特殊燃料电池。

一级电池

一级电池是一种不可循环充放电的电化学电池，其电能由不可逆化学反应产生。一级电池的阴极为**正极**，阳极为**负极**。干电池（也称勒克朗谢电池）、碱性电池和银锌电池（也称纽扣电池）是一级电池的示例。

干电池由法国化学家乔治·勒克朗谢（1839—1882）于 19 世纪 60 年代发明。该电池包含一个锌阳极、一个含 MnO_2 的惰性碳（石墨）阴极，以及由 NH_4Cl、$ZnCl_2$ 和炭黑组成的湿糊状电解质。其电池电压约 1.5 V，具有如下氧化还原半反应：

正极(+,阴极)：$2MnO_2 + 2H^+ + 2e^- \rightarrow 2MnO(OH)$

负极(−,阳极)：$Zn + 2NH_4Cl \rightarrow [Zn(NH_3)_2]Cl_2 + 2H^+ + 2e^-$

干电池制造成本低廉，但具有如能量密度低、电压降快、保质期短等缺点。

碱性电池是一种高级形式的干电池，使用碱性电解质 NaOH 或 KOH 代替 NH_4Cl。电池电压仍约 1.5 V，其氧化还原半反应为：

正极： $2MnO_2 + 2H^+ + 2e^- \rightarrow 2MnO(OH)$

负极： $Zn + 2OH^- \rightarrow Zn(OH)_2 + 2e^-$

由于锌在碱性介质中不像在酸性介质中那样易溶，碱性电池在保持电压方面做得更好，保质期更长。

尽管银锌电池已经开发为可循环充放电的电池，但其循环周期性较差，主要仍被视为一级电池。银锌电池的电池符号图为

Negative electrode: $\quad\quad\quad\quad\quad\quad Zn + 2OH^- \rightarrow Zn(OH)_2 + 2e^-$

As no solution species is involved in the cell reaction and only a very small quantity of electrolyte is needed, the silver-zinc cell is safe, portable, and with a large storage capacity.

Secondary Cells

A **secondary cell** is a rechargeable electrochemical cell in which energy is converted between chemical energy and electric energy by a reversible chemical reaction. In the secondary cells, cathode is the positive electrode on discharging and the negative electrode on charging, whereas the anode is the negative electrode on discharging and the positive electrode on charging. No matter on charging or discharging, the cathode is always the electrode at which electrons enter the battery, and the anode is always the electrode from which electrons exit the battery. Because the notations of the cathode and anode switch during charge and discharge, we usually use positive and negative electrodes to refer to the electrodes in secondary cells. The lead-acid battery (also called storage battery), nickel-cadmium cell (also called nicad battery), and lithium-ion battery are examples of secondary cells.

The lead-acid battery is commonly used in series in motor vehicles. The positive electrode is a lead grid packed with red-brown lead (IV) oxide, the negative electrode is a lead grid packed with spongy lead, and the electrolyte solution contains diluted sulfuric acid (about 35% H_2SO_4, by mass). In this strongly acidic medium, the ionization of H_2SO_4 does not go to completion. Both HSO_4^-(aq) and SO_4^{2-}(aq) are present, but HSO_4^-(aq) predominates. During discharge, the redox half-reactions are:

Positive electrode(+, cathode): $\quad\quad PbO_2 + 3H^+ + HSO_4^- + 2e^- \rightarrow PbSO_4 + 2H_2O$

Negative electrode (−, anode): $\quad\quad Pb + HSO_4^- \rightarrow PbSO_4 + H^+ + 2e^-$

After discharge, when the plates of the battery become coated with $PbSO_4$(s) and the electrolyte becomes sufficiently diluted with water, the battery must be recharged by connecting it to an external electric source. During charge, a nonspontaneous reaction is driven by the external electric source as:

Positive electrode (+, anode): $\quad\quad PbSO_4 + 2H_2O \rightarrow PbO_2 + 3H^+ + HSO_4^- + 2e^-$

Negative electrode (−, cathode): $\quad\quad PbSO_4 + H^+ + 2e^- \rightarrow Pb + HSO_4^-$

The voltage for a single lead-acid cell is about 2.0 V. Higher voltage can be obtained by joining single cells in a series fashion, positive to negative, to produce a battery. For example, a typical 12 V automobile battery consists of six single cells. The lead-acid battery is stable in voltage and cheap but quite heavy, leading to a low energy-to-weight ratio. Meanwhile, because lead is a pollution hazard, all lead-acid batteries should be disposed of properly and should not be dumped in landfills or garbage disposal sites.

The nickel-cadmium cell is one of the most common commercial batteries and is usually used in cordless electric devices. The positive electrode is the Ni(III) compound NiO(OH) supported on nickel metal, and the negative electrode is cadmium metal. The redox half-reactions are:

Positive electrode: $\quad\quad NiO(OH) + H_2O + e^- \underset{\text{charging}}{\overset{\text{discharging}}{\rightleftharpoons}} Ni(OH)_2 + OH^-$

Negative electrode: $\quad\quad Cd + 2OH^- \underset{\text{charging}}{\overset{\text{discharging}}{\rightleftharpoons}} Cd(OH)_2 + 2e^-$

This cell gives a fairly constant voltage of 1.4 V. Nicad batteries show an advantage in good cycle life, however, they suffer the memory effect, which means that they gradually lose their maximum energy capacity if repeatedly recharged after being only partially discharged, appearing to "remember" the smaller capacity.

Lithium-ion batteries are commonly used for portable electronics such as cell phones, laptop computers, etc. In this battery, lithium ions move from the negative electrode through the electrolyte to the positive electrode during discharge, and back when charging. The positive electrode consists of alithium-containing compound, such as Lithium cobaltate ($LiCoO_2$) and lithium iron phosphate ($LiFePO_4$). The negative electrode is usually highly crystallized graphite, into which lithium ions can be intercalated. The reversible insertion of a guest molecule, atom, or ion into a host layered material is called **intercalation**, and the resulting product is called an **intercalation compound**. The electrolyte is typically some lithium salts such as $LiPF_6$, $LiBF_4$, and $LiClO_4$, in some organic carbonate solvents. The illustrative redox half-reactions in a $LiCoO_2$-based battery are:

$$\text{Zn(s), ZnO(s)} | \text{KOH(饱和)} | \text{Ag}_2\text{O(s), Ag(s)}$$

电池电压为 1.8 V，放电时的氧化还原半反应为：

正极： $\text{Ag}_2\text{O} + \text{H}_2\text{O} + 2e^- \rightarrow 2\text{Ag} + 2\text{OH}^-$

负极： $\text{Zn} + 2\text{OH}^- \rightarrow \text{Zn(OH)}_2 + 2e^-$

由于电池反应中不涉及溶液，且只需极少量电解质，因此银锌电池安全、便携，具有较大的存储容量。

二级电池

二级电池是一种可循环充放电的电化学电池，其能量通过可逆化学反应在化学能和电能之间转换。在二级电池中，阴极是放电时的正极，充电时的负极，而阳极是放电时的负极，充电时的正极。不论充电还是放电，阴极总是电子进入电池的电极，而阳极总是电子离开电池的电极。由于在充放电过程中阴极和阳极的符号互换，我们通常使用正极和负极来指代二级电池中的电极。铅酸电池（也称蓄电池）、镍镉电池和锂离子电池是二级电池的示例。

铅酸电池通常在机动车中串联使用。其正极是填充有红棕色氧化铅（Ⅳ）的铅栅，负极是填充有海绵状铅的铅栅，电解质溶液含有稀硫酸（质量分数约 35% 的 H_2SO_4）。在此强酸性介质中，H_2SO_4 不会完全电离。$\text{HSO}_4^-\text{(aq)}$ 和 $\text{SO}_4^{2-}\text{(aq)}$ 同时存在，但以 $\text{HSO}_4^-\text{(aq)}$ 为主。放电时的氧化还原半反应为：

正极（+，阴极）： $\text{PbO}_2 + 3\text{H}^+ + \text{HSO}_4^- + 2e^- \rightarrow \text{PbSO}_4 + 2\text{H}_2\text{O}$

负极（−，阳极）： $\text{Pb} + \text{HSO}_4^- \rightarrow \text{PbSO}_4 + \text{H}^+ + 2e^-$

放电后，当电池板被 $\text{PbSO}_4\text{(s)}$ 覆盖、电解液被水充分稀释时，必须将电池连接到外加电源进行充电。充电时外加电源驱动的非自发反应为

正极（+，阳极）： $\text{PbSO}_4 + 2\text{H}_2\text{O} \rightarrow \text{PbO}_2 + 3\text{H}^+ + \text{HSO}_4^- + 2e^-$

负极（−，阴极）： $\text{PbSO}_4 + \text{H}^+ + 2e^- \rightarrow \text{Pb} + \text{HSO}_4^-$

单个铅酸原电池的电压约 2.0 V。通过将多个单电池从正极到负极串联起来得到电池组，可以获得更高的电压。例如，典型的 12 V 汽车电池由六个单电池组成。铅酸电池电压稳定，价格便宜，但重量较高，导致了较低的能量重量比。同时，由于铅有污染危害，所有铅酸蓄电池都应妥善处理，不得倾倒在垃圾填埋场或处理场。

镍镉电池是最常见的商用电池之一，通常用于无绳电气设备。其正极是金属镍上负载的 Ni(Ⅲ) 化合物 NiO(OH)，负极是金属镉。氧化还原半反应为：

正极： $\text{NiO(OH)} + \text{H}_2\text{O} + e^- \underset{\text{充电}}{\overset{\text{放电}}{\rightleftharpoons}} \text{Ni(OH)}_2 + \text{OH}^-$

负极： $\text{Cd} + 2\text{OH}^- \underset{\text{充电}}{\overset{\text{放电}}{\rightleftharpoons}} \text{Cd(OH)}_2 + 2e^-$

这种电池的电压相当恒定，为 1.4 V。镍镉电池在良好的循环寿命方面显示出优势，但受到记忆效应的影响，这意味着如果在部分放电之后反复充电，它们会逐渐失去最大能量容量，似乎"记住"了较小的容量。

锂离子电池常用于便携式电子产品，如手机、笔记本电脑等。在这种电池中，锂离子在放电时从负极通过电解质迁移到正极，再在充电时返回。其正极由钴酸锂（LiCoO_2）和磷酸铁锂（LiFePO_4）等含锂化合物组成。负极通常是高度结晶的石墨，锂离子可以插嵌在其中。客体分子、原子或离子可逆地插嵌入主体层状材料中，称

Positive electrode: $\text{Li}_{1-x}\text{CoO}_2(s) + x\text{Li}^+(\text{solvent}) + xe^- \underset{\text{charging}}{\overset{\text{discharging}}{\rightleftharpoons}} \text{LiCoO}_2(s)$

Negative electrode: $\text{Li}_x\text{C}_6(s) \underset{\text{charging}}{\overset{\text{discharging}}{\rightleftharpoons}} \text{C}_6(s) + x\text{Li}^+(\text{solvent}) + xe^-$

Lithium-ion batteries have advantages such as high energy density, no memory effect, and low self-discharge, etc. However, they can also be a safety hazard due to the flammability and volatility of the organic solvents used in typical electrolytes. Possible solutions include aqueous lithium-ion batteries, ceramic solid electrolytes, polymer electrolytes, etc.

Fuel Cells

A **fuel cell** is a device that converts the chemical energy from a fuel into electricity through a chemical reaction with oxygen or other oxidizing agents. Fuel cells differ from batteries in that they require a constant source of fuel and oxygen/air to sustain the chemical reaction and that they can produce electricity continually for as long as these inputs are supplied. The commonly used fuels are hydrogen, hydrocarbons, and alcohols. The redox half-reactions and the overall reaction in a H_2 fuel cell are:

Positive electrode: $O_2(g) + 2H_2O(l) + 4e^- \rightarrow 4OH^-(aq)$
Negative electrode: $H_2(g) + 2OH^-(aq) \rightarrow 2H_2O(l) + 2e^-$
Overall: $2H_2(g) + O_2(g) \rightarrow 2H_2O(l)$

and the standard cell potential is given by

$$E^\circ_{\text{cell}} = E^\circ(O_2/OH^-) - E^\circ(H_2O/H_2) = [0.401 - (-0.828)]\text{ V} = 1.229 \text{ V}$$

As ΔG° measures the theoretical maximum energy available as electric energy in any electrochemical cell, and ΔH° shows the maximum energy released when a fuel is burned, their ratio $\varepsilon = \Delta G^\circ/\Delta H^\circ$, defined as the **efficiency value**, can thus be used to evaluate the fuel cell. For the H_2-O_2 fuel cell, $\varepsilon = -474.2$ kJ mol^{-1}/(−571.6 kJ mol^{-1}) = 0.83. Another example listed below is a CH_4 fuel cell.

Positive electrode: $O_2(g) + 4H^+(aq) + 4e^- \rightarrow 2H_2O(l)$
Negative electrode: $CH_4(g) + 2H_2O(l) \rightarrow CO_2(g) + 8H^+(aq) + 8e^-$
Overall: $CH_4(g) + 2O_2(g) \rightarrow CO_2(g) + 2H_2O(l)$

$$\varepsilon = \frac{\Delta G^\circ}{\Delta H^\circ} = \frac{-818.1 \text{ kJ mol}^{-1}}{-890.5 \text{ kJ mol}^{-1}} = 0.92$$

Extended Reading Materials — Flow Batteries

A **flow battery** is a rechargeable fuel cell in which an electrolyte containing one or more dissolved redox-active materials flows on separate sides of a membrane through an electrochemical cell that reversibly converts chemical energy directly to electricity. Additional electrolytes, termed anolyte for anode electrolyte and catholyte for cathode electrolyte, are stored externally (generally in tanks), and are usually pumped through the cells of the reactor. Ion transfer inside the cell occurs through the membrane while both anolyte and catholyte circulate in their own tanks. The total amount of electricity that can be generated depends on the volume of electrolytes in the tanks. A flow battery may be used like a rechargeable battery, where regeneration of the fuel is driven by an external power source, or like a fuel cell, where rapid "recharging" can be accomplished by simply replacing the electrolytes. The typical structure of a flow battery is illustrated in **Figure 6.6**.

The typical advantages of flow batteries include: 1) flexible layout due to separation of the power and energy components; 2) long cycle life; 3) quick response time; 4) no need for "equalization" charging, which is the overcharging of a battery to ensure all cells have an equal change; 5) safety because flammable electrolytes are not involved normally and that electrolytes can be stored away from the power stack. These technical merits make redox flow batteries a well-suited option for large-scale energy storage. The major disadvantages of flow batteries compared to batteries with solid redox-active materials are low energy

为**插层**，所得生成物称为**插层化合物**。其电解质通常是一些有机碳酸盐溶剂中的锂盐（如 $LiPF_6$、$LiBF_4$ 和 $LiClO_4$ 等）。$LiCoO_2$ 电池的氧化还原半反应如下：

正极：$Li_{1-x}CoO_2(s) + xLi^+(\text{溶剂}) + xe^- \underset{\text{充电}}{\overset{\text{放电}}{\rightleftharpoons}} LiCoO_2(s)$

负极：$Li_xC_6(s) \underset{\text{充电}}{\overset{\text{放电}}{\rightleftharpoons}} C_6(s) + xLi^+(\text{溶剂}) + xe^-$

锂离子电池具有高能量密度、无记忆效应和低自放电率等优点。但由于典型电解质中使用的有机溶剂具有易燃性和挥发性，它们也存在安全隐患。可能的解决方案包括水性锂离子电池、陶瓷固态电解质、聚合物电解质等。

燃料电池

燃料电池是一种通过与氧气或其他氧化剂的化学反应，将燃料中的化学能转化为电能的装置。燃料电池与一般电池的不同之处在于，它们需要恒定的燃料和氧气/空气源来维持化学反应，并且只要提供这些原料，就能持续产生电能。常用的燃料有氢、碳氢化合物和醇。H_2 燃料电池的氧化还原半反应以及总反应为：

正极：$O_2(g) + 2H_2O(l) + 4e^- \rightarrow 4OH^-(aq)$

负极：$H_2(g) + 2OH^-(aq) \rightarrow 2H_2O(l) + 2e^-$

总反应：$2H_2(g) + O_2(g) \rightarrow 2H_2O(l)$

标准电池电压为

$$E_{\text{池}}^{\ominus} = E^{\ominus}(O_2/OH^-) - E^{\ominus}(H_2O/H_2) = [0.401-(-0.828)] \text{ V} = 1.229 \text{ V}$$

由于 $\Delta G^{\ominus}$ 是任何电化学电池理论上可用的最大电能，而 $\Delta H^{\ominus}$ 表示燃料燃烧时释放的最大能量，可将其比值 $\varepsilon = \Delta G^{\ominus}/\Delta H^{\ominus}$ 定义为**效率值**，用于评估燃料电池。对于 $H_2\text{-}O_2$ 燃料电池，$\varepsilon = -474.2 \text{ kJ mol}^{-1}/(-571.6 \text{ kJ mol}^{-1}) = 0.83$。下面给出的另一个例子是 CH_4 燃料电池：

正极：$O_2(g) + 4H^+(aq) + 4e^- \rightarrow 2H_2O(l)$

负极：$CH_4(g) + 2H_2O(l) \rightarrow CO_2(g) + 8H^+(aq) + 8e^-$

总反应：$CH_4(g) + 2O_2(g) \rightarrow CO_2(g) + 2H_2O(l)$

$$\varepsilon = \frac{\Delta G^{\ominus}}{\Delta H^{\ominus}} = \frac{-818.1 \text{ kJ mol}^{-1}}{-890.5 \text{ kJ mol}^{-1}} = 0.92$$

拓展阅读材料　液流电池

液流电池是一种可循环充放电的燃料电池，其中溶解了一种或多种氧化还原活性物质的电解液在膜的不同侧流过电化学电池，从而将化学能可逆地直接转化为电能。额外的电解液（称为**阳极电解液**和**阴极电解液**）储存在外部（一般置于存储罐中），被泵往反应器的电池中。当阳极电解液和阴极电解液在各自的存储罐中循环时，电池内的离子通过膜进行转移。可产生的总电量取决于存储罐中电解液的体积。液流电池既可以像充电电池一样使用，即在外加电源的驱动下使燃料再生，也可以像燃料电池一样使用，即通过简单地更换电解液即可实现快速"再充电"。液流电池的典型结构如**图 6.6** 所示。

液流电池的典型优势包括：1）布局灵活，因为动力组件和能量组件可分离；2）循环寿命长；3）响应时间快；4）无须"均衡"充

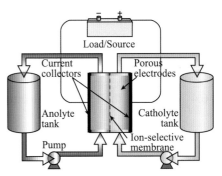

Figure 6.6 The illustrative diagram of a typical flow battery.

图 6.6 典型液流电池示意图。

density, low charge and discharge rates, and relatively low energy efficiency. Flow batteries typically have a higher energy efficiency than normal fuel cells, but lower than lithium-ion batteries.

According to whether the redox-active materials can be fully dissolved in electrolytes, flow batteries can be classified into all-liquid flow batteries and hybrid flow batteries. According to the types of redox-active materials involved, flow batteries can be classified into inorganic flow batteries (developed for decades) and organic flow batteries (first emerged in 2009). Among all types of flow batteries, the vanadium flow batteries are the most marketed ones at present. In Oct. 2022, Dalian, China began to operate a vanadium flow battery with an initial capacity of 400 MWh and an output of 100 MW, which is so far the largest flow battery in the world.

Problems

6.1 Chile saltpeter is a natural source of $NaNO_3$; it also contains $NaIO_3$, which can be used as a source of iodine. Iodine is produced from sodium iodate in a two-step process occurring under acidic conditions:

$$IO_3^-(aq) + HSO_3^-(aq) \rightarrow I^-(aq) + SO_4^{2-}(aq) \quad \text{(not balanced)}$$

$$I^-(aq) + IO_3^-(aq) \rightarrow I_2(s) + H_2O(l) \quad \text{(not balanced)}$$

A 5.00 L sample of a $NaIO_3$(aq) solution containing 5.80 g L^{-1} $NaIO_3$ is treated with the stoichiometric quantity of $NaHSO_3$. Then, a further quantity of the initial $NaIO_3$(aq) is added to the reaction mixture to bring about the second reaction.

(a) How many grams of $NaHSO_3$ are required in the first step?

(b) What additional volume of the starting solution must be added in the second step?

6.2 Blood alcohol content (BAC) is often reported in weight volume percent ($w/V\%$). For example, a BAC of 0.10% corresponds to 0.10 g CH_3CH_2OH per 100 mL of blood. Estimates of BAC can be obtained from breath samples by using a number of commercially available instruments including the Breathalyzer. The chemistry behind the Breathalyzer is described by the redox reaction below, which occurs in an acidic solution:

$$CH_3CH_2OH(g) + Cr_2O_7^{2-}(aq) \rightarrow CH_3COOH(aq) + Cr^{3+}(aq) \quad \text{(not balanced)}$$

A Breathalyzer instrument contains two ampules, each of which contains 0.75 mg $K_2Cr_2O_7$ dissolved in 3 mL of 9 mol L^{-1} H_2SO_4(aq). One of the ampules is used as a reference. When a person exhales into the tube of the Breathalyzer, the breath is directed into one of the ampules, and ethyl alcohol in the breath converts $Cr_2O_7^{2-}$ into Cr^{3+}. The instrument compares the colors of the solutions in the two ampules to determine the breath alcohol content (BrAC), and then converts this into an estimate of BAC. The conversion of BrAC into BAC rests on the assumption that 2100 mL of air exhaled from the lungs contains the same amount of alcohol as 1 mL of blood. With the theory and assumptions described in this problem, calculate the molarity of $K_2Cr_2O_7$ in the ampules before and after a breath test in which a person with a BAC of 0.05% exhales 0.500 L of his breath into a Breathalyzer instrument.

6.3 To construct a voltaic cell with E_{cell} = 0.0860 V, what [Cl^-] must be present in the cathode half-cell to achieve this result?

$$Ag(s)|Ag^+(\text{satd AgI})\|Ag^+(\text{satd AgCl}, x \text{ mol } L^{-1} \text{ } Cl^-)|Ag(s)$$

6.4 Consider the following electrochemical cell: $Pt(s)|H_2(g, 1 \text{ atm})|H^+(1 \text{ mol } L^{-1})\|Ag^+(x \text{ mol } L^{-1})|Ag(s)$

(a) Calculate E°_{cell}.

(b) Use Nernst equation to write an equation for E_{cell} when [Ag^+] = x mol L^{-1}.

(c) Now imagine titrating 50.0 mL of 0.0100 mol L^{-1} $AgNO_3$ in the cathode half-cell compartment with 0.0100 mol L^{-1} KI. The titration reaction is $Ag^+(aq) + I^-(aq) \rightarrow AgI(s)$. Calculate [$Ag^+$] and then E_{cell} after addition of the following volumes of 0.0100 mol L^{-1} KI: (i) 0.0 mL; (ii) 20.0 mL; (iii) 49.0 mL;

电（即为确保所有电池的电量变化相等而进行的过度充电）；5) 安全，因为通常不涉及易燃电解液，并且电解液可以储存在远离电源堆的地方。这些技术优势使得氧化还原液流电池非常适合大规模储能。与使用固态氧化还原活性物质的电池相比，液流电池的主要缺点有：能量密度低、充放电速率低、能量效率相对较低。液流电池的能量效率通常高于普通燃料电池，但低于锂离子电池。

根据氧化还原活性物质能否完全溶于电解液，液流电池可分为全液体液流电池和混合液流电池。根据所涉及的氧化还原活性物质的类型，液流电池可分为无机液流电池（已发展了数十年）和有机液流电池（最早出现于 2009 年）。在所有类型的液流电池中，钒液流电池是目前已上市最多的一种。2022 年 10 月，中国大连开始运营初始容量为 400 MWh、输出功率为 100 MW 的钒液流电池，这是迄今为止世界上最大的液流电池。

习题

6.1 智利硝石是 $NaNO_3$ 的天然来源；它还含有 $NaIO_3$，可用作碘源。碘由碘酸钠在酸性条件下通过两步法生产：

$$IO_3^-(aq) + HSO_3^-(aq) \rightarrow I^-(aq) + SO_4^{2-}(aq) \quad （未配平）$$

$$I^-(aq) + IO_3^-(aq) \rightarrow I_2(s) + H_2O(l) \quad （未配平）$$

用化学计量的 $NaHSO_3$ 处理含 5.80 g L^{-1} $NaIO_3$ 的 5.00 L $NaIO_3(aq)$ 溶液样品。再向反应混合物中加入更多的初始 $NaIO_3(aq)$，以进行第二步反应。

(a) 第一步需要多少克 $NaHSO_3$？

(b) 第二步中必须添加多少额外体积的初始溶液？

6.2 血液酒精含量（BAC）通常以质量体积百分比（w/V%）计。例如，0.10% 的 BAC 相当于每 100 mL 血液中含 0.10 g CH_3CH_2OH。通过使用包括呼气分析仪在内的许多商用仪器，可以从呼吸样本中获得 BAC 的估计值。呼气分析仪背后的化学反应，可由在酸性溶液中发生的以下氧化还原反应描述：

$CH_3CH_2OH(g) + Cr_2O_7^{2-}(aq) \rightarrow CH_3COOH(aq) + Cr^{3+}(aq)$（未配平）

呼气分析仪包含两个安瓿，每个均含有 0.75 mg $K_2Cr_2O_7$，溶解在 3 mL 9 mol L^{-1} $H_2SO_4(aq)$ 中。其中一个安瓿用作参比。当人的呼气进入呼气分析仪的试管时，呼气被导入其中一个安瓿，呼气中的乙醇会将 $Cr_2O_7^{2-}$ 转化为 Cr^{3+}。仪器比较两个安瓿中溶液的颜色，以确定呼气中的酒精含量（BrAC），然后将其转换为 BAC 的估计值。BrAC 转化为 BAC 的前提是，假定从肺部呼出的 2100 mL 空气中含有与 1 mL 血液相同量的酒精。根据本题中描述的理论与假定，当一个 BAC 为 0.05% 的人将 0.500 L 呼气排入呼气分析仪时，计算呼气测试前后安瓿中 $K_2Cr_2O_7$ 的摩尔浓度。

6.3 为使以下原电池的 $E_{电池}^{\ominus} = 0.0860$ V，阴极半电池的 $[Cl^-]$ 必须为多少才能实现此结果？

$$Ag(s) | Ag^+(饱和AgI) \| Ag^+(饱和AgCl, x \text{ mol L}^{-1} Cl^-) | Ag(s)$$

(iv) 50.0 mL; (v) 51.0 mL; (vi) 60.0 mL.

(d) Use the results of part (c) to sketch the titration curve of E_{cell} versus volume of titrant.

6.5 Ultimately, ΔG_f° values must be based on experimental results; in many cases, these experimental results are themselves obtained from E° values. Early in the twentieth century, G. N. Lewis conceived of an experimental approach for obtaining standard potentials of alkali metals. This approach involved using a solvent, such as ethylamine, with which the alkali metals do not react. In the following cell diagram, Na (amalg, 0.206%) represents a solution of 0.206% Na in liquid mercury.

1. $Na(s)|Na^+(\text{in ethylamine})|Na(\text{amalg}, 0.206\%)$ $E_{cell} = 0.8453$ V

Although Na(s) reacts violently with water to produce H₂(g), at least for a short time, a sodium amalgam electrode does not react with water. This makes it possible to determine E_{cell} for the following voltaic cell.

2. $Na(\text{amalg}, 0.206\%)|Na^+(1\text{ mol L}^{-1})\|H^+(1\text{ mol L}^{-1})|H_2(g, 1\text{ atm})$ $E_{cell} = 1.8673$ V

(a) Write equations for the cell reactions that occur in the voltaic cells (1) and (2).

(b) Calculate ΔG for the cell reactions written in part (a).

(c) Write the overall equation obtained by combining the equations of part (a), and calculate ΔG° for this overall reaction.

(d) Use the ΔG° value from part (c) to obtain E°_{cell} for the overall reaction. From this result, obtain $E^\circ(Na^+/Na)$. Compare your result with the value listed in **Table 6.2**.

6.6 Given that $E^\circ(BrO_4^-/BrO_3^-) = 1.025$ V, $E^\circ(BrO_3^-/Br^-) = 0.61$ V, $E^\circ(BrO^-/Br^-) = 0.761$ V, and $E^\circ(Br_2/Br^-) = 1.066$ V in a standard basic solution, construct Latimer diagram for Br by proper calculations, ignoring BrO_2^-.

6.7 Only a tiny fraction of the diffusible ions moves across a cell membrane in establishing a Nernst potential, so there is no detectable concentration change. Consider a typical cell with a volume of 10^{-8} cm³, a surface area (A) of 10^{-6} cm², and a membrane thickness (l) of 10^{-6} cm. Suppose that [K⁺] = 155 mmol L⁻¹ inside the cell and [K⁺] = 4 mmol L⁻¹ outside the cell and that the observed Nernst potential across the cell wall is 0.085 V. The membrane acts as a charge-storing device called a capacitor, with a capacitance, C, given by $C = \varepsilon_0 \varepsilon A / l$, where ε_0 is the dielectric constant of a vacuum and the product $\varepsilon_0 \varepsilon$ is the dielectric constant of the membrane, having a typical value of $3 \times 8.854 \times 10^{-12}$ C² N⁻¹ m⁻² for a biological membrane. The SI unit of capacitance is the farad (F), 1 F = 1 C V⁻¹ = 1 C² N⁻¹ m⁻¹.

(a) Determine the capacitance of the membrane for the typical cell described.

(b) What is the net charge required to maintain the observed membrane potential?

(c) How many K⁺ ions must flow through the cell membrane to produce the membrane potential?

(d) How many K⁺ ions are in the typical cell?

(e) Show that the fraction of the intracellular K⁺ ions transferred through the cell membrane to produce the membrane potential is so small that it does not change [K⁺] within the cell.

6.4 考虑如下电化学电池：

$$\text{Pt(s)}|\text{H}_2(\text{g,1 atm})|\text{H}^+(1\text{ mol L}^{-1})\|\text{Ag}^+(x\text{ mol L}^{-1})|\text{Ag(s)}$$

(a) 计算 $E_{池}^{\ominus}$。

(b) 应用能斯特方程，写出 $[\text{Ag}^+] = x$ mol L^{-1} 时 $E_{池}$ 的方程。

(c) 现在假定在阴极半电池室中用 0.0100 mol L^{-1} KI 滴定 50.0 mL 0.0100 mol L^{-1} AgNO$_3$。滴定反应为 Ag$^+$(aq) + I$^-$(aq) → AgI(s)。计算在加入以下体积 0.0100 mol L^{-1} KI 后的 $[\text{Ag}^+]$ 和 $E_{池}$：
(i) 0.0 mL；(ii) 20.0 mL；(iii) 49.0 mL；(iv) 50.0 mL；(v) 51.0 mL；(vi) 60.0 mL。

(d) 使用 (c) 的结果绘制 $E_{池}$ 对滴定剂体积的滴定曲线。

6.5 $\Delta G_f^{\ominus}$ 的最终数值必须基于实验结果；在许多情况下，这些实验结果本身就是从 $E^{\ominus}$ 中得到的。20 世纪初，G. N. 路易斯构想了一种获得碱金属标准电势的实验方法，该方法涉及使用碱金属不与之发生反应的溶剂，如乙胺。在如下电池符号图中，Na（汞齐，0.206%）表示 0.206% Na 在液汞中的溶液。

(1) $\text{Na(s)}|\text{Na}^+(\text{乙胺})|\text{Na(汞齐,0.206\%)}$ $E_{池} = 0.8453$ V

尽管 Na(s) 会与水剧烈反应生成 H$_2$(g)，但至少在短时间内，钠汞齐电极不会与水反应。这使得确定以下原电池的 $E_{池}$ 成为可能：

(2) $\text{Na(汞齐,0.206\%)}|\text{Na}^+(1\text{ mol L}^{-1})\|\text{H}^+(1\text{ mol L}^{-1})|\text{H}_2(\text{g,1 atm})$

$E_{池} = 1.8673$ V

(a) 写出原电池（1）和（2）中发生的电池反应方程式。

(b) 计算 (a) 中所述电池反应的 ΔG。

(c) 通过组合 (a) 中方程式得到总反应方程式，并计算总反应的 $\Delta G^{\ominus}$。

(d) 使用 (c) 中 $\Delta G^{\ominus}$ 得到总反应的 $E_{池}^{\ominus}$，并由该值得到 $E^{\ominus}$(Na$^+$/Na)。将结果与 **表 6.2** 中列出的相应数值进行比较。

6.6 已知在标准碱性溶液中 $E^{\ominus}$(BrO$_4^-$/BrO$_3^-$) = 1.025 V，$E^{\ominus}$(BrO$_3^-$/Br$^-$) = 0.61 V，$E^{\ominus}$(BrO$^-$/Br$^-$) = 0.761 V，$E^{\ominus}$(Br$_2$/Br$^-$) = 1.066 V，通过适当的计算构建 Br 的拉蒂默电势图（忽略 BrO$_2^-$）。

6.7 在建立能斯特电势时，只有很小一部分可扩散离子会穿过细胞膜，因此不会产生可检测的浓度变化。考虑一个体积为 10^{-8} cm^3、表面积 (A) 为 10^{-6} cm^2、膜厚 (l) 为 10^{-6} cm 的典型细胞。假定细胞内 [K$^+$] = 155 mmol L^{-1}，细胞外 [K$^+$] = 4 mmol L^{-1}，观察到的穿过细胞壁的能斯特电势为 0.085 V。细胞膜可充当电荷存储装置，即电容器，其电容 (C) 由 $C = \varepsilon_0 \varepsilon A / l$ 给出，其中 ε_0 为真空介电常数，$\varepsilon_0\varepsilon$ 的乘积为膜的介电常数，生物膜的典型值为 $3\times 8.854\times 10^{-12}$ C^2 N^{-1} m^{-2}。电容的 SI 单位是法拉（F），1 F = 1 C V^{-1} = 1 C^2 N^{-1} m^{-1}。

(a) 确定上述典型细胞膜的电容。

(b) 维持观察到的膜电势所需的净电荷是多少？

(c) 必须有多少 K$^+$ 离子流过细胞膜，才能产生此膜电势？

(d) 上述典型细胞中存在多少 K$^+$ 离子？

(e) 证明通过细胞膜转移、以产生膜电势的细胞内 K$^+$ 离子的比例非常低，因此不会改变细胞内的 [K$^+$]。

Chapter 7 Atomic Structure

The process of how we understand the structure of the microscopic world is long and tortuous. In this chapter, we will introduce the process of how we understand the atomic structure, starting from the modern picture of the nuclear atom, to the introduction of quantum theory, and then to the scientific breakthrough of quantum mechanics. We will learn the microscopic structures from the smallest atom—the hydrogen atom, to the single-electron systems—the hydrogen-like species, and then to the multielectron species. Finally, we will discuss the periodic law and the periodic table that can organize elements and atoms with periodic properties.

7.1   The Nuclear Atom

7.2   Quantum Theory

7.3   Atomic Spectra and Bohr Theory

7.4   The Nature of Microscopic Particles

7.5   Quantum Mechanical Model of Hydrogen-Like Species

7.6   Quantum Mechanical Results of Hydrogen-Like Species

7.7   Multielectron Species and Electron Configurations

7.8   The Periodic Law and the Periodic Table

7.9   Periodic Properties of the Elements

第 7 章 原子结构

我们理解微观世界结构的过程是漫长而曲折的。本章我们将介绍如何理解原子结构的过程：从核型原子的现代图像开始，到量子理论的引入，再到量子力学的科学突破。我们将按照从最小的原子（即氢原子）、到单电子体系（即类氢物种）、再到多电子物种的顺序来学习微观结构。最后，我们会讨论周期律和周期表，并通过它们将具有周期性的元素和原子组织起来。

7.1 核型原子

7.2 量子理论

7.3 原子光谱与玻尔理论

7.4 微观粒子的特性

7.5 类氢物种的量子力学模型

7.6 类氢物种的量子力学结论

7.7 多电子物种与电子组态

7.8 元素周期律与元素周期表

7.9 元素性质的周期性

7.1 The Nuclear Atom

The idea that matter is made up of discrete units appears in many ancient cultures. The word "atom", meaning "uncuttable", was coined by the Greek philosophers Leucippus and his pupil Democritus. At the end of the 18th century, scientists summarized the law of conservation of mass and the law of definite proportions from experimental results. At the beginning of the 19th century, John Dalton discovered the **law of multiple proportions**, which suggested that chemicals do not react in any arbitrary quantity, but in multiples of some basic indivisible unit of mass. In order to explain this law, Dalton proposed an atomic theory with the following features: 1) All matters are composed of minute, indivisible particles called atoms. Atoms can be neither created nor destroyed by chemical means. 2) Atoms of a given element are identical in size, mass and other properties. Atoms of different elements differ in size, mass and other properties. 3) Atoms of different elements combine in simple whole-number ratios to form chemical compounds.

It was not until the end of the 19th century that scientists realized that atoms are not indivisible. A new understanding of subatomic structures was then achieved, which will be introduced in this section.

The Discovery of Electrons and Their Properties

In ancient times, people noticed that amber attracted small objects when rubbed with fur, which was one of humanity's earliest recorded experiences with charges. Later, scientists realized that there are two kinds of charges, positive (+) and negative (−). Two similar charges repel each other, whereas two unsimilar charges attract each other. An object is considered as electrically neutral if it carries equal numbers of positive and negative charges. When charged particles travel through a magnetic field, positively and negatively charged particles are deflected in opposite directions, following the left-hand rule.

In 1859, during the study of electrical conductivity in rarefied gases, the German physicist Julius Plücker observed that phosphorescent light appeared on the tube wall near the cathode. In 1869, Plücker's student, Johann Hittorf, found that a solid body placed in between the cathode and the phosphorescence would cast a shadow upon the phosphorescent region of the tube. He inferred that there are straight invisible rays emitted from the cathode and that the phosphorescence was caused by the rays striking the tube walls. In 1876, the German physicist Eugen Goldstein showed that the rays were emitted perpendicular to the cathode surface and dubbed them **cathode rays**. During the 1870s, the English scientist William Crookes developed the first cathode-ray tube [**Figure 7.1(a)**] with a high vacuum inside. In 1874, he concluded that the cathode rays carried momentum by showing that the rays could turn a small paddle wheel when placed in their path. Furthermore, by applying electric or magnetic fields, the cathode rays can be deflected to the direction expected for negatively charged particles [**Figures 7.1(b)(c)**].

In 1897, the British physicist Joseph J. Thomson (1856—1940) established the charge-to-mass ratio (e/m) for the cathode rays by simultaneously applying an electric field and a magnetic field. The cathode-ray beam remains undeflected if the forces exerted on it by the electric and magnetic fields are counterbalanced [**Figure 7.1(d)**]. In this case, the centripetal force in a circular motion is given by

$$F_e = Ee = F_m = Bev = \frac{mv^2}{r} \tag{7.1}$$

where e, v, and r are the charge, velocity, and radius of movement of cathode rays, respectively, and E and B are the electric field strength and magnetic induction intensity (also called magnetic flux density), respectively. It is not difficult to derive from **Equation (7.1)** that

$$e/m = \frac{v}{Br} = \frac{E}{B^2 r}$$

A charge-to-mass ratio of $e/m = -1.76 \times 10^{11}$ C kg^{-1} were determined experimentally. Thomson also showed that the charge-to-mass ratio of the cathode rays was constant, independent of the cathode material or composition. He further concluded that cathode rays are negatively charged fundamental particles universally

7.1 核型原子

物质由离散的单元构成这一想法出现在许多古文明中。"原子"一词由希腊哲学家留西帕斯和他的学生德谟克利特创造，意思是"不可分割的"。18 世纪末，科学家们根据实验结果总结出质量守恒定律和定比定律。19 世纪初，约翰·道尔顿发现了**倍比定律**，即化学物质并非以任意数量发生反应，而总是以某种不可分割的基本质量单位的倍数来反应。为了解释这一定律，道尔顿提出了一种具有如下特征的原子理论：1）所有物质均由微小而不可分割的粒子（称为原子）构成，不能通过化学方法创造或破坏原子。2）同一元素的原子在大小、质量及其他性质上完全相同，不同元素的原子在大小、质量及其他性质上不同。3）不同元素的原子以简单的整数比结合形成化合物。

直到 19 世纪末，科学家们才意识到原子并非不可分割的，此后开始对亚原子结构有了新的认识，这些将在本节中介绍。

电子的发现及其性质

早在古代，人们就注意到琥珀与毛皮摩擦后会吸引轻小物体，这是人类最早记录的电荷体验之一。后来，科学家们意识到存在两种电荷：正电荷（+）与负电荷（−）。同性电荷相互排斥，而异性电荷相互吸引。如果物体带有等量的正电荷和负电荷，则该物体可视为电中性。当带电粒子在磁场中运动时，正电粒子和负电粒子依照左手定则向相反方向偏转。

1859 年，在研究稀薄气体的导电性时，德国物理学家朱利叶斯·普吕克观察到阴极附近的管壁出现磷光。1869 年，普吕克的学生约翰·希托夫发现，置于阴极和磷光之间的固体会在管的磷光区域投下阴影。他推断，有沿直线运动的不可见射线从阴极发射出，磷光是射线撞击管壁引起的。1876 年，德国物理学家尤金·戈尔茨坦证实了这些射线垂直于阴极表面发射，并将其命名为**阴极射线**。19 世纪 70 年代，英国科学家威廉·克鲁克斯研制出第一个内部具有高真空的阴极射线管 [**图 7.1(a)**]。1874 年，他根据阴极射线可以转动一个置于其路径上的小桨轮而得出结论：阴极射线具有动量。此外，当施加电场或磁场时，阴极射线会向带负电粒子所预期的方向发生偏转 [**图 7.1(b)(c)**]。

1897 年，英国物理学家约瑟夫·J. 汤姆森（1856—1940）通过同时施加电场和磁场，得到了阴极射线的荷质比（e/m）。如果电场和磁场对阴极射线所施加的力达到平衡 [**图 7.1(d)**]，则阴极射线束保持不偏转。此时，圆周运动的向心力为

$$F_{电} = Ee = F_{磁} = Bev = \frac{mv^2}{r} \tag{7.1}$$

其中 e、v 和 r 分别是阴极射线的电荷、速率及运动半径，E 和 B 分别是电场强度和磁感应强度（也称磁通密度）。由**式 (7.1)** 不难得出

$$e/m = \frac{v}{Br} = \frac{E}{B^2 r}$$

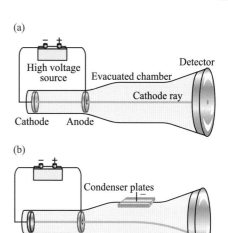

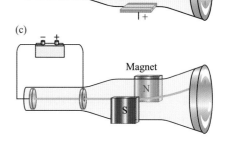

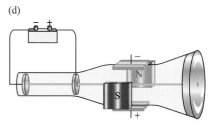

Figure 7.1 Cathode rays and the experiments to determine its charge-to-mass ratio. (a) Experimental setup of a cathode-ray tube with a high vacuum inside. Invisible rays are emitted from the cathode and are detected through the green fluorescence that they produce on the ZnS-coated screen of the detector. (b) Deflection of cathode rays in an electric field. (c) Deflection of cathode rays in a magnetic field. (d) Determination of the charge-to-mass ratio of cathode rays by simultaneously applying an electric field and a magnetic field.

图 7.1 阴极射线及确定其荷质比的实验：(a) 内部具有高真空的阴极射线管的实验装置图，不可见射线从阴极发射出，并可通过检测器的 ZnS 涂层屏幕上产生的绿色荧光来进行检测；(b) 阴极射线在电场中的偏转；(c) 阴极射线在磁场中的偏转；(d) 通过同时施加电场和磁场，可确定阴极射线的荷质比。

existing in all atoms. Subsequently, cathode rays became known as **electrons**, a term initially coined by the Irish physicist George J. Stoney in 1874.

The charge of electron was carefully measured by the American physicist Robert A. Millikan (1868—1953) through a series of oil-drop experiments **(Figure 7.2)** and the results were published in 1911. In those experiments, an electric field is used to speed up or slow down the falling of the charged oil droplets as a result of gravity. After measurement of the electronic charge, the mass of an electron can be calculated from the accurate value of e/m. The currently accepted values of electronic charge and mass are

$$e = -1.6022 \times 10^{-19} \text{ C} \quad \text{and} \quad m = 9.1094 \times 10^{-31} \text{ kg}$$

respectively.

As electrons are fundamental particles found in all atoms, scientists began to speculate on how negatively charged electrons were incorporated into electrically neutral atoms. Followed the experience of macroscopic objects, Thomson proposed the **plum-pudding model** in 1904. He suggested that electrons, like the "plum", floated in a diffuse nebulous cloud of positive charge, like "pudding". In this way, the total energy of the atom is the lowest as the attractive forces are maximized and the repulsive forces are minimized. However, this plum-pudding model was proven incorrect according to further experimental results.

X-Rays and Radioactivity

At nearly the same period of time when electrons were found, X-rays and radioactivity were also discovered. In 1895, the German physicist Wilhelm Röntgen (1845—1923) noticed that fluorescence appeared outside the cathode-ray tubes during operation, which was caused by radiation emitted by the tubes. He coined the term **X-ray** due to the unknown nature of this radiation. We now recognize that X-ray is a form of high-energy electromagnetic radiation, which will be further discussed in **Section 7.2**.

Radioactivity was first observed in 1896 by Antoine H. Becquerel, who discovered that an unexposed photographic plate carefully protected from light became exposed when salts of uranium were brought into the vicinity. Later, it was found that this effect can be produced both by salts and by pure uranium, suggesting that radioactivity was a property of the element.

In 1899, Ernest Rutherford (1871—1937) identified two types of radiation from radioactive materials: alpha (α) and beta (β) particles. When passing through an electric field, **alpha particles** are deflected slightly toward the negative plate and **beta particles** are deflected significantly toward the positive plate. It was later identified that alpha particles are He^{2+} ions and beta particles are a stream of high-speed electrons. A third beam of radiation called **gamma rays** (γ rays), which is undeflected by either electric or magnetic fields, was discovered in 1900 by Paul Villard and is recognized as electromagnetic radiation with extremely high energy. These three types of radiation from radioactive materials are illustrated in **Figure 7.3**.

α Particle Scattering Experiment

In 1909, Ernest Rutherford and Hans Geiger carried out a series of experiments to study the inner structure of atoms by bombarding thin metal foils with α particles **(Figure 7.4)**. By mounting a ZnS screen on the end of a telescope that can move in a circular track around the metal foils, α particles can be detected by the flashes of light produced when they hit the ZnS screen. The experimental results were shown as follows:

1) Most α particles penetrated the foil undeflected;
2) Some α particles experienced slight deflections;
3) Very few (about 0.005%) α particles suffered large-angle deflections;
4) Very few (about 0.005%) α particles bounced back in the incoming direction.

However, these experimental observations contradict Thomson's plum-pudding model. The plum-pudding model could not explain the large-angle deflection and bouncing back scattering of α particles, because the positive charges diffused among the entire atom would have very low charge density and thus could not exert such a large coulombic repulsive force to significantly change the trajectory of α particles. Since the observations of α particle scattering experiments were repeatable, it means that the plum-pudding model is not correct. To explain the large-angle deflection and bouncing back scattering of α particles, there

实验测定的荷质比为 $e/m = -1.76 \times 10^{11}$ C kg^{-1}。汤姆森还证明，阴极射线的荷质比恒定，与阴极的材料或组成均无关。他进一步得出结论：阴极射线是普遍存在于所有原子中的、带负电荷的基本粒子。随后，阴极射线被命名为**电子**，这一术语最初由爱尔兰物理学家乔治·J. 斯通尼于 1874 年提出。

美国物理学家罗伯特·A. 密立根（1868—1953）通过一系列油滴实验（**图 7.2**），细致地测量了电子的电荷，结果于 1911 年发表。这些实验用电场来加速或减缓带电油滴在重力下的沉降。测得电子电荷后，即可通过精确的 e/m 计算电子质量。目前普遍接受的电子的电荷及质量分别为

$$e = -1.6022 \times 10^{-19} \text{C} \quad 且 \quad m = 9.1094 \times 10^{-31} \text{ kg}$$

由于电子是存在于所有原子中的基本粒子，科学家们开始推测，带负电的电子究竟如何包含在电中性的原子中。根据宏观物体的经验，汤姆森于 1904 年提出了**葡萄干布丁模型**。他认为，电子像"葡萄干"一样，飘浮在像"布丁"一样弥散的、带正电荷的星云中。这样吸引力最大化而排斥力最小化，使得原子的总能量最低。然而，根据进一步的实验结果，这种葡萄干布丁模型被证明是不正确的。

X 射线与放射性

在发现电子的几乎同一时期，X 射线与放射性也被发现。1895 年，德国物理学家威廉·伦琴（1845—1923）注意到运行时阴极射线管外会出现荧光，这是由阴极射线管发射的辐射所引起的。由于这种辐射的本质尚属未知，他创造了 **X 射线**这一术语。我们现在认识到 X 射线是高能电磁辐射的一种形式，这将在 **7.2 节**进一步讨论。

1896 年，安托万·H. 贝克勒尔首次观察到**放射性**，他发现当靠近铀盐时，一块被小心遮挡避光的未曝光照相底片会发生曝光。后来，人们发现铀盐和纯铀均会产生这种效应，说明放射性是铀元素的一种特性。

1899 年，欧内斯特·卢瑟福（1871—1937）发现了两种来自放射性物质的辐射：阿尔法（α）和贝塔（β）粒子。通过电场时，**α 粒子**略微向负极板偏转而 **β 粒子**显著地向正极板偏转。后来发现 α 粒子是 He^{2+} 离子，β 粒子是高速电子流。1900 年，保罗·维拉德发现了称为伽马射线（**γ 射线**）的第三束辐射，它不受电场或磁场的影响，被认为是具有极高能量的电磁辐射。放射性物质的这三种辐射类型如**图 7.3** 所示。

α 粒子散射实验

1909 年，欧内斯特·卢瑟福和汉斯·盖格进行了一系列实验，通过用 α 粒子轰击薄金属箔来研究原子的内部结构（**图 7.4**）。在望远镜末端安装一个 ZnS 屏，望远镜可以沿圆形轨道围绕金属箔移动，通过撞击 ZnS 屏时产生的闪光来检测 α 粒子。该实验结果如下：

1) 绝大多数 α 粒子未偏转地穿透金属箔；
2) 一些 α 粒子发生轻微偏转；
3) 极少数（约 0.005%）α 粒子发生大角度偏转；
4) 极少数（约 0.005%）α 粒子沿入射方向反弹散射。

然而，这些实验观察结果与汤姆森的葡萄干布丁模型相矛盾。

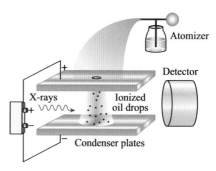

Figure 7.2 Millikan's oil-drop experiments: Ions are produced by energetic radiation (such as X-rays) and attached to oil droplets from an atomizer. An electric field is used to speed up or slow down the falling of the charged oil droplets as a result of gravity. The magnitude of the charge on a droplet is an integral multiple of a minimal charge, the charge of an electron.

图 7.2 密立根油滴实验：高能辐射（如 X 射线）产生的离子附着在雾化器喷出的油滴上，电场可加速或减缓带电油滴在重力下的沉降，油滴所带电荷的大小为一个最小电荷值（即电子电荷）的整数倍。

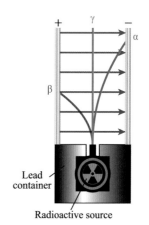

Figure 7.3 Three types of radiation from radioactive materials: α and β particles, and γ rays.

图 7.3 放射性物质的三类辐射：α 粒子、β 粒子与 γ 射线。

must be positive charges with comparable charge density to the α particles inside the atom. Due to the very small fraction of deflected α particles, the positive charge should be confined to a very small region.

The Nuclear Atom Model

In 1911, based on the α particle scattering experiment, Rutherford proposed a new atomic model with the following features:

1) The majority of the mass and all of the positive charge of an atom are concentrated in a tiny region(called the **nucleus**) surrounded by electrons. The remainder of the atom is empty space.
2) The magnitude of the positive charge differs for different atoms and is about one-half the atomic mass number of the element.
3) The numbers of electrons equal the number of units of positive charge on the nucleus (called **nuclear charge number**), counterbalanced with each other to make the whole atom electrically neutral.

Because this model suggested a nucleus centered in the atom, it is called the **nuclear atom model**. Only such an intense density of positive charge in the nucleus could produce an electric field strong enough to deflect the α particles as observed. According to the experimental results, Rutherford estimated that the size of the nucleus was less than 10^{-14} m, which was very close to the true size of about 10^{-15} m.

The Discovery of Protons and Neutrons

In 1919, in the scattering experiments of α particles by nitrogen atoms in air, Rutherford discovered positively charged fundamental particles of matter in the nuclei of atoms called **protons**. He also predicted the existence in the nucleus of electrically neutral fundamental particles, which was subsequently discovered by James Chadwick in 1932 and called **neutrons**.

The number of protons in a given atom is called the **atomic number**(Z), which is also the number of electrons in the atom. The total number of protons and neutrons in an atom is called the **mass number** (A). The number of neutrons, is then given by $A - Z$. A symbolism $^A_Z E$ is used to indicate that the atom is element E with an atomic number Z and a mass number A. Since the atomic number for the same element is always the same, the symbolism can be simplified as $^A E$. Atoms that have the same atomic number but different mass numbers are called **isotopes**.

An electron carries a unit of negative charge, a proton carries a unit of positive charge, and a neutron is electrically neutral. The **atomic mass unit** (u) is defined as exactly 1/12 of the mass of carbon-12(^{12}C). The proton and neutron have masses slightly greater than 1 u, and the mass of an electron is about 1/1823 u. The masses and charges of three subatomic fundamental particles are summarized in **Table 7.1**.

Table 7.1　Properties of Three Fundamental Particles in the Nuclear Atom
表 7.1　核型原子中三种基本粒子的性质

Particle（粒子）	Symbol（符号）	Discovery Year（发现年）	Mass（质量）		Electric Charge（电荷）	
			SI（国际单位）/kg	Atomic（原子单位）/u[a]	SI（国际单位）/C	Atomic（原子单位）/e
Electron（电子）	$^0_{-1}e$	1897	9.1094×10^{-31}	5.4858×10^{-4}	-1.6022×10^{-19}	-1
Proton（质子）	^1_1p	1919	1.6726×10^{-27}	1.0073	1.6022×10^{-19}	$+1$
Neutron（中子）	^1_0n	1932	1.6749×10^{-27}	1.0087	0	0

[a] u is the SI symbol for atomic mass unit and is defined as exactly 1/12 of ^{12}C. (u 为原子质量单位的国际制符号，精确定义为 ^{12}C 质量的 1/12。)

Limitations of the Nuclear Atom Model

Although explicating the structure of an atom, Rutherford's nuclear atom model does not indicate how electrons are arranged outside the nucleus of an atom. According to classical physics, stationary negatively charged electrons with a much lighter mass would be pulled into the much heavier positively charged nucleus, suggesting that electrons in an atom must be in motion, orbiting the nucleus just like the planets orbiting the sun. Again, according to classical physics, orbiting electrons should constantly radiate energy and emit continuous spectra. By continuously losing energy, the electrons would be drawn ever closer to the nucleus and soon spiral into it.

葡萄干布丁模型无法解释 α 粒子的大角度偏转和反弹散射，因为弥散在整个原子中的正电荷具有非常低的电荷密度，因此不能施加如此大的库仑斥力来显著改变 α 粒子的轨迹。由于 α 粒子散射实验的观测结果可重复，这意味着葡萄干布丁模型是不正确的。为了解释 α 粒子的大角度偏转和反弹散射，原子内必须存在与 α 粒子电荷密度相当的正电荷。由于发生偏转的 α 粒子比例极小，正电荷应局限在非常小的区域内。

核型原子模型

1911 年，基于 α 粒子散射实验，卢瑟福提出了一种具有以下特征的新的原子模型：
1) 原子的大部分质量和全部正电荷均集中在一个被电子包围的非常小的区域（称为**原子核**）。原子的剩余空间是空的。
2) 不同原子所带正电荷量不同，约为该元素原子质量数的一半。
3) 电子数等于原子核上单位正电荷的数量（称为**核电荷数**），二者相互抗衡，使得整个原子呈电中性。

由于该模型提出了原子中心存在一个原子核，因此被称为**核型原子模型**。只有原子核这样高密度的正电荷才能产生足够强的电场，使 α 粒子发生观察到的偏转。根据实验结果，卢瑟福估计原子核的尺寸小于 10^{-14} m，这非常接近约 10^{-15} m 的真实大小。

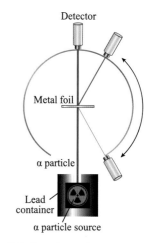

Figure 7.4 Experimental setup to study the inner structure of atoms by bombarding thin metal foils with α particle.

图 7.4 用 α 粒子轰击薄金属箔来研究原子内部结构的实验装置图。

质子与中子的发现

1919 年，通过 α 粒子对空气中氮原子的散射实验，卢瑟福发现了原子核中带正电的基本物质粒子，称为**质子**。他还预测了原子核中存在电中性的基本粒子，随后被詹姆斯·查德威克在 1932 年发现，称为**中子**。

给定原子中的质子数称为**原子序数**（Z），也等于原子中的电子数。原子中质子和中子的总数称为**质量数**（A）。中子数由 $A-Z$ 给出。符号 $^{A}_{Z}\text{E}$ 表示原子序数为 Z、质量数为 A 的元素 E 的原子。由于同一元素的原子序数始终相同，因此符号可简写为 ^{A}E。原子序数相同但质量数不同的原子，互称为**同位素**。

电子携带一个单位的负电荷，质子携带一个单位的正电荷，中子为电中性。**原子质量单位**（u）定义为碳-12（^{12}C）质量的 1/12。质子和中子的质量略大于 1 u，电子的质量约为 1/1823 u。**表 7.1** 总结了三种亚原子基本粒子的质量和电荷。

核型原子模型的局限性

尽管卢瑟福的核型原子模型阐明了原子的结构，但它并没有说明电子在原子核外是如何排布的。根据经典物理，质量轻得多的、带负电的静止电子，会被拉入重得多的、带正电的原子核，这表明原子中的电子一定在绕核运动，就像行星绕太阳运行一样。同样根据经典物理，绕核运动的电子应该不断辐射能量，并发射连续光谱。通过持续损失能量，电子将被吸引而距离原子核越来越近，并很快以螺旋形轨迹进入原子核。然而，核型原子是稳定的体系，原子光谱并非连续光谱而是线状光谱（将在 **7.3 节**进一步讨论）。这些都是核型原子模型无法解释的问题。

However, the nuclear atom is a stable system and the atomic spectra are not continuous but line spectra, which will be further discussed in **Section 7.3**. These are all problems that the nuclear atom model could not explain.

7.2 Quantum Theory

From the late 19th century to the early 20th century, a series of dilemmas in which experimental observations that could not be explained by classical physics were found. These dilemmas include but are not limited to blackbody radiation, photoelectric effect, line spectra of atoms, duality of matter, Compton effect, heat capacities of solids, etc. The line spectra of atoms and the duality of matter will be discussed in **Sections 7.3** and **7.4**, respectively. In this section, we will focus on the blackbody radiation and photoelectric effect, as well as their solutions based on Planck's quantum theory and Einstein's photon theory, respectively. As both blackbody radiation and photoelectric effect involve electromagnetic radiation, which is a type of wave, we will start with the concepts of waves and electromagnetic radiation.

The Concepts of Wave and Electromagnetic Radiation

A wave is a disturbance or oscillation that transmits energy through empty space (a vacuum) or a material medium. A prominent feature of the wave is that it is periodic in both time and space, which can be described by a 1D wave equation as a function of the spatial coordinate x and time coordinate t, as

$$y = f(x,t) = A\sin\left[\omega\left(t - \frac{x}{u}\right) + \varphi\right] \tag{7.2}$$

where A is the **amplitude**, ω is the **angular velocity**, u is the **propagation speed**, and φ is the **phase** of the wave. The durations of distance and time of one cycle in a periodic wave are called **wavelength** (λ) and **period** (T), respectively. The reciprocal of period, which is the number of periods per unit of time, is called **frequency** ($\nu = 1/T$). The reciprocal of wavelength, which is the number of wavelengths per unit of distance, is called **wavenumber** ($\tilde{\nu} = 1/\lambda$). Because the wave must propagate exactly one wavelength in one period with its propagation speed, it can be derived that

$$y = f(x,t) = f(x+\lambda, t) = f(x, t+T)$$

and

$$u = \frac{\lambda}{T} = \lambda\nu = \frac{\nu}{\tilde{\nu}} \tag{7.3}$$

The propagation speed of wave in a medium is inversely proportional to the **refractive index** (n) of the medium, given by

$$u = u_0 / n \tag{7.4}$$

where u_0 is the propagation speed of a wave in a vacuum, and $n = 1$ for a vacuum. The terminologies related to waves and their relationship are summarized in **Table 7.2**.

Table 7.2　Various Terminologies in Waves and Their Relationship
表7.2　波的各种术语及其相互关系

Terminology（术语）	Symbol（符号）	SI unit（国际制单位）	Relationship（关系）
Wavelength（波长）	λ	m	$\tilde{\nu} = 1/\lambda$
Wavenumber（波数）	$\tilde{\nu}$	m^{-1} [a]	
Period（周期）	T	s	$\nu = 1/T$
Frequency（频率）	ν	s^{-1}	
Speed（波速）	u	m s^{-1}	$u = \lambda/T = \lambda\nu = \nu/\tilde{\nu}$
Refractive Index（折射率）	n	N/A	$n = u_0/u$ [b]

[a] Although the SI unit for wavenumber is m^{-1}, a common unit of cm^{-1} is generally used instead. （尽管波数的国际制单位是m^{-1}, 通常使用常用单位cm^{-1}作为替代。）
[b] u_0 is the speed of a wave in a vacuum. （u_0为真空中的波速。）

In general, waves exhibit the following common behaviors (**Figure 7.5**) under certain circumstances:

7.2 量子理论

从 19 世纪末到 20 世纪初，出现了一系列经典物理无法解释的实验观测难题。这些难题包括但不限于黑体辐射、光电效应、原子的线状光谱、物质的二象性、康普顿效应、固体的热容等。原子的线状光谱和物质的二象性将分别在 **7.3** 和 **7.4 节**讨论。本节我们将聚焦于黑体辐射和光电效应，及其分别基于普朗克量子理论和爱因斯坦光子理论的解决方案。由于黑体辐射和光电效应均涉及电磁辐射，而电磁辐射是一种波，因此我们将从波与电磁辐射的概念讲起。

波与电磁辐射的概念

波是通过真空或物质介质传递能量的一种扰动或振荡。波的一个显著特征是，在时间和空间上均具有周期性，作为空间坐标 x 和时间坐标 t 的函数，可用如下一维波动方程描述：

$$y = f(x,t) = A\sin\left[\omega\left(t - \frac{x}{u}\right) + \varphi\right] \quad (7.2)$$

其中 A 为**振幅**，ω 为**角速度**，u 是波的**传播速率**，φ 是波的**相位**。周期性的波一个循环的持续距离和时间分别称为**波长**（λ）和**周期**（T）。周期的倒数，即单位时间内的周期数，称为**频率**（$\nu = 1/T$）。波长的倒数，即单位距离内的波长数，称为**波数**（$\tilde{\nu} = 1/\lambda$）。由于波以其传播速率在一个周期的时间内必然精确传播一个波长的距离，因此可以得出

$$y = f(x,t) = f(x+\lambda, t) = f(x, t+T)$$
$$u = \frac{\lambda}{T} = \lambda\nu = \frac{\nu}{\tilde{\nu}} \quad (7.3)$$

波在介质中的传播速率与介质的**折射率**（n）成反比，即

$$u = u_0 / n \quad (7.4)$$

其中 u_0 是波在真空中的传播速率，对于真空，$n=1$。**表 7.2** 总结了有关波的术语及其相互关系。

在某些情况下，波通常会表现出以下常见行为（**图 7.5**）：
1) **传播**：波在真空或均匀介质中通常沿直线传播。
2) **反射**：当撞击某一可反射表面时，波会改变方向，使反射角等于入射角。
3) **折射**：波从一种介质传播到另一种介质时，其方向发生改变，折射的程度取决于介质折射率的变化。
4) **衍射**：当波遇到障碍物或小孔时，它会围绕障碍物的拐角弯曲或穿过小孔，进入障碍物或小孔的几何阴影区域。当障碍物或小孔的尺寸与波的波长相当时，衍射效应变得更加明显。
5) **干涉**：两个频率相同的波相遇时，这些波可叠加形成具有不同振幅的合波。当两个波之间的相位差为 π 的偶数倍时（意味着两个波**同相**），会发生**相长干涉**，振幅变大。当相位差为 π 的奇数倍时（称为两个波**异相**），会发生**相消干涉**，振幅变小。

根据其本质，波可分为几个主要类别：机械波、电磁波（也称电磁辐射）、概率波（也称物质波，将在 **7.4 节**进一步讨论）和引

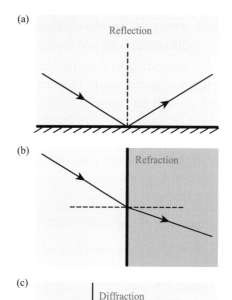

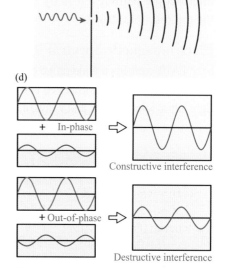

Figure 7.5 Common behaviors of wave: (a) Reflection; (b) Refraction; (c) Diffraction; (d) Interference, including constructive and destructive interference.

图 7.5 波的常见行为：(a) 反射；(b) 折射；(c) 衍射；(d) 干涉（包括相长干涉与相消干涉）。

1) **Propagation**: Waves normally propagate rectilinearly through a vacuum or a homogeneous medium.
2) **Reflection**: When a wave strikes a reflective surface, it changes direction such that the reflective angle equals the incident angle.
3) **Refraction**: The direction of a wave changes when it passes from one medium into another. The degree of refraction is dependent on the change in refractive indices of the media.
4) **Diffraction**: When a wave encounters an obstacle or an aperture, it bends around the corners of the obstacle or through the aperture into the region of geometrical shadow of the obstacle or aperture. Diffraction effects become more pronounced when the size of the obstacle or aperture is comparable to the wavelength of the wave.
5) **Interference**: When two waves of the same frequency encounter each other, the waves superpose to form a resultant wave of different amplitude. **Constructive interference** occurs and the amplitude becomes greater when the difference between the waves is an even multiple of π, which means two waves are **in-phase**. **Destructive interference** occurs and the amplitude becomes lower when the difference is an odd multiple of π, called **out-of-phase**.

According to their nature, waves can be classified into several major categories: mechanical waves, electromagnetic waves (also called electromagnetic radiation), probability waves (also called matter waves, which will be further discussed in **Section 7.4**), and gravitational waves, etc. A **mechanical wave** is a form of wave that is an oscillation of matter, such as waves on strings, water waves, and seismic waves. **Electromagnetic radiation** (EMR) is a form of wave in which oscillating electric and magnetic fields are propagated through a vacuum or a medium. The propagation speed of EMR is constant in a vacuum, often referred to as the speed of light, which is

$$c = 2.99792458 \times 10^8 \text{ m s}^{-1}$$

The propagation speed of EMR in a medium is given by $u = c/n$, where n is the refractive index of the medium.

In EMR, the direction of electric field ($\vec{E}$), the direction of magnetic field ($\vec{B}$), and the direction of wave propagation are always mutually perpendicular, which makes EMR a **transverse wave**, as shown in **Figure 7.6**. Some other waves in which the direction of oscillation is parallel to the direction of propagation are called **longitudinal waves**, such as plasma waves. EMR can interact with matter, especially with microscopic particles. It is by studying the interactions between EMR and matter that the understanding of the microscopic structures of atoms has gradually gained.

Electromagnetic Spectral Region

EMR contains a wide range of wavelengths and frequencies which can be divided into different **electromagnetic spectral regions**, as summarized in **Figure 7.7** and **Table 7.3**. These include radio waves, microwaves, infrared, visible light, ultraviolet, X-rays, and γ rays, in the order of increasing frequency or decreasing wavelength.

Table 7.3 Summary of Electromagnetic Spectral Regions
表 7.3 电磁光谱区域小结

Region（区域）	Frequency（频率）/Hz [a]	Wavelength（波长）/nm [a]	Wavenumber（波数）/cm^{-1} [a]	Type of Spectra（光谱类型）[b]
Radio waves（无线电波）	$< 10^{10}$	$> 3 \times 10^7$	< 0.3	Translational（平动）
Microwave（微波）	$3 \times 10^8 \sim 3 \times 10^{11}$	$10^6 \sim 10^9$	$0.01 \sim 10$	Pure rotational（纯转动）
Far-infrared (Far-IR, 远红外)	$3 \times 10^{11} \sim 1 \times 10^{13}$	$25\,000 \sim 10^6$	$10 \sim 400$	Rotational, weak vibrational（转动、弱振动）
Mid-infrared (Mid-IR, 中红外)	$1 \times 10^{13} \sim 1 \times 10^{14}$	$2\,500 \sim 25\,000$	$400 \sim 4\,000$	Vibrational（振动）
Near-infrared (Near-IR, 近红外)	$1 \times 10^{14} \sim 4 \times 10^{14}$	$760 \sim 2\,500$	$4\,000 \sim 13\,000$	Vibrational overtones（振动倍频）
Visible (Vis, 可见)	$4 \times 10^{14} \sim 8 \times 10^{14}$	$390 \sim 760$	$13\,000 \sim 26\,000$	Electronic（电子）
Ultraviolet (UV, 紫外)	$8 \times 10^{14} \sim 3 \times 10^{16}$	$10 \sim 390$	$26\,000 \sim 10^6$	Electronic（电子）
X-ray (X 射线)	$3 \times 10^{16} \sim 3 \times 10^{20}$	$0.01 \sim 10$	$10^6 \sim 10^{10}$	Electronic（电子）
γ ray (γ 射线)	$> 3 \times 10^{20}$	< 0.01	$> 10^{10}$	Nuclear（核）

[a] One or two significant figures are kept for the ranges of various electromagnetic spectral regions. （对各种电磁光谱区域的范围仅保留一位或两位有效数字。）
[b] The type of spectra refers to the type of microscopic energy levels between which the corresponding transitions occur. （光谱类型指发生对应跃迁的微观能级的类型。）

力波等。**机械波**是一种物质振荡的波，如绳波、水波和地震波。**电磁辐射**（EMR）是振荡的电场和磁场通过真空或介质传播的一种波的形式。电磁辐射在真空中的传播速率恒定，通常称为光速，即

$$c = 2.99792458 \times 10^8 \text{ m s}^{-1}$$

电磁辐射在介质中的传播速率为 $u=c/n$，其中 n 是介质的折射率。

在电磁辐射中，电场的方向（$\vec{E}$）、磁场的方向（$\vec{B}$）以及波的传播方向始终相互垂直，这使得电磁辐射成为**横波**，如**图 7.6** 所示。振荡方向与传播方向平行的其他一些波称为**纵波**，如等离子体波。电磁辐射可与物质相互作用，特别是与微观粒子相互作用。正是通过研究电磁辐射与物质之间的相互作用，人们才逐渐了解原子的微观结构。

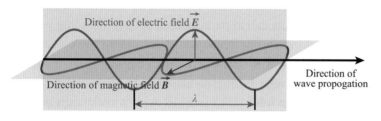

Figure 7.6 Electromagnetic radiation as a transverse wave. The direction of electric field, the direction of magnetic field, and the direction of wave propagation are mutually perpendicular.

图 7.6 作为横波的电磁辐射：电场方向、磁场方向以及波的传播方向相互垂直。

电磁光谱区域

如**图 7.7** 和**表 7.3** 所示，电磁辐射包含广泛的波长和频率，可分为不同的**电磁光谱区域**。按频率增加或波长减小的顺序排列，这些光谱区域包括无线电波、微波、红外线、可见光、紫外线、X 射线和 γ 射线。

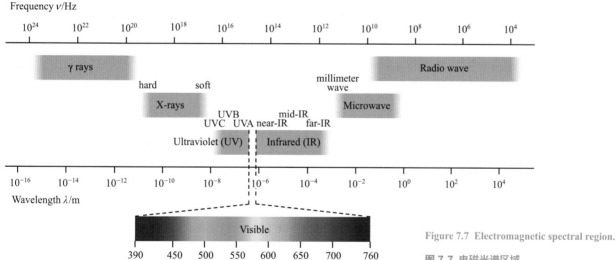

Figure 7.7 Electromagnetic spectral region.

图 7.7 电磁光谱区域。

无线电波是频率低于 10 GHz 或波长大于 0.03 m 的电磁辐射。无线电波广泛用于无线电通信、广播、雷达和无线电导航系统、通信卫星等现代技术。**微波**是波长范围从约 1 m 到 1 mm 的电磁辐射，对应于 300 MHz 到 300 GHz 之间的频率。微波的前缀"微"并不意味着其波长在微米范围内。相反，它表明与无线电波相比，微波的波长较"小"。另一方面，**毫米波**确实是波长在 1~10 mm 的电磁辐射，对应于 300~30 GHz 的频率。毫米波是微波中频率较高的一部分，也称为超高频（EHF）微波。

红外线（IR）是波长介于微波与可见光之间的电磁辐射，通常波长范围为 1 mm（频率 300 GHz）到 760 nm（400 THz）。根据与

Radio waves are EMR with frequencies below 10 GHz, or wavelengths above 0.03 m. Radio waves are widely used in modern technology for radio communication, broadcasting, radar and radio navigation systems, communication satellites, etc. **Microwaves** are EMR with wavelengths ranging from about 1 m to 1 mm, corresponding to frequencies between 300 MHz to 300 GHz. The prefix micro- in microwave is not meant to suggest a wavelength in the μm range. Rather, it indicates that microwaves are "small" in wavelength compared to radio waves. **Millimeter waves**, on the other hand, are indeed EMR with wavelengths in the 1~10 mm range, corresponding to frequencies of 300~30 GHz. Millimeter waves are the part of microwaves with high frequency, and are also called extremely high frequency (EHF) microwaves.

Infrared (IR) is EMR with a wavelength between microwaves and visible light, normally ranging from 1 mm (300 GHz) to 760 nm (400 THz). IR can be subdivided into near-IR, mid-IR, and far-IR according to how close they are from visible light. **Visible light** is the region of EMR that can be perceived by human eyes, normally ranging from 390 nm (770 THz) to 760 nm (400 THz). Visible light, shown as white light as a mixture, can be dispersed by a prism into a series of colors called pure spectral colors, including red, orange, yellow, green, blue, indigo, and violet, in the order of increasing frequency or decreasing wavelength.

As frequency increases, the damage of EMR caused to DNA and cells also increases. EMR with a frequency higher than visible light may cause considerable lasting damage to materials and tissues at the molecular level and is capable of causing cancer. **Ultraviolet** (UV) is EMR with wavelengths between 10 nm and 390 nm. According to its degree of absorption by the ozone layer, UV can be subdivided into UVA (soft UV, low frequency and long wavelength, not absorbed), UVB (intermediate UV, intermediate frequency and wavelength, mostly absorbed), and UVC (hard UV, high frequency and short wavelength, completely absorbed).Although not damage DNA directly, UVA causes indirect damage to DNA by producing reactive oxygen species. **X-rays** are a penetrating form of short-wavelength EMR ranging from 10 pm to 10 nm. X-rays with relatively longer wavelengths (0.1~10 nm) are called soft X-rays, whereas those with relatively shorter wavelengths (10 pm~0.1 nm) are called hard X-rays. **γ rays** are a penetrating form of EMR with extremely short wavelengths. There is no consensus for a definition distinguishing between X-rays and γ rays. An arbitrarily borderline between them is a wavelength of 10 pm. A common practice is to distinguish between these two types of EMR based on their source. X-rays correspond to transitions occuring between electronic energy levels, whereas γ rays to transitions between nuclear energy levels.

Blackbody Radiation and Ultraviolet Catastrophe

A so-called "blackbody" is an idealized physical body that absorbs all incidents of electromagnetic radiation, just like a black substance absorbs all colors of visible light. After absorbing the radiation, a blackbody can also emit electromagnetic radiation with continuous wavelength, called **blackbody radiation**.The intensity of blackbody radiation varies continuously with wavelength, peaking at a certain wavelength at a certain temperature. The wavelength of the peak becomes shorter at a higher temperature (shown as solid lines in **Figure 7.8**). A heated body behaves similarly to a blackbody, emitting light of different colors, from the dull red of an electric-stove heating element (about 1000 K) to the bright white of a light bulb filament (about 3000 K).

Classical physics can not provide a complete explanation of blackbody radiation. It was predicted by classical theory that the radiation intensity of an ideal blackbody would increase indefinitely with respect to decreasing wavelength (shown as dashed lines in **Figure 7.8**). The divergence between classical physics predictions and experimental observations mainly lay in the ultraviolet region of the electromagnetic radiation, and was termed **ultraviolet catastrophe**.

Planck's Quantum Theory

In 1900, in order to explain the experimental observations of blackbody radiation and to solve the problem of the ultraviolet catastrophe, Max K. Planck (1858—1947) proposed a revolutionary **quantum theory**, the main idea of which was that energy, like matter, is discontinuous. Classical physics sets no limitations on the amount of energy that a system may possess, i.e., the energy of a system can be any arbitrary value. However, quantum theory limits the energy of a system to a discrete set of specific values, which must be integer multiples of a minimal specific value, called a quantum of energy. In general, a

可见光的距离，红外可以细分为近红外、中红外和远红外。**可见光**是人眼可以感知的电磁辐射区域，通常波长范围为 390 nm（频率 770 THz）到 760 nm（400 THz）。可见光混合在一起呈白光，可以通过棱镜色散成一系列称为纯光谱色的不同颜色的单色光，按频率增加或波长减小的顺序排列，分别为红、橙、黄、绿、蓝、靛、紫色。

随着频率的增加，电磁辐射对 DNA 和细胞的损伤也会增加。频率高于可见光的电磁辐射可能在分子水平上对材料和组织造成相当大的持久损伤，并可能导致癌症。**紫外线**（UV）是波长在 10 nm 和 390 nm 之间的电磁辐射。根据其被臭氧层吸收的程度，UV 可分为 UVA（软 UV，低频长波，不被吸收）、UVB（中 UV，中频中波，大部分被吸收）和 UVC（硬 UV，高频短波，完全被吸收）。虽然 UVA 不会直接损伤 DNA，但会通过产生的活性氧物种对 DNA 造成间接损伤。**X 射线**是波长从 10 pm 到 10 nm 的具有穿透性的短波电磁辐射。波长相对较长（0.1~10 nm）的 X 射线称为软 X 射线，而波长相对较短（10 pm~0.1 nm）的称为硬 X 射线。**γ 射线**是一种具有穿透性的、波长极短的电磁辐射。如何区分 X 射线和 γ 射线，学界并没有形成共识。通常用波长为 10 pm 作为两者之间的边界。一种常见的做法是根据其来源区分这两类电磁辐射，X 射线对应于电子能级之间发生的跃迁，而 γ 射线对应于核的能级之间发生的跃迁。

黑体辐射与紫外灾难

就像黑色物质能吸收所有颜色的可见光一样，所谓"黑体"是一种理想的、能吸收全部入射电磁辐射的物体。吸收辐射之后，黑体还可以发射具有连续波长的电磁辐射，称为**黑体辐射**。黑体辐射的强度随波长不断变化，当温度一定时在特定波长处达到峰值。温度越高，峰值的波长越短（如**图 7.8** 实线所示）。受热物体的行为与黑体类似，会发射出不同颜色的光，如从电炉加热元件的暗红色（约 1000 K）到灯泡灯丝的亮白色（约 3000 K）。

经典物理无法提供对黑体辐射的完整解释。经典物理预测，理想黑体的辐射强度将随波长的减小而无限增大（如**图 7.8** 虚线所示）。经典物理预测与实验观测之间的分歧主要发生在电磁辐射的紫外区域，因此称为**紫外灾难**。

普朗克量子理论

1900 年，为解释黑体辐射的实验观测结果并解决紫外灾难问题，马克斯·K. 普朗克（1858—1947）提出了革命性的**量子理论**，其主要思想是：能量和物质一样，是不连续的。经典物理对体系可能拥有的能量没有任何限制，即体系的能量可以为任意值。而量子理论将体系的能量限制为一组离散的特定值，这些值必须是某个最小特定值（称为能量量子）的整数倍。一般来说，**量子**是任何物理实体或物理性质的最小数量。如果一个物理性质的大小只能取由一个量子的整数倍组成的离散值，则该物理实体或物理性质是"**量子化**"的。

普朗克在量子理论中使用的模型是，受热物体表面的一组原子以相同频率振荡，以发射电磁辐射。他假定能量是量子化的，因此该原子群（即振子）的所有允许的能量，必须是能量量子的整数倍，而能量与振子的频率成正比，有

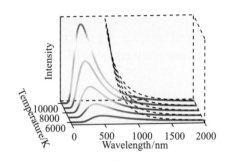

Figure 7.8 **The radiation spectra of headed bodies.** The experimental spectra (solid lines) differ significantly from those expected by classical physics (dashed lines).

图 7.8 **受热物体的辐射光谱：**实验光谱（实线）与经典物理预测的光谱（虚线）存在显著差异。

quantum is the minimum amount of any physical entity or physical property. A physical property is said to be "**quantized**" if the magnitudes of this physical entity or physical property can take on only discrete values consisting of integer multiples of one quantum.

In quantum theory, Planck used a model of a group of atoms on the surface of a heated object oscillating together at the same frequency for the emission of electromagnetic radiation. He assumed that the energy is quantized so that all allowed energies of the group of atoms, the oscillator, must be integer multiples of a quantum of energy, with the energy proportional to the oscillator frequency, given by

$$\varepsilon = nh\nu_0$$

where ε is the allowed energy, n is called a **quantum number** that can be any positive integers (n = 1,2,3,…), ν_0 is the fundamental oscillator frequency, and h is called the **Planck constant**, the value of which was determined later by experiments to be

$$h = 6.62607 \times 10^{-34} \text{ J s}$$

A more general rephrasing of Planck's postulate is that the energy (E) of a quantum of electromagnetic radiation is proportional to the frequency (ν) of the radiation, shown as

$$E = h\nu \tag{7.5}$$

where $\nu = n\nu_0$ and n = 1,2,3,…. **Equation (7.5)** is called **Planck's equation**.

As all allowed energies of a system are integer multiples of a quantum, the difference between any two allowed energies of a system must also be integer multiples of a quantum. This means that when the energy varies from one allowed value (E_1) to another (E_2), the change is also an integer multiple of a quantum, given by

$$\Delta E = E_2 - E_1 = h\nu_2 - h\nu_1 = (n_2 - n_1)h\nu_0 \tag{7.6}$$

Although Planck's quantum theory was initially designed to explain blackbody radiation, it was finally accepted to be a generally applied principle. The concept of energy quantization opened the door to understand the microscopic world. The discovery of quantum theory was a pioneering insight of modern physics and was of fundamental importance to the discovery of quantum mechanics, which was widely recognized as one of the greatest scientific findings in the 20th century. The date on which Planck read his paper at the German Physical Society, December 14, 1900, was designated as the birthdate of quantum theory.

The Photoelectric Effect

The **photoelectric effect** was first discovered by Heinrich Hertz in 1887, and further investigated by Wilhelm Hallwachs and Aleksandr Stoletov. In 1888, Hallwachs discovered that when ultraviolet light struck a freshly cleaned zinc plate connected to an electroscope, the zinc plate became positively charged, suggesting that some negatively charged particles were emitted from it. Later, these emitted negatively charged particles were found of the same nature as cathode rays, i.e., electrons, and were then named as **photoelectrons**.

In a typical apparatus for photoelectric effect measurements depicted in **Figure 7.9(a)**, two plates inside an evacuated chamber are connected to an ammeter to form a circuit. Electromagnetic radiations with adjustable intensity and frequency are allowed to shine onto the lower plate. In experiment 1, the intensity of incident light is fixed at I and the frequency varies from low to high. It is observed from the ammeter that a photocurrent fixed at I_p appears only when the frequency is greater than a certain **threshold frequency** (ν_0). In experiment 2, the intensity of incident light is fixed at $2I$ and again the frequency varies from low to high. This time the photocurrent also appears only at $\nu > \nu_0$, but its intensity is fixed at $2I_p$ [**Figure 7.9(b)**]. If $\nu \leqslant \nu_0$, no photocurrent is observed at all no matter how intense the incident light is. Therefore, we can conclude from these experiments that once the threshold condition of frequency is met, the number of photoelectrons is proportional to the intensity of incident light.

In experiment 3 [**Figure 7.9(c)**], a grid is placed between the two plates to measure the kinetic energy of the emitted photoelectrons. A capacitor is set up by connecting the lower plate to the positive electrode and the grid to the negative electrode of an external power supply with an adjustable voltage that can be measured by a voltmeter. In order for the photocurrent to flow, photoelectrons must pass through the openings in the

其中 ε 是允许的能量；n 称为**量子数**，可以是任何正整数（$n = 1, 2, 3, \cdots$）；ν_0 是振子的基频；h 称为**普朗克常数**，其值后来由实验确定为

$$h = 6.62607 \times 10^{-34} \text{ J s}$$

普朗克假定的一个更为普适的表述是：电磁辐射量子的能量（E）与辐射的频率（ν）成正比，即

$$E = h\nu \tag{7.5}$$

其中 $\nu = n\nu_0$ 且 $n = 1, 2, 3, \cdots$。**式（7.5）** 称为**普朗克方程**。

由于体系的所有允许能量均为量子的整数倍，因此体系的任何两个允许能量之差，也必须是量子的整数倍。这意味着，当能量从一个允许值（E_1）改变为另一个允许值（E_2）时，其改变量也是量子的整数倍，有

$$\Delta E = E_2 - E_1 = h\nu_2 - h\nu_1 = (n_2 - n_1)h\nu_0 \tag{7.6}$$

尽管普朗克量子理论最初只是为解释黑体辐射而创立，但最终被公认为一个普遍适用的原理。能量量子化的概念开启了认识微观世界的大门。量子理论的发现是现代物理的开拓性认识，对量子力学的发现至关重要，而量子力学被广泛认为是 20 世纪最伟大的科学发现之一。普朗克在德国物理学会宣读论文的那一天，即 1900 年 12 月 14 日，被指定为量子理论的诞生日。

光电效应

光电效应最早由海因里希·赫兹于 1887 年发现，并由威廉·哈尔瓦克斯和亚历山大·斯托列托夫进一步研究。1888 年，哈尔瓦克斯发现，当紫外线照射到一块与验电器相连、刚清洁过的锌板上时，锌板会带正电荷，这表明从其中发射出了一些带负电荷的粒子。后来，这些发射出的负电粒子被证实与阴极射线具有相同的性质，即为电子，于是被命名为**光电子**。

在如**图 7.9(a)** 所示的典型光电效应测量装置中，真空腔内的两块极板与电流表相连，以形成电路。允许强度及频率均可调的电磁辐射照射下极板。在实验 1 中，入射光的强度固定在 I，频率由低到高变化。从电流表可观察到，只有当频率大于某一**阈值频率**（ν_0）时，才会出现固定值为 I_p 的光电流。在实验 2 中，入射光的强度固定在 $2I$，频率再次由低到高变化。这一次光电流也仅出现在 $\nu > \nu_0$ 时，但其强度固定在 $2I_p$[**图 7.9(b)**]。如果 $\nu \leqslant \nu_0$，无论入射光有多强，均不能观察到光电流。因此，我们可以从这些实验得出结论：一旦满足频率的阈值条件，光电子数与入射光强度成正比。

实验 3[**图 7.9(c)**] 中，在两个极板之间放置一个栅板，以测量所发射光电子的动能。将下极板与外加电源的正极相连，栅板与负极相连，使两者之间成为一个电容器，外加电源的电压可调且可由电压表测量。为了产生流动的光电流，光电子必须穿过栅板的空隙到达上极板。这些光电子被施加在栅板上的负电势减慢。随着外加电源电压的持续增加，越来越多的光电子减速直至停止运动，不能到达上极板，导致电流表的光电流逐渐减小。总是存在一个**截止电压**（V_s），使得光电子完全无法到达上极板，光电流恰好为零。在 V_s 下，

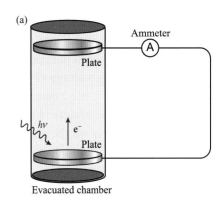

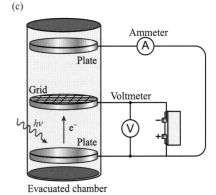

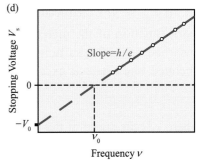

Figure 7.9 The photoelectric effect. (a) Experimental setup to measure the photoelectric current. (b) Diagram of photoelectric current versus frequency of incoming electromagnetic radiation. (c) Experimental setup to measure the stopping voltage. (d) Diagram of stopping voltage versus frequency of incoming electromagnetic radiation.

图 7.9 光电效应： (a) 测量光电流的实验装置图；(b) 光电流与入射电磁辐射频率的关系图；(c) 测量截止电压的实验装置图；(d) 截止电压与入射电磁辐射频率的关系图。

grid and reach the upper plate. These photoelectrons will be slowed down by the negative potential on the grid. As the voltage of the external power supply increases continuously, more and more photoelectrons are stopped and cannot reach the upper plate, resulting in gradually decreased photocurrent in the ammeter. There is always a **stopping voltage** (V_s) at which no photoelectron can reach the upper plate and the photocurrent is just stopped. At V_s, the kinetic energy of the emitted photoelectrons has been fully converted to potential energy. Experimental results [**Figure 7.9(d)**] show that the stopping voltage is independent of the intensity but linearly dependent on the frequency of incident light with a slope of h/e, which leads to a conclusion that the kinetic energy of photoelectrons is linearly dependent on the frequency of incident light.

In experiment 4, different metals or conductors are used as materials for the lower plate. Accordingly, different threshold frequency and stopping voltage (at the same frequency of incident light) are observed, demonstrating that the photoelectric effect is a ubiquitous phenomenon that is related to the nature of substances.

Einstein's Photon Theory

Classical theory believes that the energy of electromagnetic radiation is proportional to its intensity and the shining time, and thus photoelectrons will be eventually emitted when enough energy is accumulated in sufficient time, even for a dim light. The dependence of kinetic energy of photoelectrons on the frequency of incident light could not be explained by classical mechanics.

In 1905, Albert Einstein (1879—1955) proposed a photon theory to explain the photoelectric effects based on Planck's quantum theory. Einstein assumed that electromagnetic radiation consists of tiny packets of energy known as **photons** or light quanta, and that the energy of a photon (E_{ph}) is proportional to the frequency (ν) of the corresponding electromagnetic wave, given by

$$E_{ph} = h\nu$$

where h is the Planck constant. The intensity of electromagnetic radiation is associated with the number of photons at a point per unit time.

When a photon of energy $h\nu$ strikes a bound electron, the photon energy is absorbed. The minimum energy required to remove a bound electron from the surface of the material is called the **work function** (Φ) of the material, given by

$$\Phi = h\nu_0$$

where ν_0 is the threshold frequency. Work function is a characteristic of the material, and so is the threshold frequency, as confirmed by experiments.

If $E_{ph} \leqslant \Phi$ which leads to $\nu \leqslant \nu_0$, it means that the absorbed energy is not enough to overcome the work function of the material and no photoelectron can be liberated. Only when $E_{ph} > \Phi$ or $\nu > \nu_0$, the absorbed energy is sufficient to overcome the work function of the material so that a photoelectron is liberated, and whatever excess energy appears as the kinetic energy of the emitted photoelectron, shown as

$$E_{ph} = h\nu = \Phi + E_k = h\nu_0 + \frac{1}{2}mv^2 \tag{7.7}$$

where v is the speed of the emitted photoelectron. At higher light intensity, more photons are absorbed, and consequently, more photoelectrons are liberated, leading to a higher photocurrent. Furthermore, the kinetic energy of the emitted photoelectron can be measured by the stopping voltage as

$$E_k = \frac{1}{2}mv^2 = h(\nu - \nu_0) = eV_s \tag{7.8}$$

By rearranging **Equation (7.8)**, we have

$$V_s = \frac{h}{e}(\nu - \nu_0)$$

Extrapolation of the lines in **Figure 7.9(d)** gives a negative intercept at $-V_0$, which can be derived by

$$V_0 = -\frac{h}{e}(0 - \nu_0) = \frac{h\nu_0}{e} = \frac{\Phi}{e}$$

By rearranging **Equation (7.7)**, we also have

发射的光电子的动能完全转化为势能。实验结果 [**图 7.9(d)**] 表明，截止电压与入射光的强度无关，而与其频率呈线性关系，斜率为 h/e，由此可以得出光电子的动能与入射光频率线性相关的结论。

在实验 4 中，使用不同金属或导体作为下极板材料，相应可观察到不同的阈值频率和截止电压（在相同的入射光频率下），这表明光电效应是与物质性质相关的普遍现象。

爱因斯坦光子理论

经典力学理论认为，电磁辐射的能量与其强度和照射时间成正比，因此即使在昏暗的光线下，只要时间足够长，也能积累足够的能量，最终发射出光电子。经典力学无法解释光电子的动能对入射光频率的依赖性。

1905 年，阿尔伯特·爱因斯坦 (1879—1955) 基于普朗克量子理论，提出了光子理论来解释光电效应。爱因斯坦假定电磁辐射由称为**光子**（或光量子）的微小能量包组成，光子的能量 ($E_{光子}$) 与相应电磁波的频率 (ν) 成正比，有

$$E_{光子} = h\nu$$

其中 h 是普朗克常数。电磁辐射的强度与单位时间内某一点的光子数相关。

当能量为 $h\nu$ 的光子撞击束缚电子时，光子的能量被吸收。从材料表面移除束缚电子所需的最小能量，称为材料的**功函数** (Φ)，有

$$\Phi = h\nu_0$$

其中 ν_0 是阈值频率。实验证实，功函数是材料的特性，阈值频率也是如此。

当 $E_{光子} \leqslant \Phi$ 时，$\nu \leqslant \nu_0$，这意味着吸收的能量不足以克服材料的功函数，不能释放光电子。只有当 $E_{光子} > \Phi$ 或 $\nu > \nu_0$ 时，吸收的能量才足以克服材料的功函数，从而释放出光电子，无论有多少过剩能量，均转化为发射出的光电子的动能，即

$$E_{光子} = h\nu = \Phi + E_{k} = h\nu_0 + \frac{1}{2}mv^2 \tag{7.7}$$

其中 v 是发射出的光电子的速率。在较高的光强度下，较多的光子被吸收，相应地释放出更多光电子，从而导致了更高的光电流。此外，发射的光电子的动能可以通过截止电压来测量，有

$$E_{k} = \frac{1}{2}mv^2 = h(\nu - \nu_0) = eV_{s} \tag{7.8}$$

重排式（7.8），可得

$$V_{s} = \frac{h}{e}(\nu - \nu_0)$$

外延**图 7.9(d)** 的直线，其负截距位于 $-V_0$ 处，可推导为

$$V_0 = -\frac{h}{e}(0 - \nu_0) = \frac{h\nu_0}{e} = \frac{\Phi}{e}$$

重排式（7.7），还可得

$$h\nu = eV_0 + eV_{s}$$

即

$$V_{s} = \frac{h}{e}\nu - V_0 \tag{7.9}$$

这解释了**图 7.9(d)** 所示的截止电压与入射光频率的线性关系。

or
$$hv = eV_0 + eV_s$$
$$V_s = \frac{h}{e}v - V_0 \tag{7.9}$$

which explains the linear dependence of stopping voltage on the frequency of incident light shown in **Figure 7.9(d)**.

7.3 Atomic Spectra and Bohr Theory

In the 17th century, the word "spectrum" was introduced into optics by Isaac Newton to refer to the range of colors observed when white light was dispersed through a prism. Later, the term referred to a plot of light intensity as a function of wavelength or frequency. Spectra can be classified into continuous spectra and line spectra.

Continuous Spectra and Line Spectra

The spectrum of solar radiation on earth is illustrated in **Figure 7.10(a)** and this is said to be a **continuous spectrum** because it consists of many continuous wavelength components. The blackbody radiations shown in **Figure 7.8** are continuous spectra. On the contrary, a discrete spectrum or a **line spectrum** is a spectrum in which radiations are present only at certain discrete wavelengths with gaps between them. Atomic spectra are typical line spectra, and each element has its own distinctive spectral lines, serving as atomic "fingerprints" of the corresponding elements. The light emitted by an electric discharge through a gas, such as hydrogen gas [**Figure 7.10(b)**] produces line spectra. The flames of various ionic compounds showing distinctive colors indicative of the metal ions present also give line spectra. The reason why the flame of sodium is yellow is that the two most intense lines in the atomic spectrum of sodium are located in the yellow region of visible light.

Atomic Spectra of Hydrogen

Hydrogen is the simplest atom, which comprises only one proton in its nucleus and one electron outside the nucleus, yet it gives rise to a variety of spectral lines in its atomic spectra. Light from a hydrogen lamp appears to the naked eyes as a reddish-purple color, which contains four distinctive lines in the visible region [**Figure 7.10(b)**]: a red and most intense line at 656.3 nm (H_α), a blue line at 486.1 nm (H_β), a violet line at 434.0 nm (H_γ), and a dark violet line at 410.2 nm (H_δ). In 1885, Johann J. Balmer (1825—1898) deduced a formula for the wavelengths of these spectral lines, called the **Balmer series**, as

$$\lambda = 364.600 \text{ nm} \times \frac{n^2}{n^2 - 4}$$

Later, Johannes Rydberg (1854—1919) rearranged this formula into a simpler empirical expression as

$$\tilde{v} = \frac{1}{\lambda} = \bar{R}_H \left(\frac{1}{2^2} - \frac{1}{n^2} \right)$$

where $\tilde{v}$ is the wavenumber of light, n is a positive integer greater than 2, and $\bar{R}_H = 1.09677576 \times 10^7 \text{ m}^{-1}$ is called the **Rydberg constant** in wavenumber. It can be easily calculated that the above four spectral lines correspond to n = 3, 4, 5, and 6, respectively. More spectral lines with greater n values and higher wavenumbers are also present, but they appear in the ultraviolet region.

Later, other series in the atomic spectra of hydrogen were also discovered and listed as follows:

$$\text{Lyman series (UV)}: \quad \tilde{v} = \bar{R}_H \left(\frac{1}{1^2} - \frac{1}{n^2} \right), \quad n = 2, 3, \cdots$$

$$\text{Paschen series (near-IR)}: \quad \tilde{v} = \bar{R}_H \left(\frac{1}{3^2} - \frac{1}{n^2} \right), \quad n = 4, 5, \cdots$$

$$\text{Brackett series (near-IR)}: \quad \tilde{v} = \bar{R}_H \left(\frac{1}{4^2} - \frac{1}{n^2} \right), \quad n = 5, 6, \cdots$$

7.3 原子光谱与玻尔理论

17 世纪，艾萨克·牛顿将"光谱"一词引入光学，来表示白光通过棱镜色散时观察到的颜色范围。后来，该术语特指光的强度随波长或频率变化的曲线图。光谱可分为连续光谱和线状光谱。

连续光谱与线状光谱

地球上太阳辐射的光谱 [**图 7.10(a)**] 由许多连续的波长分量组成，因此称为**连续光谱**。**图 7.8** 所示的黑体辐射就是连续光谱。相反，离散光谱或**线状光谱**是辐射仅存在于某些离散的波长处、中间存在间隙的光谱。原子光谱是典型的线状光谱，每种元素都具有其独特的谱线，可作为相应元素的原子"指纹"。气体 [如氢气，**图 7.10(b)**] 放电发出的光会产生线状光谱。各种离子化合物的火焰呈现可指示其金属离子存在的独特颜色，也会给出线状光谱。钠的焰色之所以呈黄色，正是因为钠的原子光谱中两条强度最高的谱线位于可见光的黄光区域。

氢原子光谱

氢是最简单的原子，其原子核内仅有一个质子，核外也只有一个电子，然而氢的原子光谱中却包含大量谱线。肉眼看来，氢灯发出的光呈红紫色，在可见光区包含四条独特的谱线 [**图 7.10(b)**]：一条具有最高强度、位于 656.3 nm 的红色谱线（H_α），一条位于 486.1 nm 的蓝色谱线（H_β），一条位于 434.0 nm 的紫色谱线（H_γ），和一条位于 410.2 nm 的深紫色谱线（H_δ）。1885 年，约翰·J. 巴尔末（1825—1898）推导了这些谱线（称为**巴尔末谱系**）的波长公式，为

$$\lambda = 364.600 \text{ nm} \times \frac{n^2}{n^2 - 4}$$

后来，约翰尼斯·里德堡（1854—1919）将该公式重排为一个更简单的经验表达式，即

$$\tilde{\nu} = \frac{1}{\lambda} = \bar{R}_H \left(\frac{1}{2^2} - \frac{1}{n^2} \right)$$

其中 $\tilde{\nu}$ 是光的波数，n 为大于 2 的正整数，$\bar{R}_H = 1.09677576 \times 10^7 \text{ m}^{-1}$ 称为以波数表示的**里德堡常数**。不难算出，上述四条谱线分别对应于 $n=3$、4、5 和 6。还存在更多具有更大 n 值和更高波数的谱线，但它们出现在紫外区。

后来，氢原子光谱的其他谱系也被发现，列出如下：

莱曼谱系(紫外区)： $\tilde{\nu} = \bar{R}_H \left(\frac{1}{1^2} - \frac{1}{n^2} \right)$, $n = 2, 3, \cdots$

帕邢谱系(近红外区)： $\tilde{\nu} = \bar{R}_H \left(\frac{1}{3^2} - \frac{1}{n^2} \right)$, $n = 4, 5, \cdots$

布拉开谱系(近红外区): $\tilde{\nu} = \bar{R}_H \left(\frac{1}{4^2} - \frac{1}{n^2} \right)$, $n = 5, 6, \cdots$

普丰特谱系(中红外区): $\tilde{\nu} = \bar{R}_H \left(\frac{1}{5^2} - \frac{1}{n^2} \right)$, $n = 6, 7, \cdots$

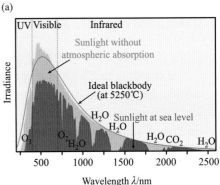
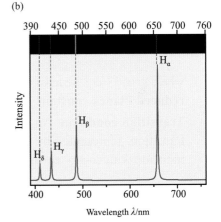

Figure 7.10 Continuous spectrum versus line spectrum. (a) The continuous spectrum of solar radiation on the earth. (b) The line spectrum of hydrogen atoms in the visible region.

图 7.10 连续光谱与线状光谱：(a) 地球上太阳辐射的连续光谱；(b) 可见光范围内氢原子的线状光谱。
(Source：http://en.wikipedia.org/wiki/Sunlight)

Pfund series (mid - IR): $\tilde{v} = \bar{R}_H \left(\dfrac{1}{5^2} - \dfrac{1}{n^2} \right)$, $n = 6, 7, \cdots$

All series in the atomic spectra of hydrogen can fit in a uniform formula, as

$$\tilde{v} = \bar{R}_H \left(\dfrac{1}{n'^2} - \dfrac{1}{n^2} \right) \tag{7.10}$$

where n and n' are two positive integers with $n > n'$.

Bohr Atomic Theory of Hydrogen

According to classical theory, electrons orbiting the nucleus should radiate energy continuously, and the atomic spectra should be continuous spectra. The line spectra of atoms contradicts the classical theory. In 1913, Niels Bohr (1885—1962) proposed a new atomic theory to explain the line spectra of hydrogen based on Planck's quantum hypothesis. In this theory, Bohr postulated three assumptions as:

1) **Stationary condition**: The electrons can only move in a set of allowed circular atomic orbits, called stationary states, around the nucleus. No energy is emitted by electrons in their stationary states.
2) **Angular momentum condition**: The angular momentum (L) of electrons is quantized when electrons are in stationary states, as

$$L = mvr = n\hbar = \dfrac{nh}{2\pi} \tag{7.11}$$

where m and v are the mass and velocity of electron, respectively, r is the orbit radius, $\hbar = h/2\pi$ is called the **reduced Planck constant**, and n is a quantum number that can be any positive integers ($n = 1, 2, 3, \ldots$).

3) **Transition condition**: An electron can pass only from one allowed orbit (n) to another (n'). In such a transition, represented by $n \to n'$, the electron can either absorb ($n < n'$) or emit ($n > n'$) electromagnetic radiation with energy equal to the difference between the two orbits, as

$$E_{ph} = hv = |\Delta E| = |E_{n'} - E_n| \tag{7.12}$$

If $E_n < E_{n'}$, $\Delta E = E_{n'} - E_n > 0$ corresponding to an absorption, and if $E_n > E_{n'}$, $\Delta E = E_{n'} - E_n < 0$ corresponding to an emission.

Based on the above assumptions, we can start the derivation. First, considering that an electron moves in a circular orbit, the electrostatic attraction between electron and nucleus serves as the centripetal force (F) of electrons in circular motion, as

$$F = \dfrac{e^2}{4\pi\varepsilon_0 r^2} = \dfrac{mv^2}{r}$$

where $\varepsilon_0 = 8.8541878128 \times 10^{-12}$ C^2 J^{-1} m^{-1} is the dielectric constant in a vacuum. We then have

$$r = \dfrac{e^2}{4\pi\varepsilon_0 mv^2}$$

Next, applying the angular momentum condition $L = mvr = n\hbar$, we can calculate that

$$v = \dfrac{e^2}{2\varepsilon_0 h} \cdot \dfrac{1}{n} = 2.187 \times 10^6 \text{ m s}^{-1} \times \dfrac{1}{n}$$

and

$$r = \dfrac{\varepsilon_0 h^2}{\pi m e^2} n^2 = a_0 n^2 \tag{7.13}$$

where $a_0 = 5.293 \times 10^{-11}$ m $= 52.93$ pm is called the **Bohr radius**. Equation (7.13) means that the allowed orbit radius (r_n) is quantized with the quantum number n, where $n = 1, 2, 3, \ldots$.

The total energy of an electron (E) equals the summation of kinetic energy (E_k) and potential energy (E_p), where

$$E_k = \dfrac{1}{2}mv^2 = \dfrac{e^2}{8\pi\varepsilon_0 r} \quad \text{and} \quad E_p = -\dfrac{e^2}{4\pi\varepsilon_0 r}$$

氢原子光谱的所有谱系均可用统一的公式拟合为

$$\tilde{\nu} = \bar{R}_H \left(\frac{1}{n'^2} - \frac{1}{n^2} \right) \quad (7.10)$$

其中 n 和 n' 是两个正整数且 $n > n'$。

玻尔氢原子理论

根据经典理论，绕核运动的电子应连续地辐射能量，原子光谱应为连续光谱。原子的线状光谱与经典理论相矛盾。1913 年，尼尔斯·玻尔（1885—1962）基于普朗克的量子假定，提出了一种新的原子理论，来解释氢原子的线状光谱。在该理论中，玻尔提出了三个假定：

1) **定态条件**：电子只能在一组允许的、圆形原子轨道（称为定态）上绕核运动，处于定态的电子不发射能量。
2) **角动量条件**：处于定态的电子，其角动量（L）是量子化的，即

$$L = mvr = n\hbar = \frac{nh}{2\pi} \quad (7.11)$$

其中 m 和 v 分别是电子的质量和速率；r 为轨道半径；$\hbar = h/2\pi$ 称为**约化普朗克常数**；n 是一个量子数，可以为任何正整数（$n = 1, 2, 3, \cdots$）。

3) **跃迁条件**：电子只能从一个允许的轨道（n）跃迁到另一个允许的轨道（n'）。在此跃迁（用 $n \to n'$ 表示）中，电子可以吸收（$n < n'$ 时）或发射（$n > n'$ 时）电磁辐射，其能量等于两个轨道的能量差，即

$$E_{ph} = h\nu = |\Delta E| = |E_{n'} - E_n| \quad (7.12)$$

如果 $E_n < E_{n'}$，$\Delta E = E_{n'} - E_n > 0$ 对应于吸收；如果 $E_n > E_{n'}$，$\Delta E = E_{n'} - E_n < 0$ 对应于发射。

基于上述假定，我们可以开始推导。首先，考虑到电子在圆形轨道上运动，电子和原子核之间的静电引力充当电子在圆周运动中的向心力（F），即

$$F = \frac{e^2}{4\pi\varepsilon_0 r^2} = \frac{mv^2}{r}$$

其中 $\varepsilon_0 = 8.8541878128 \times 10^{-12}$ C^2 J^{-1} m^{-1} 是真空中的介电常数。则有

$$r = \frac{e^2}{4\pi\varepsilon_0 mv^2}$$

接下来应用角动量条件 $L = mvr = n\hbar$，可计算得

$$v = \frac{e^2}{2\varepsilon_0 h} \cdot \frac{1}{n} = 2.187 \times 10^6 \text{ m s}^{-1} \times \frac{1}{n}$$

$$r = \frac{\varepsilon_0 h^2}{\pi m e^2} n^2 = a_0 n^2 \quad (7.13)$$

其中 $a_0 = 5.293 \times 10^{-11}$ m = 52.93 pm，称为**玻尔半径**。式（7.13）表示允许的轨道半径（r_n）是量子化的，其量子数 $n = 1, 2, 3, \cdots$。

电子的总能量（E）为其动能（E_k）与势能（E_p）之和，其中

$$E_k = \frac{1}{2}mv^2 = \frac{e^2}{8\pi\varepsilon_0 r} \quad \text{且} \quad E_p = -\frac{e^2}{4\pi\varepsilon_0 r}$$

因此

$$E = E_k + E_p = -\frac{e^2}{8\pi\varepsilon_0 r} = -\frac{me^4}{8\varepsilon_0^2 h^2} \cdot \frac{1}{n^2} = -\frac{R_H}{n^2} \quad (7.14)$$

其中 $R_H = 2.18 \times 10^{-18}$ J = 13.6 eV 是以能量表示的里德堡常数。由于能

Therefore,

$$E = E_k + E_p = -\frac{e^2}{8\pi\varepsilon_0 r} = -\frac{me^4}{8\varepsilon_0^2 h^2} \cdot \frac{1}{n^2} = -\frac{R_H}{n^2} \quad (7.14)$$

where $R_H = 2.18\times10^{-18}$ J = 13.6 eV is the Rydberg constant in energy. R_H is related to $\bar{R}_H$ by $R_H = hc\bar{R}_H$ since energy is related to wavenumber by

$$E = h\nu = \frac{hc}{\lambda} = hc\tilde{\nu} \quad (7.15)$$

Equation (7.14) means that the allowed orbit energy (E_n) is also quantized with the quantum number n, where $n = 1, 2, 3, \ldots$. The quantized orbit radii and energies for the hydrogen atom are summarized in **Table 7.4**.

Table 7.4 The Quantized Orbit Radii and Energies for the Hydrogen Atom
表 7.4 氢原子量子化的轨道半径与能量

n	r_n [a]	E_n [b]	Note（备注）
1	$r_1 = a_0$	$E_1 = -R_H$	Ground state（基态）
2	$r_2 = 4a_0$	$E_2 = -R_H/4$	First excited state（第一激发态）
3	$r_3 = 9a_0$	$E_3 = -R_H/9$	Second excited state（第二激发态）
4	$r_4 = 16a_0$	$E_4 = -R_H/16$	Third excited state（第三激发态）
...	...	...	...
n	$r_n = n^2 a_0$	$E_n = -R_H/n^2$	$(n-1)^{\text{th}}$ excited state（第 $n-1$ 激发态）

[a] $a_0 = 5.293\times10^{-11}$ m = 52.93 pm is the Bohr radius. ($a_0 = 5.293\times10^{-11}$ m = 52.93 pm 是玻尔半径。)
[b] $R_H = 2.18\times10^{-18}$ J = 13.6 eV is the Rydberg constant in energy. ($R_H = 2.18\times10^{-18}$ J = 13.6 eV 是以能量表示的里德堡常数。)

Last, from the transition condition we have

$$E_{ph} = h\nu = |\Delta E| = |E_{n'} - E_n| = R_H \left|\frac{1}{n^2} - \frac{1}{n'^2}\right|$$

Since hydrogen spectra are emission spectra from a high orbit to a low orbit, $n > n'$ and $E_n > E_{n'}$. Consequently,

$$E_{ph} = h\nu = -\Delta E = E_n - E_{n'} = R_H \left(\frac{1}{n'^2} - \frac{1}{n^2}\right)$$

and

$$\tilde{\nu} = \frac{\nu}{c} = \frac{E_{ph}}{hc} = \frac{R_H}{hc}\left(\frac{1}{n'^2} - \frac{1}{n^2}\right) = \bar{R}_H \left(\frac{1}{n'^2} - \frac{1}{n^2}\right)$$

This is in good agreement with the uniform formula for various series in hydrogen spectra shown in **Equation (7.10)**.

Strictly, there is still some very slight difference in the experimentally determined $\bar{R}_H$, which is 1.09677576×10^7 m^{-1}, and the $\bar{R}_H$ value obtained from the above derivation, which is

$$\bar{R}_H = \frac{R_H}{hc} = \frac{me^4}{8\varepsilon_0^2 h^2} \cdot \frac{1}{hc}$$

$$= \frac{(9.10938188\times10^{-31})\times(-1.602176462\times10^{-19})^4}{8\times(8.854187817\times10^{-12})^2 \times(6.62606876\times10^{-34})^3 \times(2.99792458\times10^8)} \text{ m}^{-1}$$

$$= 1.09737316\times10^7 \text{ m}^{-1}$$

This very slight difference arises from the fact that the hydrogen atom is actually a two-species system. Because the mass of an electron is much lighter than that of a proton, we normally assume that the proton does not move at all and consider only the mass of the electron. Strictly, proton and electron orbit around each other and the reduced mass (μ) of the system is given by

$$\frac{1}{\mu} = \frac{1}{m_e} + \frac{1}{m_p}$$

Replacing m with μ in **Equation (7.14)**, we have

量与波数的关系为

$$E = h\nu = \frac{hc}{\lambda} = hc\tilde{\nu} \tag{7.15}$$

R_H 与 $\bar{R}_H$ 的关系为 $R_H = hc\bar{R}_H$。**式（7.14）**表示允许的轨道能量（E_n）也是量子化的，其量子数 $n = 1, 2, 3, \cdots$。**表 7.4** 总结了氢原子量子化的轨道半径与能量。

最后，由跃迁条件可得

$$E_{ph} = h\nu = |\Delta E| = |E_{n'} - E_n| = R_H \left| \frac{1}{n^2} - \frac{1}{n'^2} \right|$$

由于氢原子光谱是从高轨道到低轨道的发射光谱，因此 $n > n'$ 且 $E_n > E_{n'}$。相应有

$$E_{ph} = h\nu = -\Delta E = E_n - E_{n'} = R_H \left(\frac{1}{n'^2} - \frac{1}{n^2} \right)$$

$$\tilde{\nu} = \frac{\nu}{c} = \frac{E_{ph}}{hc} = \frac{R_H}{hc}\left(\frac{1}{n'^2} - \frac{1}{n^2}\right) = \bar{R}_H \left(\frac{1}{n'^2} - \frac{1}{n^2} \right)$$

这与**式（7.10）**所示氢原子光谱中各种谱系的统一公式一致。

严格来说，实验测定的 $\bar{R}_H$ 为 1.09677576×10^7 m^{-1}，与从上述推导得到的 $\bar{R}_H$，即

$$\bar{R}_H = \frac{R_H}{hc} = \frac{me^4}{8\varepsilon_0^2 h^2} \cdot \frac{1}{hc}$$

$$= \frac{(9.10938188 \times 10^{-31}) \times (-1.602176462 \times 10^{-19})^4}{8 \times (8.854187817 \times 10^{-12})^2 \times (6.62606876 \times 10^{-34})^3 \times (2.99792458 \times 10^8)} \text{ m}^{-1}$$

$$= 1.09737316 \times 10^7 \text{ m}^{-1}$$

仍存在一些微小的差异。这一微小差异的来源在于，氢原子实际上是一个两物种体系。由于电子的质量比质子小得多，我们通常假定质子基本不移动，只考虑电子的质量。严格来说，质子和电子彼此环绕，体系的折合质量（μ）为

$$\frac{1}{\mu} = \frac{1}{m_e} + \frac{1}{m_p}$$

将**式（7.14）**中的 m 替换为 μ，有

$$\bar{R}_H = \frac{R_H}{hc} = \frac{\mu e^4}{8\varepsilon_0^2 h^2} \cdot \frac{1}{hc} \tag{7.16}$$

得到的 $\bar{R}_H$ 为 1.09677576×10^7 m^{-1}，与实验测定值精确匹配。

能级图与电离能

从**式（7.14）**中我们了解到允许的轨道能量是量子化的，这些允许的能量状态称为**能级**，如**图 7.11** 所示。这种图形化的表示法称为**能级图**。根据玻尔理论，量子数 n 可以是从 1 到 ∞ 的任何正整数。$n=1$ 时，$r_1=a_0$ 对应于距核最近的轨道，$E_1 = -R_H$ 是允许的最低能级，称为**基态**。电子获得能量后可以跃迁到更高的能级（$n = 2, 3, 4, \cdots$），称为**激发态**，有

$$r_2 = 4a_0, \quad r_3 = 9a_0, \quad r_4 = 16a_0, \quad \cdots$$
$$E_2 = -R_H/4, \quad E_3 = -R_H/9, \quad E_4 = -R_H/16, \quad \cdots$$

$$\bar{R}_H = \frac{R_H}{hc} = \frac{\mu e^4}{8\varepsilon_0^2 h^2} \cdot \frac{1}{hc} \tag{7.16}$$

This gives rise to an $\bar{R}_H$ of 1.09677576×10^7 m^{-1}, which matches precisely with the experimentally determined value.

Energy Level Diagram and Ionization Energy

From **Equation (7.14)**, we understand that the energy of the allowed orbits is quantized, and those allowed energy states are called **energy levels**, which are schematically represented in **Figure 7.11**. This graphic representation is called an **energy-level diagram**. According to Bohr theory, the quantum number n can be any positive integers from 1 up to ∞. When $n = 1$, $r_1 = a_0$ corresponds to the closest orbit to the nucleus and $E_1 = -R_H$ is the lowest allowed energy level, called the **ground state**. When the electron gains some energy, it jumps to a higher energy level ($n = 2, 3, 4, \ldots$), called the **excited states**, with

$$r_2 = 4a_0, \quad r_3 = 9a_0, \quad r_4 = 16a_0, \quad \cdots$$

and

$$E_2 = -R_H/4, \quad E_3 = -R_H/9, \quad E_4 = -R_H/16, \quad \cdots$$

At $n = \infty$, $r_n = a_0 n^2 = \infty$ corresponds to an orbit that is infinitely far away from the nucleus and $E_n = -R_H/n^2 = 0$ is the reference zero point in potential energy. This means that the electron is completely removed from the hydrogen atom and thus is free. In this case, the H atom is ionized into a cation H$^+$. The energy required to ionize an electron from the ground state of H atom is called the **ionization energy** (I), and is given by

$$I = E_\infty - E_1 = 0 - (-R_H) = R_H \tag{7.17}$$

More discussion about ionization energy can be found in **Section 7.9**.

The atomic spectra are **emission spectra** of atoms, obtained after the individual atoms in a collection of atoms are excited to the various possible excited states. The excited atoms then jump to the lower energy level by emitting photons of energy equal to the difference between the two corresponding energy levels. Thus, the quantization of the energy levels of atoms leads to line spectra. The emission spectrum of a sample is typically measured by dispersing the emitted light through a monochromator into a detector, which determines the intensity of light with respect to the wavelengths of individual components. On the other hand, electromagnetic radiations, such as white light, through a sample of atoms in their ground state can also be measured by dispersing through a monochromator into a detector. The intensity of light with respect to wavelengths that the atoms absorb is observed, giving rise to the **absorption spectra** of atoms.

Emission spectra are generally more complicated than absorption spectra. An excited sample contains atoms in a variety of excited states, each being able to jump to any of several lower states. An absorbing sample generally is cold and transitions are possible only from the ground state. For example, the absorption spectra from cold H atoms only contains the Lyman series ($n = 1$).

Bohr Theory for Hydrogen-Like Species

Bohr theory works not only for hydrogen atoms but also for **hydrogen-like species**, which are any atoms or ions with only one electron outside a nucleus, such as He$^+$, Li^{2+}, Be^{3+}, B^{4+}, etc. Hydrogen-like species are simple systems consisting of only two charged particles: an electron with negative charge $-e$ and a nucleus with positive charge $+Ze$, where Z is the nuclear charge or atomic number. For hydrogen-like species, the derivation process in Bohr theory is similar to that of hydrogen atoms and the results are

$$r_n = a_0 n^2 / Z \quad \text{and} \quad E_n = -R_H Z^2 / n^2 \tag{7.18}$$

Limitations of Bohr Theory

Despite the accomplishments of Bohr theory for hydrogen atoms and hydrogen-like species, Bohr theory has a number of weaknesses. The theory cannot explain the emission spectra of multielectron species, which are atoms and ions with more than one electron. In addition, the theory cannot explain the fine structure in the spectra of hydrogen atoms and hydrogen-like species, such as the splitting of emission spectra in a magnetic field. In Bohr theory, it simply assumes but does not answer the question why angular momentum

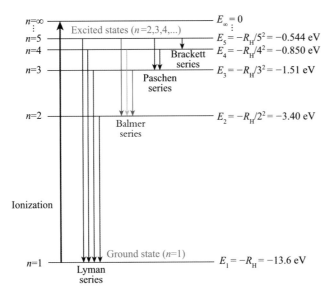

Figure 7.11 Energy-level diagram of the hydrogen atom.

图 7.11 氢原子的能级图。

$n = \infty$ 时，$r_n = a_0 n^2 = \infty$ 对应于距核无限远的轨道，$E_n = -R_H/n^2 = 0$ 是势能的参考零点。这意味着电子完全从氢原子中移除，成为自由电子。这时 H 原子被电离成 H^+ 阳离子。从基态 H 原子电离出一个电子所需的能量称为**电离能**（I），有

$$I = E_\infty - E_1 = 0 - (-R_H) = R_H \qquad (7.17)$$

关于电离能的更多讨论详见 **7.9 节**。

原子光谱是大量原子集合中的单个原子被激发到各种可能的激发态之后的原子**发射光谱**。受激原子随后通过发射能量等于两个相应能级差的光子而跃迁到较低能级。因此，原子能级的量子化导致了线状光谱。样品的发射光谱通常是将发射光经过单色器色散到检测器来进行检测，以确定光的强度相对于各个波长分量的关系。另一方面，通过基态原子样品的电磁辐射（如白光），也可以经过单色器色散到检测器来进行检测，观察到光的强度相对于原子吸收波长的关系，即为原子的**吸收光谱**。

发射光谱通常比吸收光谱更为复杂。激发的样品包含处于各种激发态的原子，每个激发态都能跃迁到几种较低能态中的任何一种。而吸收光谱的样品通常是冷的，只可能从基态开始跃迁。例如，冷的氢原子的吸收光谱仅包含莱曼谱系（$n=1$）。

类氢物种的玻尔理论

玻尔理论不仅适用于氢原子，也适用于**类氢物种**，即核外只有一个电子的任何原子或离子，如 He^+、Li^{2+}、Be^{3+}、B^{4+} 等。类氢物种是由两个带电粒子组成的简单体系：带负电荷（$-e$）的电子和带正电荷（$+Ze$）的原子核，其中 Z 是核电荷数或原子序数。对于类氢物种，玻尔理论的推导过程与氢原子类似，结论如下：

$$r_n = a_0 n^2 / Z \quad 且 \quad E_n = -R_H Z^2 / n^2 \qquad (7.18)$$

玻尔理论的局限性

尽管在解释氢原子和类氢物种方面获得了成功，玻尔理论仍有许多弱点。该理论无法解释多电子物种（即具有超过一个电子的原子和离子）的发射光谱。此外，该理论不能解释氢原子和类氢物种光谱的精细结构，如发射光谱在磁场中的分裂。玻尔理论只是假定，

is quantized. From a fundamental standpoint, Bohr theory is an uneasy mixture of classical and non-classical physics. Nevertheless, we should not underestimate the importance of Bohr theory as a scientific development because it gave the paradigm shift from classical mechanics, which is applicable to macroscopic systems, to the new quantum mechanics, which explains the nature of microscopic particles.

7.4 The Nature of Microscopic Particles

From the previous sections, we have already noticed some substantial differences between macroscopic and microscopic particles. Here in this section, we will show the differences in their natures as well as in their patterns of motion, which stimulate the new approach to quantum mechanics.

Wave-Particle Duality

Photon is a microscopic particle and let us start with the nature of photons. In history, the nature of light has been a long-debated issue since the end of the 17th century. Some scientists, represented by Isaac Newton, developed the corpuscular theory, arguing that the perfectly straight lines of reflection demonstrated the particle nature of light. Some other scientists, represented by Christiaan Huygens, mathematically refined the wave theory, demonstrating that light is an electromagnetic wave that can be described by the wave equation. In the mid-19th century, the wave theory of light began to dominate over the corpuscular theory since it could explain polarization phenomena of light that the alternatives could not. As discussed in **Section 7.2**, Einstein proposed the photon theory to explain the photoelectric effect in 1905. Later, scientists realized that separately neither corpuscular theory nor wave theory can fully explain the phenomena of light, but together they do. Electromagnetic radiation propagates following linear wave equations but can only be emitted or absorbed as discrete particles, thus acting as a wave and a particle simultaneously. In general, light shows wave-like properties in phenomena related to propagation, such as refraction, diffraction and interference, and particle-like properties when interacting with matter, such as photoelectric effect, absorption, and emission. Therefore, light appears to have a wave-particle dual nature.

In 1924, inspired by Einstein's photon theory, Louis de Broglie (1892—1987) proposed de Broglie hypothesis, which claimed that all matter exhibits wave-like behavior, called **matter wave**. By combining Einstein's mass-energy equation with the photon theory, de Broglie derived that for photons

$$E = mc^2 = h\nu$$

and

$$p = mc = \frac{h\nu}{c} = \frac{h}{\lambda}$$

In the above equations, both energy (E) and momentum (p) show particle-like properties of the photon, whereas both frequency (ν) and wavelength (λ) represent wave-like properties of the photon. They are connected to each other via the Planck constant h. De Broglie believed that these equations also apply to material particles as

$$E = h\nu \quad \text{and} \quad p = \frac{h}{\lambda} \tag{7.19}$$

where ν and λ are the frequency and wavelength of the matter wave associated with the material particles, respectively. The particle-like and wave-like properties of material particles are also connected to each other via the Planck constant h.

Figure 7.12 summaries the various relationships between E, p, ν and λ for both photons and material particles. It should be noticed that there are two velocities for a material particle: the group velocity v and the phase velocity u. The group velocity is the moving speed of the particle, so

$$E = mv^2/2 \quad \text{and} \quad p = mv$$

The phase velocity is the propagation speed of the corresponding matter wave and

$$u = \lambda\nu$$

但没有回答为什么角动量是量子化的问题。从根本上讲，玻尔理论是经典物理和非经典物理的简单混杂。然而，我们不应低估玻尔理论在科学发展中的重要性，因为它给出了从适用于宏观体系的经典力学转变为解释微观粒子特性的量子力学的科学范式。

7.4 微观粒子的特性

从前述各节中，我们已经注意到宏观和微观粒子之间的一些实质性差异。本节我们将展示它们在性质以及运动规律方面的差异，这促进了量子力学新方法的出现。

波粒二象性

光子是一种微观粒子，让我们从光的本质开始。历史上自 17 世纪末以来，光的本质一直是一个长期争论的问题。以艾萨克·牛顿为代表的一些科学家发展了微粒理论，认为光沿完美的直线反射，证明了其粒子本质。以克里斯蒂安·惠更斯为代表的其他一些科学家，从数学上完善了波动理论，证明光是可用波动方程描述的电磁波。19 世纪中叶，光的波动理论超越微粒理论开始占主导地位，因为它可以解释其他理论所不能解释的光的偏振现象。正如 **7.2 节**所述，爱因斯坦在 1905 年提出了光子理论来解释光电效应。后来科学家们意识到，单独的微粒理论和波动理论均不能完美地解释光的现象，但两者结合起来就可以解释。电磁辐射依照线性波动方程传播，但只能作为离散的粒子进行发射或吸收，因此可同时作为波和粒子。一般来说，光在与传播相关的现象（如折射、衍射和干涉）中表现出类似波的性质，而在与物质相互作用（如光电效应、吸收和发射）时表现出类似粒子的性质。因此，光具有波粒二象性。

1924 年，受爱因斯坦光子理论的启发，路易斯·德布罗意（1892—1987）提出了德布罗意假说，认为所有物质均存在类似波的行为，可称为**物质波**。通过将爱因斯坦的质能方程与光子理论相结合，德布罗意推导出光子的方程为

$$E = mc^2 = h\nu$$

$$p = mc = \frac{h\nu}{c} = \frac{h}{\lambda}$$

在上述方程中，能量（E）和动量（p）均为光子的类粒子性质，而频率（ν）和波长（λ）均为光子的类波性质，它们通过普朗克常数 h 相互关联。德布罗意认为这些方程也适用于实物粒子，有

$$E = h\nu \quad 且 \quad p = \frac{h}{\lambda} \tag{7.19}$$

其中 ν 和 λ 分别是与实物粒子相关联的物质波的频率和波长。实物粒子的类粒子性质和类波性质也可通过普朗克常数 h 相互关联。

图 7.12 总结了光子和实物粒子的 E、p、ν 和 λ 之间的各种关系。应该注意的是，实物粒子具有两种速度：群速度 v 和相速度 u。群速度是粒子的运动速度，因此

$$E = mv^2/2 \quad 且 \quad p = mv$$

相速度是对应物质波的传播速度，有

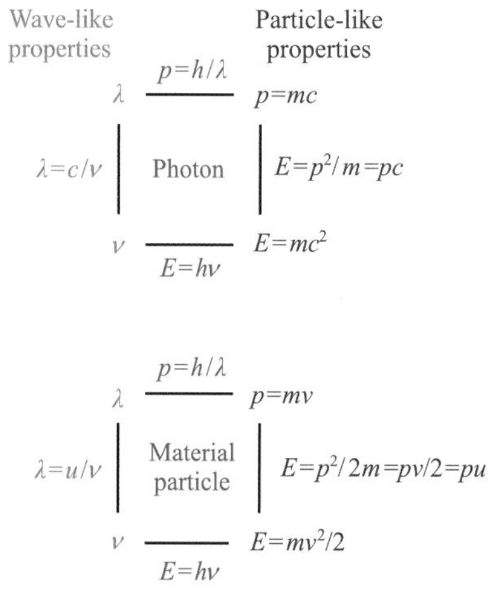

Figure 7.12 Various relationships between the wave-like properties and particle-like properties of both photons and material particles.

图 7.12 光子和实物粒子的类波性质与类粒子性质之间的各种关系。

Consequently,
$$E = \frac{p^2}{2m} = \frac{pv}{2} = \frac{hu}{\lambda} = pu$$

Therefore, $v = 2u$ for a material particle. However, $v = u = c$ for a photon.

In 1927, de Broglie hypothesis was confirmed for electrons with the observation of electron diffraction in two independent laboratories led by George Thomson and Clinton Davisson, respectively. The wavelength derived from the electron diffraction pattern matches the de Broglie wavelength of the electron calculated from **Equation (7.19)**. De Broglie was awarded the Nobel Prize in physics in 1929 and Thomson and Davisson shared the Nobel Prize in physics in 1937.

Later, diffraction patterns of other material particles such as α-particles, neutrons, some atoms and molecules were also observed, demonstrating that all material particles exhibit wave-particle duality. **Table 7.5** lists the de Broglie wavelengths of various material particles, from macroscopic to microscopic. The de Broglie wavelengths of many microscopic particles are usually comparable to those of X-rays, showing similar diffraction patterns as with X-rays (**Figure 7.13**) and exhibiting significant wave-particle duality. However, the de Broglie wavelengths of macroscopic objects, such as bullets and ping-pong balls, are too small to measure. So, macroscopic objects behave mainly like particles with negligible wave-like properties, and their motions can be described adequately by classical mechanics.

Table 7.5   De Broglie Wavelengths of Various Material Particles
表 7.5   各种实物粒子的德布罗意波长

Material Particle（实物粒子）	Mass（质量）/kg	Velocity（速度）/(m s^{-1})	Wavelength（波长）/pm
Bullet（子弹）	1.0×10^{-2}	1.0×10^{3}	6.6×10^{-23}
Ping-pong ball（乒乓球）	2.7×10^{-3}	1.0×10^{1}	2.5×10^{-20}
Xe atom（Xe 原子）	2.3×10^{-25}	2.4×10^{2}	12
He atom（He 原子）	6.6×10^{-27}	1.1×10^{3}	91
10,000 V electron（10 000 V 电子）	9.1×10^{-31}	5.9×10^{7}	12
100 V electron（100 V 电子）	9.1×10^{-31}	5.9×10^{6}	120
1 V electron（1 V 电子）	9.1×10^{-31}	5.9×10^{5}	1200

The Uncertainty Principle

According to classical mechanics, we can make precise predictions on the trajectory of an object, and there is no limit to the achieved accuracy of the predictions. All physical behavior of the object can be predicted with undoubted certainty, meaning that the uncertainty is zero. However, it was found that this is not the case when microscopic particles are concerned. In 1927, Werner Heisenberg (1901—1976) established **Heisenberg uncertainty principle** through hypothetical experiments and asserted an absolute, fundamental limit on the combined accuracy of certain pairs of physical variables of a particle known as **complementary variables**, such as the 1D position (x) - momentum (p_x), and time (t) - energy(E), as

$$\Delta x \cdot \Delta p_x \geqslant \hbar / 2 \quad (7.20)$$

and

$$\Delta t \cdot \Delta E \geqslant \hbar / 2 \quad (7.21)$$

where Δ stands for the uncertainty in the corresponding variables and $\hbar$ is the reduced Planck constant. **Equation (7.20)** demonstrates that we cannot measure position and momentum with great accuracy simultaneously. The more accurately the position of some particle is determined, the less accurately its momentum can be predicted, and vice versa. In the microscopic world, things must always be "fuzzy".

Historically, the uncertainty principle has been confused with a related effect called the observer effect, which notes that measurements of certain systems cannot be made without affecting the system. It is clear that the uncertainty principle is inherent in the properties of all wave-like systems, and that it states a fundamental property of quantum systems regardless of observers.

Table 7.6 lists the uncertainty in position with a relative uncertainty of 1% in velocity for various material particles, from macroscopic to microscopic. As can be seen, comparing to their own sizes, the uncertainty in the position of microscopic particles is so significant that it is not possible to predict their trajectories precisely. However, the uncertainty in the position of macroscopic objects is negligible, and thus

$$u = \lambda \nu$$

相应地
$$E = \frac{p^2}{2m} = \frac{pv}{2} = \frac{hu}{\lambda} = pu$$

因此，对于实物粒子，$v = 2u$；而对于光子，$v = u = c$。

1927 年，乔治·汤姆森和克林顿·戴维森领导的两个独立实验室，通过对电子衍射的观察证实了德布罗意假说。从电子衍射图案得出的波长与**式（7.19）**计算的电子的德布罗意波长相吻合。德布罗意于 1929 年获得诺贝尔物理学奖，汤姆森和戴维森于 1937 年共享诺贝尔物理学奖。

随后，其他实物粒子（如 α 粒子、中子、一些原子和分子）的衍射图案均被观察到，表明所有实物粒子均表现出波粒二象性。**表 7.5** 列出了从宏观到微观的各种实物粒子的德布罗意波长。许多微观粒子的德布罗意波长通常与 X 射线的波长相当，显示出与 X 射线类似的衍射图案（**图 7.13**），并表现出显著的波粒二象性。然而，宏观物体（如子弹和乒乓球）的德布罗意波长太小，无法测量。因此，宏观物体主要表现为类波性质可忽略不计的粒子，其运动可用经典力学充分描述。

不确定性原理

根据经典力学，我们可以对物体的运动轨迹做出精确的预测，且预测的准确度没有限制。物体的所有物理行为均可确定无疑地预测出，这意味着其不确定性为零。然而，当涉及微观粒子时，人们发现情况并非如此。1927 年，维尔纳·海森堡（1901—1976）通过假想实验建立了**海森堡不确定性原理**，断言了粒子的某些物理变量对的组合精度存在绝对的基本极限。这些物理变量对称为**互补变量**，如一维的位置 (x) - 动量 (p_x) 对，以及时间 (t) - 能量 (E) 对，有

$$\Delta x \cdot \Delta p_x \geq \hbar / 2 \quad (7.20)$$
$$\Delta t \cdot \Delta E \geq \hbar / 2 \quad (7.21)$$

其中 Δ 表示相应变量的不确定性，$\hbar$ 是约化普朗克常数。**式（7.20）** 表明，我们无法同时高准确度地测量位置和动量。某个粒子的位置确定得越准确，则其动量预测得就越不准确；反之亦然。在微观世界中，事物必须总是"模糊的"。

历史上不确定性原理常与一种称为观察者效应（即所谓的测不准原理）的相关效应相混淆。观察者效应指的是，无法在不影响体系的情况下对其进行测量。显然，不确定性原理是所有类波体系的固有特性，它陈述了量子体系的基本特性，与观察者无关。

表 7.6 列出了速度的相对不确定度为 1% 的各种实物粒子（从宏观到微观）的位置不确定度。可以看到，与其自身尺寸相比，微观粒子位置的不确定度非常大，因此不能精确预测其轨迹。而宏观物体位置的不确定度可忽略不计，因此基于经典力学对其进行精确的预测是有效的。

2003 年，小泽正直提议将海森堡不确定性不等式重新表述为

$$\Delta x \cdot \Delta p_x + \Delta x \cdot \sigma p_x + \sigma x \cdot \Delta p_x \geq \hbar / 2$$

其中 σx 和 σp_x 分别是位置和动量的标准偏差。2012 年，小泽正直与长谷川佑二通过中子自旋实验的精确测量，共同证实了上述小泽不等式，即存在 $\Delta x \cdot \Delta p_x < \hbar / 2$，但 $\Delta x \cdot \Delta p_x + \Delta x \cdot \sigma p_x + \sigma x \cdot \Delta p_x \geq \hbar / 2$ 的情况。

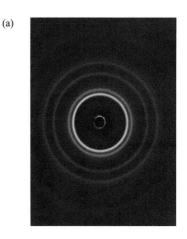

Figure 7.13 Diffraction patterns of (a) X-rays by metal foil and (b) electrons by metal foil, confirming the wave-like properties of electrons.

图 7.13 (a) X 射线与 (b) 电子通过金属箔的衍射图案，表明电子具有类波性质。

it is valid to make precise predictions based on classical mechanics.

Table 7.6   Uncertainty in Position with A Relative Uncertainty of 1% in Velocity for Various Material Particles
表 7.6   速度相对不确定度为 1% 的各种实物粒子的位置不确定度

Material Particle（实物粒子）	Mass（质量）/kg	Δv_x with 1% Relative Uncertainty in Velocity（速度相对不确定度为 1% 时的 Δv_x）/(m s^{-1})	Uncertainty in Position Δx（位置不确定度 Δx）/m
Bullet（子弹）	1.0×10^{-2}	1.0×10^{1}	5.3×10^{-34}
Ping-pong ball（乒乓球）	2.7×10^{-3}	1.0×10^{-1}	2.0×10^{-31}
Xe atom（Xe 原子）	2.3×10^{-25}	2.4	9.6×10^{-11}
He atom（He 原子）	6.6×10^{-27}	1.1×10^{1}	7.3×10^{-10}
10,000 V electron（10 000 V 电子）	9.1×10^{-31}	5.9×10^{5}	9.8×10^{-11}
100 V electron（100 V 电子）	9.1×10^{-31}	5.9×10^{4}	9.8×10^{-10}
1 V electron（1 V 电子）	9.1×10^{-31}	5.9×10^{3}	9.8×10^{-9}

In 2003, Masanao Ozawa proposed a reformulation of Heisenberg uncertainty inequality as

$$\Delta x \cdot \Delta p_x + \Delta x \cdot \sigma p_x + \sigma x \cdot \Delta p_x \geqslant \hbar/2$$

where σx and σp_x are the standard deviation for position and momentum, respectively. In 2012, together with Yuji Hasegawa, Masanao Ozawa confirmed the above Ozawa inequality by precise measurements in a neutron spin experiment that $\Delta x \cdot \Delta p_x < \hbar/2$ but $\Delta x \cdot \Delta p_x + \Delta x \cdot \sigma p_x + \sigma x \cdot \Delta p_x \geqslant \hbar/2$.

Matter Wave and Standing Wave

De Broglie hypothesis suggests that all matter displays wave-like properties and thus are matter waves with the de Broglie wavelength associated with the momentum through the Planck constant. This wave-particle duality of matter sets limited accuracy in determining the complementary variables, such as position and momentum of the particle, connected by the uncertainty principle. Matter waves are also called **probability waves** because the wave-like properties of microscopic particles behave as a wave of probability, the intensity of which corresponds to the probability of observing the given particle. Similar to the 1D wave equation shown in **Equation (7.2)**, the 3D wave equation is also a function of the 3D spatial coordinate q and time coordinate t, as $y = f(q,t)$. Therefore, there is a probability distribution in both space and time for the probability wave.

How to understand and interpret microscopic particles as probability waves? We know that the mathematical concept of probability is used to numerically describe how likely a random event is to occur based on numerous statistical data. For example, tossing coins randomly results in a probability of 50% facing up and 50% facing down. However, if you just toss a coin once, it can be either facing up or down, and the result is not affected by probability. Probability only matters when numerous statistical data are available. For microscopic particles, numerous statistical data are available in most cases, and thus we can describe their properties by probability waves. Meanwhile, probability waves also mean that we cannot predict the properties of microscopic particles with 100% certainty, but only give probabilities with uncertainty governed by the uncertainty principle.

Taking electrons as an example, the probability distribution of electrons in both space and time is related to the statistics of numerous electron motions. The numerous electron motions can be observed either with a strong electron flow over a short period of time, or with a weak electron flow over a long period of time. Experimentally, the diffraction patterns of both are consistent. For the electron diffraction pattern shown in **Figure 7.13(b)**, as an analogy to the diffraction pattern of the X-rays shown in **Figure 7.13(a)**, it is not difficult to understand that the bright and dark diffraction fringes correspond to high and low intensities of the probability wave, respectively. Provided with such an electron diffraction pattern, one cannot distinguish whether it is collected with a strong electron flow over a short period of time, or with a weak electron flow over a long period of time. If it is with a strong electron flow over a short period of time, or ultimately, with numerous electrons at some moment of time, the bright and dark fringes appear in regions with larger and smaller numbers of electrons, respectively. If it is with a weak electron flow over a long period of time, or ultimately, with a single electron over an enormously long period of time, the bright and dark fringes appear at regions where the electron has more or less chance to reach, respectively. The intensity of the probability

物质波与驻波

德布罗意假说表明，所有物质均表现出类波性质，因此是具有德布罗意波长（可通过普朗克常数与其动量相关联）的物质波。物质的波粒二象性给互补变量（由不确定性原理所关联，如粒子的位置和动量）的确定设置了有限的精度。由于微观粒子的类波性质表现为一种概率的波，其强度对应于观察到指定粒子的概率，因此物质波也称**概率波**。与**式（7.2）**所示的一维波动方程类似，三维波动方程也是三维的空间坐标 q 和时间坐标 t 的函数，$y = f(q,t)$。因此，概率波在空间和时间上均存在概率分布。

如何理解和阐释以概率波形式存在的微观粒子？我们知道，数学概念中的概率，是基于大量统计数据、以数字的形式来描述随机事件发生的可能性。例如，随机抛硬币的概率为 50% 朝上，50% 朝下。然而，如果你只抛一次硬币，它既可以朝上也可以朝下，其结果不受概率的影响。只有存在大量可用的统计数据时，概率才重要。对于微观粒子而言，在大多数情况下均存在大量可用的统计数据，因此我们可以用概率波来描述它们的性质。同时，概率波也意味着我们不能 100% 确定无疑地预测微观粒子的性质，而只能给出其不确定性（由不确定性原理控制）的概率。

以电子为例，电子在空间和时间上的概率分布，与大量电子运动的统计有关。大量电子的运动既可以短时间内用强电子流，也可以长时间内用弱电子流观察到。实验上两者的衍射图案完全一致。**图7.13(b)** 所示的电子衍射图案，作为**图 7.13(a)** 所示的 X 射线衍射图案的类比，不难理解，明、暗衍射条纹分别对应于概率波的高、低强度。如果提供一张这样的电子衍射图案，人们无法区分它是短时间内用强电子流收集到的,还是长时间内用弱电子流收集到的。如果它是短时间内的强电子流，或极限情况下在某个时刻的极其大量电子，则明、暗条纹分别出现在电子数量较多或较少的区域。如果它是长时间内的弱电子流，或极限情况下在极其长时间内的单电子，则明、暗条纹分别出现在电子有较多或较少机会到达的区域。给定区域和给定时间段内概率波的强度，对应于在该时间段和该区域内观察到给定粒子的概率。

由上述推导，我们现在应该能够认识到玻尔原子理论的一个基本错误，是将电子约束在具有固定半径的轨道上。微观粒子的运动没有明确的轨道，但在空间和时间上均表现出一定的概率分布，观察到粒子的概率与其概率波的强度呈正相关。

前面我们根据波的本质将其分类为机械波、电磁波、概率波和引力波等；我们还根据振荡方向与传播方向的关系，将波分成横波和纵波。这里，我们可以根据是否发生移动，将波分为**行波**和**驻波**。绳波或水波是行波的典型例子，其中波的每个部分均随空间和时间移动 [**图 7.14(a)**]。相反，驻波是随时间振荡，但其振幅的峰值分布不在空间中移动的波 [**图 7.14(b)**]。驻波上的不同点具有不同的振幅。振幅绝对值最小（通常为零）的位置称为**节点**，振幅绝对值最大的位置称为**波腹**（或反节点）。驻波的节点完全不发生移动。乐器的振动波是驻波的典型例子。

物质波是一种驻波，必须应用某些周期性边界条件来引入量子化。例如，玻尔推测，电子要在圆形轨道上稳定，必须表现为驻

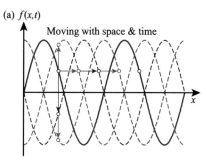

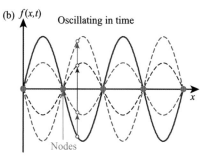

Figure 7.14 Traveling waves versus standing waves. (a) A traveling wave in which every portion of the wave moves with space and time. (b) A standing wave that oscillates in time but the peak amplitude profile of which does not move in space. Different colors indicate the oscillation of waves in time.

图 7.14 行波与驻波：(a) 波的每个部分均随空间和时间移动的行波；(b) 随时间振荡，但振幅的峰值分布不在空间中移动的驻波。不同颜色代表波随时间的振荡。

wave at a given region and over a given period of time corresponds to the probability of observing the given particle within that region at that time.

From the above deduction, we should now realize that a fundamental error in Bohr atomic theory was to constrain an electron to an orbit of a fixed radius. The motion of microscopic particles has no definite orbit but displays a certain probability distribution in both space and time, with the probability of observing the particle positively relating to the intensity of the probability wave.

Previously, we have classified waves according to their nature into mechanical waves, electromagnetic waves, probability waves, and gravitational waves, etc. We have also classified waves according to the relation between directions of oscillation and propagation into transverse waves and longitudinal waves. Here, we can classify waves according to the fact that they travel or not into **traveling waves** and **standing waves** (also called stationary waves). Waves on strings or water waves are typical examples of traveling waves, in which every portion of the wave moves with space and time [**Figure 7.14(a)**]. By contrast, standing waves are waves that oscillate in time but the peak amplitude profile of which do not move in space [**Figure 7.14(b)**]. Different points in the standing wave have different amplitude. The locations where the absolute value of the amplitude is minimum (normally zero) are called **nodes**, and the locations at which the absolute value of the amplitude is maximum are called **antinodes**. The nodes of a standing wave do not move at all. Vibrational waves in musical instruments are typical examples of standing waves.

Matter waves are a type of standing waves in which certain periodic boundary conditions must be applied to introduce quantization. For example, Bohr speculated that for an electron to be stable in a circular orbit, it has to be represented by a standing wave. In order for successive waves to reinforce one another, an integral number of wavelengths (λ) has to fit the circumference of the orbit ($2\pi r$), serving as periodic boundary conditions to introduce the quantum number n, as

$$2\pi r = n\lambda$$

Combining with the angular momentum condition

$$L = mvr = n\hbar$$

we have

$$\lambda = \frac{2\pi r}{n} = \frac{2\pi r \hbar}{mvr} = \frac{h}{mv} = \frac{h}{p}$$

This gives rise to the wavelength-momentum relationship in wave-particle duality. The total number of nodes in this case equals $2n$. If the number of wavelengths is nonintegral, successive waves tend to cancel each other. The crest in one part of the wave may overlap with a trough in another part of the wave, and there is no resultant stable wave at all.

7.5 Quantum Mechanical Model of Hydrogen-Like Species

In previous sections, we have learnt various ideas leading to the advert of quantum mechanics. In this section, we will introduce the quantum mechanical model of hydrogen-like species, starting with some basic concepts and postulates. Because the derivation of quantum mechanics is rather complicated, we will only involve some necessary parts in order to understand the results that will be discussed in the subsequent sections.

Quantum Mechanics and Wave Mechanics

Classical mechanics is the collection of theories that describes many aspects of nature at a macroscopic scale but is not sufficient for describing them at microscopic scales. **Quantum mechanics**, by contrast, is a fundamental branch of physics which deals with physical phenomena at nanoscopic or lower scales where the action is on the order of the Planck constant. Most theories in classical mechanics can be derived from quantum mechanics as an approximation valid at macroscopic scale.

Quantum mechanics was developed in the mid-1920s based on the quantum theory. It is widely recognized as one of the three most important scientific findings in the 20th century (with the other two being

的形式。为使连续波相互增强，波长（λ）的整数倍必须与轨道的周长（$2\pi r$）相匹配，作为引入量子数 n 的周期性边界条件，有

$$2\pi r = n\lambda$$

结合角动量条件

$$L = mvr = n\hbar$$

有

$$\lambda = \frac{2\pi r}{n} = \frac{2\pi r \hbar}{mvr} = \frac{h}{mv} = \frac{h}{p}$$

这就得出了波粒二象性中的波长 - 动量关系，此时节点总数等于 $2n$。如果波长数为非整数，则连续波往往会相互抵消。一部分波的波峰可能与另一部分波的波谷重叠，无法形成稳定的合波。

7.5 类氢物种的量子力学模型

在前述章节中，我们已经了解了导致量子力学产生的各种想法。本节我们将从一些基本概念和假定出发，介绍类氢物种的量子力学模型。由于量子力学的推导相当复杂，我们只涉及一些必要的部分，以便于理解后续章节中将要讨论的结果。

量子力学与波动力学

经典力学是在宏观尺度上描述自然界许多现象的理论集合，但不足以在微观尺度上进行描述。相比之下，**量子力学**是处理纳米及以下尺度物理现象的物理学基本分支，其作用通常在普朗克常数量级。经典力学的大多数理论均可从量子力学推导得出，作为在宏观尺度上的有效近似。

量子力学是在 20 世纪 20 年代中期基于量子理论发展起来的。它被广泛认为是 20 世纪最为重要的三项科学发现之一（另外两项是进化论和 DNA 的双螺旋结构）。现代量子力学可以用各种专门开发的数学公式来表述，包括但不限于矩阵力学（由维尔纳·海森堡发明）、波动力学（由欧文·薛定谔发明）和路径积分形式（由理查德·费曼发明）。在所有这些形式中，用波动力学方程来描述体系类波性质的**波动力学**，是最为常用的形式。在本教材中，我们也只讨论波动力学。

量子力学的两个基本假定

量子力学共有五个基本假定，这里我们先介绍其中两个。第一个基本假定是：一个体系的所有状态均可由**波函数** $\Psi(q,t)$ 完全描述，其中 q 为空间坐标（使用粗体字母表示它是一个多维向量），t 为时间坐标。该假定定义了体系的波函数，它可以完全描述体系的所有性质。因此，我们的最终目标就是求解波函数，因为一旦求解了波函数，就可以从中获得体系的所有性质。

波函数 $\Psi(q,t)$ 称为**含时波函数**，因为它与时间 t 相关。如果对体系如何随时间变化不感兴趣，可以使用**不含时波函数**，用 $\psi(q)$ 表示，二者的关系为

$$\Psi(q,t) = \psi(q)T(t) \tag{7.22}$$

其中 $T(t)$ 是体系的**时间演化波函数**。

theory of evolution and the double helix structure of DNA). The modern quantum mechanics is formulated in various specially developed mathematical formulations, including but not limited to matrix mechanics (invented by Werner Heisenberg), wave mechanics (invented by Erwin Schrödinger), and path integral formulation (invented by Richard Feynman). In all those formulations, the **wave mechanics**, in which a wave dynamical equation is applied to describe the wave-like properties of the system, is the most commonly used form. We will only discuss wave mechanics throughout this textbook.

Two Postulates of Quantum Mechanics

There are altogether five postulates in quantum mechanics, and here we will first introduce two of them. The very 1st postulate is that the state of a system is fully described by a **wave function** $\Psi(\boldsymbol{q},t)$, where $\boldsymbol{q}$ is the spatial coordinate (we use a bold letter to indicate that it is a multi-dimensional vector) and t is the time coordinate. This postulate defines a wave function of the system, which can fully describe all properties of the system. Therefore, our eventual goal is to solve the wave functions, because once the wave function is solved, we can obtain all the properties of the system from it.

The wave function $\Psi(\boldsymbol{q},t)$ is called the **time-dependent wave function** because it is dependent on time t. If how the system changes in time is not interested, we can use the **time-independent wave function**, denoted as $\psi(\boldsymbol{q})$, instead, with the relationship

$$\Psi(\boldsymbol{q},t) = \psi(\boldsymbol{q})T(t) \qquad (7.22)$$

where $T(t)$ is the **time evolution wave function** of the system.

The 4th postulate of quantum mechanics states that the probability that a particle will be found in a small volume $d\tau$ at the spatial point $\boldsymbol{q}$ is proportional to $|\psi(\boldsymbol{q})|^2 d\tau$. This postulate defines the physical meaning of $|\psi(\boldsymbol{q})|^2$ to be the probability density of finding a particle at the spatial point $\boldsymbol{q}$. The wave function $\psi(\boldsymbol{q})$ of a system is a complex-valued function with no direct physical meaning. Only the square of the wave function, which is the square of the modulus of the complex and is a real function, has a physical meaning as the probability density.

$$|\psi(\boldsymbol{q})|^2 = \psi^*(\boldsymbol{q})\psi(\boldsymbol{q}) = \psi^2$$

As a simple illustrative mathematical example, if $\psi = a + ib$, then $\psi^* = a - ib$, and

$$|\psi|^2 = (a-ib)(a+ib) = a^2 + b^2$$

where i is the imaginary unit with $i^2 = -1$.

Probability and Probability Density

As discussed earlier in probability wave, the mathematical concept of probability is used to numerically describe how likely a random event is to occur based on numerous statistical data. In quantum mechanics, the probability means the possibility of finding a particle inside a small volume at a given spatial point. In order to calculate this probability, we need to define a concept called **probability density**.

It may be difficult to understand the concept of probability density for a beginner. Let us take regular density (also called mass density) as an analogy. We know that the mass (m) of an object is proportional to the volume (V) as well as to the mass density (ρ) of the object, and the mass density is defined as the mass per unit volume, as

$$\rho = \frac{m}{V}$$

Similarly, the probability (dP) is proportional to the volume (dτ) as well as to the probability density ($|\psi(\boldsymbol{q})|^2$), and the probability density is defined as the probability of finding a particle per unit volume at a given point $\boldsymbol{q}$, as

$$|\psi(\boldsymbol{q})|^2 = \frac{dP}{d\tau}$$

量子力学的第四个基本假定指出，在空间某点 q 处的小体积 $\mathrm{d}\tau$ 内找到粒子的概率与 $|\psi(q)|^2 \mathrm{d}\tau$ 成正比。该假定将 $|\psi(q)|^2$ 的物理意义，定义为在空间某点 q 处找到粒子的概率密度。体系的波函数 $\psi(q)$ 是一个自身没有直接物理意义的复数函数。只有波函数的平方（即其复数模的平方）是实函数，才具有概率密度的物理意义。

$$|\psi(q)|^2 = \psi^*(q)\psi(q) = \psi^2$$

作为一个简单的数学示例，如果 $\psi = a + \mathrm{i}b$，则 $\psi^* = a - \mathrm{i}b$，且有

$$|\psi|^2 = (a-\mathrm{i}b)(a+\mathrm{i}b) = a^2 + b^2$$

其中 i 是虚数单位，$\mathrm{i}^2 = -1$。

概率与概率密度

正如前面在概率波中所讨论的，数学概念中的概率，是基于大量统计数据、用数值来描述随机事件发生的可能性。量子力学中的概率，意味着在给定空间点的小体积内找到粒子的可能性。为了计算这一概率，我们需要定义一个称为**概率密度**的概念。

初学者可能难以理解概率密度的概念。让我们以常规密度（也称质量密度）作为类比。我们知道，物体的质量（m）与体积（V）和质量密度（ρ）成正比，质量密度定义为单位体积内的质量，即

$$\rho = \frac{m}{V}$$

类似地，概率（$\mathrm{d}P$）与体积（$\mathrm{d}\tau$）和概率密度（$|\psi(q)|^2$）成正比，概率密度定义为在给定点 q 处单位体积内找到粒子的概率，即

$$|\psi(q)|^2 = \frac{\mathrm{d}P}{\mathrm{d}\tau}$$

因此
$$\mathrm{d}P = |\psi(q)|^2 \mathrm{d}\tau$$

这解释了为什么在第四个基本假定中，空间点 q 处的小体积 $\mathrm{d}\tau$ 内找到粒子的概率与 $|\psi(q)|^2 \mathrm{d}\tau$ 成正比。

对于具有等密度分布的物体，$m = \rho V$。如果物体的密度在不同部位分布不均匀，则

$$m = \sum_i \rho_i V_i$$

如果物体的密度对于每个无穷小的体积均呈不均匀分布，则

$$\mathrm{d}m = \rho \mathrm{d}V \quad \text{且} \quad m = \int \mathrm{d}m = \int \rho \mathrm{d}V$$

类似地，空间中找到粒子的概率密度对于每个无穷小的体积均呈不均匀分布，总概率应通过对全空间进行积分来计算，有

$$P = \int \mathrm{d}P = \int |\psi(q)|^2 \mathrm{d}\tau$$

由于在全空间找到粒子的总概率必为 1，有

$$\int_{-\infty}^{+\infty} |\psi(q)|^2 \mathrm{d}\tau = 1 \tag{7.23}$$

式 (7.23) 称为**归一化条件**。当满足归一化条件时，波函数称为是"归一化的"。任何量子力学体系的波函数，无论含时、不含时，还是时间演化波函数，都应该是归一化的。

Consequently,
$$dP = |\psi(q)|^2 d\tau$$
and this explains why the probability of finding a particle in a small volume $d\tau$ at the spatial point q is proportional to $|\psi(q)|^2 d\tau$ in the 4th postulate.

For an object with isodensity, $m = \rho V$. If the density of an object is unevenly distributed for different parts, then
$$m = \sum_i \rho_i V_i$$
If the density of an object is unevenly distributed for every infinitesimal volume, then
$$dm = \rho dV \quad \text{and} \quad m = \int dm = \int \rho dV$$
Similarly, the probability density of finding a particle in space is unevenly distributed for every infinitesimal volume, and the total probability should be calculated by taking integrals over the entire space, as
$$P = \int dP = \int |\psi(q)|^2 d\tau$$
As the total probability of finding a particle in the entire space must be 1, we have
$$\int_{-\infty}^{+\infty} |\psi(q)|^2 d\tau = 1 \tag{7.23}$$

Equation (7.23) is called the **normalization condition**, and the wave function is said to be "**normalized**" when the normalization condition is met. The wave functions of any quantum mechanical systems, no matter time-dependent, time-independent, or time evolution wave functions, should all be normalized.

The Hamiltonian Operator

Before going into details about the wave dynamical equation of the system, we need first to introduce a concept called operators. An **operator** is a symbol for an instruction to carry out some action (called an **operation**) on a function, and can be represented by an angle sign on top of the corresponding symbol, such as $\hat{\Omega}$. This definition seems to be quite unclear, but will become clearer after further elucidation. Here, we only consider two types of operators: the multiplication operator and the differentiation operator. Starting from a simple 1D system with a wave function $\psi(x)$, the 1D **position operator** ($\hat{x}$) is a multiplication operator given by $\hat{x}\psi(x) = x \cdot \psi(x)$. This position operator $\hat{x}$ is an instructional symbol to carry out the multiplication operation by x on any wave function of the system, no matter what wave functions the system may have. Therefore, in short, we write
$$\hat{x} = x \tag{7.24}$$
Meanwhile, the 1D **momentum operator** ($\hat{p}_x$) is a differentiation operator given by
$$\hat{p}_x = \frac{\hbar}{i} \frac{d}{dx} \tag{7.25}$$
This formula of $\hat{p}_x$ can be derived from another postulate of quantum mechanics, but since this postulate does not have significant relationship to what will be discussed in this textbook, here, we just accept the above formula as valid.

In classical mechanics, we know that the kinetic energy of a system is related to the momentum by $E_k = mv^2/2 = p^2/2m$. Similarly, the kinetic energy operator ($\hat{T}_x$) of a 1D system is given by
$$\hat{T}_x = \frac{\hat{p}_x^2}{2m} = \frac{1}{2m}\left(\frac{\hbar}{i}\frac{d}{dx}\right)^2 = -\frac{\hbar^2}{2m}\frac{d^2}{dx^2} \tag{7.26}$$

In 3D, the **kinetic energy operator** ($\hat{T}$) can be written as
$$\hat{T} = \frac{\hat{p}^2}{2m} = \frac{\hat{p}_x^2 + \hat{p}_y^2 + \hat{p}_z^2}{2m} = -\frac{\hbar^2}{2m}\left\{\frac{\partial^2}{\partial x^2} + \frac{\partial^2}{\partial y^2} + \frac{\partial^2}{\partial z^2}\right\} = -\frac{\hbar^2}{2m}\nabla^2 \tag{7.27}$$

where $\nabla^2 = \frac{\partial^2}{\partial x^2} + \frac{\partial^2}{\partial y^2} + \frac{\partial^2}{\partial z^2}$ is called the **Laplacian operator**, and partial differentials are used instead of ordinary

哈密顿算符

在详细讨论体系的波动力学方程之前，我们首先需要引入一个称为算符的概念。**算符**是对函数执行某些行为（称为**操作**）的指示符，可以用在相应符号的顶部加上角标来表示，如 $\hat{\Omega}$。这一定义看起来相当不清楚，但在进一步阐述后会变得更为明确。这里我们只考虑两类算符：乘法算符和微分算符。从简单一维体系的波函数 $\psi(x)$ 开始，**一维位置算符**（$\hat{x}$）是由 $\hat{x}\psi(x) = x \cdot \psi(x)$ 给出的乘法算符。无论体系可能具有什么样的波函数，该位置算符 $\hat{x}$ 均是一个对体系的任何波函数执行乘以 x 的乘法操作的指示符。因此可简写为

$$\hat{x} = x \tag{7.24}$$

同时，一维**动量算符**（$\hat{p}_x$）是由下式给出的微分算符：

$$\hat{p}_x = \frac{\hbar}{i}\frac{d}{dx} \tag{7.25}$$

这个 $\hat{p}_x$ 的算符形式可以从量子力学的另一个基本假定推导出，但由于该假定与本教材所讨论的内容没有显著关系，因此这里我们直接接受上述公式有效。

在经典力学中，我们知道体系的动能与动量的关系为 $E_k = mv^2/2 = p^2/2m$。类似地，一维体系的动能算符（$\hat{T}_x$）为

$$\hat{T}_x = \frac{\hat{p}_x^2}{2m} = \frac{1}{2m}\left(\frac{\hbar}{i}\frac{d}{dx}\right)^2 = -\frac{\hbar^2}{2m}\frac{d^2}{dx^2} \tag{7.26}$$

三维动能算符（$\hat{T}$）可写为

$$\hat{T} = \frac{\hat{p}^2}{2m} = \frac{\hat{p}_x^2 + \hat{p}_y^2 + \hat{p}_z^2}{2m} = -\frac{\hbar^2}{2m}\left\{\frac{\partial^2}{\partial x^2} + \frac{\partial^2}{\partial y^2} + \frac{\partial^2}{\partial z^2}\right\} = -\frac{\hbar^2}{2m}\nabla^2 \tag{7.27}$$

其中 $\nabla^2 = \frac{\partial^2}{\partial x^2} + \frac{\partial^2}{\partial y^2} + \frac{\partial^2}{\partial z^2}$ 称为**拉普拉斯算符**，由于它是三维的，因此用偏微分代替常微分。三维体系的势能算符（$\hat{V}$）可表示为

$$\hat{V} = V(\boldsymbol{q}) \tag{7.28}$$

其中 $V(\boldsymbol{q})$ 是体系的势能，是空间坐标 $\boldsymbol{q}$ 的函数。

在量子力学中，体系的**哈密顿算符**（$\hat{H}$）是对应于体系总能量的算符，包括动能和势能。因此

$$\hat{H} = \hat{T} + \hat{V} = -\frac{\hbar^2}{2m}\nabla^2 + V(\boldsymbol{q}) \tag{7.29}$$

薛定谔方程的一般形式

一般来说，如果一个算符对某个函数进行操作的结果等于同一函数乘以一个常数，即

$$\hat{\Omega}f = \omega f \tag{7.30}$$

其中 ω 为常数（必须是实数），那么称 ω 为算符 $\hat{\Omega}$ 的**本征值**，称函数 f 为算符 $\hat{\Omega}$ 的**本征函数**，称上述方程为**本征方程**。例如，令 $\hat{\Omega} = d/dx$，$f(x) = e^{mx}$，则

$$\hat{\Omega}f(x) = \frac{d}{dx}\left(e^{mx}\right) = me^{mx} = mf(x)$$

differentials because it is 3D. The **potential energy operator** ($\hat{V}$) of a 3D system can be represented by
$$\hat{V} = V(q) \tag{7.28}$$
where $V(q)$ is the potential energy of the system that is a function with respect to the spatial coordinate q.

In quantum mechanics, the **Hamiltonian operator** ($\hat{H}$) of a system is an operator corresponding to the total energy of the system, including both kinetic and potential energies. Therefore,
$$\hat{H} = \hat{T} + \hat{V} = -\frac{\hbar^2}{2m}\nabla^2 + V(q) \tag{7.29}$$

The General Schrödinger Equation

In general, if the outcome of an operator operating on a function gives rise to the same function multiplied by a constant, as
$$\hat{\Omega}f = \omega f \tag{7.30}$$
where ω is a constant (must be a real number), then we call ω the **eigenvalue** of the operator $\hat{\Omega}$, the function f the **eigenfunction** of the operator $\hat{\Omega}$, and the above equation an **eigenvalue equation**. For example, let $\hat{\Omega} = d/dx$ and $f(x) = e^{mx}$, then
$$\hat{\Omega}f(x) = \frac{d}{dx}(e^{mx}) = me^{mx} = mf(x)$$
Because m is a constant, the outcome of $\hat{\Omega}$ operating on the function $f(x)$ gives rise to the same function $f(x)$ multiplied by the constant m. Then $f(x) = e^{mx}$ is the eigenfunction of the operator $\hat{\Omega} = d/dx$ with an eigenvalue of m.

The quantum mechanical wave functions Ψ of a system should always be eigenfunctions of the Hamiltonian operator $\hat{H}$, with eigenvalues that are constants corresponding to the total energy E of the system. The corresponding eigenvalue equation can be written as $\hat{H}\Psi = E\Psi$. The more general form of the above equation was proposed by Erwin Schrödinger (1887—1961) in 1926, and was named after him as **Schrödinger equation**, given by
$$\hat{H}\Psi(q,t) = i\hbar \frac{\partial \Psi(q,t)}{\partial t} = E\Psi(q,t) \tag{7.31}$$
The above wave dynamic **equation (7.31)** is provided as the 5th postulate of quantum mechanics. Because $\Psi(q,t)$ is the time-dependent wave function, **Equation (7.31)** is also called the **general time-dependent Schrödinger equation**.

If how the system changes in time is not interested, we use the time-independent wave function $\psi(q)$ instead. The **general time-independent Schrödinger equation** can be written as
$$\hat{H}\psi(q) = E\psi(q) \tag{7.32}$$
For a single non-relativistic particle, the above general equation can be expended as
$$\hat{H}\psi(q) = \left[-\frac{\hbar^2}{2m}\nabla^2 + V(q)\right]\psi(q) = E\psi(q) \tag{7.33}$$

Although complex in expression, the above Schrödinger equation can be solved precisely for hydrogen-like species. By solving Schrödinger equation, both the wave functions and the energies of the system can be obtained. Schrödinger equation is a second-order partial differential equation, the full solutions to which require advanced mathematical skills and will not be presented here. However, we can still gain some experience by solving part of Schrödinger equation to obtain the time evolution wave function $T(t)$.

Recall that in **Equation (7.22)** we have $\Psi(q,t) = \psi(q)T(t)$. Let us apply this equation into the general time-dependent Schrödinger equation, as

由于 m 为常数，对函数 $f(x)$ 进行 $\hat{\Omega}$ 操作的结果，等于同一函数 $f(x)$ 乘以常数 m。那么 $f(x)=\mathrm{e}^{mx}$ 是算符 $\hat{\Omega}=\mathrm{d}/\mathrm{d}x$ 的本征函数，其本征值为 m。

体系的量子力学波函数 Ψ 应始终是哈密顿算符 $\hat{H}$ 的本征函数，其本征值为对应于体系总能量 E 的常数。相应的本征方程可写为 $\hat{H}\Psi=E\Psi$。上述方程更为普适的形式由欧文·薛定谔（1887—1961）于 1926 年提出，并以他的名字命名为**薛定谔方程**，即

$$\hat{H}\Psi(\boldsymbol{q},t)=\mathrm{i}\hbar\frac{\partial\Psi(\boldsymbol{q},t)}{\partial t}=E\Psi(\boldsymbol{q},t) \tag{7.31}$$

上述波动力学**方程（7.31）**即为量子力学的第五个基本假定。由于 $\Psi(\boldsymbol{q},t)$ 是含时波函数，**式（7.31）**也称**含时薛定谔方程的一般形式**。

如果对体系如何随时间变化不感兴趣，可使用不含时波函数 $\psi(\boldsymbol{q})$ 来替代。**不含时薛定谔方程的一般形式**可写为

$$\hat{H}\psi(\boldsymbol{q})=E\psi(\boldsymbol{q}) \tag{7.32}$$

对于单个非相对论粒子，上述方程的一般形式可展开为

$$\hat{H}\psi(\boldsymbol{q})=\left[-\frac{\hbar^2}{2m}\nabla^2+V(\boldsymbol{q})\right]\psi(\boldsymbol{q})=E\psi(\boldsymbol{q}) \tag{7.33}$$

尽管表达式很复杂，但上述薛定谔方程对于类氢物种可以精确求解。通过求解薛定谔方程，可以得到体系的波函数和能量。薛定谔方程是一个二阶偏微分方程，其完整求解过程需要使用较为高级的数理技巧，这里暂不介绍。但我们仍可通过部分求解薛定谔方程得到时间演化波函数 $T(t)$，来获得一些体验。

回顾**式（7.22）**，我们有 $\Psi(\boldsymbol{q},t)=\psi(\boldsymbol{q})T(t)$。将此等式应用于含时薛定谔方程的一般形式，有

$$\hat{H}\Psi(\boldsymbol{q},t)=\mathrm{i}\hbar\frac{\partial\Psi(\boldsymbol{q},t)}{\partial t}=\mathrm{i}\hbar\frac{\partial[\psi(\boldsymbol{q})T(t)]}{\partial t}=\mathrm{i}\hbar\psi(\boldsymbol{q})\frac{\mathrm{d}T(t)}{\mathrm{d}t}=E\psi(\boldsymbol{q})T(t)$$

从左到右的第三个等号有效，是因为 $\psi(\boldsymbol{q})$ 只与 $\boldsymbol{q}$ 相关而不依赖于 t，偏微分可用只含一个变量的常微分替代。在两侧除以 $\psi(\boldsymbol{q})$，可消除空间坐标 $\boldsymbol{q}$，有

$$\mathrm{i}\hbar\frac{\mathrm{d}T(t)}{\mathrm{d}t}=ET(t)$$

将所有含 $T(t)$ 的项移至左侧，所有其他项移至右侧，有

$$\frac{\mathrm{d}T(t)}{T(t)}=-\frac{\mathrm{i}E\mathrm{d}t}{\hbar}$$

两边同时积分，有

$$\int\frac{\mathrm{d}T(t)}{T(t)}=\int\left(-\frac{\mathrm{i}E\mathrm{d}t}{\hbar}\right)$$

$$\ln T(t)=-\frac{\mathrm{i}Et}{\hbar}+C$$

$$T(t)=A\exp(-\mathrm{i}Et/\hbar) \tag{7.34}$$

参数 A 可通过归一化条件 $\int T^2(t)\mathrm{d}t=1$ 求解。因此，时间演化波函数 $T(t)$ 必须采用上述形式，才能符合不含时薛定谔方程。该解法的关

$$\hat{H}\Psi(q,t) = i\hbar\frac{\partial \Psi(q,t)}{\partial t} = i\hbar\frac{\partial[\psi(q)T(t)]}{\partial t} = i\hbar\psi(q)\frac{dT(t)}{dt} = E\psi(q)T(t)$$

The third equal sign from left to right is valid because $\psi(q)$ is dependent only on q but not on t, and that the partial differential can be replaced by the ordinary differential with only one variable. Dividing $\psi(q)$ on both side, we now eliminate the spatial coordinate q, as

$$i\hbar\frac{dT(t)}{dt} = ET(t)$$

Moving all terms of $T(t)$ to the left and all other terms to the right, we have

$$\frac{dT(t)}{T(t)} = -\frac{iEdt}{\hbar}$$

Integrations on both sides gives

$$\int\frac{dT(t)}{T(t)} = \int\left(-\frac{iEdt}{\hbar}\right)$$

$$\ln T(t) = -\frac{iEt}{\hbar} + C$$

$$T(t) = A\exp(-iEt/\hbar) \tag{7.34}$$

The parameter A can be solved by the normalization condition, in which $\int T^2(t)dt = 1$. Therefore, the time evolution wave function $T(t)$ must take the above form in order to satisfy the time-independent Schrödinger equation. The key to this solution is to apply $\Psi(q,t) = \psi(q)T(t)$, which means that the function $\psi(q)$ in terms of spatial variable q and the function $T(t)$ in terms of time variable t are separable. This method of solution is called **separation of variables**, and is used to solve not only the time evolution wave function but also the time-independent wave function of hydrogen-like species.

Schrödinger Equation for Hydrogen-Like Species

The above derivations in this section are general conclusions applicable to all systems. Since the time evolution wave function $T(t)$ is solved, we only need to consider the time-independent wave function $\psi(q)$ described in **Equation (7.33)**. The kinetic energy term in the Hamiltonian operator depends only on the number of particles in the system. However, the potential energy term varies for different systems, and is the most important part in the Hamiltonian operator and Schrödinger equation.

From here to **Section 7.6**, we will focus on a specific system, hydrogen-like species, which consist of only two charged particles: an electron (m_e) with negative charge $-e$ and a nucleus (m_N) with positive charge $+Ze$. In this system, the kinetic energy operator contains two parts

$$\hat{T} = -\frac{\hbar^2}{2m_e}\nabla_e^2 - \frac{\hbar^2}{2m_N}\nabla_N^2 = -\frac{\hbar^2}{2\mu}\nabla^2$$

where μ is the reduced mass given by

$$\frac{1}{\mu} = \frac{1}{m_e} + \frac{1}{m_N}$$

In mathematics, the Laplacian operator ∇^2 can be transformed from Cartesian coordinates (x,y,z) to spherical polar coordinates (r,θ,ϕ) with relationship shown in **Figure 7.15**, as

$$\nabla^2 = \frac{\partial^2}{\partial x^2} + \frac{\partial^2}{\partial y^2} + \frac{\partial^2}{\partial z^2} = \frac{1}{r^2}\left[\frac{\partial}{\partial r}\left(r^2\frac{\partial}{\partial r}\right) + \Lambda^2\right]$$

where the **Legendre operator** Λ^2 is the angular part of the Laplacian operator given by

$$\Lambda^2 = \frac{1}{\sin\theta}\frac{\partial}{\partial \theta}\left(\sin\theta\frac{\partial}{\partial \theta}\right) + \frac{1}{\sin^2\theta}\frac{\partial^2}{\partial \phi^2}$$

键在于应用 $\Psi(\boldsymbol{q},t)=\psi(\boldsymbol{q})T(t)$，这意味着空间变量 $\boldsymbol{q}$ 的函数 $\psi(\boldsymbol{q})$ 与时间变量 t 的函数 $T(t)$ 是可分离的。这种求解方法称为**变量分离法**，不仅用于求解时间演化波函数，还可用于求解类氢物种的不含时波函数。

类氢物种的薛定谔方程

本节的上述推导，是适用于所有体系的一般性结论。由于已经求解了时间演化波函数 $T(t)$，只需要考虑**式（7.33）**描述的不含时波函数 $\psi(\boldsymbol{q})$。哈密顿算符的动能项仅取决于体系中的粒子数，但势能项对于不同体系而言是不同的，它也是哈密顿算符及薛定谔方程中最为重要的部分。

从这里到 **7.6 节**，我们将聚焦于一个特定的体系：类氢物种。它仅由两个带电粒子组成：带负电荷 $-e$ 的电子（m_e）和带正电荷 $+Ze$ 的原子核（m_N）。在该体系中，动能算符包含两部分

$$\hat{T}=-\frac{\hbar^2}{2m_e}\nabla_e^2-\frac{\hbar^2}{2m_N}\nabla_N^2=-\frac{\hbar^2}{2\mu}\nabla^2$$

其中 μ 为折合质量，有

$$\frac{1}{\mu}=\frac{1}{m_e}+\frac{1}{m_N}$$

数学上，拉普拉斯算符 ∇^2 可以从笛卡尔坐标系 (x,y,z) 转换为球极坐标系 (r,θ,ϕ)，其关系如**图 7.15** 所示，有

$$\nabla^2=\frac{\partial^2}{\partial x^2}+\frac{\partial^2}{\partial y^2}+\frac{\partial^2}{\partial z^2}=\frac{1}{r^2}\left[\frac{\partial}{\partial r}\left(r^2\frac{\partial}{\partial r}\right)+\Lambda^2\right]$$

其中**勒让德算符** Λ^2 是拉普拉斯算符的角度部分，有

$$\Lambda^2=\frac{1}{\sin\theta}\frac{\partial}{\partial\theta}\left(\sin\theta\frac{\partial}{\partial\theta}\right)+\frac{1}{\sin^2\theta}\frac{\partial^2}{\partial\phi^2}$$

在球极坐标系中，r 是径向坐标，θ 和 ϕ 是角度坐标。体系的势能为电子和原子核之间的静电势，因此势能算符可写为

$$\hat{V}=V(r)=-\frac{Ze^2}{r}$$

其中 cgs 单位制中的常数 $4\pi\varepsilon_0=1$，已省略。

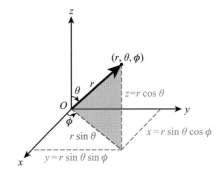

Figure 7.15 The relationship between spherical polar coordinates (r,θ,ϕ) and Cartesian coordinates (x, y, z).

图 7.15 球极坐标系 (r,θ,ϕ) 与笛卡尔坐标系 (x, y, z) 之间的关系图。

类氢物种的哈密顿算符可写为

$$\hat{H}=-\frac{\hbar^2}{2\mu}\nabla^2+V=-\frac{\hbar^2}{2\mu r^2}\left[\frac{\partial}{\partial r}\left(r^2\frac{\partial}{\partial r}\right)+\Lambda^2\right]-\frac{Ze^2}{r}$$

相应的薛定谔方程为

$$-\frac{\hbar^2}{2\mu r^2}\left[\frac{\partial}{\partial r}\left(r^2\frac{\partial\psi}{\partial r}\right)+\Lambda^2\psi\right]-\frac{Ze^2}{r}\psi=E\psi \quad (7.35)$$

$$\frac{1}{r^2}\frac{\partial}{\partial r}\left(r^2\frac{\partial\psi}{\partial r}\right)+\frac{1}{r^2\sin\theta}\frac{\partial}{\partial\theta}\left(\sin\theta\frac{\partial\psi}{\partial\theta}\right)+\frac{1}{r^2\sin^2\theta}\frac{\partial^2\psi}{\partial\phi^2}=-\frac{2\mu}{\hbar^2}\left(E+\frac{Ze^2}{r}\right)\psi$$

不含时波函数 $\psi(\boldsymbol{q})$ 在笛卡尔坐标系中表示为 $\psi(x,y,z)$，在球极坐标系中表示为 $\psi(r,\theta,\phi)$。这里你可能想知道，为什么我们采用球极坐标系而非笛卡尔坐标系的薛定谔方程。原因是类氢物种的薛定谔方程可以在球极坐标系中求解，尽管表达式看起来更为复杂。用于求解薛定谔方程的方法称为**变量分离法**，其原理与求解时间演化波

In spherical polar coordinates, r is the radial coordinate, θ and ϕ are the angular coordinates. The potential energy of the system is the electrostatic potential between the electron and the nucleus, and thus the potential energy operator can be written as

$$\hat{V} = V(r) = -\frac{Ze^2}{r}$$

where the constant $4\pi\varepsilon_0 = 1$ in cgs unit system is omitted.

The Hamiltonian operator for hydrogen-like species can be written as

$$\hat{H} = -\frac{\hbar^2}{2\mu}\nabla^2 + V = -\frac{\hbar^2}{2\mu r^2}\left[\frac{\partial}{\partial r}\left(r^2 \frac{\partial}{\partial r}\right) + \Lambda^2\right] - \frac{Ze^2}{r}$$

and the corresponding Schrödinger equation is

$$-\frac{\hbar^2}{2\mu r^2}\left[\frac{\partial}{\partial r}\left(r^2 \frac{\partial \psi}{\partial r}\right) + \Lambda^2 \psi\right] - \frac{Ze^2}{r}\psi = E\psi \tag{7.35}$$

$$\frac{1}{r^2}\frac{\partial}{\partial r}\left(r^2 \frac{\partial \psi}{\partial r}\right) + \frac{1}{r^2 \sin\theta}\frac{\partial}{\partial \theta}\left(\sin\theta \frac{\partial \psi}{\partial \theta}\right) + \frac{1}{r^2 \sin^2\theta}\frac{\partial^2 \psi}{\partial \phi^2} = -\frac{2\mu}{\hbar^2}\left(E + \frac{Ze^2}{r}\right)\psi$$

The time-independent wave function $\psi(q)$ is represented as $\psi(x,y,z)$ in Cartesian coordinates and $\psi(r,\theta,\phi)$ in spherical polar coordinates. Here, you may wonder why we use Schrödinger equation in spherical polar coordinates instead of Cartesian coordinates. The reason is that Schrödinger equation for hydrogen-like species can be solved in spherical polar coordinates, although the expression seems to be more complicated. The method used to solve this Schrödinger equation is also the separation of variables, with principles similar to what we used to solve the time evolution wave function but with more complicated details. Here, we will only provide some simplified solutions.

Previously, we assumed that the variables q and t are separable, and that $\Psi(q,t) = \psi(q)T(t)$. Here, similarly, we assume that the variables r and θ, ϕ are separable, and thus the time-independent wave function for hydrogen-like species can be separated as

$$\psi_{n,l,m_l}(r,\theta,\phi) = R_{n,l}(r)Y_{l,m_l}(\theta,\phi) \tag{7.36}$$

where $R(r)$ is called the **radial wave function**, which is dependent only on r but not on θ and ϕ, and $Y(\theta,\phi)$ is called the **angular wave function**, which is dependent only on θ and ϕ but not on r. The subscripts n, l, and m_l are three quantum numbers to which the corresponding wave functions are related. Further solution by separation of variables can be found in the extended reading materials at the end of this chapter. The solution results of Schrödinger equation for hydrogen-like species will be provided and discussed in the next section.

7.6 Quantum Mechanical Results of Hydrogen-Like Species

In the previous section, we have discussed the quantum mechanical model of hydrogen-like species, and showed some simplified solutions to Schrödinger equation. Although the full solutions to the time-independent Schrödinger **equation (7.35)** are not provided here, we must understand the full solutions contain three parts:
1) Three quantum numbers n, l, and m_l arising from certain periodic boundary conditions during the solution;
2) The quantized energy E_n, which is dependent only on n but not on l and m_l;
3) The wave functions $\psi_{n,l,m_l}(r,\theta,\phi)$ (called the **atomic orbitals**), which are the product of radial wave functions $R_{n,l}(r)$ and angular wave functions $Y_{l,m_l}(\theta,\phi)$.

Three Quantum Numbers

From various boundary conditions during the solution, three quantum numbers n, l, and m_l are

函数的原理类似，但细节更为复杂。这里我们只提供一些简化的解法。

之前我们假定变量 q 和 t 可分离，且有 $\Psi(q,t)=\psi(q)T(t)$。类似地，我们假定变量 r 和 θ、ϕ 可分离，因此类氢物种的不含时波函数可分离为

$$\psi_{n,l,m_l}(r,\theta,\phi) = R_{n,l}(r)Y_{l,m_l}(\theta,\phi) \tag{7.36}$$

其中 $R(r)$ 称为**径向波函数**，仅与 r 有关而不依赖于 θ 和 ϕ；$Y(\theta,\phi)$ 称为**角度波函数**，只与 θ 和 ϕ 有关而不依赖于 r。下标 n、l 和 m_l 是与相应波函数有关的三个量子数。进一步的变量分离解法详见章末拓展阅读材料。类氢物种薛定谔方程的求解结论将在下一节给出并进行讨论。

7.6 类氢物种的量子力学结论

上一节我们讨论了类氢物种的量子力学模型，并给出了薛定谔方程的一些简化解法。虽然这里没有提供不含时薛定谔**方程 (7.35)** 的完整解法，但我们必须理解其完整求解结果包含三个部分：
1) 三个量子数 n、l 和 m_l，由求解过程中的某些周期性边界条件产生；
2) 量子化能量 E_n，仅与 n 有关而不依赖于 l 和 m_l；
3) 波函数 $\psi_{n,l,m_l}(r,\theta,\phi)$（称为**原子轨道**），是径向波函数 $R_{n,l}(r)$ 与角度波函数 $Y_{l,m_l}(\theta,\phi)$ 的乘积。

三个量子数

表 7.7 总结了根据求解过程中的各种边界条件引入的三个量子数 n、l 和 m_l。第一个数 n 称为**主量子数**，可能的取值为正整数，即

$$n = 1, 2, 3, 4, \cdots \tag{7.37}$$

n 代表主电子层，其中 n=1, 2, 3, 4, 5, 6, 7, …分别对应于 K，L，M，N，O，P，Q，…壳层。由于能量 E_n 只与 n 有关而不依赖于其他量子数，主量子数 n 决定了原子轨道的能量。

第二个数 l 是轨道角动量量子数，简称**角量子数**。l 的取值可以是零或任何不大于 $n-1$ 的正整数，即

$$l = 0, 1, 2, \cdots, n-1 \tag{7.38}$$

其中 n 为主量子数。对于每个 n 值，都存在 n 个不同的 l 值。由于能量 E_n 与 l 无关，所有具有相同 n 值但不同 l 值的轨道均具有相同的能量，即 l 值存在 n 重简并度。回顾 **3.5 节**，简并意味着不同的排列具有相同的能量。l 代表电子亚层，其中 l=0, 1, 2, 3, …分别对应于 s, p, d, f, …亚层。角量子数 l 决定了原子轨道的形状。例如，s 轨道总是球形，p 轨道总是哑铃形。

第三个数 m_l 是**磁量子数**，可取从 $-l$ 到 l 的任何整数，包括零，即

$$m_l = -l, -l+1, \cdots, -1, 0, 1, \cdots, l-1, l \tag{7.39}$$

其中 l 为角量子数。对于每个 l 值，都存在 $2l+1$ 个不同的 m_l 值，因此 m_l 存在 $2l+1$ 重简并度。这 $2l+1$ 重简并度对应于电子亚层中 $2l+1$ 个不同的轨道，每个轨道均呈不同的空间取向。例如，总是有一个 s 轨道（l=0 时 $2l+1$=1）、三个 p 轨道（l=1 时 $2l+1$=3）和五个 d 轨道（l=2 时 $2l+1$=5）。在外加磁场下，这 $2l+1$ 重简并度被打破，意味着这 $2l+1$ 个轨道的能量不再相同，这也是 m_l 称为磁量子数的原因。

introduced and summarized in **Table 7.7**. The first number n is called the **principal quantum number**, which may have a positive integer value, as

$$n = 1, 2, 3, 4, \cdots \tag{7.37}$$

and represents the principal electronic shell, with n = 1, 2, 3, 4, 5, 6, and 7, … corresponding to the K, L, M, N, O, P, and Q, … shells, respectively. Because the energy E_n is dependent only on n but not on the other quantum numbers, the principal quantum number n determines the energy of the atomic orbitals.

Table 7.7 Three Quantum Numbers Derived from Schrödinger Equation
表 7.7 从薛定谔方程中推导得出的三个量子数

Quantum Number(量子数)	Symbol(符号)	Value（取值）	Meaning（含义）
Principal quantum number （主量子数）	n	n = 1, 2, 3, 4, ⋯	Principal electronic shell, orbital energy（对应主电子层，体现轨道能量）
Angular quantum number （角量子数）	l	l = 0, 1, 2, 3, ⋯, n−1 (s, p, d, f)	Subshell, orbital shape, n-fold degeneracy（对应电子亚层，体现轨道形状，具有 n 重简并）
Magnetic quantum number （磁量子数）	m_l	m_l = 0, ±1, ±2, ⋯, ±l	Orbital number, orbital orientation, $2l+1$-fold degeneracy, splitting in an applied magnetic field（对应轨道数量，体现轨道的空间取向，具有 $2l+1$ 重简并，在外磁场中发生裂分）

The second number l is the orbital angular momentum quantum number, or **angular quantum number** for short. l may be zero or any positive integer no larger than $n-1$, as

$$l = 0, 1, 2, \cdots, n-1 \tag{7.38}$$

where n is the principal quantum number. For each n, there are n different values of l. Since the energy E_n is independent of l, all orbitals with the same n but different l have the same energy, or there is n-fold degeneracy in l. Recall from **Section 3.5** that degeneracy means different arrangements with the same energy. l represents the electron subshells, with l = 0, 1, 2, and 3, … corresponding to the s, p, d, and f, … subshells, respectively. The angular quantum number l determines the shape of the atomic orbitals. For example, s orbitals are always spherical shape and p orbitals are always dumbbell shape.

The third number m_l is the **magnetic quantum number**, which may be any integer ranging from $-l$ to l, including zero, as

$$m_l = -l, -l+1, \cdots, -1, 0, 1, \cdots, l-1, l \tag{7.39}$$

where l is the angular quantum number. For each l, there are $2l+1$ different values of m_l and consequently, $2l+1$-fold degeneracy in m_l. This $2l+1$-fold degeneracy corresponds to $2l+1$ different orbitals in the subshell, each exhibiting different spatial orientations. For example, there is always one s ($2l+1 = 1$ for $l = 0$) orbital, three p orbitals ($2l+1 = 3$ for $l = 1$), and five d orbitals ($2l+1 = 5$ for $l = 2$). Under an applied external magnetic field, this $2l+1$-fold degeneracy is broken, meaning that the energies for these $2l+1$ orbitals are not the same any more. This is also the reason why m_l is called the magnetic quantum number.

The above three quantum numbers are closely linked with one another. **Table 7.8** lists all possible values of l and m_l for n = 1, 2, and 3. Considering both l and m_l, there is altogether n^2-fold degeneracy for each n.

Table 7.8 The Value of Quantum Numbers and the Number of Atomic Orbitals in the First Three Principal Electron Shells
表 7.8 前三个主电子层的量子数取值及原子轨道数

Orbital Symbol（轨道符号）	1s	2s	2p_z	2p_x, 2p_y	3s	3p_z	3p_x, 3p_y	3$d_{x^2-y^2}$	3d_{xz}, 3d_{yz}	3d_{xy}, 3$d_{x^2-y^2}$
n	1	2	2	2	3	3	3	3	3	3
l	0	0	1	1	0	1	1	2	2	2
m_l	0	0	0	±1	0	0	±1	0	±1	±2
Number of orbitals in subshells（亚层轨道数）	1	1	3		1	3		5		
Number of orbitals in principal shells（主层轨道数）	1	4			9					

Energy Levels for Hydrogen-Like Species

The calculated allowed orbital energies from the time-independent Schrödinger equation for hydrogen-like species are quantized, as

上述三个量子数彼此紧密联系。**表 7.8** 列出了 $n=1, 2, 3$ 时 l 和 m_l 的所有可能值。同时考虑 l 和 m_l，则每个 n 值共有 n^2 重简并度。

类氢物种的能级

从类氢物种不含时薛定谔方程中计算得的允许的轨道能量是量子化的，为

$$E_n = -R_H Z^2 / n^2$$

这与从玻尔理论推导出的轨道能量公式 **(7.18)** 一致。注意虽然中文均翻译成"轨道"，但在玻尔理论中称为"orbit"，而在量子力学中称为"orbital"，强调了这两个概念之间的区别。玻尔理论假定电子只能在具有固定半径的确定圆形轨道中运动，而量子力学的轨道没有固定半径，在空间和时间上均呈现出一定的概率分布。对于类氢物种，同一主电子层内的所有轨道都具有相同的能量，无论属于哪个亚层。**图 7.16** 给出了类氢物种的能级图及其主电子层和电子亚层的排布。

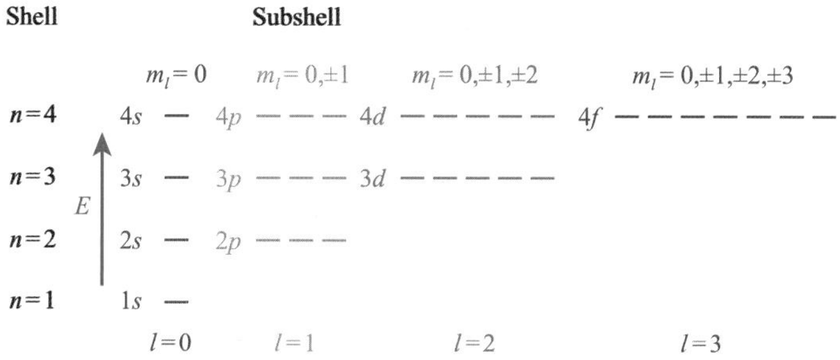

Figure 7.16 The energy-level diagram and arrangement of shells and subshells for hydrogen-like species.

图 7.16 类氢物种的能级图及其主电子层和电子亚层的排布。

类氢物种的波函数

表 7.9 列出了类氢物种归一化的径向波函数 $R_{n,l}(r)$ 和角度波函数 $Y_{l,m_l}(\theta,\phi)$ 的形式。通过将它们直接相乘，可以得到任意轨道的不含时波函数 $\psi_{n,l,m_l}(r,\theta,\phi)$。

这里可能出现的一个令人困惑的问题是，为什么**表 7.9** 中列出的所有波函数均不含虚数单位 i，而我们此前曾提到过，体系的不含时波函数 $\psi(q)$ 应为复数函数。答案是虚数单位 i 已通过采用共轭复数的线性组合加以消除了。例如，三个 $2p$ 轨道（$2p_x$、$2p_y$ 和 $2p_z$）的量子数为：$n=2$、$l=1$ 及 $m_l=-1$、0、1。它们之间的关系是 $2p_z$ 对应于 $n=2$、$l=1$ 及 $m_l=0$ 的轨道，而 $2p_x$ 和 $2p_y$ 均来自 $n=2$、$l=1$ 及 $m_l=\pm1$ 的轨道的线性组合。

从**表 7.9** 中我们看到

$$R(2p) = R_{21} = \frac{1}{2\sqrt{6}}\left(\frac{Z}{a_0}\right)^{3/2} \rho e^{-\rho/2} \quad \text{且} \quad Y(p_z) = Y_{10} = \left(\frac{3}{4\pi}\right)^{1/2} \cos\theta$$

其中 R_{21} 和 Y_{10} 是量子数形式的表示法，$\rho = 2Zr/na_0$，$a_0 = 52.93$ pm 为玻尔半径，与玻尔理论的值相同。$2p_z$ 的波函数可计算为

$$\psi(2p_z) = \psi_{210} = R(2p)Y(p_z) = \left[\frac{1}{2\sqrt{6}}\left(\frac{Z}{a_0}\right)^{3/2} \rho e^{-\rho/2}\right]\left[\left(\frac{3}{4\pi}\right)^{1/2} \cos\theta\right]$$

角度波函数 Y_{11} 和 Y_{1-1} 未在**表 7.9** 中列出，这里作为示例提供，为

$$Y_{11} = -\left(\frac{3}{8\pi}\right)^{1/2} \sin\theta e^{i\phi} \quad \text{且} \quad Y_{1-1} = \left(\frac{3}{8\pi}\right)^{1/2} \sin\theta e^{-i\phi}$$

$$E_n = -R_H Z^2 / n^2$$

This is the same as **Equation (7.18)**, the formula derived for orbit energies from Bohr atomic theory. Note that it is called "orbit" in Bohr theory and "orbital" in quantum mechanics, emphasizing on the difference in these two concepts. Bohr theory assumes that the electrons can only move in definite circular orbits with fixed radii, however, quantum mechanical orbitals have no fixed radii but exhibit a certain probability distribution in both space and time. For hydrogen-like species, all the orbitals within a principal electronic shell have the same energy, no matter to which subshell they belong. **Figure 7.16** shows an energy-level diagram and the arrangement of shells and subshells for hydrogen-like species.

Wave Functions for Hydrogen-Like Species

Table 7.9 lists the forms of normalized radial wave function $R_{n,l}(r)$ and angular wave function $Y_{l,m_l}(\theta,\phi)$ for hydrogen-like species. By directly multiplying them together, we can obtain the time-independent wave function $\psi_{n,l,m_l}(r,\theta,\phi)$ for any orbital of interest.

Table 7.9   Normalized Radial Wave Function $R_{n,l}(r)$ and Angular Wave Function $Y_{l,m_l}(\theta,\phi)$ for Hydrogen-Like Species

表 7.9   类氢物种归一化的径向波函数 $R_{n,l}(r)$ 和角度波函数 $Y_{l,m_l}(\theta,\phi)$ 的形式

Radial Wave Function（径向波函数）$R_{n,l}(r)$ [a]	Angular Wave Function（角度波函数）$Y_{l,m_l}(\theta,\phi)$
$R(1s) = R_{10} = 2\left(\dfrac{Z}{a_0}\right)^{3/2} e^{-\rho/2}$	$Y(s) = Y_{00} = \left(\dfrac{1}{4\pi}\right)^{1/2}$
$R(2s) = R_{20} = \dfrac{1}{2\sqrt{2}}\left(\dfrac{Z}{a_0}\right)^{3/2}(2-\rho)e^{-\rho/2}$	$Y(p_z) = Y_{10} = \left(\dfrac{3}{4\pi}\right)^{1/2}\cos\theta$
$R(2p) = R_{21} = \dfrac{1}{2\sqrt{6}}\left(\dfrac{Z}{a_0}\right)^{3/2}\rho e^{-\rho/2}$	$Y(p_x) = \left(\dfrac{3}{4\pi}\right)^{1/2}\sin\theta\cos\phi$
$R(3s) = R_{30} = \dfrac{1}{9\sqrt{3}}\left(\dfrac{Z}{a_0}\right)^{3/2}(6-6\rho+\rho^2)e^{-\rho/2}$	$Y(p_y) = \left(\dfrac{3}{4\pi}\right)^{1/2}\sin\theta\sin\phi$
$R(3p) = R_{31} = \dfrac{1}{9\sqrt{6}}\left(\dfrac{Z}{a_0}\right)^{3/2}(4-\rho)\rho e^{-\rho/2}$	$Y(d_{z^2}) = Y_{20} = \left(\dfrac{5}{16\pi}\right)^{1/2}(3\cos^2\theta - 1)$
$R(3d) = R_{32} = \dfrac{1}{9\sqrt{30}}\left(\dfrac{Z}{a_0}\right)^{3/2}\rho^2 e^{-\rho/2}$	$Y(d_{xz}) = \left(\dfrac{15}{16\pi}\right)^{1/2}\sin 2\theta\cos\phi$
	$Y(d_{yz}) = \left(\dfrac{15}{16\pi}\right)^{1/2}\sin 2\theta\sin\phi$
	$Y(d_{x^2-y^2}) = \left(\dfrac{15}{16\pi}\right)^{1/2}\sin^2\theta\cos 2\phi$
	$Y(d_{xy}) = \left(\dfrac{15}{16\pi}\right)^{1/2}\sin^2\theta\sin 2\phi$

[a] $\rho = 2Zr/na_0$

One confusion question that may arise here is that why all the wave functions listed in **Table 7.9** do not contain the imaginary unit i since we have mentioned previously that the time-independent wave function $\psi(q)$ of a system should be a complex-valued function. The answer is that the imaginary unit i has been eliminated by taking linear combinations of the conjugate complexes. For example, the quantum numbers for three 2p orbitals ($2p_x$, $2p_y$, and $2p_z$) are: $n = 2$, $l = 1$, and $m_l = -1$, 0, and 1. The relationship between them is that $2p_z$ corresponds to the orbital with $n = 2$, $l = 1$, and $m_l = 0$, and $2p_x$ and $2p_y$ arise from the linear combinations of the orbitals with $n = 2$, $l = 1$, and $m_l = \pm 1$.

From **Table 7.9** we see that

$$R(2p) = R_{21} = \dfrac{1}{2\sqrt{6}}\left(\dfrac{Z}{a_0}\right)^{3/2}\rho e^{-\rho/2} \quad \text{and} \quad Y(p_z) = Y_{10} = \left(\dfrac{3}{4\pi}\right)^{1/2}\cos\theta$$

$Y(p_x)$ 和 $Y(p_y)$ 可由 Y_{11} 和 Y_{1-1} 线性组合得到，有

$$Y(p_x) = \frac{1}{\sqrt{2}}(Y_{1-1} - Y_{11}) = \frac{1}{\sqrt{2}}\left(\frac{3}{8\pi}\right)^{1/2}\sin\theta\left(e^{-i\phi} + e^{i\phi}\right) = \left(\frac{3}{4\pi}\right)^{1/2}\sin\theta\cos\phi$$

$$Y(p_y) = \frac{i}{\sqrt{2}}(Y_{1-1} + Y_{11}) = \frac{i}{\sqrt{2}}\left(\frac{3}{8\pi}\right)^{1/2}\sin\theta\left(e^{-i\phi} - e^{i\phi}\right) = \left(\frac{3}{4\pi}\right)^{1/2}\sin\theta\sin\phi$$

其中 $e^{-i\phi} + e^{i\phi} = 2\cos\phi$ 且 $e^{-i\phi} - e^{i\phi} = -2i\sin\phi$。$2p_x$ 和 $2p_y$ 的波函数可计算为

$$\psi(2p_x) = R(2p)Y(p_x) = \left[\frac{1}{2\sqrt{6}}\left(\frac{Z}{a_0}\right)^{3/2}\rho e^{-\rho/2}\right]\left[\left(\frac{3}{4\pi}\right)^{1/2}\sin\theta\cos\phi\right]$$

$$\psi(2p_y) = R(2p)Y(p_y) = \left[\frac{1}{2\sqrt{6}}\left(\frac{Z}{a_0}\right)^{3/2}\rho e^{-\rho/2}\right]\left[\left(\frac{3}{4\pi}\right)^{1/2}\sin\theta\sin\phi\right]$$

类氢物种的 s、p、d 轨道

现在我们来说明三种主要轨道类型（s、p、d 轨道）以三维曲面形式表示的概率密度分布。回顾一下，轨道是波函数的线性组合，而波函数是薛定谔方程的数学解。波函数 ψ 自身没有物理意义，但 ψ^2 表示与在小体积 $d\tau$ 中找到电子的概率 $\psi^2 d\tau$ 成正比的概率密度。这里我们的目标是，基于**表 7.9** 所列波函数，获得对轨道的半定量的理解。

1) s 轨道

我们注意到**表 7.9** 中，任意 s 轨道的角度波函数 $Y(s) = \sqrt{1/4\pi}$ 总为常数，与 θ 和 ϕ 无关。这意味着所有 s 轨道都必须是球形的，因为三维空间中只有球形与角度坐标完全无关。$Y(s)$ 的值源自归一化条件，即 $\int Y^2 d\tau = 1$。

类氢物种的 $1s$ 轨道波函数可由 $n=1$ 和 $\rho = 2Zr/a_0$ 计算，为

$$\psi(1s) = R(1s)Y(s) = 2\left(\frac{Z}{a_0}\right)^{3/2}e^{-Zr/a_0} \cdot \left(\frac{1}{4\pi}\right)^{1/2} = \frac{e^{-Zr/a_0}}{\sqrt{\pi a_0^3/Z^3}}$$

概率密度为

$$\psi^2(1s) = \frac{Z^3 e^{-2Zr/a_0}}{\pi a_0^3}$$

这给出了在距离原子核 r 处找到 $1s$ 电子的概率密度。$\psi^2(1s)$ 随 r 增加呈指数单调下降，在 $r=0$（即原子核所在位置）处具有最大值 $\psi^2 = Z^3/\pi a_0^3$，在 $r=\infty$ 处具有最小值 $\psi^2 = 0$。对于 $Z=1$ 的氢原子，有

$$\psi(1s) = \frac{e^{-r/a_0}}{\sqrt{\pi a_0^3}} \quad \text{且} \quad \psi^2(1s) = \frac{e^{-2r/a_0}}{\pi a_0^3} \tag{7.40}$$

有许多方法可表示**式（7.40）** 给出的 ψ^2。一种方法是选定一个穿过原子核的平面（如 yz 平面），绘制一幅 ψ^2 作为此平面内各点的垂直高度的图。这幅图看起来像一个对称的锥形电子概率密度"山"，其峰值直接位于原子核上，如**图 7.17(a)** 所示。另一种方法是将上述三维曲面投影到二维等概率密度轮廓图上，如**图 7.17(b)** 所示。用圆形的等概率密度轮廓线将找到电子概率密度相等的点连

where R_{21} and Y_{10} are representation forms in quantum numbers, $\rho = 2Zr/na_0$, and $a_0 = 52.93$ pm is the Bohr radius, the same value as in Bohr theory. The $2p_z$ wave function can be calculated as

$$\psi(2p_z) = \psi_{210} = R(2p)Y(p_z) = \left[\frac{1}{2\sqrt{6}}\left(\frac{Z}{a_0}\right)^{3/2}\rho e^{-\rho/2}\right]\left[\left(\frac{3}{4\pi}\right)^{1/2}\cos\theta\right]$$

The angular wave functions Y_{11} and Y_{1-1} are not listed in **Table 7.9**, but provided here as an illustrative example:

$$Y_{11} = -\left(\frac{3}{8\pi}\right)^{1/2}\sin\theta e^{i\phi} \quad \text{and} \quad Y_{1-1} = \left(\frac{3}{8\pi}\right)^{1/2}\sin\theta e^{-i\phi}$$

$Y(p_x)$ and $Y(p_y)$ can be obtained from Y_{11} and Y_{1-1} by taking linear combinations as

$$Y(p_x) = \frac{1}{\sqrt{2}}(Y_{1-1} - Y_{11}) = \frac{1}{\sqrt{2}}\left(\frac{3}{8\pi}\right)^{1/2}\sin\theta(e^{-i\phi} + e^{i\phi}) = \left(\frac{3}{4\pi}\right)^{1/2}\sin\theta\cos\phi$$

$$Y(p_y) = \frac{i}{\sqrt{2}}(Y_{1-1} + Y_{11}) = \frac{i}{\sqrt{2}}\left(\frac{3}{8\pi}\right)^{1/2}\sin\theta(e^{-i\phi} - e^{i\phi}) = \left(\frac{3}{4\pi}\right)^{1/2}\sin\theta\sin\phi$$

where $e^{-i\phi} + e^{i\phi} = 2\cos\phi$ and $e^{-i\phi} - e^{i\phi} = -2i\sin\phi$. The $2p_x$ and $2p_y$ wave functions can be calculated as

$$\psi(2p_x) = R(2p)Y(p_x) = \left[\frac{1}{2\sqrt{6}}\left(\frac{Z}{a_0}\right)^{3/2}\rho e^{-\rho/2}\right]\left[\left(\frac{3}{4\pi}\right)^{1/2}\sin\theta\cos\phi\right]$$

$$\psi(2p_y) = R(2p)Y(p_y) = \left[\frac{1}{2\sqrt{6}}\left(\frac{Z}{a_0}\right)^{3/2}\rho e^{-\rho/2}\right]\left[\left(\frac{3}{4\pi}\right)^{1/2}\sin\theta\sin\phi\right]$$

The *s*, *p*, and *d* orbitals of Hydrogen-Like Species

Let us now illustrate the probability density distributions represented as 3D surfaces for the three major types of orbitals: the *s*, *p*, and *d* orbitals. Recall that orbitals are linear combinations of the wave functions, which are mathematical solutions of Schrödinger equation. The wave function ψ itself has no physical meaning, however, ψ^2 represents the probability density that is proportional to the probability $\psi^2 d\tau$ of finding an electron in the small volume $d\tau$. Here, our goal is to acquire a semi-quantitative understanding of orbitals based on wave functions listed in **Table 7.9**.

1) *s* orbitals

In **Table 7.9**, we notice that the angular wave function for any *s* orbital is always a constant, $Y(s) = \sqrt{1/4\pi}$, independent of θ and ϕ. This means that all *s* orbitals must be spherical, because only a spherical shape in 3D space is totally independent of the angular coordinates. The value of $Y(s)$ arises from the normalization condition that $\int Y^2 d\tau = 1$.

The wave function for the 1*s* orbital of hydrogen-like species can be calculated using $n = 1$, and $\rho = 2Zr/a_0$, as

$$\psi(1s) = R(1s)Y(s) = 2\left(\frac{Z}{a_0}\right)^{3/2}e^{-Zr/a_0}\cdot\left(\frac{1}{4\pi}\right)^{1/2} = \frac{e^{-Zr/a_0}}{\sqrt{\pi a_0^3/Z^3}}$$

The probability density is given by

$$\psi^2(1s) = \frac{Z^3 e^{-2Zr/a_0}}{\pi a_0^3}$$

which gives the probability density of finding a 1*s* electron at a distance *r* from the nucleus. As *r* increases, $\psi^2(1s)$ decreases exponentially, with a maximum of $\psi^2 = Z^3/\pi a_0^3$ at $r = 0$ (where the nucleus locates) and a minimum of $\psi^2 = 0$ at $r = \infty$. For a hydrogen atom with $Z = 1$, we have

$$\psi(1s) = \frac{e^{-r/a_0}}{\sqrt{\pi a_0^3}} \quad \text{and} \quad \psi^2(1s) = \frac{e^{-2r/a_0}}{\pi a_0^3} \tag{7.40}$$

接起来，靠近原子核的线连接了概率较高的点，而远离原子核的线对应于较低的概率。此外，展示电子概率的另一种简单且常用的方法，是只选定一个较大的等概率密度轮廓面，该曲面与其内所有曲面一起，包含了找到电子概率较高的区域。即只展示一个其内找到电子总概率为 95%（任意规定值）的等概率密度轮廓面。在三维图像中，此 95% 概率边界面为球面，如**图 7.17(c)** 所示。

从现在开始，我们将只关注氢原子的轨道，但其他类氢物种的轨道可用类似的方式导出，只是 Z 值不同。氢原子的 $2s$ 轨道波函数可由 $Z=1$、$n=2$ 及 $\rho = r/a_0$ 计算：

$$\psi(2s) = R(2s)Y(s) = \frac{1}{2\sqrt{2}}\left(\frac{1}{a_0}\right)^{3/2}\left(2 - \frac{r}{a_0}\right)e^{-r/2a_0} \cdot \left(\frac{1}{4\pi}\right)^{1/2}$$

$$= \frac{\left(2 - \dfrac{r}{a_0}\right)e^{-r/2a_0}}{\sqrt{32\pi a_0^3}} \tag{7.41}$$

概率密度为

$$\psi^2(2s) = \frac{\left(2 - \dfrac{r}{a_0}\right)^2 e^{-r/a_0}}{32\pi a_0^3} \tag{7.42}$$

此概率密度函数不随 r 单调变化，但在 r 较大时指数项占主导地位。与 $1s$ 轨道相比，$2s$ 轨道倾向于从原子核延伸得更远，因为指数从 $1s$ 的 $-2r/a_0$ 变成 $2s$ 的 $-r/a_0$。这意味着 $2s$ 轨道的指数项比 $1s$ 轨道衰减得更慢。

在**式 (7.41)** 中，$2s$ 波函数的符号由多项式 $(2 - r/a_0)$ 控制。若 $r/a_0 < 2$ 即 $r < 2a_0$，波函数为正；若 $r/a_0 > 2$ 即 $r > 2a_0$，波函数为负；当 $r = 2a_0$ 时，波函数为零，称为具有一个节面。**节面**是波函数改变相位（符号）的地方。由于 $\psi = RY$，不论 $R = 0$ 或 $Y = 0$ 均可使 $\psi = 0$。这里在 $r = 2a_0$ 处 $R(2s) = 0$，因此我们说 $r=2a_0$ 处的节面是**径向节面**。径向节面一定为球形，因为其与角度坐标无关。我们称 $Y=0$ 处为**角度节面**。s 轨道没有角度节面，因为 $Y(s) = \sqrt{1/4\pi} \neq 0$。注意波函数在节面处必为零，但波函数为零处不一定就是节面。例如，$r=\infty$ 时 $\psi(2s) = 0$，但在 $r=\infty$ 处 ψ 没有改变相位，因此这并不是一个节面。

$2s$ 轨道的 ψ^2 比 $1s$ 轨道延伸得离核更远，再加上节面的存在，意味着 $2s$ 轨道的 95% 概率边界球面大于 $1s$ 轨道，且在其内包含一个位于 $r=2a_0$ 处的零概率密度球面。这些特征如**图 7.18** 所示，该图比较了 $1s$、$2s$ 和 $3s$ 轨道。注意，$3s$ 轨道在 $6 - 6\rho + \rho^2 = 0$ 处有两个径向节面，且大于 $1s$ 和 $2s$ 轨道。为了突出轨道从原子核向外推进时的相位变化，我们使用了不同的颜色来表示相位交替。

2) p 轨道

氢原子 $2p$ 轨道的径向函数为

$$R(2p) = \frac{1}{2\sqrt{6}}\left(\frac{1}{a_0}\right)^{3/2}\frac{r}{a_0}e^{-r/2a_0}$$

在 $r=0$ 及 $r=\infty$ 处，$R(2p) = 0$。然而由于其在 $r=0$ 或 $r=\infty$ 处均没有改变相位，$2p$ 轨道没有径向节面。当 r 从 0 增加到 ∞ 时，$R(2p)$ 从

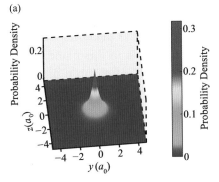

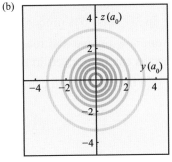

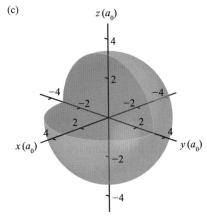

Figure 7.17 Three representations of the electron probability density for the 1s orbital of hydrogen atom. (a) The probability density is represented by the height above an arbitrarily chosen *yz* plane. (b) 2D contour map of the probability density in the *yz* plane, with the overall probability within the outermost contour corresponding to 95%. (c) 3D representation of the 95% probability boundary surface, with 1/4 of the surface being sliced out to show its inner structure.

图 7.17 氢原子 1s 轨道电子概率密度的三种表示法：(a) 用 *yz* 平面（任意选定）上的高度表示的概率密度；(b) *yz* 平面上的二维等概率密度轮廓图，最外圈轮廓以内的总概率对应于 95%；(c) 95% 概率边界曲面的三维表示图，其中 1/4 曲面被切去以展示其内部结构。

There are many ways to represent ψ^2 given in **Equation (7.40)**. One way is to select a plane (for example, the yz plane) passing through the nucleus and plot a graph of ψ^2 as perpendicular heights above various points in this plane. The resultant graph, seen in **Figure 7.17(a)**, looks like a symmetrical, cone-shaped "hill" of electron probability densities with its peak directly above the nucleus. Another way is to project the above 3D surface onto a 2D contour map, shown in **Figure 7.17(b)**. The circular contour lines join points of equal probability density of finding the electron. The contours close to the nucleus join points of high probability, and those farther away correspond to a lower probability. In addition, a simple and commonly used way to display the electron probability is to select just one large contour surface that, together with all the contours within it, encompasses an area of high probability of finding the electron. That is, only a contour surface within which the chance of finding electron is 95% (an arbitrarily chosen value) is shown. In 3D, this 95% probability boundary surface is a sphere, as shown in **Figure 7.17(c)**.

From now on, we will focus only on the orbitals of a hydrogen atom, but the orbitals of other hydrogen-like species can be derived in a similar manner, just with different values of Z. The wave function for the $2s$ orbital of a hydrogen atom can be calculated using $Z = 1$, $n = 2$, and $\rho = r/a_0$.

$$\psi(2s) = R(2s)Y(s) = \frac{1}{2\sqrt{2}}\left(\frac{1}{a_0}\right)^{3/2}\left(2 - \frac{r}{a_0}\right)e^{-r/2a_0} \cdot \left(\frac{1}{4\pi}\right)^{1/2}$$

$$= \frac{\left(2 - \dfrac{r}{a_0}\right)e^{-r/2a_0}}{\sqrt{32\pi a_0^3}} \tag{7.41}$$

The probability density is given by

$$\psi^2(2s) = \frac{\left(2 - \dfrac{r}{a_0}\right)^2 e^{-r/a_0}}{32\pi a_0^3} \tag{7.42}$$

This probability density function does not vary monotonically with respect to r, but the exponential term dominates at large r. When compared to the $1s$ orbital, the $2s$ orbital tends to extend further from the nucleus because the exponential has changed from $-2r/a_0$ for the $1s$ to $-r/a_0$ for the $2s$ orbital. This means that the exponential term of the $2s$ orbital decays more slowly than that of the $1s$.

In **Equation (7.41)**, the sign of the $2s$ wave function is controlled by the polynomial term $(2-r/a_0)$. If $r/a_0 < 2$ or $r < 2a_0$, the wave function is positive, and if $r/a_0 > 2$ or $r > 2a_0$, the wave function is negative. At $r = 2a_0$, the wave function is zero and is said to have a **node**, where the wave function changes phase (sign). Since $\psi = RY$, $\psi = 0$ can be achieved by either $R = 0$ or $Y = 0$. Here, $R(2s) = 0$ at $r = 2a_0$, so we say that the node at $r = 2a_0$ is a **radial node**. Radial nodes must always be spherical because they are independent of the angular coordinates. We call it an **angular node** at $Y = 0$. There is no angular node in s orbitals as $Y(s) = \sqrt{1/4\pi} \neq 0$. Note that the wave function must be zero at the node, however, it is not necessarily a node where the wave function is zero. For example, $\psi(2s) = 0$ at $r = \infty$, but since ψ does not change phase at $r = \infty$, this is not a node.

The fact that ψ^2 of the $2s$ orbital extends farther from the nucleus than that of the $1s$ orbital, together with the presence of the node, means that the 95% probability boundary sphere of a $2s$ orbital is bigger than that of a $1s$ orbital and contains a sphere of zero probability density at $r = 2a_0$. These features are illustrated in **Figure 7.18**, which compares the $1s$, $2s$, and $3s$ orbitals. Note that the $3s$ orbital exhibits two radial nodes at $6 - 6\rho + \rho^2 = 0$, and is larger than both the $1s$ and $2s$ orbitals. To highlight the change in phase of an orbital progressing outward from the nucleus, we have used different colors to represent phase alternation.

2) p orbitals

The radial function for the $2p$ orbital of a hydrogen atom is

$$R(2p) = \frac{1}{2\sqrt{6}}\left(\frac{1}{a_0}\right)^{3/2}\frac{r}{a_0}e^{-r/2a_0}$$

0 开始先增加到最大值，再减小回 0。

与 s 轨道恒定的角度函数相反，p 轨道的角度函数不是常数，而是 θ 和 ϕ 的函数，这意味着 p 轨道的电子概率密度分布不是球形。p_x、p_y 和 p_z 三个角度函数的形状相似，但空间取向不同。这里我们仅以最简单的 $Y(p_z)$ 为例进行说明。如**表 7.9** 所列，$Y(p_z)$ 与 $\cos\theta$ 成正比。以函数 $\cos\theta$ 对 θ 作图，如**图 7.19(a)** 所示，其结果为左右两个相切的圆，其中右圆为正相位，左圆为负相位。我们将在后续 **8.4~8.5 节**中看到，轨道的相位对于理解化学键至关重要。由于 $0 \leq \cos\theta \leq 1$，函数 $\cos^2\theta$ 呈更窄的双泪滴形状，如**图 7.19(b)** 所示。

将角度函数与径向函数相乘，可得总波函数 $\psi(2p_z) = R(2p)Y(p_z)$，以及概率密度 $\psi^2(2p_z) = R^2(2p)Y^2(p_z)$。$\psi^2(2p_z)$ 的等概率密度轮廓图如**图 7.19(c)** 所示。可以看到，最大概率密度的取向（称为**波瓣**）位于 z 轴的正向和负向，分别对应 $\theta=0$ 和 π，均使 $\cos^2\theta=1$。这也正是将其命名为 p_z 的原因。由于在 xy 平面的任意位置均有 $\theta=\pi/2$ 且 $\cos\theta=0$，故 xy 平面始终是 p_z 轨道的角度节面。

氢原子三个 2p 轨道的 95% 概率边界面及其角度节面如**图 7.20**所示。这里我们总结一下轨道的三维曲面表示法的特点：

① 形状：轨道曲面的形状源自等概率密度轮廓面，这意味着该轨道曲面上的每个点都具有相同的概率密度。
② 大小：轨道边界面的大小由 95%（任意规定值）概率得出，这是沿等概率密度面从内向外的概率积分。
③ 颜色：轨道曲面的颜色表示原始波函数 ψ 的相位交替，尽管概率密度 ψ^2 本身总是非负值，且缺乏相位信息。

3) d 轨道

我们将不讨论 d 轨道波函数的数学形式，而只关注其三维轨道曲面及其各种截面。**图 7.21** 显示了五个 3d 轨道的三维曲面：d_{z^2} ($m_l=0$)、d_{xz} 和 d_{yz} ($m_l = \pm 1$)、d_{xy} 和 $d_{x^2-y^2}$ ($m_l = \pm 2$)。其中 d_{z^2} 具有独特的形状，沿 z 轴有两个正波瓣，在 xy 平面存在环绕的负波瓣。除了相对于轴的取向不同之外，所有其他四个 d 轨道均具有相同的基本形状。$d_{x^2-y^2}$ 的波瓣指向 x 轴和 y 轴，而其余三个指向相应的笛卡尔轴之间。每个 d 轨道总是有两个角度节面，这可以从 $Y(d)=0$ 得到。五个 d 轨道的截面以及相应的角度节面如**图 7.22** 所示。

4) n 值较高的轨道：2s、3p 和 4d

由于角度波函数与 n 无关，当考虑 n 值较高的轨道时，我们可以乘上不同的径向波函数。定性地说，我们可以通过加上额外的径向节面来比较它们。任何轨道的角度节面数均恰好等于其 l 值：s 轨道没有角度节面，p 轨道有 1 个角度节面，d 轨道有 2 个角度节面。任何轨道的总节面数均等于 $n-1$，因此径向节面数为 $n-l-1$。也就是说，1s、2p 和 3d 轨道没有径向节面，2s、3p 和 4d 轨道各有 1 个径向节面，以此类推。径向节面数也可由 $R=0$ 得到，或者令 R 中的多项式为 0。

图 7.23 比较了 (a) 1s 和 2s、(b) $2p_z$ 和 $3p_z$ 以及 (c) $3d_{z^2}$ 和 $4d_{z^2}$ 的等概率密度轮廓图。由于具有相同的角度波函数和角度节面，较高 n 值与较低 n 值的对应轨道具有相似的一般形状，而额外的径向

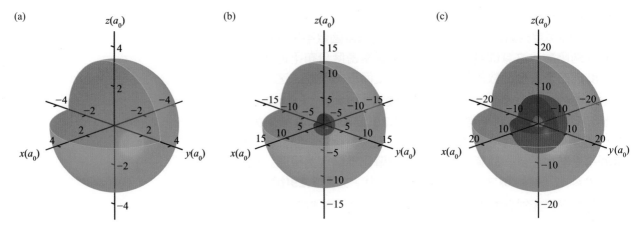

Figure 7.18 3D surface representation of the 95% probability boundary surface for the (a) 1s, (b) 2s, and (c) 3s orbitals of hydrogen atom. 1/4 of the surfaces are sliced out to show their inner structures. Different colors indicate the phase alternation of the wave function.

图 7.18 氢原子 (a) 1s、(b) 2s 和 (c) 3s 轨道 95% 概率边界曲面的三维表示图：其中 1/4 曲面被切去以展示其内部结构，不同颜色表示波函数的相位交替。

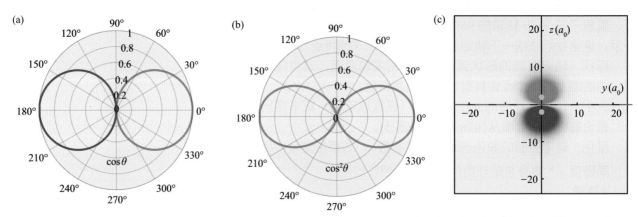

Figure 7.19 Various representations of the 2p orbital wave functions of hydrogen atom. (a) A plot of $\cos\theta$ as a function of θ, representing the angular part of the p_z wave function, as $Y(p_z) \propto \cos\theta$. Different colors indicate the phase alternation of the wave function. (b) A plot of $\cos^2\theta$ as a function of θ, representing the angular probability density of p_z, as $Y^2(p_z) \propto \cos^2\theta$. (c) The contour map of $\psi^2(2p_z) = R^2(2p)Y^2(p_z)$. Red dashed line indicates its angular node at the xy plane.

图 7.19 氢原子 2p 轨道波函数的各种表示法：(a) $\cos\theta$ 对 θ 的函数图，代表了 p_z 波函数的角度部分，因为 $Y(p_z) \propto \cos\theta$，不同颜色表示波函数的相位交替；(b) $\cos^2\theta$ 对 θ 的函数图，代表了 p_z 的角度概率密度，因为 $Y^2(p_z) \propto \cos^2\theta$；(c) $\psi^2(2p_z) = R^2(2p)Y^2(p_z)$ 的等概率密度轮廓图。红色虚线表示其在 xy 平面的角度节面。

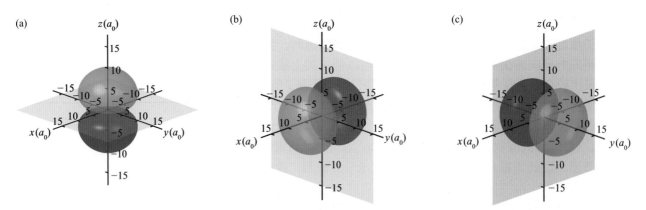

Figure 7.20 3D surface representation of the 95% probability boundary surface and the corresponding angular nodes (gray shaded surface) for the three 2p orbitals of hydrogen atom: (a) $2p_z$, (b) $2p_x$, and (c) $2p_y$.

图 7.20 氢原子三个 2p 轨道 95% 概率边界曲面的三维表示图以及相应的角度节面（灰色阴影面）：(a) $2p_z$、(b) $2p_x$ 和 (c) $2p_y$。

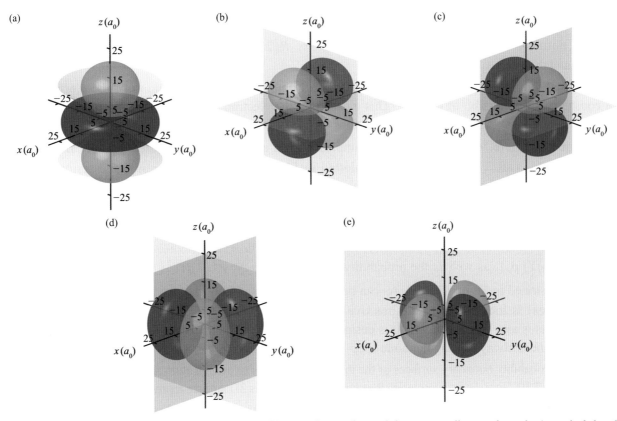

Figure 7.21 3D surface representation of the 95% probability boundary surface and the corresponding angular nodes (gray shaded surface) for the five 3d orbitals of hydrogen atom: (a) $3d_{z^2}$, (b) $3d_{xz}$, (c) $3d_{yz}$, (d) $3d_{xy}$, and (e) $3d_{x^2-y^2}$.

图 7.21 氢原子五个 3d 轨道 95% 概率边界曲面的三维表示图以及相应的角度节面（灰色阴影面）：(a) $3d_{z^2}$、(b) $3d_{xz}$、(c) $3d_{yz}$、(d) $3d_{xy}$ 和 (e) $3d_{x^2-y^2}$。

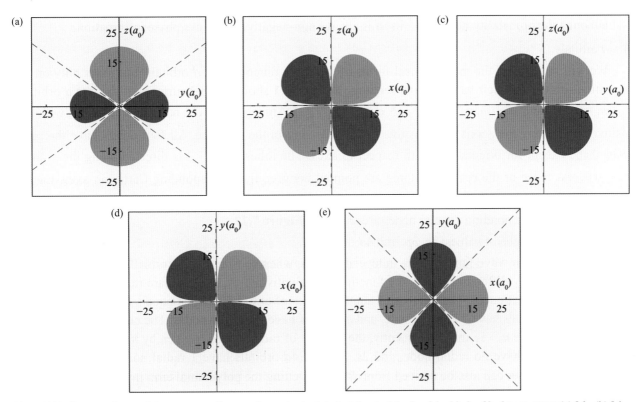

Figure 7.22 Cross sections and the corresponding angular nodes (red dashed lines) of the five 3d orbitals of hydrogen atom: (a) $3d_{z^2}$, (b) $3d_{xz}$, (c) $3d_{yz}$, (d) $3d_{xy}$, and (e) $3d_{x^2-y^2}$.

图 7.22 氢原子五个 3d 轨道的截面以及相应的角度节面（红色虚线）：(a) $3d_{z^2}$、(b) $3d_{xz}$、(c) $3d_{yz}$、(d) $3d_{xy}$ 和 (e) $3d_{x^2-y^2}$。

At $r = 0$ and $r = \infty$, $R(2p) = 0$. However, since it does not change phase at either $r = 0$ or $r = \infty$, the $2p$ orbital has no radial nodes. As r increases from 0 to ∞, $R(2p)$ increases from a value of 0, reaches a maximum, and then decreases back to 0.

In contrast to the constant angular function of the s orbitals, the angular functions of the p orbitals are not a constant, but a function of θ and ϕ. This means that the electron probability density distribution of p orbitals is not spherical. The three angular functions of p_x, p_y, and p_z are similar in shape, but with different spatial orientations. Here, we only illustrate the simplest $Y(p_z)$ as an example. As listed in **Table 7.9**, $Y(p_z)$ is proportional to $\cos\theta$. The function $\cos\theta$ is plotted in **Figure 7.19(a)** in terms of θ and results in two tangential circles, with a positive phase on the right circle and a negative phase on the left circle. We will see later in **Sections 8.4~8.5** that the phase of orbitals is important in understanding chemical bonding. Because $0 \leqslant \cos\theta \leqslant 1$, the function $\cos^2\theta$ results in a narrower double teardrop shape, shown in **Figure 7.19(b)**.

By multiplying the angular function with the radial function, we can obtain the total wave function $\psi(2p_z) = R(2p)Y(p_z)$, as well as the probability density $\psi^2(2p_z) = R^2(2p)Y^2(p_z)$. The contour map of $\psi^2(2p_z)$ is illustrated in **Figure 7.19(c)**. As can be seen, the orientation of the maximum probability density (called the **lobes**) is located along the positive and negative z axis, where $\cos^2\theta = 1$ at $\theta = 0$ and π, respectively. This is also the reason why it is designated p_z. As $\theta = \pi/2$ and $\cos\theta = 0$ at everywhere in the xy plane, the xy plane is always an angular node for p_z orbitals.

The 95% probability boundary surfaces of the three $2p$ orbitals of hydrogen atom and their angular nodes are shown in **Figure 7.20**. Here, let us summarize the characteristics of this 3D surface representation of orbitals:

① Shape: The shape of the orbital surface arises from the contour with an equal probability density, which means that every point on this orbital surface has the same probability density.

② Size: The size of the orbital boundary surface results from the arbitrarily chosen 95% probability, which is the integral of probability from the inner outwards along the contours.

③ Color: The colors of the orbital surface represent the phase alternation of the original wave function ψ, although the probability density ψ^2 itself is always non-negative and lacks phase information.

3) d orbitals

We will not discuss the mathematical forms of the wave functions for d orbitals, but focus only on the 3D orbital surface and their various cross sections. **Figure 7.21** shows the 3D surfaces of the five d orbitals: d_{z^2} ($m_l = 0$), d_{xz} and d_{yz} ($m_l = \pm 1$), d_{xy} and $d_{x^2-y^2}$ ($m_l = \pm 2$). Among them, d_{z^2} has a unique shape, with two positive lobes along the z axis and negative lobes encircling in the xy plane. All four others have the same basic shape except for orientations with respect to axes. The lobes of $d_{x^2-y^2}$ is directed along the x and y axes, whereas those of the remaining three are pointed between the corresponding Cartesian axes. Each d orbital always has two angular nodes that can be derived from $Y(d) = 0$. The cross sections of the five d orbitals and the corresponding angular nodes are shown in **Figure 7.22**.

4) Orbitals with higher n values: 2s, 3p, and 4d

Since the angular wave functions are independent of n, when considering the orbitals with higher values of n, we can multiply different radial wave functions. Qualitatively, we can compare them by including the extra radial nodes that occur. The number of angular nodes of any orbitals just equals l: s orbitals have no angular node, p orbitals have 1 angular node, and d orbitals have 2 angular nodes. The number of total nodes of any orbitals equals $n - 1$, and consequently, the number of radial nodes is given by $n - l - 1$. That is, 1s, 2p, and 3d orbitals have no radial node, and 2s, 3p, and 4d orbitals have 1 radial nodes, and so on. The number of radial nodes can also be derived from $R = 0$, or letting the polynomial term in R to be 0.

Figure 7.23 compares the contour maps between (a) 1s and 2s, (b) $2p_z$ and $3p_z$, and (c) $3d_{z^2}$ and $4d_{z^2}$. The orbitals with higher n values have similar general shapes as the corresponding orbitals with lower n values due to the same angular wave functions and angular nodes, but the extra radial node appears as a

节面显示为圆形（如**图 7.23** 的红色虚线所示）。较高 n 值的轨道显示为，在较大的轨道内出现较小的轨道。内轨道的等概率密度轮廓线比外轨道的更密集，这是因为内轨道具有更高的概率密度。

径向概率分布

除了上面讨论的以三维曲面表示的概率密度分布之外，我们还可以使用一种称为**径向概率分布** $D(r)$ 的简化分布来描述不同的轨道。如前述 **7.5 节**所述，找到电子的总概率 P 可以通过对整个三维空间进行积分来计算，为

$$P = \int |\psi(\mathbf{q})|^2 \, d\tau = \iiint \psi^2(r,\theta,\phi) \, d\tau$$

已知 $\psi(r,\theta,\phi) = R(r)\Theta(\theta)\Phi(\phi)$ 和 $d\tau = r^2 \sin\theta \, dr d\theta d\phi$，有

$$P = \int_{r=0}^{\infty}\int_{\theta=0}^{\pi}\int_{\phi=0}^{2\pi} \left[R(r)\Theta(\theta)\Phi(\phi)\right]^2 r^2 \sin\theta \, dr d\theta d\phi$$

将此三重积分分离为三个单重积分，有

$$P = \int_0^{\infty} r^2 R^2(r) dr \int_0^{\pi} \Theta^2(\theta) \sin\theta \, d\theta \int_0^{2\pi} \Phi^2(\phi) d\phi$$

由于角度积分 $\int_0^{\pi} \Theta^2(\theta) \sin\theta \, d\theta \int_0^{2\pi} \Phi^2(\phi) d\phi = 1$，有

$$P = \int_0^{\infty} r^2 R^2(r) dr$$

通过定义

$$D(r) = r^2 R^2(r) \tag{7.43}$$

我们可以将原始的三重积分简化为 $D(r)$ 的单重积分，即

$$P = \iiint \psi^2 d\tau = \int D(r) dr \tag{7.44}$$

这就是使用径向概率分布 $D(r)$ 的优势，它给出了在半径为 r、厚度为无穷小的球壳内任意位置找到电子的概率，而不管方向如何，因为 $D(r)$ 与 θ 和 ϕ 无关。

图 7.24 比较了各种轨道的径向概率密度 R^2 与径向概率分布 $D=r^2R^2$ 对 r 的关系图。例如，1s 轨道的 R^2 在原子核处 ($r=0$) 具有最大值，但由于 r^2 非常小，在原子核处 $D=0$。同时，**最概然半径**定义为最可能找到电子的半径，这是 D 达到最大值的地方，可以由下式计算：

$$\frac{dD(r)}{dr} = 0 \tag{7.45}$$

对于类氢物种的 1s 轨道

$$\frac{dD(r)}{dr} = \frac{d}{dr}\left(\frac{4Z^3 r^2}{a_0^3} e^{-2Zr/a_0}\right) = \frac{4Z^3}{a_0^3} e^{-2Zr/a_0}\left(2r - \frac{2Zr^2}{a_0}\right) = 0$$

$$\therefore r = \frac{a_0}{Z}$$

氢原子 1s 轨道的最概然半径为 a_0，与第一玻尔轨道的半径相同。氢原子 1s 轨道 95% 概率边界面的半径也可通过 $D(r)$ 计算，为

$$P = 0.95 = \int_0^r D(r) dr = \int_0^r \frac{4r^2}{a_0^3} e^{-2r/a_0} dr$$

所得 r 约为 141 pm，远大于 a_0=53 pm。

从径向概率分布，我们可以直接可视化某个轨道的径向节面数 $(n-l-1)$ 及峰的数目 $(n-l)$。在比较具有相同 $n-l$ 的轨道（如 1s、2p 和 3d）时，最概然半径可分别计算为 a_0、$4a_0$ 和 $9a_0$。在比较

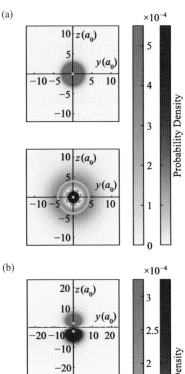

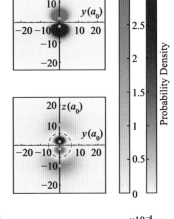

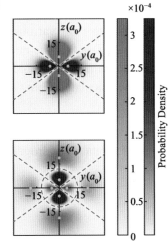

Figure 7.23 Comparison of the contour maps of hydrogen atom between (a) 1s and 2s, (b) $2p_z$ and $3p_z$, and (c) $3d_{z^2}$ and $4d_{z^2}$. Red dashed lines represent the nodes and cyan solid dots or lines indicate the locations of the local maximal electron probability density in each lobes.

图 7.23 氢原子 (a) 1s 和 2s、(b) $2p_z$ 和 $3p_z$ 以及 (c) $3d_{z^2}$ 和 $4d_{z^2}$ 的等概率密度轮廓图的比较。红色虚线代表节面，青色点或实线表示每个波瓣电子概率密度极大值所处的位置。

circle (shown as the dashed lines in **Figure 7.23**). The appearance of higher n orbitals is that of a smaller orbital inside a larger one. The fact that the contour lines are denser in the inner orbital than in the outer one is due to higher probability density in the inner orbital.

Radial Probability Distributions

Other than probability density distributions represented as 3D surfaces discussed above, we can also use a kind of simplified distribution called **radial probability distribution** $D(r)$ to describe different orbitals. As shown previously in **Section 7.5**, the total probability P of finding an electron can be calculated by taking integrals over the entire 3D space, as

$$P = \int |\psi(q)|^2 \, d\tau = \iiint \psi^2(r,\theta,\phi) \, d\tau$$

Given that $\psi(r,\theta,\phi) = R(r)\Theta(\theta)\Phi(\phi)$ and $d\tau = r^2 \sin\theta \, dr\, d\theta\, d\phi$, we have

$$P = \int_{r=0}^{\infty} \int_{\theta=0}^{\pi} \int_{\phi=0}^{2\pi} \left[R(r)\Theta(\theta)\Phi(\phi) \right]^2 r^2 \sin\theta \, dr\, d\theta\, d\phi$$

Separation of this triple integral into three single integrals gives rise to

$$P = \int_0^{\infty} r^2 R^2(r) \, dr \int_0^{\pi} \Theta^2(\theta) \sin\theta \, d\theta \int_0^{2\pi} \Phi^2(\phi) \, d\phi$$

As the angular integration $\int_0^{\pi} \Theta^2(\theta) \sin\theta \, d\theta \int_0^{2\pi} \Phi^2(\phi) \, d\phi = 1$, we can derive

$$P = \int_0^{\infty} r^2 R^2(r) \, dr$$

By defining
$$D(r) = r^2 R^2(r) \tag{7.43}$$

we can then simplify the original triple integral into a single integral of $D(r)$, as

$$P = \iiint \psi^2 \, d\tau = \int D(r) \, dr \tag{7.44}$$

This is the advantage of using the radial probability distribution $D(r)$, which gives the probability of finding the electron anywhere in a spherical shell of radius r and an infinitesimal thickness dr, regardless of the direction since $D(r)$ is independent of θ and ϕ.

Figure 7.24 compares the radial probability densities R^2 and radial probability distributions $D = r^2 R^2$ with respect to r for various orbitals. For example, R^2 for a 1s orbital has a maximum at the nucleus ($r = 0$), however, $D = 0$ at the nucleus because r^2 is vanishingly small. Meanwhile, the **most probable radius** is defined as the radius at which the electron is most likely to be found. This is where D reaches a maximum and can be calculated from

$$\frac{dD(r)}{dr} = 0 \tag{7.45}$$

For the 1s orbital of hydrogen-like species

$$\frac{dD(r)}{dr} = \frac{d}{dr}\left(\frac{4Z^3 r^2}{a_0^3} e^{-2Zr/a_0} \right) = \frac{4Z^3}{a_0^3} e^{-2Zr/a_0} \left(2r - \frac{2Zr^2}{a_0} \right) = 0$$

$$\therefore r = \frac{a_0}{Z}$$

For the 1s orbital of a hydrogen atom, the most probable radius is a_0, the same as that of the first Bohr orbit. The radius for the 95% probability boundary surface of the hydrogen 1s orbital can also be calculated using $D(r)$, as

$$P = 0.95 = \int_0^r D(r) \, dr = \int_0^r \frac{4r^2}{a_0^3} e^{-2r/a_0} \, dr$$

The resultant r is about 141 pm, much larger than $a_0 = 53$ pm.

From the radial probability distribution, we can directly visualize the number of radial nodes ($n-l-1$) and the number of peaks ($n-l$) for an orbital. In comparing orbitals with the same $n-l$ value, such as 1s, 2p, and 3d, the most probable radius can be calculated as a_0, $4a_0$, and $9a_0$, respectively. In comparing orbitals

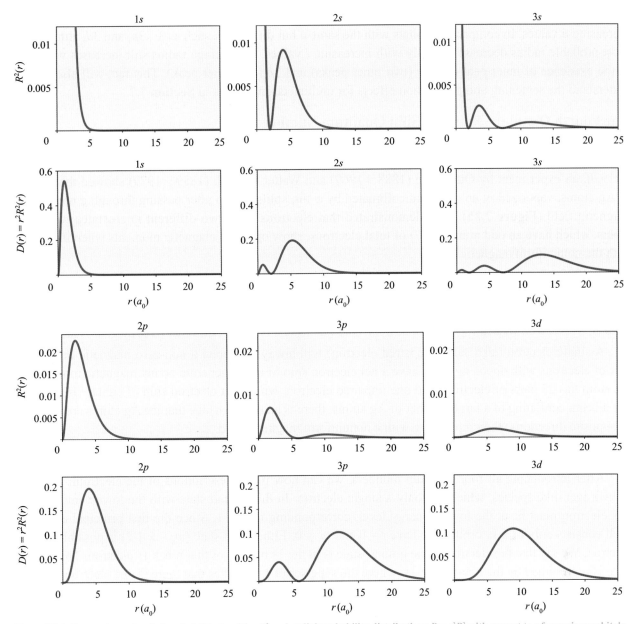

Figure 7.24 Comparison of radial probability densities R^2 and radial probability distributions $D = r^2R^2$ with respect to r for various orbitals.

图 7.24 各种不同轨道的径向概率密度 R^2 与径向概率分布 $D = r^2R^2$ 对 r 的关系比较图。

具有相同 l 但不同 n 的轨道（如 1s、2s 和 3s）时，最概然半径随 n 的增加而显著增加。在比较具有相同 n 但不同 l 的轨道（如 3s、3p 和 3d）时，尽管最概然半径随 l 的增加而略微减小，但由于 3s 存在两个内峰，3p 存在一个内峰，其平均半径仍随 l 的增加而增加。这一事实也将有助于我们理解 **7.7 节**中多电子物种的屏蔽和钻穿效应。

第四个量子数：自旋量子数

如前所述，在薛定谔方程的求解过程中引入了三个量子数（n、l 和 m_l），来描述每个主电子层具有 n^2 重简并度的各种轨道。1920 年，奥托·斯特恩（1888—1969）和沃尔特·格拉赫（1889—1979）的实验表明，从炉源中蒸发并通过狭缝准直的一束 Ag 原子，在通过非均匀磁场后会裂分成两束（**图 7.25**）。这表明电子具有两种不同的微观状态，因此在与非均匀磁场相互作用时，总电子数为奇数（47）的 Ag 原子会

with the same *l* but different *n*, such as 1s, 2s, and 3s, the most probable radius increases significantly with increasing *n* values. In comparing orbitals with the same *n* but different *l*, such as 3s, 3p, and 3d, although the most probable radius decreases slightly with increasing *l* values, the average radius still increases with *l* due to the existence of inner peaks for 3s (two inner peaks) and 3p (one inner peak). This fact will also help us understand the screening and penetration effects for multielectron species in **Section 7.7**.

The Fourth Quantum Number: Spin Quantum Number

As mentioned earlier, during the solution to Schrödinger equation, three quantum numbers (n, l, and m_l) are introduced to describe various orbitals with a n^2-fold degeneracy in each principal electronic shell. In 1920, an experiment by Otto Stern (1888—1969) and Walter Gerlach (1889—1979) showed that a beam of Ag atoms, vaporized in an oven and collimated by a slit, splits in two after passing through a nonuniform magnetic field (**Figure 7.25**). This demonstrated that electrons have two different microstates so that Ag atoms, which have an odd number (47) of total electrons, show opposite magnetic moments when interacting with the nonuniform magnetic field.

In 1925, George Uhlenbeck and Samuel Goudsmit assumed that an electron acts as if it "spins", i.e., an electron may have two different spin angular moment. This leads to the fourth quantum number: the electron **spin quantum number** m_s, which may have a value of either +1/2 (also denoted by α or an up arrow ↑) or −1/2 (also denoted by β or a down arrow ↓). The value of m_s does not depend on any of the other three quantum numbers.

A single electron (also called unpaired electron) will always generate a non-zero magnetic moment. A pair of electrons with opposing spins have a net electron spin of 0 and generate no net magnetic moment. An Ag atom has 23 pairs of electrons and one unpaired electron, with a net electron spin of either +1/2 or −1/2. For a beam consisting of a large number of Ag atoms, there is an equal chance that the Ag atoms are deflected in opposite directions when interacting with a nonuniform magnetic field.

Electronic Structure of Hydrogen-Like Species

After introducing all four quantum numbers, we can now make descriptions of the electronic structure of hydrogen-like species, which has only a single electron. In the ground state with the lowest total energy, this electron must be at the lowest energy level, corresponding to $n = 1$. Since the first principal electronic shell consists only of a 1s orbital, we have $l = 0$ and $m_l = 0$. Either spin state ($m_s = \pm 1/2$) is possible for this electron. We say that the electron in the ground-state is in the 1s orbital, or that it is a 1s electron. This ground state is represented by the notation of $1s^1$, where the superscript 1 indicates that there is one electron in the 1s orbital. Thus, the ground-state electronic structure is:

Ground state ($1s^1$): $n = 1$, $l = 0$, $m_l = 0$, $m_s = +1/2$ or $-1/2$

In excited states with higher energies, the electron may occupy orbitals with higher values of n. For example, with $n = 2$, the electron can occupy either the 2s or one of the 2p orbitals, both with the same energy, represented by:

Excited state ($2s^1$): $n = 2$, $l = 0$, $m_l = 0$, $m_s = +1/2$ or $-1/2$

Excited state ($2p^1$): $n = 2$, $l = 1$, $m_l = -1, 0,$ or $+1$, $m_s = +1/2$ or $-1/2$

Other excited states with even higher values of n can be represented in a similar manner. Because the probability density extends farther from the nucleus in higher n orbitals than in the 1s orbital, the excited-state atom (or ion) is larger than the ground-state.

7.7 Multielectron Species and Electron Configurations

Previously, we have discussed the quantum mechanical model and results for single-electron systems: the hydrogen-like species. Now we will extend to systems with more than one electron: the multielectron species.

显示出相反的磁矩。

1925 年，乔治·乌伦贝克和塞缪尔·古兹米特假定电子的行为就像它在"自旋"一样，即电子可能具有两种不同的自旋角动量。这引入了第四个量子数：电子的**自旋量子数** m_s，其值可以为 +1/2（也用 α 或向上的箭头↑表示）或 −1/2（也用 β 或向下的箭头↓表示）。m_s 的值与其他三个量子数均无关。

单个电子（也称未成对电子）将始终产生非零磁矩。一对自旋相反的电子的净电子自旋为 0，不产生净磁矩。Ag 原子有 23 对电子和一个未成对电子，净电子自旋为 +1/2 或 −1/2。当一束由大量 Ag 原子组成的原子束与非均匀磁场相互作用时，Ag 原子向相反方向偏转的概率相等。

类氢物种的电子结构

在引入了所有四个量子数之后，我们现在可以描述类氢物种（只有一个电子）的电子结构。在总能量最低的基态中，该电子必须处于最低能级，对应于 $n=1$。由于第一主电子层仅由 $1s$ 轨道组成，因此 $l=0$、$m_l=0$。对于该电子，任意自旋态（$m_s=\pm 1/2$）均可能。我们称处于基态的电子在 $1s$ 轨道上，或者说它是 $1s$ 电子。这个基态用 $1s^1$ 表示，其中上标 1 表示 $1s$ 轨道上有一个电子。因此，基态的电子结构为：

基态（$1s^1$）：$n=1$，$l=0$，$m_l=0$，$m_s=+1/2$ 或 $-1/2$

在具有更高能量的激发态中，电子可能占据具有更高 n 值的轨道。例如，当 $n=2$ 时，电子可以占据 $2s$ 或 $2p$ 轨道之一，两者均具有相同能量，可表示为：

激发态（$2s^1$）：$n=2$，$l=0$，$m_l=0$，$m_s=+1/2$ 或 $-1/2$

激发态（$2p^1$）：$n=2$，$l=1$，$m_l=-1$、0 或 +1，$m_s=+1/2$ 或 $-1/2$

具有更高 n 值的其他激发态可用类似的方式表示。由于高 n 值轨道的概率密度比 $1s$ 轨道延伸得离核更远，因此激发态的原子（或离子）比基态的更大。

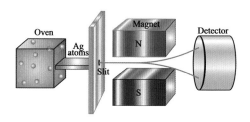

Figure 7.25 An experimental setup demonstrates that electrons have two different microstates (or spins).

图 7.25 表明电子具有两种不同的微观状态（或自旋）的实验装置图。

7.7 多电子物种与电子组态

此前我们讨论了单电子体系（即类氢物种）的量子力学模型及其结论。现在我们扩展到具有一个以上电子的体系：多电子物种。

多体效应

薛定谔方程的精确求解只能在单电子体系中得到。即便只是再多一个电子，薛定谔方程也无法精确求解。原因在于薛定谔方程的势能项：多电子物种的电子之间存在相互排斥（称为**多体效应**）。电子之间的排斥倾向于使它们彼此远离，从而导致了彼此纠缠的电子运动。

虽然多电子物种的薛定谔方程不能精确求解，但我们仍然可以得到一些近似解。最常用的方法之一称为**中心力场模型**。其要点在于，认为电子可以逐一加入由原子核与其他电子所建立的中心力场中。例如，原子核与第一个电子形成第一个单电子体系，其薛定谔方程

Many-Body Problem

The precise solutions to Schrödinger equation can only be obtained for single-electron systems. With even just one more electron, Schrödinger equation cannot be solved precisely. The reason lies in the potential energy term of Schrödinger equation: the existence of mutual repulsion between electrons in multielectron species, which is called **many-body problem**. The repulsion between electrons tends to keep them away from one another, leading to inextricably entangled electron motions.

Although Schrödinger equation cannot be solved precisely for multielectron species, we can still obtain some approximate solutions. One of the most commonly used methods is called the **central force-field model**. In this model, the key point is that the electrons are considered to be added in, one by one, in the central force-field established by the nucleus and the other electrons. For example, the nucleus and the very first electron form a first single-electron system, Schrödinger equation of which can be solved precisely. Then, the second electron is introduced and is regarded as a second single-electron system together with the first system, which is considered as a "new nucleus" with a smaller nuclear charge than the original nucleus due to the existence of the first electron. The second electron experiences a central force-field established by the first system. Schrödinger equation of the second system can also be solved. After that, the third electron is added in, and a similar process continues until all electrons are introduced. In this method, approximate solutions can be obtained for multielectron species, but the repulsions between electrons are neglected.

Compared with those for hydrogen-like species, the resultant approximate solutions in central force-field model for multielectron species also contain three parts:

1) The three quantum numbers n, l, and m_l are the same as those for hydrogen-like species.
2) The quantized energy E depends not only on n, but also on l. The orbitals with the same n but different l are not degenerate, with $E_{ns}<E_{np}<E_{nd}$.
3) The wave functions (called the **hydrogen-like orbitals**, sometimes still called the orbitals for short) have the same angular parts but different radial parts as those for hydrogen-like species. Note that hydrogen-like orbitals are approximate solutions for multielectron species, not for hydrogen-like species.

Screening and Penetration Effects

In central force-field model, when a new electron is introduced, we assume that it experiences a central force-field established by the nucleus and all other electrons that have already been added in. The existence of the other electrons screens or shields the nucleus and reduces the effectiveness of the nucleus in attracting the newly introduced electron as if the nuclear charge is reduced to some extent. This is called **screening effect** or shielding effect.

The magnitude of screening effect depends on both n and l. In general, electrons in orbitals closer to the nucleus are more effective in screening the nucleus from outer electrons than electrons farther away. This can be related to the radial probability distribution of various orbitals. For example, we have discussed previously that for orbitals with the same n but different l, such as ns, np, and nd, the average radius of the orbitals, which is also the mean distance between the corresponding electron and the nucleus, increases with l. Therefore, electrons in ns orbitals are generally closer to the nucleus, and thus more effective at screening the nucleus from outer electrons than are electrons in np or nd orbitals. The stronger attraction between the ns electrons and the nucleus leads to lower energy in ns orbitals than in np or nd orbitals, resulting in $E_{ns}<E_{np}<E_{nd}$. The ability of electrons in ns orbitals that allows them to get closer to the nucleus than electrons in np or nd orbitals is called **penetration effect**. An electron in an orbital with good penetration is better at screening, and consequently, has a lower energy than one with low penetration.

The nuclear charge that an electron actually experiences, defined as the **effective nuclear charge** (Z_{eff}), is less than the original nuclear charge (Z, or atomic number), as

$$Z_{eff} = Z - \sigma \tag{7.46}$$

where σ is the total **screening constant**, and can be calculated based on a set of empirical rules, called **Slater rules** (see **Problem 7.11**), which was originally devised by John C. Slater (1900—1976) in 1930. The orbital

可以精确求解。之后引入第二个电子，并将其与第一个体系一起视为第二个单电子体系。而第一个体系则被视为一个"新的原子核"，由于第一个电子的存在，其核电荷小于初始的原子核。第二个电子处于由第一个体系所建立的中心力场中。第二个体系的薛定谔方程也可以求解。随后加入第三个电子，继续类似的过程，直到引入所有电子。采用这种方法可以获得多电子物种的近似解，但电子之间的排斥力被忽略。

与类氢物种相比，多电子物种在中心力场模型中的近似求解结果也包含三个部分：

1) 三个量子数 n、l 和 m_l，与类氢物种相同。
2) 量子化的能量 E，不仅与 n 有关，还与 l 有关。具有相同 n 但不同 l 的轨道不再简并，有 $E_{ns}<E_{np}<E_{nd}$。
3) 波函数（称为**类氢轨道**，有时仍简称轨道）具有与类氢物种相同的角度部分，但径向部分不同。注意，类氢轨道是多电子物种而非类氢物种的近似解。

屏蔽效应与钻穿效应

在中心力场模型中，当引入一个新电子时，假定其处于由原子核与所有其他已加入的电子所建立的中心力场中。其他电子的存在部分屏蔽了原子核，并降低了原子核吸引新引入电子的有效性，就好像在某种程度上减少了核电荷一样，这种效应称为**屏蔽效应**。

屏蔽效应的大小取决于 n 和 l。一般来说，距核较近的轨道中的电子比离核更远的电子，能更有效地屏蔽原子核对核外电子的吸引。这与各种轨道的径向概率分布有关。例如，前面已经讨论过，对于具有相同 n 但不同 l 的轨道（如 ns、np 和 nd），其平均半径（即相应电子与原子核之间的平均距离）随 l 的增加而增加。因此，ns 轨道中的电子通常更靠近原子核，比 np 或 nd 轨道中的电子能更有效地屏蔽原子核对核外电子的吸引。ns 电子与原子核之间较强的吸引力，导致 ns 轨道的能量低于 np 或 nd 轨道，使得 $E_{ns}<E_{np}<E_{nd}$。ns 轨道中的电子比 np 或 nd 轨道中的电子更接近原子核的能力，称为**钻穿效应**。钻穿效应较强的轨道中的电子具有较强的屏蔽效应，因此其能量低于钻穿效应较弱的轨道中的电子。

有效核电荷（$Z_{有效}$）是电子实际感受的核电荷，其值小于初始的核电荷（Z，或原子序数），有

$$Z_{有效} = Z - \sigma \tag{7.46}$$

其中 σ 是总**屏蔽常数**，可根据一组称为**斯莱特规则**（见**习题 7.11**）的经验规则进行计算，此规则最初由约翰·C. 斯莱特（1900—1976）于 1930 年提出。轨道能量可近似写为

$$E_n = -R_H \frac{Z_{有效}^2}{n^2} = -R_H \frac{(Z-\sigma)^2}{n^2} \tag{7.47}$$

看上去**式（7.47）**中只出现了 n，但由于 σ 与 l 有关，轨道能量与 n 和 l 均有关。$Z_{有效}$ 越大，原子核对电子的吸引力就越大，因此电子所在轨道的能量就越低。

轨道能级图

由于不同轨道的屏蔽和钻穿效应不同，主电子层的能级分裂为

energy is then approximated by

$$E_n = -R_H \frac{Z_{eff}^2}{n^2} = -R_H \frac{(Z-\sigma)^2}{n^2} \tag{7.47}$$

It seems that only n appears in **Equation (7.47)**, however, since σ is dependent on l, the orbital energy depends on both n and l. The larger the value of Z_{eff}, the larger is the attraction of the nucleus to the electron, and hence the lower is the energy of the orbital in which the electron is found.

Orbital Energy-Level Diagram

Due to the different screening and penetration effects for various orbitals, the energy level of a principal shell is split into separate levels for its subshells. There is no further splitting of energies within a subshell, however, because all the orbitals in the subshell have the same radial characteristics and thereby experience the same Z_{eff}. The combined effect of the decreased spacing between successive energy levels at higher n (because of $E_n \propto 1/n^2$) and the splitting of subshell energy levels (because of screening and penetration) causes some energy levels to overlap.

However, the order in energy levels is not routine for all elements. For example, because of the extra penetration of a 4s electron over that of a 3d electron, the 4s energy level is generally below the 3d level for some elements such as K and Ca, but not for other elements such as Li, Na, Sc and Ti. **Figure 7.26** compares the orbital energy-level diagram for the first several principal shells for H, Li, Na, and K. Note that the steady decrease in all orbital energies with increasing atomic number is caused by the general increase in Z_{eff}.

Electron Configuration

The electron configuration of an atom is a designation of how electrons are distributed among various orbitals in principal shells and subshells. In later sections, we will find that many of the physical and chemical properties of elements can be correlated with electron configurations. Here, we will first discuss two fundamental principles of electron configurations.

1) **Pauli exclusion principle**: No two electrons in an atom can have all four quantum numbers alike. The first three quantum numbers (n, l, and m_l) determine a specific orbital, and the last quantum number m_s can only have a value of either +1/2 or −1/2. If three electrons occupy the same orbital, it means that the third electron must have all four quantum numbers the same as either the first or the second electron, which violates Pauli exclusion principle. Therefore, this principle can also be expressed as each atomic orbital can only accommodate two electrons, and these electrons must have opposing spins. Because each principal shell has n^2 orbitals, it can accommodate $2n^2$ electrons at most.

2) **Lowest energy principle**: Electrons occupy orbitals in a way that minimizes the total energy of the atom. The exact order of filling the orbitals has been established by spectroscopic and magnetic experiments. With only a few exceptions, the general order of filling of orbitals is

$$1s, 2s, 2p, 3s, 3p, 4s, 3d, 4p, 5s, 4d, 5p, 6s, 4f, 5d, 6p, 7s, 5f, 6d, 7p, \cdots \tag{7.48}$$

The general order of filling the electronic subshells is illustrated in **Figure 7.27** and can also be demonstrated based on the periodic table. To determine which configuration has the lowest energy, a so-called **Hund's rule** applies in some cases. It states that when degenerate orbitals are available, a configuration in which electrons singly occupy these orbitals with the same spins has the lowest energy. Consequently, an atom tends to have as many unpaired electrons as possible, and a fully filled or half-filled subshell leads to a more stable configuration.

In assigning electron configurations, Pauli exclusion principle is a fundamental principle that must always be valid. However, the lowest energy principle only defines the ground-state electron configuration, which is the most stable or the most energetically favorable configuration for isolated atoms. Other configurations that are less stable and with higher energy, may still exist, as far as they do not violate Pauli exclusion principle, and are said to be in excited states.

不同的亚层能级。而亚层并没有进一步的能级分裂，这是由于亚层的所有轨道均具有相同的径向特征，因此感受的 $Z_{有效}$ 相同。n 值较高时相邻能级之间间距减小（因为 $E_n \propto 1/n^2$）以及亚层能级分裂（由于屏蔽和钻穿效应）的组合效应，使得一些能级发生交错。

然而，能级的顺序对于所有元素而言并非一成不变。例如，由于 $4s$ 电子相对于 $3d$ 电子存在额外的钻穿效应，某些元素（如 K 和 Ca）的 $4s$ 能级通常低于 $3d$ 能级，但其他元素（如 Li、Na、Sc 和 Ti）则相反。**图 7.26** 比较了 H、Li、Na 和 K 的前几个主电子层的轨道能级图。注意，随着原子序数的增加，所有轨道能量均稳定下降，这是由 $Z_{有效}$ 的逐渐增加所导致的。

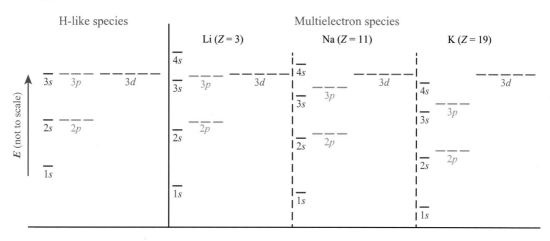

Figure 7.26 Comparison of the orbital energy-level diagrams for hydrogen-like species and multielectron species.

图 7.26 类氢物种与多电子物种轨道能级图的比较。

电子组态

原子的**电子组态**（也称**电子构型**）指的是电子在主层和亚层的各种轨道上的排布。在后续章节里，我们会发现元素的许多物理和化学性质均与其电子组态相关。这里我们将首先讨论电子组态的两个基本原理。

1) **泡利不相容原理**：原子中没有两个电子的四个量子数完全相同。前三个量子数（n、l 和 m_l）决定了某个特定的轨道，而最后一个量子数 m_s 的值只能为 $+1/2$ 或 $-1/2$。如果在同一轨道内存在三个电子，这意味着第三个电子的所有四个量子数，必须与第一个或第二个电子完全相同，而这违反了泡利不相容原理。因此该原理也可表述为：每个原子轨道仅能容纳两个电子，且这两个电子必须自旋相反。由于每个主电子层有 n^2 个轨道，因此最多可以容纳 $2n^2$ 个电子。

2) **能量最低原理**：电子以使原子总能量最低的方式占据轨道。由光谱和磁学实验可确定填充轨道的确切顺序。除了少数例外，轨道填充的一般顺序为

$$1s, 2s, 2p, 3s, 3p, 4s, 3d, 4p, 5s, 4d, 5p, \\ 6s, 4f, 5d, 6p, 7s, 5f, 6d, 7p, \cdots \tag{7.48}$$

填充电子亚层的一般顺序如**图 7.27** 所示，也可基于元素周期表证实。为了确定哪种组态具有最低能量，在某些情况下可采用所谓的**洪特规则**。该规则指出，当存在多个可用的简并轨道时，电子以自旋相同的方式分占不同轨道的组态具有最低能量。因

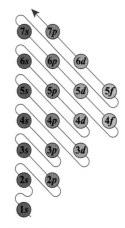

Figure 7.27 The general order of filling the electronic subshells.

图 7.27 填充电子亚层的一般顺序。

Table 7.10 Ground-State Electron Configurations for A Series of Elements by the Aufbau Process
表 7.10 通过构造过程得到的一系列元素的基态电子组态

No. of Period（周期序号）	Atomic No.（原子序数）	Element Symbol（元素符号）	Ground-State Electron Configurations（基态电子组态）
1	1	H	$1s^1$
1	2	He	$1s^2$
2	3	Li	[He] $2s^1$
2	4	Be	[He] $2s^2$
2	5	B	[He] $2s^2 2p^1$
2	6	C	[He] $2s^2 2p^2$
2	7	N	[He] $2s^2 2p^3$
2	8	O	[He] $2s^2 2p^4$
2	9	F	[He] $2s^2 2p^5$
2	10	Ne	[He] $2s^2 2p^6$
3	11	Na	[Ne] $3s^1$
3	12	Mg	[Ne] $3s^2$
3	13	Al	[Ne] $3s^2 3p^1$
3	14	Si	[Ne] $3s^2 3p^2$
3	15	P	[Ne] $3s^2 3p^3$
3	16	S	[Ne] $3s^2 3p^4$
3	17	Cl	[Ne] $3s^2 3p^5$
3	18	Ar	[Ne] $3s^2 3p^6$
4	19	K	[Ar] $4s^1$
4	20	Ca	[Ar] $4s^2$
4	21	Sc	[Ar] $3d^1 4s^2$
4	22	Ti	[Ar] $3d^2 4s^2$
4	23	V	[Ar] $3d^3 4s^2$
4	24	Cr	[Ar] $3d^5 4s^1$
4	25	Mn	[Ar] $3d^5 4s^2$
4	26	Fe	[Ar] $3d^6 4s^2$
4	27	Co	[Ar] $3d^7 4s^2$
4	28	Ni	[Ar] $3d^8 4s^2$
4	29	Cu	[Ar] $3d^{10} 4s^1$
4	30	Zn	[Ar] $3d^{10} 4s^2$
4	31	Ga	[Ar] $3d^{10} 4s^2 4p^1$
4	32	Ge	[Ar] $3d^{10} 4s^2 4p^2$
4	33	As	[Ar] $3d^{10} 4s^2 4p^3$
4	34	Se	[Ar] $3d^{10} 4s^2 4p^4$
4	35	Br	[Ar] $3d^{10} 4s^2 4p^5$
4	36	Kr	[Ar] $3d^{10} 4s^2 4p^6$

There are many ways to represent the electron configurations. Here, we give an example of representing the ground-state electron configuration of a nitrogen atom in three different ways:

Condensed *spdf* notation:　　　N　　$1s^2 2s^2 2p^3$

Expanded *spdf* notation:　　　N　　$1s^2 2s^2 2p_x^1 2p_y^1 2p_z^1$

Orbital diagram:　　　　　　　N　　↑↓　↑ ↑ ↑
　　　　　　　　　　　　　　　　　　2s　　2p

Two electrons in the same orbital must have opposing spins, and are called **paired electrons**. Electrons that singly occupy the degenerate orbitals with the same (or parallel) spins are called **parallel-spin electrons**. According to Hund's rule, the ground-state electron configuration of a nitrogen atom has three parallel-spin electrons in three $2p$ orbitals.

The principal shell with the highest n in an atom is called the **valence shell**. Electrons in the valence shell are called **valence electrons**. The nucleus and the inner electrons that have a noble-gas configuration

此，原子倾向于拥有尽可能多的未成对电子，全充满或半充满的亚层对应于更稳定的组态。

在分配电子组态时，泡利不相容原理是一个必须始终满足的基本原理。而能量最低原理仅定义了基态的电子组态，这是孤立原子最稳定或能量上最有利的组态。只要不违反泡利不相容原理，其他不太稳定、能量较高的组态仍可存在，称为处于激发态。

有许多方法可用于表示电子组态，这里我们以氮原子的基态电子组态为例，给出三种不同的表示方法：

简版 spdf 表示法：　　　　N　　$1s^2 2s^2 2p^3$

扩展 spdf 表示法：　　　　N　　$1s^2 2s^2 2p_x^1 2p_y^1 2p_z^1$

电子轨道图表示法：　　　　N　　(↑↓) (↑)(↑)(↑)
　　　　　　　　　　　　　　　　2s　　2p

同一轨道上的两个电子必须具有相反的自旋，称为**成对电子**。以自旋相同（或平行）的方式分占不同简并轨道的电子，称为**自旋平行电子**。根据洪特规则，氮原子的基态电子组态在三个 2p 轨道上存在三个自旋平行电子。

原子中 n 值最高的主电子层称为**价层**，价层中的电子称为**价电子**。原子核以及具有稀有气体组态的内层电子共同组成**芯**。电子组态也可用芯价表示法来表示，由稀有气体芯和价电子组成，如：

芯价表示法：　　　　　　　N　　[He] $2s^2 2p^3$

构造过程

写出电子组态的方法称为**构造 (Aufbau) 过程**（或构造原理），其中 Aufbau 是一个德语单词，意为"构造"。这是一个通过逐渐向低能轨道添加电子来"构造"原子、以获得最低总能量的假想过程。**表 7.10** 总结了通过构造过程得到的一系列元素的基态电子组态。

7.8 元素周期律与元素周期表

元素周期表是将化学元素按照原子序数、电子组态以及重复出现的化学性质排列的表格形式。元素周期表的结构显示了周期性的趋势（即**元素周期律**）。

元素周期律与门捷列夫元素周期表

元素周期律由德米特里·门捷列夫（1834—1907）和洛萨·迈耶（1830—1895）于 1869 年独立提出。他们指出，如果按其原子质量的顺序排列，元素可被归类为周期性重复的具有相似化学和物理性质的族。迈耶发现，当用摩尔体积（即元素的摩尔质量除以其固体形式的密度）与原子质量作图时，折线图在 Li、Na、K、Rb 和 Cs 上出现了一系列极大值。后来，迈耶还观察到元素及其化合物的其他性质（如硬度和沸点等）均具有类似的周期性趋势。

在迈耶之前的几个月，门捷列夫制定了周期律，并创建了一个富有远见的元素周期表版本。他不仅修正了一些已知元素在当时已被接受的性质，还预测了八种未发现元素的性质。例如，门捷列夫将铀的价从 3 修正为 6，原子质量从 120 修正为 240（接近现代的质

form a **core**. The electron configuration can also be represented by a core-valence notation, which consists of a noble-gas core and the valence electrons, as:

Core-valence notation:　　　　　　N　　[He] $2s^22p^3$

The Aufbau Process

The method for writing electron configurations is called the **Aufbau process**, where *Aufbau* is a German word that means "building up". This is a hypothetical process in which an atom is "built up" by progressively adding electrons to the low-lying orbitals to achieve the lowest total energy. **Table 7.10** summarizes the ground-state electron configurations for a series of elements by the Aufbau process.

7.8 The Periodic Law and the Periodic Table

The **periodic table** of elements is a tabular display of the chemical elements, arranged by atomic number, electron configuration, and recurring chemical properties. The structure of the periodic table shows periodic trend (or **periodic law**).

The Periodic Law and Mendeleev's Periodic Table

The periodic law was independently proposed by Dmitri Mendeleev (1834—1907) and Lothar Meyer (1830—1895) in 1869, stating that if the elements are arranged in the order of their atomic masses, they can be classified into groups of similar chemical and physical properties repeated periodically. Meyer found that when the molar volume, which is the molar mass of an element divided by the density of its solid form, was plotted against the atomic mass, the line chart shows a series of maxima at Li, Na, K, Rb, and Cs. Later, Meyer also observed similar periodic trends with other properties of the elements and their compounds, such as hardness and boiling points.

Several months before Meyer, Mendeleev formulated the periodic law and created a far sighted version of the periodic table of elements. He not only corrected the then-accepted properties of some known elements, but also predicted the properties of eight undiscovered elements. For example, Mendeleev corrected the valence of uranium from 3 to 6, and the atomic mass from 120 to 240 (close to the modern value of 238). He also accurately predicted the properties of new elements he called ekaaluminium and ekasilicon, which were now known as gallium (Ga) and germanium (Ge), found in 1875 and 1886, respectively.

In his periodic table, Mendeleev arranged the elements into eight groups and twelve rows. He put formulas of their oxides and hydrides on top of each group. For example, in group I, Mendeleev listed those elements at the maximum in the line chart of molar volumes in the order of decreasing melting points as

Li (174°C) > Na (97.8°C) > K (63.7°C) >Rb (38.9°C) > Cs (28.5°C)

He also noticed that those elements all exhibit an oxidation state of +1 and form oxides with the formula R_2O. Those are the elements we now know as alkali metals.

In his periodic table, Mendeleev left no blanks for the noble gas because he did not anticipate them. The discoverer of the noble gas, William Ramsay (1852—1916), proposed to place them in a separate group called group 0.

Modern Periodic Table

When creating his periodic table, Mendeleev placed several elements out of the order of increasing atomic mass in order to put them into the proper groups. He attributed this to the errors in the experimental measurements of atomic masses. For example, Mendeleev placed tellurium (Te) before iodine, but noted that tellurium has a higher atomic mass than iodine. So, he incorrectly predicted that the measured atomic mass of tellurium was at fault. However, it became clear that a few elements might always remain "out of order" in atomic mass. Later, based on the research led by Henry Moseley (1887—1915) on the X-ray spectra of the elements, it was clear that the periodic law depends on the atomic number rather than atomic mass.

量数 238）。他还准确预测了称为类铝和类硅的新元素的性质，而这两种元素现在称为镓（Ga）和锗（Ge），分别发现于 1875 年和 1886 年。

在其周期表中，门捷列夫将元素分成八族十二行，将元素的氧化物和氢化物的化学式放在每族顶部。例如，在第 1 族（IA）中，门捷列夫将摩尔体积折线图中极大值所对应的元素按照熔点降低的顺序排列为

Li (174 °C) > Na (97.8 °C) > K (63.7 °C) > Rb (38.9 °C) > Cs (28.5 °C)

他还注意到，这些元素均表现出 +1 的氧化态，并形成具有式 R_2O 的氧化物。这些元素即我们现在所知道的碱金属。

由于没有预料到稀有气体（惰性气体），门捷列夫并没有在其周期表中留下它们的位置。稀有气体的发现者威廉·拉姆齐（1852—1916）提议，将它们置于单独一族，称为第 0 族。

现代元素周期表

在创建周期表时，为了将一些元素归入适当的族中，门捷列夫并没有将其按照原子质量递增的顺序排列。他把这归因于原子质量实验测量中的误差。例如，门捷列夫将碲（Te）置于碘之前，但注意到碲的原子质量高于碘。因此，他错误地预测了碲的测量原子质量有误。然而很明显，一些元素的原子质量可能总是处于"不按顺序排列"中。后来，根据亨利·莫斯利（1887—1915）对元素 X 射线光谱的研究，显然周期律取决于原子序数而非原子质量。

量子理论建立后，人们很快意识到周期表与量子理论之间的主要联系在于电子组态：周期表中同一族元素的电子组态相似。之所以具有类似物理和化学性质的元素会周期性重复出现，直接源于类似的原子外层电子组态的周期性重复出现。这带来了现代元素周期表，其中的元素按照 18 个族排列。例如，第 17 族为卤素，所有元素均具有 7 个外层（价）电子，组态为 ns^2np^5。

现代元素周期表（**图 7.28**）可分为五个近矩形区域（称为**区**），与最后一个价电子填充的亚层相关：

1) s 区：最后一个价电子填充在最高 n 值的 s 轨道中，组态为

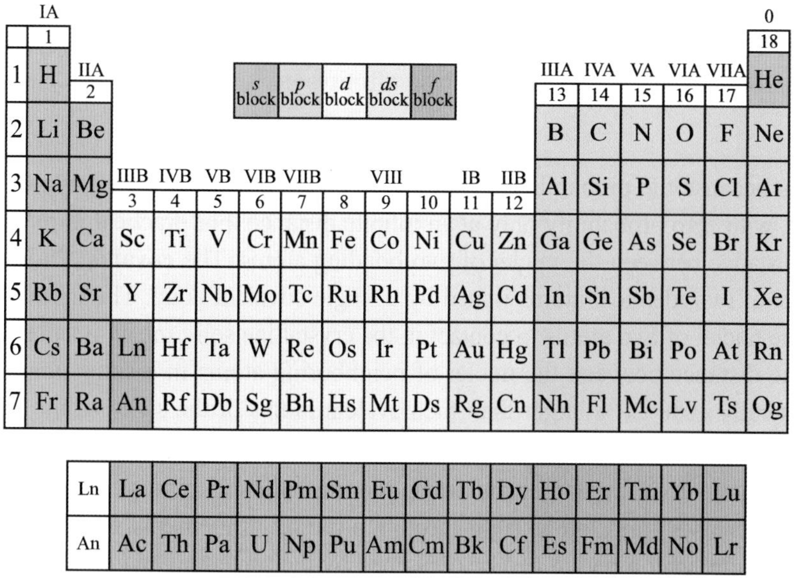

Figure 7.28 Different blocks in a modern periodic table.

图 7.28 现代元素周期表的不同分区。

After the quantum theory was established, it was soon realized that the main connection between the periodic table and quantum theory lies in electron configurations: Elements in the same group of the periodic table show similar electron configurations. The periodic recurrence of elements with similar physical and chemical properties results directly from the periodic recurrence of similar electronic configurations in the outer shells of the atoms. This leads to the modern periodic table, in which elements are arranged in 18 groups. For example, group 17 elements are halogens, all with 7 outer-shell (valence) electrons in the configuration ns^2np^5.

The modern periodic table (**Figure 7.28**) can be divided into five near-rectangular areas (called the **blocks**), associated with the subshells where the last valence electron fills:

1) *s* block: The last valence electron fills in the *s* orbital of highest *n*, with a configuration of $ns^{1\sim2}$. The *s* block consists of group 1 (alkali metals) and group 2 (alkaline earth metals) as well as hydrogen and helium.

2) *p* block: The last valence electron fills in the *p* orbitals of highest *n*, with a configuration of $ns^2np^{1\sim6}$. The *p* block comprises groups 13, 14, 15 (pnictogens), 16 (chalcogens), 17 (halogens), and 18 (noble gases, except helium).

3) *d* block: The last valence electron fills in the *d* orbitals of second highest *n*, or *n*−1, with a configuration of $(n-1)d^{1\sim9}ns^{1\sim2}$ (with a few exceptions). The *d* block consists of groups 3~10.

4) *ds* block: The last valence electron fills in the *s* orbital of highest *n*, but the inner (*n*−1)*d* orbitals are fully filled, unlike *s* block with empty (*n*−1)*d* orbitals. The electron configuration is $(n-1)d^{10}ns^{1\sim2}$. The *ds* block comprises groups 11 and 12. In some textbooks, the *ds* block is incorporated into the *d* block.

5) *f* block: The last valence electron fills in the *f* orbitals of third highest *n*, or *n*−2, with a configuration of $(n-2)f^{1\sim14}(n-1)d^{0\sim2}ns^2$. The *f* block are the lanthanides (Ln, $Z = 57\sim71$) and the actinides (An, $Z = 89\sim103$).

Both *s* and *p* blocks have their last valence electrons fill in the outermost shell, and constitute the **main-group elements**. Those lie between the *s* and *p* blocks are known as **transition elements**. IUPAC defines a transition element as an element whose atom has an incomplete *d* subshell, or which can give rise to cations with an incomplete *d* subshell. By this definition, transition elements comprise all *d* blocks as well as group 11 elements (Cu, Ag, and Au), but do not include group 12 elements (Zn, Cd, Hg). The *f* block elements are also known as **inner transition elements**, which are often offset below the rest of the periodic table and have no group numbers. Since transition and inner transition elements are all metals, they are also called transition metals and inner transition metals. A hypothetical *g* block is expected to begin around element 121, only a few elements away from what is currently known.

7.9 Periodic Properties of the Elements

As the basis of the periodic table, electron configurations of the elements also affect their various properties, such as atomic radius, ionic radius, ionization energy, electron affinity, etc.

Atomic Radius

It is not easy to define atomic radius because the probability of finding an electron decreases continuously with increasing distance from the nucleus, with zero probability only at an infinite distance. For practical purpose, atomic radius is defined in terms of the distance between the nuclei of two bonding atoms. The **covalent radius** is one-half the distance between the nuclei of two identical atoms joined by a single covalent bond. The **metallic radius** is one-half the distance between the nuclei of two atoms in contact in the crystalline solid metal. Similarly, the **van der Waals radius** is one-half the distance between the nuclei of neighboring atoms in a solid sample of a noble gas. The metallic radius is generally 10%~20% larger than the covalent radius of the same atom. For example, the metallic and covalent radii of sodium are 186 and 157 pm, respectively.

In general, the more electronic shells in an atom, the larger the atom. Atomic radius increases from top to bottom through a group of elements and decreases from left to right through a period of elements. The general decreasing trend in atomic radii from left to right over a period can be explained by the screening and penetration

$ns^{1~2}$。s 区包括第 1 族（碱金属）和第 2 族（碱土金属）以及氢和氦。

2) p 区：最后一个价电子填充在最高 n 值的 p 轨道中，组态为 $ns^2np^{1~6}$。p 区包括第 13、14、15 族（氮族）、16 族（硫族）、17 族（卤素）和 18 族（稀有气体，氦除外）。

3) d 区：最后一个价电子填充在第二高 n 值（即 $n-1$）的 d 轨道中，组态为 $(n-1)d^{1~9}ns^{1~2}$（存在少数例外）。d 区包括第 3~10 族。

4) ds 区：最后一个价电子填充在最高 n 值的 s 轨道，但内层的 $(n-1)d$ 轨道全充满，不像 s 区具有空的 $(n-1)d$ 轨道。组态为 $(n-1)d^{10}ns^{1~2}$。ds 区包括第 11 族和第 12 族。在一些教材中，ds 区被并入 d 区。

5) f 区：最后一个价电子填充在第三高 n 值（即 $n-2$）的 f 轨道中，组态为 $(n-2)f^{1~14}(n-1)d^{0~2}ns^2$。$f$ 区包括镧系元素（Ln，$Z=57~71$）和锕系元素（An，$Z=89~103$）。

s 区和 p 区的最后一个价电子填充在最外层，构成了**主族元素**。位于 s 区和 p 区之间的元素称为**过渡元素**。IUPAC 将过渡元素定义为原子具有不完整的 d 亚层，或者可以产生具有不完整 d 亚层的阳离子的元素。根据该定义，过渡元素包括所有 d 区元素以及第 11 族元素（Cu、Ag 和 Au），但不包括第 12 族元素（Zn、Cd、Hg）。f 区元素也称**内过渡元素**，通常移至周期表其余部分之下，且没有族号。由于过渡元素和内过渡元素均为金属，也称过渡金属和内过渡金属。假想的 g 区预计开始于 121 号元素附近，仅与当前已知的元素相距几种元素。

7.9 元素性质的周期性

元素的电子组态是元素周期表的基础，它同样也会影响元素的各种性质，如原子半径、离子半径、电离能、电子亲和能等。

原子半径

定义原子半径并不容易，因为找到电子的概率随着与原子核距离的增加而不断降低，只有在无限远处才为零。出于实际目的，原子半径是根据两个成键原子的核间距来定义的。**共价半径**是两个相同原子以共价单键连接时核间距的一半。**金属半径**是晶态固体金属中相接触的两个原子的核间距的一半。类似地，**范德华半径**是惰性气体的固体样品中，相邻原子的核间距的一半。金属半径通常比同一原子的共价半径大 10% ~ 20%。例如，钠的金属半径和共价半径分别为 186 pm 和 157 pm。

一般来说，原子中的电子壳层越多，原子就越大。原子半径在同族元素中从上至下逐渐增大，在同周期元素中从左至右逐渐减小。原子半径在同一周期内从左至右呈总体下降的趋势，可用 **7.7 节**讨论过的屏蔽和钻穿效应来解释。对于主族元素，最后一个价电子总是填充在最外层，同一壳层内屏蔽效应的微小差异，导致了原子半径从左至右连续下降。对于过渡元素，最后一个价电子填充在 $(n-1)d$ 轨道，是具有更强屏蔽效应的内层。尽管 $Z_{有效}$ 仍随 Z 的增加而增加，但原子半径的减小幅度低于主族元素的减小幅度。对于内过渡元素，最后一个价电子填充在 $(n-2)f$ 轨道，具有更显著的屏蔽效应，导致

effects that we have discussed in **Section 7.7**. For main-group elements, the last valence electron always fills in the outermost shell, and the little difference in screening within the same shell leads to a continuous decrease in atomic radii from left to right. For transition elements, the last valence electron fills in the $(n-1)d$ orbital, which is in an inner shell with stronger screening. Although Z_{eff} still increases with increasing Z, the magnitude of decrease in atomic radii is less than that for the main-group elements. For inner transition elements, the last valence electron fills in the $(n-2)f$ orbital that has even more significant screening, causing a very slight decrease in atomic radii. The average decrease in atomic radii for neighboring elements through a period is

$$\text{Main-group (~10 pm)} > \text{transition (~5 pm)} > \text{inner transition (less than 1 pm)}$$

The lanthanides (Ln, Z = 57~71) mainly differ from each other in their $4f$ electron configurations. The so-called **lanthanide contraction** effect refers to the slight decrease in their atomic radii due to the stronger screening effect of $4f$ electrons than $5d$ electrons. Lanthanide contraction normally has two meanings:

1) Influence on the Ln series itself: the decrease in radii of Ln is smaller than otherwise expected. If there were no $4f$ orbitals, the valence electrons would fill in $5d$ orbitals instead, and the decrease in atomic radii would be more significant.

2) Influence on the post-Ln elements: the radii of the post-Ln elements are smaller than otherwise expected. The total decrease in radius from La to Lu is about 11 pm, which compensates the increase in radius from period 5 to 6, causing the radii as well as properties of the post-Ln elements to be very close to those of the previous elements, for example, Zr vs. Hf, Nb vs. Ta, and Mo vs. W, in the same group. This is the more important meaning of lanthanide contraction.

Meanwhile, we can see two irregular peaks at Europium (Eu) and Ytterbium (Yb) in the atomic radii of lanthanides. This irregularity can be explained by Hund's rule. Eu ($4f^7 6s^2$) has a half-filled $4f$ subshell and Yb ($4f^{14} 6s^2$) has a fully filled $4f$ subshell, both are more symmetric and stable configurations with relatively larger atomic radii.

Ionic Radius

The **ionic radius** is also defined based on the distance between the nuclei of ions joined by an ionic bond. However, because ionic bonds cannot be formed between identical ions, this distance must be properly apportioned between the cation and anion. Assigning O^{2-} an ionic radius of 140 pm as the reference, the ionic radii of all other ions can be inferred accordingly.

Cations are formed by losing electrons from the atoms. Thus, cations are smaller than the corresponding atoms. For **isoelectronic** (species with equal number of electrons in identical configurations) cations, the more positive the ionic charge, the higher the nuclear charge, and the smaller the ionic radius. For example, K^+ and Ca^{2+} are isoelectronic cations with the same configuration as Ar. The ionic radii $r(Ca^{2+})$ = 100 pm is smaller than $r(K^+)$ = 138 pm.

Similarly, anions are formed by gaining electrons to the atoms. Thus, anions are larger than the corresponding atoms. For isoelectronic anions, the more negative the ionic charge, the lower the nuclear charge, and the larger the ionic radius. For example, F^- and O^{2-} are isoelectronic anions with the same configuration as Ne. The ionic radius $r(O^{2-})$ = 140 pm is larger than $r(F^-)$ = 133 pm.

Knowledge of atomic and ionic radii can help to tune certain properties of elements. One example concerns strengthening glass. Normal glass containing Na^+ and Ca^{2+} ions is brittle and shatters easily when struck by a hard blow. One way to strengthen the glass is to replace the surface Na^+ ions with larger K^+ ions to fill up the surface sites. This replacement results in a shatter-resistant glass that has less opportunity for cracking. Another example is the striking results when Cr^{3+} ions replace about 1% of the Al^{3+} in Al_2O_3. This substitution is possible because Cr^{3+} ions are only slightly larger (by 9 pm and by 17%) than Al^{3+} ions. This substitution makes corundum into ruby. Ruby and other gemstones can be made artificially and are used as jewelry as well as in devices such as lasers.

Ionization Energy

In an atom, negatively charged electrons are attracted to the positively charged nucleus and additional energy

了原子半径的减小幅度非常轻微。同一周期内相邻元素原子半径的平均减少量为：

主族元素（≈10 pm）> 过渡元素（≈5 pm）> 内过渡元素（小于 1 pm）

镧系元素（Ln，$Z=57\sim71$）彼此的主要不同，在于其 $4f$ 电子组态。所谓的**镧系收缩**指的是，由于 $4f$ 电子比 $5d$ 电子具有更强的屏蔽效应，导致其原子半径略微减小的现象。镧系收缩通常具有两层含义：

1) 对镧系元素自身的影响：镧系元素半径的减小低于预期。如果没有 $4f$ 轨道，价电子将填充在 $5d$ 轨道，原子半径的减小将更加显著。
2) 对镧系之后元素的影响：镧系之后元素的半径低于预期。从 La 到 Lu 的总半径减小约为 11 pm，这补偿了从第五周期到第六周期的半径增加，导致镧系之后元素的半径和性质，非常接近同一族中上一元素的半径和性质（如 Zr 与 Hf、Nb 与 Ta、Mo 与 W）。这是镧系收缩更为重要的意义所在。

同时，镧系元素的原子半径在铕（Eu）和镱（Yb）处可以看到两个不规则的峰值，这种不规则性可用洪特规则来解释。Eu（$4f^7 6s^2$）具有半充满的 $4f$ 亚层，Yb（$4f^{14} 6s^2$）具有全充满的 $4f$ 亚层，两者均为更对称、更稳定的组态，具有相对较大的原子半径。

离子半径

离子半径也是基于以离子键连接的离子的核间距来定义的。然而，由于同种离子之间不能形成离子键，因此该核间距必须在阳离子和阴离子之间适当分配。指定 O^{2-} 的离子半径为 140 pm 作为参考值，据此可推断出所有其他离子的离子半径。

阳离子由原子失去电子而形成，因此阳离子半径小于相应的原子。对于**等电子体**（指电子数和电子组态均相同的物种）阳离子，离子的正电荷越多，核电荷就越高，离子半径就越小。例如，K^+ 和 Ca^{2+} 是具有与 Ar 相同组态的等电子体阳离子，离子半径 $r(Ca^{2+})$=100 pm 小于 $r(K^+)$ = 138 pm。

类似地，阴离子由原子获得电子而形成，因此阴离子半径大于相应的原子。对于等电子体阴离子，离子的负电荷越多，核电荷就越低，离子半径就越大。例如，F^- 和 O^{2-} 是具有与 Ne 相同组态的等电子体阴离子，离子半径 $r(O^{2-})$ = 140 pm 大于 $r(F^-)$ = 133 pm。

了解原子半径和离子半径的知识有助于调控元素的某些性质。以强化玻璃为例，含有 Na^+ 和 Ca^{2+} 离子的普通玻璃比较脆，受到重击时容易破碎。一种强化玻璃的方法是采用更大的 K^+ 离子取代表面的 Na^+ 离子以填充表面位点。这种替换得到了破碎可能性较小的抗碎玻璃。另一个例子是，用 Cr^{3+} 离子取代 Al_2O_3 中约 1% 的 Al^{3+} 时会出现惊人的结果。由于 Cr^{3+} 离子半径仅比 Al^{3+} 离子稍大（大 9 pm，约 17%），这种替代是可能的，使得刚玉变成了红宝石。红宝石及其他宝石均可人工制造，用于珠宝制作以及激光等设备。

电离能

原子中带负电的电子与带正电的原子核相互吸引，需要额外的能量才能使原子中的电子电离。如果提供足够的能量，电子将逐一电离，最松散的电子首先被释放。第一**电离能**（I_1）是基态的气态原子释放出第一个电子所需吸收的能量。释放出第二个电子的相应能

is necessary to ionize electrons from an atom. If sufficient energy is provided, the electrons will be ionized one by one, with the most loosely held electron being expelled first. The first **ionization energy** (I_1) is the quantity of energy a ground-state gaseous atom must absorb to expel the first electron. The corresponding energy to expel the second electron is called the second ionization energy (I_2). Further ionization energies are I_3, I_4, and so on. Each succeeding ionization energy is invariably larger than the preceding one, i.e., $I_1<I_2<I_3<…$, because it becomes more difficult to ionize negatively charged electrons from a cation than from a neutral atom, and so on. The lower the ionization energy, the more easily the electron is lost, and the more metallic a species is considered to be.

Ionization energies are usually measured through experiments based on the photoelectric effect, in which gaseous atoms at low pressures are bombarded with photons of sufficient energy to eject an electron. For example

$$Mg(g) \rightarrow Mg^+(g) + e^- \quad I_1 = 737.7 \text{ kJ mol}^{-1}$$
$$Mg^+(g) \rightarrow Mg^{2+}(g) + e^- \quad I_2 = 1451 \text{ kJ mol}^{-1}$$
$$Mg^{2+}(g) \rightarrow Mg^{3+}(g) + e^- \quad I_3 = 7733 \text{ kJ mol}^{-1}$$

We can see that $I_1<I_2\ll I_3$. Therefore, ionization energies can be used to distinguish valence electrons from core electrons. A large break in ionization energy appears when the first core electron is removed.

As discussed previously in **Section 7.3**, the ionization energy is the energy required to ionize an electron from the corresponding orbital to infinity, as

$$I = E_\infty - E_n = 0 - \left(-R_H \frac{Z_{eff}^2}{n^2}\right) = R_H \frac{Z_{eff}^2}{n^2} \tag{7.49}$$

where n is the principal quantum number of the electron to be ionized. **Equation (7.49)** suggests a linear relationship between ionization energy and Z_{eff}^2. In fact, one of the earliest estimates of Z_{eff} was obtained by analyzing the data of ionization energies.

From **Equation (7.49)**, we understand that ionization energy tends to decrease down a group, due to increased n and slightly increased Z_{eff}. Thus, atoms lose electrons more easily and become more metallic down a group. Meanwhile, ionization energy tends to increase from left to right across a period, since n remains constant and Z_{eff} increases. **Table 7.11** lists the stepwise ionization energies for the third-period elements. With minor exceptions, the increase trend from left to right across the period is clear. For I_1, the reversal occurs at Al and S. The reason that $I_1(Al) < I_1(Mg)$ is because Mg loses an electron from a lower-energy 3s orbital whereas Al loses an electron from a higher-energy 3p orbital. $I_1(S) < I_1(P)$ can be explained by Hund's rule, since it is easier to remove an electron from a less stable S atom ([Ne]$3s^2 3p^4$) and become a more stable half-filled S$^+$ ion ([Ne]$3s^2 3p^3$) than to remove an electron from a more stable half-filled P atom ([Ne]$3s^2 3p^3$) and become a less stable P$^+$ ion ([Ne]$3s^2 3p^2$).

Table 7.11   Stepwise Ionization Energies for the Third-Period Elements (Unit: kJ mol^{-1})
表 7.11   第三周期元素的逐级电离能（单位：kJ mol^{-1}）

I_n	Na	Mg	Al	Si	P	S	Cl	Ar
I_1	495.8	737.7	577.5	786.5	1 012	999.6	1 251	1 521
I_2	4 562	1 451	1 817	1 577	1 907	2 252	2 298	2 666
I_3	6 910	7 733	2 745	3 232	2 914	3 357	3 822	3 931
I_4	9 543	10 543	11 577	4 356	4 964	4 556	5 159	5 771
I_5	13 354	13 630	14 842	16 091	6 274	7 004	6 542	7 238
I_6	16 613	18 020	18 379	19 805	21 267	8 496	9 362	8 781
I_7	20 117	21 711	23 326	23 780	25 431	27 107	11 018	11 995
I_8	25 496	25 661	27 465	29 287	29 872	31 719	33 604	13 842

If we plot the square root of the first ionization energy (in kJ mol^{-1}) for a series of isoelectronic species Na, Mg$^+$, Al^{2+}, Si^{3+}, P^{4+}, S^{5+}, Cl^{6+}, and Ar^{7+}, which are the ionization energies shown above the zigzag diagonal

量称为第二电离能（I_2）。进一步的电离能为 I_3、I_4 等。每个后续的电离能总是大于前一个电离能，即有 $I_1<I_2<I_3<\cdots$，因为从阳离子电离出带负电的电子比从中性原子电离更为困难，以此类推。电离能越低，电子越容易失去，该物种被认为更具金属性。

电离能一般通过光电效应的实验来测量，在该实验中，低压气态原子被具有足够能量的光子轰击以释放出电子。例如

$$Mg(g) \rightarrow Mg^+(g) + e^- \quad I_1 = 737.7 \text{ kJ mol}^{-1}$$
$$Mg^+(g) \rightarrow Mg^{2+}(g) + e^- \quad I_2 = 1451 \text{ kJ mol}^{-1}$$
$$Mg^{2+}(g) \rightarrow Mg^{3+}(g) + e^- \quad I_3 = 7733 \text{ kJ mol}^{-1}$$

可以看到 $I_1<I_2\ll I_3$。因此，电离能可用来区分价电子和芯电子。当第一个芯电子被移除时，电离能值会出现较大的突跃。

如 **7.3 节**所述，电离能是将电子从相应轨道电离至无穷远所需的能量，即

$$I = E_\infty - E_n = 0 - \left(-R_H \frac{Z_{有效}^2}{n^2}\right) = R_H \frac{Z_{有效}^2}{n^2} \quad (7.49)$$

其中 n 为待电离电子的主量子数。**式（7.49）**表明电离能与 $Z_{有效}^2$ 之间存在线性关系。事实上，$Z_{有效}$ 的最早估算方法之一即为通过分析电离能数据获得。

从**式（7.49）**可知，同一族从上至下，由于 n 增加而 $Z_{有效}$ 仅略微增加，故电离能趋于降低。因此同一族从上至下，原子更容易失去电子，金属性更强。同时，同一周期从左至右，由于 n 保持不变而 $Z_{有效}$ 增加，电离能趋于增加。**表 7.11** 列出了第三周期元素的逐级电离能。除了少数例外，该周期从左至右增加的趋势很明显。I_1 在 Al 和 S 处发生了反转。I_1(Al) $<I_1$(Mg) 的原因在于，Mg 从较低能量的 $3s$ 轨道失去电子，而 Al 从较高能量的 $3p$ 轨道失去电子。I_1(S) $<I_1$(P) 的原因可用洪特规则解释，与从更稳定的半充满 P 原子（[Ne]$3s^23p^3$）中移除电子、变为较不稳定的 P$^+$ 离子（[Ne]$3s^23p^2$）相比，从较不稳定的 S 原子（[Ne]$3s^24p^4$）中移除电子、变为更稳定的半充满 S$^+$ 离子（[Ne]$3s^23p^3$）要更加容易。

如果我们将一系列等电子体物种 Na、Mg$^+$、Al^{2+}、Si^{3+}、P^{4+}、S^{5+}、Cl^{6+} 和 Ar^{7+} 的第一电离能（位于**表 7.11** 中锯齿形对角线上方，单位为 kJ mol^{-1}）的平方根，相对于原子序数（Z）绘图，其结果呈线性关系 [**图 7.29(b)**]，拟合公式为

$$\sqrt{I} = 13.5Z - 124$$

与对应原子的 I_1 相对于 Z 的折线图 [**图 7.29(a)**] 相比，线性关系更为简明，且具有科学意义。让我们来尝试推导该线性关系。

上述等电子体物种的电子组态为 $1s^22s^22p^63s^1$，即在氖芯（$1s^22s^22p^6$）之外存在一个价电子（$3s^1$）。由于芯电子均匀地屏蔽了原子核对单个价电子的吸引，这些物种与具有相应 $Z_{有效}$ 的玻尔原子类似。作为近似，我们可用**式（7.49）**计算其电离能，有

$$I = R_H \frac{Z_{有效}^2}{n^2} = R_H \frac{(Z-\sigma)^2}{n^2}$$

$$\sqrt{I} = \sqrt{R_H}\left(\frac{Z-\sigma}{n}\right)$$

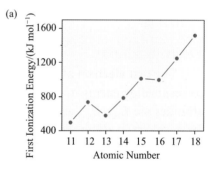

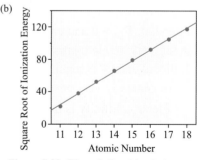

Figure 7.29 The relationships between ionization energy and atomic number. (a) First ionization energies (in kJ mol^{-1}) as a function of atomic number for Na, Mg, Al, Si, P, S, Cl, and Ar. (b) Square root of first ionization energies (in kJ mol^{-1}) as a function of atomic number for Na, Mg$^+$, Al^{2+}, Si^{3+}, P^{4+}, S^{5+}, Cl^{6+}, and Ar^{7+}.

图 7.29 电离能与原子序数的关系图：(a) Na、Mg、Al、Si、P、S、Cl 和 Ar 的第一电离能（单位：kJ mol^{-1}）对原子序数的函数关系图；(b) Na、Mg$^+$、Al^{2+}、Si^{3+}、P^{4+}、S^{5+}、Cl^{6+} 和 Ar^{7+} 的第一电离能（单位：kJ mol^{-1}）的平方根对原子序数的函数关系图。

line in **Table 7.11**, with respect to their atomic number (Z), a linear relationship [**Figure 7.29(b)**] is observed with the following fitting formula:

$$\sqrt{I} = 13.5Z - 124$$

Compared to the line chart of I_1 of the corresponding atoms with respect to Z [**Figure 7.29(a)**], the linear relationship is more concise and of scientific significance. Let us try to derive the origin of this linear relationship.

The electron configuration for the above isoelectronic species is $1s^2 2s^2 2p^6 3s^1$, i.e., a single valence electron ($3s^1$) beyond the neon core ($1s^2 2s^2 2p^6$). As core electrons evenly screen the nucleus from the single valence electron, these species are analogues of the Bohr atom with the corresponding Z_{eff}. As an approximation, we can use **Equation (7.49)** to calculate the ionization energy, as

$$I = R_{\text{H}} \frac{Z_{\text{eff}}^2}{n^2} = R_{\text{H}} \frac{(Z-\sigma)^2}{n^2}$$

$$\sqrt{I} = \sqrt{R_{\text{H}}} \left(\frac{Z-\sigma}{n} \right)$$

Taking $n = 3$ for the $3s^1$ electron to be ionized, $R_{\text{H}} = 1311.6$ kJ mol^{-1}, and assuming that the core electrons screen the valence electron perfectly and $\sigma = 10$, we have

$$\sqrt{I} = \sqrt{1311.6} \left(\frac{Z-10}{3} \right) = 12.07Z - 120.7$$

This explains the linear relationship of $\sqrt{I}$ with respect to Z and agrees with the experimentally determined expression.

Electron Affinity

The first **electron affinity** (E_{ea1}) is the energy released when a gaseous atom gains the first electron. The corresponding energy to gain the second electron is called the second electron affinity (E_{ea2}). Further electron affinities are E_{ea3}, E_{ea4}, and so on. By definition, electron affinities equal the negative of the enthalpy change for the electron affinity reaction, such as

$$O(g) + e^- \rightarrow O^-(g) \quad E_{\text{ea1}} = -\Delta H_1^\circ = 141.0 \text{ kJ mol}^{-1}$$

$$O^-(g) + e^- \rightarrow O^{2-}(g) \quad E_{\text{ea2}} = -\Delta H_2^\circ = -780 \text{ kJ mol}^{-1}$$

In general, $E_{\text{ea1}} > 0$ and energy is given off when a neutral atom gains an electron for most elements, except for the very stable configurations such as noble gases and N, Mg, etc. Normally, $\ldots < E_{\text{ea3}} < E_{\text{ea2}} < 0$ because additional energy is required for negatively charged anions to gain electrons. Gaseous anions such as O^{2-} and S^{2-} are generally very unstable.

Electron affinity is usually measured as the ionization energy of the corresponding anion, also based on the photoelectric effect.

$$X^-(g) \rightarrow X(g) + e^- \quad I = E_{\text{ph}}$$

$$X(g) + e^- \rightarrow X^-(g) \quad \Delta H = -I \quad \text{and} \quad E_{\text{ea}} = -\Delta H = I$$

Electronegativity

The ionization energy shows the ability for a gaseous atom to lose electrons to form cations, and electron affinity represents the ability for a gaseous atom to gain electrons to become anions. When two gaseous atoms combine to form a molecule, the concept of **electronegativity** is introduced to measure the tendency of an atom to attract a bonding electron pair.

Although the concept was studied by many chemists including Avogadro, the term "electronegativity" was introduced by Jöns J. Berzelius in 1811. However, an accurate scale of electronegativity was not developed until 1932 when Linus Pauling (1901—1994) proposed an electronegativity scale dependent on bond energies. If the bond energies of A—A, B—B, and A—B are E_{AA}, E_{BB}, and E_{AB}, respectively, Pauling

对于待电离的 $3s^1$ 电子，$n=3$，且 $R_H=1311.6 \text{ kJ mol}^{-1}$，假定芯电子完美地屏蔽了价电子，故 $\sigma=10$，有

$$\sqrt{I} = \sqrt{1311.6}\left(\frac{Z-10}{3}\right) = 12.07Z - 120.7$$

这解释了 $\sqrt{I}$ 对 Z 的线性关系，且与实验确定的表达式较为一致。

电子亲和能

第一电子亲和能（E_{ea1}）是气态原子获得第一个电子时释放的能量。获得第二个电子的相应能量称为第二电子亲和能（E_{ea2}）。进一步的电子亲和能为 E_{ea3}、E_{ea4} 等。根据定义，电子亲和能等于电子亲和反应的焓变的负值，例如

$$O(g) + e^- \to O^-(g) \quad E_{ea1} = -\Delta H_1^\ominus = 141.0 \text{ kJ mol}^{-1}$$
$$O^-(g) + e^- \to O^{2-}(g) \quad E_{ea2} = -\Delta H_2^\ominus = -780 \text{ kJ mol}^{-1}$$

一般而言，$E_{ea1}>0$，大多数元素的中性原子获得一个电子时会释放能量，但非常稳定的组态如稀有气体和 N、Mg 等除外。通常有 $\cdots < E_{ea3} < E_{ea2} < 0$，因为带负电的阴离子需要额外的能量才能获得电子。气态阴离子（如 O^{2-} 和 S^{2-}）通常很不稳定。

电子亲和能常通过对应阴离子的电离能来测量，也基于光电效应。

$$X^-(g) \to X(g) + e^- \quad I = E_{光子}$$
$$X(g) + e^- \to X^-(g) \quad \Delta H = -I \quad \text{且} \quad E_{ea} = -\Delta H = I$$

电负性

电离能反映了一个气态原子失去电子形成阳离子的能力，而电子亲和能反映了一个气态原子获得电子形成阴离子的能力。当两个气态原子结合形成一个分子时，我们引入**电负性**的概念来衡量原子吸引成键电子对的趋势。

尽管包括阿伏伽德罗在内的许多化学家均对这个概念进行了研究，"电负性"一词由约翰斯·J. 贝泽利乌斯于 1811 年才提出。然而直到 1932 年莱纳斯·鲍林（1901—1994）提出了依赖于键能的电负性标度之后，电负性的精确标度才得以发展。如果化学键 A—A、B—B 和 A—B 的键能分别为 E_{AA}、E_{BB} 和 E_{AB}，鲍林发现 E_{AB} 值总是高于 $\sqrt{E_{AA} \cdot E_{BB}}$。他将这种异核键的额外稳定性归因于离子性成分对成键的贡献，并认为此额外稳定性可作为电负性的标度，即

$$\Delta = E_{AB} - \sqrt{E_{AA} \cdot E_{BB}}$$

鲍林采用一个量纲为 1 的量 χ 来表示电负性。通过实验数据拟合，他提出了以下半经验公式：

$$\Delta = 1.3 \text{ eV} \cdot (\chi_A - \chi_B)^2$$

通过令 $\chi_F = 4.0$ 作为基准值，可以相应地导出其他原子的电负性。在所有元素中，氟（$\chi_F = 4.0$）和铯（$\chi_{Cs} = 0.8$）分别是电负性最大和最小的元素。

1934 年，罗伯特·S. 穆利肯（1896—1986）提出，第一电离能和第一电子亲和能的算术平均值可用于衡量原子吸引成键电子对的趋势，有

found that E_{AB} is always higher than $\sqrt{E_{AA} \cdot E_{BB}}$. He attributed this additional stabilization of the heteronuclear bond to the contribution of ionic canonical forms to the bonding, and believed that this additional stability can serve as the scale to electronegativity, as

$$\Delta = E_{AB} - \sqrt{E_{AA} \cdot E_{BB}}$$

Pauling used a dimensionless quantity χ to represent the electronegativity. By experimental data fitting, he proposed the following semi-empirical formula, as

$$\Delta = 1.3\,\text{eV} \cdot (\chi_A - \chi_B)^2$$

By setting $\chi_F = 4.0$ as a reference, the electronegativity of other atoms can be derived accordingly. Among all elements, fluorine ($\chi_F = 4.0$) and cesium ($\chi_{Cs} = 0.8$) are the most and least electronegative elements, respectively.

In 1934, Robert S. Mulliken (1896—1986) proposed that the arithmetic mean of the first ionization energy and the first electron affinity can be used as a measure of the tendency of an atom to attract a bonding electron pair, as

$$\chi = \frac{1}{2}(I_1 + E_{ea1}) \tag{7.50}$$

As this definition is not dependent on an arbitrary relative scale, it has also been termed absolute electronegativity. However, it is conventional to linearly transform the absolute electronegativity into values that resemble the more familiar Pauling electronegativity values. For I_1 and E_{ea1} in eV, the linear relationship is given by

$$\chi = 0.187(I_1 + E_{ea1}) + 0.17$$

Later, the Allred-Rochow electronegativity, Sanderson electronegativity, and Allen electronegativity were also defined based on different sets of data.

We need always remember that electronegativity is not a property of an atom alone, but rather a property of an atom in a molecule. It is to be expected that the electronegativity of an element will vary with its chemical environment, but it is usually considered to be a transferable property, i.e., similar values will be valid in a variety of situations.

Other Atomic Properties

In this section, we have learned how some atomic properties periodically recur within groups and periods of elements. The metallic and nonmetallic characteristics of atoms can also be associated with a set of atomic properties, especially with electronegativity. The metallic character relates to the ability to lose electrons, and large atomic radii and low electronegativity are generally associated with metals. The metallic character increases across a group, and decreases from left to right over a period. Therefore, the metallic character generally increases in the direction from top-right to bottom-left in the periodic table. Meanwhile, the nonmetallic character relates to the ability to gain electrons, and small atomic radii and high electronegativity are associated with nonmetal. The nonmetallic character decreases down a group, and increases from left to right over a period. Therefore, the nonmetallic character generally increases in the direction from bottom-left to top-right in the periodic table. These general trends in some atomic properties are summarized in **Figure 7.30**.

Metals and nonmetals are often separated by a stairstep diagonal line, and several elements near this line are called metalloids. **Metalloids** are some elements that have a preponderance of properties in between those of metals and nonmetals, such as silicon, germanium, arsenic.

Metals usually form oxides that give basic solutions, whereas nonmetal oxides form acidic solutions. Some nonmetal oxides are called acidic oxides or acid anhydrides. The term **anhydride** means "without water". An acid "without water", i.e., an acid anhydride, becomes an acid when water is added. At the break between clearly basic and acidic properties, some metals and metalloids exhibit **amphoteric** behavior because they react with both acids and bases.

$$\chi = \frac{1}{2}(I_1 + E_{ea1}) \tag{7.50}$$

由于该定义不依赖于任意规定的相对标度，因此也被称为绝对电负性。然而，传统上仍会将这些绝对电负性线性地转换为与更为熟悉的鲍林电负性类似的值。对于以 eV 为单位的 I_1 和 E_{ea1}，其线性关系为

$$\chi = 0.187(I_1 + E_{ea1}) + 0.17$$

后来，基于不同数据集还分别定义了奥尔瑞德 - 罗肖电负性、桑德逊电负性和艾伦电负性。

我们需要始终记住，电负性不仅是原子的性质，更是分子中原子的性质。可以预期，元素的电负性将随其化学环境而变化，但通常视为一种可调用性质，即类似的值在各种情况下均基本有效。

原子的其他性质

本节我们学习了一些原子性质是如何在元素的同族和同周期内周期性重复出现的。原子的金属性和非金属性也可与一些原子性质（特别是电负性）相关联。金属性与失去电子的能力有关，金属通常具有较大的原子半径和较低的电负性。金属性在同一族内从上至下逐渐增加，在同一周期内从左至右逐渐降低。因此，在周期表中金属性通常按照从右上到左下的方向增加。同时，非金属性与得到电子的能力有关，非金属通常具有较小的原子半径和较高的电负性。非金属性在同一族内从上至下逐渐降低，在同一周期内从左至右逐渐增加。因此，在周期表中非金属性通常按照从左下到右上的方向增加。**图 7.30** 总结了原子性质的总体趋势。

金属和非金属通常由一条阶梯形对角线隔开，这条线附近的几种元素称为半金属。**半金属**是具有介于金属和非金属之间的优异性质的一些元素，如硅、锗、砷等。

金属形成的氧化物通常能产生碱性溶液，而非金属的氧化物则会形成酸性溶液。一些非金属氧化物称为酸性氧化物或酸酐。术语**酐**的意思是"无水"。当添加水时，"无水"酸（即酸酐）就变成了酸。在明显的碱性和酸性之间，一些金属和半金属表现出**两性**行为，因为它们与酸和碱均能发生反应。

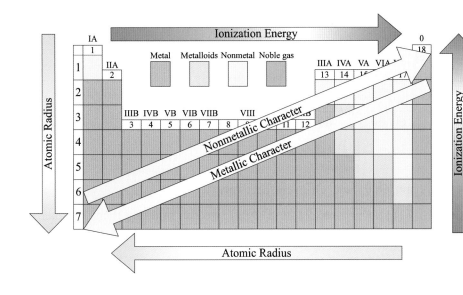

Figure 7.30 The general trends in some atomic properties of elements in the periodic table. Atomic radius refers to the metallic radius for metals and the covalent radius for nonmetals. Ionization energies refer to first ionization energies. Metallic and nonmetallic characters relate generally to the abilities to lose and gain electrons, respectively.

图 7.30 周期表中元素的一些原子性质的总体趋势：原子半径对金属为金属半径，对非金属为共价半径；电离能指第一电离能；金属/非金属性通常分别与失/得电子的能力相关联。

Extended Reading Materials — Separation of Variables for Hydrogen-Like Species

Assuming that the spatial variables r and θ, ϕ are separable, we can separate the time-independent wave function $\psi(r,\theta,\phi)$ for hydrogen-like species into the radial wave function $R(r)$ and the angular wave function $Y(\theta,\phi)$, as

$$\psi_{n,l,m_l}(r,\theta,\phi) = R_{n,l}(r) Y_{l,m_l}(\theta,\phi)$$

Since $R(r)$ is dependent only on r but not on θ and ϕ, and $Y(\theta,\phi)$ is dependent only on θ and ϕ but not on r, we have

$$\frac{\partial \psi}{\partial r} = Y \frac{dR}{dr}, \quad \frac{\partial \psi}{\partial \theta} = R \frac{\partial Y}{\partial \theta} \quad \text{and} \quad \frac{\partial \psi}{\partial \phi} = R \frac{\partial Y}{\partial \phi}$$

For hydrogen-like species, Schrödinger equation is given by

$$-\frac{\hbar^2}{2\mu r^2}\left[\frac{\partial}{\partial r}\left(r^2 \frac{\partial \psi}{\partial r}\right) + \Lambda^2 \psi\right] - \frac{Ze^2}{r}\psi = E\psi$$

$$\frac{1}{r^2}\frac{\partial}{\partial r}\left(r^2 \frac{\partial \psi}{\partial r}\right) + \frac{1}{r^2 \sin\theta}\frac{\partial}{\partial \theta}\left(\sin\theta \frac{\partial \psi}{\partial \theta}\right) + \frac{1}{r^2 \sin^2\theta}\frac{\partial^2 \psi}{\partial \phi^2} = -\frac{2\mu}{\hbar^2}\left(E + \frac{Ze^2}{r}\right)\psi$$

Applying $\psi = RY$ and multiplying $r^2/(RY)$ on both sides, we can derive

$$\frac{1}{R}\frac{d}{dr}\left(r^2 \frac{dR}{dr}\right) + \frac{2\mu r^2}{\hbar^2}\left(E + \frac{Ze^2}{r}\right) = -\frac{1}{Y\sin\theta}\frac{\partial}{\partial \theta}\left(\sin\theta \frac{\partial Y}{\partial \theta}\right) - \frac{1}{Y\sin^2\theta}\frac{\partial^2 Y}{\partial \phi^2}$$

Since the left side is dependent only on r and the right side only on θ and ϕ, both sides must be equal to a constant. This is because any change in r alone will not cause changes on the right, and consequently, the left side will remain unchanged. So will changes in θ and ϕ make the right side unchanged. Therefore, both sides will keep unchanged no matter which variables change, leading to a constant value. Therefore,

$$\frac{1}{R}\frac{d}{dr}\left(r^2 \frac{dR}{dr}\right) + \frac{2\mu r^2}{\hbar^2}\left(E + \frac{Ze^2}{r}\right) = C$$

The above equation is dependent only on r and can be solved to obtain the radial wave function $R(r)$.

$$-\frac{1}{Y\sin\theta}\frac{\partial}{\partial \theta}\left(\sin\theta \frac{\partial Y}{\partial \theta}\right) - \frac{1}{Y\sin^2\theta}\frac{\partial^2 Y}{\partial \phi^2} = C$$

Further assuming that $Y(\theta,\phi) = \Theta(\theta)\Phi(\phi)$ and multiplying $\sin^2\theta$ on both sides, we have

$$\frac{\sin\theta}{\Theta}\frac{d}{d\theta}\left(\sin\theta \frac{d\Theta}{d\theta}\right) + C\sin^2\theta = -\frac{1}{\Phi}\frac{d^2\Phi}{d\phi^2}$$

Similarly, both sides of the above equation must also equal a constant, giving

$$\frac{\sin\theta}{\Theta}\frac{d}{d\theta}\left(\sin\theta \frac{d\Theta}{d\theta}\right) + C\sin^2\theta = N$$

$$-\frac{1}{\Phi}\frac{d^2\Phi}{d\phi^2} = N$$

The above equations can be solved to obtain $\Theta(\theta)$ and $\Phi(\phi)$, respectively.

After the independent solutions of $T(t)$, $R(r)$, $\Theta(\theta)$, and $\Phi(\phi)$, we can multiply them together to obtain the time-independent wave function $\psi(r,\theta,\phi)$, as

$$\psi(r,\theta,\phi) = R(r)\Theta(\theta)\Phi(\phi)$$

拓展阅读材料　类氢物种的变量分离解法

假定空间变量 r 和 θ、ϕ 可分离，我们可将类氢物种的不含时波函数 $\psi(r,\theta,\phi)$ 分离为径向波函数 $R(r)$ 和角度波函数 $Y(\theta,\phi)$，有

$$\psi_{n,l,m_l}(r,\theta,\phi) = R_{n,l}(r) Y_{l,m_l}(\theta,\phi)$$

由于 $R(r)$ 仅与 r 有关而不依赖于 θ 和 ϕ，而 $Y(\theta,\phi)$ 只与 θ 和 ϕ 有关而不依赖于 r，有

$$\frac{\partial \psi}{\partial r} = Y \frac{dR}{dr}, \quad \frac{\partial \psi}{\partial \theta} = R \frac{\partial Y}{\partial \theta} \quad \text{且} \quad \frac{\partial \psi}{\partial \phi} = R \frac{\partial Y}{\partial \phi}$$

类氢物种的薛定谔方程为

$$-\frac{\hbar^2}{2\mu r^2}\left[\frac{\partial}{\partial r}\left(r^2 \frac{\partial \psi}{\partial r}\right) + \Lambda^2 \psi\right] - \frac{Ze^2}{r}\psi = E\psi$$

$$\frac{1}{r^2}\frac{\partial}{\partial r}\left(r^2 \frac{\partial \psi}{\partial r}\right) + \frac{1}{r^2 \sin\theta}\frac{\partial}{\partial \theta}\left(\sin\theta \frac{\partial \psi}{\partial \theta}\right) + \frac{1}{r^2 \sin^2\theta}\frac{\partial^2 \psi}{\partial \phi^2} = -\frac{2\mu}{\hbar^2}\left(E + \frac{Ze^2}{r}\right)\psi$$

应用 $\psi = RY$ 并在两侧同乘以 $r^2/(RY)$，有

$$\frac{1}{R}\frac{d}{dr}\left(r^2 \frac{dR}{dr}\right) + \frac{2\mu r^2}{\hbar^2}\left(E + \frac{Ze^2}{r}\right) = -\frac{1}{Y\sin\theta}\frac{\partial}{\partial \theta}\left(\sin\theta \frac{\partial Y}{\partial \theta}\right) - \frac{1}{Y\sin^2\theta}\frac{\partial^2 Y}{\partial \phi^2}$$

由于左侧仅依赖于 r 而右侧仅依赖于 θ 和 ϕ，因此两侧必须等于一个常数。这是因为 r 单独发生的任何变化均不会导致右侧变化，因此左侧也将保持不变。同样，θ 和 ϕ 的变化也会使得右侧不变。因此，无论哪个变量发生变化，两侧都将保持不变，等于一个恒定值，故

$$\frac{1}{R}\frac{d}{dr}\left(r^2 \frac{dR}{dr}\right) + \frac{2\mu r^2}{\hbar^2}\left(E + \frac{Ze^2}{r}\right) = C$$

上式仅依赖于 r，可求解以获得径向波函数 $R(r)$。

$$-\frac{1}{Y\sin\theta}\frac{\partial}{\partial \theta}\left(\sin\theta \frac{\partial Y}{\partial \theta}\right) - \frac{1}{Y\sin^2\theta}\frac{\partial^2 Y}{\partial \phi^2} = C$$

进一步假定 $Y(\theta,\phi) = \Theta(\theta)\Phi(\phi)$，并在两侧同乘以 $\sin^2\theta$，有

$$\frac{\sin\theta}{\Theta}\frac{d}{d\theta}\left(\sin\theta \frac{d\Theta}{d\theta}\right) + C\sin^2\theta = -\frac{1}{\Phi}\frac{d^2\Phi}{d\phi^2}$$

类似地，上式的两侧也必须等于一个常数，有

$$\frac{\sin\theta}{\Theta}\frac{d}{d\theta}\left(\sin\theta \frac{d\Theta}{d\theta}\right) + C\sin^2\theta = N$$

$$-\frac{1}{\Phi}\frac{d^2\Phi}{d\phi^2} = N$$

求解上述方程，可分别获得 $\Theta(\theta)$ 和 $\Phi(\phi)$。

在求得 $T(t)$、$R(r)$、$\Theta(\theta)$ 和 $\Phi(\phi)$ 的独立解之后，可以将其相乘，得到不含时波函数 $\psi(r,\theta,\phi)$，为

$$\psi(r,\theta,\phi) = R(r)\Theta(\theta)\Phi(\phi)$$

和含时波函数 $\Psi(r,\theta,\phi,t)$，为

$$\Psi(r,\theta,\phi,t) = R(r)\Theta(\theta)\Phi(\phi)T(t)$$

and the time-dependent wave function $\Psi(r,\theta,\phi,t)$, as

$$\Psi(r,\theta,\phi,t) = R(r)\Theta(\theta)\Phi(\phi)T(t)$$

Problems

7.1 Mass spectrometry is one of the most versatile and powerful tools in chemical analysis because of its capacity to discriminate between atoms of different masses. When a sample containing a mixture of isotopes is introduced into a mass spectrometer, the ratio of the peaks observed reflects the ratio of the percent natural abundances of the isotope atoms. This ratio provides an internal standard from which the amount of a certain isotope present in a sample can be determined. This is accomplished by deliberately introducing a known quantity of a particular isotope into the sample to be analyzed. A comparison of the new isotope ratio to the first ratio allows the determination of the amount of the isotope present in the original sample.

An analysis was done on a rock sample to determine its rubidium content. The rubidium content of a portion of rock weighing 0.350 g was extracted, and to the extracted sample was added an additional 19.45 μg of ^{87}Rb. The mass spectrum of this spiked sample showed a ^{87}Rb peak that was 1.12 times as high as the peak for ^{85}Rb. Assuming that the two isotopes react identically, calculate the Rb content of the rock (expressed in parts per million by mass)? The percent natural (molar) abundances and isotopic masses are shown in **Table P7.1**.

Table P7.1 （表 P7.1）

Isotope （同位素）	Natural Abundance （天然丰度）/ (%)	Atomic Mass （原子质量）/u
^{87}Rb	27.83	86.909
^{85}Rb	72.17	84.912

7.2 The work function of mercury is equivalent to 435 kJ mol^{-1} photons.

(a) Can the photoelectric effect be obtained with mercury by using visible light? Explain.

(b) What is the kinetic energy, in joules, of the ejected photoelectrons when light of 215 nm strikes a mercury surface?

(c) What is the velocity, in m s^{-1}, of the ejected photoelectrons in part (b)?

7.3 Determine the wavelength of the line in the emission spectrum of He$^+$ produced by an electron transition from $n=5$ to $n=2$.

7.4 If all other rules governing electron configurations were valid, what would be the electron configuration of Cs if (a) there were three possibilities for electron spin; (b) the quantum number l could have the value n?

7.5 Do the following calculation based on the relationships given in **Table 7.9**:

(a) Calculate the value of r, in terms of a_0, of the nodes for a 3s orbital of the hydrogen atom.

(b) Calculate the value of r, in terms of a_0, at which the probability of finding a 2s electron is maximum for He$^+$.

*7.6 In the ground state of a hydrogen atom, what is the overall probability of finding an electron anywhere in a sphere of radius (a) a_0, or (b) $2a_0$?

*7.7 Emission and absorption spectra of the hydrogen atom exhibit line spectra characteristic of quantized systems. In an absorption spectrum experiment, a sample of hydrogen atoms is irradiated with light at wavelengths ranging from 100 to 1000 nm. In an emission spectrum experiment, the hydrogen atoms are excited through an energy source that provides a range of energies from 1230 to 1240 kJ mol^{-1} to the atoms. Assume that the absorption spectrum is obtained at room temperature, when all atoms are in the ground state.

(a) Calculate the position of the lines in the absorption spectrum.

(b) Calculate the position of the lines in the emission spectrum.

(c) Compare the line spectra observed in the two experiments. In particular, will the overall number of

习题

7.1 质谱是化学分析中最通用且强大的工具之一，因为它能够区分不同质量的原子。当含有同位素混合物的样品被引入质谱仪时，观察到的峰面积的比值，反映了同位素原子的天然丰度的比值。该比值提供了一个可以从中确定样品中存在的某种同位素的量的内标。这是通过在待分析样品中故意引入已知量的特定同位素来实现的。将新的同位素比值与前一比值进行比较，即可确定原始样品中存在的同位素的量。

对某岩石样品进行分析，以确定其中铷的含量。提取重为 0.350 g 的岩石中的铷，并向提取的样品中额外添加 19.45 μg 的 ^{87}Rb。该加标样品的质谱显示，^{87}Rb 峰是 ^{85}Rb 峰的 1.12 倍。假定这两种同位素的反应相同，计算岩石中的 Rb 含量（以质量的百万分率 ppm 表示）。铷的天然（摩尔）丰度及同位素质量如**表 P7.1** 所示。

7.2 汞的功函数相当于 435 kJ mol^{-1} 的光子。
   (a) 用可见光能使汞产生光电效应吗？解释之。
   (b) 当 215 nm 的光照射到汞表面时，产生的光电子的动能是多少焦?
   (c) (b) 中产生的光电子的速度是多少 m s^{-1}?

7.3 在 He$^+$ 的发射光谱中，确定从 $n = 5$ 到 $n = 2$ 的电子跃迁产生的谱线的波长。

7.4 如果其他所有关于电子组态的规则均有效，但 (a) 电子自旋存在三种可能性；(b) 量子数 l 可以具有 n 的值，Cs 的电子组态分别是什么？

7.5 根据**表 7.9** 给出的关系进行以下计算：
   (a) 计算氢原子 $3s$ 轨道节面的 r 值（用 a_0 表示）。
   (b) 计算 He$^+$ 的 $2s$ 电子概率最大处的 r 值（用 a_0 表示）。

*7.6 基态氢原子中，在半径为 (a) a_0 或 (b) $2a_0$ 的球体内找到电子的总概率是多少？

*7.7 氢原子的发射光谱和吸收光谱表现出量子化体系的线状光谱特征。在吸收光谱实验中，用波长范围为 100~1000 nm 的光照射氢原子样品。在发射光谱实验中，能量源提供范围为 1230~1240 kJ mol^{-1} 的能量给氢原子使其激发。假定吸收光谱在室温下获得，此时所有原子均处于基态。
   (a) 计算吸收光谱中各谱线的位置。
   (b) 计算发射光谱中各谱线的位置。
   (c) 比较两个实验中观察到的线状光谱。尤其是，观察到的谱线总数是否相同？

*7.8 类氢物种的不含时薛定谔方程为

$$-\frac{\hbar^2}{2\mu r^2}\left[\frac{\partial}{\partial r}\left(r^2\frac{\partial \psi}{\partial r}\right) + \Lambda^2\psi\right] - \frac{Ze^2}{4\pi\varepsilon_0 r}\psi = E\psi$$

其中 $\Lambda^2 = \dfrac{1}{\sin^2\theta}\dfrac{\partial^2}{\partial \phi^2} + \dfrac{1}{\sin\theta}\dfrac{\partial}{\partial \theta}\left(\sin\theta\dfrac{\partial}{\partial \theta}\right)$。已知 $a_0 = \dfrac{\varepsilon_0 h^2}{\pi\mu e^2}$ 且

lines observed be the same?

*7.8 The time-independent Schrödinger equation for hydrogen-like species is given by:

$$-\frac{\hbar^2}{2\mu r^2}\left[\frac{\partial}{\partial r}\left(r^2\frac{\partial \psi}{\partial r}\right)+\Lambda^2\psi\right]-\frac{Ze^2}{4\pi\varepsilon_0 r}\psi = E\psi$$

where $\Lambda^2 = \frac{1}{\sin^2\theta}\frac{\partial^2}{\partial\phi^2}+\frac{1}{\sin\theta}\frac{\partial}{\partial\theta}\left(\sin\theta\frac{\partial}{\partial\theta}\right)$. Given that $a_0 = \frac{\varepsilon_0 h^2}{\pi\mu e^2}$ and $R_H = \frac{\mu e^4}{8\varepsilon_0^2 h^2}$.

(a) Calculate the energy of 1s orbital (E_{1s}) in the hydrogen atom in terms of the reduced mass (μ) and the Bohr radius (a_0) by applying the above time-independent Schrödinger equation.

(b) Prove that for the hydrogen atom $E_{1s} = -R_H$.

7.9 Calculate the second ionization energy for the He atom based on Z_{eff}. Compare your result with the tabulated value of 5251 kJ mol^{-1}.

7.10 Arrange the following ionization energies in order of increasing value: I_1 for F; I_2 for Ba; I_3 for Sc; I_2 for Na; I_3 for Mg. Explain your answer as well as the basis of any uncertainty.

*7.11 The effective nuclear charge (Z_{eff}) can be modified from the original nuclear charge (Z), as $Z_{eff} = Z - \sigma$, to account for the phenomenon of screening. In 1930, John C. Slater devised the following set of empirical rules to calculate a screening constant (σ) for a designated electron in the orbital ns or np:

(i) Write the electron configuration of the element, and group the subshells as follows: (1s), (2s, 2p), (3s, 3p), (3d), (4s, 4p), (4d), (4f), (5s, 5p), etc.

(ii) Electrons in groups to the right of the (ns, np) group contribute nothing to the screening constant for the designated electron.

(iii) All other electrons in the (ns, np) group screen the designated electron to the extent of 0.35 each (use 0.30 instead if both electrons are in 1s orbital).

(iv) All electrons in the n–1 shell screen to the extent of 0.85 each.

(v) All electrons in the n–2 shell, or lower, screen completely: their contributions to the screening constant are 1.00 each.

When the designated electron being screened is in an nd or nf group, rules (ii) and (iii) remain the same but rules (iv) and (v) are replaced by:

(vi) Each electron in a group lying to the left of the nd or nf group contributes 1.00 to the screening constant.

These rules are a simplified generalization based on the average behavior of different types of electrons. Use these rules to do the following:

(a) Calculate Z_{eff} for a valence electron of O.

(b) Calculate Z_{eff} for the 4s electron in Cu.

(c) Calculate Z_{eff} for a 3d electron in Cu.

(d) Evaluate Z_{eff} for the valence electrons in the group 1 elements (including H), and show that the ionization energies observed for this group are accounted for by using Slater rules.

(e) Evaluate Z_{eff} for a valence electron in the elements Li through Ne, and use the results to explain the observed trend in first ionization energies for these elements.

(f) Using the radial functions given in **Table 7.9** and Z_{eff} estimated with Slater rules, compare plots of the radial probability for the 3s, 3p, and 3d orbitals for the H atom and the Na atom. What do you observe from these plots regarding the effect of screening on radial probability distributions?

$$R_\mathrm{H} = \frac{\mu e^4}{8\varepsilon_0^2 h^2}。$$

(a) 通过应用以上不含时薛定谔方程，计算氢原子 1s 轨道的能量 E_{1s}，以折合质量（μ）和玻尔半径（a_0）的形式表示。

(b) 证明 $E_{1s} = -R_\mathrm{H}$ 对于氢原子成立。

7.9 基于 $Z_{\text{有效}}$ 计算 He 原子的第二电离能。将结果与列表值 5251 kJ mol^{-1} 进行比较。

7.10 按照值增加的顺序排列以下电离能：F 的 I_1；Ba 的 I_2；Sc 的 I_3；Na 的 I_2；Mg 的 I_3。解释你的答案以及其中存在任何不确定性的依据。

*7.11 为解释屏蔽现象，有效核电荷数（$Z_{\text{有效}}$）可从原始核电荷数（Z）修正为：$Z_{\text{有效}} = Z - \sigma$。1930 年，约翰·C. 斯莱特设计了如下一组经验规则，来计算 ns 或 np 轨道中指定电子的屏蔽常数（σ）：

(i) 写出元素的电子组态，并将电子亚层按如下分组：(1s), (2s, 2p), (3s, 3p), (3d), (4s, 4p), (4d), (4f), (5s, 5p) 等。

(ii) (ns, np) 组右侧组中的电子，对指定电子的屏蔽常数没有任何贡献。

(iii) (ns, np) 组内所有其他电子，对指定电子的屏蔽为每个 0.35（如同在 1s 轨道则为 0.30）。

(iv) $n-1$ 主层内所有电子，对指定电子的屏蔽为每个 0.85。

(v) $n-2$ 或更低的主层内所有电子，对指定电子完全屏蔽：它们对屏蔽常数的贡献为每个 1.00。

当被屏蔽的指定电子在 nd 或 nf 组时，上述规则 (ii) 和 (iii) 仍然成立，但规则 (iv) 和 (v) 被如下规则替代：

(vi) 在 nd 或 nf 组左侧组内的每个电子，对屏蔽常数的贡献均为 1.00。

这些规则是基于不同类型电子平均行为的简化概括。使用这些规则完成以下内容：

(a) 计算 O 的价电子的 $Z_{\text{有效}}$。

(b) 计算 Cu 的 4s 电子的 $Z_{\text{有效}}$。

(c) 计算 Cu 的 3d 电子的 $Z_{\text{有效}}$。

(d) 评估第 1 族元素（包括 H）中价电子的 $Z_{\text{有效}}$，并表明采用斯莱特规则可以解释该族元素观察到的电离能。

(e) 评估从元素 Li 到 Ne 中价电子的 $Z_{\text{有效}}$，并用该结果解释这些元素观察到的第一电离能的趋势。

(f) 用**表 7.9** 给出的径向函数和用斯莱特规则估算的 $Z_{\text{有效}}$，比较 H 原子和 Na 原子的 3s、3p 和 3d 轨道的径向概率图。从这些图中，你观察到屏蔽效应对径向概率分布有哪些影响？

Chapter 8 Molecular Structure and Crystal Structure

Based on the atomic structure we have learnt in **Chapter 7**, we will discuss in this chapter the interactions between atoms when they combine to form molecules as well as the nature of chemical bonds, in the order of covalent bonds, metallic bonds, intermolecular forces, and ionic bonds. We will also introduce the structure of crystalline solids in this chapter.

8.1　Lewis Theory

8.2　Shape of Molecules

8.3　Polarity of Molecules

8.4　Valence-Bond Theory

8.5　Molecular Orbital Theory

8.6　Bonding in Metals and Band Theory

8.7　Intermolecular Forces

8.8　Crystal Structure

8.9　Various Types of Crystals and Their Structures

第 8 章　分子结构与晶体结构

基于**第 7 章**学过的原子结构，我们将在本章讨论原子结合成分子时的相互作用以及化学键的本质，依次介绍共价键、金属键、分子间作用力及离子键。本章我们还将介绍晶态固体的结构。

8.1　路易斯理论

8.2　分子的形状

8.3　分子的极性

8.4　价键理论

8.5　分子轨道理论

8.6　金属键与能带理论

8.7　分子间作用力

8.8　晶体结构

8.9　各种晶体类型及其结构

8.1 Lewis Theory

In 1916, Gilbert N. Lewis proposed **Lewis theory**, which provides one of the simplest methods of representing chemical bonds. By noticing that the electron configurations of noble gas atoms account for their inertness, Lewis believed that atoms of other elements could combine with one another to acquire similar stable configurations as noble gases. Although Lewis's work dealt mostly with covalent bonds, its ideas also apply to ionic bonds.

Lewis Symbols and Lewis Structures

Lewis developed a special set of notations called Lewis symbol for his theory. A **Lewis symbol** consists of 1) a chemical symbol that represents the nucleus and the core electrons of an atom; 2) dots around the symbol that represent the valence electrons. Lewis symbols for several main-group elements are written as follows:

$$\cdot \ddot{S}i \cdot \quad \cdot \ddot{N} \cdot \quad \cdot \ddot{A}l \quad \cdot \ddot{S}e \cdot \quad : \ddot{I} \cdot \quad : \ddot{A}r :$$

A **Lewis structure** is a structural schematic diagram that represents the electron configurations in a chemical bond by a combination of Lewis symbols. **Ionic bonds** are formed when electrons are transferred between atoms, whereas **covalent bonds** are formed by the sharing of electron pairs between atoms. For example

Ionic Bonds (Transfer of electrons) $\quad Na \times + \cdot \ddot{C}l: \longrightarrow [Na]^+[:\ddot{C}l:]^-$

Covalent Bonds (Sharing of electrons) $\quad H \times + \cdot \ddot{C}l: \longrightarrow H:\ddot{C}l:$

Here, we designate the valence electrons from one atom as (×) and from the other as (·) to indicate the difference. However, it is impossible to distinguish between electrons, and henceforth we will use only dots (·) to represent electrons in Lewis structures. In ionic bonds, square brackets are used to identify ions, and the charge on the ion is put in the superscript.

In covalent bonds, the sharing of a single pair of electrons between bonded atoms produces a **single covalent bond**. This pair of electrons in a single covalent bond is called a **bond pair**, and is customarily replaced by a short line (—) in Lewis structures. The sharing of multiple pairs of electrons between the bonded atoms produces **multiple covalent bonds**, including **double covalent bond** (═) and **triple covalent bond** (≡). The pair of electrons that is not involved in covalent bonds is called a **lone pair**.

Although covalent bonding describes the sharing of an electron pair between two bonded atoms, it does not necessarily mean that each atom contributes one electron to the bond. If a single atom contributes both of the electrons to a shared pair, a **coordinate covalent bond** is formed. The coordinate covalent bond is sometimes represented by an arrow (→) from the electron-pair donor to the acceptor. The electron-pair donor is the Lewis base and the acceptor is the Lewis acid, which was discussed previously in Lewis theory of acids and bases in **Section 5.1**. More discussion about coordination covalent bonds can be found in **Section 9.1**.

Octet Rule

The rule to write the Lewis structure is also referred to as the **octet rule**, because the stable configurations of noble gas atoms are usually with eight electrons in the outermost shell, forming an **octet**. The octet rule is summarized as follows:

1) Calculate the total number of valence electrons, including the charges of ions. Make sure all valence electrons appear in the Lewis structure.
2) Sketch the **skeletal structure**. In a skeletal structure of more than two atoms, the atom that is bonded to more than one atom is called a **central atom**, whereas the atom that is bonded to just one other atom is called a **terminal atom**. Connect central and terminal atoms with short lines to represent single covalent bonds. In general, central atoms are those with lower electronegativity and terminal atoms are those with higher electronegativity. Specifically, carbon atoms are always central atoms, and halogen, hydrogen (except in borane), and oxygen [except in a peroxo linkage (—O—O—) or in a hydroxy group (—O—H)]

8.1 路易斯理论

1916年，吉尔伯特·N.路易斯提出了**路易斯理论**，这是表示化学键的最简单方法之一。路易斯注意到稀有气体原子的电子组态与其化学惰性之间的联系，认为其他元素的原子也可以相互结合，获得与稀有气体类似的稳定组态。虽然路易斯的工作主要涉及共价键，但其思想同样适用于离子键。

路易斯符号与路易斯结构

路易斯为其理论发展了一套特殊的符号，称为**路易斯符号**，它包括：1）一种化学元素符号，代表原子核及芯电子；2）环绕在符号外的黑点，代表价电子。几种主族元素的路易斯符号举例如下：

$$\cdot \overset{\cdot}{Si} \cdot \quad \cdot \overset{\cdot}{N} \cdot \quad \cdot \overset{\cdot}{Al} \quad \cdot \overset{\cdot}{Se} \cdot \quad : \overset{\cdot}{I} : \quad : \overset{\cdot \cdot}{Ar} :$$

路易斯结构是用路易斯符号的组合来表示化学键中电子组态的结构示意图。在原子之间转移电子可形成**离子键**，在原子之间共用电子对可形成**共价键**。例如

离子键（转移电子）　　$Na \times + \cdot \overset{\cdot \cdot}{Cl} : \longrightarrow [Na]^+ [: \overset{\cdot \cdot}{Cl} :]^-$

共价键（共用电子）　　$H \times + \cdot \overset{\cdot \cdot}{Cl} : \longrightarrow H : \overset{\cdot \cdot}{Cl} :$

这里我们将一个原子的价电子指定为点（·），另一个原子的价电子指定为叉（×），以示区别。然而电子实际上不可区分，此后我们将仅使用点（·）来表示路易斯结构中的电子。离子键中用方括号来识别离子，将离子的电荷置于上标。

在共价键中，成键原子之间共用一对电子形成**共价单键**，这对电子称为**成键电子对**（简称键对），在路易斯结构中通常用一条短线（—）代替。成键原子之间共用多对电子形成**多重共价键**，包括**共价双键**（═）和**共价叁键**（≡）。不参与共价成键的电子对称为**孤电子对**（简称孤对）。

尽管共价键描述了两个成键原子之间共用电子对，但这并不一定意味着每个原子均为成键贡献一个电子。如果成键电子对的两个电子均由一个原子贡献，可形成**配位共价键**。配位共价键有时用从电子对给体到受体的箭头（→）表示。电子对给体是路易斯碱，电子对受体是路易斯酸，这在前述**5.1节**的路易斯酸碱理论中已经讨论过。关于配位共价键的更多讨论详见**9.1节**。

八隅律

书写路易斯结构的规则也称**八隅律**，这是因为惰性气体原子的稳定组态通常在最外层有八个电子，形成**八隅体**。八隅律总结如下：
1) 计算价电子总数（包括离子的电荷数）。保证所有价电子都出现在路易斯结构中。
2) 画出**骨架结构**。含有两个以上原子的骨架结构中，与一个以上原子成键的原子称为**中心原子**，而仅与另一个原子成键的原子称为**端基原子**。用短线连接中心原子和端基原子，来表示共价单键。一般来说，中心原子具有较低电负性，端基原子具有较

atoms are usually terminal atoms. Inorganic molecules and ions generally have compact, symmetrical structures, whereas organic molecules and ions can form chain-like structure. For each single covalent bond in the skeletal structure, subtract two from the total number of valence electrons.

3) Assign the remaining electrons. First complete the octets of the terminal atoms (except two electrons for hydrogen). Then, to the extent possible, complete the octets of the central atom(s). If one or more central atoms are left with an incomplete octet, move lone-pair electrons from one or more terminal atoms to form multiple covalent bonds to central atoms.

4) Check the structure to make sure that all atoms are complete octets.

Formal Charge

The so-written Lewis structure following the above octet rule is called a plausible Lewis structure. For many molecules and ions, there might be more than one plausible Lewis structures, possibly some more probable than others. To ascertain which plausible Lewis structure is more satisfactory, we need take into consideration the formal charges. **Formal charges** (FC) are hypothetical charges on the atoms in a Lewis structure if all bonds were 100% covalent, meaning that bonding electron pairs were divided 100% equally between the bonded atoms.

The formal charge on an atom is calculated by the number of valence electrons in the free (uncombined) atom minus the sum of the number of lone-pair electrons and one-half the number of bond-pair electrons, as

$$FC = N(\text{valence e}^-) - N(\text{lone-pair e}^-) - N(\text{bond-pair e}^-)/2$$

Formal charges in a Lewis structure can be shown by using small, encircled numbers under the corresponding atoms. For example

$$\underset{\underset{\text{Plausible}}{\circledcirc \ \circledcirc \ \circledcirc}}{H-\ddot{O}-H} \quad \underset{\underset{\text{Plausible}}{\circledcirc \ \oplus \ \circledcirc}}{[\ddot{O}=N=\ddot{O}]^+} \quad \underset{\underset{\text{Improbable}}{\ominus \ \oplus \ \oplus}}{[:\ddot{O}-N\equiv O:]^+}$$

The general rules that can help to determine the plausibility of a Lewis structure based on its formal charges are listed as follows:

1) The sum of formal charges must equal the overall charge of molecules or ions.
2) The formal charges should be as small as possible.
3) Negative formal charges usually appear on the most electronegative elements, and positive formal charges on the least electronegative elements.
4) Structures with formal charges of the same sign on adjacent atoms are unlikely.

Note that formal charges are not the actual charges on atoms. Recall from **Section 6.1** that the oxidation states are hypothetical charges when assuming all bonds are 100% ionic. Here, the formal charges are hypothetical charges when assuming all bonds were 100% covalent. Therefore, the concept of the oxidation state tends to exaggerate the ionic character of the bonds, whereas the concept of formal charge tends to exaggerate the covalent character of the bonding.

Limitations of Lewis Theory

Despite its simplicity, Lewis theory captures many key features of the electronic structure of a variety of molecular systems, and is especially useful in the field of organic chemistry. However, it also has many limitations as follows:

1) Lewis theory does not state the nature of chemical bonds. It does not explain why sharing electron pairs can bind two atoms firmly.
2) The octet rule is a chemical rule of thumb that arises from observations mainly in the compounds of main-group elements. It may not be applicable to transition elements and to complex compounds.
3) Lewis theory focuses on electron pairs, and thus cannot explain the structures of species with unpaired electrons. For example, the octet rule is not applicable to radicals, which are highly reactive molecular fragments with one or more unpaired electrons, such as ·CH_3, :CH_2, ·OH, etc., and species with three-

高电负性。具体而言，碳原子总是中心原子，卤素、氢（硼烷除外）和氧 [过氧键（—O—O—）或羟基（—O—H）除外] 原子通常为端基原子。无机分子和离子通常采用紧凑对称的结构，而有机分子和离子可以形成链状结构。对于骨架结构中的每个共价单键，从价电子总数中减去 2。

3) 分配剩余电子。首先使端基原子形成八隅体（氢为两电子除外），再尽可能地使中心原子形成八隅体。如果一个或多个中心原子无法成为八隅体，则从一个或多个端基原子上移动孤电子对，与中心原子形成多重共价键。

4) 检查结构以确保所有原子均为完整的八隅体。

形式电荷

按照上述八隅律书写的路易斯结构称为合理的路易斯结构。对于许多分子和离子，可能存在不止一种合理的路易斯结构，而有些结构比其他结构更为合理。为了确定哪种路易斯结构更为合理，我们需要考虑形式电荷。**形式电荷**（FC）是假定所有键均为 100% 共价性（意味着成键电子对 100% 均等分布在成键原子之间）时路易斯结构中原子的假想电荷。

原子的形式电荷可通过自由原子（即未成键原子）的价电子数减去孤对电子数和键对电子数的一半来计算，即

$$FC = N(\text{价电子}) - N(\text{孤对电子}) - N(\text{键对电子})/2$$

路易斯结构的形式电荷可用相应原子下方小的绕圈数字来表示。例如

$$\text{H}-\ddot{\text{O}}-\text{H} \quad [\ddot{\text{O}}=\text{N}=\ddot{\text{O}}]^+ \quad [:\!\ddot{\text{O}}-\text{N}\equiv\text{O}:]^+$$
$$\text{合理} \qquad \text{合理} \qquad \text{不太可能}$$

根据形式电荷确定路易斯结构合理性的一般规则如下：

1) 形式电荷之和必须等于分子或离子的总电荷数。
2) 形式电荷应尽可能小。
3) 形式负电荷通常出现在电负性最强的元素上，而形式正电荷出现在电负性最弱的元素上。
4) 在相邻原子上出现同号形式电荷的结构不太可能。

注意，形式电荷并非原子的实际电荷。回顾 **6.1 节**，氧化态是假定所有键均为 100% 离子性时的假想电荷。这里，形式电荷是假定所有键均为 100% 共价性时的假想电荷。因此，氧化态的概念倾向于夸大了成键的离子性，而形式电荷的概念则倾向于夸大了成键的共价性。

路易斯理论的局限性

尽管简单，路易斯理论抓住了各种分子体系电子结构的许多关键特征，在有机化学领域尤其有用。但它也存在以下许多局限性：

1) 路易斯理论没有阐明化学键的本质，不能解释为什么共用电子对可使两个原子牢固地结合。
2) 八隅律是一个最初从主族元素化合物中观察到的化学经验规则，可能不适用于过渡元素和配位化合物。
3) 路易斯理论强调电子对，因此无法解释具有未成对电子的物种结构。例如，八隅律不适用于自由基（即存在一个或多个未成对电子、具有高反应性的分子碎片），如 $\cdot \text{CH}_3$、$:\text{CH}_2$、$\cdot \text{OH}$ 等，

electron bonds, such as NO, O₂, etc.

4) The octet rule cannot explicate the structures of species with incomplete (less than eight electrons, such as BF₃, BeF₂, etc.) or expanded (more than eight electrons, such as PCl₅, SO₃, etc.) valence shell.

8.2 Shape of Molecules

In addition to the above limitations, Lewis theory tells nothing about the shapes of molecules, which are very important structural information directly relating to their properties, such as the polarity of molecules that will be discussed in the next section. To determine the shape of a molecule requires tremendous experimental data such as spectroscopic data. The molecular shape can also be determined by quantum mechanical calculations confirmed by experiments. However, a relatively simple model called **valence-shell electron-pair repulsion (VSEPR) theory** can give rise to quite good agreement with the experiments and calculations in predicting the approximate shape of a molecule.

VSEPR Theory

The VSEPR theory was developed by Ronald J. Gillespie (1924—2021) and Ronald S. Nyholm (1917—1971) in 1957. The key point is that the valence-shell electron pairs surrounding an atom tend to repel each other and will, therefore, adopt an arrangement that minimizes this repulsion, which in turn decreases the total energy of the molecule and determines the molecular shape.

The VSEPR theory focuses on groups of electrons in the valence shell of a central atom in a structure. A group of electrons can be either a single-, double-, or triple-bond pair (all denoted as X) or a lone pair (denoted as E). Each of the above is treated as one electron group, and in some rare cases a single unpaired electron on an atom can also be treated as one electron group. We need to count the total number of electron groups around a central atom (denoted as A). For example, the Lewis structures for SF_4 and CO_3^{2-} are shown below. The central S atom in SF_4 has five electron groups, including four bond pairs and one lone pair, denoted as AX_4E. The central C atom in CO_3^{2-} has three electron groups, including one double-bond pair and two single-bond pairs, denoted as AX_3.

The repulsions between electron groups force them to be as far apart in space as possible. Thus, the total number of electron groups around a central atom determines directly the geometric distribution of these electron groups, called the **electron-group geometry**. For 2, 3, 4, 5, and 6 electron groups, the electron-group geometries are linear, trigonal planar, tetrahedral, trigonal bipyramidal, and octahedral, respectively.

Rules in Determining Molecular Geometry

By noticing the difference between bond pairs (with an atomic nucleus attaching to the other end of the bond) and lone pairs, the **molecular geometry** can be determined based on the corresponding electron-group geometry. The molecular geometry is the same as the electron-group geometry only when all electron groups are bond pairs. If one or more electron groups are lone pairs, the molecular geometry is different from the electron-group geometry, although still derived from it. The relationship between electron-group geometry and molecular geometry is summarized in **Table 8.1**.

To determine the molecular geometry, two rules in comparison of the repulsion between electron pairs are necessary:

1) Lone-pair electrons spread out more than do bond-pair electrons. The repulsion forces are in the order of

$$\text{Lone pair-lone pair} > \text{lone pair-bond pair} > \text{bond pair-bond pair}$$

2) The closer together two groups of electrons are forced, the stronger the repulsion between them. The repulsion forces are in the order of

$$\text{Angle of } 90° > \text{angle of } 120° > \text{angle of } 180°$$

For example, the central atoms in CH_4, NH_3, and H_2O all have 4 electron groups and a tetrahedral

以及具有三电子键的物种，如 NO、O_2 等。
4) 八隅律无法解释具有不完整价层（少于八个电子，如 BF_3、BeF_2 等）或超价价层（多于八个电子，如 PCl_5、SO_3 等）的物种结构。

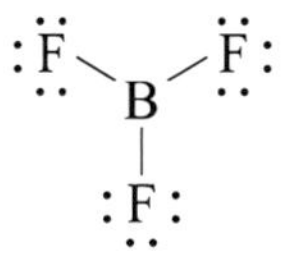

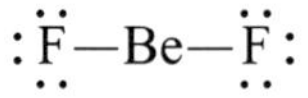

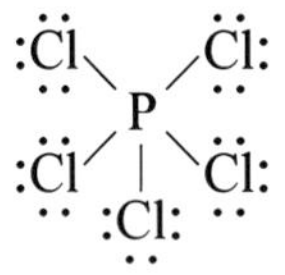

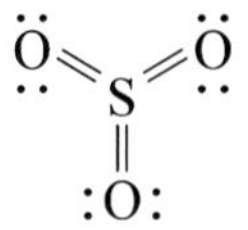

8.2 分子的形状

除上述局限性之外，路易斯理论不能给出分子的形状，而分子的形状是与其性质（如下一节即将讨论的分子极性）直接相关的至关重要的结构信息。确定分子的形状需要大量诸如光谱等实验数据，也可通过经由实验证实的量子力学计算来确定。而一个相对简单的称为**价层电子对互斥（VSEPR）理论**的模型，在预测分子的大致形状时可给出与实验和计算相当一致的结论。

VSEPR 理论

VSEPR 理论由罗纳德·J. 吉莱斯皮（1924—2021）和罗纳德·S. 尼霍姆（1917—1971）于 1957 年提出。其要点在于，原子周围的价层电子对倾向于相互排斥，因此将采用一种将此排斥力最小化的排布形式，而这反过来降低了分子的总能量，并决定了分子的形状。

VSEPR 理论重点关注某一结构的中心原子价层的电子组。电子组可以是成键电子对（单键、双键或叁键，均用 X 表示）或孤电子对（用 E 表示）。上述每一个均视为一个电子组，而在一些罕见的情况下，原子的单个未成对电子也可视为一个电子组。我们需要计算中心原子（用 A 表示）周围的电子组总数。例如，SF_4 和 CO_3^{2-} 的路易斯结构如右所示。SF_4 的中心 S 原子有五个电子组，包括四个成键电子对和一个孤电子对，可表示为 AX_4E。CO_3^{2-} 的中心 C 原子有三个电子组，包括一个双键电子对和两个单键电子对，可表示为 AX_3。

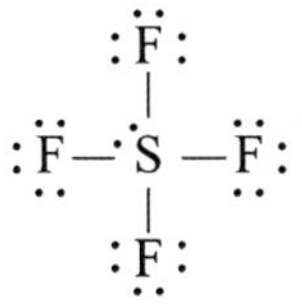

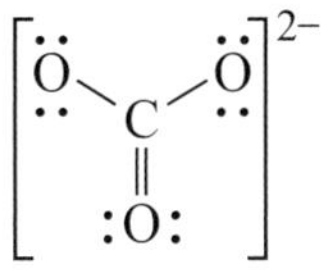

电子组之间的排斥力使其在空间上尽可能彼此远离。因此，中心原子周围的电子组总数，直接决定了这些电子组的几何分布，称为**电子组几何构型**。对于 2、3、4、5 和 6 个电子组，其电子组几何构型分别为直线形、平面三角形、四面体形、三角双锥形和八面体形。

决定分子几何构型的规则

考虑到成键电子对（有一个原子核附在键的另一端）与孤电子对之间的差异，可以根据相应的电子组几何构型来确定**分子几何构型**。只有当所有电子组均为成键电子对时，分子几何构型才与电子组几何构型相同。如果一个或多个电子组为孤电子对，尽管仍由其衍生，分子几何构型与电子组几何构型存在差异。**表 8.1** 总结了电子组几何构型与分子几何构型之间的关系。

为确定分子几何构型，比较电子对之间排斥力需要的两个规则是：
1) 孤电子对比成键电子对在空间中更为铺展，排斥力的大小顺序为
  孤电子对 - 孤电子对 > 孤电子对 - 成键电子对 > 成键电子对 - 成键电子对
2) 两个电子组距离越近，排斥力越强。排斥力的大小顺序为

$$90° 键角 > 120° 键角 > 180° 键角$$

例如，CH_4、NH_3 和 H_2O 的中心原子均有 4 个电子组，其电子组几何构型均为四面体形。CH_4 具有四个相同的成键电子对，可表

electron-group geometry. CH$_4$ has four identical bond pairs, denoted as AX$_4$, and thus a tetrahedral molecular geometry. NH$_3$ has three bond pairs and one lone pair, denoted as AX$_3$E, and the molecular geometry is trigonal pyramidal. H$_2$O has two bond pairs and two lone pairs, denoted as AX$_2$E$_2$, and the molecular geometry is V-shaped, or bent (**Figure 8.1**).

For tetrahedral electron-group geometry, the bond angles should be 109.5°, and this is the case for CH$_4$. According to rule 1, the repulsion between the two lone pairs in H$_2$O and the lone pair-bond pair repulsion in NH$_3$ are greater than the repulsion between bond pairs in CH$_4$. The greater repulsions force the bond-pair electrons closer together and reduce the bond angles: 107° for the H—N—H bond angle in NH$_3$ and 104.5° for the H—O—H bond angle in H$_2$O.

The examples listed in **Table 8.1** are straight forward except for the cases of trigonal bipyramidal electron-group geometry. For the notation AX$_4$E of trigonal bipyramidal, the lone pair can be either in the central plane [**Figure 8.2(a)**] or at the pole [**Figure 8.2(b)**]. As no lone pair-lone pair repulsion available, we only consider the different bond angles for the lone pair-bond pair repulsion according to rule 2. The former has two repulsions of 90° and two of 120°. The latter, however, has three repulsions of 90° and one of 180°. To minimize the repulsion, **Figure 8.2(a)** show the more favorable arrangement.

Table 8.1 Relationship between Electron-Group Geometry and Molecular Geometry
表 8.1 电子组几何构型与分子几何构型的关系

No. of Electron Groups (电子组数目)	Electron-Group Geometry (电子组几何构型)	VSEPR Notation (VSEPR表示法)	Molecular Geometry (分子几何构型)	Ideal Bond Angles (理想键角)	Example (示例)
2	Linear (直线形)	AX$_2$	Linear (直线形)	180°	BeCl$_2$
3	Trigonal planar (平面三角形)	AX$_3$	Trigonal planar (平面三角形)	120°	BF$_3$
		AX$_2$E	Bent (折叠形)	120°	SO$_2$
4	Tetrahedral (四面体形)	AX$_4$	Tetrahedral (四面体形)	109.5°	CH$_4$
		AX$_3$E	Trigonal pyramidal (三角锥形)	109.5°	NH$_3$
		AX$_2$E$_2$	Bent (折叠形)	109.5°	H$_2$O
5	Trigonal bipyramidal (三角双锥形)	AX$_5$	Trigonal bipyramidal (三角双锥形)	90°,120°	PCl$_5$
		AX$_4$E	Seesaw (跷跷板形)	90°,120°	SF$_4$
		AX$_3$E$_2$	T-shaped (T字形)	90°	ClF$_3$
		AX$_2$E$_3$	Linear (直线形)	180°	XeF$_2$
6	Octahedral (八面体形)	AX$_6$	Octahedral (八面体形)	90°	SF$_6$
		AX$_5$E	Square pyramidal (四方锥形)	90°	BrF$_5$
		AX$_4$E$_2$	Square planar (平面四边形)	90°	XeF$_4$

Strategy for Predicting Shape of Molecules

For predicting the shapes of molecules, the following general strategy can be used:
1) Draw a plausible Lewis structure of the species.
2) Determine the number of electron groups around the central atom, and identify them as being either bond-pair electron groups (single, double, or triple bonds, X) or lone-pair electron groups (E).
3) Establish the electron-group geometry around the central atom: linear (2), trigonal planar (3), tetrahedral (4), trigonal bipyramidal (5), or octahedral (6).
4) Determine the molecular geometry from the numbers of bond-pair and lone-pair electron groups.
5) If the species contains more than one central atoms, the geometric distribution of terminal atoms around each central atom must be determined. The results then combine into a single description of the molecular shape.

示为 AX_4，因此具有四面体形分子几何构型。NH_3 有三个成键电子对和一个孤电子对，可表示为 AX_3E，分子几何构型为三角锥形。H_2O 有两个成键电子对和两个孤电子对，可表示为 AX_2E_2，分子几何构型为 V 形或折叠形（**图 8.1**）。

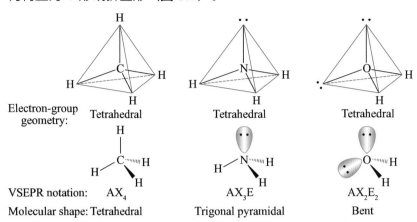

Figure 8.1 Molecular shapes of CH_4 (tetrahedral), NH_3 (trigonal pyramidal), and H_2O (bent), based on tetrahedral electron-group geometry.

图 8.1 基于四面体形电子组几何构型的分子形状：CH_4（四面体形）；NH_3（三角锥形）；H_2O（折叠形）。

对于四面体形电子组几何构型，键角应为 109.5°，这是 CH_4 的情况。根据规则 1，H_2O 中两个孤电子对之间的排斥力，以及 NH_3 中孤电子对 - 成键电子对的排斥力，均大于 CH_4 中成键电子对之间的排斥力。更大的排斥力迫使成键电子更为靠近，具有更小的键角：NH_3 中 H—N—H 的键角为 107°，H_2O 中 H—O—H 的键角为 104.5°。

除三角双锥形电子组几何构型外，**表 8.1** 中列出的例子均较为直观。对于表示为 AX_4E 的三角双锥形，孤电子对可位于中心平面 [**图 8.2(a)**] 或极点 [**图 8.2(b)**] 处。由于没有孤电子对之间的排斥力，我们仅需根据规则 2 考虑不同键角时孤电子对 - 成键电子对的排斥力。前者存在两个 90° 和两个 120° 键角的排斥力。而后者存在三个 90° 和一个 180° 键角的排斥力。为使排斥力最小化，**图 8.2(a)** 给出了更为有利的排布形式。

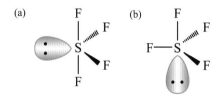

Figure 8.2 Two possible structures for the notation AX_4E of trigonal bipyramidal electron-group geometry: the lone pair is located (a) in the central plane and (b) at the pole.

图 8.2 用 AX_4E 表示的三角双锥形电子组几何构型的两种可能结构：孤电子对位于 (a) 中心平面；(b) 极点。

预测分子形状的方法

预测分子的形状可使用以下一般方法：
1) 画出该物种合理的路易斯结构。
2) 确定中心原子周围的电子组数目，并将其识别为键对电子组（单键、双键或叁键，X）或孤对电子组（E）。
3) 确定围绕中心原子的电子组几何构型：直线形（2）、平面三角形（3）、四面体形（4）、三角双锥形（5）或八面体形（6）。
4) 根据键对电子组和孤对电子组数目，确定分子几何构型。
5) 如果该物种包含多个中心原子，必须确定每个中心原子周围端基原子的几何分布。再将这些结果合并为对分子形状的单一描述。

8.3 分子的极性

分子的极性是决定许多其他性质的一种重要性质。例如，我们都知道水是极性分子，这种极性可以解释其多种性质，包括溶解各种物质的能力、相对较高的沸点、反常密度等。分子的极性取决于分子的形状以及分子中键的极性。一个分子究竟是极性分子还是非极性分子，

8.3 Polarity of Molecules

The polarity of molecules is an important property that determines many other properties of the molecules. For example, we all know that water is a polar molecule, and this polarity accounts for many properties including its ability to dissolve various substances, its relatively high boiling point, its abnormal density, etc. The polarity of a molecule depends on both the shape of the molecule and the polarity of bonds in the molecule. Whether a molecule is polar or nonpolar can be directly visualized in an electrostatic potential map, which is based on the concepts of electron density and electrostatic potential.

Electron density and Electrostatic Potential

Recall that in **Section 7.6** we have shown the 3D surface representations of the electron probability density distributions for various atomic orbitals. The combination of the electron probability density distributions for all the atomic orbitals with filled electrons can give rise to the total electron probability density of an atom. When atoms combine to form molecules, the total electron probability density throughout a molecule [**Figure 8.3(a)**] can also be derived. From now on, we use **electron density** as short for electron probability density, which is a measure of the probability density distributions of all electrons in a molecule. Similar to the commonly used 95% probability boundary surface representations of the atomic orbitals, the electron density of a molecule can also be represented as 3D surfaces, with characteristics as:

1) Shape: The shape of the electron density surface arises from the contour with equal electron density, which means that every point on this 3D surface has the same electron density.
2) Size: The size of the electron density surface either results from some arbitrarily chosen probability (such as 95%) of the total electrons in the molecule, or simply is some isodensity surface with arbitrarily chosen electron density. The latter is used in the figures in this chapter.
3) Color: An arbitrary uniform gray color is used on the electron density surface. It serves as a canvas upon which other electronic properties can be displayed in various colors.

The electron density only measures the density distribution of electrons outside the nuclei in a molecule. When all the charges in a molecule are considered, we must also take into account the nuclei. In this case, we use electrostatic potential to measure the density distribution of all the charges in a molecule. The **electrostatic potential** is the work done in moving a unit of positive point charge at a constant speed from infinity to a specific location. The value of electrostatic potential depends on both the attractive force between the positive point charge and the electrons, and the repulsive force between the positive point charge and the nuclei.

Electrostatic Potential Map

The electrostatic potential is an important electronic property and can be displayed in rainbow colors upon the electron density surface in an **electrostatic potential map** [EPM, **Figure 8.3(b)**]. The EPM is obtained by hypothetically probing an electron density surface with a unit of positive point charge. That is, the value of electrostatic potential on each point on the electron density surface is obtained as the work done in moving a unit of positive point charge from infinity to the corresponding point. For an electron-rich region where excess negative charge is available, the work is negative and so is the electrostatic potential. Conversely, for an electron-poor region with excess positive charge, both the work and the electrostatic potential are positive. The EPM is usually displayed in arbitrary rainbow colors upon the electron density surface, with red indicating the most negative electrostatic potential, blue the most positive electrostatic potential, and intermediate colors representing intermediate values of electrostatic potential.

Figure 8.4 shows the computed EPMs of several molecules in rainbow colors, which give information about the charge density distributions in those molecules. We see that Cl_2 has a uniform color distribution, indicating a uniform charge density distribution. This is typical for all diatomic molecules containing identical atoms. The NaCl molecule, conversely, exhibits a highly nonuniform charge density distribution. The Na atom is almost exclusively

可以在基于电子密度和静电势概念的静电势图中直接可视化。

电子密度与静电势

回顾 **7.6** 节，我们展示了以三维曲面表示的各种原子轨道的电子概率密度分布。由所有填充了电子的原子轨道电子概率密度分布的组合，可以得到一个原子的总电子概率密度。当原子结合形成分子时，也可得出整个分子的总电子概率密度 [**图 8.3(a)**]。从现在起，我们使用**电子密度**作为电子概率密度的简称，它是对分子中所有电子的概率密度分布的度量。与常用的原子轨道 95% 概率边界面的表示法类似，分子的电子密度也可用三维曲面来表示，其特征如下：

1) 形状：电子密度曲面的形状源自等电子密度轮廓面，这意味着三维曲面上的每个点都具有相同的电子密度。
2) 大小：电子密度曲面的大小可以由分子中所有电子的某个任意选定的概率值（如 95%）得出，也可以就是具有某个任意选定电子密度值的等密度曲面。本章的所有图片均采用后者。
3) 颜色：电子密度曲面采用均匀的灰色（任意选定的颜色）。它就像一块画布，可以在上面用各种颜色来显示其他的电子性质。

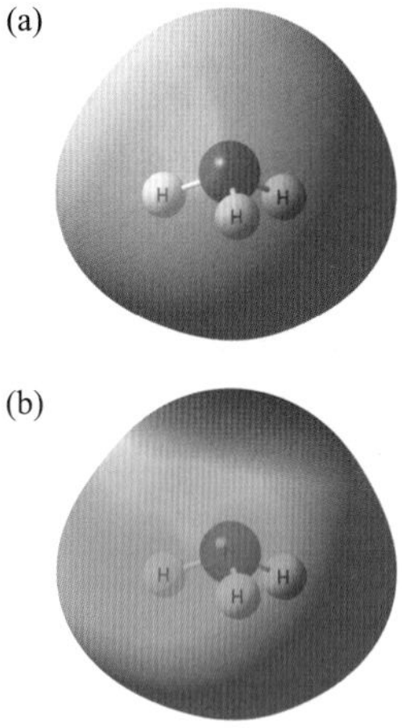

电子密度仅度量分子中核外电子的密度分布。当考虑一个分子的所有电荷时，我们还必须考虑原子核。在这种情况下，我们使用静电势来度量分子中所有电荷的密度分布。**静电势**是以恒定速度将单位正点电荷从无穷远处移至特定位置所做的功。静电势的值取决于此正点电荷与电子之间的吸引力以及与原子核之间的排斥力。

静电势图

静电势是一种重要的电子性质，可以在**静电势图** [EPM，**图 8.3(b)**] 中以彩虹色显示在电子密度曲面上。用单位正点电荷探测电子密度曲面，可得静电势图。也就是说，电子密度曲面上每个点的静电势，等于将单位正点电荷从无穷远处移至对应点所做的功。对于存在过量负电荷的富电子区域，功为负值，静电势也为负值。相反，对于存在过量正电荷的缺电子区域，功和静电势均为正值。静电势图通常以任意规定的彩虹色显示在电子密度曲面上，红色代表最负的静电势值，蓝色代表最正的静电势值，中间色代表静电势的中间值。

图 8.4 展示了用彩虹色表示的几个分子的计算静电势图，给出了这些分子中电荷密度分布的信息。我们看到 Cl_2 具有均匀的颜色分布，代表均匀的电荷密度分布。这是所有同核双原子分子的典型情况。相反，NaCl 分子表现出高度不均匀的电荷密度分布。Na 原子几乎完全处于正电荷的蓝色极点，而 Cl 原子处于负电荷的红色极点。这种静电势图为典型的离子键，但从图可以清楚地看出，从 Na 原子到 Cl 原子的电子密度转移并不完全。HCl 分子也具有不对称的电荷密度分布，Cl 原子上带有部分负电荷（橙红色区域），而 H 原子上带有部分正电荷（淡蓝色区域）。静电势图清楚地描述了 HCl 中键的极性本质。

有趣的是，所有分子的电子密度曲面上的总静电势并不为零，而是某个正值。这是因为电子密度曲面是具有某个值的概率边界面，比如 95%。这意味着在该曲面内找到所有电子的总概率为 95%。加上原子核，则曲面内的总电荷为正 5%。因此，将正点电荷从无穷远处移至带正电荷的曲面，所做的总功必为正值，总静电势也为正值。

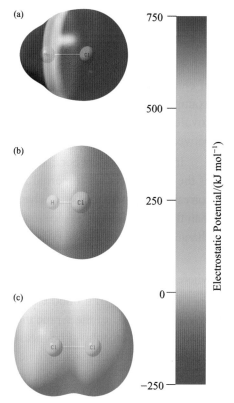

Figure 8.3 3D surface representations of the (a) electron density map and (b) electrostatic potential map of NH_3.

图 8.3 以三维曲面表示的 NH_3 的 (a) 电子密度图和 (b) 静电势图。

Figure 8.4 The electrostatic potential maps of (a) NaCl, (b) HCl, and (c) Cl_2.

图 8.4 (a) NaCl、(b) HCl 和 (c) Cl_2 的静电势图。

at the blue extreme of positive charge and the Cl atom at the red extreme of negative charge. This EPM is typical of an ionic bond, yet it is clear from the map that the transfer of electron density from the Na atom to the Cl atom is not complete. The HCl molecule also has an unsymmetrical charge density distribution, with a partial negative charge (orange-red region) on the Cl atom and a partial positive charge (pale blue region) on the H atom. The EPM clearly depicts the polar nature of the bond in HCl.

It is interesting to note that the overall electrostatic potentials on the electron density surface for all molecules are not zero, but some positive values. This is because the electron density surface is the boundary surface of some probability, for example, 95%. This means that the total probability of finding all electrons inside the surface is 95%. Together with the nuclei, the total charge inside the surface is 5% positive. Therefore, the total work done in moving a positive point charge from infinity to a surface of positive charge must be positive, and so is the overall electrostatic potential.

Dipole Moment

We can clearly see from **Figure 8.4** that there is a separation between positive and negative charges (called charge displacement) in HCl so that HCl is a polar molecule. The extent of the charge displacement in a bond can be represented by the **dipole moment** (μ) of the bond. An electric **dipole** consists of two equal and opposite point charges. If not specified, dipole moment usually refers to an electric dipole moment, which is a vector quantity with a magnitude equal to the strength of each charge (δ) times the separation between them (d), as

$$\mu = \delta d \tag{8.1}$$

in scalar form. By convention in chemistry, the dipole moment points from positive charge to negative charge, although it should point from negative charge to positive charge by physical definition. Its vector form is

$$\boldsymbol{\mu} = \delta \boldsymbol{d} \tag{8.2}$$

where $\boldsymbol{d}$ is the corresponding displacement vector.

The SI unit for dipole moment is C m. However, this is too large to be practical on the molecular scale. Dipole moments are commonly measured in debyes (D), where $1\text{ D} = 3.33564 \times 10^{-30}$ C m.

Polarity of Bonds

Ideally, ionic bonds involve a complete transfer of electrons and covalent bonds involve an equal sharing of electron pairs. In reality, no bond is 100% ionic and all ionic bonds have some covalent character. For example, experiments show that the ionic bond between Na and Cl is only about 80% ionic. Although the covalent bonds formed between two identical atoms are indeed 100% covalent, the covalent bonds formed between heterogeneous atoms are always **polar covalent bonds** in which electrons are not shared equally.

In a polar covalent bond, electrons are displaced toward the more electronegative atom. The unequal sharing of the electrons leads to partial negative charge on the more electronegative atom, designated by $\delta-$, and a corresponding partial positive charge on the less electronegative atom, signified by $\delta+$. Here, partial means less than the unit charge of an electron and falls between 100% covalent (with a charge of 0) and 100% ionic (with a charge of ±1 in a unit of e). For example, the polar covalent bond in HCl can be represented as

$$^{\delta+}\text{H} \longrightarrow \text{Cl}^{\delta-}$$

where the partial charge $\delta+$ on H and $\delta-$ on Cl indicate that the bond pair of electrons lies closer to Cl than to H.

The polarity of a bond can be represented by its percent ionic character based on the experimentally determined value of the dipole moment of the bond (called **bond dipole moment**). For example, the H—Cl bond dipole moment can be measured with an electrical capacitor to be 1.11 D, and the H—Cl bond length can be derived from spectroscopic data to be 127.5 pm. From **Equation (8.1)**, we can calculate that

$$\delta = \frac{\mu}{d} = \frac{1.11\text{ D} \times 3.34 \times 10^{-30}\text{ C m/D}}{127.5 \times 10^{-12}\text{ m}} = 2.91 \times 10^{-20}\text{ C} = 0.181e$$

This means that the partial charge is about 18% of the charge on an electron and suggests that HCl is about 18% ionic.

Meanwhile, the polarity of a bond can also be related to the difference in electronegativity (ΔEN) between

偶极矩

从**图 8.4**中可以清楚地看到，HCl 中存在正负电荷的分离（称为电荷位移），因此 HCl 是极性分子。化学键中电荷位移的程度，可用键的**偶极矩**（μ）表示。一个电偶极由两个带等量相反电量的点电荷组成。如果未加说明，偶极矩通常指电偶极矩，它是一个矢量，大小等于每个电荷的电量（δ）与它们之间距离（d）的乘积。其标量形式为

$$\mu = \delta d \tag{8.1}$$

按化学中的惯例，偶极矩从正电荷指向负电荷（尽管根据物理定义应从负电荷指向正电荷）。其矢量形式为

$$\boldsymbol{\mu} = \delta \boldsymbol{d} \tag{8.2}$$

其中 $\boldsymbol{d}$ 为相应的位移矢量。

偶极矩的 SI 单位是 C m。但该单位太大，在分子尺度上并不实用。偶极矩的常用单位是德拜（D），1 D = 3.33564×10^{-30} C m。

键的极性

理想情况下，离子键涉及电子的完全转移，共价键涉及电子对的均等共用。实际上没有任何键具有 100% 离子性，所有离子键都具有一定的共价性。例如，实验表明 Na 和 Cl 之间的离子键仅为约 80% 离子性。尽管两个相同原子之间形成的共价键确实是 100% 共价性，但不同原子之间形成的共价键始终是**极性共价键**，其中电子的共用是不均等的。

在极性共价键中，电子会偏向电负性更强的原子。电子的不均等共用，导致了电负性较强的原子上带有部分负电荷（用 $\delta-$ 表示），而电负性较弱的原子上带有相应的部分正电荷（用 $\delta+$ 表示）。这里的"部分"指的是电荷小于电子所带的单位电荷，介于 100% 共价性（电荷为 0）和 100% 离子性（以 e 为单位，电荷为 ±1）之间。例如，HCl 中的极性共价键可表示为

$$^{\delta+}\text{H} \longleftrightarrow \text{Cl}^{\delta-}$$

其中 H 上的部分电荷 $\delta+$ 和 Cl 上的部分电荷 $\delta-$，表明成键电子对距离 Cl 比距离 H 更近。

键的极性可以基于实验确定的键偶极矩（称为**键矩**），通过其离子性百分数来表示。例如，用电容器测得 H—Cl 的键矩为 1.11 D，从光谱数据得到 H—Cl 的键长为 127.5 pm。由**式 (8.1)** 可算得

$$\delta = \frac{\mu}{d} = \frac{1.11 \text{ D} \times 3.34 \times 10^{-30} \text{ C m/D}}{127.5 \times 10^{-12} \text{ m}} = 2.91 \times 10^{-20} \text{ C} = 0.181 e$$

这意味着部分电荷约为电子电荷的 18%，表明 HCl 具有 18% 的离子性。

同时，键的极性也与两个成键原子的电负性之差（ΔEN）有关。如果 ΔEN 很小（<0.5，鲍林标度），它们之间的键基本上是非极性共价键；如果 ΔEN 较大（>1.7），则基本上是离子键；对于 ΔEN 的中间值（0.5~1.7），通常可视为极性共价键。**图 8.5** 绘制了离子性百分数与 ΔEN 之间的有用的近似关系。我们可以看到，ΔEN=1.7 对应于 50% 的离子性。

分子的极性

根据极性，分子可分为极性分子和非极性分子。极性分子具有非

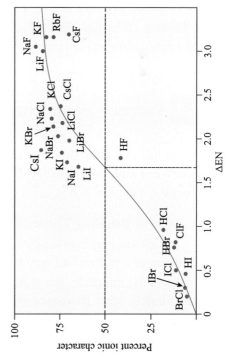

Figure 8.5 Percent ionic character of a chemical bond as a function of electronegativity difference (ΔEN).

图 8.5 化学键的离子性百分数与电负性差值（ΔEN）的函数关系图。

the two bonded atoms. If ΔEN is very small (< 0.5, Pauling scale), the bond between them is essentially nonpolar covalent. If ΔEN is large (> 1.7), the bond is essentially ionic. For intermediate values of ΔEN (0.5~1.7), the bond is generally regarded as polar covalent. **Figure 8.5** plots a useful approximate relationship between the percent ionic character and ΔEN. We can see that ΔEN = 1.7 corresponds to 50% ionic character.

Polarity of Molecules

According to polarity, molecules can be classified as polar and nonpolar molecules. A polar molecule has a non-zero net molecular dipole moment, or $\mu_{net} \neq 0$, resulting from asymmetric arrangement of partial charges from polar bonds. Conversely, a nonpolar molecule has a net molecule dipole moment of 0, or $\mu_{net} = 0$. A molecule may be nonpolar either because of the equal sharing of electrons in all the bonds, or due to the symmetric arrangement of polar bonds in the molecule.

For diatomic molecules, there is only one (either single or multiple) bond so the polarity of the molecule is determined by the polarity of the bond, and the molecular dipole moment equals the bond dipole moment, with typical values in the range of 0~11 D. For example, homogeneous diatomic molecule Cl_2 is a nonpolar molecule with a dipole moment of 0, and heterogeneous diatomic KBr, which is a molecule in gaseous phase, is a polar molecule with a dipole moment of 10.6 D.

For polyatomic molecules, there is more than one bond. The total molecular dipole moment equals the vector sum of the individual bond dipole moments. A molecule is always nonpolar if the shape of the molecule is totally symmetric in 3D, no matter the bonds are polar or not, because the bond dipole moments always cancel out with one another, resulting in a non-zero net dipole moment, such as P_4, CO_2, BF_3, CCl_4, etc. A molecule is polar if the shape of the molecule is not totally symmetric in 3D, and the net dipole moment can be affected by its structure. For example, H_2O contains two polar O—H bonds in a bent shape, which is not totally symmetric. Thus, water is a typical polar molecule that is generally able to dissolve other polar molecules. The dipole moment of H_2O is higher in solid (3.09 D) and liquid (2.95 D) states than in gaseous state (1.85 D) due to structural differences caused by hydrogen-bonded environments.

In practice, the bond dipole moments are obtained from the molecular dipole moments, which can be derived from experimental data, by vector decomposition. As the bond dipole moments generally do not vary significantly in different molecules, their values are transferrable among molecules. Then, the vector sum of the transferred bond dipole moments can give rise to an estimate for the total dipole moments of other molecules.

8.4 Valence-Bond Theory

In this and the following sections, we will introduce two basic bonding theories—**valence-bond (VB) theory** and molecular orbital (MO) theory—that were developed to use the methods of quantum mechanics to explain chemical bonds, especially covalent bonds.

Purpose of Bonding Theories

As mentioned previously in **Section 7.5**, the potential energy term is the most important part in Schrödinger equation and may vary from system to system. For a covalent bonding system, such as the H—H bond in a H_2 molecule, the typical potential energy curve is shown in **Figure 8.6(a)**. When two H atoms are infinitely apart from each other (condition 1), there is no interaction between them, and the potential energy of the system is set to be 0 by convention. As they approach each other, additional interactions (other than the attractions of N_1-e_1 and N_2-e_2 in the free H atoms) occur: the repulsion between electrons e_1-e_2, the repulsion between nuclei N_1-N_2, and the attractions of N_1-e_2 and N_2-e_1, as shown in **Figure 8.6(b)**. The attractions predominate over the repulsions and the potential energy becomes negative at intermediate distances (condition 2). However, the repulsions exceed the attractions, and the potential energy turns positive at very close distances (condition 4). Consequently, there must be a particular distance (0.74 Å, condition 3), at which the potential energy reaches its minimum (−436 kJ mol^{-1}). This is the condition in

零的净分子偶极矩（即 $\mu_{净} \neq 0$），来自极性键中部分电荷的不对称排布。相反，非极性分子的净分子偶极矩为 0（即 $\mu_{净} = 0$）。非极性分子可能因为所有键都均等共用电子，也可能由于分子的极性键均呈对称排布。

双原子分子只有一个（单重或多重）键，因此分子的极性由键的极性决定，分子的偶极矩等于键矩，典型值在 0~11D 范围内。例如，同核双原子分子 Cl_2 是偶极矩为 0 的非极性分子；而气相的异核双原子分子 KBr，是偶极矩为 10.6 D 的极性分子。

多原子分子存在不止一个键。分子的总偶极矩等于单个键矩的矢量和。如果分子的形状在三维空间中完全对称，不论键是否为极性，该分子始终为非极性，因为其键矩总是相互抵消而得到非零的净偶极矩，如 P_4、CO_2、BF_3、CCl_4 等。如果分子的形状在三维空间中不完全对称，则该分子是极性分子，且净偶极矩受其结构影响。例如，H_2O 包含两个呈折叠形的极性 O—H 键，不是完全对称的。因此，水是典型的极性分子，通常能溶解其他极性分子。由于氢键环境所引起的结构差异，固态（3.09 D）和液态（2.95 D）H_2O 的偶极矩高于气态（1.85 D）。

实际上，键矩可通过矢量分解从分子的偶极矩中得到，而分子的偶极矩可由实验数据获得。由于键矩在不同分子中通常不会显著变化，因此其值可在分子之间调用。可用调用的键矩的矢量和，来估算其他分子的总偶极矩。

8.4 价键理论

在本节和下一节中，我们将介绍两种基本成键理论：**价键（VB）理论**和**分子轨道（MO）理论**，这两种理论均采用量子力学的方法来解释化学键（特别是共价键）的形成。

成键理论的目标

如 **7.5 节**所述，势能项是薛定谔方程中最重要的部分，但可能因体系而异。对于共价键体系，如 H_2 分子中的 H—H 键，典型的势能曲线如**图 8.6(a)** 所示。当两个 H 原子彼此距离无限远（条件 1）时，它们之间没有相互作用，体系的势能按惯例设为 0。当它们彼此接近时，会发生额外的相互作用（除了自由 H 原子中 N_1-e_1 和 N_2-e_2 的吸引力之外）：电子 e_1-e_2 之间的排斥力、核 N_1-N_2 之间的排斥力，以及 N_1-e_2 和 N_2-e_1 的吸引力，如**图 8.6(b)** 所示。在中间距离处，吸引力大于排斥力，占主导地位，势能为负值（条件 2）。然而，在非常近的距离处，排斥力超过吸引力，势能变为正值（条件 4）。因此，必然存在一个势能达到其最小值（–436 kJ mol^{-1}）的特定距离（0.74 Å，条件 3）。这就是形成具有 H—H 共价键的 H_2 分子的条件。该特定距离称为**平衡核间距**或**键长**。平衡核间距处势能的负值称为键的**解离能**。

成键理论的目标在于，其应帮助我们理解共价键的性质（如键参数和键的解离能等），以及解释为什么势能曲线的形状如**图 8.6(a)** 所示。为此，仅有路易斯理论和 VSEPR 理论是不够的。

价键理论的要点

价键理论最初由沃尔特·海特勒（1904—1981）和弗里茨·伦敦

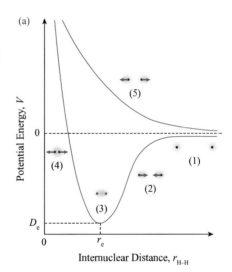

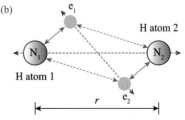

Figure 8.6 Interactions in a system comprising of two H atoms. (a) Potential energy of the system plotted with respect to the internuclear distance. (b) Attractive and repulsive interactions between two H atoms. Additional interactions during the bonding of an H_2 molecule are shown as dashed lines.

图 8.6 由两个氢原子组成的体系的相互作用：(a) 体系势能与核间距的关系图；(b) 两个氢原子之间的吸引力与排斥力，其中虚线表示形成氢分子时额外的相互作用。

which a H₂ molecule with a H—H covalent bond is formed. This particular distance is called the **equilibrium internuclear distance** or **bond length**. The negative of the potential energy at the equilibrium internuclear distance is called the bond **dissociation energy**.

The purpose of bonding theories should help us in understanding the properties of covalent bonds, such as bond parameters and bond dissociation energy, as well as in explaining why the shape of potential energy curve looks like that shown in **Figure 8.6(a)**. For this purpose, Lewis theory and VSEPR theory are not enough.

Key Points of Valence-Bond Theory

The VB theory was first developed by Walter Heitler (1904—1981) and Fritz London (1900—1954) in 1927, and later modified by Linus Pauling (1901—1994). In VB theory, as the two atoms that are about to bond approach each other, the regions with high electron density in their atomic orbitals begin to interpenetrate (called **overlap**). The electron density between the two atoms increases, and the increased electron density attracts the two positively charged nuclei, leading to a decreased total energy of the system and forming a covalent bond between the two atoms.

Again, let us take the H—H bond in a H₂ molecule as an example. Recall that in **Section 7.6** and **Figure 7.17** we have shown the area of high probability of finding the electron in a 1s orbital. As the two H atoms approach each other, the interpenetration between the two half-filled 1s orbitals occurs, and there is interference between the two 1s atomic wave functions. The constructive interference, which occurs when the two atomic wave functions are in phase, leads to increased electron density between the two H atoms and lower total energy. The energy becomes the lowest at the equilibrium internuclear distance, where a covalent H—H bond is formed. This lowest energy state is called the ground state of H₂ molecule. Meanwhile, the destructive interference when the two atomic wave functions are out of phase, leads to decreased electron density between the two H atoms and higher total energy. This corresponds to an excited state of the H₂ molecule [condition 5 in **Figure 8.6(a)**].

Explanation of the Characteristics of Covalent Bonds

The two major characteristics of covalent bonds are saturability and directivity, both can be explained by the VB theory. We say that the covalent bonds are saturable since the total number of covalent bonds formed by an atom is limited. For example, in general, a hydrogen atom can form only one covalent bond but a carbon atom can form four covalent bonds. This is because the total number of bonding orbitals as well as the total number of unpaired electrons of an atom are limited. Once paired with another unpaired electron to form a covalent bond, the electron cannot pair again unless the bond is broken. The saturability of covalent bonds determines the quantitative relationship between various atoms when they combine to form molecules.

All atomic orbitals other than the spherical s orbitals show certain directivity in 3D space, meaning that their lobes (the direction of the maximum probability density) are located at certain space orientations. In order for the maximum overlap between two orbitals, the covalent bond must be formed in certain directions. The directivity of covalent bonds determines the configuration of the molecule.

Classification of Covalent Bonds

There are two major types of covalent bonds in VB theory: **σ (sigma) bond** and **π (pi) bond** (**Figure 8.7**). A σ bond is formed when orbitals overlap in a head-to-head fashion, such as the bonds formed between two 1s orbitals in H₂, and between the 3p orbital of Cl and 1s orbital of H in HCl. The shape and signs of the orbitals forming a σ bond remain unchanged when rotating around the bonding axis, whatever the rotation angle is. A π bond is formed when two parallel orbitals overlap in side-to-side fashion. When rotating 180° around the bonding axis, the shape of the orbitals forming a π bond do not change but the signs change (with a nodal plane along the bonding axis).

For multiple covalent bonds, a double bond comprises a σ bond and a π bond whereas a triple bond consists of one σ bond and two π bonds. The bond energy of a π bond is slightly lower than that of the corresponding σ bond. For molecules with multiple covalent bonds, the shape of the molecule is determined only by the σ bonds (called σ-bond framework). Rotation of a single bond is allowed, since the σ bond

(1900—1954）于 1927 年提出，后经莱纳斯·鲍林（1901—1994）修正。在价键理论中，当即将成键的两个原子彼此接近时，原子轨道中具有较高电子密度的区域开始相互钻穿（称为**重叠**）。两个原子之间的电子密度增加，而增加的电子密度吸引两个带正电的原子核，使得体系的总能量降低，并在两个原子之间形成共价键。

让我们再次以 H_2 分子中的 H—H 键为例。回顾 **7.6 节**和**图 7.17**，我们展示了在 $1s$ 轨道中找到电子概率较高的区域。当两个 H 原子彼此接近时，两个半充满的 $1s$ 轨道之间发生相互钻穿，两个 $1s$ 原子波函数之间发生干涉。当两个原子波函数同相时，会发生相长干涉，使得两个 H 原子之间的电子密度增加、总能量降低。在形成共价 H—H 键的平衡核间距处，能量变得最低。这种最低能量的状态称为 H_2 分子的基态。同时，当两个原子波函数异相时，相消干涉使得两个 H 原子之间的电子密度降低、总能量增加。这对应于 H_2 分子的激发态 [**图 8.6(a)** 中的条件 5]。

对共价键特性的解释

共价键的两大特性是饱和性与方向性，均可用价键理论解释。我们说共价键具有饱和性，指的是一个原子所能形成的共价键总数是有限的。例如，通常而言，氢原子只能形成一个共价键，而碳原子可形成四个共价键。这是由于原子的成键轨道总数以及未成对电子总数均是有限的，一旦与另一未成对电子配对形成共价键之后，除非共价键断裂，该电子不能再与其他电子配对。共价键的饱和性决定了各种原子结合形成分子时的定量关系。

除球形的 s 轨道外，所有原子轨道在三维空间中都具有一定的方向性，即其波瓣（最大概率密度所在的方向）具有特定的空间取向。为实现两个轨道之间的最大重叠，共价键必须在特定方向上形成。共价键的方向性决定了分子的几何构型。

共价键的分类

价键理论中有两类主要的共价键：**σ 键**和 **π 键**（**图 8.7**）。当轨道以"头碰头"的方式重叠时，形成 σ 键，如 H_2 中两个 $1s$ 轨道之间形成的键，以及 HCl 中 Cl 的 $3p$ 轨道与 H 的 $1s$ 轨道之间形成的键。当围绕键轴旋转时，无论旋转角度如何，形成 σ 键的轨道的形状和符号均保持不变。当两个平行轨道以"肩并肩"的方式重叠时，形成 π 键。围绕键轴旋转 180°，形成 π 键的轨道形状不变，但符号会发生改变（在沿键轴的方向上存在一个节面）。

对于多重共价键，双键包括一个 σ 键和一个 π 键，而叁键包括一个 σ 键和两个 π 键。π 键的键能略低于相应的 σ 键。对于具有多重共价键的分子，其形状仅由 σ 键决定（称为 σ 键骨架）。围绕单键的旋转是允许的，因为 σ 键在旋转过程中保持不变。而围绕多重键的旋转严格受限，因为任何扭曲或旋转均会减少成键轨道的重叠量并削弱 π 键。因此，我们说双键和叁键都是刚性的。

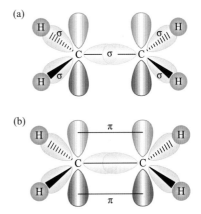

Figure 8.7 (a) σ and (b) π bonds in C_2H_4.

图 8.7 C_2H_4 中的 (a) σ 键和 (b) π 键。

半定量分析

价键理论是基于量子力学的定量理论，但这里我们将以半定量的方式来介绍它。前述 **7.5 节**介绍了量子力学和波函数。我们已经理解，类氢物种的不含时波函数 $\psi(q)$ 是空间坐标 q 的函数。严格

remains unchanged during rotation. However, rotation of multiple bonds is severely restricted because any twisting or rotation would reduce the amount of overlap of the bonding orbitals and weaken the π bond. Therefore, we say that double and triple bonds are rigid.

Semi-Quantitative Analysis

The VB theory is a quantitative theory based on quantum mechanics, but here we will introduce it in a semi-quantitative way. Previously in **Section 7.5**, we have introduced quantum mechanics and wave function. We have understood that the time-independent wave function of the hydrogen-like species $\psi(q)$ is a function in terms of the spatial coordinate q. Strictly, q is the spatial coordinate of the system consisting of an electron and a nucleus. However, since the nucleus is much heavier than the electron, the reduced mass of the system is approximately equal to that of the electron, and q of the system is also approximately equal to that of the electron. Similarly, the wave function of a molecular system can also be considered as a function in terms of the spatial coordinates q of its electrons, approximately.

For two individual H atoms where atom **A** has an unpaired electron **1** in its valance atomic orbital ψ_a, and **B** has an unpaired electron **2** in ψ_b, shown in **Figure 8.8(a)**, we have

$$\psi_A(1) = \psi_a(1) \quad \text{and} \quad \psi_B(2) = \psi_b(2) \tag{8.3}$$

For conciseness, here we use **1** and **2** as abbreviations for q_1 and q_2, respectively. If both electrons are in their ground state, then $\psi_A = \psi_{1s}(1)$ and $\psi_B = \psi_{1s}(2)$. But this may not necessarily be the case, so we use ψ_a and ψ_b instead of ψ_{1s}.

Now let us consider a system comprising these two H atoms. When these two H atoms are infinitely apart, they are simply two individual free H atoms without any interaction between them. Thus, the wave function of the system equals the multiplication of its individual wave functions, as

$$\psi_{AB}(1,2) = \psi_A(1)\psi_B(2) = \psi_a(1)\psi_b(2) \tag{8.4}$$

If these two H atoms approach each other, interaction between them happens and interpenetration of wave functions occurs. Because electrons are indistinguishable, there are possibilities that **1** in ψ_a and **2** in ψ_b [**Figure 8.8(b)**], as well as that **2** in ψ_a and **1** in ψ_b [**Figure 8.8(c)**]. Then, the wave functions of the system equal the linear combinations of both possibilities, as

$$\text{Constructive(bonding)}: \psi_{AB}(1,2) = \psi_a(1)\psi_b(2) + \psi_a(2)\psi_b(1) \tag{8.5}$$

and
$$\text{Destructive(anti-bonding)}: \psi_{AB}^*(1,2) = \psi_a(1)\psi_b(2) - \psi_a(2)\psi_b(1) \tag{8.6}$$

where ψ_{AB} is the constructive interference between $\psi_a(1)\psi_b(2)$ and $\psi_a(2)\psi_b(1)$, corresponding to the bonding orbital, and ψ_{AB}^* with a superscript asterisk (*) is the destructive interference between $\psi_a(1)\psi_b(2)$ and $\psi_a(2)\psi_b(1)$, corresponding to the anti-bonding orbital with higher energy than ψ_{AB}. As a semi-quantitative analysis, the coefficients in linear combination and the normalization coefficients of the wave functions are omitted for conciseness.

Similar to the atomic system, the square of the wave function, ψ_{AB}^2, also gives the electron density. The square of **Equations (8.5)** and **(8.6)** gives that

$$\psi_{AB}^2(1,2) = \psi_a^2(1)\psi_b^2(2) + \psi_a^2(2)\psi_b^2(1) + 2\psi_a(1)\psi_a(2)\psi_b(1)\psi_b(2)$$

$$\psi_{AB}^{*2}(1,2) = \psi_a^2(1)\psi_b^2(2) + \psi_a^2(2)\psi_b^2(1) - 2\psi_a(1)\psi_a(2)\psi_b(1)\psi_b(2)$$

As we can see, compared to ψ_{AB} in **Equation (8.4)**, ψ_{AB} in **Equation (8.5)** has an additional term $\psi_a(2)\psi_b(1)$, corresponding to additional possible electron arrangement. This additional electron arrangement results in an additional cross term $2\psi_a(1)\psi_a(2)\psi_b(1)\psi_b(2)$ in electron density, with a positive sign in the bonding orbital and a negative sign in the anti-bonding orbital. This cross term corresponds to the electron density between the two H atoms. Thus, the bonding orbital has increased electron density between the two H atoms and lower total energy, whereas the anti-bonding orbital has decreased electron density between the two H atoms and higher total energy.

地说，q 是由一个电子和一个原子核组成的体系的空间坐标。然而，由于原子核比电子重得多，体系的折合质量约等于电子的质量，体系的 q 也约等于电子的 q。类似地，分子体系的波函数也可近似认为是其电子的空间坐标 q 的函数。

对于两个单独的 H 原子，其中原子 **A** 在其价原子轨道 ψ_a 中有一个未成对电子 **1**，而原子 **B** 在 ψ_b 中有一个未成对电子 **2**，如**图 8.8(a)** 所示，有

$$\psi_A(1) = \psi_a(1) \quad 且 \quad \psi_B(2) = \psi_b(2) \tag{8.3}$$

为简明起见，这里我们分别使用 **1** 和 **2** 作为 q_1 和 q_2 的缩写。如果两个电子均处于基态，则 $\psi_A = \psi_{1s}(1)$ 且 $\psi_B = \psi_{1s}(2)$。但情况未必如此，所以我们使用 ψ_a 和 ψ_b 代替 ψ_{1s}。

现在让我们考虑包含这两个氢原子的体系。当这两个氢原子相距无限远时，它们只是两个单独的自由氢原子，之间没有任何相互作用。因此，体系的波函数等于其各自波函数的乘积，即

$$\psi_{AB}(1,2) = \psi_A(1)\psi_B(2) = \psi_a(1)\psi_b(2) \tag{8.4}$$

如果这两个氢原子彼此接近，它们之间会发生相互作用，波函数也会相互钻穿。由于电子是不可区分的，因此 **1** 在 ψ_a、**2** 在 ψ_b [**图 8.8(b)**] 和 **2** 在 ψ_a、**1** 在 ψ_b [**图 8.8(c)**] 的可能性均存在。故体系的波函数等于这两种可能性的线性组合，即

相长干涉（成键）：$\psi_{AB}(1,2) = \psi_a(1)\psi_b(2) + \psi_a(2)\psi_b(1)$ (8.5)

相消干涉（反键）：$\psi_{AB}^*(1,2) = \psi_a(1)\psi_b(2) - \psi_a(2)\psi_b(1)$ (8.6)

其中，ψ_{AB} 是 $\psi_a(1)\psi_b(2)$ 和 $\psi_a(2)\psi_b(1)$ 之间的相长干涉，对应于成键轨道；而带星号（*）上标的 ψ_{AB}^* 是 $\psi_a(1)\psi_b(2)$ 和 $\psi_a(2)\psi_b(1)$ 之间的相消干涉，对应于反键轨道。作为半定量分析，为简明起见，我们省略了线性组合的系数以及波函数的归一化系数。

与原子体系类似，波函数的平方（ψ_{AB}^2）也给出了电子密度。**式 (8.5)** 和 **(8.6)** 的平方为

$$\psi_{AB}^2(1,2) = \psi_a^2(1)\psi_b^2(2) + \psi_a^2(2)\psi_b^2(1) + 2\psi_a(1)\psi_a(2)\psi_b(1)\psi_b(2)$$

$$\psi_{AB}^{*2}(1,2) = \psi_a^2(1)\psi_b^2(2) + \psi_a^2(2)\psi_b^2(1) - 2\psi_a(1)\psi_a(2)\psi_b(1)\psi_b(2)$$

如我们所见，与**式 (8.4)** 的 ψ_{AB} 相比，**式 (8.5)** 的 ψ_{AB} 具有额外项 $\psi_a(2)\psi_b(1)$，对应于额外可能的电子排布。这种额外的电子排布导致了电子密度中的额外交叉项 $2\psi_a(1)\psi_a(2)\psi_b(1)\psi_b(2)$，其中成键轨道的交叉项为正号，而反键轨道的交叉项为负号。该交叉项对应于两个 H 原子之间的电子密度。因此，成键轨道增加了两个 H 原子之间的电子密度，总能量较低，而反键轨道降低了两个 H 原子之间的电子密度，总能量较高。

上述价键理论可以解释一些简单分子（如 H_2）的共价成键及结构。然而，从未修正的原子轨道的简单重叠推导出的分子几何构型，在许多情况下与观察到的结果并不一致。历史上，价键理论存在两个修正：杂化理论和共振理论。

修正一：杂化理论

杂化理论最早由莱纳斯·鲍林于 1931 年提出，作为价键理论的修正，用于解释 CH_4 的结构。碳的基态价电子组态为

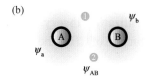
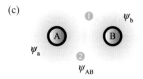

Figure 8.8 Schematic diagrams for the wave functions of a system comprising of two H atoms. (a) The wave functions ψ_A and ψ_B of two individual H atoms. (b) The wave function ψ_{AB} with **1** in ψ_a and **2** in ψ_b. (c) The wave function ψ_{AB} with **2** in ψ_a and **1** in ψ_b.

图 8.8 由两个氢原子组成的体系的波函数示意图：(a) 两个单独的氢原子的波函数 ψ_A 和 ψ_B；(b) **1** 在 ψ_a、**2** 在 ψ_b 的波函数 ψ_{AB}；(c) **2** 在 ψ_a、**1** 在 ψ_b 的波函数 ψ_{AB}。

The VB theory described above can explain the covalent bonding and the structure of some simple molecules such as H$_2$. However, the molecular geometry resulting from the simple overlap of unmodified atomic orbitals does not agree with the observed measurements in many cases. Historically, two modifications of the VB theory were proposed: the hybridization theory and the resonance theory.

Modification 1: Hybridization Theory

The **hybridization theory** was first developed by Linus Pauling in 1931 as a modification of VB theory to explain the structure of CH$_4$. The ground-state electron configuration of the valence shell for C is

$$C \quad \underset{2s}{(\uparrow\downarrow)} \quad \underset{2p}{(\uparrow)(\uparrow)()}$$

With two half-filled 2p orbitals, we expect the existence of the molecule CH$_2$ with a bond angle of 90°. The molecule CH$_2$ does exist under specially designed circumstances and is highly reactive. In order to combine with 4 H atoms to form CH$_4$, an excited-state electron configuration of carbon, with one of the 2s electrons promoting to the empty 2p orbital is needed, as

$$C \quad \underset{2s}{(\uparrow)} \quad \underset{2p}{(\uparrow)(\uparrow)(\uparrow)}$$

This excited-state electron configuration suggests a CH$_4$ molecule with three mutually perpendicular C—H bonds based on the three 2p orbitals of C (90° bond angles) and a fourth bond in some arbitrary direction based on the spherical 2s orbital. However, the actual CH$_4$ molecule is a tetrahedron with four identical H—C—H bond angles of 109.5°, which does not agree with the above description predicted from VB theory.

VB theory describes the orbitals in bonded atoms as though they are the same as those in isolated, nonbonded atoms. In many cases, those unmodified pure atomic orbitals need to be modified. Recall that atomic orbitals (AOs) are mathematical wave functions of the electrons in an atom. Algebraic combinations of the wave functions of different orbitals can produce a new set of orbitals. The mathematical process of replacing pure AOs with reformulated AOs for bonded atoms is called **hybridization**, and the reformulated orbitals are called **hybrid orbitals**.

The general rules for orbital hybridization are listed as follows:

1) Only AOs with similar energy in the same atom can hybridize. The total energy is conserved.
2) The number of hybrid orbitals equals the total number of AOs that are combined, although the shape and orientation may be different.
3) It is possible that not all AOs are involved in hybridization. Not used hybrid orbitals should be avoided.

Taking gaseous BeCl$_2$ as an example, the ground-state valence electron configuration of Be is $(2s)^2(2p)^0$. In the hybridization scheme, the 2s orbital and one of the 2p orbitals of Be are hybridized into two ***sp* hybrid orbitals**, and the remaining two 2p orbitals are left unhybridized (**Figure 8.9**). In order for the total energy to be conserved, the energy of *sp* hybrid orbitals lies at the midpoint of the energies of *s* and *p* orbitals. The two electrons occupy the two *sp* hybrid orbitals, with parallel spins. A linear electron-group geometry and a 180° bond angle are present in the *sp* hybridization scheme, as in BeCl$_2$ (g).

In quantum mechanics, the wave functions of the *sp* hybrid orbitals can be obtained from the linear combination of the wave functions of the corresponding orbitals, such as the 2s and 2p_z orbitals [**Figures 8.10(a)** and **8.10(b)**], as

$$\begin{cases} \psi_1(sp) = \dfrac{1}{\sqrt{2}}\left[\psi(2s) + \psi(2p_z)\right] \\ \psi_2(sp) = \dfrac{1}{\sqrt{2}}\left[\psi(2s) - \psi(2p_z)\right] \end{cases} \tag{8.7}$$

where $1/\sqrt{2}$ is the normalization coefficient. The 2D contour maps and 3D probability density maps of the *sp* hybrid orbitals are shown in **Figures 8.10(c)(d)** and **8.10(e)(f)**, respectively.

Three ***sp*^{2} hybrid orbitals** are formed from the hybridization of one *s* orbital and two *p* orbitals of the same atom. The energy of *sp*^{2} hybrid orbitals lies in 1/3 of the energy difference between the *s* and *p* orbitals,

由于具有两个半充满的 $2p$ 轨道,我们预期存在 CH_2 分子,其键角为 $90°$。在特殊设计的环境下 CH_2 分子确实存在,且具有高反应活性。为与 4 个 H 原子结合形成 CH_4,需要碳的激发态电子组态,其中有一个 $2s$ 电子激发到空 $2p$ 轨道,即

这种激发态电子组态表明,CH_4 分子应该具有三个互相垂直的、基于 C 的三个 $2p$ 轨道的 C—H 键($90°$ 键角),而第四个键基于球形 $2s$ 轨道,可在任意方向。然而实际的 CH_4 分子是四面体,具有四个完全相同的 H—C—H 键,键角为 $109.5°$。这与从价键理论预测的上述描述不一致。

价键理论对成键原子轨道的描述,与对孤立、未成键原子轨道的描述完全相同。在许多情况下,需要修正这些未经修正的纯原子轨道。回顾原子轨道(AOs)是原子中电子的数学形式的波函数。不同轨道波函数的代数组合,可以产生一组新的轨道。将纯原子轨道替换为重新生成的原子轨道的数学过程,称为**杂化**,重新生成的轨道称为**杂化轨道**。

轨道杂化的一般规则如下:
1) 只有同一原子中能量相近的原子轨道才能杂化,总能量守恒。
2) 杂化轨道的总数等于参与杂化的原子轨道总数,尽管轨道形状和空间取向可能不同。
3) 并非所有原子轨道都必须参与杂化,应尽量避免杂化后的轨道不被利用。

以气态 $BeCl_2$ 为例,Be 的基态价电子组态为 $(2s)^2(2p)^0$。在杂化方案中,Be 的 $2s$ 轨道和一个 $2p$ 轨道发生杂化,形成两个 **sp 杂化轨道**,剩余的两个 $2p$ 轨道不参与杂化(**图 8.9**)。为使总能量守恒,sp 杂化轨道的能量位于 s 和 p 轨道能量的中点。两个电子以自旋平行的方式,占据两个 sp 杂化轨道。sp 杂化方案对应于直线形电子组几何构型和 $180°$ 键角,如 $BeCl_2(g)$。

量子力学中,sp 杂化轨道的波函数可由相应轨道(如 $2s$ 和 $2p_z$ 轨道)波函数的线性组合获得 [**图 8.10(a)** 和 **8.10(b)**],有

$$\begin{cases} \psi_1(sp) = \dfrac{1}{\sqrt{2}}\left[\psi(2s) + \psi(2p_z)\right] \\ \psi_2(sp) = \dfrac{1}{\sqrt{2}}\left[\psi(2s) - \psi(2p_z)\right] \end{cases} \tag{8.7}$$

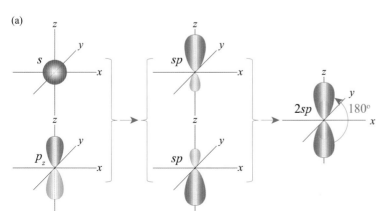

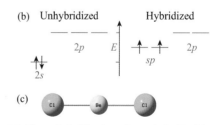

Figure 8.9 The sp hybridization scheme in $BeCl_2$. (a) sp hybrid orbital diagrams of Be. (b) Energy-level diagrams of Be before and after hybridization. (c) Ball-and-stick model of $BeCl_2$.

图 8.9 $BeCl_2$ 的 sp 杂化方案:(a) Be 的 sp 杂化轨道示意图;(b) 杂化前后 Be 的能级图;(c) $BeCl_2$ 的球棍模型。

downward from the energy of *p* orbital. A trigonal-planar electron-group geometry and 120° bond angles are present in the *sp*² hybridization scheme, as in BF₃ (**Figure 8.11**). The wave functions of the *sp*² hybrid orbitals in the *xz* plane are

$$\begin{cases} \psi_1(sp^2) = \dfrac{1}{\sqrt{3}}\psi(2s) + \dfrac{\sqrt{2}}{\sqrt{3}}\psi(2p_x) \\ \psi_2(sp^2) = \dfrac{1}{\sqrt{3}}\psi(2s) - \dfrac{1}{\sqrt{6}}\psi(2p_x) + \dfrac{1}{\sqrt{2}}\psi(2p_z) \\ \psi_3(sp^2) = \dfrac{1}{\sqrt{3}}\psi(2s) - \dfrac{1}{\sqrt{6}}\psi(2p_x) - \dfrac{1}{\sqrt{2}}\psi(2p_z) \end{cases} \quad (8.8)$$

Four ***sp*³ hybrid orbitals** are formed from the hybridization of one *s* orbital and three *p* orbitals of the same atom. The energy of *sp*³ hybrid orbitals lies in 1/4 of the energy difference between the *s* and *p* orbitals, downward from the energy of *p* orbital. A tetrahedral electron-group geometry and 109.5° bond angles are present in the *sp*³ hybridization scheme, as in CH₄ (**Figure 8.12**). The wave functions of the *sp*³ hybrid orbitals are

$$\begin{cases} \psi_1(sp^3) = \dfrac{1}{2}\psi(2s) + \dfrac{1}{2}\psi(2p_x) + \dfrac{1}{2}\psi(2p_y) + \dfrac{1}{2}\psi(2p_z) \\ \psi_2(sp^3) = \dfrac{1}{2}\psi(2s) - \dfrac{1}{2}\psi(2p_x) + \dfrac{1}{2}\psi(2p_y) - \dfrac{1}{2}\psi(2p_z) \\ \psi_3(sp^3) = \dfrac{1}{2}\psi(2s) - \dfrac{1}{2}\psi(2p_x) - \dfrac{1}{2}\psi(2p_y) + \dfrac{1}{2}\psi(2p_z) \\ \psi_4(sp^3) = \dfrac{1}{2}\psi(2s) + \dfrac{1}{2}\psi(2p_x) - \dfrac{1}{2}\psi(2p_y) - \dfrac{1}{2}\psi(2p_z) \end{cases} \quad (8.9)$$

In the cases of NH₃ and H₂O, the hybridization scheme for the central atoms is still *sp*³, so the electron-group geometry is still tetrahedral. Because N has 5 valence electrons, one of the *sp*³ orbitals is occupied by a lone pair of electrons and its energy is slightly lower than those of the other three half-filled *sp*³ orbitals involved in bond formation, leading to a trigonal-pyramidal molecular geometry of 107.3° bond angles for NH₃. Similarly, two of the *sp*³ orbitals are occupied by lone-pair electrons, resulting in a bent structure with a 104.5° bond angle for H₂O.

More complicated cases involve *d*-orbital contributions. Five ***sp*³*d* hybrid orbitals** are combined from one *s*, three *p*, and one *d* orbitals of the valence shell to form a trigonal-bipyramidal molecular geometry, as in PCl₅ [**Figures 8.13(a)(b)**]. Six ***sp*³*d*² hybrid orbitals** are combined from one *s*, three *p*, and two *d* orbitals of the valence shell to form an octahedral molecular geometry, as in SF₆ [**Figures 8.13(c)(d)**].

Modification 2: Resonance Theory

VB theory is a localized bonding theory—the charge density of the bonding electrons is concentrated in the region of orbital overlap, and thus cannot explain delocalized electron systems, such as benzene (C₆H₆). The **resonance theory** was first developed by Linus Pauling in 1928 as a modification of VB theory to explain the structure of benzene. Based on the hybridization theory, a flat, hexagonal ring of six C atoms with *sp*² hybridization, joined by alternating single and double covalent bonds (the **Kekulé structure**), is expected for benzene. However, spectroscopic data show that all the C—C bonds are actually identical in benzene.

The resonance theory proposes that the true structure of a molecule is a resonance hybrid of all plausible contributing Lewis structures. The skeletal structures of all acceptable contributing structures to a resonance hybrid must be the same, but how the electrons are distributed within the structure can be different. In this way, electrons are delocalized to some extent. The bond order is the weighted average of all plausible contributing structures, and the weight is higher for the more plausible contributing structure.

For example, we can write two plausible contributing Lewis structures for O₃, each suggesting that one O—O bond is single and the other is double. Yet experimental evidence indicates that the two O—O bonds are

其中 $1/\sqrt{2}$ 是归一化系数。**图 8.10(c)(d)** 和 **图 8.10(e)(f)** 分别给出了 sp 杂化轨道的二维等概率密度轮廓线图和三维概率密度图。

三个 ***sp^2 杂化轨道***由同一原子的一个 s 轨道和两个 p 轨道杂化形成。sp^2 杂化轨道的能量位于 s 轨道和 p 轨道之间、从 p 轨道能量向下 1/3 处。sp^2 杂化方案对应于平面三角形电子组几何构型和 120° 键角，如 BF_3（**图 8.11**）。xz 平面中 sp^2 杂化轨道的波函数为

$$\begin{cases} \psi_1(sp^2) = \dfrac{1}{\sqrt{3}}\psi(2s) + \dfrac{\sqrt{2}}{\sqrt{3}}\psi(2p_x) \\ \psi_2(sp^2) = \dfrac{1}{\sqrt{3}}\psi(2s) - \dfrac{1}{\sqrt{6}}\psi(2p_x) + \dfrac{1}{\sqrt{2}}\psi(2p_z) \\ \psi_3(sp^2) = \dfrac{1}{\sqrt{3}}\psi(2s) - \dfrac{1}{\sqrt{6}}\psi(2p_x) - \dfrac{1}{\sqrt{2}}\psi(2p_z) \end{cases} \quad (8.8)$$

四个 ***sp^3 杂化轨道***由同一原子的一个 s 轨道和三个 p 轨道杂化形成。sp^3 杂化轨道的能量位于 s 轨道和 p 轨道之间、从 p 轨道能量向下 1/4 处。sp^3 杂化方案对应于四面体形电子组几何构型和 109.5° 键角，如 CH_4（**图 8.12**）。sp^3 杂化轨道的波函数为

$$\begin{cases} \psi_1(sp^3) = \dfrac{1}{2}\psi(2s) + \dfrac{1}{2}\psi(2p_x) + \dfrac{1}{2}\psi(2p_y) + \dfrac{1}{2}\psi(2p_z) \\ \psi_2(sp^3) = \dfrac{1}{2}\psi(2s) - \dfrac{1}{2}\psi(2p_x) + \dfrac{1}{2}\psi(2p_y) - \dfrac{1}{2}\psi(2p_z) \\ \psi_3(sp^3) = \dfrac{1}{2}\psi(2s) - \dfrac{1}{2}\psi(2p_x) - \dfrac{1}{2}\psi(2p_y) + \dfrac{1}{2}\psi(2p_z) \\ \psi_4(sp^3) = \dfrac{1}{2}\psi(2s) + \dfrac{1}{2}\psi(2p_x) - \dfrac{1}{2}\psi(2p_y) - \dfrac{1}{2}\psi(2p_z) \end{cases} \quad (8.9)$$

对于 NH_3 和 H_2O，其中心原子的杂化方案仍为 sp^3，因此电子组几何构型仍为四面体形。由于 N 有 5 个价电子，其中一个 sp^3 轨道被一对孤电子对占据，其能量略低于参与成键的其他三个半充满的 sp^3 轨道，导致 NH_3 的分子几何构型为三角锥形，键角为 107.3°。类似地，H_2O 的两个 sp^3 轨道被孤电子对占据，导致了键角为 104.5° 的折叠形结构。

涉及 d 轨道的情况更为复杂。五个 ***sp^3d 杂化轨道***由价层的一个 s、三个 p 和一个 d 轨道组合而成，形成三角双锥形分子几何构型，如 PCl_5[**图 8.13(a)(b)**]。六个 ***sp^3d^2 杂化轨道***由价层的一个 s、三个 p 和两个 d 轨道组合而成，形成八面体形分子几何构型，如 SF_6[**图 8.13(c)(d)**]。

修正二：共振理论

价键理论是一种定域键理论（即成键电子的电荷密度集中在轨道重叠的区域），因此无法解释离域电子体系，如苯（C_6H_6）。**共振理论**最早由莱纳斯·鲍林于 1928 年提出，作为价键理论的修正，用于解释苯的结构。杂化理论预期，苯的六个 C 原子由 sp^2 杂化形成平面六边形环，通过交替的共价单 / 双键连接（即**凯库勒结构**）。然而光谱数据表明，苯的所有 C—C 键实际上完全相同。

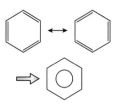

共振理论提出，分子的真实结构是所有合理的路易斯共振体的共振混合体。同一共振混合体的所有可接受的共振体，其骨架结构必须相同，但电子在其中的分布方式可能不同，这样电子在某种程度上是离域的。键级是所有合理的共振体的加权平均值，而更为合

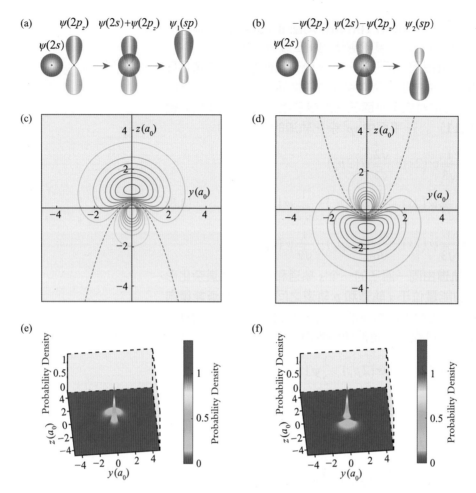

Figure 8.10 The formation of *sp* hybrid orbitals. Linear combination of $\psi(2s)$ and $\psi(2p_z)$ to form (a) $\psi_1(sp)$ and (b) $\psi_2(sp)$. 2D contour maps of (c) $\psi_1(sp)$ and (d) $\psi_2(sp)$. 3D probability density maps of (e) $\psi_1(sp)$ and (f) $\psi_2(sp)$.

图 8.10 *sp* 杂化轨道的形成：$\psi(2s)$ 与 $\psi(2p_z)$ 线性组合，形成 (a) $\psi_1(sp)$ 与 (b) $\psi_2(sp)$；(c) $\psi_1(sp)$ 与 (d) $\psi_2(sp)$ 的二维等概率密度轮廓线图；(e) $\psi_1(sp)$ 与 (f) $\psi_2(sp)$ 的三维概率密度图。

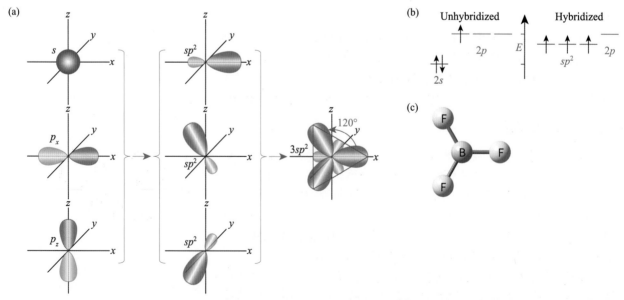

Figure 8.11 The *sp*² hybridization scheme in BF₃. (a) sp^2 hybrid orbital diagrams of B. (b) Energy-level diagrams of B before and after hybridization. (c) Ball-and-stick model of BeF₃.

图 8.11 BF₃ 的 sp^2 杂化方案：(a) B 的 sp^2 杂化轨道示意图；(b) 杂化前后 B 的能级图；(c) BF₃ 的球棍模型。

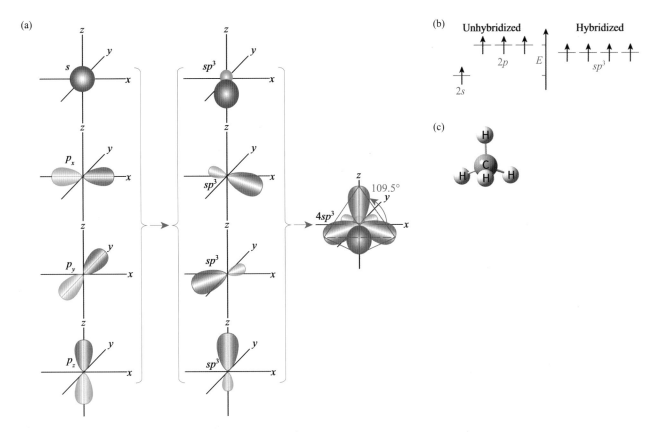

Figure 8.12 The sp^3 hybridization scheme in CH_4. (a) sp^3 hybrid orbital diagrams of C. (b) Energy-level diagrams of C before and after hybridization. (c) Ball-and-stick model of CH_4.

图 8.12 CH_4 的 sp^3 杂化方案：(a) C 的 sp^3 杂化轨道示意图；(b) 杂化前后 C 的能级图；(c) CH_4 的球棍模型。

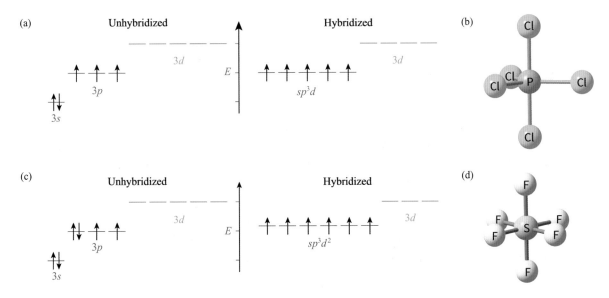

Figure 8.13 The sp^3d and sp^3d^2 hybridization schemes. (a) Energy-level diagrams of P before and after sp^3d hybridization in PCl_5. (b) Ball-and-stick model of PCl_5. (c) Energy-level diagrams of S before and after sp^3d^2 hybridization in SF_6. (d) Ball-and-stick model of SF_6.

图 8.13 sp^3d 和 sp^3d^2 杂化方案：(a) sp^3d 杂化前后 PCl_5 中 P 的能级图；(b) PCl_5 的球棍模型；(c) sp^3d^2 杂化前后 SF_6 中 S 的能级图；(d) SF_6 的球棍模型。

identical, with a length of 128 pm. This bond length is shorter than the normal O—O single bond length of ~145 pm but longer than the normal O=O double bond length of ~121 pm. The bonds in O_3 are intermediate between a single and a double bond. Therefore, the true structure of O_3 is neither of the previously proposed structures but a resonance hybrid of the two. Since both structures are equivalent, they are equally weighted, that is, they contribute equally to the structure of the resonance hybrid. The average O—O bond order is, then, (1+2)/2 = 1.5.

$$:\ddot{\underset{\ominus}{O}}-\underset{\oplus}{\ddot{O}}=\underset{0}{\ddot{O}} \leftrightarrow \underset{0}{\ddot{O}}-\underset{\oplus}{\ddot{O}}=\underset{\ominus}{\ddot{O}}:$$

Another example is the azide anion (N_3^-), for which three contributing structures are shown below:

$$\left[:\underset{\ominus}{\ddot{N}}-\underset{\oplus}{N}\equiv\underset{0}{N}:\right]^- \leftrightarrow \left[\underset{\ominus}{\ddot{N}}=\underset{\oplus}{N}=\underset{\ominus}{\ddot{N}}\right]^- \leftrightarrow \left[:\underset{0}{N}\equiv\underset{\oplus}{N}-\underset{\ominus}{\ddot{N}}:\right]^-$$

By applying the general rules for formal charges, we understand that the central contributing structure is more plausible since it avoids the unlikely large formal charge of −2 found on an N atom in the other two structures. Consequently, we expect that the central structure contributes most to the resonance hybrid of N_3^-. The average N—N bond order, no matter the weights, equals 2.

Back to the case of benzene, the two equivalent Kekulé structures are the contributing structures to a resonance hybrid, symbolized by a hexagonal ring with a circle inside. The average C—C bond order is (1+2)/2 = 1.5.

Limitations of Valence-Bond Theory

VB theory is an after-the-fact theory, which is good for explanation but poor for prediction of the molecular shape and structure. VB theory uses the concept of electron-pairs from Lewis theory, and thus cannot explain the structure and properties of some odd-electron species, such as the paramagnetism of O_2, and the existences of H_2^+ and He_2^+. Meanwhile, hybridization is NOT an actual physical phenomenon. Electron charge distributions changing from those of pure AOs to those of hybrid orbitals have not been experimentally observed. For some covalent bonds no single hybridization scheme works well. Resonance theory does not suggest that the molecule has one structure part of the time and the other structure the rest of the time. The molecule has the same structure all the time. Nevertheless, the concepts of hybridization and resonance work very well for carbon-containing molecules, and are therefore used a great deal in organic chemistry.

8.5 Molecular Orbital Theory

The **molecular orbital (MO) theory** was developed primarily by Friedrich Hund (1896—1997) and Robert S. Mulliken (1896—1986) around 1932.

Key Points of Molecular-Orbital Theory

In MO theory, electrons in a molecule are not constrained to individual chemical bonds but are considered as moving in an average potential field arising from the nuclei and all other electrons (called **single electron approximation**). The status of the electrons in the molecule can be described by **molecular orbitals**, which are mathematical wave functions related to the probability of finding electrons in certain regions of a molecule. MOs can be considered as linear combinations of atomic orbitals (LCAOs). The number of MOs formed must equal the number of AOs that are combined.

For example, as two H atoms approach each other, the two 1s orbitals combine by interfering constructively to form a **bonding molecular orbital**, as

$$\psi_{\sigma_{1s}} = \psi_{1s_A} + \psi_{1s_B} \tag{8.10}$$

with increased electron density between the two nuclei and lower energy than the original 1s orbitals, as

$$\psi_{\sigma_{1s}}^2 = \psi_{1s_A}^2 + \psi_{1s_B}^2 + 2\psi_{1s_A}\psi_{1s_B}$$

and destructively to form an **anti-bonding molecule orbital**, as

理的共振体具有更高的权重。

例如，我们可以写出 O_3 的两个合理的路易斯共振体，每个共振体都具有一个 O—O 单键和一个双键。然而实验表明，两个 O—O 键完全相同，键长为 128 pm。该键长比常规 O—O 单键的键长（约 145 pm）更短，但比常规 O═O 双键的键长（约 121 pm）更长，即 O_3 的键介于单键和双键之间。因此 O_3 的真正结构并非前面提到的两个结构，而是二者的共振混合体。由于这两种结构对等，因此其权重相等，即它们对共振混合体的贡献相等。O—O 键的平均键级为 (1+2)/2=1.5。

$$:\ddot{\text{O}}—\ddot{\text{O}}=\ddot{\text{O}} \leftrightarrow \ddot{\text{O}}—\ddot{\text{O}}=\ddot{\text{O}}:$$

另一个例子是叠氮阴离子（N_3^-），其三种共振体如下：

$$[:\ddot{\text{N}}—\text{N}\equiv\text{N}:]^- \leftrightarrow [\ddot{\text{N}}=\text{N}=\ddot{\text{N}}]^- \leftrightarrow [:\text{N}\equiv\text{N}—\ddot{\text{N}}:]^-$$

应用形式电荷的一般规则，我们知道中间的共振体更为合理，因为它避免了其他两个共振体中 N 原子上出现的较大的形式电荷 −2。因此，我们预期中间的共振体对 N_3^- 的共振混合体贡献最大。无论权重如何，N—N 键的平均键级均为 2。

回到苯的情况，两个对等的凯库勒结构是同一共振混合体的共振体，用内含圆圈的六边形环表示。C—C 的平均键级为 (1+2)/2 = 1.5。

价键理论的局限性

价键理论是一个后验理论，即它可以很好地解释分子形状和结构，但对分子形状和结构的预测性较差。价键理论使用了路易斯理论中的电子对概念，因此无法解释某些存在未成对电子的物种的结构与性质，如 O_2 的顺磁性，以及 H_2^+ 和 He_2^+ 的存在。同时，杂化并非一种实际的物理现象，实验上并没有观察到电子电荷分布从纯原子轨道到杂化轨道的变化。对于某些共价键，没有单一的杂化方案能很好地奏效。共振理论并非认为分子在某些时间具有一种结构，而在其余时间具有另一种结构；分子一直都具有相同的结构。尽管如此，杂化和共振的概念对含碳分子非常有效，因此被大量应用于有机化学中。

Electrostatic potential map of O_3.
O_3 的静电势图

Electrostatic potential map of N_3^-.
N_3^- 的静电势图

8.5 分子轨道理论

分子轨道（MO）理论主要由弗里德里希·洪特（1896—1997）和罗伯特·S. 穆利肯（1896—1986）在 1932 年左右提出。

分子轨道理论的要点

在分子轨道理论中，分子中的电子不受单个化学键的约束，而被认为在由原子核和所有其他电子产生的平均势场中运动（称为**单电子近似**）。分子中电子的状态可用**分子轨道**来描述，分子轨道是与在分子的某些区域找到电子的概率相关的数学形式的波函数，可视为原子轨道的线性组合（LCAO）。形成的分子轨道数目必须等于参与组合的原子轨道数目。

例如，当两个 H 原子彼此接近时，两个 $1s$ 轨道通过相长干涉形成**成键分子轨道**，即

$$\psi^*_{\sigma_{1s}} = \psi_{1s_A} - \psi_{1s_B} \tag{8.11}$$

with decreased electron density between the two nuclei and higher energy than the original 1s orbitals, as

$$\psi^{*2}_{\sigma_{1s}} = \psi^2_{1s_A} + \psi^2_{1s_B} - 2\psi_{1s_A}\psi_{1s_B}$$

The combination of two 1s orbitals of H atoms into two MOs in a H_2 molecule is summarized in **Figure 8.14**.

There are three main requirements for the effective combination of AOs into MOs:

1) The AOs to be combined must have the correct symmetry. The σ and π bonding MOs are formed by the overlap between the regions of the two corresponding AOs with the same signs, i.e., + to + and − to −. The σ* and π* anti-bonding MOs are formed by the overlap between the regions of the two corresponding AOs with the opposite signs, i.e., + to −.

2) The AOs to be combined must be at similar energy levels. AOs with large energy difference cannot be combined to form effective MOs.

3) Maximum overlap should occur between two AOs to form effective MOs. In this way, the total energy of the system is minimized.

In ground-state configurations, the arrangements of electrons in MOs also follow the same rules as those in AOs, that is, Pauli exclusion principle (each MO can only accommodate two electrons with opposing spins), lowest energy principle (electrons occupy MOs in a way that minimizes the total energy of the molecule), and Hund's rule (electrons enter degenerate MOs singly before they pair up).

After electrons are assigned to MOs, a stable molecular species should have more electrons in bonding MOs than in anti-bonding MOs. Since two electrons correspond to a single covalent bond in Lewis theory, the bond order in MO theory can be calculated as one-half the difference between the number of bonding and anti-bonding electrons.

Homonuclear Diatomics of the First-Period Elements

There are only two elements in the first row of the periodic table: H and He. Here, we consider four homonuclear diatomic species of the first-period elements: H_2, H_2^+, He_2, and He_2^+. Two 1s AOs are combined to form a bonding σ_{1s} MO and an anti-bonding σ_{1s}^* MO. The available electrons are then distributed in these two MOs. The details are listed below and summarized in **Figure 8.15**.

1) H_2: This molecule has two electrons. According to Pauli exclusion and lowest energy principles, these two electrons are distributed in the bonding σ_{1s} MO in the ground-state configuration, with opposing spins. The bond order is $(2 − 0)/2 = 1$, corresponding to a single covalent H—H bond.

2) H_2^+: This molecular ion has a single electron, and it enters the bonding σ_{1s} MO. The bond order is $(1 − 0)/2 = 1/2$, equivalent to a half bond. Therefore, H_2^+ is a stable ion.

3) He_2: This molecule has four electrons, two in σ_{1s} and two in σ_{1s}^*. The bond order is $(2 − 2)/2 = 0$. No bond is produced—He_2 is not a stable species.

4) He_2^+: This molecular ion has three electrons, two in σ_{1s} and one in σ_{1s}^*. This species exists as a stable ion with a bond order of $(2 − 1)/2 = 1/2$.

Homonuclear Diatomics of the Second-Period Elements

The situation for homonuclear diatomic species of the second-period elements is more interesting because both 2s and 2p orbitals are involved. The MOs formed by combining 2s orbitals are similar to those from 1s orbitals, except for with higher energies. The situation for combining 2p orbitals, however, is different. As illustratively shown in O_2, there are two possible ways to combine 2p orbitals into MOs:

1) An end-to-end combination produces a pair of σ-type MOs [**Figures 8.16(a)(b)**] along the bonding axis: the bonding σ_{2p} (overlap between the lobes of the 2p orbitals with the same phase) and anti-bonding σ_{2p}^* (overlap between the lobes of the 2p orbitals with opposite phase) MOs. There is no nodal plane in σ_{2p} MO but one nodal plane midway between the nuclei in σ_{2p}^* MO. Only one pair of p orbitals in the bonding axis direction can combine to form σ-type MOs.

$$\psi_{\sigma_{1s}} = \psi_{1s_A} + \psi_{1s_B} \tag{8.10}$$

两个原子核之间的电子密度增加，能量低于原 1s 轨道，有

$$\psi_{\sigma_{1s}}^2 = \psi_{1s_A}^2 + \psi_{1s_B}^2 + 2\psi_{1s_A}\psi_{1s_B}$$

通过相消干涉形成**反键分子轨道**，即

$$\psi_{\sigma_{1s}}^* = \psi_{1s_A} - \psi_{1s_B} \tag{8.11}$$

两个原子核之间的电子密度降低，能量高于原 1s 轨道，有

$$\psi_{\sigma_{1s}}^{*2} = \psi_{1s_A}^2 + \psi_{1s_B}^2 - 2\psi_{1s_A}\psi_{1s_B}$$

图 8.14 总结了 H 原子的两个 1s 轨道组合形成 H$_2$ 分子的两个分子轨道的情况。

将原子轨道有效组合形成分子轨道，有三个主要要求：
1) 待组合的原子轨道必须具有正确的对称性。σ 和 π 的成键分子轨道由两个符号相同的对应原子轨道的区域重叠形成，即 + 与 + 重叠，− 与 − 重叠。σ* 和 π* 的反键分子轨道由两个符号相反的对应原子轨道的区域重叠形成，即 + 与 − 重叠。
2) 待组合的原子轨道的能级必须接近。具有较大能量差的原子轨道，不能组合形成有效的分子轨道。
3) 两个原子轨道之间应出现最大重叠，以形成有效的分子轨道，这样可使体系的总能量最小化。

基态组态中，电子在分子轨道内的排布也遵循与在原子轨道内排布相同的规则，即泡利不相容原理（每个分子轨道只能容纳两个自旋相反的电子）、能量最低原理（电子以使分子总能量最小化的方式占据分子轨道）和洪特规则（电子在配对前优先分占简并的分子轨道）。

将电子排布在分子轨道上之后，稳定的分子物种在成键分子轨道中的电子应比在反键分子轨道中更多。由于两个电子对应于路易斯理论的共价单键，因此分子轨道理论中的键级，可用成键和反键电子数之差的一半来计算。

第一周期元素同核双原子物种

元素周期表的第一周期只有两种元素：H 和 He。这里我们考虑第一周期元素的四种同核双原子物种：H$_2$、H$_2^+$、He$_2$ 和 He$_2^+$。两个 1s 原子轨道组合形成一个成键 σ$_{1s}$ 和一个反键 σ$_{1s}^*$ 分子轨道。可用的电子排布在这两个分子轨道中，详情如下所示，并总结于**图 8.15**。

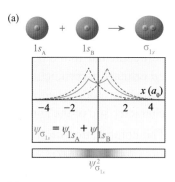

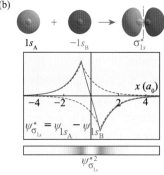

Figure 8.14 The combination of two 1s orbitals of H atoms into two molecular orbitals in a H$_2$ molecule. (a) Constructive interference of the two H 1s orbitals to form a σ$_{1s}$ bonding molecular orbital. (b) Destructive interference of the two H 1s orbitals to form a σ$_{1s}^*$ anti-bonding molecular orbital.

图 8.14 由 H 原子的两个 1s 轨道组合形成 H$_2$ 分子的两个分子轨道：(a) 两个 H 1s 轨道相长干涉形成 σ$_{1s}$ 成键分子轨道；(b) 两个 H 1s 轨道相消干涉形成 σ$_{1s}^*$ 反键分子轨道。

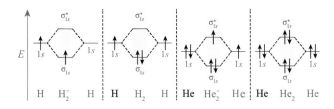

Figure 8.15 Molecular orbital energy-level diagram for the homonuclear diatomics of the first-period elements.

图 8.15 第一周期元素同核双原子物种的分子轨道能级图。

1) H$_2$：该分子有两个电子。根据泡利不相容原理和能量最低原理，基态组态时，这两个电子以自旋相反的方式排布在成键 σ$_{1s}$ 分子轨道中。键级为 $(2-0)/2=1$，对应于 H—H 共价单键。
2) H$_2^+$：该分子离子有一个单电子，进入成键 σ$_{1s}$ 分子轨道。键级为 $(1-0)/2=1/2$，相当于半键。因此，H$_2^+$ 是一种稳定的离子。
3) He$_2$：该分子有四个电子，两个在 σ$_{1s}$，两个在 σ$_{1s}^*$。键级为 $(2-2)/2=0$。没有成键，故 He$_2$ 不是稳定的物种。

2) A side-to-side combination produces two pairs of π-type MOs [**Figures 8.16(c)(d)**]: two bonding π_{2p} and two anti-bonding π_{2p}^* MOs. The bonding π_{2p} orbital is formed by the addition of the two $2p$ orbitals perpendicular to the bonding axis. The anti-bonding π_{2p}^* orbital is formed by the subtraction of the two $2p$ orbitals perpendicular to the bonding axes. In addition to the atomic node that contains the nuclei, a molecular node where the electron charge density falls to 0 is formed between the nuclei, and this is a characteristic of the anti-bonding orbitals.

The combination of $2p$ orbitals forms six MOs. If we designate the bonding axis as in the z direction, then the two σ-type MOs are σ_{2p_z} and $\sigma_{2p_z}^*$, and the four π-type MOs are π_{2p_x} and π_{2p_y} (degenerate), and $\pi_{2p_x}^*$ and $\pi_{2p_y}^*$ (degenerate), as shown in **Figure 8.17**. In addition to the above set of symbols of MOs that signify the AOs from which the MOs are formed, there is another set of symbols that signify the symmetry of the formed MOs in homonuclear diatomics. If the formed MO is symmetric / antisymmetric with respect to the center of inversion, it is designated as g (gerade) / u (ungerade) symmetry, respectively. Therefore, the σ_{2p}, σ_{2p}^*, π_{2p}, and π_{2p}^* orbitals can be designated as σ_g, σ_u, π_u, and π_g, respectively. Meanwhile, the corresponding symbols for σ_{2s}, σ_{2s}^* are σ_g and σ_u, respectively. Note that all bonding σ orbitals are always σ_g, and all anti-bonding σ* orbitals are σ_u, no matter formed from s or p orbitals. Conversely, all bonding π orbitals are always π_u, and all anti-bonding π* orbitals are π_g. This is determined by symmetry of the orbitals.

The energy-level diagram for the MOs formed from AOs is related to the corresponding AO energy levels. In general, the MOs formed from $2s$ orbitals are lower in energy than those formed from $2p$ orbitals—the same relationship as between the $2s$ and $2p$ AOs. The σ-type bonding MOs should have lower energies than π-type because end-to-end overlap of p orbitals should be more extensive than side-to-side overlap.

There are eight elements in the second row of the periodic table, and consequently, eight homonuclear diatomic molecules are formed. These eight molecules can be classified into two categories according to their MO energy-level diagram. One is for O_2, F_2, and Ne_2 ($Z \geq 8$), the energy-level diagram of which are shown in **Figure 8.18(a)** and the electron assignments in **Figure 8.18(b)**. In this category, the valence-shell MOs are in a normal order of increasing energy, as σ_{2s}, σ_{2s}^*, σ_{2p}, π_{2p}, π_{2p}^*, and σ_{2p}^*. If symmetry symbols are used, the corresponding MOs are labeled as $1\sigma_g$, $1\sigma_u$, $2\sigma_g$, $1\pi_u$, $1\pi_g$, and $2\sigma_u$. The number before the symmetry symbol indicates the order of the MO with the same bond type and symmetry in the valence shell, from low to high in energy. For example, σ_{2s} is the first σ_g and σ_{2p} is the second σ_g in the valence shell. The other category is for Li_2, Be_2, B_2, C_2, and N_2 ($Z < 8$), the modified energy-level diagram and electron assignments of which are given in **Figure 8.19**. The difference lies in the switch of σ_{2p} and π_{2p} orbitals in the order. The reason for this switch is due to the more similar energy in $2s$ and $2p$ orbitals and the consequent stronger $2s$-$2p$ interactions (called $2s$-$2p$ mixing) for $Li_2 \sim N_2$ than for $O_2 \sim Ne_2$, as shown in **Figure 8.20**.

The modified energy-level diagrams are derived from experimental evidences. For example, if we assign the 8 valence electrons of the molecule C_2 to the unmodified diagram in **Figure 8.18(b)**, we obtain

$$C_2 \quad \underset{\sigma_{2s}}{(\uparrow\downarrow)} \; \underset{\sigma_{2s}^*}{(\uparrow\downarrow)} \; \underset{\sigma_{2p}}{(\uparrow\downarrow)} \; \underset{\pi_{2p}}{(\uparrow)(\uparrow)} \; \underset{\pi_{2p}^*}{(\;)(\;)} \; \underset{\sigma_{2p}^*}{(\;)}$$

With two unpaired electrons, the unmodified diagram suggests that C_2 is paramagnetic. However, experiment shows that C_2 is actually diamagnetic. Assigning 8 valence electrons to the modified diagram in **Figure 8.19(b)**, we have

$$C_2 \quad \underset{\sigma_{2s}}{(\uparrow\downarrow)} \; \underset{\sigma_{2s}^*}{(\uparrow\downarrow)} \; \underset{\pi_{2p}}{(\uparrow\downarrow)(\uparrow\downarrow)} \; \underset{\sigma_{2p}}{(\;)} \; \underset{\pi_{2p}^*}{(\;)(\;)} \; \underset{\sigma_{2p}^*}{(\;)}$$

This is consistent with the fact that C_2 is diamagnetic.

Another example is O_2. Assigning 12 valence electrons into the valence-shell MOs of O_2, we have

$$O_2 \quad \underset{\sigma_{2s}}{(\uparrow\downarrow)} \; \underset{\sigma_{2s}^*}{(\uparrow\downarrow)} \; \underset{\sigma_{2p}}{(\uparrow\downarrow)} \; \underset{\pi_{2p}}{(\uparrow\downarrow)(\uparrow\downarrow)} \; \underset{\pi_{2p}^*}{(\uparrow)(\uparrow)} \; \underset{\sigma_{2p}^*}{(\;)}$$

This demonstrates that O_2 has two unpaired electrons and explains the paramagnetism of O_2. The bond order is $(8 - 4)/2 = 2$, corresponding to a double covalent bond. More discussion about magnetism can be found in **Section 9.4**.

4) He_2^+: 该分子离子有三个电子，两个在 σ_{1s}，一个在 σ_{1s}^*。该物种以稳定的离子形式存在，键级为 $(2-1)/2 = 1/2$。

第二周期元素同核双原子物种

第二周期元素的同核双原子物种的情况更为有趣，因为涉及了 $2s$ 与 $2p$ 轨道。由 $2s$ 轨道组合而成的分子轨道类似于 $1s$ 轨道形成的分子轨道（除了具有更高能量）。然而，$2p$ 轨道组合的情况不同。以 O_2 分子为例，有两种可能的方式将 $2p$ 轨道组合形成分子轨道：

1) "头碰头"方式的组合会沿键轴形成一对 σ 型分子轨道 [**图 8.16(a)(b)**]：成键 σ_{2p}（同相的 $2p$ 轨道波瓣的重叠）和反键 σ_{2p}^*（反相的 $2p$ 轨道波瓣的重叠）分子轨道。σ_{2p} 分子轨道没有节面，而 σ_{2p}^* 分子轨道在两个原子核的中点处有一个节面。只有沿键轴方向的一对 p 轨道可以组合形成 σ 型分子轨道。

2) "肩并肩"方式的组合会形成两对 π 型分子轨道 [**图 8.16(c)(d)**]：两个成键 π_{2p} 和两个反键 π_{2p}^* 分子轨道。成键 π_{2p} 轨道由垂直于键轴的两个 $2p$ 轨道相加形成，反键 π_{2p}^* 轨道由垂直于键轴的两个 $2p$ 轨道相减形成。除了包含原子核的原子节面外，在原子核之间还会形成电子电荷密度降至0的分子节面，这是反键轨道的一个特征。$2p$ 轨道的组合可形成六个分子轨道。如果将键轴指定为在 z 方向，则两个 σ 型分子轨道为 σ_{2p_z} 和 $\sigma_{2p_z}^*$，四个 π 型分子轨道分别为 π_{2p_x} 和 π_{2p_y}（简并），以及 $\pi_{2p_x}^*$ 和 $\pi_{2p_y}^*$（简并），如**图 8.17** 所示。除了上述强调从哪些原子轨道形成分子轨道的一套符号之外，还有另一套强调同核双原子物种中形成分子轨道的对称性的符号。如果形成的分子轨道相对于对称中心是对称或反对称的，则分别表示为 g (gerade) 或 u (ungerade) 对称性。因此，σ_{2p}、σ_{2p}^*、π_{2p} 和 π_{2p}^* 轨道可分别表示为 σ_g、σ_u、π_u 和 π_g。同时，σ_{2s} 和 σ_{2s}^* 的对应符号分别为 σ_g 和 σ_u。注意，所有成键 σ 轨道总是 σ_g，所有反键 σ* 轨道均为 σ_u，无论由 s 轨道还是 p 轨道形成。相反，所有成键 π 轨道总是 π_u，所有反键 π* 轨道均为 π_g。这是由轨道对称性决定的。

由原子轨道形成分子轨道的能级图，与相应的原子轨道的能级相关。一般来说，由 $2s$ 轨道形成的分子轨道，比由 $2p$ 轨道形成的能量更低，这与 $2s$ 和 $2p$ 原子轨道之间的关系相同。σ 型成键分子轨道的能量应低于 π 型，因为 p 轨道"头碰头"的重叠应比"肩并肩"的重叠在空间中延伸更广。

周期表的第二行有八种元素，因此可形成八个同核双原子分子。根据其分子轨道能级图，可将这八个分子分为两类。一类为 O_2、F_2 和 Ne_2 ($Z \geq 8$)，其能级图如**图 8.18(a)** 所示，电子排布如**图 8.18(b)** 所示。这一类的价层分子轨道按能量增加的正常顺序排列，即 σ_{2s}、σ_{2s}^*、σ_{2p}、π_{2p}、π_{2p}^* 和 σ_{2p}^*。如果使用对称性符号，相应的分子轨道可标记为 $1\sigma_g$、$1\sigma_u$、$2\sigma_g$、$1\pi_u$、$1\pi_g$ 和 $2\sigma_u$。对称性符号前的数字，表示价层中具有相同键型和对称性的分子轨道从低到高能量的顺序。例如，σ_{2s} 是价层中的第一个 σ_g，σ_{2p} 是第二个 σ_g。另一类为 Li_2、Be_2、B_2、C_2 和 N_2 ($Z<8$)，**图 8.19** 给出了它们的修正后的能级图及电子排布。不同之处在于 σ_{2p} 和 π_{2p} 轨道的顺序交换。这种顺序交换的原因在于，由于 $Li_2 \sim N_2$ 的 $2s$ 和 $2p$ 轨道的能量更接近，因此 $2s$-$2p$ 相

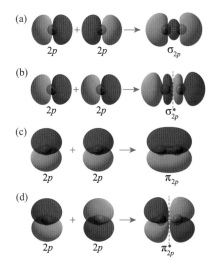

Figure 8.16 Formation of bonding and antibonding molecular orbitals from $2p$ orbitals. (a) The end-to-end addition of two in-phase $2p$ orbitals to form a σ_{2p} bonding molecular orbital. (b) The end-to-end addition of two out-of-phase $2p$ orbitals to form a σ_{2p}^* anti-bonding molecular orbital. (c) The side-to-side addition of two in-phase $2p$ orbitals to form a π_{2p} bonding molecular orbital. (d) The side-to-side addition of two out-of-phase $2p$ orbitals to form a π_{2p}^* anti-bonding molecular orbital.

图 8.16 由 $2p$ 轨道形成的成键和反键分子轨道：(a) 两个同相 $2p$ 轨道以"头碰头"方式相加，形成 σ_{2p} 成键分子轨道；(b) 两个反相 $2p$ 轨道以"头碰头"方式相加，形成 σ_{2p}^* 反键分子轨道；(c) 两个同相 $2p$ 轨道以"肩并肩"方式相加，形成 π_{2p} 成键分子轨道；(d) 两个反相 $2p$ 轨道以"肩并肩"方式相加，形成 π_{2p}^* 反键分子轨道。

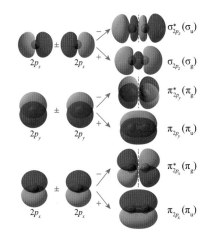

Figure 8.17 Geometric orientation and symmetry of the molecular are orbitals formed from $2p$ orbitals.

图 8.17 由 $2p$ 轨道形成的分子轨道的空间取向及对称性。

Heteronuclear Diatomic Species

Similar ideas can be extended, with some care, to heteronuclear diatomic species. The first note needs to be taken is that now the two atoms are different and so are the energy levels of these two atoms. Based on previous content, we know that Z_{eff} increases continuously and the energy levels ($E_n = -R_H Z_{eff}^2 / n^2$) decrease across a period. Let us take the molecule CO as an example. The 2s and 2p orbitals of C are higher in energy than those of O. This means that the contributions by each atom to the MOs are unequal. The next care is to decide whether or not there is 2s-2p mixing. The 2s-2p separation in atoms with $3 \leq Z < 8$ is narrow enough, and so we expect 2s-2p mixing to occur in the σ orbitals in CO.

The MO energy-level diagram for CO is shown in **Figure 8.21(a)**. Since there is no center of inversion in heteronuclear diatomics, the g and u symmetries are not applicable. We label the MOs in increasing energy as 1σ, 2σ, 1π, 3σ, 2π, and 4σ. Here, we do not use the notations such as $σ_{2s}$ and $π_{2p}$ because there is additional mixing between the 2s orbital in C and 2p orbital in O due to similar energy. Assigning 10 valence electrons to the MOs, the configuration is

$$\text{CO: } 1σ^2 2σ^2 1π^4 3σ^2$$

The bond order is (8 − 2)/2 = 3, corresponding to a triple covalent bond. Among all MOs, the **highest occupied molecular orbital** (HOMO) is 3σ and the **lowest unoccupied molecular orbital** (LUMO) is 2π, as shown in **Figure 8.21(b)**. HOMO and LUMO are collectively called the **frontier molecular orbitals** (FMO).

Figures 8.22(a) shows the MO diagrams for NO. The valence electron configuration is

$$\text{NO: } 1σ^2 2σ^2 1π^4 3σ^2 2π^1$$

The bond order is (8 − 3)/2 = 2.5. We expect that the bond energy in NO is less than that in CO and that NO is paramagnetic. **Figures 8.22(b)** shows the MO diagrams for HF. The orbitals of HF can be constructed by the overlap of the 1s orbital of H with the $2p_z$ orbitals of F, with z being the bonding axis. The $2p_x$ and $2p_y$ orbitals of F, however, are left unaffected, called the **non-bonding molecular orbital**, as they have π symmetry and there is no valence H orbital of that symmetry. The valence electron configuration is

$$\text{HF: } 1σ^2 2σ^2 1π^4$$

Here, both 1σ and 1π are largely non-bonding MOs and are confined mainly to the F atom. The 2σ is a bonding MO (contributed mainly from F $2p_z$ orbital) and 3σ is an anti-bonding MO (contributed mainly from H 1s orbital). The bond order is (2 − 0)/2 = 1. In general, the atom with higher electronegativity in a heteronuclear diatomic species contributes more to bonding MOs, and the atom with lower electronegativity contributes more to anti-bonding MOs.

Delocalized Electrons in Molecular Orbitals

MO theory focuses on a picture of delocalized electrons in delocalized MOs. In benzene, six 2p orbitals of C combine to form six MOs of the π type: three bonding and three anti-bonding MOs [**Figure 8.23(a)**]. What do these π MOs are like? Recall that as the energy of MO increases, the number of molecular nodes also increases. So, we expect that the lowest π MO has no molecular node: all six 2p orbitals are in phase as indicated by the fact that all the + lobes are on one side of the σ framework. The next two π-bonding MOs each have one molecular node [dashed line in **Figure 8.23(b)**] and are degenerate. The next pair of orbitals, which are degenerate anti-bonding π orbitals, have two molecular nodes, and the final orbital has three molecular nodes. The computed π MOs of benzene are given in **Figures 8.23(c)(d)**. The overall π bond order is (6 − 0)/2 = 3, distributed among six C atoms to result in a half-bond between each pair of C atoms. With an additional σ bond order of 1, the total bond order is 1.5 for each C—C bond in benzene. Because the π-type MOs are spread out among all six C atoms instead of being confined to regions between pairs of C atoms, these MOs are called **delocalized molecular orbitals**. A similar example is O_3 (**Figure 8.24**).

MO theory can also be used to explain the colors of plants. Two pigment molecules typically isolated from vegetables are β-carotene found in carrots, and lycopene present in tomatoes. The common feature of these molecules is the π conjugate system. The many p orbitals on the trigonal-planar C atoms contribute to many π MOs. As a consequence, the molecules have many π molecular energy levels that become very

互作用（称为 2s-2p 混杂）比 O_2~Ne_2 更强，如**图 8.20** 所示。

对能级图的修正来自实验证据。例如，如果我们将 C_2 分子的 8 个价电子排布在**图 8.18(b)** 的未修正能级图中，有

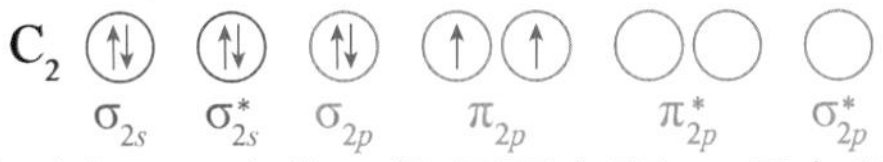

$$C_2 \quad (\uparrow\downarrow)\,(\uparrow\downarrow)\,(\uparrow\downarrow)(\uparrow\downarrow)\,(\uparrow)(\uparrow)\,(\,)(\,)\,(\,)$$
$$\quad \sigma_{2s} \quad \sigma_{2s}^* \quad \sigma_{2p} \quad \pi_{2p} \quad \pi_{2p}^* \quad \sigma_{2p}^*$$

由于存在两个未成对电子，未修正能级图表明 C_2 具有顺磁性。然而，实验表明 C_2 实际上是反磁性的。将 8 个价电子排布在**图 8.19(b)** 的修正能级图中，有

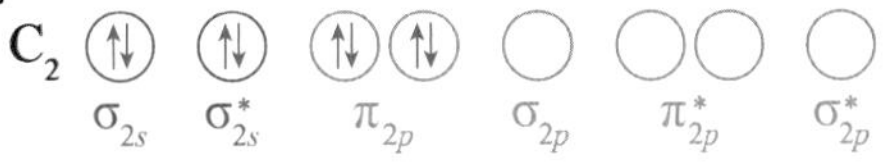

$$C_2 \quad (\uparrow\downarrow)\,(\uparrow\downarrow)\,(\uparrow\downarrow)(\uparrow\downarrow)\,(\,)\,(\,)(\,)\,(\,)$$
$$\quad \sigma_{2s} \quad \sigma_{2s}^* \quad \pi_{2p} \quad \sigma_{2p} \quad \pi_{2p}^* \quad \sigma_{2p}^*$$

这与 C_2 为反磁性的事实一致。

另一个例子是 O_2。将 12 个价电子排布在 O_2 的价层分子轨道中，可得

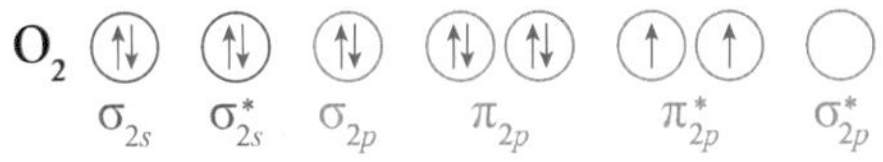

$$O_2 \quad (\uparrow\downarrow)\,(\uparrow\downarrow)\,(\uparrow\downarrow)\,(\uparrow\downarrow)(\uparrow\downarrow)\,(\uparrow)(\uparrow)\,(\,)$$
$$\quad \sigma_{2s} \quad \sigma_{2s}^* \quad \sigma_{2p} \quad \pi_{2p} \quad \pi_{2p}^* \quad \sigma_{2p}^*$$

这表明 O_2 存在两个未成对电子，解释了 O_2 的顺磁性。其键级为 $(8-4)/2 = 2$，对应于一个共价双键。关于磁性的更多讨论详见 **9.4 节**。

异核双原子物种

类似的方法可以谨慎地扩展到异核双原子物种。首先需要注意的是，现在这两个原子是不同的，其能级也是不同的。基于前述内容，我们知道在同一周期内 $Z_{有效}$ 持续增加，能级（$E_n = -R_H Z_{有效}^2 / n^2$）持续降低。以 CO 分子为例，C 的 $2s$ 和 $2p$ 轨道能量比 O 的高，这意味着每个原子对分子轨道的贡献不等。下一个注意事项是，确定是否存在 2s-2p 混杂。$3 \leqslant Z < 8$ 的原子中 2s-2p 能级差足够窄，因此我们预期 CO 的 σ 轨道中会发生 2s-2p 混杂。

CO 的分子轨道能级图如**图 8.21(a)** 所示。由于异核双原子分子没有对称中心，g 和 u 对称性不适用。我们将其分子轨道按能量递增的顺序标记为 1σ、2σ、1π、3σ、2π 和 4σ。这里我们也不使用 σ_{2s} 和 π_{2p} 等符号，因为由于能量相近，C 的 $2s$ 轨道和 O 的 $2p$ 轨道之间存在额外的混杂。将 10 个价电子排布在分子轨道上，其组态为

$$\text{CO}: 1\sigma^2 2\sigma^2 1\pi^4 3\sigma^2$$

键级为 $(8-2)/2 = 3$，对应于一个共价叁键。在所有分子轨道中，**最高占据轨道**（HOMO）为 3σ，**最低未占据轨道**（LUMO）为 2π，如**图 8.21(b)** 所示。HOMO 和 LUMO 统称为**前线轨道**（FMO）。

图 8.22(a) 给出了 NO 的分子轨道能级图，其价电子组态为

$$\text{NO}: 1\sigma^2 2\sigma^2 1\pi^4 3\sigma^2 2\pi^1$$

键级为 $(8-3)/2 = 2.5$。我们预期 NO 的键能小于 CO 的键能，且 NO 具有顺磁性。**图 8.22(b)** 给出了 HF 的分子轨道能级图。HF 的轨道可通过 H 的 $1s$ 轨道与 F 的 $2p_z$ 轨道的重叠来构建，其中 z 为键轴。F 的 $2p_x$ 和 $2p_y$ 轨道不受影响，称为**非键分子轨道**，因为它们具有 π 对称性，而 H 没有这种对称性的价轨道。其价电子组态为

$$\text{HF}: 1\sigma^2 2\sigma^2 1\pi^4$$

这里 1σ 和 1π 基本上都是非键分子轨道，主要局域在 F 原子周围。2σ 是成键分子轨道（主要来自 F 的 $2p_z$ 轨道），3σ 是反键分子轨道（主要来自 H 的 $1s$ 轨道）。键级为 $(2-0)/2 = 1$。一般来说，异核双原子物种中电负性较高的原子对成键分子轨道的贡献较大，而电负性

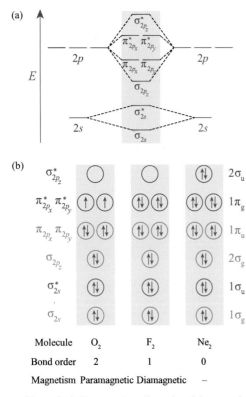

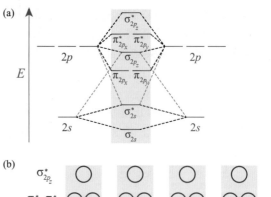

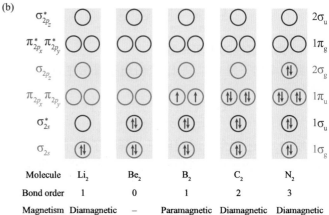

Figure 8.18 Homonuclear diatomics of the second-period elements with $Z \geqslant 8$. (a) Molecular orbital energy level diagrams with the normally expected ordering in which σ_{2p} lies below π_{2p}. (b) Electron assignment diagrams.

图 8.18 $Z \geqslant 8$ 的第二周期元素同核双原子分子：(a) 分子轨道能级图，其中 σ_{2p} 位于 π_{2p} 之下，为正常预期顺序；(b) 电子排布图。

Figure 8.19 Homonuclear diatomics of the second-period elements with $Z < 8$. (a) Molecular orbital energy level diagrams with the modified ordering in which σ_{2p} lies above π_{2p} due to 2s-2p mixing. (b) Electron assignment diagrams.

图 8.19 $Z < 8$ 的第二周期元素同核双原子分子：(a) 分子轨道能级图，其中 σ_{2p} 位于 π_{2p} 之上，是由于 2s-2p 混杂而修正的顺序；(b) 电子排布图。

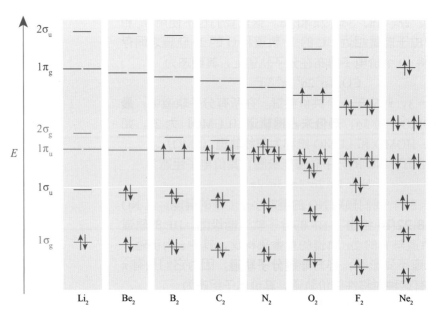

Figure 8.20 The variation of molecular orbital energies for the homonuclear diatomics of the second-period elements.

图 8.20 第二周期同核双原子分子的分子轨道能量变化图。

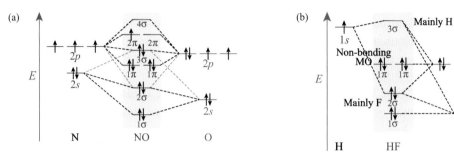

Figure 8.21 (a) Molecular orbital energy-level diagram and (b) frontier molecular orbital (HOMO and LUMO) diagrams of CO.

图 8.21 CO 的 (a) 分子轨道能级图及 (b) 前线轨道（HOMO 和 LUMO）示意图。

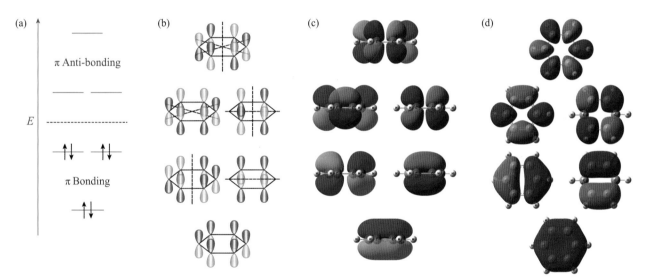

Figure 8.22 Molecular orbital energy-level diagrams of (a) NO and (b) HF.

图 8.22 (a) NO 和 (b) HF 的分子轨道能级图。

Figure 8.23 π molecular orbital diagrams for benzene (C_6H_6). (a) π molecular orbital energy level diagram with three bonding and three anti-bonding molecular orbitals. (b) Individual 2p orbital diagrams forming π molecular orbitals. (c) Side view and (d) top view of the computed diagrams for π molecular orbitals.

图 8.23 苯（C_6H_6）的 π 分子轨道图：(a) π 分子轨道能级图，其中有三个成键分子轨道和三个反键分子轨道；(b) 形成 π 分子轨道的单个 2p 轨道图；计算得的 π 分子轨道的 (c) 侧视图与 (d) 俯视图。

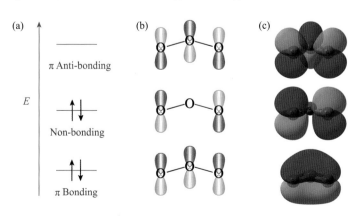

Figure 8.24 π molecular orbital diagrams for ozone (O_3). (a) π molecular orbital energy level diagram with one bonding, one non-bonding, and one anti-bonding molecular orbitals. (b) Individual 2p orbital diagrams forming π molecular orbitals. (c) Computed diagrams for π molecular orbitals.

图 8.24 臭氧（O_3）的 π 分子轨道图：(a) π 分子轨道能级图，其中有一个成键分子轨道、一个非键分子轨道和一个反键分子轨道；(b) 形成 π 分子轨道的单个 2p 轨道图；(c) 计算得的 π 分子轨道图。

closely spaced (**Figure 8.25**). In molecules with large extended π systems, the HOMO is very close in energy to the LUMO. As a result, it takes very little energy to excite an electron from HOMO to LUMO. Photons of visible light have enough energy to excite the electrons across the energy gap between HOMO and LUMO, and the absorption of these photons is responsible for the colors of plants that we see.

8.6 Bonding in Metals and Band Theory

In this section, we will introduce the typical bonding in metals: metallic bonding, and a theory that can explain the properties of various solids including metals: the band theory.

Bonding in Metals and the Electron Sea Model

In general, nonmetals have more valence electrons than valence orbitals, whereas metals have less valence electrons than valence orbitals. For example, both F and Li are second-period elements with four valence orbitals. A F atom has seven valence electrons and a F_2 molecule can be formed, in either solid, liquid, or gaseous states, by sharing a pair of electrons in the F—F single covalent bond. However, a Li atom has only one valence electron. This may account for the formation of the gaseous molecule Li_2, but in the solid metal, each Li atom is somehow bonded to eight neighbors. The challenge to a bonding theory for metals is to explain how so much bonding can occur with so few electrons. The theory should also account for the various distinctive properties of metals.

A simple theory that can explain some of the properties of metals is called the **electron sea model**. In this model, a solid metal is considered as a network of positive ions immersed in a "sea of electrons". Electrons in the sea are free, meaning that they are not attached to any particular ion, and mobile. Metallic bonding arises from the electrostatic attraction between positive ions and free electrons. The interactions between electrons are neglected. Various properties and characteristics of metals can be explained by the electron sea model as:

1) Opacity: Electrons can absorb the photons of visible light.
2) Lustrous appearance: Surface electrons of a metal can reradiate light.
3) Good thermal and electrical conductivity: Free electrons in a metal can conduct heat and electricity.
4) Good ductility: The sea of electrons can rapidly adjust to the new shape of metal in response to applied force, while the internal structure of the metal remains essentially unchanged.

Key Points of Band Theory

Compared to the electron sea model that describes the properties of metals in a simple and qualitative manner, a more sophisticated theory called **band theory** can explain various properties of solids based on quantum mechanics. Again, let us take the Li metal as an example. In MO theory, a Li_2 molecule is formed by each Li atom contributing one $2s$ orbital to produce a bonding σ_{2s} MO and an anti-bonding σ_{2s}^* MO. The two valence electrons of Li_2 fill the σ_{2s} and leave the σ_{2s}^* empty, as given in **Figure 8.19**. Extending this combination of Li atoms from Li_2 to Li_N, where N stands for the total number of atoms in a crystal of Li and is an enormously large number ($N \approx 10^{22}$), is similar to the case from C_2 to C_{40} in **Figure 8.25**. As more and more Li atoms are added to the growing "Li molecule", additional energy levels are also added and the spacing between levels becomes increasingly smaller. In an entire crystal of N atoms, the energy levels merge into a band of N closely spaced, almost continuous levels. The lower $N/2$ levels are filled with electrons, and the upper $N/2$ levels are empty, as shown in **Figure 8.26**.

In solid-state physics, the **electronic band structure** of a solid describes the range of energy levels that an electron may have (called **energy bands**), and the ranges of energy levels that it may not have (called **band gaps** or **forbidden bands**). Band theory derives these bands and band gaps by examining the allowed quantum mechanical wave functions for an electron in a large, periodic lattice of atoms or molecules. Band

较低的原子对反键分子轨道的贡献较大。

分子轨道中的离域电子

分子轨道理论聚焦于离域分子轨道中离域电子的图像。在苯中，C 的六个 $2p$ 轨道组合形成六个 π 型分子轨道：三个成键和三个反键分子轨道 [**图 8.23(a)**]。这些 π 型分子轨道是什么样的？回想一下，随着分子轨道能量的增加，其分子节面的数量也增加。因此我们预期能量最低的 π 型分子轨道没有分子节面：所有六个 $2p$ 轨道同相，即所有正的波瓣均位于 σ 骨架的一侧。接下来的两个 π 型成键分子轨道均具有一个分子节面 [**图 8.23(b)** 的虚线]，能量简并。下一对轨道是简并的反键 π 轨道，具有两个分子节面，最后一个轨道具有三个分子节面。**图 8.23(c)(d)** 给出了计算得到的苯的 π 型分子轨道。π 键的总键级为 $(6-0)/2 = 3$，分布在六个 C 原子之间，使得每对 C 原子之间形成半键。加上一个额外的键级为 1 的 σ 键，苯的每个 C—C 键的总键级为 1.5。由于 π 型分子轨道分布在所有六个 C 原子之间，而并非局限于 C 原子对之间的区域，因此这些分子轨道称为**离域分子轨道**。类似的例子有 O_3（**图 8.24**）等。

分子轨道理论还可以用来解释植物的颜色。胡萝卜中的 β- 胡萝卜素和番茄中的番茄红素是两种从蔬菜中分离出的典型色素分子。这些分子的共同特征是存在大 π 共轭体系。平面三角形 C 原子上的 p 轨道贡献了许多 π 型分子轨道，因此这些分子具有许多紧密排列的 π 分子能级（**图 8.25**）。在具有大的扩展 π 体系的分子中，其 HOMO 能量非常接近 LUMO。因此，只需较低的能量即可将一个电子从 HOMO 激发到 LUMO。可见光光子具有足够的能量激发电子穿过 HOMO 和 LUMO 之间的能隙，而这些光子的吸收就是我们看到植物颜色的原因。

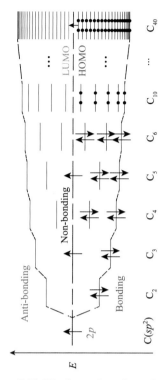

Figure 8.25 The formation of π molecular orbitals in π conjugate systems.

图 8.25 π 共轭体系中 π 分子轨道的形成。

8.6 金属键与能带理论

本节我们将介绍金属中的典型成键：金属键，以及可以解释包括金属在内的各种固体性质的理论：能带理论。

金属键与电子海模型

一般来说，非金属的价电子比价轨道多，而金属的价电子比价轨道少。例如，F 和 Li 都是具有四个价轨道的第二周期元素。F 原子有七个价电子，通过共用 F—F 共价单键中的一对电子，无论在固态、液态还是气态中均可形成 F_2 分子。而 Li 原子只有一个价电子，这可以解释气态 Li_2 分子的形成，但在固态金属中，每个 Li 原子以某种方式与八个相邻原子成键。金属的成键理论所面临的挑战是，如何解释这么少的电子却能形成这么多的键。该理论还应解释金属的各种独特性质。

电子海模型是一个可以解释金属的一些性质的简单理论。在该模型中，固态金属可视为浸泡在"电子海"中的正离子网格。电子海中的电子是自由的，意味着它们不依附于任何特定的离子，而且可以迁移。金属键由正离子与自由电子之间的静电引力产生，电子之间的相互作用可忽略不计。金属的各种性质和特征可用电子海模型解释为：

1) 不透明：电子可以吸收可见光光子；

theory can be viewed as an extended MO theory.

Classification of Bands and Solids in Band Theory

According to how many electrons are filled in, energy bands can be classified into full bands (which are completely filled with electrons), empty bands (which do not have any electrons in at all), and partially filled bands (which are partially filled with electrons). Meanwhile, the energy band that valence electrons are found within it is called a **valence band**. The energy band that is responsible for the conductivity of the material is called a **conduction band**. A partially filled band is both a valence and a conduction band, because the energy differences between the occupied and unoccupied levels within it are so small that mobile electrons that can conduct heat and electricity can be easily produced by applying just a small heat or electric potential difference across the crystal.

In band theory, solids can be classified into three major types according to their thermal and electrical conductivity:

1) Metals or conductors: Heat and electricity can be conducted by the directional movement of free electrons in their conduction bands in metals. For example, the valence band of the alkaline metals, such as the $2s$ band of Li_N and the $3s$ band of Na_N, serves as the conduction band [**Figure 8.27(a)**]. In the case of the alkaline earth metals such as Be_N, the valence $2s$ band itself is a full band, with N orbitals and $2N$ electrons. However, due to the fact that the full $2s$ band overlaps with the empty $2p$ band of Be_N, which means that the lowest levels of the $2p$ band are at a lower energy than the highest levels of the $2s$ band, the overlapped $2s$&$2p$ bands then serve as a large partially filled band and thus as a conduction band [**Figure 8.27(b)**]. Although both Li_N and Be_N are considered as metals, the conductivity of Be_N is poorer than that of Li_N. The conductivity of metals generally decreases with increasing temperature, because the more vigorous vibrations of positive ions hinder the movement of free electrons.

2) Insulators: The atoms have tightly bound electrons that cannot readily move, so that the conductivities of insulators are very poor. In general, the valence bands of an insulator are full but there is a large band gap (normally > 4 eV) between the valence band and the conduction band [**Figure 8.27(d)**]. Materials such as glass, diamond, and silica (SiO_2) are examples of insulators.

3) Semiconductors: The conductivity of semiconductors lies between that of conductors and insulators and is sensitive to light or heat. The electronic band structure of semiconductors is similar to that of insulators except that the band gap is smaller, normally ⩽ 4 eV [**Figure 8.27(c)**]. Electrons can be excited from the valence band to the conduction band upon visible light, some heat or electric potential difference. Both the excited electrons in the conduction band and the positively charged holes left in the valence band can conduct heat and electricity. The greater the thermal energy, the more electrons can be excited. Therefore, the conductivity of semiconductors generally increases with temperature.

Semiconductors and Photovoltaic Cells

Semiconductors are the foundation of modern electronics, including transistors, photovoltaic/solar cells, light-emitting diodes (LEDs), quantum dots, digital and analog integrated circuits, etc. Pure semiconductors that have a fixed energy band gap are called **intrinsic semiconductors**, or i-type semiconductors [**Figure 8.28(a)**], such as CdS and GaAs. In an intrinsic semiconductor, the concentration of excited electrons (n) and that of remaining holes (p) are equal: $n = p$. The band gap of an intrinsic semiconductor is dependent only on the material itself.

The conductivity of a semiconductor can be easily modified by introducing a small amount of impurities, a process called **doping**, into its crystal structure. The introduced impurities are called **dopants**, and the resulting semiconductors are called **extrinsic semiconductors**, or **doped semiconductors**. The band gap as well as the conductivity of an extrinsic semiconductor can be tuned by the dopant concentrations.

The materials chosen as suitable dopants depend on the atomic properties of both the dopant and

2) 具有金属光泽：金属表面的电子可以重新辐射光；
3) 良好的导热性和导电性：金属中的自由电子可以传热导电；
4) 良好的延展性：电子海可以根据施加的外力迅速适应金属的新形状，而保持金属的内部结构基本不变。

能带理论的要点

与以简单、定性方式描述金属性质的电子海模型相比，**能带理论**是一个基于量子力学解释固体各种性质的更为复杂的理论。让我们仍以金属 Li 为例。在分子轨道理论中，Li_2 分子中每个 Li 原子贡献一个 $2s$ 轨道，形成成键 σ_{2s} 和反键 σ_{2s}^* 分子轨道，Li_2 的两个价电子填充在 σ_{2s}，而 σ_{2s}^* 为空轨道，如**图 8.19**所示。将 Li 原子的组合由 Li_2 扩展到 Li_N，其中 N 代表 Li 晶体中的原子总数，是一个非常大的值（$N \approx 10^{22}$），与**图 8.25** 中从 C_2 到 C_{40} 的情况类似。随着越来越多的 Li 原子被添加到不断增长的"Li 分子"中，额外的能级也增加了，能级之间的间距越来越小。在由 N 个原子组成的整个晶体中，能级合并成一个由 N 个紧密间隔、几乎连续的能级组成的带。如**图 8.26** 所示，较低的 $N/2$ 个能级充满电子，而较高的 $N/2$ 个能级为空。

在固体物理中，固体的**电子能带结构**描述了电子可能具有的能级范围（称为**能带**），以及不可能具有的能级范围（称为**带隙**或**禁带**）。能带理论通过原子或分子中的电子在一个大的周期性晶格中所允许的量子力学波函数，来推导这些能带和带隙。能带理论可视为一种扩展的分子轨道理论。

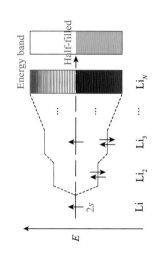

Figure 8.26 Formation of the 2s energy band in lithium metal.

图 8.26 锂金属中 $2s$ 能带的形成。

能带理论对能带和固体的分类

根据电子的填充量，能带可分为满带（完全填充电子）、空带（完全没有填充电子）和部分填充带（部分填充电子）。同时，价电子填充的能带称为**价带**，与材料导热、导电性相关的能带称为**导带**。部分填充带既是价带又是导带，因为其中已占据能级和未占据能级之间的能量差非常小，因此只要在晶体上施加一个很小的热或电势差，即可很容易地产生能够传热导电的可迁移电子。

根据热导率和电导率（统称传导率），能带理论中的固体可分为三种主要类型：

1) 金属或导体：热和电可通过金属中的自由电子在导带中的定向迁移来传导。例如，碱金属的价带（如 Li_N 的 $2s$ 带和 Na_N 的 $3s$ 带）可作为导带 [**图 8.27(a)**]。对于碱土金属（如 Be_N），其 $2s$ 价带本身是满带，具有 N 个轨道和 $2N$ 个电子。但由于 Be_N 的 $2s$ 满带与 $2p$ 空带重叠，即 $2p$ 带最低能级的能量低于 $2s$ 带最高能级，因此重叠的 $2s$ 和 $2p$ 带可作为一个大的部分填充带，故为导带 [**图 8.27(b)**]。尽管 Li_N 和 Be_N 均视为金属，但 Be_N 的导电性比 Li_N 差。金属的传导率通常随温度的升高而降低，因为正离子更为剧烈的振动阻碍了自由电子的迁移。

2) 绝缘体：原子具有紧密结合、不易迁移的电子，因此绝缘体的传导率非常小。一般来说，绝缘体的价带是满带，但价带和导带之间存在较大的带隙 [通常 > 4 eV，**图 8.27(d)**]。玻璃、金刚石和二氧化硅（SiO_2）等材料是绝缘体的示例。

3) 半导体：半导体的传导率介于导体和绝缘体之间，对光或热敏

the material to be doped. In general, dopants that produce the desired controlled changes are classified as either electron donors or acceptors. Semiconductors doped with donor impurities ($n>p$) are called **n-type semiconductors**, whereas those doped with acceptor impurities ($n<p$) are known as **p-type semiconductors**. The *n* and *p* type designations indicate that the majority charge carrier of the material is negative (electron) or positive (hole), respectively.

Let us take the widely used element in the current semiconductor industry—silicon—as an example. Pure Si has four valence electrons bonding with other four neighboring Si atoms to form a network covalent crystal. The most common dopants of Si are group III and group V elements. Group V elements such as P have five valence electrons, which can form bonds to four neighboring Si atoms. With an extra valence electron remaining, these atoms thus serve as donor atoms. The migration of the extra electrons [with immobile P^+ ion left, **Figure 8.29(a)**] creates directional current, and is responsible for the conductivity of the material. From the perspective of energy level, the energy level of the donor atoms generally lies just below the conduction band of Si [**Figure 8.28(b)**]. If visible light, or some thermal or electrical energy is provided, the extra valence electrons in the donor energy level can be promoted to the conduction band of Si. The promoted electrons in the conduction band are the majority of charge carriers, and thus this type of semiconductor doped with donor atoms is *n*-type.

When Si is doped with group III elements such as Al, these dopants can form regular electron-pair bonds with three neighboring Si atoms but only a single-electron bond with a fourth Si atom. Another electron can migrate to this single-electron bond to form an immobile Al^- ion and create another single-electron bond called a "hole" with positive charge [**Figure 8.29(b)**]. Although holes themselves do not move but electrons elsewhere can migrate to fill the hole, it seems that holes also "migrate" and can be considered as a type of positive charge carrier. Thus, these group III atoms function as acceptor dopants for Si. Meanwhile, the energy level of the acceptor atoms usually lies just above the valence band of Si [**Figure 8.28(c)**]. If additional energy is introduced, electrons in the valence band of Si can be promoted to the acceptor energy level. The remaining holes in the valence band then become the majority of charge carriers, and this material doped with acceptor atoms is a *p*-type semiconductor.

When a *p*-type and *n*-type semiconductors are in close contact with each other inside a single crystal of semiconductor, a so-called **p-n junction** is formed at the interface [**Figure 8.30(a)**]. Because the concentration of electrons is greater in *n*-type than in *p*-type, some electrons will migrate across the junction from *n*-type into *p*-type and combine with holes. The positively charged donor atoms left in the *n*-type are immobile, and thus cause a region near the junction in the *n*-type to be positively charged. Similarly, a region near the junction in the *p*-type becomes negatively charged due to the migration of holes. In this way, the regions near the *p-n* junction lose their neutrality and most of their mobile charge carriers, forming a "depletion zone" that acts as a non-conductive barrier. The *p-n* junctions are elementary building blocks of semiconductor electronic devices.

A single *p-n* junction is commonly used as a diode that serves as a one-way valve. When a positive voltage is applied to the *p*-type and negative to the *n*-type [called forward bias, **Figure 8.30(b)**], it shrinks the depletion zone, causing the current to flow from *p*-type to *n*-type. When a negative voltage is applied to the *p*-type and positive to the *n*-type [called reverse bias, **Figure 8.30(c)**], it increases the depletion zone and prevents current from flowing.

A **photovoltaic/solar cell** is an electrical device that generates voltage and electric current in a semiconducting material upon exposure to light. The simplest structure of a photovoltaic cell (**Figure 8.31**) contains a *p-n* junction and wires that connect the *p*-type end to the *n*-type end through an external load. Upon light exposure, for example, to the *p*-type, electrons in the valence band absorb photons and are promoted to the acceptor energy level. Unlike holes, these photo-excited electrons in the *p*-type can easily cross the junction into the *n*-type, migrate through the external load, and eventually return to the *p*-type to recombine with holes. This directional flow of electrons sets up an electric current from the *p*-type end to the *n*-type end in the external circuit, as long as light shines on the solar cell.

感。半导体的电子能带结构与绝缘体类似，但带隙通常较小 [≤ 4 eV，**图 8.27(c)**]。在可见光、热或电势差的作用下，电子可由价带激发至导带。导带中的激发电子和价带中留下的带正电的空穴，均可传热导电。热能越大，激发的电子越多。因此，半导体的传导率通常随温度的升高而增大。

半导体与光伏器件

半导体是包括晶体管、光伏/太阳能电池、发光二极管（LED）、量子点、数字和模拟集成电路等在内的现代电子学的基础。带隙具有固定能量的纯半导体称为**本征半导体**，或 i 型半导体 [**图 8.28(a)**]，如 CdS 和 GaAs。本征半导体中，激发的电子（n）和剩余的空穴（p）浓度相等：$n=p$。本征半导体的带隙仅取决于材料自身。

半导体的传导率可通过在其晶体结构中引入少量杂质（该过程称为**掺杂**）来改变。引入的杂质称为**掺杂剂**，得到的半导体称为**非本征半导体**或**掺杂半导体**。非本征半导体的带隙和传导率可通过掺杂剂的浓度来调控。

合适掺杂剂材料的选取，取决于掺杂剂和待掺杂材料的原子性质。通常，产生所需可控变化的掺杂剂分为电子给体和受体。掺杂给体杂质（$n>p$）的半导体称为 **n 型半导体**，而掺杂受体杂质（$n<p$）的半导体称为 **p 型半导体**。n 型或 p 型分别表示材料的主要载流子带负电（电子）或正电（空穴）。

让我们以当前半导体工业中广泛使用的硅元素为例。纯硅有四个价电子，可与其他四个相邻的硅原子成键，形成网状共价晶体。Si 的最为常见的掺杂剂是Ⅲ族和Ⅴ族元素。Ⅴ族元素如 P 有五个价电子，可与四个相邻的 Si 原子成键。由于剩余一个额外的价电子，这些原子可充当给体原子。额外电子的迁移 [留下固定不动的 P^+ 离子，**图 8.29(a)**] 产生定向电流，使材料具有导电性。从能级的角度来看，给体原子的能级通常恰好位于硅的导带下方 [**图 8.28(b)**]。如果提供可见光、一些热能或电能，给体能级中额外的价电子可以跃迁到 Si 的导带。导带中的激发电子是多数载流子，因此这种掺杂了给体原子的半导体为 n 型。

当 Si 掺杂了Ⅲ族元素如 Al 时，这些掺杂剂可以与三个相邻的 Si 原子形成常规的电子对键，但与第四个 Si 原子仅形成单电子键。另一个电子可以迁移到这个单电子键上，形成一个固定不动的 Al^- 离子，并产生另一个单电子键 [称为"空穴"，带正电荷，**图 8.29(b)**]。虽然空穴自身不能移动，但其他地方的电子可以迁移过来填充空穴，就仿佛空穴在"迁移"一样，可视为一种带正电荷的载流子。因此，这些Ⅲ族原子可以充当 Si 的受体掺杂剂。同时，受体原子的能级通常恰好位于 Si 的价带上方 [**图 8.28(c)**]。如果提供额外的能量，Si 价带中的电子可以跃迁到受体能级。价带中剩余的空穴成为多数载流子，这种掺杂了受体原子的材料是 p 型半导体。

当 p 型和 n 型半导体在一块半导体单晶内彼此紧密接触时，界面处会形成所谓的 **p-n 结** [**图 8.30(a)**]。由于 n 型的电子浓度大于 p 型，一些电子会穿过结从 n 型迁移到 p 型，并与空穴结合。留在 n 型中的带正电的给体原子不能移动，因此导致结附近的 n 型区域带正电。类似地，由于空穴的迁移，结附近的 p 型区域带负电。这样，p-n 结附近

8.7 Intermolecular Forces

As discussed previously in **Section 2.6**, **intermolecular forces** are the attractive or repulsive interactions between molecular species. Intermolecular forces are generally weak relative to **intramolecular forces**, which are the forces binding a molecular species together, such as chemical bonds. The intermolecular forces include but are not limited to van der Waals forces and hydrogen bonds. The **van der Waals forces**, named after Johannes van der Waals and described as a combination of London force, Keesom force, and Debye force, originate from the interactions between various molecular dipoles.

Three Types of Dipoles

We have introduced the definitions of dipole and dipole moment in **Section 8.3**. There are three types of dipoles between molecules:

1) **Permanent dipole**: the dipole of a polar molecule due to the non-superimposition of the positive and negative charge centers. If not specified, a dipole usually refers to a permanent dipole. Only polar molecules have permanent dipole, and the permanent dipole moment of a nonpolar molecule must be zero.

2) **Instantaneous dipole**: the momentary dipole of a molecule caused by displacement of electrons at some particular instant [**Figure 8.32(a)**]. For all molecules, either polar or nonpolar, there is always some instant, purely by chance, when the electrons happen to be more concentrated in one region than another in the molecule, resulting in a momentary dipole.

3) **Induced dipole**: the newly formed dipole induced by the dipole of a neighboring species. The existence of a dipole, either permanent or instantaneous, may have interactions with the electrons of another nearby molecule, inducing a dipole in that molecule. According to the original dipole, induced dipole can be classified into (permanent) dipole-induced dipole and instantaneous dipole-induced dipole [**Figure 8.32(b)**].

London Force and Polarizability

The **London force**, also called **dispersion force**, is the attraction between an instantaneous dipole and its induced dipole. It was named after Fritz London for his theoretical explanation of these forces in 1928. Since instantaneous dipole is present in all molecules and it may cause induced dipoles in all neighboring molecules, London force universally exists between all molecules, either polar or nonpolar, and is the dominant component in van der Waals forces.

The London force is positively related to a term called **polarizability**, which is the relative tendency for the electron cloud of a species to be distorted from its normal shape by an external electric field. In practical, the polarizability (α) of a species is defined as the ratio of its induced dipole moment to the local electric field, as

$$\alpha = \frac{\mu_{\text{ind}}}{E} \tag{8.12}$$

where E is the electric field strength that produces the induced dipole moment μ_{ind}.

In general, polarizability is inversely related to charge density, which is the ratio of charge to volume of the electron cloud, of a species. Larger molecules usually have more loosely held electrons farther away from the nuclei than smaller molecules, and polarizability increases with molecular size and molecular mass. Therefore, the London force as well as the melting and boiling points of molecular substances generally increase with molecular mass. For example, the melting and boiling points of the halogens increase in the order of F_2, Cl_2, Br_2, and I_2. The boiling point of He (4.22 K) is lower than that of Rn (211.5 K).

的区域不再呈电中性,且失去了大部分可移动的载流子,形成了一个可充当非导电势垒的"耗尽区"。p-n 结是半导体电子器件的基本构造块。

单个 p-n 结通常可用作二极管,充当单向阀。当对 p 型施加正电压、对 n 型施加负电压时[称为正向偏压,**图 8.30(b)**],耗尽区会缩小,导致电流从 p 型流向 n 型。当对 p 型施加负电压、对 n 型施加正电压时[称为反向偏压,**图 8.30(c)**],耗尽区会扩大,阻碍产生电流。

光伏/太阳能电池是一种暴露在光照下能在半导体材料中产生电压和电流的电气设备。光伏电池最简单的结构(**图 8.31**)包含一个 p-n 结以及通过外加负载将 p 型端连接到 n 型端的导线。例如,当 p 型端暴露在光照下,价带中的电子吸收光子并跃迁到受体能级。与空穴不同,p 型中的这些光激发电子可以很容易地穿过结进入 n 型,再通过外加负载迁移,最终返回 p 型与空穴复合。只要有光照在太阳能电池上,这种定向电子流就会在外加电路中形成从 p 型端到 n 型端的电流。

8.7 分子间作用力

如 **2.6 节**所述,**分子间作用力**是分子物种之间的吸引或排斥相互作用。分子间作用力通常弱于**分子内作用力**(即将一个分子物种结合在一起的力,如化学键)。分子间作用力包括但不限于范德华力和氢键。**范德华力**命名自约翰尼斯·范德华,为色散力、取向力和诱导力的组合,范德华力源于分子的各种偶极之间的相互作用。

三类偶极

我们在 **8.3 节**中介绍了偶极和偶极矩的定义。分子之间存在三种类型的偶极:

1) **永久偶极**:极性分子由于正、负电荷中心不重合所产生的偶极。如果未加说明,偶极通常指的就是永久偶极。只有极性分子具有永久偶极,非极性分子的永久偶极矩必为零。
2) **瞬时偶极**:由于分子的电子在某一特定时刻的位移而引起的瞬时存在的偶极[**图 8.32(a)**]。所有分子(不论极性还是非极性)都会有一些纯偶然的瞬间,电子碰巧集中在分子中的一些区域而非另一些区域,从而产生瞬时偶极。
3) **诱导偶极**:在邻近物种偶极的诱导下新形成的偶极。不论永久偶极还是瞬时偶极的存在,均可能与附近另一个分子的电子发生相互作用,在该分子中诱导形成一个偶极。根据原偶极的类型,诱导偶极可分为(永久)偶极的诱导偶极和瞬时偶极的诱导偶极[**图 8.32(b)**]。

色散力与极化率

色散力,也称**伦敦力**,是瞬时偶极与其诱导偶极之间的吸引力。它是根据弗里茨·伦敦在 1928 年对这些作用力的理论解释而命名的。由于瞬时偶极存在于所有分子中,并可能在所有邻近分子中产生诱导偶极,色散力普遍存在于所有分子之间(不论极性还是非极性分子),是范德华力最主要的组成部分。

色散力与极化率正相关,**极化率**是一个物种的电子云在外加电场作用下从正常形状发生扭曲的相对趋势。具体而言,物种的极化率(α)

The strength of London force also depends on its molecular shape. The elongated molecules are more polarizable than small, compact, symmetrical molecules. Isomers with identical molecular mass but different molecular shapes may have different properties. For instance, the boiling point of neopentane (9.5 °C, compact) is lower than that of pentane (36.1 °C, elongated).

Keesom Force

The **Keesom force**, named after Willem H. Keesom (1876—1956) and also called dipole-dipole interaction, originates from the attractive or repulsive electrostatic interaction between permanent dipoles. As only polar molecules have permanent dipoles, Keesom force only exists between polar molecules. Due to Keesom force, polar molecules tend to line up, with the negative end of one dipole directed toward the positive ends of the neighboring dipoles, to reduce the overall potential energy.

For substances with comparable molecular masses, the magnitude of London force is comparable, and then Keesom force produces significant differences in their melting or boiling points. For example, by comparing the boiling points of N_2, O_2, and NO, we know that both N_2 and O_2 are nonpolar and that the boiling point of N_2 (77.34 K) should be lower than that of O_2 (90.19 K). Considering only London forces, we would expect the boiling point of NO to be intermediate to those of N_2 and O_2. With additional Keesom force, the slightly polar molecule NO has the highest boiling point (121.39 K) of the three.

Debye Force

The **Debye force**, named after Peter J. W. Debye (1884—1966), results from the attraction between a permanent dipole and its induced dipole. The Debye force is also called dipole-induced dipole interaction. Debye force exists either between polar molecules or between polar and nonpolar molecules. Both Debye and Keesom forces belong to polar interactions because polar molecules with permanent dipoles are required. The magnitude of Debye force is generally weaker than that of the corresponding Keesom force.

An example of Debye force is the interaction between HCl and Ar. The permanent dipole in HCl repels the electrons in a nearby Ar atom, causing its electrons to be attracted to the H side of HCl or repelled from the Cl side of HCl. The Debye force is then the attraction between this permanent dipole-induced dipole in Ar and the original permanent dipole in HCl.

Summary on van der Waals Forces

The main characteristics of van der Waals forces are:

1) Van der Waals forces are a combination of London force, Keesom force, and Debye force, they are orders of magnitude weaker than normal chemical bonds. The general energy scales for London, Keesom, and Debye forces are 0.05~40, 5~25, and 2~10 kJ mol^{-1}, respectively, whereas those for covalent, ionic, and metallic bonds are 100~1000, 400~4000, and 100~1000 kJ mol^{-1}, respectively. For substances of widely different molecular masses, London forces are usually more significant than Keesom and Debye forces. For substances of comparable molecular masses, the van der Waals forces are greater for polar molecules than for nonpolar molecules.

2) Van der Waals forces cannot be saturated and have no directional characteristic. The nature of van der Waals forces are electrostatic forces between molecular dipoles. Like ionic bonds but unlike covalent bonds, they are with no saturability or directivity.

3) Van der Waals forces are short-range forces and hence only interactions between the nearest particles instead of all the particles need to be considered. Van der Waals forces are inversely proportional to the sixth power of the distance, which decreases significantly with increasing distance. As a comparison, the electrostatic Coulomb force is regarded as relatively long-range because it is inversely proportional to the square of distance.

4) Keesom force is inversely proportional to temperature. Both London and Debye forces are independent of temperature.

定义为其诱导偶极矩与局域电场的比值，即

$$\alpha = \frac{\mu_{诱导}}{E} \tag{8.12}$$

其中 E 是产生诱导偶极矩 $\mu_{诱导}$ 的电场强度。

一般来说，极化率与物种的电荷密度反相关，而电荷密度是该物种电子云的电荷与体积之比。较大的分子通常比较小的分子具有离核更远、更为松散的电子，极化率随分子大小和分子质量的增大而增大。因此，色散力以及由分子组成的物质的熔沸点通常随分子质量的增大而增大。例如，卤素的熔点和沸点按 F_2、Cl_2、Br_2 和 I_2 的顺序增大；He 的沸点（4.22 K）低于 Rn（211.5 K）。

色散力的强度也取决于分子的形状。细长型的分子比小的、紧凑对称型的分子更易极化。相对分子质量相同但分子形状不同的异构体可能具有不同的性质。例如，新戊烷的沸点（9.5°C、紧凑型）低于正戊烷（36.1°C、细长型）。

取向力

取向力（即基索姆力）命名自威廉·H. 基索姆（1876—1956），也称偶极-偶极作用力，源于永久偶极之间的吸引或排斥的静电相互作用。由于只有极性分子具有永久偶极，因此取向力只存在于极性分子之间。由于取向力，极性分子倾向于以一个偶极的负端对准相邻偶极正端的形式排列，以降低总势能。

对于相对分子质量相当的分子，色散力的大小差不多，取向力会导致其熔沸点产生显著差异。例如，比较 N_2、O_2 和 NO 的沸点，我们知道 N_2 和 O_2 均为非极性，N_2 的沸点（77.34 K）应低于 O_2（90.19 K）。若仅考虑色散力，我们预期 NO 的沸点会介于 N_2 和 O_2 之间。加上额外存在的取向力，略带极性的 NO 分子具有三者中最高的沸点（121.39 K）。

诱导力

诱导力（即德拜力）命名自彼得·J. W. 德拜（1884—1966），源于永久偶极与其诱导偶极之间的吸引力，也称偶极-诱导偶极作用力。诱导力存在于极性分子之间或极性与非极性分子之间。诱导力和取向力均属于极性相互作用，因为需要具有永久偶极的极性分子。诱导力的强度通常弱于相应的取向力。

HCl 和 Ar 之间的相互作用是诱导力的一个例子。HCl 的永久偶极会排斥附近 Ar 原子的电子，使其电子被吸引到 HCl 的 H 原子一侧，或被排斥远离 HCl 的 Cl 原子一侧。诱导力即为 Ar 的由永久偶极诱导的偶极与 HCl 的原永久偶极之间的吸引力。

范德华力小结

范德华力的主要特征有：
1) 范德华力为色散力、取向力和诱导力的组合，比正常化学键弱几个数量级。色散力、取向力和诱导力的一般能量范围分别为 0.05~40、5~25 和 2~10 kJ mol^{-1}，而共价键、离子键和金属键分别为 100~1000、400~4000 和 100~1000 kJ mol^{-1}。对于相对分子质量差异较大的分子，色散力通常比取向力和诱导力更重要。对于相对分子质量相当的分子，极性分子的范德华力大于非极性分子。
2) 范德华力没有饱和性和方向性。范德华力的本质是分子偶极之

Hydrogen Bond

A hydrogen bond is the electrostatic attraction between a hydrogen atom that is covalently bonded to a highly electronegative atom X, such as F, O, or N, and the lone pair of another highly electronegative atom Y. This can be denoted as X—H⋯Y, where the hydrogen bond is represented by a dotted line (⋯). The energy of a hydrogen bond depends on the geometry, the environment, and the nature of X and Y atoms, and can vary on the scale of 10~40 kJ mol^{-1}. This makes the hydrogen bond somewhat stronger than van der Waals forces but weaker than normal chemical bonds.

Hydrogen bonds can be either intramolecular (occurring among parts of the same molecule) or intermolecular (occurring between separate molecules), as shown in **Figures 8.33(a)** and **8.33(b)**, respectively. Another example of intermolecular hydrogen bonds is the pairing between adenine (A) and thymine (T) in DNA [**Figure 8.33(c)**].

The hydrogen bond is responsible for many abnormal physical and chemical properties of compounds of F, O, and N. In particular, intermolecular hydrogen bond is responsible for the higher boiling points of HF, H_2O, and NH_3 than those of any other hydride in their groups. CH_4 with no hydrogen bond, on the other hand, has the lowest boiling point in the hydrides of the group 14 elements, as normally expected.

Other Intermolecular Forces

In addition to van der Waals forces and hydrogen bonds, there are other types of intermolecular forces such as ion-dipole force (40~600 kJ mol^{-1}) and ion-induced dipole force (3~15 kJ mol^{-1}). Those ion-involving forces are similar to dipole-dipole and dipole-induced dipole forces but stronger, because the charge of any ion is greater than the partial charge of a dipole.

An ion-dipole force is the electrostatic interaction between an ion and a polar molecule. An important example of ion-dipole interaction is the hydration of ions in water. The ions in water are surrounded by polar H_2O molecules and the energy released during this process is known as hydration enthalpy. The interaction has its immense importance in justifying the stability of various ions (like Cu^{2+}) in water.

8.8 Crystal Structure

In crystallography, crystal structure is a description of the ordered arrangement of atoms, ions, or molecules in a crystalline material. In this and the next sections, we will briefly introduce some basic definitions and concepts in crystallography, as well as various types of crystals and their structures.

Definition of Crystal

The word *crystal* derives from the Ancient Greek word *krustallos*, meaning both "ice" and "rock crystal". The definition of crystal has evolved with time. Traditionally, a **crystal** or a **crystalline solid** is defined as a solid material whose constituents (such as atoms, molecules, or ions) are arranged in a highly ordered microscopic structure, forming a crystal lattice that extends in all directions. The scientific study of crystals and crystal formation is known as **crystallography**. The traditional characteristics of crystals include regular geometric shapes, sharp melting point, anisotropy, periodicity, sharp X-ray diffraction (XRD) peaks, etc.

In 1982, a special type of crystal called a **quasicrystal** or a quasiperiodic crystal, which consists of arrays of constituents that are ordered but not strictly periodic, was discovered by Dan Shechtman (1941—). Quasicrystals have many attributes in common with ordinary crystals, such as displaying discrete XRD patterns, and the ability to form shapes with smooth, flat faces. A quasicrystalline pattern can continuously fill all available space, but it lacks translational symmetry. Therefore, quasicrystals are not included in the traditional definition of crystal. Later, the International Union

间的静电作用力，与离子键类似，但与共价键不同，因此没有饱和性和方向性。
3) 范德华力是短程力，因此只需要考虑最邻近粒子之间的相互作用，而不用考虑所有粒子。范德华力与距离的六次方成反比，随着距离的增加而显著减小。作为比较，静电库仑力被认为是相对长程的作用力，因为其与距离的平方成反比。
4) 取向力与温度成反比，色散力和诱导力均与温度无关。

氢键

氢键是与高电负性原子 X（如 F、O 或 N）以共价键相连的氢原子，和另一高电负性原子 Y 的孤电子对之间的静电吸引作用，可表示为 X—H⋯Y，其中的氢键用虚线（⋯）表示。氢键的能量取决于 X 和 Y 原子的几何形状、化学环境和性质，可以在 10~40 kJ mol^{-1} 的范围内变化。这使得氢键通常强于范德华力，但弱于正常的化学键。

氢键既可以在分子内（发生在同一分子的不同部分之间），也可以在分子间（发生在不同的分子之间），分别如**图 8.33(a)** 和 **8.33(b)** 所示。DNA 中腺嘌呤（A）与胸腺嘧啶（T）之间的配对 [**图 8.33(c)**] 是分子间氢键的另一个例子。

氢键导致含 F、O 和 N 的化合物具有许多异常的物理和化学性质。特别是分子间氢键导致 HF、H_2O 和 NH_3 的沸点高于其同族中任何其他氢化物的沸点。另一方面，正如通常预期，不存在氢键的 CH_4 在第 14 族元素的氢化物中具有最低沸点。

其他分子间作用力

除了范德华力和氢键之外，还存在其他类型的分子间作用力，如离子-偶极作用力（40~600 kJ mol^{-1}）、离子-诱导偶极作用力（3~15 kJ mol^{-1}）等。这些涉及离子的作用力类似于偶极-偶极作用力和偶极-诱导偶极作用力，但是更强，因为任何离子所带的电荷均大于偶极所带的部分电荷。

离子-偶极作用力是离子与极性分子之间的静电相互作用。离子-偶极作用力的一个重要例子是水溶液中离子的水合作用。水中的离子被极性 H_2O 分子包围，在此过程中释放的能量称为水合焓。这种相互作用对于水中各种离子（如 Cu^{2+}）的稳定性，具有极其重要的意义。

8.8 晶体结构

在晶体学中，晶体结构是对晶态材料中原子、离子或分子有序排列的描述。在本节和下一节中，我们将简要介绍晶体学的一些基本定义和概念，以及各种类型的晶体及其结构。

晶体的定义

晶体一词源于古希腊语"krustallos"，同时包含"冰"和"水晶"的意思。晶体的定义随着时间而演变。传统上，**晶体**或**晶态固体**定义为其成分（如原子、分子或离子）以高度有序的微观结构排列、形成向所有方向延伸的晶格的固体材料。研究晶体以及晶体形成的科学，称为**晶体学**。晶体的传统特征包括规则的几何外形、固定的

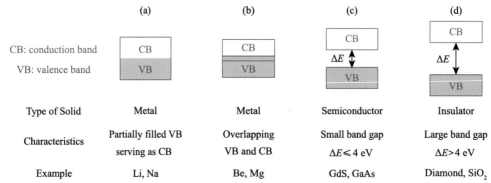

Figure 8.27 Classification of solids according to band theory. (a) Metals with a partially filled valence band that can serve as a conduction band. (b) Metals with overlapping valence band (normally full) and conduction band (normally empty). (c) Semiconductors with a relatively small band gap (normally $\Delta E \leqslant 4$ eV) between the full valence band and empty conduction band. (d) Insulators with relatively large band gap (normally $\Delta E > 4$ eV) between the full valence band and empty conduction band.

图 8.27 能带理论对固体的分类：(a) 金属，其部分填充的价带可作为导带；(b) 金属，其价带（通常为满带）与导带（通常为空带）相互重叠；(c) 半导体，其满的价带与空的导带之间存在一个较小的带隙（通常 $\Delta E \leqslant 4$ eV）；(d) 绝缘体，其满的价带与空的导带之间存在一个较大的带隙（通常 $\Delta E > 4$ eV）。

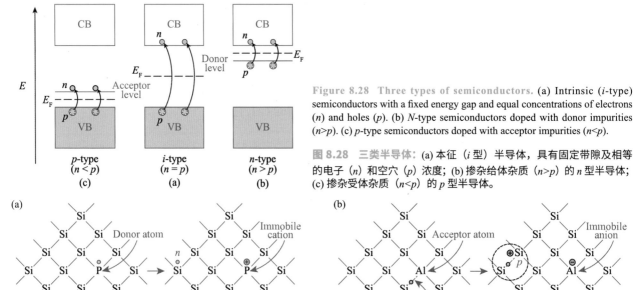

Figure 8.28 Three types of semiconductors. (a) Intrinsic (i-type) semiconductors with a fixed energy gap and equal concentrations of electrons (n) and holes (p). (b) N-type semiconductors doped with donor impurities (n>p). (c) p-type semiconductors doped with acceptor impurities (n<p).

图 8.28 三类半导体：(a) 本征（i 型）半导体，具有固定带隙及相等的电子（n）和空穴（p）浓度；(b) 掺杂给体杂质（n>p）的 n 型半导体；(c) 掺杂受体杂质（n<p）的 p 型半导体。

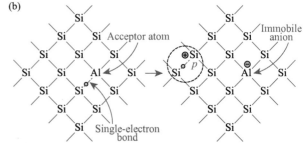

Figure 8.29 Illustrative structural diagrams of (a) n-type semiconductor with doped donor atoms such as P and (b) p-type semiconductor with doped acceptor atoms such as Al in silicon.

图 8.29 在硅中掺杂了 (a) 给体原子（如 P）的 n 型半导体与 (b) 受体原子（如 Al）的 p 型半导体的结构示意图。

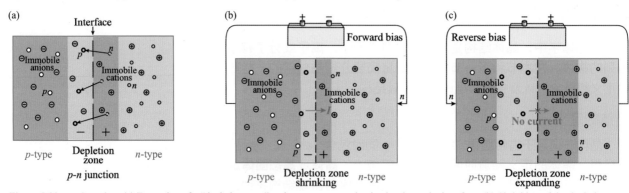

Figure 8.30 p-n junction. (a) Formation of a "depletion zone" acting as a non-conductive barrier at the interface. (b) Shrinking of the depletion zone at a forward bias. (c) Expanding of the depletion zone at a reverse bias.

图 8.30 p-n 结：(a) 在界面处形成可充当非导电势垒的"耗尽区"；(b) 正向偏压下耗尽区缩小；(c) 反向偏压下耗尽区扩大。

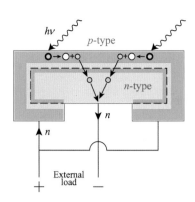

Figure 8.31 A photovoltaic/solar cell based on a *p-n* junction.

图 8.31 基于 *p-n* 结的光伏/太阳能电池。

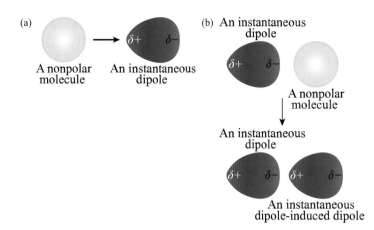

Figure 8.32 An instantaneous dipole and its induced dipole. (a) At some instantaneous moment, the displacement of electronic charge produces an instantaneous dipole with a charge separation represented as $\delta+$ and $\delta-$. (b) The instantaneous dipole on the left molecule induces a charge separation on the right molecule, resulting in an instantaneous dipole-induced dipole, as well as the attraction between these two dipoles (London force).

图 8.32 瞬时偶极及其诱导偶极：(a) 在某一瞬间，电子电荷的位移会产生瞬时偶极，其电荷分别用 $\delta+$ 和 $\delta-$ 表示。(b) 左侧分子的瞬时偶极诱导了右侧分子的电荷分离，产生了瞬时偶极的诱导偶极，以及这两个偶极之间的吸引力（即色散力）。

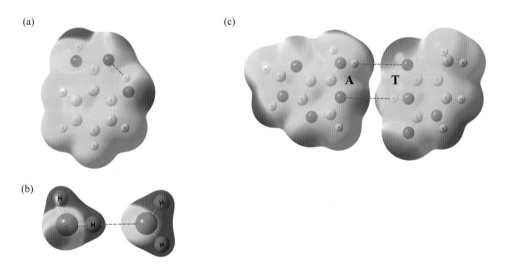

Figure 8.33 Illustrative diagrams for hydrogen bonds. (a) Intramolecular hydrogen bonds inside salicylic acid. (b) Intermolecular hydrogen bonds between two water molecules. (c) Intermolecular hydrogen bonds between adenine (A) and thymine (T) in DNA. Red dashed lines indicated the corresponding hydrogen bonds.

图 8.33 氢键示意图：(a) 水杨酸的分子内氢键；(b) 两个水分子之间的分子间氢键；(c) DNA 中腺嘌呤（A）与胸腺嘧啶（T）之间的分子间氢键。红色虚线表示相应的氢键。

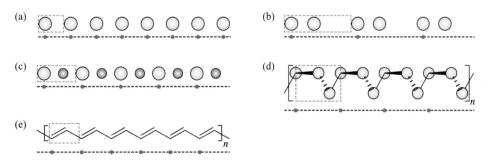

Figure 8.34 1D structure and the lattices of (a) Cu, (b) graphite, (c) NaCl, (d) Se, and (e) polyacetylene. Red dots indicate the lattice points and red dashed rectangles represent the corresponding structure motifs.

图 8.34 (a) Cu、(b) 石墨、(c) NaCl、(d) Se 和 (e) 聚乙炔的一维结构及晶格。红点代表晶格点，红色虚线框表示相应的结构基元。

of Crystallography has redefined the term *crystal* to include both ordinary periodic crystals and quasicrystals as "any solid having an essentially discrete diffraction diagram". Shechtman was awarded the 2011 Nobel Prize in chemistry for "the discovery of quasicrystals".

Lattice, Structure Motif and Unit Cell

To understand the crystal structure, a number of basic concepts are necessary:

1) The smallest unit of repeated arrangement, including the type and number of atoms or molecules, as well as their chemical environment, in a crystal is called a **structural motif**.
2) Signifying the structural motif as a point, an infinite array of discrete points that can well represent and reflect the arrangement of crystals in space is called a **lattice** or a Bravais lattice. The discrete point in a lattice is called a **lattice point**. A lattice can be generated by applying an infinite set of discrete translation operations on a single lattice point.
3) The smallest unit that contains all of the structural and symmetry information to build up the macroscopic structure of a crystal by translation is called a **unit cell**. The entire crystal can be generated by placing the structural motif on each lattice point on the lattice, as

$$\text{Crystal} = \text{Structural motif} @ \text{Lattice}$$

Lattice Types and Lattice Parameters

Lattices can be classified into different lattice types and described by the corresponding lattice parameters. Here, we will introduce those lattice types and lattice parameters from simple to complex, in the order of 0D, 1D, 2D, and 3D lattices.

To become 0D, the lattice has only one lattice point, and thus there is only one type of 0D lattice. For 1D lattice, there is also one type. An infinite array of discrete points to build up a 1D crystal by translation can be described by only one lattice parameter: the distance between neighboring points a. **Figure 8.34** shows the 1D structure and lattices of Cu, graphite, NaCl, Se, and polyacetylene. Although the values of a are different in these 1D crystals, the lattice type is all the same. However, the structure motifs are different, represented by the dashed rectangles in **Figure 8.34**.

For 2D lattices, there are five different Bravais lattice types, as shown in **Figure 8.35**, with the unit cell being represented by the filled parallelogram. To describe a 2D lattice, three parameters are needed: a_1, a_2, and φ. These five types of 2D lattice and the corresponding parameters are:

1) Oblique: $a_1 \neq a_2$ and $\varphi \neq 90°$;
2) Primitive rectangular: $a_1 \neq a_2$ and $\varphi = 90°$;
3) Centered rectangular: $a_1 \neq a_2$ and $\varphi = 90°$;
4) Hexagonal: $a_1 = a_2$ and $\varphi = 120°$;
5) Square: $a_1 = a_2$ and $\varphi = 90°$.

Although both primitive and centered rectangular lattices correspond to $a_1 \neq a_2$ and $\varphi = 90°$, their types are different. Both primitive rectangular and centered rectangular belong to the rectangular lattice system. Therefore, these five 2D Bravais lattice types can be further classified into four lattice systems.

For 3D lattices, there are altogether 14 types of Bravais lattices, which can be classified into 7 lattice systems, as summarized in **Table 8.2**. Describing a 3D lattice requires six parameters: a, b, c, α, β, and γ, and the basic unit cell is a parallel epiped. Taking the cubic lattice system as an example (**Figure 8.36**), the parameters are $a=b=c$ and $\alpha = \beta = \gamma = 90°$. Only one independent parameter is necessary, which means that if a is given, the rest parameters are all determined. The cubic lattice system contains three different lattice types: **primitive cubic** (pcc, a lattice point at each corner of the cube), **body-centered cubic** (bcc, a lattice point at the center of the cube as well as at each corner), and **face-centered cubic** (fcc, a lattice point at the center of each face as well as at each corner). The corresponding examples of cubic lattice are CsCl (pcc), K (bcc), and NaCl (fcc).

熔点、各向异性、周期性、尖锐的 X 射线衍射（XRD）峰等。

1982 年，一种特殊类型的晶体被丹·舍特曼（1941— ）发现。它由有序但不具有严格周期性的成分阵列组成，称为**准晶**，又称准周期性晶体。准晶具有许多与常规晶体相同的属性，如离散的 XRD 图案、形成光滑平坦的表面形状的能力等。准晶图案可以连续地填充所有可用空间，但缺乏平移对称性，因此并未包含在传统的晶体定义中。随后，国际晶体学联合会将"晶体"一词重新定义为"具有基本离散的衍射图案的任何固体"，这样就将常规周期性晶体和准晶都包含在内。因准晶的发现，舍特曼获得 2011 年诺贝尔化学奖。

晶格、结构基元与晶胞

为了理解晶体结构，需要了解一些基本概念：
1) 晶体中重复排列的最小单位（包括原子或分子的类型、数目及其化学环境），称为**结构基元**。
2) 将结构基元抽象为点，可以很好地表示和反映空间中晶体排列规律的、由离散点组成的无限阵列，称为**晶格**或布拉维晶格。晶格中的离散点称为**晶格点**。可以通过在单个晶格点上应用无限组离散的平移操作，来生成晶格。
3) 包含所有结构和对称性信息、可通过平移操作建立晶体宏观结构的最小单位，称为**晶胞**。将结构基元置于晶格的每个晶格点上，就可以得到整个晶体，即

$$晶体 = 结构基元 @ 晶格$$

晶格类型与晶格参数

晶格可分为不同的晶格类型，并用相应的晶格参数来描述。这里我们将按零维、一维、二维和三维晶格的顺序，由简至繁地介绍这些晶格类型和晶格参数。

零维晶格只能有一个晶格点，因此只存在一种类型的零维晶格。一维晶格也只有一种类型。通过平移操作可建立一维晶体的、由离散点组成的无限阵列，只需一个晶格参数即可描述相邻晶格点之间的距离 a。**图 8.34** 给出了 Cu、石墨、NaCl、Se 和聚乙炔的一维结构和晶格。尽管这些一维晶体的 a 不同，但其晶格类型完全相同，而结构基元（如**图 8.34** 的虚线框所示）各不相同。

对于二维晶格，共有五种不同的布拉维晶格类型，如**图 8.35** 所示，其晶胞用填充的平行四边形表示。为了描述二维晶格，需要使用三个参数：a_1、a_2 和 φ。这五种二维晶格类型及其相应参数为：
1) 斜方：$a_1 \neq a_2$ 且 $\varphi \neq 90°$；
2) 简单长方：$a_1 \neq a_2$ 且 $\varphi = 90°$；
3) 面心长方：$a_1 \neq a_2$ 且 $\varphi = 90°$；
4) 六角：$a_1 = a_2$ 且 $\varphi = 120°$；
5) 正方：$a_1 = a_2$ 且 $\varphi = 90°$。

尽管简单长方和面心长方的晶格都对应于 $a_1 \neq a_2$ 且 $\varphi = 90°$，但它们的类型不同。简单长方和面心长方均属于长方晶系，因此这五种二维布拉维晶格类型可进一步划归为四种晶系。

对于三维晶格，共有 14 种布拉维晶格类型，可划归为 7 种晶系，如**表 8.2** 所示。描述三维晶格需要使用六个参数：a、b、c、α、β 和

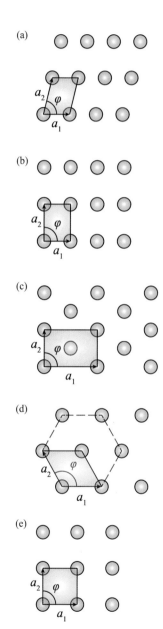

Figure 8.35 The five fundamental 2D Bravais lattices: (a) oblique; (b) rectangular; (c) centered rectangular; (d) hexagonal; (e) square, the unit cell of which is represented by the filled parallelogram.

图 8.35 五种二维晶格基本类型 (a) 斜方；(b) 简单长方；(c) 面心长方；(d) 六角；(e) 正方，其晶胞用填充的平行四边形表示。

X-Ray Diffraction

The experimental measurements of the structure of a crystal usually exploit the interactions between light and crystal, which requires the wavelength of light to be comparable to the repeated arrangement units in the crystal. Since the size of atoms, molecules, or ions in a crystal is normally in angstrom to nanometer scale, the light best interacting with crystals is X-rays. When a beam of X-ray encounters a crystal, the structure of the crystal causes the beam to diffract in many specific directions that are related to the structure. By measuring the angles and intensities of these diffracted beams, a series of diffraction patterns can be obtained. These diffraction patterns are then used to infer the microscopic structure of the crystal, with the aid of high-speed computers to process vast amounts of X-ray diffraction data. This method is called **X-ray diffraction** (XRD) or X-ray crystallography.

Table 8.2 14 Types of Bravais Lattices in 7 Lattice Systems for 3D Lattices

表 8.2 三维晶格的 7 种晶系和 14 种布拉维晶格类型

Lattice Group (晶族)	Lattice System (晶系)	Lattice Parameter (晶格参数)	Bravais Lattice Type (布拉维晶格类型)[a]			
Anorthic (a, 三斜)	Triclinic (三斜)	$a \neq b \neq c$ $\alpha \neq \beta \neq \gamma \neq 90°$	aP			
Monoclinic (m, 单斜)	Monoclinic (单斜)	$a \neq b \neq c$ $\alpha = \gamma = 90° \neq \beta$	mP	mC		
Orthorhombic (o, 正交)	Orthorhombic (正交)	$a \neq b \neq c$ $\alpha = \beta = \gamma = 90°$	oP	oC	oI	oF
Tetragonal (t, 四方)	Tetragonal (四方)	$a = b \neq c$ $\alpha = \beta = \gamma = 90°$	tP	tI		
Hexagonal (h, 六方)	Trigonal (三方)	$a = b \neq c$ $\alpha = \beta = 90°$ $\gamma = 120°$	hP	hR		
	Hexagonal (六方)	$a = b \neq c$ $\alpha = \beta = 90°$ $\gamma = 120°$	hP			
Cubic (c, 立方)	Cubic (立方)	$a = b = c$ $\alpha = \beta = \gamma = 90°$	cP (pcc)	cI (bcc)	cF (fcc)	

[a] P: primitive; C: base-centered; I: body-centered; F: face-centered; R: Rhombohedral. (P: 简单；C: 底心；I: 体心；F: 面心；R: R 心。)

γ，基本晶胞为平行六面体。以立方晶系为例（**图 8.36**），其参数为 $a=b=c$ 且 $\alpha=\beta=\gamma=90°$。其中只有一个独立的参数，这意味着如果 a 值给定，则其余参数均已确定。立方晶系包括三种不同的晶格类型：**简单立方**（pcc，立方体的每个角上各有一个晶格点）、**体心立方**（bcc，立方体的中心以及每个角上各有一个晶格点）和**面心立方**（fcc，立方体每个面的中心以及每个角上各有一个晶格点）。立方晶格的相应例子有 CsCl（pcc）、K（bcc）和 NaCl（fcc）。

X 射线衍射

晶体结构的实验测量通常利用光与晶体之间的相互作用，这要求光的波长与晶体中重复排列的单元相当。由于晶体中的原子、分子或离子的尺寸通常在从埃（Å）到纳米（nm）的量级，因此与晶体相互作用最好的光是 X 射线。当一束 X 射线照射在晶体上，晶体的结构会使 X 射线束衍射到与其结构相关的许多特定方向。通过测量这些衍射光束的角度和强度，可以获得一系列衍射图案。借助高速计算机处理大量 X 射线衍射数据，这些衍射图案可用来推断晶体的微观结构。这种方法称为 **X 射线衍射**（XRD）或 X 射线晶体学。

当 X 射线照射到晶体上，它主要与晶体中的电子发生相互作用。就像海浪撞击灯塔会产生从灯塔发出的次级圆形波一样，X 射线撞击电子也会产生从电子发出的次级球形波，使得原光束向所有方向散射。尽管这些波在大多数方向上通过相消干涉会相互抵消，但在几个特定方向上它们会发生相长叠加，这些方向可由**布拉格定律**确定，有

$$n\lambda = 2d\sin\theta \tag{8.13}$$

其中 d 是衍射平面之间的间距，θ 是入射角，n 为任意整数，λ 是入射光束的波长。布拉格定律由威廉·H. 布拉格（1862—1942）和威廉·L. 布拉格（1890—1971）于 1913 年提出，如**图 8.37** 所示。

8.9 各种晶体类型及其结构

本节我们将介绍四类典型晶体的结构：金属晶体、离子晶体、分子晶体和网状共价晶体。

金属晶体结构

金属晶体可视为由大量具有相同金属半径的硬球（离子或原子）堆砌而成，让我们先搭建等径圆球的堆砌模型。如果我们在一个平面上排列等径圆球，最密堆积的排列如**图 8.38(a)** 所示，并在**图 8.38(b)** 中标记为 A 层（蓝色），其中六个相邻球体以六边形的方式与中心球体相接触。任意三个相邻球体中间类似三角形的空隙，称为三角形空隙 [**图 8.38(a)**]。为搭建最密堆积排列，下一层 B 层（红色）的球体必须位于 A 层的凹坑中。在 A、B 两层之间存在两种不同类型的空隙：位于 A 层球体正上方的四面体空隙和位于 A 层三角形空隙正上方的八面体空隙 [**图 8.38(b)(c)**]。第三层（黄色）存在两种可能性。一种是第三层的球体恰好位于四面体空隙的正上方，

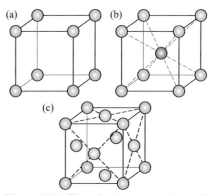

Figure 8.36 Three lattice types in the cubic lattice system: (a) primitive cubic (pcc); (b) body-centered cubic (bcc); (c) face-centered cubic (fcc).

图 8.36 立方晶系的三种晶格类型：(a) 简单立方 (pcc)；(b) 体心立方 (bcc)；(c) 面心立方 (fcc)。

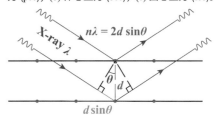

Figure 8.37 Derivation of Bragg's law that determines the crystal structure by X-ray diffraction.

图 8.37 通过 X 射线衍射（XRD）确定晶体结构的布拉格定律的推导。

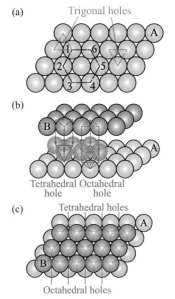

Figure 8.38 Various holes in closest packed structure of congruent spheres. (a) Closest arrangement on a plane where trigonal holes (orange) located among any three neighboring spheres. (b) Side view and (c) top view of the tetrahedral (red) and octahedral (green) holes between two neighboring layers in the closest arrangement.

图 8.38 等径圆球最密堆积结构中的各种空隙：(a) 平面最密堆积中位于任意三个相邻球体之间的三角形空隙（橙色）；最密堆积中相邻两层之间四面体空隙（红色）和八面体空隙（绿色）的 (b) 侧视图与 (c) 俯视图。

When encountering a crystal, an X-ray primarily interacts with the electrons in the crystal. Just like an ocean wave striking a lighthouse produces secondary circular waves emanating from the lighthouse, an X-ray striking an electron also produces secondary spherical waves emanating from the electron, causing the original beam to scatter in all directions. Although these waves cancel one another out in most directions through destructive interference, they add constructively in a few specific directions, determined by **Bragg's law**, as

$$n\lambda = 2d\sin\theta \tag{8.13}$$

where d is the spacing between diffracting planes, θ is the incident angle, n is any integer, and λ is the wavelength of the incident beam. Bragg's law was proposed by William H. Bragg (1862—1942) and William L. Bragg (1890—1971) in 1913 and illustrated in **Figure 8.37**.

8.9 Various Types of Crystals and Their Structures

In this section, we will introduce the structures of four typical types of crystals: metallic, ionic, molecular, and network covalent crystals.

Metallic Crystal Structure

A metallic crystal can be viewed as being packed up by a large number of hard spheres (ions or atoms) with equal metallic radius. Let us first build up the arrangement model of congruent spheres. If we arrange congruent spheres on a plane, the closest arrangement is shown in **Figure 8.38(a)** and labelled as layer A (blue) in **Figure 8.38(b)** with six neighboring spheres in contact with a central sphere in a hexagonal fashion. A hole among any three neighboring spheres resembling a triangle is called a trigonal hole [**Figure 8.38(a)**]. To build up the closest arrangement, the spheres in the next layer, layer B (red), must be located in the dimple of layer A. Two different types of holes exist between layers A and B: tetrahedral holes sitting directly above spheres in layer A, and octahedral holes falling directly over trigonal holes in layer A [**Figure 8.38(b)**]. There are two possibilities for the third layer (yellow). In one arrangement, called **hexagonal closest packed** [hcp, **Figures 8.39(a)(b)**], the spheres in the third layer are located exactly on top of the tetrahedral holes. The third layer is identical to layer A, and the structure is A B A B A B …. In the other arrangement, called **cubic closest packed** [ccp, **Figures 8.39(c)(d)**], the spheres in the third layer are directly above the octahedral holes. The third layer (layer C) is different than layers A and B, and the structure is A B C A B C ….

Take a close look at the ccp structure, and you will find that it has a fcc unit cell [**Figures 8.40(a)(b)**]. The **coordination number**, which is the number of atoms with which a given atom is in contact, is 12 for fcc [**Figure 8.39(d)**]: six in the same layer, three in the layer above, and three in the layer below. The total number of atoms in a fcc unit cell is four [**Figure 8.40(c)**]: eight corner atoms, each shared among eight adjoining unit cells, account for $8 \times 1/8 = 1$ atom, and six atoms in the center of faces, each shared between two adjoining unit cells, account for $6 \times 1/2 = 3$ atoms. The occupied volume percentage, which is the percentage of the volume occupied by the atoms with respect to the total volume in the unit cell, can be calculated as

$$\sqrt{2}a = 4r$$

$$\eta_{ccp} = \eta_{fcc} = \frac{4 \times 4\pi r^3/3}{\left(4r/\sqrt{2}\right)^3} = \frac{\sqrt{2}\pi}{6} = 74.05\% \tag{8.14}$$

where a is the edge of the unit cell and r is the radius of the atom.

For hcp [**Figures 8.41(a)(b)**], the unit cell is a parallel epiped with a rhombic base, and there are two atoms per unit cell. The same as ccp, the coordination number is $3 + 6 + 3 = 12$ [**Figure 8.39(b)**], and the occupied volume percentage is 74.05%. Both ccp and hcp have the greatest occupied volume percentage, and this is why they are called "closest" packed.

For bcc [**Figure 8.41(c)**], the coordination number is 8, and the number of atoms per unit cell is 2. The occupied volume percentage can be calculated as

称为**六方最密堆积** [hcp，**图 8.39(a)(b)**]，其第三层与 A 层完全相同，结构为 A B A B A B…。另一种是第三层的球体直接位于八面体空隙的正上方，称为**立方最密堆积** [ccp，**图 8.39(c)(d)**]，其第三层（C 层）与 A、B 层均不同，结构为 A B C A B C…。

仔细观察 ccp 结构，会发现它具有 fcc 晶胞 [**图 8.40(a)(b)**]。**配位数**是与给定原子相接触的原子数目，fcc 的配位数为 12 [**图 8.39(d)**]：同一层 6，上一层 3，下一层 3。fcc 晶胞内的原子总数为四个 [**图 8.40(c)**]：八个角原子，每个在八个相邻晶胞之间共享，算作 8 × 1/8 = 1 个原子；六个面心原子，每个在两个相邻晶胞之间共享，算作 6 × 1/2 = 3 个原子。体积占有百分比，即原子所占体积相对于晶胞总体积的百分比，可计算为

$$\sqrt{2}a = 4r$$
$$\eta_{\mathrm{ccp}} = \eta_{\mathrm{fcc}} = \frac{4 \times 4\pi r^3 / 3}{\left(4r/\sqrt{2}\right)^3} = \frac{\sqrt{2}\pi}{6} = 74.05\% \tag{8.14}$$

其中 a 为晶胞的边长，r 是原子的半径。

hcp 晶胞 [**图 8.41(a)(b)**] 是一个具有菱形底面的平行六面体，每个晶胞中有两个原子。其配位数为 3 + 6 + 3 = 12 [**图 8.39(b)**]，体积占有百分比为 74.05%，与 ccp 相同。ccp 和 hcp 均具有最大的体积占有百分比，这就是它们被称为"最密"堆积的原因。

bcc [**图 8.41(c)**] 的配位数为 8，晶胞内的原子总数为 2。体积占有百分比可计算为

$$\sqrt{3}a = 4r$$
$$\eta_{\mathrm{bcc}} = \frac{2 \times 4\pi r^3/3}{\left(4r/\sqrt{3}\right)^3} = \frac{\sqrt{3}\pi}{8} = 68.02\% \tag{8.15}$$

pcc [**图 8.41(d)**] 的配位数为 6，晶胞内的原子总数为 1。体积占有百分比可计算为

$$a = 2r$$
$$\eta_{\mathrm{pcc}} = \frac{4\pi r^3/3}{(2r)^3} = \frac{\pi}{6} = 52.36\% \tag{8.16}$$

bcc 和 pcc 被认为是密堆积，但并非最密堆积，因为它们的体积占有百分比小于 ccp 和 hcp。

金属晶体是以金属键结合形成的晶体，而金属键源于金属离子与自由电子之间的静电吸引力。由于其静电力本质，金属键没有饱

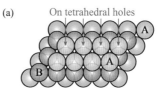
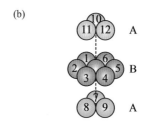

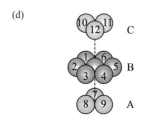

Figure 8.39 Closest packed structures of congruent spheres. (a) Top view and (b) coordination number of the hexagonal closest packed (hcp) structure. The spheres in the third layer (layer A, yellow spheres) are located exactly on top of the tetrahedral holes between layers A (blue spheres) and B (red spheres). (c) Top view and (d) coordination number of the cubic closest packed (ccp) structure. The spheres in the third layer (layer C, yellow spheres) are directly above the octahedral holes between layers A (blue spheres) and B (red spheres).

图 8.39 等径圆球最密堆积结构：六方最密堆积（hcp）结构的 (a) 俯视图与 (b) 配位数，第三层球体（A 层，黄球）恰好位于 A（蓝球）、B（红球）两层之间四面体空隙正上方；立方最密堆积（ccp）结构的 (c) 俯视图与 (d) 配位数，第三层球体（C 层，黄球）直接位于 A（蓝球）、B（红球）两层之间八面体空隙正上方。

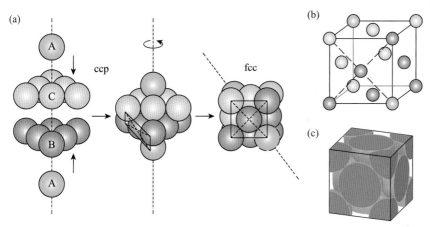

Figure 8.40 Relationship between cubic closest packed (ccp) structure and face-centered cubic (fcc) unit cell. (a) Rotation of 14 spheres extracted from a ccp structural array reveals (b) the fcc unit cell. (c) The fcc unit cell showing a total number of four atoms.

图 8.40 立方最密堆积（ccp）结构与面心立方（fcc）晶胞之间的关系。(a) 旋转从 ccp 结构阵列中提取的 14 个圆球，显示其具有 (b) fcc 晶胞。(c) fcc 晶胞内共有 4 个原子。

$$\sqrt{3}a = 4r$$

$$\eta_{bcc} = \frac{2 \times 4\pi r^3 / 3}{\left(4r/\sqrt{3}\right)^3} = \frac{\sqrt{3}\pi}{8} = 68.02\% \tag{8.15}$$

For pcc [**Figure 8.41(d)**], the coordination number is 6, and the number of atoms per unit cell is 1. The occupied volume percentage can be calculated as

$$a = 2r$$

$$\eta_{pcc} = \frac{4\pi r^3 / 3}{(2r)^3} = \frac{\pi}{6} = 52.36\% \tag{8.16}$$

Both bcc and pcc are considered as close packed but not the closest, because they have a smaller occupied volume percentage than ccp and hcp.

A **metallic crystal** is a crystal formed by metallic bonds, which arise from the electrostatic attraction between metal ions and free electrons. Due to its electrostatic nature, the metallic bond lacks saturability and directivity. There is no single molecule in metallic crystals. In order to minimize the overall energy, metal ions tend to pack up around each other, leading to a high coordination number. XRD data demonstrate that more than half of the metals adopt the closest packed structures in their stable forms at room temperature. **Table 8.3** lists the crystal structures of many metals in the periodic table.

Table 8.3 Four Special Ways of Packing Congruent Spheres
表 8.3 等径圆球堆砌的四种特殊方式

Unit Cell (晶胞)	Coordination Number (配位数)	Atoms per Unit Cell (晶胞内原子数)	Volume Occupied/（%） (体积占有百分比)	Example (示例)
Simple cubic (pcc) (简单立方)	6	1	52	Po
Body-centered cubic (bcc) (体心立方)	8	2	68	Fe, Na, K, W
Face-centered cubic (fcc or ccp) (面心立方)	12	4	74	Ag, Cu, Pb
Hexagonal closest packed (hcp) (六方最密堆积)	12	2	74	Cd, Mg, Ti, Zn

Ionic Crystal Structure

An **ionic crystal** is a crystal in which ions are bound together by ionic bonds, which is a type of chemical bond that involves the electrostatic interactions between charged ions. Due to its electrostatic nature, the ionic bond also lacks saturability and directivity. Typical properties and characteristics of ionic crystals include high melting and boiling points, high hardness, good electrical conductivity in aqueous solution or molten state, lack of ductility, good solubility in polar solvents such as water.

To understand the structure of ionic crystals, we can also apply the previous packing-of-sphere model with the following two considerations:

1) Ions are either positively or negatively charged;
2) Cations and anions are of different sizes.

To minimize the overall energy, we should maximize the attractive forces and minimize the repulsive forces. This can be realized by keeping the oppositely charged ions in close proximity and preventing the like-charged ions from direct contact. Since anions are generally larger than cations, the electrostatic repulsion between anions is relatively weak. In most cases, we can think of ionic crystals as a fairly closely packed arrangement of anions, with holes filled by cations of smaller size. The rule of filling is that cations should be fitted into holes slightly smaller than their actual size, so the anions are pushed slightly apart while the anion and cation are in contact.

For a binary ionic crystal with anions in fcc arrangement, the cations may occupy one of the three types of holes: trigonal, tetrahedral, and octahedral, in the order of increasing size [**Figures 8.42(a)~(c)**]. If the

和性和方向性。金属晶体中不存在单个分子，为使总能量最小化，尽可能多的金属离子倾向于彼此聚集，从而导致高配位数。XRD 数据表明，室温下超过一半的金属最稳定的形式均采用最密堆积结构。**表 8.3** 列出了元素周期表中许多金属的晶体结构。

离子晶体结构

离子晶体是离子之间通过离子键结合形成的晶体，而离子键是一种涉及带电离子之间静电相互作用的化学键。由于其静电力本质，离子键也没有饱和性和方向性。离子晶体的典型性质和特征包括：高熔沸点、高硬度、在水溶液或熔融状态下良好的导电性、缺乏延展性、在极性溶剂（如水）中良好的溶解性。

为了理解离子晶体的结构，我们仍可应用前述硬球堆积模型，需考虑如下两个因素：
1) 阴、阳离子带异种电荷；
2) 阴、阳离子的大小不等。

为使总能量最低，我们应该使吸引力最大化而排斥力最小化。这可以通过使带异种电荷的离子保持互相接近，并防止带同种电荷的离子直接接触来实现。由于阴离子通常大于阳离子，阴离子之间的静电排斥力相对较弱。在大多数情况下，我们可以将离子晶体看作一种阴离子的相当密实的排列，而具有较小尺寸的阳离子则填充在阴离子的空隙中。填隙规则是，将阳离子填入比其自身实际尺寸略小的空隙中，使得阴离子被略微推开，而阴、阳离子之间直接接触。

对于阴离子以 fcc 排布的二元离子晶体，阳离子可占据三种类型的空隙之一：三角形、四面体形和八面体形空隙，其大小依次增加 [**图 8.42(a)~(c)**]。如果阴离子和阳离子的半径分别用 R_- 和 r_+ 表示，不难计算出使得阳离子恰好填入阴离子的三角形、四面体形和八面体形空隙的填充比 r_+/R_-，分别为 0.155、0.225 和 0.414。同时，阴离子以 pcc 排布的二元离子晶体，在其晶胞中心还具有一个立方体形空隙 [**图 8.42(d)**]，填充比 $r_+/R_- = 0.732$。应用填隙规则，阳离子应填入以下空隙：

三角形 (fcc)：$r_+/R_- < 0.225$

四面体形 (fcc)：$0.225 < r_+/R_- < 0.414$

八面体形 (fcc)：$0.414 < r_+/R_- < 0.732$

立方体形 (pcc)：$0.732 < r_+/R_- < 1$

如果 $r_+/R_- \geq 1$，这意味着阳离子甚至大于阴离子。这时应该使用阳离子进行排布，而用阴离子来填充空隙。

Na^+、Cs^+ 和 Cl^- 的离子半径分别为 102、174 和 181 pm，相应的填充比分别为

$$\frac{r_+}{R_-}(NaCl) = \frac{r_+(Na^+)}{R_-(Cl^-)} = \frac{102}{181} = 0.564$$

$$\frac{r_+}{R_-}(CsCl) = \frac{r_+(Cs^+)}{R_-(Cl^-)} = \frac{174}{181} = 0.961$$

我们预期 NaCl 中 Na^+ 会占据 Cl^- 的 fcc 阵列的八面体形空隙 [**图**

radii of the anion and cation are denoted as R_- and r_+, respectively, it is not difficult to calculate the ratio of r_+/R_- that makes the cation just fit into the trigonal, tetrahedral, and octahedral holes of the anion to be 0.155, 0.225, and 0.414, respectively. Meanwhile, a binary ionic crystal with anions in pcc arrangement has a cubic hole at the center of the unit cell [**Figure 8.42(d)**]. The fitting ratio is $r_+/R_- = 0.732$. By applying the rule of filling, a cation should accommodate the following holes:

$$\text{Trigonal(fcc)}: r_+/R_- < 0.225$$

$$\text{Tetrahedral(fcc)}: 0.225 < r_+/R_- < 0.414$$

$$\text{Octahedral(fcc)}: 0.414 < r_+/R_- < 0.732$$

$$\text{Cubic(pcc)}: 0.732 < r_+/R_- < 1$$

If $r_+/R_- \geq 1$, it means that the cation is even bigger than the anion. Then we should use cations in the arrangement and anions to fill in the holes.

The ionic radii for Na^+, Cs^+, and Cl^- are 102, 174, and 181 pm, respectively. The corresponding fitting ratios are

$$\frac{r_+}{R_-}(NaCl) = \frac{r_+(Na^+)}{R_-(Cl^-)} = \frac{102}{181} = 0.564$$

$$\frac{r_+}{R_-}(CsCl) = \frac{r_+(Cs^+)}{R_-(Cl^-)} = \frac{174}{181} = 0.961$$

We expect Na^+ to occupy the octahedral holes of the fcc arrays of Cl^- in NaCl [**Figure 8.43(a)**] and Cs^+ to occupy the cubic holes of the pcc arrays of Cl^- in CsCl [**Figure 8.43(b)**]. In the NaCl unit cell, there are $8 \times 1/8 + 6 \times 1/2 = 4$ Cl^- ions and $1 + 12 \times 1/4 = 4$ Na^+ ions. The ratio of Na^+ to Cl^- is 4:4, in accord with the formula NaCl. To establish the coordination numbers in an ionic crystal, we take the ratio of the number of nearest neighboring anions of any cations to the number of nearest neighboring cations of anions. In NaCl, each Na^+ is surrounded by six Cl^- and each Cl^- is also surrounded by six Na^+. Thus, the ratio of coordination numbers is 6:6, corresponding to the formula NaCl. In the CsCl unit cell, the ratio of Cs^+ to Cl^- is 1:1, and the ratio of coordination numbers is 8:8.

Ionic compounds of the type $M^{2+}X^{2-}$, such as MgO, BaS, CaO, may form crystals of the NaCl type. However, if the cation is small enough so that $r_+/R_- < 0.414$, as in the case of ZnS in which

$$\frac{r_+}{R_-}(ZnS) = \frac{r_+(Zn^{2+})}{R_-(S^{2-})} = \frac{60}{184} = 0.326$$

it can occupy the tetrahedral holes. To satisfy the stoichiometry, only half of the eight tetrahedral holes are occupied, to correspond to the four S^{2-} forming fcc array [**Figure 8.43(c)**]. In the ZnS unit cell, the ratio of Zn^{2+} to S^{2-} is 4:4, and the ratio of coordination numbers is also 4:4. The structure of CaF_2 is similar to that of ZnS, except that eight F^- ions occupy eight tetrahedral holes [**Figure 8.43(d)**], leading to an ion ratio of 4:8 and a coordination number ratio of 8:4.

Ionic bonds are usually formed between atoms with relatively large differences in their electronegativity, normally $\Delta\chi > 1.7$. The strength of an ionic bond is determined by its **lattice energy**, which is the energy given off when separated gaseous ions come together to form 1 mol of a solid ionic crystal. Lattice energy can help predict the melting points and water solubilities of ionic compounds. The lattice energy generally increases with increasing ion charges and with decreasing ionic sizes.

To calculate the lattice energy quantitatively, it is quite difficult to do a direct calculation because both attractive and repulsive interactions must be considered at the same time. The indirect calculation of the lattice energy can be accomplished through an application of Hess's law, known as the Born-Fajans-Haber cycle, named after its originators Max Born, Kasimir Fajans, and Fritz Haber. **Figure 8.44** illustrates an example to indirectly calculate the lattice energy of NaCl by the cycle.

Molecular Crystal Structure

A **molecular crystal** is a crystal consisting of a lattice array of molecules bound by intermolecular

8.43(a)]，而 CsCl 中 Cs⁺ 会占据 Cl⁻ 的 pcc 阵列的立方体形空隙 [**图 8.43(b)**]。在 NaCl 晶胞中，共有 8 × 1/8 + 6 × 1/2 = 4 个 Cl⁻ 和 1 + 12 × 1/4 = 4 个 Na⁺ 离子。Na⁺ 与 Cl⁻ 的比率为 4:4，与化学式 NaCl 一致。为了确定离子晶体中的配位数，我们取任意阳离子周围最近邻阴离子的数量与任意阴离子周围最近邻阳离子的数量之比。在 NaCl 中，每个 Na⁺ 被六个 Cl⁻ 包围，每个 Cl⁻ 也被六个 Na⁺ 包围。因此配位数之比为 6:6，也对应于化学式 NaCl。在 CsCl 晶胞中，Cs⁺ 与 Cl⁻ 的比率为 1:1，配位数之比为 8:8。

$M^{2+}X^{2-}$ 型离子化合物（如 MgO、BaS、CaO 等）可形成 NaCl 型晶体。但如果阳离子足够小，满足填充比 r_+/R_-<0.414，如对于 ZnS：

$$\frac{r_+}{R_-}(\text{ZnS}) = \frac{r_+(\text{Zn}^{2+})}{R_-(\text{S}^{2-})} = \frac{60}{184} = 0.326$$

它可以占据四面体形空隙。为满足化学计量数，在对应于 4 个 S²⁻ 形成的 fcc 阵列中，8 个四面体形空隙中只有一半被占据 [**图 8.43(c)**]。在 ZnS 晶胞中，Zn²⁺ 与 S²⁻ 的比率为 4:4，配位数之比也是 4:4。CaF₂ 的结构与 ZnS 相似，只是八个 F⁻ 离子占据八个四面体形空隙 [**图 8.43(d)**]，导致其离子比为 4:8，配位数之比为 8:4。

离子键通常在电负性差异较大（通常 $\Delta\chi$>1.7）的原子之间形成。离子键的强度由其**晶格能**决定，晶格能是分离的气态离子聚集在一起，形成 1 mol 固态离子晶体时释放的能量。晶格能有助于预测离子化合物的熔点及水溶性。晶格能通常随离子电荷的增加和离子尺寸的减小而增大。

晶格能的定量计算通常较难直接进行，因为必须同时考虑吸引力和排斥力。晶格能的间接计算可以通过应用盖斯定律来完成，称为玻恩-法扬斯-哈伯循环，命名自其创始人麦克斯·玻恩、卡西米尔·法扬斯和弗里茨·哈伯。**图 8.44** 给出了通过该循环间接计算 NaCl 晶格能的示例。

分子晶体结构

分子晶体是由以分子间作用力结合的分子晶格阵列组成的晶体。在分子晶体中，分子之间存在弱的分子间作用力，而在分子内存在强的共价键。分子晶体的熔化或沸腾只需克服相对较弱的分子间作用力。分子晶体的典型特征有：熔沸点低、硬度低、导电导热性差等。

由球形或近球形分子构成的分子晶体通常采用最密堆积结构，配位数为 12。所有稀有气体的分子晶体均为 ccp（fcc）或 hcp 结构。球形 CH₄ 分子晶体采用 fcc 结构，近球形 H₂ 分子晶体采用 hcp 结构。

网状共价晶体结构

网状共价晶体是原子以共价键网格连接在一起组成的晶体。整个晶体可视为一个巨大的单分子。破坏网状共价晶体需要断裂共价键。网状共价晶体的典型特征有：熔沸点高、硬度高、配位数低、密度低、导电导热性差、易碎、通常不溶于大多数溶剂等。

金刚石是一种典型的网状共价晶体。每个 C 原子可通过四个 sp^3 杂化轨道，与四个相邻 C 原子形成共价键，进而形成四面体形网状

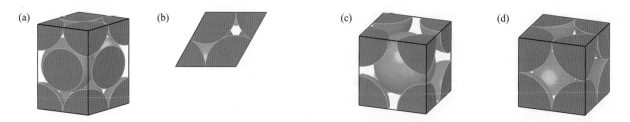

Figure 8.41 The total number of atoms in various unit cells of congruent spheres. (a) The hcp unit cell showing a total number of two atoms. (b) Top view of hcp unit cell. (c) The bcc unit cell showing a total number of two atoms. (d) The pcc unit cell showing a total number of one atom.

图 8.41 各种晶胞（等径圆球）内的原子总数： (a) hcp 晶胞内共有 2 个原子；(b) hcp 晶胞的俯视图；(c) bcc 晶胞内共有 2 个原子；(d) pcc 晶胞内共有 1 个原子。

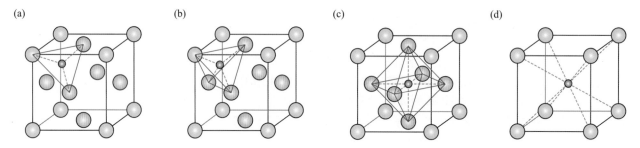

Figure 8.42 Various holes in a face-centered cubic (fcc) and a primitive cubic (pcc) unit cells. (a) Trigonal, (b) tetrahedral, and (c) octahedral holes in a fcc unit cell, and (d) a cubic hole in a pcc unit cell.

图 8.42 面心立方（fcc）和简单立方（pcc）晶胞中的各种空隙： fcc 晶胞中的 (a) 三角形、(b) 四面体形、(c) 八面体形空隙，以及 pcc 晶胞中的 (d) 立方体形空隙。

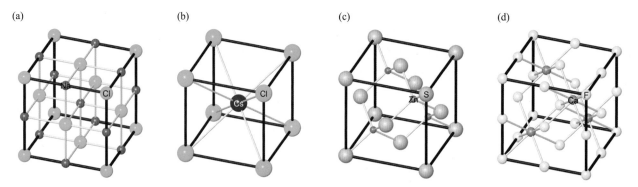

Figure 8.43 Illustrative examples of crystal structures: (a) NaCl; (b) CsCl; (c) ZnS; (d) CaF_2.

图 8.43 晶体结构示例：(a) NaCl；(b) CsCl；(c) ZnS；(d) CaF_2。

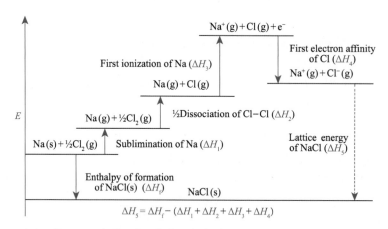

Figure 8.44 An illustrative enthalpy diagram to indirectly calculate the lattice energy of NaCl by the Born-Fajans-Haber cycle.

图 8.44 通过玻恩 - 法扬斯 - 哈伯循环间接计算 NaCl 晶格能的焓图示意。

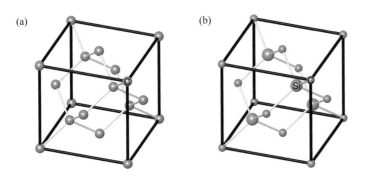

Figure 8.45 The crystal structures of (a) diamond and (b) SiC.

图 8.45 (a) 金刚石与 (b) SiC 的晶体结构。

Figure 8.46 A cartoon representing four types of chemical bonds or forces. (a) Covalent bonds. (b) Metallic bonds. (c) Intermolecular forces. (d) Ionic bonds.

图 8.46 代表四种化学键或作用力的卡通图：(a) 共价键；(b) 金属键；(c) 分子间作用力；(d) 离子键。
(Picture redrawn by Junyu Yang and idea from Ball, P. *Nature*, **2011**, *469*, 26–28.)

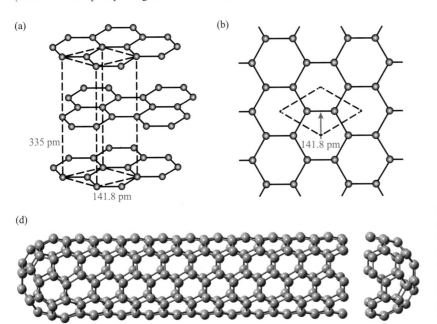

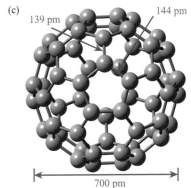

Figure 8.47 Structures of (a) graphite, (b) single-layer graphene, (c) fullerene (a buckyball C_{60}), and (d) single-walled carbon nanotube [a (5,5) armchair nanotube with ends being closed by a half C_{60} cap].

图 8.47 (a) 石墨、(b) 单层石墨烯、(c) 富勒烯（足球烯 C_{60}）和 (d) 单壁碳纳米管 [由半个 C_{60} 加帽封端的扶手椅型 (5,5) 管] 的结构。

forces. In a molecular crystal, weak intermolecular forces are present between the molecules, and strong covalent bonds exist within those molecules. The melting or boiling of a molecular crystal requires only to overcome the relatively weak intermolecular forces. Typical characteristics of molecular crystals are low melting and boiling points, low hardness, poor electrical and thermal conductivities, etc.

Molecular crystals of spherical or near-spherical molecules generally adopt the closest packed structure with a coordination number of 12. All noble gas molecular crystals are ccp (fcc) or hcp structures. The spherical CH_4 molecular crystal adopts fcc structure, and the near-spherical H_2 molecular crystal adopts hcp structure.

Network Covalent Crystal Structure

A **network covalent crystal** consists of atoms held together by a network of covalent bonds. The entire crystal can be regarded as a large, single molecule. To break a network covalent crystal requires to break covalent bonds. Typical characteristics of network covalent crystals include high melting and boiling points, high hardness, low coordination number, low density, poor electrical and thermal conductivities, brittle, generally insoluble in most solvents, etc.

Diamond is a typical network covalent crystal. With four sp^3 hybrid orbitals, each C atom can form covalent bonds with four neighboring C atoms, leading to a tetrahedral network. In the unit cell of diamond shown in **Figure 8.45(a)**, C atoms are located at the fcc lattice points as well as at the center of half of the tetrahedral holes. The coordination number is 4, and the total number of C atoms in the unit cell is 4 + 4 = 8. The occupied volume percentage is only 34%, much lower than those of metallic or ionic crystals. Diamond is the hardest substance known. It is an electric insulator, and the melting point of diamond is very high (above 3500 ℃). Replacing the four C atoms at the center of tetrahedral holes in the diamond with Si atoms results in the structure of SiC [**Figure 8.45(b)**].

Figure 8.46 shows a cartoon representing the four types of chemical bonds or forces, i.e., covalent bonds, ionic bonds, metallic bonds, and intermolecular forces, that we have discussed in this chapter.

Extended Reading Materials — Structures of the Elements and Compounds of Carbon

In addition to some gaseous low-carbon molecules that may exist instantaneously, such as C_1, C_2, C_3, C_4, C_5, etc., the allotropies of carbon elements mainly exist in the forms of diamond, graphite, graphene, fullerene, carbon nanotube, and a morphous carbon. The structure of the diamond is given in **Figure 8.45(a)**. The graphite crystals [**Figure 8.47(a)**] are stacked up from the infinitely extended planar layers of hexagonal carbon rings, with relatively weak van der Waals interactions between the layers and covalent bonds within the layer. Each C atom is bonded to three neighboring C atoms via sp^2 hybridization to form three equidistance σ bonds within the layer, whereas the overlap involving the unhybridized perpendicular p_z orbital on every C atom generates delocalized π bonds. The interlayer distance is 335 pm, and the distance between neighboring C atoms is 141.8 pm, which is between the C—C single bond length and C=C double bond length. Graphene is the common name for single-layer [**Figure 8.47(b)**] and oligolayer graphite as well as their derivatives. Fullerenes are a series of near-spherical discrete molecules of a closed cage structure made up of tens to hundreds of C atoms, including C_{60} [**Figure 8.47(c)**], C_{70}, C_{84}, etc. C_{60} is also called a buckyball and has icosahedral (I_h) symmetry with 60 vertices, 90 edges, 12 pentagons, and 20 hexagons. The C—C bond lengths are 139 pm for 6/6 bonds (fusion of two six-membered rings) and 144 pm for 6/5 bonds (fusion between six- and five-membered rings). All C atoms lie within a shell with a mean diameter of 700 pm. Each C atom forms three σ bonds with three neighbors, and the remaining orbitals and electrons form delocalized π bonds. The mean C—C—C angle is 116°. The π orbital lies normal to the convex surface, with an angle of 101.64° between the σ and π orbitals. Carbon nanotubes can be classified into single-walled [**Figure 8.47(d)**] and multi-walled carbon nanotubes. A single-walled carbon nanotube can be visualized as a seamless hollow cylinder rolling up from a single layer of graphene. It can also be considered as a largely elongated tubular fullerene, with a network of fused six-membered rings and both ends being closed by a half-fullerene cap. A morphous carbon is a general term that covers non-crystalline forms of carbon such as coal, coke, charcoal, carbon black

结构。在**图 8.45(a)** 所示的金刚石晶胞中，C 原子位于所有 fcc 晶格点以及一半的四面体形空隙中心。配位数为 4，晶胞中 C 原子的总数为 4+4=8，体积占有百分比仅为 34%，远低于金属晶体或离子晶体。金刚石是已知最坚硬的物质，是一种电绝缘体，具有非常高的熔点（3500℃ 以上）。用 Si 原子替换金刚石四面体形空隙中心的四个 C 原子，可以得到 SiC 结构 [**图 8.45(b)**]。

图 8.46 给出了一幅代表在本章中讨论过的四种化学键或作用力（共价键、离子键、金属键和分子间作用力）的卡通图。

拓展阅读材料　碳的单质及化合物的结构

除少量瞬间存在的气态低碳分子，如 C_1、C_2、C_3、C_4、C_5 等外，碳单质的同素异形体有金刚石、石墨、石墨烯、富勒烯（也称球碳）、碳纳米管和无定形碳等主要存在形式。金刚石的结构如**图 8.45(a)** 所示。石墨晶体 [**图 8.47(a)**] 由无限伸展的六角碳环平面层堆积而成，层间作用力为较弱的范德华力，层内为共价键，其中每个 C 原子以 sp^2 杂化与三个相邻 C 原子形成等距离的三个 σ 键，而每个 C 原子垂直于该平面、未参与杂化的 p_z 轨道互相叠加，形成离域 π 键。层间距离为 335 pm，层内相邻 C 原子间距为 141.8 pm，键长介于 C—C 单键和 C═C 双键之间。石墨烯是单层石墨 [**图 8.47(b)**] 和寡层石墨及其衍生物的通用名称。富勒烯是由几十至数百个 C 原子组成的一系列具有封闭笼状结构的近球形离散分子，包括 C_{60}[**图 8.47(c)**]、C_{70}、C_{84} 等。C_{60} 也称足球烯，具有 I_h 对称性，有 60 个顶点、90 条边、12 个五边形面和 20 个六边形面。其 6/6（即两个六元环汇合处）的 C—C 键长为 139 pm，6/5（即六元环与五元环汇合处）的 C—C 键长为 144 pm。所有 C 原子均位于一个平均直径为 700 pm 的壳层内。每个 C 原子和周围三个 C 原子形成三个 σ 键，剩余的轨道和电子则共同组成离域 π 键。C—C—C 键角的平均值为 116°，π 轨道垂直于球面，σ 和 π 轨道间夹角为 101.64°。碳纳米管包括单壁碳纳米管和多壁碳纳米管。单壁碳纳米管 [**图 8.47(d)**] 可视为由单层石墨烯卷曲而成的无缝中空管状结构，也可视为延伸很长的管状富勒烯，中间圆柱形管具有稠合六元环网格，两端各由半个富勒烯笼加帽封闭。无定形碳是碳的各种非晶态类型的总称，如煤、焦炭、木炭、炭黑（烟炱）、活性炭、碳纤维等。

从结构化学的角度来看，一种元素在其单质的同素异形体中的键合模式和键参数等可以进一步扩展到其化合物。因此，可按碳的三种同素异形体将碳的化合物(其中大多数是含有 C—C 键的有机物)方便地分为三族：来自金刚石的脂肪族化合物、来自石墨烯的芳香族化合物和来自富勒烯即球碳的球碳族化合物。脂肪族化合物包括烃及其衍生物，其分子骨架由四面体的 C 原子通过 C—C 单键组成。这些四面体的 C 原子可以排列成链状、环形或有限的三维骨架，且常在不同位置上连接官能团阵列作为取代基。在芳香族化合物中，平面的 C 原子骨架可看作石墨烯碎片，平面骨架中的 C 原子彼此通过 sp^2 杂化形成 σ 键，剩下平行的 p_z 轨道重叠形成离域 π 键。球碳的衍生物称为球碳族化合物。

(soot), activated carbon, carbon fiber, etc.

From the perspective of structural chemistry, the modes of bonding and the bond parameters of a particular element in its allotropes may be further extended to its compounds. Thus, the compounds of carbon, most of which are organic compounds containing C—C bonds, can be conveniently divided into three families that originate from three allotropies: aliphatic compounds from diamond, aromatic compounds from graphene, and fullerenic compounds from fullerenes. Aliphatic compounds comprise hydrocarbons and their derivatives in which the molecular skeletons consist of tetrahedral C atoms connected by C—C single bonds. These tetrahedral C atoms can be arranged as chains, rings, or finite 3D frame works, and often with an array of functional groups as substituents on various sites. In aromatic compounds, planar carbon skeletons can be considered as fragments of graphene, each consisting of C atoms which form σ bonds to one another via sp^2 hybridization and generate delocalized π bonds through overlap between the remaining parallel p_z orbitals. The derivatives of fullerenes are called fullerenic compounds.

Problems

8.1 Two electrostatic potential maps are shown in **Figure P8.1**, one corresponding to a molecule containing only Cl and F, the other P and F. Give the molecular formulas of these two compounds, and match the electrostatic potential maps with the formulas.

8.2 For LiBr, the dipole moment (measured in the gas phase) and the bond length (measured in the solid state) are 7.268 D and 217.0 pm, respectively. For NaCl, the corresponding values are 9.001 D and 236.1 pm.

(a) Calculate the percent of ionic character for each bond.

(b) Compare these values with the expected ionic character based on differences in electronegativity given in **Figure 8.5**.

8.3 The bond angle in the H_2O molecule is 104.5° and the resultant dipole moment is μ = 1.85 D.

(a) By an appropriate geometric calculation, determine the value of the H—O bond dipole in H_2O.

(b) Use the same method as in part (a) to estimate the bond angle in H_2S, given that the H—S bond dipole is 0.67 D and that the resultant dipole moment is μ = 0.93 D.

(c) Given the bond dipoles 1.87 D for the C—Cl bond and 0.30 D for the C—H bond, together with μ = 1.04 D, estimate the H—C—Cl bond angle in $CHCl_3$.

8.4 A group of spectroscopists believe that they have detected one of the following species: NeF, NeF^+, or NeF^-. Assume that the unmodified energy-level diagrams (**Figure 8.18**) apply, and describe bonding in these species. Which of these species would you expect the spectroscopists to have observed? Why?

8.5 Methyl nitrate, CH_3NO_3, is used as a rocket propellant. The skeletal structure of the molecule is CH_3ONO_2. The N and three O atoms all lie in the same plane, but the CH_3 group is not in the same plane as the group. The bond angle C—O—N is 105°, and the bond angle O—N—O is 125°. One nitrogen-to-oxygen bond length is 136 pm, and the other two are 126 pm.

(a) Draw a sketch of the molecule showing its geometric structue.

(b) Label all the bonds in the molecule as σ or π, and indicate the probable orbital overlaps involved.

(c) Explain why all three nitrogen-to-oxygen bond lengths are not the same.

8.6 Think of the reaction shown below as involving the transfer of a F^- ion from ClF_3 to AsF_5 to form the ions ClF_2^+ and AsF_6^-. As a result, the hybridization scheme of each central atom must change. For each reactant molecule and product ion, indicate (a) its geometric structure and (b) the hybridization scheme for its central atom.

$$ClF_3 + AsF_5 \rightarrow \left(ClF_2^+\right)\left(AsF_6^-\right)$$

8.7 The anion I_4^{2-} is linear, and the anion I_5^- is V-shaped, with a 95° angle between the two arms of the V. For the central atoms in these ions, propose hybridization schemes that are consistent with the above observations.

习题

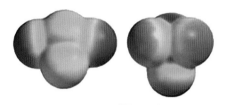

Figure P8.1(图 P8.1)

8.1 **图 P8.1** 给出了两个静电势图,一个对应于仅含 Cl 和 F 的分子,另一个对应于仅含 P 和 F 的分子。给出这两个化合物的分子式,并将静电势图与分子式相匹配。

8.2 LiBr 的偶极矩(在气相中测量)和键长(在固相中测量)分别为 7.268 D 和 217.0 pm。NaCl 的相应值为 9.001 D 和 236.1 pm。
(a) 计算每个键的离子性百分比。
(b) 将这些值与**图 8.5** 中基于电负性差异所预期的离子性进行比较。

8.3 H_2O 分子的键角为 104.5°,合偶极矩 $\mu=1.85$ D。
(a) 通过适当的几何计算,确定 H_2O 中 H—O 键的键矩。
(b) 使用与 (a) 中相同的方法估算 H_2S 的键角,已知 H—S 键的键矩为 0.67 D,合偶极矩 $\mu=0.93$ D。
(c) 已知 C—Cl 键的键矩为 1.87 D,C—H 键的键矩为 0.30 D,且有 $\mu=1.04$ D,估算 $CHCl_3$ 的 H—C—Cl 键角。

8.4 某些光谱学家认为,他们检测到了以下物种之一:NeF、NeF^+ 或 NeF^-。假定未修正的能级图(**图 8.18**)适用,且可描述这些物种的成键情况。你认为光谱学家观察到的物种是哪一种?为什么?

8.5 硝酸甲酯 CH_3NO_3 可用作火箭推进剂,其分子的骨架结构为 CH_3ONO_2。N 和三个 O 原子共面,但与 CH_3 基团不共面。已知 C—O—N 键角为 105°,O—N—O 键角为 125°。一个氮氧键键长为 136 pm,另两个为 126 pm。
(a) 绘制一张显示其几何构型的分子草图。
(b) 用 σ 或 π 标记分子中的所有键,并指出可能涉及的轨道重叠。
(c) 解释为什么三个氮氧键键长不同。

8.6 将下述反应视为从 ClF_3 向 AsF_5 转移 F^- 离子,以形成 ClF_2^+ 和 AsF_6^- 离子。因此,每个中心原子的杂化方案必须改变。指出所有反应物分子和产物离子的 (a) 几何构型;(b) 中心原子的杂化方案。

8.7 I_4^{2-} 阴离子结构为线性,I_5^- 阴离子为 V 形,其中 V 形的两臂间角为 95°。对于这些离子的中心原子,提出与上述观察结果一致的杂化方案。

8.8 共轭烃具有交替的单、双键。绘制 1,3,5-己三烯的 π 体系的分子轨道。如果将电子从其 HOMO 激发到 LUMO,所需的能量对应于 256 nm 波长,你预期 1,3,5,7-辛四烯中相应激发的波长会更长还是更短?为什么?

*8.9 sp 杂化轨道是 s 轨道与 p 轨道的代数组合。$2s$ 与 $2p_z$ 轨道的所需组合为

$$\begin{cases} \psi_1(sp) = \dfrac{1}{\sqrt{2}}\left[\psi(2s)+\psi(2p_z)\right] \\ \psi_2(sp) = \dfrac{1}{\sqrt{2}}\left[\psi(2s)-\psi(2p_z)\right] \end{cases}$$

通过**表 7.9** 中适当函数的组合,构建 H 原子的上述函数在 xz 平面上的极坐标图。在极坐标图中,r/a_0 设为定值,这里你可以将其设为 $2p$ 轨道径向概率分布的相应峰值处。

8.8 A conjugated hydrocarbon has an alternation of double and single bonds. Draw the molecular orbitals of the π system of 1,3,5-hexatriene. If the energy required to excite an electron from its HOMO to LUMO corresponds to a wavelength of 256 nm, do you expect the wavelength for the corresponding excitation in 1,3,5,7-octatetraene to be a longer or shorter wavelength? Why?

*8.9 The *sp* hybrid orbitals are algebraic combinations of the *s* and *p* orbitals. The required combinations of $2s$ and $2p_z$ orbitals are

$$\begin{cases} \psi_1(sp) = \dfrac{1}{\sqrt{2}}[\psi(2s) + \psi(2p_z)] \\ \psi_2(sp) = \dfrac{1}{\sqrt{2}}[\psi(2s) - \psi(2p_z)] \end{cases}$$

By combining the appropriate functions given in **Table 7.9**, construct a polar plot for each of the above functions in the xz plane for H atom. In a polar plot, the value of r/a_0 is set at a fixed value. Here, you may set it at the corresponding peak position of the radial probability distribution of the $2p$ orbital.

8.10 In some barbecue grills the electric lighter consists of a small hammer-like device striking a small crystal, which generates voltage and causes a spark between wires that are attached to opposite surfaces of the crystal. The phenomenon of causing an electric potential through mechanical stress is known as the piezoelectric effect. One type of crystal that exhibits the piezoelectric effect is lead zirconate titanate. In this perovskite crystal structure, a titanium (IV) ion sits in the middle of a tetragonal unit cell with dimensions of 0.403 nm × 0.398 nm × 0.398 nm. At each corner is a lead (II) ion, and at the center of each face is an oxygen anion. Some of the Ti(IV) are replaced by Zr(IV). This substitution, along with Pb(II), results in the piezoelectric behavior.

(a) How many oxygen ions are in the unit cell?

(b) How many lead(II) ions are in the unit cell?

(c) How many titanium(IV) ions are in the unit cell?

(d) What is the density of the unit cell?

*8.11 In an ionic crystal lattice each cation will be attracted by anions next to it and repulsed by cations near it. Consequently the Coulomb potential leading to the lattice energy depends on the type of crystal. To get the total lattice energy you must sum all of the electrostatic interactions on a given ion. The general form of the electrostatic potential is

$$V = -\dfrac{Q_1 Q_2 e^2}{d_{12}}$$

where Q_1 and Q_2 are the charges on ions 1 and 2, d_{12} is the distance between them in the crystal lattice, and e is the charge on the electron.

(a) Consider a uniformly spaced infinite 1D linear "crystal" shown in **Figure P8.2**. The distance between the centers of adjacent spheres is d. Assume that the blue spheres are cations and that the red spheres are anions, all with a charge of Q. Prove that the total electrostatic potential for this 1D "crystal" is

$$V = -\dfrac{2\ln(2Q^2 e^2)}{d}$$

(b) In general, the electrostatic potential in a crystal can be written as

$$V = -k_M \dfrac{Q^2 e^2}{d}$$

where k_M is a geometric constant, called the Madelung constant, for a particular crystal system under consideration. For the NaCl crystal structure, let d be the distance between the centers of sodium and chloride ions. Show that by considering four layers of nearest neighbors to a central chloride ion, k_M is given by

$$k_M = 6 - \dfrac{12}{\sqrt{2}} + \dfrac{8}{\sqrt{3}} - \dfrac{6}{\sqrt{4}} + \cdots$$

(c) What is the k_M for CsCl? Give the first four terms.

8.10 某些烧烤炉的电打火机由一个小锤状装置组成，该装置通过敲击小晶体产生电压，并在连接晶体相对表面的电线之间产生火花。这种通过机械应力产生电势的现象，称为压电效应。锆钛酸铅是一种表现出压电效应的晶体。在这种钙钛矿晶体结构中，钛（Ⅳ）离子位于尺寸为 0.403 nm×0.398 nm×0.398 nm 的四方晶胞中心。每个角落都有一个铅（Ⅱ）离子，每个面的中心都有一个氧负离子。一些 Ti（Ⅳ）被 Zr（Ⅳ）取代，这种取代与 Pb（Ⅱ）一起导致了压电行为。

(a) 晶胞中有多少氧离子？

(b) 晶胞中有多少铅（Ⅱ）离子？

(c) 晶胞中有多少钛（Ⅳ）离子？

(d) 晶胞的密度是多少？

*8.11 在离子晶格中，每个阳离子都会被附近的阴离子吸引，并被附近的阳离子排斥。因此，导致晶格能的库仑势取决于晶体的类型。为获得总晶格能，必须对给定离子上的所有静电相互作用求和。静电势的一般形式为

$$V = -\frac{Q_1 Q_2 e^2}{d_{12}}$$

其中 Q_1 和 Q_2 是离子 1 和 2 的电荷，d_{12} 是它们在晶格中的距离，e 是电子的电荷。

(a) 考虑如**图 P8.2** 所示均匀间隔的无限一维线性"晶体"。相邻球体中心之间的距离为 d。假定蓝色球体为阳离子，红色球体为阴离子，均带电荷 Q。证明该一维"晶体"的总静电势为

$$V = -\frac{2\ln(2Q^2 e^2)}{d}$$

(b) 晶体中的静电势通常可写为

$$V = -k_M \frac{Q^2 e^2}{d}$$

其中 k_M 是所考虑的特定晶体体系的几何常数，称为马德隆常数。对于 NaCl 晶体结构，令 d 为钠离子和氯离子中心之间的距离。考虑中心氯离子最近邻的四层，证明 k_M 可由下式给出：

$$k_M = 6 - \frac{12}{\sqrt{2}} + \frac{8}{\sqrt{3}} - \frac{6}{\sqrt{4}} + \cdots$$

(c) CsCl 的 k_M 是多少？给出公式前四项。

Figure P8.2（图 P8.2）

Chapter 9 Complex-Ion Equilibria and Coordination Compounds

In this chapter, we will introduce the last of four general kinds of chemical equilibria: complex-ion equilibria. We will focus on coordination compounds, which are the compounds produced in the complex-ion equilibria. The basic concepts, structures, bonding, properties, and applications of coordination compounds will be discussed.

9.1 Basic Concepts in Coordination Compound
9.2 Isomerism
9.3 Crystal Field Theory
9.4 Properties of Coordination Compounds
9.5 Complex-Ion Equilibria
9.6 Applications of Coordination Chemistry

第 9 章 配位解离平衡与配合物

本章我们将介绍四大化学平衡的最后一类：配位解离平衡。我们将聚焦于配位化合物（简称配合物，又称络合物），即配位解离平衡中生成的化合物。我们将讨论配合物的基本概念、结构、成键、性质及其应用。

9.1 配合物的基本概念

9.2 异构现象

9.3 晶体场理论

9.4 配合物的性质

9.5 配位解离平衡

9.6 配位化学的应用

9.1 Basic Concepts in Coordination Compound

In this section, we will learn some basic concepts related to coordination compound, such as complex, donor atom, coordination number, and ligand. We will also briefly introduce the nomenclature for naming and writing the formula of coordination compounds.

Discovery of Coordination Compounds

The first known coordination compound, Prussian blue, was accidentally discovered in the early 18th century. However, it was not until nearly a century later that the unique structure of the coordination compounds came to be understood.

In the late 18th century, it was discovered that compounds of different colors could be produced when different amounts of $NH_3(aq)$ are mixed with the same amount of $CoCl_3(aq)$. For example, when the amount ratio of $CoCl_3$ and NH_3 was 1:6, a yellow compound (formula $CoCl_3 \cdot 6NH_3$) was formed, while a purple compound (formula $CoCl_3 \cdot 5NH_3$) was produced if the ratio was 1:5, as

$$CoCl_3 + 6NH_3 \rightarrow CoCl_3 \cdot 6NH_3$$
$$CoCl_3 + 5NH_3 \rightarrow CoCl_3 \cdot 5NH_3$$

Meanwhile, when treated with excessive $AgNO_3(aq)$, 1 mol $CoCl_3 \cdot 6NH_3$ formed 3 mol $AgCl(s)$ whereas 1 mol $CoCl_3 \cdot 5NH_3$ formed only 2 mol $AgCl(s)$. This means that 3 and 2 mol Cl^- were ionized for 1 mol $CoCl_3 \cdot 6NH_3$ and $CoCl_3 \cdot 5NH_3$, respectively, as

$$CoCl_3 \cdot 6NH_3 + 3AgNO_3 \rightarrow ? + 3AgCl \downarrow$$
$$CoCl_3 \cdot 5NH_3 + 2AgNO_3 \rightarrow ? + 2AgCl \downarrow$$

These experimental observations troubled scientists at that time. The main questions they could not understand were:

1) Both $CoCl_3$ and NH_3 are stable and capable of independent existence. Why do they further combine to form other stable compounds?
2) The status of three Cl atoms in $CoCl_3 \cdot 5NH_3$ is clearly different. What structure in the compound can account for this difference?

Werner's Theory

The above questions were answered by **Werner's theory** that forms the fundamental of coordination chemistry. This theory was named after the Swiss chemist Alfred Werner (1866—1919), who proposed it in 1893 and won the Nobel Prize in chemistry in 1913 for it. In this theory, Werner proposed that certain metal atoms have two types of valence or bonding capacity. The primary valence is the number of electrons the metal atom loses in forming a central ion, called the **coordination center**. The secondary valence is the number of bonds that coordinate a surrounding array of molecules or ions, called **ligands**, to the central ion.

For example, $CoCl_3 \cdot 6NH_3$ should be written as $[Co(NH_3)_6]Cl_3$. The Co atom loses three electrons to form the coordination center Co(III), so the primary valence is three. Co(III) forms six bonds with the surrounding NH_3 ligands, so the secondary valence is six. In $[Co(NH_3)_6]Cl_3$, the central Co(III) ion is surrounded by six NH_3 ligands, and the three Cl^- ions can be ionized freely. Similar, $CoCl_3 \cdot 5NH_3$ should be written as $[CoCl(NH_3)_5]Cl_2$, and the primary and secondary valences of Co are also three and six, respectively. In $[CoCl(NH_3)_5]Cl_2$, the central Co(III) ion is surrounded by six ligands (five NH_3 molecules and one Cl^- ion), and the other two Cl^- can be ionized freely. Thus, the previous equations can be rewritten as

$$CoCl_3 + 6NH_3 \rightarrow \left[Co(NH_3)_6\right]Cl_3 \tag{9.1}$$

$$CoCl_3 + 5NH_3 \rightarrow \left[CoCl(NH_3)_5\right]Cl_2 \tag{9.2}$$

9.1 配合物的基本概念

本节我们将学习有关配合物的一些基本概念，如配位单元、配位原子、配位数和配体等。我们还将简要介绍配合物的命名法及其化学式的书写法。

配合物的发现

第一例已知的配合物普鲁士蓝在 18 世纪初被意外发现，然而直到近一个世纪后，配合物的独特结构才被人们所理解。

18 世纪后期人们发现，当不同量的 $NH_3(aq)$ 与等量 $CoCl_3(aq)$ 混合时，可以生成不同颜色的化合物。例如，当 $CoCl_3$ 与 NH_3 的物质的量之比为 1:6 时，会生成黄色化合物（化学式为 $CoCl_3 \cdot 6NH_3$）；而当比例为 1:5 时，则会生成紫色化合物（化学式为 $CoCl_3 \cdot 5NH_3$），即

$$CoCl_3 + 6NH_3 \rightarrow CoCl_3 \cdot 6NH_3$$
$$CoCl_3 + 5NH_3 \rightarrow CoCl_3 \cdot 5NH_3$$

同时，当加入过量 $AgNO_3(aq)$ 时，1 mol $CoCl_3 \cdot 6NH_3$ 可生成 3 mol $AgCl(s)$，而 1 mol $CoCl_3 \cdot 5NH_3$ 仅生成 2 mol $AgCl(s)$。这意味着 1 mol $CoCl_3 \cdot 6NH_3$ 和 $CoCl_3 \cdot 5NH_3$ 可分别电离出 3 mol 和 2 mol Cl^-，即

$$CoCl_3 \cdot 6NH_3 + 3AgNO_3 \rightarrow ? + 3AgCl\downarrow$$
$$CoCl_3 \cdot 5NH_3 + 2AgNO_3 \rightarrow ? + 2AgCl\downarrow$$

这些实验现象令当时的科学家们感到困惑。他们无法理解的主要问题有：

1) $CoCl_3$ 和 NH_3 均能单独稳定存在，为什么两者还能进一步结合形成其他稳定的化合物？
2) $CoCl_3 \cdot 5NH_3$ 中三个 Cl 原子的状态明显不同。怎样的化合物结构才能解释这种差异？

维尔纳理论

上述问题可用**维尔纳理论**来解答，该理论是配位化学的基础，命名自瑞士化学家阿尔弗雷德·维尔纳（1866—1919）。他于 1893 年提出该理论，并因此获得 1913 年诺贝尔化学奖。维尔纳在其理论中提出，某些金属原子具有两重价态或成键能力。第一价态是金属原子形成中心离子（称为**配位中心**）所失去的电子数。第二价态是周围一系列分子或离子（称为**配体**）与中心离子配位所形成的键数。

例如，$CoCl_3 \cdot 6NH_3$ 应写为 $[Co(NH_3)_6]Cl_3$。Co 原子失去三个电子形成配位中心 Co(Ⅲ)，因此第一价态为 3。Co(Ⅲ) 与周围的 NH_3 配体形成六个键，因此第二价态为 6。在 $[Co(NH_3)_6]Cl_3$ 中，中心的 Co(Ⅲ) 离子被六个 NH_3 配体包围，三个 Cl^- 离子均可自由电离。类似地，$CoCl_3 \cdot 5NH_3$ 应写为 $[CoCl(NH_3)_5]Cl_2$，Co 的第一和第二价态也分别为 3 和 6。在 $[CoCl(NH_3)_5]Cl_2$ 中，中心的 Co(Ⅲ) 离子被六个配体（五个 NH_3 分子和一个 Cl^- 离子）包围，另两个 Cl^- 离子可自由电离。因此，前述方程式可改写为

$$CoCl_3 + 6NH_3 \rightarrow \left[Co(NH_3)_6\right]Cl_3 \tag{9.1}$$

$$[Co(NH_3)_6]Cl_3 + 3AgNO_3 \rightarrow [Co(NH_3)_6](NO_3)_3 + 3AgCl\downarrow \qquad (9.3)$$

$$[CoCl(NH_3)_5]Cl_2 + 2AgNO_3 \rightarrow [CoCl(NH_3)_5](NO_3)_2 + 2AgCl\downarrow \qquad (9.4)$$

Werner confirmed his proposed structure of coordination compounds by measuring the electrical conductivity of these compounds in aqueous solutions. Experiments showed that the electrical conductivity of $[CoCl(NH_3)_5]Cl_2$ (producing three ions per formula unit) is poorer than that of $[Co(NH_3)_6]Cl_3$ (producing four ions per formula unit). Experiments also revealed that $[CoCl_2(NH_3)_4]Cl$ is an even poorer conductor, and that $[CoCl_3(NH_3)_3]$ is a nonelectrolyte.

Complex and Coordination Compound

A **complex** is any species involving the coordination of ligands to a coordination center, where ligands act as Lewis bases, the coordination center acts as a Lewis acid, and coordinated covalent bonds (also called **coordination bonds**) are formed by ligands donating electron pairs to the coordination center. The complex can be either a cation, an anion, or a neutral molecule, set off by square brackets [ ] in the chemical formula. In aqueous solutions, ionization only happens outside the square brackets. **Coordination compounds** are compounds containing complex ions or neutral complexes. $[CoCl(NH_3)_5]^{2+}$, $[CoCl_5(NH_3)]^{2-}$, $[CoCl_3(NH_3)_3]$, and $[Co(NH_3)_6]Cl_3$ are examples of a complex cation, a complex anion, a neutral complex, and a coordination compound, respectively.

The atom in a ligand that donates an electron pair to form a coordinated covalent bond with the coordination center is called the **donor atom**. Some ligands may have more than one donor atom. The number of donor atoms around the coordination center is called the **coordination number** of a complex. Coordination numbers generally range from 2 to 12, and 6 is by far the most common number, followed by 4. A coordination center with high charge and large diameter tend to form complexes with high coordination number. For example, coordination numbers greater than 6 are not often observed in members of the first transition series but are more common in those of the second and third series. The coordination number 2 is limited mostly to complexes of Cu(I), Ag(I), and Au(I). Ligands with high charge and large diameter tends to form complexes with low coordination number. The common coordination numbers of some metal ions are listed in **Table 9.1**. The four most commonly observed geometric shapes of complexes are linear, tetrahedral, square planar, and octahedral, corresponding to a coordination number of 2, 4, 4, and 6, respectively.

Ligand

Virtually all anions and molecules with lone-pair electrons that act as Lewis bases can serve as ligands. For example, halide and pseudohalide ions are important anionic ligands whereas ammonia, carbon monoxide, and water are common charge-neutral ligands. Some organic species such as pyridine and methylamine are also common ligands.

According to the number of times bonded to the coordination center through noncontiguous donor atoms, ligands can be classified into monodentate and polydentate ligands. A ligand that bonds to the coordination center through only one donor atom is called a **monodentate** ligand. Some common monodentate ligands are listed in **Table 9.2**. Some ligands may have more than one potential donor atom, but because these atoms are contiguous, only one of them can bond to the coordination center. These ligands are termed **ambidentate ligands**. For example, both N and O atoms in the ligand NO_2^- have lone pair electrons and can serve as the donor atoms. However, these two atoms cannot simultaneously coordinate to the coordination center, otherwise a three-membered ring with extremely high tension will be formed. Thus, NO_2^- is a monodentate and ambidentate ligand. The donor atom in such ligands should be specified. We name the ligand as nitrito-N- and write it as NO_2^- if the donor atom is N, whereas nitrito-O- and ONO^- if the donor atom is O.

A ligand with more than one noncontiguous donor atom that can simultaneously attach to the coordination center is called a **polydentate** ligand. For example, the two N atoms in the molecule ethylenediamine (en) can donate two pairs of electrons and attach simultaneously to a coordination center at two points. Thus, en is a **bidentate** ligand. A five-membered ring is formed when en is bound to a

$$CoCl_3 + 5NH_3 \rightarrow \left[CoCl(NH_3)_5\right]Cl_2 \quad (9.2)$$

$$\left[Co(NH_3)_6\right]Cl_3 + 3AgNO_3 \rightarrow \left[Co(NH_3)_6\right](NO_3)_3 + 3AgCl\downarrow \quad (9.3)$$

$$\left[CoCl(NH_3)_5\right]Cl_2 + 2AgNO_3 \rightarrow \left[CoCl(NH_3)_5\right](NO_3)_2 + 2AgCl\downarrow \quad (9.4)$$

维尔纳通过测量这些配合物在水溶液中的电导率，证实了他所提出的配合物结构。实验表明，$[CoCl(NH_3)_5]Cl_2$（每化学式可生成三个离子）的电导率低于 $[Co(NH_3)_6]Cl_3$（每化学式可生成四个离子）。实验还表明，$[CoCl_2(NH_3)_4]Cl$ 是一种导电性更差的导体，而 $[CoCl_3(NH_3)_3]$ 为非电解质。

配位单元与配合物

配位单元是包含配体与配位中心相配位的任何物种，其中配体充当路易斯碱，配位中心充当路易斯酸，由配体向配位中心提供电子对以形成配位共价键（简称**配位键**）。配位单元可以是阳离子、阴离子或中性分子，在化学式中用方括号 [ ] 隔开。水溶液中的电离只发生在方括号外。**配合物**是含有配位单元离子（称为配离子）或中性配位单元（称为配分子）的化合物。$[CoCl(NH_3)_5]^{2+}$、$[CoCl_5(NH_3)]^{2-}$、$[CoCl_3(NH_3)_3]$ 和 $[Co(NH_3)_6]Cl_3$ 分别是配阳离子、配阴离子、配分子和配合物的示例。

配体中提供电子对与配位中心形成配位键的原子，称为**配位原子**。一些配体可能具有多个配位原子。配位中心周围的配位原子数，称为配位单元的**配位数**。配位数一般在 2~12 之间，6 是最常见的配位数，其次是 4。具有高电荷和大直径的配位中心，倾向于形成高配位数的配位单元。例如，大于 6 的配位数在第一过渡系元素中不常见，但在第二和第三过渡系元素中更为常见。配位数 2 主要限于 Cu（Ⅰ）、Ag（Ⅰ）和 Au（Ⅰ）。具有高电荷和大直径的配体，倾向于形成低配位数的配位单元。**表 9.1** 列出了一些金属离子的常见配位数。四种最为常见的配位单元的几何构型是直线形、四面体形、平面正方形和八面体形，分别对应于配位数 2、4、4 和 6。

配体

事实上，所有具有可用作路易斯碱的孤电子对的阴离子和分子均可作为配体。例如，卤离子和类卤离子是重要的阴离子配体，而氨、一氧化碳和水是常见的中性配体。一些有机物种（如吡啶和甲胺）也是常见的配体。

根据通过非邻接配位原子与配位中心结合的次数，配体可分为单齿配体和多齿配体。仅通过一个配位原子与配位中心结合的配体称为**单齿配体**，**表 9.2** 列出了一些常见的单齿配体。一些配体可能存在一个以上潜在的配位原子，但由于这些原子相互邻接，因此只有其中一个可以与配位中心结合，这类配体称为**两可配体**。例如，配体 NO_2^- 中的 N 原子和 O 原子均具有孤电子对，都可用作配位原子。然而这两个原子不能同时与配位中心配位，否则将形成具有极高张力的三元环。因此 NO_2^- 是单齿配体和两可配体。这类配体需要指明其中的配位原子：如果配位原子为 N，此配体命名为 nitrito-*N*- 并写作 NO_2^-；如果配位原子为 O，则命名为 nitrito-*O*- 并写作 ONO^-。

具有一个以上、可同时连接到配位中心的非邻接配位原子的配体，称为**多齿配体**。例如，乙二胺（en）分子中的两个 N 原子可提供两对

coordination center (**Figure 9.1**). The complex in which a ring is produced by bonding a polydentate ligand to a coordination center is called a **chelate**. The most stable and common rings formed in chelates are five- or six-membered rings. Some common polydentate ligands are listed in **Table 9.3**.

Nomenclature in Coordination Compounds

The nomenclature in naming complexes originated with Werner, but it has been modified several times over the years. There is a great deal of detail in the nomenclature. Here, we only introduce some general rules, as:

1) Anions as ligands are named by using the ending *-o*, with the *-ide* endings normally changing to *-o*, *-ite* to *-ito*, and *-ate* to *-ato*. Neutral molecules as ligands generally carry an unmodified name. Aqua, ammine, carbonyl, and nitrosyl are important exceptions (**Table 9.2**).

2) The number of ligands of a given type is denoted by a prefix. The usual prefixes are *mono* = 1, *di* = 2, *tri* = 3, *tetra* = 4, *penta* =5, and *hexa* = 6. The prefix *mono-* is often omitted. If the ligand name is a composite name that itself contains a numerical prefix, such as ethylene*di*amine, place parentheses around the name and precede it with *bis* = 2, *tris* = 3, *tetrakis* = 4, and so on. Thus, *tetra*aqua signifies four H_2O molecules as ligands, and *tris*(ethylenediamine) signifies three en ligands.

3) To name a complex, ligands are named first, in alphabetical order, followed by the name of the coordination center, with the oxidation state denoted by a Roman numeral in a parenthesis. For a complex anion, *-ate* is attached to the name of the metal. Prefixes (*di*, *tri*, *bis*, *tris*,...) are ignored in establishing the alphabetical order. For some specific metals, the Latin name (**Table 9.4**) is used instead of the English name. For example,

$[CoCl_2(NH_3)_4]^+$ tetraamminedichlorocobalt(III) ion
$[CrCl_2(en)_2]^+$ dichlorobis(ethylenediamine)chromium(III) ion
$[CuCl_4]^{2-}$ tetrachlorocuprate(II) ion

4) To write the formula of a complex, the chemical symbol of the coordination center is written first, followed by the formulas of an ion ligands and then neutral molecule ligands. If there are two or more ligands of the same type, write them in alphabetical order according to the first chemical symbols of their formulas. For example, the formula of tetraamminechloronitrito-*N*-platinum(IV) ion is $[PtCl(NO_2)(NH_3)_4]^{2+}$.

5) In the names and formulas of coordination compounds, cations come first followed by anions. For example, the formula $[Pt(NH_3)_4][PtCl_4]$ represents the coordination compound tetraammineplatinum(II) tetrachloroplatinate(II).

Although most complexes are named in the manner shown above, some common, or trivial, names are still in use. Two such trivial names are ferrocyanide for $[Fe(CN)_6]^{4-}$ and ferricyanide for $[Fe(CN)_6]^{3-}$. Their systematic names are hexacyanoferrate(II) and hexacyanoferrate(III), respectively.

9.2 Isomerism

One of the distinct properties of coordination compounds is that many of them have different isomers. It was also Werner who made the key breakthrough in reconciling the isomers of coordination compounds.

Isomerism and Isomers

Isomers are compounds that have the same formula but differ in their structures and properties. The phenomenon of the existence of isomers is called **isomerism**. The isomerism of coordination compounds can be classified into the following two broad categories:

1) **Structural isomerism**: the isomerism arising from difference in basic structure or bond type.
2) **Stereoisomerism**: the isomerism in which isomers have the same basic structure and bond type, but differ in the way how ligands occupy the space around the coordination center. This category generally

电子，并通过两个位点同时与配位中心相连，因此 en 是**双齿配体**。当 en 与配位中心结合时，可形成一个五元环（**图 9.1**）。多齿配体与配位中心结合成环的配合物，称为**螯合物**。螯合物中形成的最稳定且最常见的环是五元环或六元环。**表 9.3** 列出了一些常见的多齿配体。

配合物的命名法

配合物的命名法源自维尔纳，但多年来已历经多次修改。命名法中存在大量细节，这里我们只介绍如下一般规则：

1) 阴离子配体的命名以 -o 结尾，通常将 -ide 词尾改为 -o，-ite 改为 -ito，-ate 改为 -ato。中性分子配体通常采用未修改的名称，水、氨、羰基和亚硝基是重要的例外（见**表 9.2**）。
2) 给定类型配体的数量用前缀表示，常见前缀有 mono = 1、di = 2、tri = 3、tetra = 4、penta = 5 以及 hexa = 6，前缀 mono- 经常省略。如果配体名称是一个自身包含数字前缀的复合名称，如乙二胺，则在名称左右加上括号，并在前面加上 bis=2、tris=3、tetrakis=4 等。因此，tetraaqua 表示四个 H_2O 分子作为配体，tris(ethylenediamine) 表示三个乙二胺配体。
3) 命名配位单元时，首先按字母顺序命名配体，然后是配位中心的名称，其氧化态用括号中的罗马数字表示。对于配阴离子，在金属名称上附加 -ate。当按字母顺序排列时，前缀（di、tri、bis、tris…）均忽略。对于某些特定金属，使用其拉丁名称（**表 9.4**）代替英文名称。例如：

　　$[CoCl_2(NH_3)_4]^+$　　　　二氯·四氨合钴（Ⅲ）离子
　　$[CrCl_2(en)_2]^+$　　　　二氯·二（乙二胺）合铬（Ⅲ）离子
　　$[CuCl_4]^{2-}$　　　　　　四氯合铜（Ⅱ）离子

4) 书写配位单元的化学式时，首先写配位中心的化学符号，接着是阴离子配体的化学式，然后是中性分子配体的化学式。如果存在两个及以上同类配体，根据其化学式里首个化学符号的字母顺序来书写。如氯·硝基·四氨合铂（Ⅳ）离子的化学式为 $[PtCl(NO_2)(NH_3)_4]^{2+}$。
5) 在配合物的命名以及化学式的书写中，先阳离子后阴离子。例如，化学式 $[Pt(NH_3)_4][PtCl_4]$ 表示配合物四氯合铂（Ⅱ）酸四氨合铂（Ⅱ）。

尽管绝大多数配位单元均按上述方式命名，但一些常用名或俗名仍在使用。两个常见俗名是：$[Fe(CN)_6]^{4-}$ 称为亚铁氰酸根，$[Fe(CN)_6]^{3-}$ 称为铁氰酸根，其系统命名法的名称分别为六氰合铁（Ⅱ）酸根和六氰合铁（Ⅲ）酸根。

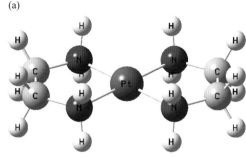

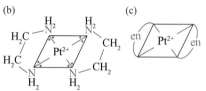

Figure 9.1 Three representations of the chelate $[Pt(en)_2]^{2+}$.

图 9.1 螯合物 $[Pt(en)_2]^{2+}$ 的三种表示法。

注： 正文为对应英文的中文翻译，说明的是配合物的英文命名法。以下为配合物的中文命名法规则简介：

1) 命名配合物时，先阴离子后阳离子。阴离子为简单离子时，在阴阳离子之间加"化"字；阴离子是含氧酸根或配阴离子时，在阴阳离子之间加"酸"字。
2) 命名配位单元时，先配体后配位中心。配体与配位中心之间加"合"字。配体前面用二、三、四……表示该配体的个数，没有数字则表示该配体为一个。几种不同的配体之间用"·"隔开。配位中心后加圆括号，内写罗马数字表示配位中心的氧化态。

9.2 异构现象

许多配合物都具有不同的异构体，这是配合物的独特性质之一。维尔纳在解释配合物的异构体方面也取得了重大突破。

异构现象与异构体

异构体是具有相同化学式但结构和性质不同的化合物，存在异构体的现象称为**异构现象**。配合物的异构现象可分为以下两大类：

Table 9.1 Some Common Coordination Numbers of Metal Ions (Roman Number: Oxidation State; Arabic Number: Coordination Number)
表 9.1 一些金属离子的常见配位数（罗马数字：氧化态；阿拉伯数字：配位数）

Highest Coord No. (最高配位数)	No. of Period (周期数)																	
2	1	H I 2																
4	2	Li I 4	Be II 4										B III 4	C IV 4	N	O	F
	3	Na I 6	Mg II 4,6										Al III 4,6	Si IV 6	P V 6	S	Cl
6	4	K I 6,8	Ca II 6,8	Sc III 6	Ti IV 6	V III 6 IV 5,6	Cr II 6 III 6	Mn I 6 II 4,6 III 6	Fe II 4,6 III 4,6	Co II 4,6 III 6	Ni 0 4 II 6	Cu I 2,3 II 4,6	Zn II 4,6	Ga III 4,6	Ge IV 6	As III 4 V 6	Se	Br
	5	Rb I 8	Sr II 6,8	Y III 6	Zr IV 6,8	Nb V 6~8	Mo III 6 IV 6,8 V 8	Tc	Ru II 6 III 6	Rh III 6	Pd II 4,6 IV 6	Ag I 2,3 II 4	Cd II 4,6	In III 4,6	Sn II 4 IV 6	Sb III 6 V 6	Te	I
8	6	Cs I 8	Ba II 6,8	La	Hf IV 6,8	Ta V 6~8	W V 6,8	Re IV 4,6	Os III 6	Ir III 6	Pt II 4,6 IV 6	Au I 2,3 III 4	Hg II 4,6	Tl I 2,4	Pb II 4 IV 6	Bi III 4-6 V 6	Po	At

Table 9.2　Some Common Monodentate Ligands
表 9.2　一些常见单齿配体

Neutral Molecule as Ligand（中性分子配体）		Anion as Ligand（阴离子配体）			
Formula（化学式）	Name（名称）	Formula（化学式）	Name（名称）	Formula（化学式）	Name（名称）
H_2O	Aqua（水）	F^-	Fluoro（氟）	SO_4^{2-}	Sulfato（硫酸根）
NH_3	Ammine（氨）	Cl^-	Chloro（氯）	$S_2O_3^{2-}$	Thiosulfato（硫代硫酸根）
CO	Carbonyl（羰基）	Br^-	Bromo（溴）	NO_2^-	Nitrito-N-[a]（硝基）
NO	Nitrosyl（亚硝酰基）	I^-	Iodo（碘）	ONO^-	Nitrito-O-[a]（亚硝酸根）
CH_3NH_2	Methylamine（甲胺）	OH^-	Hydroxo（羟基）	SCN^-	Thiocyanato-S-[b]（硫氰酸根）
C_5H_5N	Pyridine（py, 吡啶）	CN^-	Cyano（氰）	NCS^-	Thiocyanato-N-[b]（异硫氰酸根）

[a] Nitrito-N- signifies that the donor atom is N in NO_2^- whereas Nitrito-O- means that the donor atom is O in ONO^-. （Nitrito-N- 表示 NO_2^- 中的配位原子为 N，而 Nitrito-O- 则表示 ONO^- 中的配位原子为 O。）
[b] Thiocyanato-S- means that the donor atom is S in SCN^- whereas Thiocyanato-N- signifies that the donor atom is N in NCS^-. （Thiocyanato-S- 表示 SCN^- 中的配位原子为 S，而 Thiocyanato-N- 则表示 NCS^- 中的配位原子为 N。）

Table 9.3　Some Common Polydentate Ligands
表 9.3　一些常见多齿配体

Abbreviation（简写）	Name（名称）	No. of Coordination Bonds（配位键数）	Formula（化学式）
en	Ethylenediamine（乙二胺）	2	$H_2N-CH_2-CH_2-NH_2$
phen	o-phenanthroline（邻菲咯啉）	2	
ox^{2-} [a]	Oxalate（草酸根）	2	
$EDTA^{4-}$ [b]	Ethylenediaminetetraacetato（乙二胺四乙酸根）	6	

[a] Oxalic acid is a diprotic acid denoted H_2ox. It is the ox anion that binds as a bidentate ligand. （草酸是一种可表示为 H_2ox 的二元酸，草酸根阴离子可作为双齿配体进行配位。）
[b] Ethylenediaminetetraacetic acid is a tetraprotic acid denoted H_4EDTA. （乙二胺四乙酸是一种可表示为 H_4EDTA 的四元酸。）

Table 9.4　Names for Some Metals in Complex Anion
表 9.4　配阴离子中一些金属的命名

Element（元素）	Name（名称）	Element（元素）	Name（名称）
Fe	Ferrate	Sn	Stannate
Cu	Cuprate	Au	Aurate
Ag	Argentate	Pb	Plumbate

consists of geometric and optical isomerism.

We will discuss those types of isomerism with some illustrative examples in the following contents.

Structural Isomerism

The structural isomerism generally includes ionization, coordination, linkage, and ligand isomerism:

1) **Ionization isomerism**: The isomers give different ions in the solution. The ionization isomerism occurs when the counter ion of the complex ion is also a potential ligand. For example, pentaamminebromocobalt(III) sulphate $[CoBr(NH_3)_5]SO_4$ gives a precipitate with $BaCl_2$, confirming the presence of SO_4^{2-} ion, whereas pentaamminesulphatecobalt(III) bromide $[Co(SO_4)(NH_3)_5]Br$ tests negative for SO_4^{2-} ion in solution, but instead gives a precipitate of AgBr with $AgNO_3$.

2) **Coordination isomerism**: This isomerism occurs when a coordination compound is composed of both complex cations and complex anions. The ligands can be exchanged between the two complex ions. For example, hexaamminecobalt(III) hexacyanochromate(III) $[Co(NH_3)_6][Cr(CN)_6]$ and hexaamminechromium(III) hexacyanocobaltate(III) $[Cr(NH_3)_6][Co(CN)_6]$.

3) **Linkage isomerism**: This isomerism occurs with ambidentate ligands that can attach to the coordination center via different donor atoms. For example, pentaamminenitrito-*N*-chromium(III) ion $[Cr(NO_2)(NH_3)_5]^{2+}$ and pentaammienitrito-*O*-chromium(III) ion $[Cr(ONO)(NH_3)_5]^{2+}$.

4) **Ligand isomerism**: This isomerism arises when the ligands themselves have isomers. For example, $[CoCl_2(NH_2CH_2CH(NH_2)CH_3)_2]$ and $[CoCl_2(NH_2CH_2CH_2CH_2NH_2)_2]$.

Geometric Isomerism

Geometric isomerism is the isomerism in which different isomers are formed by different geometric positions of the ligands. It usually occurs in square-planar (a coordination number of 4) and octahedral (a coordination number of 6) complexes. In describing geometric isomerism, we usually denote the coordination center as M, and the different ligands as A, B, C, and so on.

In a square-planar complex with four monodentate ligands, denoted as MA_4, it is obvious that this complex shows no geometric isomerism, that is, the number of geometric isomers is only 1 [**Figure 9.2(a)**]. If one ligand A is substituted by another monodentate ligand B, the complex is denoted as MA_3B and still shows no geometric isomerism [**Figure 9.2(b)**] because all four possibilities are alike no matter at which corner of the square this substitution is made. If another A is substituted by another B, there are two distinct possibilities [**Figure 9.2(c)**] for the resultant complex, MA_2B_2. The two remaining A ligands can be either along the same edge of the square (*cis*) or on opposite corners, across from each other (*trans*). If a third and a fourth A are further substituted by B, geometric isomerism disappears for both MAB_3 and MB_4 [**Figures 9.2(d)(e)**].

For an octahedral complex with a formula of MA_2B_4, *cis* and *trans* isomers result. The *cis* isomer has two A ligands along the same edge of the octahedron [**Figure 9.3(a)**], whereas the *trans* isomer has two A on opposite corners, that is, at opposite ends of a line drawn through M [**Figure 9.3(b)**]. For an octahedral complex with a formula of MA_3B_3, there are also two possibilities. If the three A ligands occupy one face of the octahedron, the isomer is said to be facial, or *fac* [**Figure 9.4(a)**]. If the three A ligands are in a plane together with M, the isomer is said to be meridional, or *mer* [**Figure 9.4(b)**]. In a *fac* isomer, any two identical ligands, either A or B, are adjacent or *cis* to each other. A *mer* isomer can be considered as a combination of a *trans* and two *cis*, since it contains one pair of *trans* and two pairs of *cis* of identical ligands.

For an octahedral complex with a formula of $MA_2B_2C_2$, the situation is a bit more complicated. This complex has five geometric isomers: one with all *cis* pairs of identical ligands [**Figure 9.5(a)**], one with all *trans* pairs of identical ligands [**Figure 9.5(b)**], three with one *trans* pair of identical ligands, either A, B, or C, and two other pairs of identical ligands in *cis* [**Figures 9.5(c)**]. The number of geometric isomers for various complexes is summarized in **Table 9.5**.

1) **结构异构**：由于基本结构或成键类型不同而产生的异构现象。
2) **立体异构**（又称**空间异构**）：异构体具有相同的基本结构和成键类型，但由于配体占据配位中心周围空间的方式不同而导致的异构现象。通常包括几何异构和光学异构。

接下来我们将通过一些示例来讨论这些异构类型。

结构异构

结构异构一般包括电离异构、配位异构、键合异构和配体异构等：

1) **电离异构**：异构体在溶液中可电离出不同的离子。当配离子的抗衡离子同时也是潜在配体时，可发生电离异构。例如，硫酸溴·五氨合钴（Ⅲ）$[CoBr(NH_3)_5]SO_4$ 可与 $BaCl_2$ 形成沉淀，证实了 SO_4^{2-} 离子的存在，而溴化硫酸·五氨合钴（Ⅲ）$[Co(SO_4)(NH_3)_5]Br$ 对 SO_4^{2-} 离子的测试显阴性，但会与 $AgNO_3$ 生成 $AgBr$ 沉淀。

2) **配位异构**：当配合物由配阳离子和配阴离子组成时，会发生这种异构。配体可以在两个配离子之间交换。例如，六氰合铬（Ⅲ）酸六氨合钴（Ⅲ）$[Co(NH_3)_6][Cr(CN)_6]$ 和六氰合钴（Ⅲ）酸六氨合铬（Ⅲ）$[Cr(NH_3)_6][Co(CN)_6]$。

3) **键合异构**：此异构发生在含有的两可配体通过不同配位原子连接到配位中心时。例如，硝基·五氨合铬（Ⅲ）离子 $[Cr(NO_2)(NH_3)_5]^{2+}$ 和亚硝酸根·五氨合铬（Ⅲ）离子 $[Cr(ONO)(NH_3)_5]^{2+}$。

4) **配体异构**：当配体自身具有异构体时会出现这种异构。例如，$[CoCl_2(NH_2CH_2CH(NH_2)CH_3)_2]$ 和 $[CoCl_2(NH_2CH_2CH_2CH_2NH_2)_2]$。

几何异构

几何异构是指因配体的不同几何位置而形成不同异构体的异构现象。它通常出现在平面正方形（配位数为 4）和八面体形（配位数为 6）配位单元中。描述几何异构时，我们通常将配位中心表示为 M，将不同配体表示为 A、B、C 等。

具有四个单齿配体的平面正方形配位单元（表示为 MA_4）显然没有几何异构，即其几何异构体数仅为 1[**图 9.2(a)**]。如果一个 A 配体被另一个单齿配体 B 取代，该配位单元可表示为 MA_3B，仍然没有几何异构 [**图 9.2(b)**]，因为无论该取代位于正方形哪个角落，所有四种可能性均相同。如果再有一个 A 被 B 取代，则生成的配位单元 MA_2B_2 存在两种不同的可能性 [**图 9.2(c)**]。余下的两个 A 配体可以在正方形的同一条边上（顺式），或者位于对角上呈交叉分布（反式）。如果第三和第四个 A 进一步被 B 取代，MAB_3 和 MB_4 的几何异构均消失 [**图 9.2(d)(e)**]。

化学式为 MA_2B_4 的八面体形配位单元可产生顺式和反式异构体。顺式异构体的两个 A 配体位于八面体的同一条边上 [**图 9.3(a)**]；而反式异构体的两个 A 在对角上，即位于一条通过 M 的线的两端 [**图 9.3(b)**]。对于化学式为 MA_3B_3 的八面体形配位单元，也存在两种可能性。如果三个 A 配体占据八面体的一个面，则该异构体称为面式异构体或 *fac* [**图 9.4(a)**]。如果三个 A 配体与 M 位于同一平面，则该异构体称为经式异构体或 *mer* [**图 9.4(b)**]。在面式异构体中，任何两个相同的配体（A 或 B）彼此相邻或处于顺式。经式异构体可视为一个反式和两个顺式的组合，因为它包含相同配体的一对反式和两对顺式。

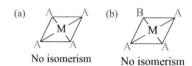

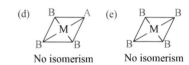

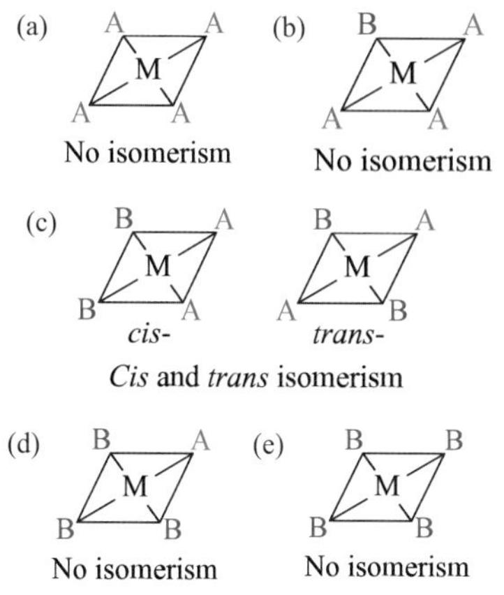

Figure 9.2 Geometric isomerism of various square-planar complexes. (a) MA_4: no isomerism. (b) MA_3B: no isomerism. (c) MA_2B_2: *cis* and *trans* isomerism. (d) MAB_3: no isomerism. (e) MB_4: no isomerism.

图 9.2 各种平面正方形配位单元的几何异构：(a) MA_4：无几何异构；(b) MA_3B：无几何异构；(c) MA_2B_2：顺反异构；(d) MAB_3：无几何异构；(e) MB_4：无几何异构。

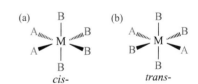

Figure 9.3 Geometric isomerism for an octahedral complex with a formula of MA_2B_4. (a) *Cis* isomer. (b) *Trans* isomer.

图 9.3 化学式为 MA_2B_4 的八面体形配位单元的几何异构：(a) 顺式异构体；(b) 反式异构体。

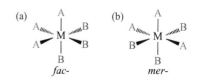

Figure 9.4 Geometric isomerism for an octahedral complex with a formula of MA_3B_3. (a) *Fac* isomer. (b) *Mer* isomer.

图 9.4 化学式为 MA_3B_3 的八面体形配位单元的几何异构：(a) 面式异构体；(b) 经式异构体。

Table 9.5 Summary on the Number of Geometric Isomers for Various Complexes
表 9.5 各种配位单元的几何异构体数小结

Coordination No. (配位数)	Geometric Shape (几何形状)	Formula (化学式)	No. of Geometric Isomers (几何异构体数)
2	Linear (直线形)	MA_2, MAB	1
4	Tetrahedral (四面体形)	MA_4, MAB_3, MA_2B_2, MA_2BC, $MABCD$	1
4	Square-planar (平面正方形)	MA_4, MAB_3, $MABCD$	1
		MA_2B_2, MA_2BC	2 (*cis-trans*, 顺反异构)
6	Octahedral (八面体形)	MA_6, MAB_5, etc.	1
		MA_2B_4, etc.	2 (*cis-trans*, 顺反异构)
		MA_3B_3, etc.	2 (*fac-mer*, 面经异构)
		$MA_2B_2C_2$, etc.	5 (refer to **Fig. 9.5**, 参考图 9.5)

Optical Isomerism

To understand optical isomerism requires to first understand the relationship between an object and its mirror image. Features on the left side of the object appear on the right side of its image in a mirror, and vice versa. **Optical isomerism** occurs when a complex is *nonsuperimposable* to its mirror image, where nonsuperimposable means that no matter how you rotate an object in 3-D space, it will not be identical to its mirror image. For example, an open-top box can be superimposed on its mirror image [**Figure 9.6(a)**]. However, with a sticker (no matter symmetric or not) placed at a corner of one side of the box, there is no way that the box and its mirror image can be superimposed [**Figure 9.6(b)**]. Moreover, if a symmetric sticker is placed at a corner of one side of a regular box, not open-top, it is again possible to rotate the box to make it superimposed on its mirror image [**Figure 9.6(c)**]. A left hand is always nonsuperimposable on its mirror image (a right hand).

The two structures of $[Cr(en)_3]^{3+}$ (**Figure 9.7**) are related to each other as are an object and its mirror image. These two structures are nonsuperimposable, like a left and a right hand. These two structures represent a pair of different optical isomers. A pair of isomers that are mirror images but cannot be superimposed on each other are called **enantiomers**. Both enantiomers are said to be chiral, whereas structures that are superimposable are said to be achiral. The geometric relationship between enantiomers is called **chirality**, or **handedness**. Unlike other types of isomers that may differ significantly in their physical and chemical properties, enantiomers have identical properties except in a few specialized situations, especially in optical activity. This is also why this type of isomerism is called optical isomerism. Molecules with either a plane of symmetry or a center of inversion must be optically inactive.

Optical activity is a property of enantiomers that causes the plane of polarization of the incident light to rotate due to the interactions between the polarized light and the electrons in the enantiomers. A **polarized light** is a light with its electric vector being oriented in a predictable fashion with respect to the direction of propagation. By default, a polarized light usually refers to a linearly polarized light, also called a plane polarized light, which has a fixed direction of electric vector, not varying with time. The plane of polarization is then defined by this fixed direction of the electric vector together with the propagation direction. The electromagnetic radiation shown in **Figure 7.6** is a linearly polarized light. In an unpolarized light, the electric vector is oriented in a random, unpredictable fashion. Lasers are typical sources of polarized lights, whereas sunlight as well as ordinary sources such as incandescent lamps, tungsten lamps, and mercury lamps are sources of unpolarized lights.

The experimental setup for detecting the optical activity is pictured in **Figure 9.8**. The unpolarized light from an ordinary source is passed through a polarizer, which screens out all electromagnetic radiations except those polarized on a particular plane. The resulting polarized light is passed through a chamber containing an optically active sample. The plane of polarization of the transmitted light deflects from the original plane at an angle. This angle can be determined by rotating an analyzer (a second polarizer) to the extent that all

对于化学式为 $MA_2B_2C_2$ 的八面体形配位单元,情况要复杂一些。该配位单元共有五种几何异构体:一种所有相同配体均处于顺式[**图 9.5(a)**],一种所有相同配体均处于反式[**图 9.5(b)**],三种只有一对相同配体(A、B 或 C 之一)处于反式、另两对相同配体处于顺式[**图 9.5(c)**]。**表 9.5** 总结了各种配位单元的几何异构体数。

光学异构

为理解光学异构,需要先理解物体与其镜像之间的关系。物体左侧的特征将出现在其镜像的右侧,反之亦然。当配位单元与其镜像不可重合时,即存在**光学异构**,其中不可重合意味着无论怎样在三维空间中旋转一个物体,它都不会与其镜像完全相同。例如,一个顶部开口的盒子与其镜像可完全重合[**图 9.6(a)**]。但如果在盒子某一面的角落处贴一张标签(不论是否对称),则盒子与其镜像就无法重合了[**图 9.6(b)**]。此外,如果将一张对称的标签贴在普通盒子(顶部未开口)某一面的角落,则通过旋转使其与镜像重合又变得可能了[**图 9.6(c)**]。左手与其镜像(右手)总是不可重合。

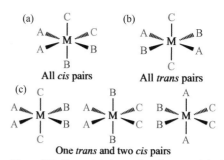

Figure 9.5 Geometric isomerism for an octahedral complex with a formula of $MA_2B_2C_2$. (a) An isomer with all *cis* pairs of identical ligands. (b) An isomer with all *trans* pairs of identical ligands. (c) Three isomers with one *trans* pair and two *cis* pairs of identical ligands.

图 9.5 化学式为 $MA_2B_2C_2$ 的八面体形配位单元的几何异构:(a) 所有相同配体均处于顺式的异构体;(b) 所有相同配体均处于反式的异构体;(c) 相同配体一对处于反式、两对处于顺式的三种异构体。

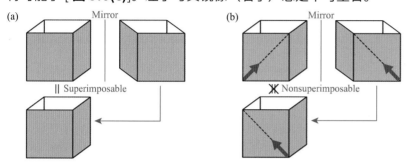

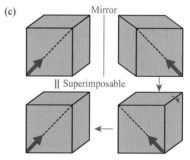

$[Cr(en)_3]^{3+}$ 的两种结构(**图 9.7**)相互关联,就像物体与其镜像一样。这两种结构不可重合,正如左手与右手,它们代表了一对不同的光学异构体。一对互为镜像但不能完全重合的异构体,称为**对映体**。两种对映体均认为是手性的,而可以重合的结构则认为是非手性的。对映体之间的几何关系称为**手性**。不像其他类型的异构体具有显著不同的物理和化学性质,对映体的其他所有性质完全相同,只有少数特殊性质(特别是光学活性)不同,这也是为什么将这种异构称为光学异构的原因。存在对称面或对称中心的分子,一定没有光学活性。

光学活性是一种通过偏振光与对映体的电子之间的相互作用而导致入射光的偏振平面发生旋转的对映体性质。**偏振光**是其电场矢量方向相对于传播方向以可预测的方式定向的光。默认情况下,偏振光通常指线偏振光,也称平面偏振光,它具有固定的、不随时间变化的电场矢量方向,该方向与传播方向共同定义了偏振平面。**图 7.6** 所示的电磁辐射为线偏振光。非偏振光的电场矢量方向随机变化,不可预测。激光是典型的偏振光源,太阳光以及常规光源(如白炽灯、钨灯和汞灯等)均为非偏振光源。

检测光学活性的实验装置如**图 9.8** 所示。来自常规光源的非偏振光先通过起偏器,屏蔽了除偏振方向位于特定平面之外的所有其他电磁辐射。产生的偏振光再通过装载光学活性样品的腔室。透射光的偏振平面从原平面偏转了一定角度,该角度可通过旋转检偏器(第二个偏振器),使所有偏振光均被吸收来检测。绝对偏转角取决于腔室的长度、样品的浓度和样品的性质。在单位腔室长度和单位样品浓度的

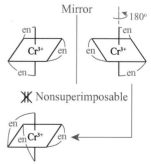

Figure 9.7 The two optical isomers of $[Cr(en)_3]^{3+}$ are nonsuperimposable mirror images of each other.

图 9.7 $[Cr(en)_3]^{3+}$ 的两个光学异构体是彼此的不可重合镜像。

the polarized light is absorbed. The absolute deflection angle depends on the length of the chamber, the concentration of the sample, and the nature of the sample. With a unit of chamber length and a unit of sample concentration, the one enantiomer that rotates the plane of polarization to the right (clockwise) is said to be **dextrorotatory** (designated + or *d*). The other enantiomer that rotates the plane of polarization to the same extent, but to the left (counter clockwise) is said to be **levorotatory** (designated − or *l*).

A homogeneous mixture with equal amounts of both enantiomers of a chiral species is called a **racemic mixture**. Because the rotation of the polarization plane by one optical isomer cancels that by the other, the racemic mixture produces no net rotation of the polarization plane, behaving like optically inactive. The separation of enantiomers from a racemic mixture is called **resolution**. The component enantiomers often have different chemical reactivity from other enantiomer substances. This may lead to chirality-controlled chemical reactions.

Chirality is generally observed in many organic compounds in vivo. For example, all amino acids are chiral except for glycine (NH_2CH_2COOH). Almost all natural amino acids found in organisms on earth are left-handed. Humans are also living organisms made of left-handed amino acids and cannot well metabolize to right-handed species. The component enantiomers of a racemic drug may differ significantly in their pharmaceutical performance.

Although chirality was first observed in and once thought to be restricted to organic compounds, Werner overthrew this misconception by the resolution of a purely inorganic compound $\{[Co(NH_3)_4(OH)_2]_3Co\}(SO_4)_3$ in 1911. He demonstrated that the nature of chirality arises from geometric structure instead of carbon atoms.

9.3 Crystal Field Theory

The theory of chemical bonding in explaining the characteristics and properties of complexes, called **crystal field theory** (CFT), was developed by Hans Bethe (1906—2005) and John H. van Vleck (1899—1980) in the 1930s.

Key Points of Crystal Field Theory

In CFT, bonding in a complex is regarded to arise from the electrostatic interactions between the ligands and the central ion, including:
1) The attractions between the negatively charged non-bonding electrons of the ligands and the positively charged nucleus of the central ion.
2) The repulsions between the ligand electrons and the electrons in the central ion.

In brief, CFT is a model that describes the splitting of degeneracies of the electron orbital energy levels, usually the energy levels of *d* or *f* orbitals in the central metal, due to the electrostatic field produced by the surrounding ligands.

To understand the above information, let us first recall the degeneracy and orientations of the *d* orbitals introduced in **Figure 7.21** in **Section 7.6**. All the five *d* orbitals degenerate in energy in an isolated atom or ion, but their spatial orientations are different: d_{z^2} has lobes along the *z* axis, $d_{x^2-y^2}$ is directed along the *x* and *y* axes, and the remaining three are pointed between the corresponding Cartesian axes.

Next, we consider the non-bonding electrons of the ligands as point charges. As the ligands approach the central ion in a non-spherical manner, for example, creating an octahedral (with a coordination number of 6), or a tetrahedral or square-planar (with a coordination number of 4) electrostatic field, the electrons of the ligands are closer to some of the *d* orbitals and farther away from others. The different repulsions between ligand electrons and *d* electrons cause the *d* orbital energy levels of the central ion to be raised to different extent. The *d* orbitals closer to the ligands are raised to a higher energy than those farther away, breaking the original five-fold degeneracy of the *d* orbitals. Meanwhile, the attractions between the ligand electrons and the nucleus of the central ion lower the total energy of the five *d* orbitals to the same extent, leading to the bonding in a stable complex.

条件下，将使偏振平面向右（顺时针方向）旋转的一种对映体称为**右旋**对映体（记为 + 或 d）；将使偏振平面旋转到相同程度，但向左（逆时针方向）旋转的另一种对映体称为**左旋**对映体（记为 − 或 l）。

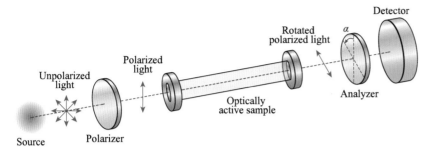

Figure 9.8 The experimental setup to detect the optical activity of a sample.

图 9.8 检测样品光学活性的实验装置图。

手性物质的两种对映体含量相等的均匀混合物，称为**外消旋体**。由于一种光学异构体对偏振平面的旋转抵消了另一种的旋转，外消旋体不会产生偏振平面的净旋转，表现为非光学活性。从外消旋体中分离出对映体的过程，称为**手性拆分**。组分对映体通常会对其他的对映体物质表现出不同的化学反应活性，这可能导致手性可控的化学反应。

手性在生物体内的许多有机化合物中可普遍观察到。例如，氨基酸除甘氨酸（NH_2CH_2COOH）之外均具有手性。地球上生物体内所发现的天然氨基酸几乎全为左旋，人类也是由左旋氨基酸组成的活生物体，不能很好地代谢右旋物种。外消旋药物的组分对映体在药物性能上可能存在显著差异。

虽然手性最初是在有机物中观察到的，并且一度认为仅限于有机物，但 1911 年维尔纳通过对纯无机物 $\{[Co(NH_3)_4(OH)_2]_3Co\}(SO_4)_3$ 的手性拆分，推翻了这一误解。他证明了手性的本质来源于几何构型，而非碳原子。

9.3 晶体场理论

用于解释配位单元特征和性质的化学键理论，称为**晶体场理论**（CFT），由汉斯·贝特（1906—2005）和约翰·H. 范弗莱克（1899—1980）在 20 世纪 30 年代提出。

晶体场理论的要点

晶体场理论认为，配位单元中的化学键由配体与中心离子之间的静电相互作用产生，包括：
1) 配体带负电的非键电子与中心离子带正电的原子核之间的吸引力。
2) 配体电子与中心离子的电子之间的排斥力。

简言之，晶体场理论是一个描述在周围配体产生的静电场下，简并的电子轨道能级（通常是中心金属的 d 或 f 轨道能级）发生分裂的模型。

为理解以上内容，让我们先回顾 **7.6 节图 7.21** 中介绍的 d 轨道的简并性及空间取向。在孤立的原子或离子中，所有五个 d 轨道的能量都是简并的，但它们的空间取向各不相同：d_{z^2} 的波瓣位于 z 轴，$d_{x^2-y^2}$ 的取向沿 x 和 y 轴方向，其余三个指向相应的坐标轴之间。

接下来，我们将配体的非键电子视为点电荷。当配体以非球形的方式接近中心离子时，例如产生八面体形（配位数为 6）或四面体形

Octahedral Complexes

In an octahedral complex, six ligands approach the central ion along the x, y, and z axes. The d_{z^2} and $d_{x^2-y^2}$ orbitals with lobes in these directions are raised to a higher energy than the other three orbitals with lobes avoiding these directions (**Figure 9.9**). Calculations show that the original five-fold degeneracy of the d orbitals breaks into two groups: a two-fold degeneracy group of d_{z^2} and $d_{x^2-y^2}$, and a three-fold degeneracy group of d_{xy}, d_{xz}, and d_{yz}. The energy difference between these two groups of d orbitals is called **crystal field splitting** and is denoted as Δ_o, where the subscript o emphasizes that it is for an octahedral complex (**Figure 9.10**). In order for the total energy to be conserved, it is not difficult to calculate that the energy of d_{z^2} and $d_{x^2-y^2}$ group is $0.6\Delta_o$ higher than the average energy of d orbitals and that of d_{xy}, d_{xz}, and d_{yz} group is $0.4\Delta_o$ lower.

In ground-state configurations, the electron arrangements in the orbitals of complexes also follow three basic rules: Pauli exclusion principle, lowest energy principle, and Hund's rule. In the cases of octahedral complexes, the electron configurations with 4 to 7 d electrons are particularly interesting. Taking Cr^{2+} with a d^4 configuration as an example, the first three d electrons singly occupy the three-fold degenerate d_{xy}, d_{xz}, and d_{yz} orbitals according to Hund's rule. There are two possibilities for the fourth electron: it may either pair up with any one of the three electrons in the d_{xy}, d_{xz}, and d_{yz} orbitals, or occupy one of the d_{z^2} and $d_{x^2-y^2}$ orbitals (**Figure 9.11**). In the former configuration, the additional energy required to force another electron into an orbital that is already occupied by an electron, is called the **pairing energy** (P). The latter requires an additional energy of Δ_o. Which one of the two possibilities the fourth electron may take depends on the relative magnitude of P and Δ_o. The value of P is quite constant, usually within the range of 20,000~30,000 cm^{-1}, that is, about 2.5~3.7 eV. If $\Delta_o > P$, the crystal field is said to be a **strong field**, and the fourth electron tends to pair with one at the lower level. This situation corresponds to the minimum number of unpaired electrons (with Hund's rule fulfilled) and is referred to as **low spin**. If $\Delta_o < P$, the crystal field is said to be a **weak field**, and it is more stable that the fourth electron occupies a higher level. This corresponds to a situation with the maximum number of unpaired electrons and is referred to as **high spin**. Therefore, a strong field generally leads to producing low-spin complexes, whereas a weak field usually results in high-spin complexes.

As discussed previously in **Section 7.6**, the electron spin quantum number m_s may have a value of either $+1/2$ or $-1/2$. For a multielectron species, the total electron spin quantum number, denoted by M_S, is the summation of the m_s values of all individual electrons. For paired electrons, which is a pair of electrons on the same orbital with opposing spins, $M_S = 1/2 - 1/2 = 0$. Thus, paired electrons make no contribution to the total spin of the species. M_S can be simply calculated by the summation of the m_s values of all the unpaired electrons. High spin or low spin refers to a high M_S value or a low M_S value, respectively. For example, the Cr(II) complex in a strong field has two unpaired electrons with $M_S = 2 \times 1/2 = 1$, whereas that in a weak field has four unpaired electrons with $M_S = 4 \times 1/2 = 2$. Consequently, the former corresponds to a low-spin complex, and the latter is a high-spin complex (**Figure 9.11**).

Whether the crystal field that a complex produces is considered as a strong field or a weak field depends on both the ligands and the central ion. Spectroscopic data of the complexes is used to arrange a series, called the **spectrochemical series**, in the order of crystal field splitting. The spectrochemical series of central ions is given by (strong field, large Δ_o, low spin) $Pt^{4+} > Ir^{3+} > Pd^{4+} > Ru^{3+} > Rh^{3+} > Mo^{3+} > Co^{3+} > Fe^{3+} > V^{2+} > Fe^{2+} > Co^{2+} > Ni^{2+} > Mn^{2+}$ (weak field, small Δ_o, high spin)

The general trend is that Δ_o increases with the Lewis acidity. The spectrochemical series of ligands is given by (strong field, large Δ_o, low spin) $CO > \underline{C}N^- > PPh_3 > \underline{N}O_2^- > phen > bpy > en > NH_3 > py > CH_3CN > \underline{N}CS^- > H_2O > O^{2-} > C_2O_4^{2-} > OH^- > F^- > N_3^- > \underline{N}O_2^- > Cl^- > \underline{S}CN^- > S^{2-} > Br^- > I^-$ (weak field, small Δ_o, high spin)

where the donor atom in an ambidentate ligand is underlined. PPh_3, phen, bpy, and py stand for triphenylphosphine, o-phenanthroline, 2,2'-bipyridine, and pyridine, respectively. The general trend is that Δ_o

平面正方形（配位数为 4）静电场时，配体的电子距离某些 d 轨道更近，而距离其他 d 轨道更远。配体电子与 d 电子之间不同程度的排斥力，导致中心离子的 d 轨道能级发生不同程度的提升。距离配体较近的 d 轨道比距离较远的 d 轨道提升的能量更高，打破了 d 轨道原有的五重简并。同时，配体电子与中心离子原子核之间的吸引力，使五个 d 轨道的总能量同等程度地降低，从而导致了配位单元的稳定成键。

八面体形配位单元

在八面体形配位单元中，六个配体沿 x、y 和 z 轴方向接近中心离子。波瓣在这些方向的 d_{z^2} 和 $d_{x^2-y^2}$ 轨道，比波瓣避开这些方向的其他三个轨道提升的能量更高（**图 9.9**）。计算表明，d 轨道原有的五重简并分裂成两组：具有两重简并的 d_{z^2} 和 $d_{x^2-y^2}$ 组，和具有三重简并的 d_{xy}、d_{xz} 和 d_{yz} 组。这两组 d 轨道之间的能量差，称为**晶体场分裂能**，记为 Δ_o，其中下标 o 强调其为八面体形配位单元（**图 9.10**）。为使总能量守恒，不难计算出 d_{z^2} 和 $d_{x^2-y^2}$ 组能量比 d 轨道的平均能量高 $0.6\Delta_o$，而 d_{xy}、d_{xz} 和 d_{yz} 组能量比平均能量低 $0.4\Delta_o$。

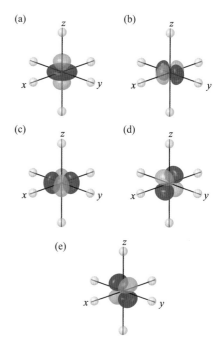

Figure 9.9 Interactions between six approaching ligands along the x, y, and z axes and various d orbitals of the central ion in an octahedral complex: (a) d_{z^2}; (b) $d_{x^2-y^2}$; (c) d_{xy}; (d) d_{xz}; (e) d_{yz}.

图 9.9 八面体形配位单元的六个沿 x、y 和 z 轴方向靠近的配体与中心离子各 d 轨道之间的相互作用：(a) d_{z^2}；(b) $d_{x^2-y^2}$；(c) d_{xy}；(d) d_{xz}；(e) d_{yz}。

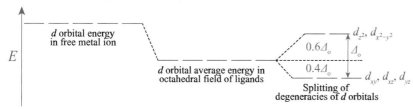

Figure 9.10 Splitting of the d orbital degeneracies in an octahedral crystal field of the ligands.

图 9.10 简并的 d 轨道在配体的八面体形晶体场中的分裂。

配位单元基态组态的电子排布也遵循三个基本规则：泡利不相容原理、能量最低原理和洪特规则。在八面体形配位单元中，具有 4~7 个 d 电子的电子组态特别有趣。以具有 d^4 组态的 Cr^{2+} 为例，根据洪特规则，前三个 d 电子单独占据三重简并的 d_{xy}、d_{xz} 和 d_{yz} 轨道。第四个电子则存在两种可能：它可以与 d_{xy}、d_{xz} 和 d_{yz} 轨道中三个电子的任何一个配对，也可以占据 d_{z^2} 和 $d_{x^2-y^2}$ 轨道中的一个（**图 9.11**）。在前一种组态中，迫使另一个电子进入已经被电子占据的轨道所需的额外能量，称为**电子成对能**（P）。后一种组态要求的额外能量即为 Δ_o。第四个电子采取哪一种可能性，取决于 P 和 Δ_o 的相对大小。P 的大小相对恒定，通常在 20 000~30 000 cm^{-1} 范围内，即约 2.5~3.7 eV。如果 $\Delta_o>P$，此晶体场称为**强场**，第四个电子倾向于与较低能级电子成对。这种情况对应的未成对电子数最小（满足洪特规则时），称为**低自旋**。如果 $\Delta_o<P$，此晶体场称为**弱场**，第四个电子占据较高能级时更为稳定。这对应于未成对电子数最大的情况，称为**高自旋**。因此，强场通常产生低自旋配位单元，而弱场通常导致高自旋配位单元。

如 **7.6 节**所述，电子的自旋量子数 m_s 取值可为 +1/2 或 −1/2。多电子物种的总电子自旋量子数，记为 M_S，是其所有单个电子的 m_s 之和。对于成对电子，即在同一轨道上具有相反自旋的一对电子，$M_S=1/2-1/2 = 0$，因此成对电子对物种的总自旋没有贡献。M_S 可通过所有未成对电子 m_s 之和来简单计算。所谓高自旋或低自旋，分别

increases with decreasing electronegativity of the donor atom. H₂O usually serves as the boundary between strong-field ligands and weak-field ligands.

Crystal Field Stabilization Energy

Let us illustrate with $[FeF_6]^{3-}$ and $[Fe(CN)_6]^{3-}$ as examples. According to the spectrochemical series of ligands, F^- is a weak-field ligand whereas CN^- is a strong-field ligand. Experimental data show that Δ_o = 13,700 cm^{-1}<P for $[FeF_6]^{3-}$ and Δ_o = 34,250 cm^{-1}>P for $[Fe(CN)_6]^{3-}$. Therefore, $[FeF_6]^{3-}$ is more stable to form a high-spin complex, with five unpaired electrons singly occupying the five d orbitals [M_S = 5/2, **Figure 9.12(a)**]. $[Fe(CN)_6]^{3-}$ forms a low-spin complex, with two pairs of electrons and one unpaired electron occupying the three lower energy levels [M_S = 1/2, **Figure 9.12(b)**].

The total energy (E_o) of assigning electrons into the split crystal field is generally not higher than that (E_d) of placing electrons into the original non-split d orbitals. This energy difference accounts for the stability of the complex, and is termed **crystal field stabilization energy** (CFSE). For $[FeF_6]^{3-}$:

$$CFSE = E_d - E_o = 0 - (2 \times 0.6\Delta_o - 3 \times 0.4\Delta_o) = 0$$

For $[Fe(CN)_6]^{3-}$:

$$CFSE = E_d - E_o = 0 - (-5 \times 0.4\Delta_o + 2P) = 2\Delta_o - 2P$$

Note that the corresponding pairing energies should be included in the calculation of CFSE.

Tetrahedral and Square-Planar Complexes

In a tetrahedral complex, four ligands approach the central ion in directions avoiding the x, y, and z axes (**Figure 9.13**). The tetrahedral crystal field (**Figure 9.14**) differs from the octahedral crystal field mainly in two aspects:

1) The five-fold degeneracy of the d orbitals breaks into two groups: a two-fold degeneracy group of d_{z^2} and $d_{x^2-y^2}$ with lower energy and a three-fold degeneracy group of d_{xy}, d_{xz}, and d_{yz} with higher energy.

2) The tetrahedral crystal field splitting, denoted as Δ_t, where t stands for tetrahedral, is smaller than the corresponding octahedral splitting Δ_o with the same central ion and ligands (Δ_t = 0.44 Δ_o). Δ_t<P generally applies and almost all tetrahedral complexes are high spin.

In a square-planar complex, four ligands approach the central ion along the x and y axes, and no ligand approaches along the z axis. Compared to the octahedral crystal field, the energy levels of $d_{x^2-y^2}$ and d_{xy} are raised to higher energies, but their splitting remains unchanged, still equal to Δ_o. The energy of d_{z^2} is substantially lower than that of $d_{x^2-y^2}$ due to the lack of ligands along the z axis. Similarly, the energy of d_{xz} and d_{yz}, which remains degenerate, is also considerably lower than that of d_{xy}. The crystal field splitting in a square-planar complex (**Figure 9.15**) is the maximum possible energy difference, that is, the energy difference between $d_{x^2-y^2}$ and d_{xz} or d_{yz}, in this crystal field. Δ_{sp} = 1.74 Δ_o, where Δ_o is the corresponding octahedral splitting with the same central ion and ligands.

Limitations of Crystal Field Theory

CFT considers only the electrostatic interactions between the central ion and ligands and treats ligands as point charges. Thus, CFT cannot account for the spectrochemical series of ligands. Later experimental evidence confirmed that there are overlap between the orbitals of the central ion and the ligand orbitals. This suggested that there is some covalent character in the bonding between the central ion and ligands. By combining CFT together with molecular orbital theory, a more sophisticated theory called **ligand field theory** (LFT) was developed to concentrate on the overlap between the d orbitals of the central ion and the valence orbitals of the ligands. LFT provides a more substantial framework to understand the origins of energy level splitting in complexes and can explain the spectrochemical series of ligands.

指高的 M_S 值或低的 M_S 值。例如，强场中 Cr(Ⅱ) 配位单元存在两个未成对电子，$M_S=2\times1/2=1$；而弱场中的 Cr(Ⅱ) 配位单元存在四个未成对电子，$M_S=4\times1/2=2$。因此，前者对应于低自旋配位单元，后者对应于高自旋配位单元（**图 9.11**）。

配位单元产生的晶体场究竟是强场还是弱场，取决于配体和中心离子。基于配位单元的光谱数据，按照晶体场分裂能的大小顺序进行排列的序列，称为**光谱化学序列**。中心离子的光谱化学序列为（强场、大 Δ_o 值、低自旋）$Pt^{4+}> Ir^{3+}> Pd^{4+}> Ru^{3+}> Rh^{3+}> Mo^{3+}> Co^{3+}> Fe^{3+}> V^{2+}> Fe^{2+}> Co^{2+}> Ni^{2+}> Mn^{2+}$（弱场、小 Δ_o 值、高自旋）总体趋势是，Δ_o 随路易斯酸性增加而增大。配体的光谱化学序列为（强场、大 Δ_o 值、低自旋）$CO >\underline{C}N^-> PPh_3>\underline{N}O_2^->phen>bpy>en>NH_3>py> CH_3CN >\underline{N}CS^->H_2O> O^{2-}> C_2O_4^{2-}> OH^-> F^-> N_3^-> N\underline{O}_2^-> Cl^->\underline{S}CN^-> S^{2-}> Br^-> I^-$（弱场、小 Δ_o 值、高自旋）其中下划线表示两可配体的配位原子。PPh_3、phen、bpy 和 py 分别代表三苯基膦、邻菲咯啉、2,2′-联吡啶和吡啶。总体趋势是，Δ_o 随配位原子电负性降低而增大。H_2O 通常作为强场配体和弱场配体的分界线。

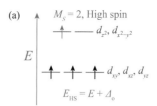

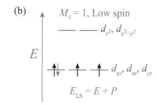

Figure 9.11 The (a) high-spin and (b) low-spin complexes with a d^4 configuration.

图 9.11 具有 d^4 组态的 (a) 高自旋和 (b) 低自旋配位单元。

晶体场稳定化能

让我们以 $[FeF_6]^{3-}$ 和 $[Fe(CN)_6]^{3-}$ 为例进行说明。根据配体的光谱化学序列，F^- 是弱场配体而 CN^- 是强场配体。实验数据表明，$[FeF_6]^{3-}$ 的 $\Delta_o=13\ 700\ cm^{-1}<P$，$[Fe(CN)_6]^{3-}$ 的 $\Delta_o=34\ 250\ cm^{-1}>P$。因此，$[FeF_6]^{3-}$ 可更稳定地形成高自旋配位单元，五个未成对电子单独占据五个 d 轨道 [$M_S=5/2$，**图 9.12(a)**]。$[Fe(CN)_6]^{3-}$ 可形成低自旋配位单元，两对电子和一个未成对电子占据三个较低能级 [$M_S=1/2$，**图 9.12(b)**]。

将电子排布到分裂的晶体场中的总能量（E_o），通常不高于将电子排布到原有未分裂的 d 轨道中的总能量（E_d）。这种能量差称为**晶体场稳定化能**（CFSE），解释了配位单元的稳定性。对于 $[FeF_6]^{3-}$：

$$CFSE = E_d - E_o = 0 - (2\times0.6\Delta_o - 3\times0.4\Delta_o) = 0$$

对于 $[Fe(CN)_6]^{3-}$：

$$CFSE = E_d - E_o = 0 - (-5\times0.4\Delta_o + 2P) = 2\Delta_o - 2P$$

注意计算 CFSE 时应考虑相应的电子成对能。

四面体形和平面正方形配位单元

在四面体形配位单元中，四个配体以避开 x、y 和 z 轴的方向接近中心离子（**图 9.13**）。四面体形晶体场（**图 9.14**）与八面体形晶体场主要存在两个方面的不同：

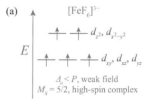

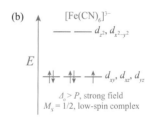

Figure 9.12 Examples of high-spin and low-spin complexes. (a) The high-spin complex $[FeF_6]^{3-}$ with a weak crystal field ($\Delta_o<P$). (b) The low-spin complex $[Fe(CN)_6]^{3-}$ with a strong crystal field ($\Delta_o>P$).

图 9.12 高自旋与低自旋配位单元举例：(a) 弱晶体场（$\Delta_o<P$）的高自旋配位单元 $[FeF_6]^{3-}$；(b) 强晶体场（$\Delta_o>P$）的低自旋配位单元 $[Fe(CN)_6]^{3-}$。

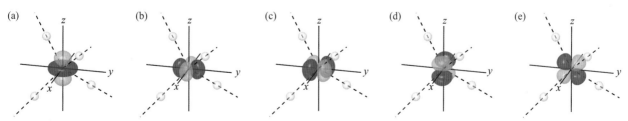

Figure 9.13 Interactions between four approaching ligands avoiding the x, y, and z axes and various d orbitals of the central ion in a tetrahedral complex: (a) d_{z^2} ; (b) $d_{x^2-y^2}$; (c) d_{xy} ; (d) d_{xz} ; (e) d_{yz} .

图 9.13 四面体形配位单元的四个沿避开 x、y 和 z 轴方向靠近的配体与中心离子各 d 轨道之间的相互作用：(a) d_{z^2} ；(b) $d_{x^2-y^2}$ ；(c) d_{xy} ；(d) d_{xz} ；(e) d_{yz} 。

9.4 Properties of Coordination Compounds

In this section, we will discuss various properties of coordination compounds, especially the optical and magnetic properties, and how these properties can be related to their electronic structures.

Optical Properties

The optical properties of a substance are properties related to light or electromagnetic radiation passing through it, such as absorption, scattering, refraction, polarization, etc. The previously discussed optical activity of chiral complexes to rotate the plane of polarization of the incident light is also an optical property. Here, we will focus on the explanation of the colors of coordination compounds via absorption spectroscopy.

When a non-luminescent substance is irradiated by a white light, the substance is colorless and transparent if all the visible light is transmitted. The substance becomes white if all the visible light is reflected and black if all absorbed. If only part of the visible light is absorbed by the substance, the remaining light will be either transmitted (for transparent substances) or reflected (for opaque substances), showing the complementary color of the absorbed light. The complementary relationship of colors is shown in **Figure 9.16**. If there is no absorption in the visible region, all the visible light will be transmitted or reflected.

The absorption of photons of light occurs when the energies of those photons match the differences between the electronic energy levels of the substance. The photons are then absorbed by the substance to promote transitions of electrons between the corresponding energy levels. For coordination compounds, these transitions normally occur between the split d energy levels with partially filled electrons ($d^{1\sim 9}$), called the **d-d transitions**. For example, the electron configuration of Ti(III) in $[Ti(H_2O)_6]^{3+}$ is $3d^1$ [**Figure 9.17(a)**], and $\Delta_o \approx 19,500$ cm^{-1}, corresponding to the energy of green light at $\lambda = 510$ nm. When a beam of white light passes through a $[Ti(H_2O)_6]^{3+}$ solution, the green light is absorbed, and the transmitted light becomes reddish purple. This explains the reddish-purple color of the $[Ti(H_2O)_6]^{3+}$ solution.

Absorbance is a measure of the proportion of monochromatic (single-color) light that is absorbed as the light passes through a substance. According to Beer's law,

$$A = -\log T = -\log \frac{I}{I_0} = \log \frac{I_0}{I} \tag{9.5}$$

where A is the absorbance, T is called transmittance, and I_0 and I are the intensities of light before and after the absorption by the substance, respectively. A high absorbance results in a low transmittance, meaning that a large proportion of the light entering the substance is absorbed. The absorbance of a substance is wavelength dependent. An **absorption spectrum** is a graph of absorbance plotted as a function of wavelength. The absorption spectrum of $[Ti(H_2O)_6]^{3+}$(aq) is shown in **Figure 9.17(b)**.

Ions with full orbital configurations have large HOMO-LUMO transition energies that usually do not locate in the visible region. White light passes through these substances without being absorbed. These ions, such as the alkali and alkaline earth metal ions, the halide ions, Ag^+, Zn^{2+}, Al^{3+}, Bi^{3+}, and Ti^{4+}, etc. are normally colorless in solution.

For coordination compounds with d^0 or d^{10} configuration, although d-d transitions are not possible to occur, another type of transitions called **charge transfer transitions** is still possible. A charge transfer band results from the shift of charge density between a metal-based orbital and a ligand-based orbital. If the transfer occurs from the ligand-based orbital to the metal-based orbital, the transition is called a ligand-to-metal charge transfer (LMCT) transition. If the electronic charge shifts from the metal-based orbital to the ligand-based orbital, the resulting band is called a metal-to-ligand charge transfer (MLCT) band. Thus, a LMCT results in oxidation of the ligand and reduction of the metal center, whereas a MLCT leads to oxidation of the metal center and reduction of the ligand. Charge transfer bands are typically more intense than d-d transition bands.

For instance, the colors of the tetraoxides of d^0 metal ions of the first transition series, such as yellow for

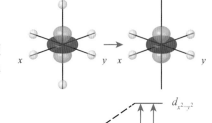

1) d 轨道的五重简并分裂成两组：能量较低的两重简并 d_{z^2} 和 $d_{x^2-y^2}$ 组，和能量较高的三重简并 d_{xy}、d_{xz} 和 d_{yz} 组。
2) 四面体形晶体场分裂能 Δ_t（t 代表四面体形）小于具有相同中心离子和配体的八面体场分裂能 Δ_o（$\Delta_t = 0.44\Delta_o$）。$\Delta_t < P$ 通常成立，几乎所有四面体形配位单元均为高自旋。

在平面正方形配位单元中，四个配体沿 x 和 y 轴接近中心离子，没有配体沿 z 轴接近。与八面体形晶体场相比，$d_{x^2-y^2}$ 和 d_{xy} 能级被提升至更高能量，但它们之间的分裂能保持不变，仍等于 Δ_o。由于 z 轴方向没有配体，d_{z^2} 的能量显著低于 $d_{x^2-y^2}$。类似地，d_{xz} 和 d_{yz} 的能量（保持简并）也远低于 d_{xy}。平面正方形配位单元的晶体场分裂能（**图 9.15**）是该晶体场中可能的最大能量差，即 $d_{x^2-y^2}$ 与 d_{xz} 或 d_{yz} 之间的能量差。$\Delta_{sp} = 1.74\Delta_o$，其中 Δ_o 是具有相同中心离子和配体的八面体场分裂能。

Figure 9.14 Splitting of the d orbital degeneracies in a tetrahedral crystal field of the ligands.

图 9.14 简并的 d 轨道在配体的四面体形晶体场中的分裂。

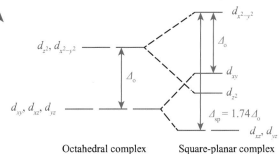

Figure 9.15 Comparison of crystal field spitting in a square-planar and an octahedral complex.

图 9.15 平面正方形与八面体形配位单元的晶体场分裂能对比。

晶体场理论的局限性

晶体场理论只考虑了中心离子和配体之间的静电相互作用，并将配体视为点电荷。因此，晶体场理论不能解释配体的光谱化学序列。后续实验证据证实，中心离子的轨道与配体轨道之间存在重叠。这表明中心离子与配体之间的成键，具有一定程度的共价性质。通过将晶体场理论与分子轨道理论相结合，发展了一种更为精细的理论，称为**配体场理论**（LFT）。该理论聚焦于中心离子的 d 轨道与配体的价层轨道之间的重叠。配体场理论为理解配位单元能级分裂的起源提供了更为实质性的框架，并可以解释配体的光谱化学序列。

9.4 配合物的性质

本节我们将讨论配合物的各种性质，特别是光学和磁学性质，以及这些性质如何与其电子结构相关联。

光学性质

物质的光学性质是指与通过该物质的光或电磁辐射相关的性质，如吸收、散射、折射、偏振等。前述手性配位单元具有可旋转入射光偏振平面的光学活性，这也是一种光学性质。这里我们将重点放在通过吸收光谱来解释配合物的颜色。

当白光照射一个自身不发光的物质时，如果所有可见光均被透射，则该物质呈无色透明。如果所有可见光均被反射，则该物质显

CrO_4^{2-} and purple for MnO_4^-, can be attributed to LMCT, involving the transfer of non-bonding electrons on the oxo ligands to the empty d orbital of the metal. Since the charge transfer is stronger in MnO_4^- than CrO_4^{2-} due to the higher oxidation state of Mn(VII) than Cr(VI), the charge transfer band locates in the yellow-green region for MnO_4^-, lower in energy than that in the dark blue region for CrO_4^{2-}. The complementary colors are purple for MnO_4^- and yellow for CrO_4^{2-}. The corresponding charge transfer transitions for heavier metals occur in the UV region, and thus ReO_4^-, MoO_4^{2-}, and WO_4^{2-} are colorless. These transition energies accord to the order of the spectrochemical series.

Magnetic Properties

The magnetic properties of a substance, also called **magnetism**, are the attributes mediated by a magnetic field. When placed in an external magnetic field, a material will be magnetized, and an induced magnetic field will be generated consequently. The magnetism of a material is related to the number of unpaired electrons in the atoms or molecules of the material. The major types of magnetism include:

1) **Paramagnetism**: the general magnetic property observed in materials with unpaired electrons. The unpaired electrons, like individual magnetic needles, will be aligned in the same direction as the external magnetic field, resulting in enhanced magnetic field strength. The more unpaired electrons present, the stronger the enhancement. For metals in the first transition series, the number of unpaired electrons n is related to the magnetic moment μ by

$$\mu = \sqrt{n(n+2)}\mu_B \quad (9.6)$$

where μ_B is called the Bohr magneton, which is the unit of magnetic moment generated by a single free electron.

2) **Diamagnetism**: the tendency of all materials to generate an induced magnetic field in the opposite direction to the external magnetic field, and thus to be repelled by the field. Any paired electrons, required by Pauli exclusion principle, must have their intrinsic magnetic moments pointing in opposite directions and thus cancel out, showing diamagnetic behavior. Although diamagnetism appears in all materials due to the general existence of paired electrons, diamagnetic behavior is only observed in purely diamagnetic materials with no unpaired electrons. Once there is any unpaired electron, the paramagnetic behavior dominates.

3) **Ferromagnetism**: a special magnetism found in materials containing Fe, Co, and Ni. In the solid state of ferromagnetic materials, there is a tendency for the electron magnetic moments in small regions called domains, to orient parallel to each other even in the absence of an external magnetic field. In an unmagnetized material, the domains are oriented in random directions and thus their magnetic moments cancel out. When placed in an external magnetic field, however, the domains line up and a strong net magnetic moment is produced. The ordering of domains can persist when the material is removed from the magnetic field, and thus permanent magnetism results. As temperature increases, thermal motion competes with the ferromagnetic tendency to align. Above a certain critical point, called the **Curie temperature**, ferromagnetic materials become paramagnetic. Paramagnetism and ferromagnetism are compared in **Figure 9.18**.

The magnetism of coordination compounds can be measured using a Gouy magnetic balance. A sample is first weighed in the absence of a magnetic field. When the field is turned on, the balanced condition is upset. A paramagnetic substance is pulled into the magnetic field and weighs more within the field under the experimental setup shown in **Figure 9.19**. Meanwhile, a diamagnetic substance is pushed out of the field and weighs less within the field under the same experimental setup.

The degree to which a paramagnetic substance weights more in the magnetic field depends on the number of unpaired electrons. We know that a high-spin d^n complex has more unpaired electrons than a low-spin d^n complex. Thus, measuring the change in weight of the coordination compounds in a magnetic field allows us to determine whether it is high or low spin. The results of measuring the magnetic properties of coordination compounds can be interpreted from CFT.

For example, experiments show that the complex anion $[Ni(CN)_4]^{2-}$ is diamagnetic, and we can speculate its probable structure from this based on CFT. With a coordination number of four, the choice is between

白色；如果全部被吸收则显黑色。如果只有一部分可见光被物质吸收，剩余的光将被透射（对于透明物质）或反射（对于不透明物质），从而呈现出吸收光的互补色。颜色的互补关系如**图 9.16** 所示。如果在可见光区域没有吸收，则所有可见光均会被透射或反射。

当光子的能量与物质的电子能级差相匹配时，就会发生光子的吸收。光子被物质吸收后，会使电子在相应能级之间进行跃迁。对配合物而言，这些跃迁通常发生在电子部分填充（d^{1-9}）的分裂的 d 能级之间，称为 **d-d 跃迁**。例如，$[Ti(H_2O)_6]^{3+}$ 中 Ti（Ⅲ）的电子组态为 $3d^1$[**图 9.17(a)**]，$\Delta_o \approx 19\,500\ cm^{-1}$，对应于 $\lambda=510\ nm$ 的绿光能量。当一束白光通过 $[Ti(H_2O)_6]^{3+}$ 溶液时，绿色光被吸收，透射光变为红紫色。这解释了 $[Ti(H_2O)_6]^{3+}$ 溶液的红紫色。

吸光度是光通过物质时单色光被吸收比例的量度。根据比尔定律

$$A = -\lg T = -\lg \frac{I}{I_0} = \lg \frac{I_0}{I} \tag{9.5}$$

其中 A 是吸光度，T 称为透光率，I_0 和 I 分别是被物质吸收前后的光强度。高吸光度导致低透光率，这意味着进入物质的大部分光被吸收。物质的吸光度与波长相关。**吸收光谱**是吸光度对波长的函数关系图。$[Ti(H_2O)_6]^{3+}(aq)$ 的吸收光谱如**图 9.17(b)** 所示。

满轨道组态的离子具有较大的 HOMO-LUMO 跃迁能，通常位于可见光区域之外，白光通过这些物质时不被吸收。这些离子，如碱金属和碱土金属离子、卤素离子、Ag^+、Zn^{2+}、Al^{3+}、Bi^{3+} 和 Ti^{4+} 等，在溶液中通常呈无色。

对于具有 d^0 或 d^{10} 组态的配合物，尽管不可能发生 d-d 跃迁，但称为**荷移跃迁**的另一类跃迁仍然可能发生。荷移吸收谱带是由于电荷密度在基于金属的轨道与基于配体的轨道之间转移而产生的。如果从基于配体的轨道荷移至基于金属的轨道，这种跃迁称为配体 - 金属荷移（LMCT）跃迁。如果电子电荷从基于金属的轨道转移至基于配体的轨道，对应的跃迁称为金属 - 配体荷移（MLCT）跃迁。因此，LMCT 会导致配体氧化和金属中心还原，而 MLCT 则会导致金属中心氧化和配体还原。荷移吸收谱带通常比 d-d 跃迁吸收谱带强度更高。

例如，第一过渡系的 d^0 金属离子四氧化物的颜色（如 CrO_4^{2-} 呈黄色，MnO_4^- 呈紫色），可以归因于 LMCT，涉及氧配体的非键电子转移到金属的空 d 轨道。由于 Mn（Ⅶ）的氧化态高于 Cr（Ⅵ），MnO_4^- 的荷移强于 CrO_4^{2-}，因此 MnO_4^- 的荷移吸收谱带位于黄绿色区，能量低于 CrO_4^{2-} 吸收谱带的深蓝色区。对应的互补色为紫色（MnO_4^-）和黄色（CrO_4^{2-}）。较重金属的相应荷移跃迁发生在紫外区，因此 ReO_4^-、MoO_4^{2-} 和 WO_4^{2-} 均无色。这些跃迁能与光谱化学序列的顺序一致。

磁学性质

物质的磁学性质，也称**磁性**，是由磁场介导的属性。置于外磁场中的材料会被磁化，从而产生感应磁场。材料的磁性与其原子或分子中的未成对电子数相关。磁性的主要类型有：

1) **顺磁性**：是在含有未成对电子的材料中所观察到的一般磁性。未成对电子就像单个小磁针一样，将顺着与外磁场一致的方向排布，从而增强磁场强度。未成对电子数越多，磁场增强就越大。对于第一过渡系金属，未成对电子数 n 与磁矩 μ 的关系为

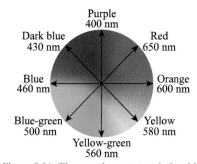

Figure 9.16 The complementary relationship of colors.

图 9.16 颜色的互补关系图。

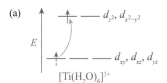

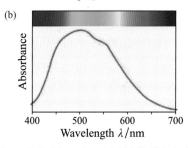

Figure 9.17 The color of $[Ti(H_2O)_6]^{3+}(aq)$. (a) The $3d^1$ electron configuration of Ti(III) and the corresponding electronic transition with energy in the green light region. (b) The absorption spectrum of $[Ti(H_2O)_6]^{3+}(aq)$ with the maximum absorbance at $\lambda = 510\ nm$.

图 9.17 $[Ti(H_2O)_6]^{3+}(aq)$ 的颜色：(a) 电子组态为 $3d^1$ 的 Ti（Ⅲ）及其对应于绿光能量范围的电子跃迁；(b) $[Ti(H_2O)_6]^{3+}(aq)$ 的吸收光谱，最大吸光度位于 $\lambda= 510\ nm$。

tetrahedral and square-planar geometries. The electron configuration of Ni is $[Ar]3d^8 4s^2$, and that of Ni(II) is $[Ar]3d^8$. All these eight $3d$ electrons must be paired due to the diamagnetic character. If the structure were tetrahedral, we would first place two pairs of electrons in the two lower d levels and then distribute the remaining four electrons among the three higher d levels. Two of the electrons would be unpaired [**Figure 9.20(a)**], and this complex anion would be paramagnetic. In a square-planar structure, we would place four pairs of electrons in the four lower d orbitals and leave the highest $d_{x^2-y^2}$ orbital empty [**Figure 9.20(b)**]. This corresponds to a diamagnetic complex anion. Therefore, the structure of $[Ni(CN)_4]^{2-}$ is square-planar according to its magnetic property.

9.5 Complex-Ion Equilibria

When a coordination compound is dissolved in water to form an aqueous solution, it will be ionized into complex ions and their counter ions. Meanwhile, the complex ions will be partially dissociated into central ions and ligands, among which a dynamic equilibrium, called **complex-ion equilibrium**, is established.

Formation/Stability Constant

The constant of the dynamic equilibrium for the formation of a complex in solution of a central ion and the ligands is called the **formation constant** (K_f), also referred to as the **stability constant**, of the complex. For example, the complex cation $[Zn(NH_3)_4]^{2+}$ can be formed from $Zn^{2+}(aq)$ and $NH_3(aq)$, as

$$Zn^{2+}(aq) + 4NH_3(aq) \rightleftharpoons [Zn(NH_3)_4]^{2+}(aq) \tag{9.7}$$

The formation constant is written as

$$K_f = \frac{\left[[Zn(NH_3)_4]^{2+}\right]}{[Zn^{2+}][NH_3]^4} = 2.9 \times 10^9$$

In fact, cations in aqueous solution exist mostly in hydrated form. That is, $Zn^{2+}(aq)$ is actually $[Zn(H_2O)_4]^{2+}$, which is a complex cation with H_2O molecules the ligands bonding to Zn^{2+}. As a result, when NH_3 molecules are introduced, these NH_3 molecules do not enter the empty coordination sphere of Zn^{2+}, but rather substitute for the H_2O ligands to bond to Zn^{2+}. This substitution occurs in a stepwise fashion. The first-step reaction

$$[Zn(H_2O)_4]^{2+} + NH_3 \rightleftharpoons [Zn(H_2O)_3(NH_3)]^{2+} + H_2O \tag{9.8}$$

for which

$$K_1 = \frac{\left[[Zn(H_2O)_3(NH_3)]^{2+}\right]}{\left[[Zn(H_2O)_4]^{2+}\right][NH_3]} = 2.3 \times 10^2$$

is followed by the second-step reaction

$$[Zn(H_2O)_3(NH_3)]^{2+} + NH_3 \rightleftharpoons [Zn(H_2O)_2(NH_3)_2]^{2+} + H_2O \tag{9.9}$$

for which

$$K_2 = \frac{\left[[Zn(H_2O)_2(NH_3)_2]^{2+}\right]}{\left[[Zn(H_2O)_3(NH_3)]^{2+}\right][NH_3]} = 2.8 \times 10^2$$

and so on. K_1, K_2, K_3, and K_4 are called the **stepwise formation constants**, referring to the formation of the complexes one step at a time.

The formation of $[Zn(H_2O)_2(NH_3)_2]^{2+}$ can also be represented by the sum of **Equations (9.8)** and **(9.9)**, as

$$\mu = \sqrt{n(n+2)}\mu_B \qquad (9.6)$$

其中 μ_B 称为玻尔磁子,是单个自由电子产生的磁矩单位。

2) **抗磁性**或**反磁性**:所有材料中均会产生与外磁场方向相反的感应磁场,从而具有被外磁场排斥的趋势。泡利不相容原理要求所有成对电子的固有磁矩必须指向相反的方向,因此相互抵消,表现出抗磁行为。尽管由于成对电子普遍存在,所有材料都具有抗磁性,但只有在没有未成对电子的纯抗磁性材料中,才能观察到抗磁行为。一旦存在任何未成对电子,顺磁行为将占主导地位。

3) **铁磁性**:是在含有铁、钴和镍的材料中所发现的一种特殊磁性。在固态铁磁性材料中,即使没有外磁场时,被称为磁畴的小区域内的电子磁矩也具有平行排布的倾向。在未磁化的材料中,磁畴的取向随机,因此磁矩相互抵消。然而,当置于外磁场中时,磁畴整齐排列,产生强的净磁矩。将材料从磁场中移除后,磁畴的有序性得以保持,从而产生永久磁性。随着温度的升高,热运动与铁磁排列的趋势相竞争。当超过一个称为**居里温度**的临界点后,铁磁性材料会变成顺磁性。**图 9.18** 比较了顺磁性和铁磁性。

配合物的磁性可用古埃磁天平测量。首先在没有磁场时称量样品的质量。开启磁场之后,原平衡状态被破坏。顺磁性物质将被拉入磁场,在如**图 9.19** 所示的实验装置下,其在磁场中的质量会增大。同时,反磁性物质则被推出磁场,在同一装置下其在磁场中的质量会减小。

顺磁性物质在磁场中质量增加的程度,取决于其未成对电子数。我们知道一个高自旋 d^n 配位单元,比低自旋 d^n 配位单元具有更多的未成对电子。因此,测量配合物在磁场中的质量变化,可以确定其是高自旋还是低自旋。配合物磁性的测量结果可用晶体场理论解释。

例如,实验表明配阴离子 $[Ni(CN)_4]^{2-}$ 为抗磁性,我们可以根据晶体场理论推测其可能的结构。配位数为 4 时,可以是四面体形或平面正方形几何构型。Ni 的电子组态为 $[Ar]3d^8 4s^2$,Ni(Ⅱ) 的电子组态为 $[Ar]3d^8$。由于呈抗磁性,所有八个 $3d$ 电子必须成对。如果结构是四面体形,我们会首先在两个较低的 d 能级中排布两对电子,再将剩余的四个电子分配到三个较高的 d 能级。其中有两个电子不成对 [**图 9.20(a)**],该配阴离子应呈顺磁性。在平面正方形结构中,我们将在四个较低的 d 轨道中排布四对电子,并使最高的 $d_{x^2-y^2}$ 轨道为空 [**图 9.20(b)**]。这对应于抗磁性配阴离子。因此根据其磁性,$[Ni(CN)_4]^{2-}$ 的结构应为平面正方形。

9.5 配位解离平衡

当配合物溶于水形成水溶液时,它会电离成配离子及其抗衡离子。同时,配离子又将部分解离成中心离子和配体,在它们之间建立的动态平衡,称为**配位解离平衡**(或配离子平衡)。

生成 / 稳定常数

由中心离子和配体在溶液中生成配位单元的动态平衡常数,称

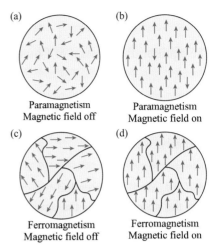

Figure 9.18 Comparison of paramagnetism and ferromagnetism. (a) A paramagnetic material with a magnetic field off. (b) A paramagnetic material with a magnetic field on. (c) A ferromagnetic material with a magnetic field off. (b) A ferromagnetic material with a magnetic field on.

图 9.18 顺磁性与铁磁性的比较:(a) 无磁场下的顺磁材料;(b) 磁场下的顺磁材料;(c) 无磁场下的铁磁材料;(d) 磁场下的铁磁材料。

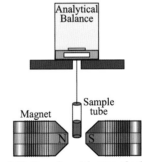

Figure 9.19 Experimental setup of a Gouy magnetic balance to measure the magnetism of a substance.

图 9.19 测量物质磁性的古埃磁天平装置图。

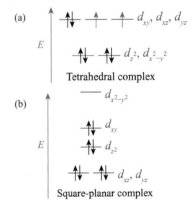

Figure 9.20 Prediction of the structure of $[Ni(CN)_4]^{2-}$ from its magnetic properties based on crystal field theory. (a) Two unpaired electrons would be present if the structure were tetrahedral, contradicting its diamagnetic nature. (b) All electrons are paired in a square-planar structure, in accordance to its diamagnetic nature.

图 9.20 基于晶体场理论由磁性预测 $[Ni(CN)_4]^{2-}$ 的结构:(a) 四面体形结构存在两个未成对电子,与其抗磁性相矛盾;(b) 平面正方形结构所有电子均成对,符合其抗磁性。

$$[Zn(H_2O)_4]^{2+} + 2NH_3 \rightleftharpoons [Zn(H_2O)_2(NH_3)_2]^{2+} + 2H_2O \tag{9.10}$$

The equilibrium constant of **Equation (9.10)**, represented by β_2, is given by

$$\beta_2 = \frac{[[Zn(H_2O)_2(NH_3)_2]^{2+}]}{[[Zn(H_2O)_4]^{2+}][NH_3]^2} = K_1 \cdot K_2 = 6.5 \times 10^4$$

Similarly, we have

$$\beta_1 = K_1, \quad \beta_3 = K_1 \cdot K_2 \cdot K_3, \quad \text{and} \quad \beta_4 = K_1 \cdot K_2 \cdot K_3 \cdot K_4$$

β_1, β_2, β_3, and β_4 are called the **cumulative formation constants**. The final cumulative formation constant, β_4 in the above case, is the equilibrium constant corresponding to the overall **Equation (9.7)**. Therefore, $\beta_4 = K_f$ and this is, by default, the overall formation constant of $[Zn(NH_3)_4]^{2+}$. The stepwise and overall formation constants for several complex ions are given in **Table 9.6**.

Table 9.6　Stepwise and Overall Formation/Stability Constants for Several Complex Ions
表 9.6　一些配离子的逐级与累积生成 / 稳定常数

Metal Ion (金属离子)	Ligand (配体)	K_1	K_2	K_3	K_4	K_5	K_6	K_f (or β_n)[a]
Ag^+	NH_3	1.7×10^3	6.5×10^3					1.1×10^7
Zn^{2+}	NH_3	2.3×10^2	2.8×10^2	3.2×10^2	1.4×10^2			2.9×10^9
Cu^{2+}	NH_3	2.0×10^4	4.7×10^3	1.1×10^3	2.0×10^2			2.1×10^{13}
Ni^{2+}	NH_3	6.3×10^2	1.7×10^2	5.4×10^1	1.5×10^1	5.6	1.1	5.5×10^8
Cu^{2+}	en	4.7×10^{10}	2.1×10^9					1.0×10^{20}
Ni^{2+}	en	3.3×10^7	2.1×10^6	3.1×10^4				2.1×10^{18}
Ni^{2+}	EDTA	3.6×10^{18}						3.6×10^{18}

[a] The β_n listed is for the number of steps shown: e.g., $K_f = \beta_2 = K_1 \cdot K_2$ for $[Ag(NH_3)_2]^+$; $K_f = \beta_3 = K_1 \cdot K_2 \cdot K_3$ for $[Ni(en)_3]^{2+}$; $K_f = \beta_1 = K_1$ for $[Ni(EDTA)]^{2+}$. （列出的 β_n 对应于如下步骤：例如，对于 $[Ag(NH_3)_2]^+$，$K_f = \beta_2 = K_1 \cdot K_2$；对于 $[Ni(en)_3]^{2+}$，$K_f = \beta_3 = K_1 \cdot K_2 \cdot K_3$；对于 $[Ni(EDTA)]^{2+}$，$K_f = \beta_1 = K_1$。）

The large numerical value of K_1 for **reaction (9.8)** indicates that Zn^{2+} has a greater affinity for NH_3, which is a stronger Lewis base, than it does for H_2O. The substitution of NH_3 for H_2O occurs even if the concentration of NH_3 is much smaller than that of H_2O, as in dilute $NH_3(aq)$. The fact that the successive K values decrease regularly in the substitution process, at least for most substitutions involving neutral molecules as ligands, is due in part to statistical factors. That is, an NH_3 molecule has a better chance of replacing a H_2O molecule in $[Zn(H_2O)_4]^{2+}$, in which each coordination position is occupied by H_2O, than in $[Zn(H_2O)_3(NH_3)]^{2+}$, in which one of the positions is already occupied by NH_3. Meanwhile, once the degree of substitution of NH_3 for H_2O has become large, the chance of H_2O to replace NH_3 in a reverse reaction increases, which also tends to reduce the value of K. If irregularities arise in the succession of K values, it is often because of a change in structure of the complex at some point in the series of substitution reactions.

Polydentate ligands substitute as many H_2O molecules as there are points of attachment. Thus, en replaces H_2O molecules in $[Ni(H_2O)_6]^{2+}$ two at a time, in three steps. Complex ions with polydentate ligands usually have much larger K_f than do those with monodentate ligands. This is known as the **chelation effect**, which can be partly attributed to the increase in entropy because of the increased number of particles after chelation.

Multiple Equilibria

In general, the complex-ion equilibrium between the metal ion (M^{n+}), the ligand (L^-), and the complex ion $ML_x^{(n-x)+}$ can be written as

$$M^{n+} + xL^- \rightleftharpoons ML_x^{(n-x)+} \tag{9.11}$$

If some other agents which can react with M^{n+} or L^- are introduced, these reactions may couple with the

为该配位单元的**生成常数**，也称**稳定常数**（K_f）。例如，配阳离子 $[Zn(NH_3)_4]^{2+}$ 可由 $Zn^{2+}(aq)$ 和 $NH_3(aq)$ 生成，有

$$Zn^{2+}(aq) + 4NH_3(aq) \rightleftharpoons [Zn(NH_3)_4]^{2+}(aq) \quad (9.7)$$

其生成常数可写为

$$K_f = \frac{[[Zn(NH_3)_4]^{2+}]}{[Zn^{2+}][NH_3]^4} = 2.9 \times 10^9$$

事实上，水溶液中的阳离子大多以水合形式存在。也就是说，$Zn^{2+}(aq)$ 实际上是 $[Zn(H_2O)_4]^{2+}$，这是一种以 H_2O 分子作为配体与 Zn^{2+} 键合的配阳离子。因此当引入 NH_3 分子时，这些 NH_3 分子不是进入了一个 Zn^{2+} 的空配位球，而是取代配体 H_2O 与 Zn^{2+} 键合。这种取代以分步的方式进行。第一步反应

$$[Zn(H_2O)_4]^{2+} + NH_3 \rightleftharpoons [Zn(H_2O)_3(NH_3)]^{2+} + H_2O \quad (9.8)$$

$$K_1 = \frac{[[Zn(H_2O)_3(NH_3)]^{2+}]}{[[Zn(H_2O)_4]^{2+}][NH_3]} = 2.3 \times 10^2$$

之后进行第二步反应

$$[Zn(H_2O)_3(NH_3)]^{2+} + NH_3 \rightleftharpoons [Zn(H_2O)_2(NH_3)_2]^{2+} + H_2O$$
$$(9.9)$$

$$K_2 = \frac{[[Zn(H_2O)_2(NH_3)_2]^{2+}]}{[[Zn(H_2O)_3(NH_3)]^{2+}][NH_3]} = 2.8 \times 10^2$$

以此类推。K_1、K_2、K_3 和 K_4 称为**逐级生成常数**，指的是分步逐级地生成配位单元。

$[Zn(H_2O)_2(NH_3)_2]^{2+}$ 的生成也可由**式（9.8）**和**（9.9）**之和得到，有

$$[Zn(H_2O)_4]^{2+} + 2NH_3 \rightleftharpoons [Zn(H_2O)_2(NH_3)_2]^{2+} + 2H_2O \quad (9.10)$$

式（9.10）的平衡常数用 β_2 表示，有

$$\beta_2 = \frac{[[Zn(H_2O)_2(NH_3)_2]^{2+}]}{[[Zn(H_2O)_4]^{2+}][NH_3]^2} = K_1 \cdot K_2 = 6.5 \times 10^4$$

类似地，有

$$\beta_1 = K_1, \quad \beta_3 = K_1 \cdot K_2 \cdot K_3, \quad \text{且} \quad \beta_4 = K_1 \cdot K_2 \cdot K_3 \cdot K_4$$

β_1、β_2、β_3 和 β_4 称为**累积生成常数**。最终的累积生成常数（上述情况下为 β_4），即为总方程[**式（9.7）**]对应的平衡常数。因此，$\beta_4 = K_f$，即为 $[Zn(H_2O)_4]^{2+}$ 默认的总生成常数。**表 9.6** 给出了几种配离子的逐级和累积生成常数。

反应（9.8）的 K_1 较大，这表明 Zn^{2+} 对 NH_3（更强的路易斯碱）的亲和力大于对 H_2O 的亲和力。即使 NH_3 的浓度比 H_2O 的浓度小得多，如在稀 NH_3（aq）中，也会发生 NH_3 对 H_2O 的取代。在取代过程中（至少对于大多数涉及中性分子配体的取代过程而言），K 逐级有规律地减小，其部分原因来自统计因素。也就是说，相对于一

above complex-ion equilibrium to form multiple equilibria. Here, we give some illustrative examples of how complex-ion equilibrium is coupled with other equilibria.

1) Complex-ion equilibrium coupled with acid-base equilibrium

Some complex ions may act as Brønsted-Lowry acids or bases to donate or accept protons. For example, Fe^{3+}(aq), which is actually $[Fe(H_2O)_6]^{3+}$, may undergo the following hydrolysis reaction

$$[Fe(H_2O)_6]^{3+} + H_2O \rightleftharpoons [Fe(OH)(H_2O)_5]^{3+} + H_3O^+ \tag{9.12}$$

in which the hexaaquairon(III) ion serves as an acid to donate a proton from one of its ligand water molecules to a solvent water molecule with an acid ionization constant $K_{a1} = 9 \times 10^{-4}$. From this K_{a1} value, we see that Fe^{3+}(aq) is even more acidic than acetic acid with $K_a = 1.8 \times 10^{-5}$. To suppress the hydrolysis of $[Fe(H_2O)_6]^{3+}$, a low pH by the addition of strong acids such as HNO_3 or $HClO_4$ is needed. The $[Fe(H_2O)_6]^{3+}$ ion is violet in color, but the Fe^{3+}(aq) solution are generally yellow due to the presence of hydroxo complex ions. The acid strength of aqua complex ions depends on the charge-to-radius ratio of the central metal ion. A small and highly charged central ion attracts electrons away from an O—H bond in a ligand water molecule strongly. Thus, $[Fe(H_2O)_6]^{3+}$ is a stronger acid than $[Fe(H_2O)_6]^{2+}$ ($K_{a1} = 1 \times 10^{-7}$).

Some ligands such as F^-, CN^-, and CO_3^{2-} can form weak acids with introduced H_3O^+ and cause the complex-ion equilibrium to shift. For example, when H_3O^+ is introduced to a FeF_3 solution

$$F^- + H_3O^+ \rightleftharpoons HF + H_2O \tag{9.13}$$

$$Fe^{3+} + 3F^- \rightleftharpoons FeF_3 \tag{9.14}$$

the acid-base equilibrium **(9.13)** shifts to the right and $[F^-]$ is reduced. Then, the complex-ion equilibrium **(9.14)** shifts to the left and $[FeF_3]$ is reduced. For the overall reaction from $3 \times$ **(9.13)**−**(9.14)**

$$FeF_3 + 3H_3O^+ \rightleftharpoons Fe^{3+} + 3HF + 3H_2O \tag{9.15}$$

the equilibrium constant is given by

$$K = \frac{1}{K_f(FeF_3) \cdot K_a^3(HF)}$$

These complexes are more easily dissolved by the introduced H_3O^+ at a smaller K_f (meaning that the complex is less stable) and a smaller K_a (meaning that the acid of the ligand is weaker).

2) Complex-ion equilibrium coupled with solubility equilibrium

A metal ion can simultaneously form a complex ion equilibrium with a ligand and a solubility equilibrium with a precipitation agent. These two equilibria compete with each other. We can tell the final product by comparing the corresponding K_f and K_{sp}. For example

$$Ag^+ + 2NH_3 \rightleftharpoons [Ag(NH_3)_2]^+ \tag{9.16}$$

$$Ag^+ + Cl^- \rightleftharpoons AgCl \tag{9.17}$$

By applying **(9.16)**−**(9.17)**, we have

$$AgCl + 2NH_3 \rightleftharpoons [Ag(NH_3)_2]^+ + Cl^- \tag{9.18}$$

with an equilibrium constant of

$$K = K_f([Ag(NH_3)_2]^+) \cdot K_{sp}(AgCl) = 2.0 \times 10^{-3}$$

Although this K is not very large, the precipitate AgCl(s) can be dissolved to form the complex ion $[Ag(NH_3)_2]^+$ in relatively concentrated NH_3(aq).

3) Complex-ion equilibrium coupled with redox equilibrium

The complex-ion equilibrium and redox equilibrium can also be coupled. For example, we know that gold cannot be dissolved in either concentrated nitric acid or concentrated hydrochloric acid but can react with their 1:3 (volume ratio) mixture called aqua regia. This is due to the enhanced oxidation ability of NO_3^-

个位置已被 NH_3 占据的 $[Zn(H_2O)_3(NH_3)]^{2+}$ 而言,在所有配位点均为 H_2O 占据的 $[Zn(H_2O)_4]^{2+}$ 中,NH_3 分子取代 H_2O 分子的机会更大。同时,一旦 NH_3 对 H_2O 的取代程度变大,H_2O 在逆反应中取代 NH_3 的机会就会增加,这也倾向于降低 K 的值。如果逐级 K 中出现不规则的情况,通常是因为在一系列取代反应的某个点上,配位单元的结构发生了改变。

多齿配体取代 H_2O 分子的数量与其配位点数相同。因此,乙二胺可以通过每步两个、总共三步取代掉 $[Ni(H_2O)_6]^{2+}$ 的 H_2O 分子。与单齿配体的配离子相比,多齿配体的配离子通常具有更大的 K_f。这就是所谓的**螯合效应**,可部分归因于螯合后粒子数量增加的熵增效应。

多重平衡

总体而言,金属离子(M^{n+})、配体(L^-)以及配离子 $ML_x^{(n-x)+}$ 之间的配位解离平衡可写为

$$M^{n+} + xL^- \rightleftharpoons ML_x^{(n-x)+} \qquad (9.11)$$

如果加入可与 M^{n+} 或 L^- 发生反应的一些其他试剂,这些反应会与上述配位解离平衡耦合,形成多重平衡。这里,我们给出配位解离平衡与其他平衡发生耦合的一些示例。

1) 配位解离平衡与酸碱电离平衡相耦合

一些配离子可作为布朗斯特 - 劳莱酸或碱,来提供或接受质子。例如,$Fe^{3+}(aq)$ 实际上是 $[Fe(H_2O)_6]^{3+}$,可以经历如下水解反应:

$$[Fe(H_2O)_6]^{3+} + H_2O \rightleftharpoons [Fe(OH)(H_2O)_5]^{3+} + H_3O^+ \qquad (9.12)$$

其中六水合铁(Ⅲ)离子作为酸,从其一个水分子配体中提供一个质子给一个溶剂水分子,其酸式电离常数 $K_{a1} = 9 \times 10^{-4}$。从这个 K_{a1} 我们可以看出,$Fe^{3+}(aq)$ 的酸性甚至比醋酸($K_a = 1.75 \times 10^{-5}$)更强。为抑制 $[Fe(H_2O)_6]^{3+}$ 的水解,需要添加如 HNO_3 或 $HClO_4$ 等强酸来降低 pH。$[Fe(H_2O)_6]^{3+}$ 离子呈紫色,但由于羟基配离子的存在,Fe^{3+} 溶液通常呈黄色。含水配体的配离子的酸性,取决于中心金属离子的电荷半径比。半径小而电荷高的中心离子,会强烈地吸引电子远离配体水分子的 O—H 键。因此,$[Fe(H_2O)_6]^{3+}$ 是比 $[Fe(H_2O)_6]^{2+}$($K_{a1} = 1 \times 10^{-7}$)更强的酸。

一些配体(如 F^-、CN^- 和 CO_3^{2-})可以与加入的 H_3O^+ 生成弱酸,并导致配位解离平衡发生移动。例如,在 FeF_3 溶液中加入 H_3O^+ 时

$$F^- + H_3O^+ \rightleftharpoons HF + H_2O \qquad (9.13)$$

$$Fe^{3+} + 3F^- \rightleftharpoons FeF_3 \qquad (9.14)$$

酸碱平衡**式(9.13)** 向右移动,使得 $[F^-]$ 减小。然后配位解离平衡**式(9.14)** 向左移动,使得 $[FeF_3]$ 减小。对于 $3 \times$ **(9.13)** − **(9.14)** 的总反应

$$FeF_3 + 3H_3O^+ \rightleftharpoons Fe^{3+} + 3HF + 3H_2O \qquad (9.15)$$

平衡常数为

$$K = \frac{1}{K_f(FeF_3) \cdot K_a^3(HF)}$$

当 K_f 较小(意味着配位单元更不稳定)和 K_a 较小(意味着配体形成的酸更弱)时,这些配位单元更容易被引入的 H_3O^+ 溶解。

in the presence of the complex ion [AuCl$_4$]$^-$. Given that

$$Au^{3+} + 3e^- \rightleftharpoons Au \quad E_1^\circ = 1.498 \text{ V} \quad (9.19)$$

$$4H^+ + NO_3^- + 3e^- \rightleftharpoons NO + 2H_2O \quad E_2^\circ = 0.96 \text{ V} \quad (9.20)$$

$$Au^{3+} + 4Cl^- \rightleftharpoons [AuCl_4]^- \quad K_f = 1.4 \times 10^{25} \quad (9.21)$$

By applying **(9.20)− (9.19)**, we have

$$Au + 4H^+ + NO_3^- \rightleftharpoons Au^{3+} + NO + 2H_2O \quad (9.22)$$

for which

$$K_4^\circ = 10^{\frac{3 \times (E_2^\circ - E_1^\circ)}{0.0592}} = 5 \times 10^{-28}$$

Using **(9.20)− (9.19) + (9.21)**, we obtain

$$Au + 4H^+ + NO_3^- + 4Cl^- \rightleftharpoons [AuCl_4]^- + NO + 2H_2O \quad (9.23)$$

for which

$$K_5^\circ = K_4^\circ \cdot K_f = 7 \times 10^{-3}$$

K_5° is rather small, indicating that it is very difficult for Au to react with HNO$_3$ alone. In the presence of [AuCl$_4$]$^-$, however, K_5° becomes considerable at standard conditions with [H$^+$] = 1 mol L^{-1}. In practice, [H$^+$] in aqua regia is much greater than 1 mol L^{-1}, and thus gold can be dissolved in aqua regia.

4) Coupled complex-ion equilibria

In the presence of more than one ligand, a metal ion can form different complexes via different complex-ion equilibria. These equilibria also compete with one another. For example,

$$Fe^{3+} + 3F^- \rightleftharpoons FeF_3 \quad (9.24)$$

$$Fe^{3+} + SCN^- \rightleftharpoons [Fe(NCS)]^{2+} \quad (9.25)$$

By apply **(9.24) − (9.25)**, we have

$$[Fe(NCS)]^{2+} + 3F^- \rightleftharpoons FeF_3 + SCN^- \quad (9.26)$$

for which

$$K = \frac{K_f(FeF_3)}{K_f([Fe(NCS)]^{2+})} = 4.8 \times 10^8$$

This K is very large, and thus almost all Fe^{3+} will form FeF$_3$ instead of [Fe(NCS)]$^{2+}$ at comparable concentrations of F$^-$ and SCN$^-$.

9.6 Applications of Coordination Chemistry

Coordination chemistry is the subbranch of inorganic chemistry that studies the characteristics, bonding, structure, properties, preparation, and reactions of coordination compounds. The applications of coordination chemistry are widely distributed in chemical analysis, catalysis, biomedicine, etc. Here, we will give several illustrative examples.

Cisplatin: A Cancer-Fighting Drug

Chemotherapy is a type of cancer treatment that utilizes one or more anti-cancer drugs to destroy cancer cells. Cisplatin is a commonly used anti-cancer drug to treat testicular, ovarian, cervical, bladder, esophageal, and lung cancers. It is generally given by injection into a vein.

2) 配位解离平衡与沉淀溶解平衡相耦合

金属离子可同时与配体形成配位解离平衡，与沉淀剂形成沉淀溶解平衡，这两种平衡相互竞争。我们可以通过比较相应的 K_f 和 K_{sp} 来判断最终产物。例如

$$Ag^+ + 2NH_3 \rightleftharpoons [Ag(NH_3)_2]^+ \quad (9.16)$$

$$Ag^+ + Cl^- \rightleftharpoons AgCl \quad (9.17)$$

(9.16) − (9.17)，可得

$$AgCl + 2NH_3 \rightleftharpoons [Ag(NH_3)_2]^+ + Cl^- \quad (9.18)$$

平衡常数为

$$K = K_f\left([Ag(NH_3)_2]^+\right) \cdot K_{sp}(AgCl) = 2.0 \times 10^{-3}$$

尽管这个 K 不是很大，但 AgCl(s) 沉淀可以溶解在较浓的 NH_3(aq) 中，生成 $[Ag(NH_3)_2]^+$ 配离子。

3) 配位解离平衡与氧化还原平衡相耦合

配位解离平衡与氧化还原平衡也可以耦合。例如，我们知道金不溶于浓硝酸或浓盐酸，但可以与它们的体积比为 1:3 的混合物（称为王水）反应。这是由于配离子 $[AuCl_4]^-$ 的存在增强了 NO_3^- 的氧化能力。已知

$$Au^{3+} + 3e^- \rightleftharpoons Au \quad E_1^\ominus = 1.498V \quad (9.19)$$

$$4H^+ + NO_3^- + 3e^- \rightleftharpoons NO + 2H_2O \quad E_2^\ominus = 0.96V \quad (9.20)$$

$$Au^{3+} + 4Cl^- \rightleftharpoons [AuCl_4]^- \quad K_f = 1.4 \times 10^{25} \quad (9.21)$$

(9.20) − (9.19)，可得

$$Au + 4H^+ + NO_3^- \rightleftharpoons Au^{3+} + NO + 2H_2O \quad (9.22)$$

$$K_4^\ominus = 10^{\frac{3 \times (E_2^\ominus - E_1^\ominus)}{0.0592}} = 5 \times 10^{-28}$$

(9.20) − (9.19) + (9.21)，可得

$$Au + 4H^+ + NO_3^- + 4Cl^- \rightleftharpoons [AuCl_4]^- + NO + 2H_2O \quad (9.23)$$

$$K_5^\ominus = K_4^\ominus \cdot K_f = 7 \times 10^{-3}$$

$K_4^\ominus$ 值相当小，说明 Au 很难单独与 HNO_3 反应。然而当 $[AuCl_4]^-$ 存在时，在 $[H^+] = 1\ mol\ L^{-1}$ 的标准条件下，$K_5^\ominus$ 变得相当可观。实际上，王水中的 $[H^+]$ 远大于 $1\ mol\ L^{-1}$，因此金可以溶解在王水中。

4) 耦合的配位解离平衡

当多个配体存在时，金属离子可以通过不同的配位解离平衡生成不同的配位单元，这些平衡也存在相互竞争。例如

$$Fe^{3+} + 3F^- \rightleftharpoons FeF_3 \quad (9.24)$$

$$Fe^{3+} + SCN^- \rightleftharpoons [Fe(NCS)]^{2+} \quad (9.25)$$

(9.24) − (9.25)，可得

$$[Fe(NCS)]^{2+} + 3F^- \rightleftharpoons FeF_3 + SCN^- \quad (9.26)$$

$$K = \frac{K_f(FeF_3)}{K_f\left([Fe(NCS)]^{2+}\right)} = 4.8 \times 10^8$$

这个 K 非常大，因此当 F^- 和 SCN^- 浓度相当时，几乎所有 Fe^{3+} 都将形成 FeF_3 而非 $[Fe(NCS)]^{2+}$。

Cisplatin was first synthesized by the Italian chemist Michele Peyrone in 1845 and has been known for a long time as Peyrone's salt. Its structure was elucidated by Alfred Werner in 1893. In 1965, Barnett Rosenberg (1926—2009) discovered the anti-cancer activity of cisplatin, and reported that transplatin, the *trans*-stereoisomer of cisplatin, was ineffective. Cisplatin was licensed for medical use in the USA in 1978 and in the UK and several other European countries in 1979.

Cisplatin is the common name of the square planar complex *cis*-$[PtCl_2(NH_3)_2]$. One method to synthesize cisplatin is shown in **Figure 9.21**. One obstacle is the facile formation of Magnus's green salt (MGS, $[Pt(NH_3)_4][PtCl_4]$), which has the same empirical formula as cisplatin. The traditional way to avoid MGS involves the conversion of $K_2[PtCl_4]$ to $K_2[PtI_4]$. The key step (3) to form the *cis* isomer is governed by the *trans* effect. Because NH_3 is a weaker *trans* director than I^-, the second NH_3 molecule is preferentially directed to a position that is *trans* to I^-, rather than being directed to a position that is *trans* to NH_3.

The anti-cancer activity of cisplatin is achieved by interfering with DNA replication, which kills the fastest proliferating cells, presumably, cancerous. Cisplatin enters cancer cells mainly by diffusion. Once inside the cell, cisplatin undergoes an aquation process, in which one Cl^- ligand is replaced by a H_2O molecule. Dissociation of Cl^- is favored inside the cell because the intracellular chloride concentration is much lower than the chloride concentration in the extracellular fluid. The water molecule in *cis*-$[PtCl(NH_3)_2(H_2O)]^+$ can be easily displaced by the *N*-heterocyclic based on DNA. The subsequent crosslinking can occur via displacement of the other Cl^-. Cisplatin can crosslink DNA in several different ways, interfering with the normal division of cells by mitosis. The damaged DNA elicits DNA repair mechanisms, which in turn activate apoptosis when repair proves impossible.

Transplatin does not exhibit a comparably useful pharmacological effect. Two mechanisms have been suggested to explain the reduced anti-cancer effect of transplatins. One is that the *trans* arrangement of the Cl^- ligands is thought to confer transplatin with greater chemical reactivity, causing transplatin to become deactivated before it reaches the DNA, where cisplatin exerts its pharmacological action. The other is that the stereo-conformation of transplatin is such that it is unable to form the characteristic adducts formed by cisplatin in abundance.

Despite a number of side effects of cisplatin, such as kidney damage, nerve damage, nausea and vomiting, hearing loss, and electrolyte disturbance, cisplatin combination chemotherapy is the cornerstone of treatment of many cancers. To date, thousands of Pt-containing compounds have been investigated as potential chemotherapy drugs. Worldwide annual sales of Pt-based anti-cancer drugs are currently in excess of $2 billion.

Porphins and Porphyrins

Porphyrins [**Figure 9.22(a)**] are a group of heterocyclic macrocycle organic compounds, composed of four modified pyrrole subunits interconnected at their α carbon atoms via methine bridges (=CH—). If all the R groups are H atoms, the molecule is called **porphin** [$C_{20}H_{14}N_4$, **Figure 9.22(b)**]. A porphin is the simplest porphyrin and a porphyrin is a substituted porphin.

A porphyrin has an inner cycle with 18 π electrons and an outer cycle with 26 π electrons. Both cycles meet the $4n + 2$ rule and are aromatic. The large conjugated system in porphyrins leads to strong absorption in the visible region of light, and thus porphyrins are deeply colored. With the two H atoms on the central N being removed, porphyrin can serve as a tetradentate ligand with the four N atoms lying on a plane and simultaneously coordinating to a metal ion. The resulted complexes, called **metalporphyrins** [**Figure 9.22(c)**], are commonly found in animals and plants.

Heme is an octahedral iron-porphyrin complex, consisting of an Fe(II) ion coordinated to a porphyrin acting as a planar tetradentate ligand, and to one or two axial ligands. Heme is a substance precursive to hemoglobin, which is the oxygen-transport protein in the red blood cells. Heme is biosynthesized in both the bone marrow and the liver. In an oxygen enriched environment, such as in the lungs or gills, the O_2 molecule serves as one of the axial ligands to bind with the central Fe(II) ion. After being transported to the rest of the body, which is an oxygen poor environment, O_2 is released to permit aerobic respiration and to provide energy via metabolism. The coordination bond between heme and the ligand O_2 molecule is medium

9.6 配位化学的应用

配位化学是研究配合物的特征、成键、结构、性质、制备以及反应的无机化学分支。配位化学的应用广泛分布于化学分析、催化、生物医药等领域。这里我们将给出几个示例。

顺铂：一种抗癌药

化疗是利用一种或多种抗癌药物摧毁癌细胞的癌症治疗方法。顺铂是一种常用的抗癌药物，可用于治疗睾丸癌、卵巢癌、宫颈癌、膀胱癌、食管癌和肺癌等，一般通过静脉注射给药。

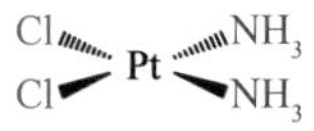

cis-$[Pt(NH_3)_2Cl_2]$
cisplatin(顺铂)

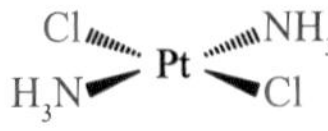

trans-$[Pt(NH_3)_2Cl_2]$
transplatin(反铂)

顺铂最早由意大利化学家米歇尔·佩龙于 1845 年合成，并长期称为佩龙盐。1893 年，其结构由阿尔弗雷德·维尔纳阐明。1965 年，巴内特·罗森堡（1926—2009）发现了顺铂的抗癌活性，并报告称顺铂的反式立体异构体反铂没有活性。顺铂于 1978 年在美国获得医疗许可，1979 年在英国和其他几个欧洲国家获得医疗许可。

顺铂是平面正方形配位单元 *cis*-$[PtCl_2(NH_3)_2]$ 的俗名。**图 9.21** 给出了顺铂的一种合成方法。该方法的一个阻碍是容易形成马格努斯绿盐（MGS，$[Pt(NH_3)_4][PtCl_4]$），其经验式与顺铂相同。避免生成 MGS 的传统方法，是将 $K_2[PtCl_4]$ 转换为 $K_2[PtI_4]$。形成顺式异构体的关键步骤（3）由反式效应控制。由于 NH_3 是比 I^- 更弱的反式导向剂，第二个 NH_3 分子会优先导向至 I^- 的反位，而非 NH_3 的反位。

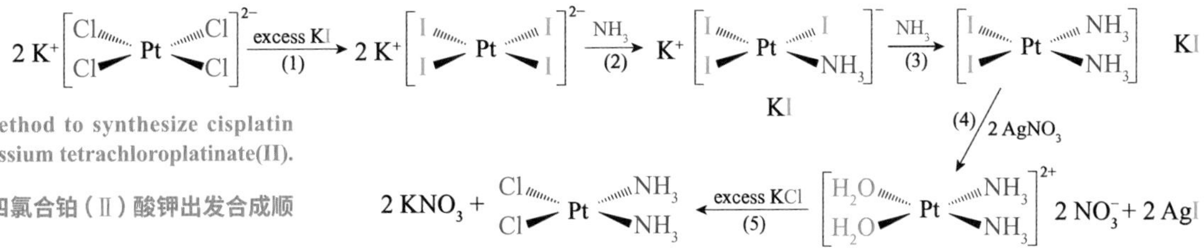

Figure 9.21 A method to synthesize cisplatin starting from potassium tetrachloroplatinate(II).

图 9.21　一种从四氯合铂（Ⅱ）酸钾出发合成顺铂的方法。

顺铂的抗癌活性通过干扰 DNA 的复制来实现，从而杀死增殖最快的细胞，而这些细胞最有可能是癌细胞。顺铂主要通过扩散进入癌细胞。一旦进入细胞，顺铂会发生水合，其中一个 Cl^- 配体被 H_2O 分子取代。由于细胞内氯化物的浓度远低于细胞外液中氯化物的浓度，因此细胞内有利于 Cl^- 的解离。*cis*-$[PtCl(NH_3)_2(H_2O)]^+$ 中的水分子很容易被 DNA 上的 *N*- 杂环碱基取代。随后的交联可以通过取代其他 Cl^- 发生。顺铂可以采用几种不同的方式交联 DNA，通过有丝分裂干扰细胞的正常分裂。受损的 DNA 会引发 DNA 修复机制，当修复被证明不可能时，DNA 修复机制反过来会激活细胞凋亡。

反铂没有显示出同等疗效的药理作用，学界提出了两种机制来解释反铂降低的抗癌作用。一种是 Cl^- 配体的反式结构使反铂具有更高的化学反应性，导致反铂在到达顺铂能发挥药理作用的 DNA 之前就已经失活。另一种是反铂的立体构象使其无法大量形成顺铂所能形成的特征加合物。

尽管顺铂存在许多副作用，如肾损伤、神经损伤、恶心呕吐、听力下降和电解质紊乱等，但顺铂联合化疗是许多癌症治疗的基石。迄今为止，已有数千种含铂化合物作为潜在的化疗药物被研究。目前，铂类抗癌药物的全球年销售额已超过 20 亿美元。

in strength, enabling the shift in the complex-ion equilibrium at different O_2 concentrations. Meanwhile, CO binds to the Fe(II) ion in hemoglobin much more strongly than does O_2. Thus, the toxicity of CO arises because it prevents hemoglobin from binding with O_2.

Chlorophyll and Photosynthesis Reactions

Chlorophyll is any of several related green pigments found in the chloroplasts of algae and plants. It can absorb sunlight and catalyze the photosynthesis reaction to form carbohydrates from CO_2 and H_2O, as

$$nCO_2 + nH_2O \xrightarrow[\text{chlorophyll}]{\text{sunlight}} (CH_2O)_n + nO_2 \tag{9.27}$$

Figure 9.23(a) shows the structure of the most widely distributed chlorophyll in terrestrial plants: chlorophyll *a*, which is a magnesium porphyrin complex. The structure of chlorophyll *b* differs from that of chlorophyll *a* only in that one of the methyl (—CH_3) groups in chlorophyll *a* is replaced by a formyl (—CHO) group. This difference causes a considerable change in the absorption spectrum [**Figure 9.23(b)**]. As can be seen, both chlorophyll *a* and *b* have strong absorption bands in the red and purple regions of visible light, which is the complementary color of green. This explains why chlorophyll is green and also suggests that green plants should grow more readily in red light. Some experimental evidence indicates that this is the case. For example, the maximum rate of formation of $O_2(g)$ by **reaction (9.27)** occurs with a red light.

Extended Reading Materials Antiferromagnetism and Ferrimagnetism

In addition to paramagnetism, diamagnetism, and ferromagnetism, some materials can exhibit other types of magnetization properties such as antiferromagnetism and ferrimagnetism. Similar to ferromagnetism, both antiferromagnetism and ferrimagnetism are manifestations of ordered magnetism.

The phenomenon of **antiferromagnetism** was first introduced by Lev Landau (1908—1968) in 1933. In antiferromagnetic materials, the electron magnetic moments of atoms or molecules align in a regular pattern with neighboring spins pointing in opposite directions and cancel out, resulting in a zero net magnetic moment [**Figures 9.24(a)(b)**]. In general, antiferromagnetic orders may exist at sufficiently low temperatures, but vanish at and above the Néel temperature, which was named after Louis Néel (1904—2000), who had first identified this type of magnetic ordering. Above the Néel temperature, antiferromagnetic materials are typically paramagnetic.

Néel also discovered **ferrimagnetism** in 1948 and was awarded the Nobel Prize in physics in 1970 for his contribution on magnetism. Ferrimagnetism is a type of cooperative magnetism similar to antiferromagnetism, except that the electron magnetic moment in one direction is larger than that in the opposite direction, so magnetization is retained in the absence of an external magnetic field [**Figures 9.24(c)(d)**]. Like ferromagnetic materials, ferrimagnetic materials also have a critical Curie temperature above which they become paramagnetic.

Problems

9.1 Following are the names of five coordination compounds containing complexes with platinum(II) as the central metal ion and ammonia molecules and/or chloride ions as ligands: (a) potassium amminetrichloroplatinate(II), (b) diamminedichloroplatinum(II), (c) triamminechloroplatinum(II) chloride, (d) tetraammineplatinum(II) chloride, (e) potassium tetrachloroplatinate(II). Sketch the expected graph of the electric conductivity versus the chlorine content of the compounds.

9.2 A Cu electrode is immersed in a solution that is 1.00 mol L^{-1} NH_3 and 1.00 mol L^{-1} $[Cu(NH_3)_4]^{2+}$. If a standard hydrogen electrode is the cathode, E_{cell} is +0.08 V. What is the value obtained by this method for the formation constant, K_f, of $[Cu(NH_3)_4]^{2+}$?

9.3 The amino acid glycine (NH_2CH_2COOH, denoted Hgly) binds to an anion and is a bidentate ligand. Draw and name all possible isomers of $[Co(gly)_3]$. How many isomers are possible for the compound $[Co(gly)_2Cl(NH_3)]$?

卟吩与卟啉

卟啉 [**图 9.22(a)**] 是一类由四个修饰的吡咯亚基的 α-碳原子通过次甲基桥（═CH—）互联而成的杂环大环有机化合物。如果所有 R 基团均为 H 原子，该分子称为**卟吩** [$C_{20}H_{14}N_4$，**图 9.22(b)**]。卟吩是最简单的卟啉，而卟啉是具有取代基的卟吩。

卟啉的内环有 18 个 π 电子，外环有 26 个 π 电子。两个环均符合 $4n+2$ 规则，具有芳香性。卟啉的大共轭体系使其在可见光区存在强吸收，因此显深色。如果中心 N 上的两个 H 原子被去除，卟啉可作为四齿配体，其中四个 N 原子位于同一平面，并同时与金属离子配位。生成的配位单元称为**金属卟啉** [**图 9.22(c)**]，常在动植物体内发现。

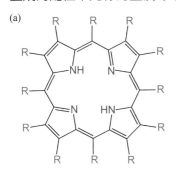
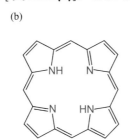
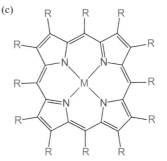

Figure 9.22 The structures of (a) porphyrins, (b) porphin, and (c) metal porphyrins.

图 9.22 (a) 卟啉、(b) 卟吩与 (c) 金属卟啉的结构。

血红素是一种八面体形铁卟啉配位单元，包含一个与平面四齿配体卟啉以及一个或两个轴向配体配位的 Fe(Ⅱ) 离子。血红素是血红蛋白的前体，而血红蛋白是红细胞中的氧转运蛋白。血红素通常在骨髓和肝脏中生物合成。在富氧环境中，如在肺或鳃中，O_2 分子可作为轴向配体之一与中心 Fe(Ⅱ) 离子结合。当被输送到身体其他部位（缺氧环境）后，可释放出氧气，进行有氧呼吸，并通过新陈代谢提供能量。血红素与配体 O_2 分子之间配位键的强度适中，能够在不同 O_2 浓度下使其配位解离平衡发生移动。而 CO 与血红蛋白中 Fe(Ⅱ) 离子的结合比 O_2 强得多，因此 CO 的毒性来源于其阻止了血红蛋白与 O_2 的结合。

叶绿素与光合反应

叶绿素是在藻类和植物的叶绿体中发现的几种相关绿色色素之一。它可以吸收阳光并催化光合反应，从 CO_2 和 H_2O 合成碳水化合物，如

$$n\mathrm{CO}_2 + n\mathrm{H}_2\mathrm{O} \xrightarrow[\text{叶绿素}]{\text{光照}} (\mathrm{CH}_2\mathrm{O})_n + n\mathrm{O}_2 \quad (9.27)$$

图 9.23(a) 给出了陆地植物中分布最广的叶绿素 a 的结构，它是一种镁卟啉配位单元。叶绿素 b 与叶绿素 a 的结构不同之处在于，叶绿素 a 中的一个甲基（—CH_3）被甲酰基（—CHO）取代。这种差异导致其吸收光谱发生显著变化 [**图 9.23(b)**]。可以看到，叶绿素 a 和 b 在可见光的红色和紫色区域均存在很强的吸收峰，而它们正是绿色的互补色。这解释了为什么叶绿素是绿色的，也说明绿色植物在红光下应该更容易生长。一些实验证据表明情况确实如此。例如，由**反应 (9.27)** 生成 $O_2(g)$ 的最大速率发生在红光下。

拓展阅读材料 反铁磁性与亚铁磁性

除了顺磁性、抗磁性和铁磁性外，一些材料还可以表现出其他类型的磁性，如反铁磁性和亚铁磁性。与铁磁性类似，反铁磁性和

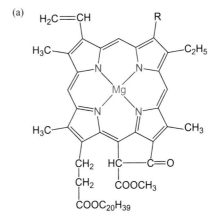

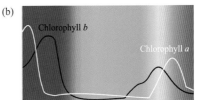

Figure 9.23 Chlorophyll and the absorption spectra. (a) The structures of chlorophyll a (R = CH_3) and chlorophyll b (R = CHO). (b) The absorption spectra of chlorophyll a and b, both with strong absorption bands in the red and purple regions.

图 9.23 叶绿素及其吸收光谱：(a) 叶绿素 a（R = CH_3）与叶绿素 b（R = CHO）的结构；(b) 叶绿素 a 与 b 的吸收光谱，二者在红光和紫光区均存在强吸收峰。

9.4 A structure that Werner examined as a possible alternative to the octahedron is the trigonal prism.

(a) Does this structure predict the correct number of isomers for the complex ion $[CoCl_2(NH_3)_4]^+$? Why or why not?

(b) Does this structure account for optical isomerism in $[Co(en)_3]^{3+}$? Why or why not?

*9.5 **Table P9.1** contains the enthalpy of hydration for the reaction

$$M^{2+}(g) + 6H_2O(l) \rightarrow [M(H_2O)_6]^{2+}(aq)$$

(a) Plot the hydration enthalpies as a function of the atomic number of the metals shown.

Table P9.1 (表 P9.1)

Divalent Metal (二价金属)	Hydration Enthalpy (水合焓) /(kJ mol^{-1})
Ca	−2468
Sc	−2673
Ti	−2750
V	−2814
Cr	−2799
Mn	−2743
Fe	−2843
Co	−2904
Ni	−2986
Cu	−2989
Zn	−2939

(b) Assuming that all the hexaaqua complexes are high spin, which ions have zero CFSE?

(c) If lines are drawn between those ions with CFSE = 0, a line of negative slope is obtained. Explain.

(d) The ions with CFSE ≠ 0 have hydration enthalpies that are more negative than the lines drawn in part (c). What is the explanation for this?

(e) Estimate the value of Δ_o for the Fe(II) ion in an octahedral field of water molecules.

(f) What wavelength of light would the $[Fe(H_2O)_6]^{2+}$ ion absorb?

亚铁磁性均是有序磁性的表现。

反铁磁现象最早由列夫·朗道（1908—1968）于 1933 年提出。在反铁磁性材料中，原子或分子的电子磁矩以相邻自旋指向相反方向的规律模式排列，并且相互抵消，从而导致零净磁矩 [**图 9.24(a)(b)**]。一般而言，反铁磁有序性只存在于足够低的温度下，但在奈尔温度及以上时消失。该温度命名自路易·奈尔（1904—2000），他首先发现了这种磁有序性。在奈尔温度以上，反铁磁性材料通常变为顺磁性。

奈尔在 1948 年还发现了亚铁磁性，并因其在磁性方面的贡献获得 1970 年诺贝尔物理学奖。亚铁磁性是一种与反铁磁性类似的协同磁性，只是在一个方向的电子磁矩大于在相反方向的电子磁矩，因此在没有外磁场时仍可保持磁化 [**图 9.24(c)(d)**]。像铁磁性材料一样，亚铁磁性材料也具有临界居里温度，超过该温度即会变成顺磁性。

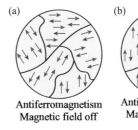

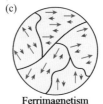
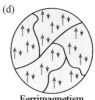

Figure 9.24 Comparison of antiferromagnetism and ferrimagnetism. (a) An antiferromagnetic material with a magnetic field off. (b) An antiferromagnetic material with a magnetic field on. (c) A ferrimagnetic material with a magnetic field off. (b) A ferrimagnetic material with a magnetic field on.

图 9.24 反铁磁性与亚铁磁性的比较：(a) 无磁场下的反铁磁材料；(b) 磁场下的反铁磁材料；(c) 无磁场下的亚铁磁材料；(d) 磁场下的亚铁磁材料。

习题

9.1 以下是含有以铂（Ⅱ）为中心金属离子、氨分子和/或氯离子为配体的配位单元的五种配位化合物的名称：(a) 三氯·氨合铂（Ⅱ）酸钾；(b) 二氯·二氨合铂（Ⅱ）；(c) 氯化氯·三氨合铂（Ⅱ）；(d) 氯化四氨合铂（Ⅱ）；(e) 四氯合铂（Ⅱ）酸钾。绘制预期的化合物电导率对氯含量的关系图。

9.2 将 Cu 电极浸入 1.00 mol L^{-1} NH$_3$ 和 1.00 mol L^{-1} [Cu(NH$_3$)$_4$]$^{2+}$ 溶液中。如果以标准氢电极为阴极，$E_{池}$ 为 +0.08 V。通过该方法获得的 [Cu(NH$_3$)$_4$]$^{2+}$ 的生成常数 K_f 是多少？

9.3 甘氨酸（NH$_2$CH$_2$COOH，表示为 Hgly）可作为阴离子进行配位，是一个双齿配体。绘制出 [Co(gly)$_3$] 所有可能的异构体，并命名之。化合物 [Co(gly)$_2$Cl(NH$_3$)] 有多少种可能的异构体？

9.4 维尔纳研究过的一种可能替代八面体的结构是三棱柱形。
(a) 该结构是否能正确地预测配离子 [CoCl$_2$(NH$_3$)$_4$]$^+$ 的异构体数量？为什么？
(b) 该结构是否能解释 [Co(en)$_3$]$^{3+}$ 的光学异构？为什么？

*9.5 **表 P9.1** 给出了如下反应的水合焓：

$$M^{2+}(g) + 6H_2O(l) \rightarrow [M(H_2O)_6]^{2+}(aq)$$

(a) 绘制水合焓与所列金属原子序数的函数关系图。
(b) 假定所有六水配合物均为高自旋，哪些离子的 CFSE 为零？
(c) 如果在 CFSE = 0 的离子之间绘制直线，将获得一条斜率为负的直线。解释之。
(d) CFSE ≠ 0 的离子的水合焓值低于 (c) 中绘制的直线，对此应如何解释？
(e) 估算在水分子的八面体场中 Fe（Ⅱ）离子的 Δ_o。
(f) [Fe(H$_2$O)$_6$]$^{2+}$ 离子会吸收什么波长的光？

Appendix

Appendix A Names of Chemical Compounds

Appendix B Table of Mathematical and Physical Constants

Appendix C Tables of Chemical Data

附录

附录 A　化合物的命名

附录 B　数理常数表

附录 C　化学数据表

Appendix A Names of Chemical Compounds

When we look up the properties of a compound in a handbook or locate a chemical in a stockroom, knowing the name of a compound would be very helpful. Some compounds have common or trivial names, such as water (H_2O), ammonia (NH_3), or glucose ($C_6H_{12}O_6$). However, it is impossible to refer to all compounds by their common or trivial names, because then we would have to learn and remember billions of unrelated names. In this appendix, we will briefly introduce the system of **nomenclature**, which is a systematic method of assigning names of various chemical compounds.

Compounds formed by carbon and hydrogen or carbon and hydrogen together with oxygen, nitrogen, and a few other elements are **organic compounds**. The rest are **inorganic compounds**. We will introduce the naming of inorganic compounds first, and then the naming of organic compounds.

Simple Ions

When an atom loses n electrons it becomes a simple positive ion, or **cation**, with a positive oxidation state $+n$. When an atom gains n electrons it becomes a simple negative ion, or **anion**, with a negative oxidation state $-n$. More discussion about oxidation states can be found in **Section 6.1**. To name a cation, we use the name of the element followed by the word *ion*. Because some metal elements may form more than one cation, we can distinguish them by putting a Roman numeral, indicating the oxidation state or simply the charge on the ion, in parenthesis immediately following the name of the metal. To name an anion, we write the name of the element, modified with an *-ide* ending (or suffix), then followed by the word *ion*. For example,

Cu^+ = copper(I) ion

Cu^{2+} = copper(II) ion

H^- = hydride ion

Table A.1 lists the names and symbols of some simple ions formed by metals and nonmetals.

An earlier system of nomenclature that is still used to some extent applies two different word endings to distinguish metals with two common oxidation states: the *-ic* ending for the higher oxidation state and the *-ous* ending for the lower oxidation state. Iron (Fe) and copper (Cu) are two typical metals that use this *-ic/-ous* system. Fe^{3+} is called the ferric ion and Fe^{2+} is called the ferrous ion, whereas Cu^{2+}/Cu^+ are the cupric/cuprous ions. There are many inadequacies in this *-ic/-ous* system, and we will not use it in this textbook.

Table A.1 The Names and Symbols of Some Simple Ions
表 A.1 一些简单离子的名称及符号

Name（名称）	Symbol（符号）	Name（名称）	Symbol（符号）
Positive ions /cations（正离子 / 阳离子）			
Lithium ion（锂离子）	Li^+	Chromium(II) ion（铬(II)离子）	Cr^{2+}
Sodium ion（钠离子）	Na^+	Chromium(III) ion（铬(III)离子）	Cr^{3+}
Potassium ion（钾离子）	K^+	Iron(II) ion（铁(II)离子）	Fe^{2+}
Rubidium ion（铷离子）	Rb^+	Iron(III) ion（铁(III)离子）	Fe^{3+}
Cesium ion（铯离子）	Cs^+	Cobalt(II) ion（钴(II)离子）	Co^{2+}
Magnesium ion（镁离子）	Mg^{2+}	Cobalt(III) ion（钴(III)离子）	Co^{3+}
Calcium ion（钙离子）	Ca^{2+}	Copper(I) ion（铜(I)离子）	Cu^+
Strontium ion（锶离子）	Sr^{2+}	Copper(II) ion（铜(II)离子）	Cu^{2+}
Barium ion（钡离子）	Ba^{2+}	Mercury(I) ion（汞(I)离子）	Hg_2^{2+}
Aluminum ion（铝离子）	Al^{3+}	Mercury(II) ion（汞(II)离子）	Hg^{2+}
Zinc ion（锌离子）	Zn^{2+}	Tin(II) ion（锡(II)离子）	Sn^{2+}
Silver ion（银离子）	Ag^+	Lead(II) ion（铅(II)离子）	Pb^{2+}
Negative ions/anions（负离子 / 阴离子）			
Hydride ion（氢负离子）	H^-	Iodide ion（碘离子）	I^-
Fluoride ion（氟离子）	F^-	Oxide ion（氧离子）	O^{2-}
Chloride ion（氯离子）	Cl^-	Sulfide ion（硫离子）	S^{2-}
Bromide ion（溴离子）	Br^-	Nitride ion（氮离子）	N^{3-}

附录 A　化合物的命名

当我们在手册中查找一种化合物的性质，或者从仓库寻找一种化学品时，知道化合物的名称将非常有用。一些化合物有常用名或俗名，如水（H_2O）、氨（NH_3）或葡萄糖（$C_6H_{12}O_6$）等。然而，我们不可能用常用名或俗名来指代所有化合物，因为那样我们就必须学习和记住数十亿个不相关的名字。本附录将简要介绍**命名法**系统，这是一种系统地给各种化合物确定名称的方法。

由碳、氢或碳、氢、氧、氮和其他一些元素组成的化合物是**有机化合物**，其余的是**无机化合物**。我们将首先介绍无机化合物的命名，然后介绍有机化合物的命名。

简单离子

当一个原子失去 n 个电子时，它变成一个简单正离子或**阳离子**，具有正氧化态 $+n$；当一个原子获得 n 个电子时，它变成一个简单负离子或**阴离子**，具有负氧化态 $-n$。关于氧化态的更多讨论详见 **6.1 节**。命名阳离子时，采用在元素的名称之后跟 ion（离子）这个词。由于一些金属元素可形成不止一种阳离子，通过在金属名称后紧跟的括号里插入一个罗马数字来加以区分，该罗马数字表示氧化态或离子上的电荷。命名阴离子时，先写元素名称，将词尾（或后缀）修改为 -ide，再加上 ion（离子）这个词。如

Cu^+ = copper(I) ion = 铜（Ⅰ）离子

Cu^{2+} = copper(II) ion = 铜（Ⅱ）离子

H^- = hydride ion = 氢负离子

表 A.1 列出了一些由金属和非金属组成的简单离子的名称及符号。

一种仍在使用的早期命名系统采用两种不同词尾来区分具有两种常见氧化态的金属：较高氧化态采用 -ic 词尾，较低氧化态采用 -ous 词尾。铁（Fe）和铜（Cu）是两种典型的采用 -ic/-ous 系统的金属。Fe^{3+} 称为 ferric ion（铁离子），Fe^{2+} 称为 ferrous ion（亚铁离子），而 Cu^{2+}/Cu^+ 是 cupric/cuprous ions（铜 / 亚铜离子）。此 -ic/-ous 系统有很多不足之处，在本教材中我们基本不使用。

由金属和非金属组成的二元化合物

二元化合物是两种元素之间形成的化合物。如果其中一种元素为金属而另一种为非金属，此二元化合物通常由离子构成，即所谓的二元**离子化合物**。命名由一种金属和一种非金属组成的二元化合物时，先写金属的未修改名称，再写非金属以 -ide 结尾的修改名称。如

$NaCl$ = sodium chloride = 氯化钠

Al_2O_3 = aluminum oxide = 氧化铝

$FeCl_2$ = iron(II) chloride = 氯化铁（Ⅱ）

$FeCl_3$ = iron(III) chloride = 氯化铁（Ⅲ）

由两种非金属组成的二元化合物

如果二元化合物中的两种元素均为非金属，则该化合物通常为

Binary Compounds of Metals and Nonmetals

Binary compounds are compounds formed between two elements. If one element is a metal and the other a nonmetal, the binary compound is usually made up of ions, a so-called binary **ionic compound**. To name a binary compound of a metal and a nonmetal, we first write the unmodified name of the metal, and then write the name of the nonmetal, modified to end in *-ide*. For example,

NaCl = sodium chloride

Al_2O_3 = aluminum oxide

$FeCl_2$ = iron(II) chloride

$FeCl_3$ = iron(III) chloride

Binary Compounds of Two Nonmetals

If the two elements in a binary compound are both nonmetals, the compound is usually a **molecular compound**. To name a binary compound of two nonmetals, we first write the unmodified name of the element with the positive oxidation state, and then write the name of the element with the negative oxidation state, modified to end in *-ide*. For example,

HCl = hydrogen chloride

Many pairs of nonmetals form more than one binary molecular compound, and we can distinguish among them by indicating the relative numbers of atoms through prefixes: *mono* = 1, *di* = 2, *tri* = 3, *tetra* = 4, *penta* = 5, *hexa* = 6, *hepta* = 7, *octa* = 8, *nona* = 9, *deca* = 10, and so on. Thus,

NO = nitrogen *mono*oxide

NO_2 = nitrogen *di*oxide

B_2Br_4 = *di*boron*tetra*bromide

Table A.2 gives the names and formulas of some binary compounds formed by two nonmetals. Note that the prefix *mono-* is treated in a special way that generally it is not used for the first named element. Thus, CO is called carbon monoxide, not *mono*carbon monoxide. Finally, several substances have common or trivial names that are so well recognized that their systematic names are almost never used. For example,

H_2O = water (*di*hydrogen *mon*oxide)

NH_3 = ammonia (*tri*hydrogen *mono*nitride)

Binary Acids

An **acid** is a substance that ionizes in water to produce hydrogen ions (H^+) and anions. A binary acid is an acid formed by hydrogen and another nonmetal. For example, HCl ionized into hydrogen ions (H^+) and chloride ions (Cl^-) in water, and thus it is a binary acid. More discussion about acids can be found in **Section 5.1**.

To name a binary acid, we use the prefix *hydro-* followed by the name of the other nonmetal modified with an *-ic* ending. The most commonly used binary acids are given as:

HF(aq) = *hydro*fluor*ic* acid

HCl(aq) = *hydro*chlor*ic* acid

HBr(aq) = *hydro*brom*ic* acid

HI(aq) = *hydro*iod*ic* acid

H_2S(aq) = *hydro*sulfur*ic* acid

The symbol (aq) signifies an aqueous (water) solution. Note that although the formulas of both are HCl, HCl alone indicates the pure binary molecular compound hydrogen chloride, and HCl(aq) emphasizes the binary hydrochloric acid in aqueous solution.

Polyatomic Ions

All the ions listed in **Table A.1** except for Hg_2^{2+} are monatomic ions, each consisting of a single atom.

分子化合物。命名由两种非金属组成的二元化合物时，先写具有正氧化态的元素的未修改名称，再写具有负氧化态的元素以 -ide 结尾的修改名称。如

HCl = hydrogen chloride = 氯化氢

许多非金属对之间可以形成不止一种二元分子化合物，可通过表示原子相对数量的前缀来加以区分：*mono*=1、*di*=2、*tri*=3、*tetra*=4、*penta*=5、*hexa*=6、*hepta*=7、*octa*=8、*nona*=9、*deca*=10 等等。故

NO = nitrogen *mon*oxide = 一氧化氮

NO_2 = nitrogen *di*oxide = 二氧化氮

B_2Br_4 = *di*boron*tetra*bromide = 四溴化二硼

表 A.2 给出了一些由两种非金属组成的二元化合物的名称及化学式。注意前缀 *mono-* 的处理方式较为特殊，它通常不用于命名中的第一种元素。因此，CO 称为 carbon monoxide（一氧化碳）而不是 *mono*carbon monoxide（一氧化一碳）。最后，有几种物质的通用名或俗名太常用了，因而其系统名称几乎从未使用。如

H_2O = water (*di*hydrogen *mon*oxide) = 水（一氧化二氢）

NH_3 = ammonia (*tri*hydrogen *mon*onitride) = 氨（三氢化一氮）

二元素酸

酸是一种在水中电离产生氢离子（H^+）和阴离子的物质。二元素酸是由氢和另一种非金属组成的酸。例如，HCl 在水中电离成氢离子（H^+）和氯离子（Cl^-），因此它是一种二元素酸。关于酸的更多讨论详见 **5.1 节**。

命名二元素酸时，使用前缀 *hydro-* 后跟另一种非金属以 *-ic* 结尾的修改名称。最为常用的二元素酸有

HF(aq) = *hydro*fluor*ic* acid = 氢氟酸

HCl(aq) = *hydro*chlor*ic* acid = 氢氯酸

HBr(aq) = *hydro*brom*ic* acid = 氢溴酸

HI(aq) = *hydro*iod*ic* acid = 氢碘酸

H_2S(aq) = *hydro*sulfur*ic* acid = 氢硫酸

符号（aq）表示水溶液。注意虽然化学式均为 HCl，但单独的 HCl 表示纯二元分子化合物氯化氢，而 HCl（aq）强调是水溶液中的二元素酸氢氯酸。

多原子离子

表 A.1 中列出的所有离子除 Hg_2^{2+} 外均为由一个原子组成的单原子离子。在多原子离子中，两个或多个原子通过共价键连接在一起。**表 A.3** 列出了一些常用的多原子离子及含有这些离子的化合物。

多原子阴离子比多原子阳离子更为常见。通式为 $A_xO_y^{z-}$（其中 A 代表一种元素）的含氧阴离子是典型的多原子阴离子。某些非金属（如 Cl、N、P 和 S）可形成一系列含有不同数量氧原子的含氧阴离子。其命名与 O 原子所键合的非金属原子的氧化态相关，可依据以下方案命名：

In polyatomic ions, two or more atoms are joined together by covalent bonds. **Table A.3** lists a number of commonly used polyatomic ions and the compounds containing these ions.

Polyatomic anions are more common than polyatomic cations. Oxoanions, the anions with the generic formula $A_xO_y^{z-}$ where A represents an element, are typical polyatomic anions. Certain nonmetals (such as Cl, N, P, and S) form a series of oxoanions containing different numbers of oxygen atoms. Their names are related to the oxidation state of the nonmetal atom to which the O atoms are bonded, according to the following scheme:

$$\text{Increasing oxidation state of the nonmetal atom} \longrightarrow$$
$$\text{hypo\_\_\_ite} \quad \text{\_\_\_ite} \quad \text{\_\_\_ate} \quad \text{per\_\_\_ate}$$
$$\text{Increasing number of oxygen atoms} \longrightarrow$$

Oxoacids

Oxoacids are ternary acids formed by hydrogen ions (H^+) and oxoanions with the generic formula $H_zA_xO_y$ where A is an element. Their naming scheme is similar to that for oxoanions, except that the ending *-ous* is used instead of *-ite* and *-ic* instead of *-ate*. **Table A.4** lists the names and formulas of several oxoacids and their salts.

Hydrocarbons

Carbon is the central element in organic chemistry. The great diversity of organic compounds arises from the ability of carbon atoms to combine with each other or with other atoms to form a framework of chains or rings. There are millions of organic compounds. A systematic approach to naming these compounds is crucial and can be found in some organic chemistry textbooks. In this textbook, we only need to recognize some organic compounds and use their common names together with an occasional systematic name.

Hydrocarbons are compounds containing only carbon and hydrogen. The simplest hydrocarbon is methane (CH_4), which contains one carbon atom and four hydrogen atoms. The complexity of organic compounds arises as the number of carbon atoms increases, because carbon atoms can form chains and rings, and the nature of the chemical bonds between the carbon atoms can vary.

Hydrocarbons that contain only single bonds are called **alkanes**, the generic formula of which is C_nH_{2n+2}, $n = 1,2,\ldots$. The names of the alkanes consist of two parts: a word stem (or prefix) and the ending *-ane* indicating that it is an alkane. **Table A.5** lists the word stem indicating the number of carbon atoms in simple organic molecules. For example, C_4H_{10} is butane and C_8H_{18} is octane.

Hydrocarbons with at least one C=C double bonds are called **alkenes**. The generic formula of alkenes with only one C=C double bond is C_nH_{2n}, $n = 1,2,\ldots$. The names of the alkenes consist of a word stem and the ending *-ene*. Benzene (C_6H_6) is a molecule with six carbon atoms arranged in a hexagonal ring. Hydrocarbons with at least one C≡C triple bonds are called **alkynes** with the ending *-yne* in their names. The generic formula of alkynes with only one C≡C triple bond is C_nH_{2n-2}, $n = 1,2,\ldots$.

Functional Groups

Carbon chains provide the framework for organic compounds. Other atoms or groups of atoms, the so-called **functional groups**, can replace one or more of the hydrogen atoms of hydrocarbons to form different compounds with characteristic properties and chemical reactivity. Compounds with the same functional group generally have similar chemical properties and reactivity regardless of the molecule size. This allows for systematic predictions of chemical reactions and behavior of organic compounds as well as the rational design of chemical syntheses.

If one or more hydrogen atoms are replaced by the functional group —OH, a class of organic molecules called **alcohols** is formed. The systematic name of alcohols is derived from the name of the corresponding hydrocarbons with the final *-e* replaced by the ending *-ol*. For example, the name of the molecule with one of the H atoms of ethane replaced by an —OH group is ethanol (CH_3CH_2OH), or ethyl alcohol as the common name. The ending *-yl* in replacing the final *-ane* of the name of the hydrocarbon indicates the attachment of the —OH group to the corresponding hydrocarbon chain. It is often the case that the common name of one compound will provide the generic name for a complete class of compounds. In this case, all alcohols contain

非金属原子的氧化态增加⟶
hypo___ite   ___ite   ___ate   per___ate
次___酸   亚___酸   ___酸   高___酸
氧原子数增加⟶

含氧酸

含氧酸是由氢离子（H^+）和含氧阴离子组成的三元素酸，通式为 $H_zA_xO_y$（其中 A 代表一种元素）。其命名方案与含氧阴离子类似，不同之处在于词尾用 -ous 代替 -ite、用 -ic 代替 -ate。**表 A.4** 列出了几种含氧酸及其盐的名称及化学式。

碳氢化合物

碳是有机化学的中心元素。有机化合物的多样性源于碳原子之间或碳原子与其他原子结合形成链状或环状骨架的能力，存在数以百万计的有机化合物。命名这些化合物的系统方法至关重要，可以在一些有机化学教材中找到。本教材中，我们仅需识别一些有机化合物并使用其常用名，偶尔需要使用其系统名称。

碳氢化合物是只含有碳和氢的化合物。甲烷（CH_4）是最简单的碳氢化合物，含有一个碳原子和四个氢原子。有机化合物的复杂性随碳原子数增加而增加，因为碳原子可以成链和成环，且碳原子之间的化学键性质多变。

只含单键的碳氢化合物称为**烷烃**，通式为 C_nH_{2n+2}，$n = 1, 2, \cdots$。烷烃的名称由两部分组成：词干（或前缀）以及表示其为烷烃的词尾 -ane。**表 A.5** 列出了表示简单有机分子中碳原子数目的词干。如 C_4H_{10} 是丁烷，C_8H_{18} 是辛烷。

至少含有一个 C=C 双键的碳氢化合物称为**烯烃**。只有一个 C=C 双键的烯烃的通式为 C_nH_{2n}，$n = 1, 2, \cdots$。烯烃的名称由词干和词尾 -ene 组成。苯（C_6H_6）是六个碳原子排列成六角形的环状分子。至少含有一个 C≡C 叁键的碳氢化合物称为**炔烃**，其名称以 -yne 结尾。只有一个 C≡C 叁键的炔烃的通式为 C_nH_{2n-2}，$n = 1, 2, \cdots$。

官能团

碳链提供了有机化合物的骨架，其他原子或原子团（即所谓的**官能团**）可以取代碳氢化合物的一个或多个氢原子，形成具有独特性质及化学活性的不同化合物。无论分子大小如何，具有相同官能团的化合物通常具有相似的化学性质及反应活性。这使得我们可以对化学反应和有机化合物的行为进行系统预测，并且对化学合成进行合理设计。

如果一个或多个氢原子被官能团—OH 取代，会形成一类称为**醇**的有机分子。醇的系统名称源于相应碳氢化合物的名称，将词尾 -e 用 -ol 取代。如 ethane（乙烷）的一个 H 原子被—OH 基团取代的分子，名称为 ethanol（乙醇，CH_3CH_2OH），常用名为 ethyl alcohol。将碳氢化合物名称的词尾 -ane 替换为 -yl，表示有—OH 基团接在对应的碳氢链上。一种化合物的常用名通常将为一整类化合物提供通用名，如所有醇类都至少含有一个—OH 基团；官能团—OH 称为**羟基**。

羧基—COOH 是另一类重要的官能团，它使分子显酸性。接在

Table A.2　The Names and Formulas of Some Binary Molecular Compounds
表 A.2　一些二元分子化合物的名称及化学式

Formula（化学式）	Name（名称）[a]	Formula（化学式）	Name（名称）[a]
BCl_3	Boron trichloride（三氯化硼）	N_2O_3	Dinitrogen trioxide（三氧化二氮）
B_2Br_4	Diboron tetrabromide（四溴化二硼）	N_2O_4	Dinitrogen tetroxide（四氧化二氮）
CCl_4	Carbon tetrachloride（四氯化碳）	N_2O_5	Dinitrogen pentoxide（五氧化二氮）
CO	Carbon monoxide（一氧化碳）	PCl_3	Phosphorus trichloride（三氯化磷）
CO_2	Carbon dioxide（二氧化碳）	PCl_5	Phosphorus pentachloride（五氯化磷）
SiO_2	Silicon dioxide（二氧化硅）	SO_2	Sulfur dioxide（二氧化硫）
NO	Nitrogen monoxide（一氧化氮）	SO_3	Sulfur trioxide（三氧化硫）
NO_2	Nitrogen dioxide（二氧化氮）	SF_6	Sulfur hexafluoride（六氟化硫）
N_2O	Dinitrogen monoxide（一氧化二氮）		

[a] When the prefix ends in a or o and the element name begins with a or o, the final vowel of the prefix is dropped for ease of pronunciation. For example, carbon monoxide, not carbon monooxide, and dinitrogen tetroxide, not dinitrogen tetraoxide. However, PI_3 is phosphorus triiodide, not phosphorus triodide. （当前缀以 a 或 o 结尾且元素名称以 a 或 o 开头时，前缀的最后一个元音将被删除，以便于发音。例如，carbon monoxide 而非 carbon monooxide，dinitrogen tetroxide 而非 dinitrogen tetraoxide。但 PI_3 是 phosphorus triiodide 而非 phosphorus triodide。）

Table A.3　The Names and Formulas of Some Polyatomic Ions
表 A.3　一些多原子离子的名称及化学式

Name（名称）	Formula（化学式）	Typical Compound（典型化合物）
Positive ions /cations（正离子 / 阳离子）		
Ammonium ion（铵离子）	NH_4^+	NH_4Cl
Negative ions/anions（负离子 / 阴离子）		
Acetate ion（醋酸根离子）	CH_3COO^-	KCH_3COO
Carbonate ion（碳酸根离子）	CO_3^{2-}	K_2CO_3
Hydrogen carbonate ion (or bicarbonate ion)（碳酸氢根离子）[a]	HCO_3^-	$KHCO_3$
Hypochlorite ion（次氯酸根离子）	ClO^-	$KClO$
Chlorite ion（亚氯酸根离子）	ClO_2^-	$KClO_2$
Chlorate ion（氯酸根离子）	ClO_3^-	$KClO_3$
Perchlorate ion（高氯酸根离子）	ClO_4^-	$KClO_4$
Chromate ion（铬酸根离子）	CrO_4^{2-}	K_2CrO_4
Dichromate ion（重铬酸根离子）	$Cr_2O_7^{2-}$	$K_2Cr_2O_7$
Cyanide ion（氰根离子）	CN^-	KCN
Hydroxide ion（氢氧根离子）	OH^-	KOH
Nitrite ion（亚硝酸根离子）	NO_2^-	KNO_2
Nitrate ion（硝酸根离子）	NO_3^-	KNO_3
Oxalate ion（草酸根离子）	$C_2O_4^{2-}$	$K_2C_2O_4$
Manganate ion（锰酸根离子）	MnO_4^{2-}	K_2MnO_4
Permanganate ion（高锰酸根离子）	MnO_4^-	$KMnO_4$
Phosphate ion（磷酸根离子）	PO_4^{3-}	K_3PO_4
Hydrogen phosphate ion（磷酸氢根离子）[a]	HPO_4^{2-}	K_2HPO_4
Dihydrogen phosphate ion（磷酸二氢根离子）[a]	$H_2PO_4^-$	KH_2PO_4
Sulfite ion（亚硫酸根离子）	SO_3^{2-}	K_2SO_3
Hydrogen sulfite ion (or bisulfite ion)（亚硫酸氢根离子）[a]	HSO_3^-	$KHSO_3$
Sulfate ion（硫酸根离子）	SO_4^{2-}	K_2SO_4
Hydrogen sulfate ion (or bisulfate ion)（硫酸氢根离子）[a]	HSO_4^-	$KHSO_4$
Thiosulfate ion（硫代硫酸根离子）	$S_2O_3^{2-}$	$K_2S_2O_3$

[a] These anion names are sometimes written as a single word. For example, hydrogencarbonate, dihydrogenphosphate, and so on. （这些阴离子的英文名称有时被写成一个单词，如 hydrogencarbonate、dihydrogenphosphate 等。）

Table A.4 Nomenclature of Some Oxoacids and Their Salts
表 A.4 一些含氧酸及其盐的系统命名

Oxidation State (氧化态)	Formula of Acid (酸的化学式)[a]	Name of Acid (酸的名称)[b]	Formula of Salt (盐的化学式)[b]	Name of Salt (盐的名称)
Cl: +1	$HClO$	Hypochlorous acid（次氯酸）	$KClO$	Potassium hypochlorite（次氯酸钾）
Cl: +3	$HClO_2$	Chlorous acid（亚氯酸）	$KClO_2$	Potassium chlorite（亚氯酸钾）
Cl: +5	$HClO_3$	Chloric acid（氯酸）	$KClO_3$	Potassium chlorate（氯酸钾）
Cl: +7	$HClO_4$	Perchloric acid（高氯酸）	$KClO_4$	Potassium perchlorate（高氯酸钾）
S: +4	H_2SO_3	Sulfurous acid（亚硫酸）	K_2SO_3	Potassium sulfite（亚硫酸钾）
S: +6	H_2SO_4	Sulfuric acid（硫酸）	K_2SO_4	Potassium sulfate（硫酸钾）
N: +3	HNO_2	Nitrous acid（亚硝酸）	KNO_2	Potassium nitrite（亚硝酸钾）
N: +5	HNO_3	Nitric acid（硝酸）	KNO_3	Potassium nitrate（硝酸钾）
P: +5	H_3PO_4	Phosphoric acid（磷酸）	K_3PO_4	Potassium phosphate（磷酸钾）
C: +3	$H_2C_2O_4$	Oxalic acid（草酸）	$K_2C_2O_4$	Potassium oxalate（草酸钾）
C: +4	H_2CO_3	Carbonic acid（碳酸）	K_2CO_3	Potassium carbonate（碳酸钾）

[a] In all these acids, H atoms are bonded to O atoms, not the central nonmetal atom. Formulas are often written to reflect this fact, for instance, HOCl instead of HClO and HOClO instead of HClO₂.（在所有这些酸中，H 原子与 O 原子而非中心非金属原子键连。化学式的写法通常反映了这一事实，如 HOCl 而非 HClO，HOClO 而非 HClO₂。）

[b] In general, the -ic and -ate names are assigned to compounds in which the central nonmetal atom has an oxidation state equal to the periodic table group number minus 10. Halogen compounds are exceptional in that the -ic and -ate names are assigned to compounds in which the halogen has an oxidation state of +5 (even though the group number is 17).[一般来说，-ic 和 -ate 名称指的是中心非金属原子的氧化态等于周期表族数减 10 的化合物。卤化物的特殊性在于 -ic 和 -ate 名称被分配给卤素氧化态为 +5 的化合物（尽管其族数为 17）。]

Table A.5 Word Stem (or Prefix) Indicating the Number of Carbon Atoms in Simple Organic Compounds
表 A.5 表示简单有机化合物中碳原子数量的词干（或前缀）

Stem or Prefix（词干或前缀）	Number of C Atoms（碳原子数）	Stem or Prefix（词干或前缀）	Number of C Atoms（碳原子数）
Meth- (甲-)	1	Hex- (己-)	6
Eth- (乙-)	2	Hept- (庚-)	7
Prop- (丙-)	3	Oct- (辛-)	8
But- (丁-)	4	Non- (壬-)	9
Pent- (戊-)	5	Dec- (癸-)	10

Table A.6 Some Common Functional Groups in Organic Compounds
表 A.6 一些有机化合物中的常见官能团

Functional Group（官能团）	Formula（化学式）	Example（示例）
Hydroxyl（羟基）	—OH	CH_3CH_2OH
Carboxyl（羧基）	—COOH	CH_3COOH
Halogen（卤基）	—F, —Cl, —Br, —I	CH_3CH_2Cl
Aldehyde（醛基）	—CHO	CH_3CHO
Phenyl（苯基）	—C_6H_5	$C_6H_5CH_3$
Ester（酯基）	—COO—	$CH_3COOCH_2CH_3$

at least one —OH group; the functional group —OH is called the **hydroxyl group**.

Another important functional group is the **carboxyl group**, —COOH, which confers acidic properties on a molecule. The hydrogen attached to one of the oxygen atoms in a carboxyl group is ionizable and accounts for the acidity. Compounds containing the carboxyl group are called **carboxylic acids**. The simplest carboxylic acid is methanoic acid (HCOOH). In the systematic name, methan- indicates one carbon atom and the *-oic* acid indicates a carboxylic acid. The common name for methanoic acid is formic acid. The carboxylic acid based on ethane is ethanoic acid (CH_3COOH), more commonly known as acetic acid. Vinegar is a solution of acetic acid in water.

Halogen group is a functional group that can substitute for one or more H atoms in hydrocarbons. When present as functional groups, the halogens carry the names of fluoro-, chloro-, bromo-, and iodo-. Hydrocarbon dismissing one or more H atoms can also be viewed as a functional group called **hydrocarbyl group**. Table A.6 lists some common functional groups in organic compounds.

羧基的一个氧原子上的氢可电离，这就是酸性的来源。含有羧基的化合物称为**羧酸**，甲酸（HCOOH）是最简单的羧酸。在系统名称中，methan-（甲）表示一个碳原子，-*oic* acid（酸）表示一种羧酸。甲酸的常用名是 formic acid（蚁酸）。乙酸（CH_3COOH）是源于乙烷的羧酸，通常称为醋酸，醋是醋酸的水溶液。

卤基是一类可以取代碳氢化合物中一个或多个氢原子的官能团。当卤素以官能团形式存在时，其名称为 fluoro-（氟）、chloro-（氯）、bromo-（溴）和 iodo-（碘）。碳氢化合物除去一个或几个氢原子也可视为一个称为**烃基**的官能团。**表 A.6** 列出了一些有机化合物中的常见官能团。

Appendix B Table of Mathematical and Physical Constants
附录 B 数理常数表

Table B.1 Commonly Used Mathematical and Physical Constants
表 B.1 常用数理常数

Name (名称)	Symbol (符号)	Value (数值)
Circular constant (圆周率)	π	3.141 592 653 589 793
Base of natural logarithm (自然对数的底数)	e	2.718 281 828 459 045
Molar gas constant (摩尔气体常数)	R	8.314 462 618 153 24 J mol^{-1} K^{-1}
Avogadro constant (阿伏伽德罗常数)	N_A	6.022 140 76×10^{23} mol^{-1}
Boltzmann constant (玻尔兹曼常数)	k_B	1.380 649×10^{-23} J mol^{-1} K^{-1}
Faraday constant (法拉第常数)	F	9.648 533 212 331 002 ×10^4 C mol^{-1}
Speed of light in a vacuum (真空中的光速)	c	2.997 924 58 ×10^8 m s^{-1}
Vacuum electric permittivity (真空介电常数)	ε_0	8.854 187 812 8×10^{-12} F m^{-1}
Elementary charge (基本电荷)	e	1.602 176 634×10^{-19} C
Electron mass (电子质量)	m_e	9.109 383 701 5×10^{-31} kg
Planck constant (普朗克常数)	h	6.626 070 15×10^{-34} J s
Rydberg constant (里德堡常数)	R_H	1.097 373 156 816 0 ×10^7 m^{-1}
Bohr radius (玻尔半径)	a_0	5.291 772 109 04×10^{-11} m
Bohr magneton (玻尔磁子)	μ_B	9.274 010 078 3×10^{-24} J T^{-1}

Appendix C Tables of Chemical Data
附录 C 化学数据表

Source of Data（数据来源）：
CRC Handbook of Chemistry and Physics(95th Ed.), CRC Press (2014).
Lange's Handbook of Chemistry (13th Ed.), McGraw-Hill Book Co. (1992).

Table C.1 Thermodynamic Properties of Selected Substances at Standard States and 298.15 K
表 C.1 选定物质在标准状态和 298.15 K 下的热力学性质

Formula (化学式)	Name (名称)	ΔH_f° /(kJ mol^{-1})	ΔG_f° /(kJ mol^{-1})	S° /(J mol^{-1} K^{-1})	C_p /(J mol^{-1} K^{-1})
Substances not containing carbon（不含碳物质）					
Silver（银）					
Ag(s)	Silver（银）	0	0	42.6	25.4
Ag$^+$(aq)	Silver ion（银离子）	105.6	77.1	72.7	21.8
AgBr(s)	Silver(I) bromide（溴化银）	−100.4	−96.9	107.1	52.4
AgCl(s)	Silver(I) chloride（氯化银）	−127.0	−109.8	96.3	50.8
AgF(s)	Silver(I) fluoride（氟化银）	−204.6	——	——	——
AgI(s)	Silver(I) iodide（碘化银）	−61.8	−66.2	115.5	56.8
AgNO$_3$(s)	Silver(I) nitrate（硝酸银）	−124.4	−33.4	140.9	93.1
Ag$_2$CrO$_4$(s)	Silver(I) chromate（铬酸银）	−731.7	−641.8	217.6	142.3
Ag$_2$O(s)	Silver(I) oxide（氧化银）	−31.1	−11.2	121.3	65.9
Ag$_2$SO$_4$(s)	Silver(I) sulfate（硫酸银）	−715.9	−618.4	200.4	131.4
Aluminum（铝）					
Al(s)	Aluminum（铝）	0	0	28.3	24.2
Al^{3+}(aq)	Aluminum ion（铝离子）	−531.0	−485.0	−321.7	——
AlCl$_3$(s)	Aluminum chloride（氯化铝）	−704.2	−628.8	109.3	91.1
AlF$_3$(s)	Aluminum fluoride（氟化铝）	−1510.4	−1431.1	66.5	75.1
Al$_2$Cl$_6$(g)	Aluminum hexachloride（六氯化铝）	−1290.8	−1220.4	490.0	——
Al$_2$O$_3$(α solid)	Aluminum oxide（氧化铝）	−1675.7	−1582.3	50.9	79.0
Boron（硼）					
B(s)	Boron (β-rhombohedral)（β-三方-硼）	0	0	5.9	11.1
BBr$_3$(l)	Boron tribromide（三溴化硼）	−239.7	−238.5	229.7	——
BCl$_3$(l)	Boron trichloride（三氯化硼）	−427.2	−387.4	206.3	106.7
BF$_3$(g)	Boron trifluoride（三氟化硼）	−1136.0	−1119.4	254.4	——
BF$_4^-$(aq)	Tetrafluoroborate ion（四氟合硼阴离子）	−1574.9	−1486.9	180.0	——
BO$_2^-$(aq)	Metaborate ion（偏硼酸根离子）	−772.4	−678.9	−37.2	——
B$_2$H$_6$(g)	Diborane（乙硼烷）	36.4	87.6	232.1	56.7
B$_2$O$_3$(s)	Boron oxide（氧化硼）	−1273.5	−1194.3	54.0	62.8
B$_2$S$_3$(s)	Boron sulfide（硫化硼）	−240.6	——	100.0	111.7
Barium（钡）					
Ba(s)	Barium（钡）	0	0	62.5	28.1
Ba^{2+}(aq)	Barium ion（钡离子）	−537.6	−560.8	9.6	——
BaBr$_2$(s)	Barium bromide（溴化钡）	−757.3	−736.8	146.0	——
BaCl$_2$(s)	Barium chloride（氯化钡）	−855.0	−806.7	123.7	75.1
BaF$_2$(s)	Barium fluoride（氟化钡）	−1207.1	——	−1156.8	96.4
BaI$_2$(s)	Barium iodide（碘化钡）	−602.1	——	——	——
BaO(s)	Barium oxide（氧化钡）	−548.0	−520.3	72.1	47.3
Ba(OH)$_2$(s)	Barium hydroxide（氢氧化钡）	−944.7	——	——	——
BaS(s)	Barium sulfide（硫化钡）	−460.0	−456.0	78.2	49.4

续表

Formula (化学式)	Name (名称)	ΔH_f° /(kJ mol^{-1})	ΔG_f° /(kJ mol^{-1})	S° /(J mol^{-1} K^{-1})	C_p /(J mol^{-1} K^{-1})
BaSO$_4$(s)	Barium sulfate (硫酸钡)	−1473.2	−1362.2	132.2	101.8
	Beryllium (铍)				
Be(s)	Beryllium (铍)	0	0	9.5	16.4
Be(g)	Beryllium (铍)	324.0	286.6	136.3	20.8
Be^{2+}(aq)	Beryllium ion (铍离子)	−382.8	−379.7	−129.7	——
BeBr$_2$(s)	Beryllium bromide (溴化铍)	−353.5	——	108.0	69.4
BeCl$_2$(s)	Beryllium chloride (氯化铍)	−490.4	−445.6	75.8	62.4
BeF$_2$(s)	Beryllium fluoride (氟化铍)	−1026.8	−979.4	53.4	51.8
BeI$_2$(s)	Beryllium iodide (碘化铍)	−192.5	——	121.0	71.1
BeO(s)	Beryllium oxide (氧化铍)	−609.4	−580.1	13.8	25.6
Be(OH)$_2$(s)	Beryllium hydroxide (氢氧化铍)	−902.5	−815.0	45.5	62.1
BeS(s)	Beryllium sulfide (硫化铍)	−234.3		34.0	34.0
BeSO$_4$(s)	Beryllium sulfate (硫酸铍)	−1205.2	−1093.8	77.9	85.7
	Bismuth (铋)				
Bi(s)	Bismuth (铋)	0	0	56.7	25.5
Bi^{3+}(aq)	Bismuth ion (铋离子)	——	82.8	——	
BiCl$_3$(s)	Bismuth trichloride (三氯化铋)	−379.1	−315.0	177.0	105.0
Bi(OH)$_3$(s)	Bismuth hydroxide (氢氧化铋)	−711.3			
Bi$_2$O$_3$(s)	Bismuth oxide (三氧化二铋)	−573.9	−493.7	151.5	113.5
	Bromide (溴)				
Br(g)	Bromine (atomic) (溴原子)	111.9	82.4	175.0	20.8
Br$^-$(aq)	Bromine ion (溴离子)	−121.6	−104.0	82.4	−141.8
BrCl(g)	Bromine chloride (氯化溴)	14.6	−1.0	240.1	35.0
BrF$_3$(l)	Bromine trifluoride (三氟化溴)	−300.8	−240.5	178.2	124.6
BrF$_3$(g)	Bromine trifluoride (三氟化溴)	−255.6	−229.4	292.5	66.6
Br$_2$(l)	Bromine (溴)	0.0	——	152.2	75.7
Br$_2$(g)	Bromine (溴)	30.9	3.1	245.5	36.0
	Calcium (钙)				
Ca(s)	Calcium (钙)	0	0	41.6	25.9
Ca^{2+}(aq)	Calcium ion (钙离子)	−542.8	−553.6	−53.1	——
CaCl$_2$(s)	Calcium chloride (氯化钙)	−795.4	−748.8	108.4	72.9
CaF$_2$(s)	Calcium fluoride (氟化钙)	−1228.0	−1175.6	68.5	67.0
CaH$_2$(s)	Calcium hydride (氢化钙)	−181.5	−142.5	41.4	41.0
CaO(s)	Calcium oxide (氧化钙)	−634.9	−603.3	38.1	42.0
Ca(OH)$_2$(s)	Calcium hydroxide (氢氧化钙)	−985.2	−897.5	83.4	87.5
CaS(s)	Calcium sulfide (硫化钙)	−482.4	−477.4	56.5	47.4
CaSO$_4$(s)	Calcium sulfate (硫酸钙)	−1434.5	−1322.0	106.5	99.7
	Cadmium (镉)				
Cd(s)	Cadmium (镉)	0	0	——	51.8
Cd^{2+}(aq)	Cadmium ion (镉离子)	−75.9	−77.6	−73.2	——
CdCl$_2$(s)	Cadmium chloride (氯化镉)	−391.5	−343.9	115.3	74.7
CdF$_2$(s)	Cadmium fluoride (氟化镉)	−700.4	−647.7	77.4	
CdO(s)	Cadmium oxide (氧化镉)	−258.4	−228.7	54.8	43.4
Cd(OH)$_2$(s)	Cadmium hydroxide (氢氧化镉)	−560.7	−473.6	96.0	
CdS(s)	Cadmium sulfide (硫化镉)	−161.9	−156.5	64.9	——
CdSO$_4$(s)	Cadmium sulfate (硫酸镉)	−933.3	−822.7	123.0	99.6
	Chlorine (氯)				
Cl(g)	Chlorine (atomic) (氯原子)	121.3	105.3	165.2	21.8
Cl$^-$(aq)	Chlorine ion (氯离子)	−167.2	−131.2	56.5	−136.4
ClF(g)	Chlorine fluoride (氟化氯)	−50.3	−51.8	217.9	32.1
ClF$_3$(g)	Chlorine trifluoride (三氟化氯)	−163.2	−123.0	281.6	63.9
ClO$_2$(g)	Chlorine dioxide (二氧化氯)	102.5	120.5	256.8	42.0
Cl$_2$(g)	Chlorine (氯气)	0	0	223.1	33.9
Cl$_2$O(g)	Chlorine monoxide (一氧化二氯)	80.3	97.9	266.2	45.4

Formula（化学式）	Name（名称）	ΔH_f° /(kJ mol^{-1})	ΔG_f° /(kJ mol^{-1})	S° /(J mol^{-1} K^{-1})	C_p /(J mol^{-1} K^{-1})
		Cobalt（钴）			
Co(s)	Cobalt（钴）	0	0	30.0	24.8
Co^{2+}(aq)	Cobalt(II) ion（亚钴离子）	−58.2	−54.4	−113.0	——
Co^{3+}(aq)	Cobalt(III) ion（钴离子）	92.0	134.0	−305.0	——
CoF$_2$(s)	Cobalt(II) fluoride（氟化亚钴）	−692.0	−647.2	82.0	68.8
CoO(s)	Cobalt(II) oxide（氧化亚钴）	−237.9	−214.2	53.0	55.2
Co(OH)$_2$(s)	Cobalt(II) hydroxide（氢氧化亚钴）	−539.7	−454.3	79.0	
CoS(s)	Cobalt(II) sulfide（硫化亚钴）	−82.8	——		
CoSO$_4$(s)	Cobalt(II) sulfate（硫酸亚钴）	−888.3	−782.3	118.0	
Co$_2$S$_3$(s)	Cobalt(III) sulfide（三硫化二钴）	−147.3	——		
Co$_3$O$_4$(s)	Cobalt(II,III) oxide（四氧化三钴）	−891.0	−774.0	102.5	123.4
		Chromium（铬）			
Cr(s)	Chromium（铬）	0	0	23.8	23.4
CrO$_4^{2-}$(aq)	Chromate ion（铬酸根离子）	−881.2	−727.8	50.2	——
Cr$_2$O$_3$(s)	Chromium(III) oxide（三氧化二铬）	−1139.7	−1058.1	81.2	118.7
Cr$_2$O$_7^{2-}$(aq)	Dichromate ion（重铬酸根离子）	−1490.3	−1301.3	261.9	——
		Copper（铜）			
Cu(s)	Copper（铜）	0	0	33.2	24.4
Cu^{2+}(aq)	Copper(II) ion（铜离子）	64.8	65.5	−99.6	——
CuO(s)	Copper(II) oxide（氧化铜）	−157.3	−129.7	42.6	42.3
Cu(OH)$_2$(s)	Copper(II) hydroxide（氢氧化铜）	−449.8			
CuS(s)	Copper(II) sulfide（硫化铜）	−53.1	−53.6	66.5	47.8
CuSO$_4$(s)	Copper(II) sulfate（硫酸铜）	−771.4	−662.2	109.2	
Cu$_2$O(s)	Copper(I) oxide（氧化亚铜）	−168.6	−146.0	93.1	63.6
Cu$_2$S(s)	Copper(I) sulfide（硫化亚铜）	−79.5	−86.2	120.9	76.3
		Fluorine（氟）			
F(g)	Fluorine (atomic)（氟原子）	79.4	62.3	158.8	22.7
F$^-$(aq)	Fluorine ion（氟离子）	−332.6	−278.8	−13.8	−106.7
F$_2$(g)	Fluorine（氟气）	0.0	——	202.8	31.3
		Iron（铁）			
Fe(s)	Iron（铁）	0	0	27.3	25.1
Fe^{2+}(aq)	Iron(II) ion（亚铁离子）	−89.1	−78.9	−137.7	
Fe^{3+}(aq)	Iron(III) ion（铁离子）	−48.5	−4.7	−315.9	
FeO(s)	Iron(II) oxide（氧化亚铁）	−272.0			
FeS(s)	Iron(II) sulfide（硫化亚铁）	−100.0	−100.4	60.3	50.5
FeSO$_4$(s)	Iron(II) sulfate（硫酸亚铁）	−928.4	−820.8	107.5	100.6
Fe$_2$O$_3$(s)	Iron(III) oxide（三氧化二铁）	−824.2	−742.2	87.4	103.9
Fe$_3$O$_4$(s)	Iron(II,III) oxide（四氧化三铁）	−1118.4	−1015.4	146.4	143.4
		Hydrogen（氢）			
H(g)	Hydrogen (atomic)（氢原子）	218.0	203.3	114.7	20.8
H$^+$(aq)	Hydron（氢离子）	0	0	0	——
HBr(g)	Hydrogen bromide（溴化氢）	−36.3	−53.4	198.7	29.1
HCl(g)	Hydrogen chloride（氯化氢）	−92.3	−95.3	186.9	29.1
HCl(aq)	Hydrochloric acid（盐酸）	−167.15	−131.25	56.5	−136.4
HF(g)	Hydrogen fluoride（氟化氢）	−273.3	−275.4	173.8	——
HI(g)	Hydrogen iodide（碘化氢）	26.5	1.7	206.6	29.2
HNO$_3$(l)	Nitric acid（硝酸）	−174.1	−80.7	155.6	109.9
HNO$_3$(aq)	Nitric acid（硝酸）	−207.36	−111.34	146.4	−86.6
H$_2$(g)	Hydrogen（氢气）	0	0	130.7	28.8
H$_2$O(l)	Water（水）	−285.8	−237.1	70.0	75.3
H$_2$O(g)	Water（水）	−241.8	−228.6	188.8	33.6
H$_2$O$_2$(l)	Hydrogen peroxide（过氧化氢）	−187.8	−120.4	109.6	89.1
H$_2$O$_2$(g)	Hydrogen peroxide（过氧化氢）	−136.3	−105.6	232.7	43.1
H$_2$S(g)	Hydrogen sulfide（硫化氢）	−20.6	−33.4	205.8	34.2

续表

Formula （化学式）	Name （名称）	ΔH_f° /(kJ mol^{-1})	ΔG_f° /(kJ mol^{-1})	S° /(J mol^{-1} K^{-1})	C_p /(J mol^{-1} K^{-1})
H_2SO_4(l)	Sulfuric acid（硫酸）	−814.0	−690.0	156.9	138.9
H_2SO_4(aq)	Sulfuric acid（硫酸）	−909.27	−744.63	20.1	293
	Helium（氦）				
He(g)	Helium（氦）	0	0	126.2	20.8
	Mercury（汞）				
Hg(l)	Mercury（汞）	0	0	75.9	28.0
Hg(g)	Mercury（汞）	61.4	31.8	175.0	20.8
HgO(s)	Mercury(II) oxide（氧化汞）	−90.8	−58.5	70.3	44.1
HgS(s)	Mercury(II) sulfide (red)（硫化汞-红）	−58.2	−50.6	82.4	48.4
	Iodine（碘）				
I(g)	Iodine (atomic)（碘原子）	106.8	70.2	180.8	20.8
I$^-$(aq)	Iodine ion（碘离子）	−55.2	−51.6	111.3	−142.3
IBr(g)	Iodine bromide（溴化碘）	40.8	3.7	258.8	36.4
ICl(g)	Iodine chloride（氯化碘）	17.8	−5.5	247.6	35.6
ICl(l)	Iodine chloride（氯化碘）	−23.9	−13.6	135.1	——
I_2(s)	Iodine (rhombic)（碘）	0	0	116.1	54.4
I_2(g)	Iodine (rhombic)（碘）	62.4	19.3	260.7	36.9
	Potassium（钾）				
K(s)	Potassium（钾）	0	0	64.7	29.6
K(g)	Potassium（钾）	89.0	60.5	160.3	20.8
K$^+$(aq)	Potassium ion（钾离子）	−252.4	−283.3	102.5	21.8
KBr(s)	Potassium bromide（溴化钾）	−393.8	−380.7	95.9	52.3
KCl(s)	Potassium chloride（氯化钾）	−436.5	−408.5	82.6	51.3
$KClO_3$(s)	Potassium chlorate（氯酸钾）	−397.7	−296.3	143.1	100.3
$KClO_4$(s)	Potassium perchlorate（高氯酸钾）	−432.8	−303.1	151.0	112.4
KF(s)	Potassium fluoride（氟化钾）	−567.3	−537.8	66.6	49.0
KI(s)	Potassium iodide（碘化钾）	−327.9	−324.9	106.3	52.9
KNO_3(s)	Potassium nitrate（硝酸钾）	−494.6	−394.9	133.1	96.4
KOH(s)	Potassium hydroxide（氢氧化钾）	−424.6	−379.4	81.2	68.9
KOH(aq)	Potassium hydroxide（氢氧化钾）	−482.37	−440.53	91.6	−126.8
K_2O(s)	Potassium oxide（氧化钾）	−361.5	——	——	——
K_2O_2(s)	Potassium peroxide（过氧化钾）	−494.1	−425.1	102.1	——
K_2SO_4(s)	Potassium sulfate（硫酸钾）	−1437.8	−1321.4	175.6	131.5
	Lithium（锂）				
Li(s)	Lithium（锂）	0	0	29.1	24.8
Li(g)	Lithium（锂）	159.3	126.6	138.8	20.8
Li$^+$(aq)	Lithium ion（锂离子）	−278.5	−293.3	13.4	68.6
LiCl(s)	Lithium chloride（氯化锂）	−408.6	−384.4	59.3	48.0
$LiNO_3$(s)	Lithium nitrate（硝酸锂）	−483.1	−381.1	90.0	——
LiOH(s)	Lithium hydroxide（氢氧化锂）	−487.5	−441.5	42.8	49.6
Li_2O(s)	Lithium oxide（氧化锂）	−597.9	−561.2	37.6	54.1
	Magnesium（镁）				
Mg(s)	Magnesium（镁）	0	0	32.7	24.9
Mg^{2+}(aq)	Magnesium ion（镁离子）	−466.9	−454.8	−138.1	——
$MgCl_2$(s)	Magnesium chloride（氯化镁）	−641.3	−591.8	89.6	71.4
MgF_2(s)	Magnesium fluoride（氟化镁）	−1124.2	−1071.1	57.2	61.6
MgO(s)	Magnesium oxide（氧化镁）	−601.6	−569.3	27.0	37.2
$Mg(OH)_2$(s)	Magnesium hydroxide（氢氧化镁）	−924.5	−833.5	63.2	77.0
MgS(s)	Magnesium sulfide（硫化镁）	−346.0	−341.8	50.3	45.6
$MgSO_4$(s)	Magnesium sulfate（硫酸镁）	−1284.9	−1170.6	91.6	96.5
	Manganese（锰）				
Mn(s)	Manganese（锰）	0	0	32.0	26.3
Mn^{2+}(aq)	Manganese ion（锰离子）	−220.8	−228.1	−73.6	50.0
MnO_2(s)	Manganese(IV) oxide（二氧化锰）	−520.0	−465.1	53.1	54.1

Formula (化学式)	Name (名称)	ΔH_f° /(kJ mol^{-1})	ΔG_f° /(kJ mol^{-1})	S° /(J mol^{-1} K^{-1})	C_p /(J mol^{-1} K^{-1})
MnO$_4^-$(aq)	Permanganate ion (高锰酸根离子)	−541.4	−447.2	191.2	−82.0
	Nitrogen (氮)				
N(g)	Nitrogen (atomic) (氮原子)	472.7	455.5	153.3	20.8
NF$_3$(g)	Nitrogen trifluoride (三氟化氮)	−132.1	−90.6	260.8	53.4
NH$_3$(g)	Ammonia (氨)	−45.9	−16.4	192.8	35.1
NH$_3$(aq)	Aqueous ammonia (氨水)	−80.29	−26.57	111.3	——
NH$_4^+$(aq)	Ammonium ion (铵离子)	−132.5	−79.3	113.4	79.9
NH$_4$Br(s)	Ammonium bromide (溴化铵)	−270.8	−175.2	113.0	96.0
NH$_4$Cl(s)	Ammonium chloride (氯化铵)	−314.4	−202.9	94.6	84.1
NH$_4$F(s)	Ammonium fluoride (氟化铵)	−464.0	−348.7	72.0	65.3
NH$_4$I(s)	Ammonium iodide (碘化铵)	−201.4	−112.5	117.0	——
NH$_4$NO$_3$(s)	Ammonium nitrate (硝酸铵)	−365.6	−183.9	151.1	139.3
NH$_4$NO$_3$(aq)	Ammonium nitrate (硝酸铵)	−339.87	−190.71	259.8	−6.7
(NH$_4$)$_2$SO$_4$(s)	Ammonium sulfate (硫酸铵)	−1180.9	−901.7	220.1	187.5
NO(g)	Nitric oxide (一氧化氮)	91.3	87.6	210.8	29.9
NO$_2$(g)	Nitrogen dioxide (二氧化氮)	33.2	51.3	240.1	37.2
NO$_2^-$(aq)	Nitrite ion (亚硝酸根离子)	−104.6	−32.2	123.0	−97.5
NO$_3^-$(aq)	Nitrate ion (硝酸根离子)	−207.4	−111.3	146.4	−86.6
NOBr(g)	Nitrosyl bromide (亚硝酰溴)	82.2	82.4	273.7	45.5
NOCl(g)	Nitrosyl chloride (亚硝酰氯)	51.7	66.1	261.7	44.7
N$_2$(g)	Nitrogen (氮气)	0	0	191.6	29.1
N$_2$H$_4$(l)	Hydrazine (肼)	50.6	149.3	121.2	98.9
N$_2$H$_4$(g)	Hydrazine (肼)	95.4	159.4	238.5	48.4
N$_2$O(g)	Nitrous oxide (一氧化二氮)	81.6	103.7	220.0	38.6
N$_2$O$_4$(l)	Nitrogen tetroxide (四氧化二氮)	−19.5	97.5	209.2	142.7
N$_2$O$_4$(g)	Nitrogen tetroxide (四氧化二氮)	11.1	99.8	304.4	79.2
N$_2$O$_5$(g)	Nitrogen pentoxide (五氧化二氮)	13.3	117.1	355.7	95.3
	Sodium (钠)				
Na(s)	Sodium (钠)	0	0	51.3	28.2
Na(g)	Sodium (钠)	107.5	77.0	153.7	20.8
Na$^+$(aq)	Sodium ion (钠离子)	−240.1	−261.9	59.0	46.4
NaBr(s)	Sodium bromide (溴化钠)	−361.1	−349.0	86.8	51.4
NaCl(s)	Sodium chloride (氯化钠)	−411.2	−384.1	72.1	50.5
NaCl(aq)	Sodium chloride (氯化钠)	−407.27	−393.17	115.5	−90.0
NaClO$_3$(s)	Sodium chlorate (氯酸钠)	−365.8	−262.3	123.4	——
NaClO$_4$(s)	Sodium perchlorate (高氯酸钠)	−383.3	−254.9	142.3	——
NaF(s)	Sodium fluoride (氟化钠)	−576.6	−546.3	51.1	46.9
NaH(s)	Sodium hydride (氢化钠)	−56.3	−33.5	40.0	36.4
NaI(s)	Sodium iodide (碘化钠)	−287.8	−286.1	98.5	52.1
NaNO$_3$(s)	Sodium nitrate (硝酸钠)	−467.9	−367.0	116.5	92.9
NaNO$_3$(aq)	Sodium nitrate (硝酸钠)	−447.48	−373.21	205.4	−40.2
NaOH(s)	Sodium hydroxide (氢氧化钠)	−425.8	−379.7	64.4	59.5
NaOH(aq)	Sodium hydroxide (氢氧化钠)	−469.15	−419.20	48.1	−102.1
NaHSO$_4$(s)	Sodium hydrogen sulfate (硫酸氢钠)	−1125.5	−992.8	113.0	——
Na$_2$(g)	Disodium (二钠)	142.1	103.9	230.2	37.6
Na$_2$O(s)	Sodium oxide (氧化钠)	−414.2	−375.5	75.1	69.1
Na$_2$O$_2$(s)	Sodium peroxide (过氧化钠)	−510.9	−447.7	95.0	89.2
Na$_2$S(s)	Sodium sulfide (硫化钠)	−364.8	−349.8	83.7	——
Na$_2$SO$_4$(s)	Sodium sulfate (硫酸钠)	−1387.1	−1270.2	149.6	128.2
	Oxygen (氧)				
O(g)	Oxygen (atomic) (氧原子)	249.2	231.7	161.1	21.9
OH$^-$(aq)	Hydroxide ion (氢氧根离子)	−230.0	−157.2	−10.8	−148.5
OF$_2$(g)	Fluorine monoxide (二氟化氧)	24.5	41.8	247.5	43.3
O$_2$(g)	Oxygen (氧气)	0	0	205.2	29.4

续表

Formula (化学式)	Name (名称)	ΔH_f° /(kJ mol^{-1})	ΔG_f° /(kJ mol^{-1})	S° /(J mol^{-1} K^{-1})	C_p /(J mol^{-1} K^{-1})
O_3(g)	Ozone (臭氧)	142.7	163.2	238.9	39.2
	Phosphorus (磷)				
P(s)	Phosphorus (white) (白磷)	0	0	41.1	23.8
P(s)	Phosphorus (red) (红磷)	−17.6	——	22.8	21.2
P(s)	Phosphorus (black) (黑磷)	−39.3	——		
PCl_3(g)	Phosphorus(III) chloride (三氯化磷)	−287.0	−267.8	311.8	71.8
PCl_5(g)	Phosphorus(V) chloride (五氯化磷)	−374.9	−305.0	364.6	112.8
PH_3(g)	Phosphine (膦)	5.4	13.5	210.2	37.1
PO_4^{3-}(aq)	Phosphate ion (磷酸根离子)	−1277.4	−1018.7	−220.5	
P_4(g)	Tetraphosphorus (四磷)	58.9	24.4	280.0	67.2
	Lead (铅)				
Pb(s)	Lead (铅)	0	0	64.8	26.4
Pb^{2+}(aq)	Lead ion (铅离子)	−1.7	−24.4	10.5	
PbI_2(s)	Lead(II) iodide (二碘化铅)	−175.5	−173.6	174.9	77.4
PbO_2(s)	Lead(IV) oxide (氧化铅(IV))	−277.4	−217.3	68.6	64.6
PbS(s)	Lead(II) sulfide (硫化铅(II))	−100.4	−98.7	91.2	49.5
$PbSO_4$(s)	Lead(II) sulfate (硫酸铅)	−920.0	−813.0	148.5	103.2
	Platinum (铂)				
Pt(s)	Platinum (铂)	0	0	41.6	25.9
	Sulfur (硫)				
S(s)	Sulfur (rhombic) (正交硫)	0	0	32.1	22.6
S(g)	Sulfur (rhombic) (正交硫)	277.2	236.7	167.8	23.7
S(s)	Sulfur (monoclinic) (单斜硫)	0.3	——		
S^{2-}(aq)	Sulfur ion (硫离子)	33.1	85.8	−14.6	
SF_6(g)	Sulfur hexafluoride (六氟化硫)	−1220.5	−1116.5	291.5	97.0
SO_2(g)	Sulfur dioxide (二氧化硫)	−296.8	−300.1	248.2	39.9
SO_2Cl_2(g)	Sulfuryl chloride (二氯二氧化硫)	−364.0	−320.0	311.9	77.0
SO_3(g)	Sulfur trioxide (三氧化硫)	−395.7	−371.1	256.8	50.7
SO_3^{2-}(aq)	Sulfite ion (亚硫酸根离子)	−635.5	−486.5	−29.0	
SO_4^{2-}(aq)	Sulfate ion (硫酸根离子)	−909.3	−744.5	20.1	−293.0
$S_2O_3^{2-}$(aq)	Thiosulfate ion (硫代硫酸根离子)	−652.3	−522.5	67.0	
	Silicon (硅)				
Si(s)	Silicon (硅)	0	0	18.8	20.0
SiH_4(g)	Silane (硅烷)	34.3	56.9	204.6	42.8
SiO_2(s)	Silicon dioxide (α-quartz) (α-石英)	−910.7	−856.3	41.5	44.4
Si_2H_6(g)	Disilane (乙硅烷)	80.3	127.3	272.7	80.8
	Tin (锡)				
Sn(s)	Tin (white) (白锡)	0	0	51.2	27.0
Sn(s)	Tin (gray) (灰锡)	−2.1	0.1	44.1	25.8
Sn^{2+}(aq)	Tin(II) ion (锡离子)	−8.8	−27.2	−17.0	
$SnCl_4$(l)	Tin(IV) chloride (四氯化锡)	−511.3	−440.1	258.6	165.3
SnO(s)	Tin(II) oxide (氧化亚锡)	−280.7	−251.9	57.2	44.3
SnO_2(s)	Tin(IV) oxide (氧化锡)	−577.6	−515.8	49.0	52.6
	Titanium (钛)				
Ti(s)	Titanium (钛)	0	0	30.7	25.0
$TiCl_4$(l)	Titanium(IV) chloride (四氯化钛)	−804.2	−737.2	252.3	145.2
$TiCl_4$(g)	Titanium(IV) chloride (四氯化钛)	−763.2	−726.3	353.2	95.4
TiO_2(s)	Titanium(IV) oxide (二氧化钛)	−944.0	−888.8	50.6	55.0
	Uranium (铀)				
U(s)	Uranium (铀)	0	0	50.2	27.7
UF_6(s)	Uranium(VI) fluoride (六氟化铀)	−2197.0	−2068.5	227.6	166.8
UF_6(g)	Uranium(VI) fluoride (六氟化铀)	−2147.4	−2063.7	377.9	129.6
UO_2(s)	Uranium(IV) oxide (二氧化铀)	−1085.0	−1031.8	77.0	63.6
	Zinc (锌)				

Formula（化学式）	Name（名称）	ΔH_f° /(kJ mol^{-1})	ΔG_f° /(kJ mol^{-1})	S° /(J mol^{-1} K^{-1})	C_p /(J mol^{-1} K^{-1})
Zn(s)	Zinc（锌）	0	0	41.6	25.4
Zn^{2+}(aq)	Zinc ion（锌离子）	−153.9	−147.1	−112.1	46.0
ZnO(s)	Zinc oxide（氧化锌）	−350.5	−320.5	43.7	40.3
Substances containing carbon（含碳物质）					
C(s)	Carbon, graphite（碳、石墨）	0	0	5.7	8.5
C(s)	Carbon, diamond（碳、金刚石）	1.9	2.9	2.4	6.1
C(g)	Carbon（碳）	716.7	671.3	158.1	20.8
CCl$_4$(l)	Tetrachloromethane（四氯化碳）	−128.2	——	——	130.7
CCl$_4$(g)	Tetrachloromethane（四氯化碳）	−95.7	——	——	83.3
CH$_4$(g)	Methane（甲烷）	−74.6	−50.5	186.3	35.7
C$_2$H$_2$(g)	Acetylene（乙炔）	227.4	209.9	200.9	44.0
C$_2$H$_4$(g)	Ethylene（乙烯）	52.4	68.4	219.3	42.9
C$_2$H$_6$(g)	Ethane（乙烷）	−84.0	−32.0	229.2	52.5
C$_3$H$_8$(g)	Propane（丙烷）	−103.8	−23.4	270.3	73.6
C$_4$H$_{10}$(g)	Butane（丁烷）	−125.7	——	——	——
C$_5$H$_{12}$(g)	Pentane（戊烷）	−146.9	——	——	——
C$_6$H$_6$(l)	Benzene（苯）	49.1	124.5	173.4	136.0
C$_6$H$_6$(g)	Benzene（苯）	82.9	129.7	269.2	82.4
C$_6$H$_{12}$(l)	Cyclohexane（环己烷）	−156.4	——	——	154.9
C$_6$H$_{12}$(g)	Cyclohexane（环己烷）	−123.4	——	——	——
C$_{10}$H$_8$(s)	Naphthalene（萘）	78.5	201.6	167.4	165.7
C$_{10}$H$_8$(g)	Naphthalene（萘）	150.6	224.1	333.1	131.9
CH$_2$O(g)	Formaldehyde（甲醛）	−108.6	−102.5	218.8	35.4
CH$_3$CHO(l)	Acetaldehyde（乙醛）	−192.2	−127.6	160.2	89.0
CH$_3$CHO(g)	Acetaldehyde（乙醛）	−166.2	−133.0	263.8	55.3
CH$_3$OH(l)	Methanol（甲醇）	−239.2	−166.6	126.8	81.1
CH$_3$OH(g)	Methanol（甲醇）	−201.0	−162.3	239.9	44.1
CH$_3$CH$_2$OH(l)	Ethanol（乙醇）	−277.6	−174.8	160.7	112.3
CH$_3$CH$_2$OH(g)	Ethanol（乙醇）	−234.8	−167.9	281.6	65.6
C$_6$H$_5$OH(s)	Phenol（苯酚）	−165.1	——	144.0	127.4
(CH$_3$)$_2$CO(l)	Acetone（丙酮）	−248.4	——	199.8	126.3
(CH$_3$)$_2$CO(g)	Acetone（丙酮）	−217.1	−152.7	295.3	74.5
CH$_3$COOH(l)	Acetic acid（乙酸）	−484.3	−389.9	159.8	123.3
CH$_3$COOH(g)	Acetic acid（乙酸）	−432.2	−374.2	283.5	63.4
CH$_3$COO$^-$(aq)	Acetate ion（乙酸根离子）	−486.0	−369.3	86.6	−6.3
C$_6$H$_5$COOH(s)	Benzoic acid（苯甲酸）	−385.2	——	167.6	146.8
CH$_3$NH$_2$(g)	Methylamine（甲胺）	−22.5	32.7	242.9	50.1
C$_6$H$_5$NH$_2$(l)	Aniline（苯胺）	31.6	——	——	191.9
C$_6$H$_5$NH$_2$(g)	Aniline（苯胺）	87.5	−7.0	317.9	107.9
C$_2$N$_2$(g)	Cyanogen（氰）	306.7	——	241.9	56.8
CO(g)	Carbon monoxide（一氧化碳）	−110.5	−137.2	197.7	29.1
CO$_2$(g)	Carbon dioxide（二氧化碳）	−393.5	−394.4	213.8	37.1
CO$_3^{2-}$(aq)	Carbonate ion（碳酸根离子）	−677.1	−527.8	−56.9	——
COCl$_2$(g)	Carbonyl chloride（碳酰氯）	−219.1	−204.9	283.5	57.7
COS(g)	Carbon oxysulfide（氧硫化碳）	−142.0	−169.2	231.6	41.5
CS$_2$(l)	Carbon disulfide（二硫化碳）	89.0	64.6	151.3	76.4
CS$_2$(g)	Carbon disulfide（二硫化碳）	116.7	67.1	237.8	45.4
BaCO$_3$(s)	Barium carbonate（碳酸钡）	−1213.0	−1134.4	112.1	86.0
CaCO$_3$(s)	Cadmium carbonate（碳酸钙）	−750.6	−669.4	92.5	——
MgCO$_3$(s)	Magnesium carbonate（碳酸镁）	−1095.8	−1012.1	65.7	75.5
Na$_2$CO$_3$(s)	Sodium carbonate（碳酸钠）	−1130.7	−1044.4	135.0	112.3
NaHCO$_3$(s)	Sodium hydrogen carbonate（碳酸氢钠）	−950.8	−851.0	101.7	87.6
PbCO$_3$(s)	Lead(II) carbonate（碳酸铅）	−699.1	−625.5	131.0	87.4
ZnCO$_3$(s)	Zinc carbonate（碳酸锌）	−812.8	−731.5	82.4	79.7

Table C.2 Various Equilibrium Constants
表 C.2 各种平衡常数

A. Ionization Constants of Weak Acids at 298.15 K（298.15 K 时弱酸的电离常数）

Formula（化学式）	Name（名称）	K_a	Formula（化学式）	Name（名称）	K_a
H_3AsO_4	Arsenic acid (砷酸)	5.5×10^{-3} (K_{a1}) 1.7×10^{-7} (K_{a2}) 5.1×10^{-12} (K_{a3})	H_2SeO_3	Selenous acid (亚硒酸)	2.4×10^{-3} (K_{a1}) 4.8×10^{-9} (K_{a2})
H_3AsO_3	Arsenous acid (亚砷酸)	5.1×10^{-10}	H_2SO_4	Sulfuric acid (硫酸)	1.0×10^{-2} (K_{a2})
H_2CO_3	Carbonic acid (碳酸)	4.5×10^{-7} (K_{a1}) 4.7×10^{-11} (K_{a2})	H_2SO_3	Sulfurous acid (亚硫酸)	1.4×10^{-2} (K_{a1}) 6×10^{-8} (K_{a2})
$HClO_2$	Chlorous acid (亚氯酸)	1.1×10^{-2}	HCOOH	Formic acid (甲酸)	1.8×10^{-4}
HCNO	Cyanic acid (氰酸)	3.5×10^{-4}	CCl_3COOH	Trichloroacetic acid (三氯乙酸)	2.2×10^{-1}
HN_3	Hydrazoic acid (叠氮酸)	3×10^{-5}	$CHCl_2COOH$	Dichloroacetic acid (二氯乙酸)	4.5×10^{-2}
HCN	Hydrocyanic acid (氢氰酸)	6.2×10^{-10}	$H_2C_2O_4$	Oxalic acid (草酸)	5.6×10^{-2} (K_{a1}) 1.5×10^{-4} (K_{a2})
HF	Hydrofluoric acid (氢氟酸)	6.3×10^{-4}	$CH_2BrCOOH$	Bromoacetic acid (溴乙酸)	1.3×10^{-3}
H_2O_2	Hydrogen peroxide (过氧化氢)	2.4×10^{-12}	$CH_2ClCOOH$	Chloroacetic acid (氯乙酸)	1.3×10^{-3}
H_2Se	Hydrogen selenide (硒化氢)	1.3×10^{-4} (K_{a1}) 1.0×10^{-11} (K_{a2})	CH_2FCOOH	Fluoroacetic acid (氟乙酸)	2.6×10^{-3}
H_2S	Hydrogen sulfide (硫化氢)	8.9×10^{-8} (K_{a1}) 1×10^{-19} (K_{a2})	CH_2ICOOH	Iodoacetic acid (碘乙酸)	6.6×10^{-4}
H_2Te	Hydrogen telluride (碲化氢)	3×10^{-3} (K_{a1}) 1×10^{-11} (K_{a2})	CH_3COOH	Acetic acid (乙酸)	1.75×10^{-5}
HOBr	Hypobromous acid (次溴酸)	2.8×10^{-9}	$HC_3H_3O_2$	Acrylic acid (丙烯酸)	5.6×10^{-5}
HOCl	Hypochlorous acid (次氯酸)	4.0×10^{-8}	$H_2C_3H_2O_4$	Malonic acid (丙二酸)	1.4×10^{-3} (K_{a1}) 2.0×10^{-6} (K_{a2})
HOI	Hypoiodous acid (次碘酸)	3×10^{-11}	CH_3CH_2COOH	Propionic acid (丙酸)	1.3×10^{-5}
HIO_3	Iodic acid (碘酸)	1.7×10^{-1}	$H_2C_4H_4O_4$	Succinic acid (丁二酸)	6.2×10^{-5} (K_{a1}) 2.3×10^{-6} (K_{a2})
HNO_2	Nitrous acid (亚硝酸)	5.6×10^{-4}	$HC_4H_7O_2$	Butyric acid (丁酸)	1.5×10^{-5}
H_3PO_4	Phosphoric acid (磷酸)	6.9×10^{-3} (K_{a1}) 6.2×10^{-8} (K_{a2}) 4.8×10^{-13} (K_{a3})	C_6H_5OH	Phenol (苯酚)	1.0×10^{-10}
			C_6H_5SH	Benzenethiol (苯硫酚)	2.4×10^{-7}
H_3PO_3	Phosphorous acid (亚磷酸)	5×10^{-2} (K_{a1}) 2.0×10^{-7} (K_{a2})	$H_3C_6H_5O_7$	Citric acid (柠檬酸)	7.4×10^{-4} (K_{a1}) 1.7×10^{-5} (K_{a2}) 4.0×10^{-7} (K_{a3})
$H_4P_2O_7$	Pyrophosphoric acid (焦磷酸)	1.2×10^{-1} (K_{a1}) 7.9×10^{-3} (K_{a2}) 2.0×10^{-7} (K_{a3}) 4.8×10^{-10} (K_{a4})	C_6H_5COOH	Benzoic acid (苯甲酸)	6.25×10^{-5}
			$HC_8H_7O_2$	Phenylacetic acid (苯乙酸)	4.9×10^{-5}

B. Ionization Constants of Weak Bases at 298.15 K（298.15 K 时弱碱的电离常数）

Formula（化学式）	Name（名称）	K_b	Formula（化学式）	Name（名称）	K_b
NH_3	Ammonia（氨）	1.8×10^{-5}	C_5H_5N	Pyridine（吡啶）	1.7×10^{-9}
NH_2NH_2	Hydrazine（联氨）	1.3×10^{-6}	$C_5H_{11}N$	Piperdine（哌啶）	1.33×10^{-3}
NH_2OH	Hydroxylamine（羟胺）	8.7×10^{-9}	$C_6H_5NH_2$	Aniline（苯胺）	7.4×10^{-10}
CH_3NH_2	Methylamine（甲胺）	4.6×10^{-4}	$(C_2H_5)_3N$	Triethylamine（三乙基胺）	5.6×10^{-4}
$C_2H_5NH_2$	Ethylamine（乙胺）	4.5×10^{-4}	$C_6H_{15}O_3N$	Triethanolamine（三乙醇胺）	5.8×10^{-7}
$(CH_3)_2NH$	Dimethylamine（二甲胺）	5.4×10^{-4}	C_9H_7N	Quinoline（喹啉）	7.9×10^{-10}
$(CH_3)_3N$	Trimethylamine（三甲胺）	6.3×10^{-5}	C_9H_7N	Isoquinoline（异喹啉）	2.5×10^{-9}
$(C_2H_5)_2NH$	Diethylamine（二乙胺）	6.9×10^{-4}	$C_{17}H_{19}O_3N$	Morphine（吗啡）	$1.6 \times 10^{-6}(K_{b1})$ $7.1 \times 10^{-5}(K_{b2})$

C. Solubility Product Constants（溶度积常数）

Formula（化学式）	Name（名称）	K_{sp}	Formula（化学式）	Name（名称）	K_{sp}
AgBr	Silver bromide（溴化银）	5.35×10^{-13}	$Bi(OH)_3$	Bismuth hydroxide（氢氧化铋）	6.0×10^{-31}
AgCl	Silver chloride（氯化银）	1.77×10^{-10}	$CaCO_3$	Calcium carbonate（碳酸钙）	2.8×10^{-9}
Ag_2CO_3	Silver carbonate（碳酸银）	8.46×10^{-12}	CaC_2O_4	Calcium oxalate（草酸钙）	2.32×10^{-9}
Ag_2CrO_4	Silver chromate（铬酸银）	1.12×10^{-12}	CaF_2	Calcium fluoride（氟化钙）	5.3×10^{-9}
AgI	Silver iodide（碘化银）	8.52×10^{-17}	$Ca(OH)_2$	Calcium hydroxide（氢氧化钙）	5.02×10^{-6}
Ag_2S	Silver(I) sulfide（硫化银）	6.3×10^{-50}	$CaSO_4$	Calcium sulfate（硫酸钙）	4.93×10^{-5}
$Al(OH)_3$	Aluminum hydroxide（氢氧化铝）	1.3×10^{-33}	CdS	Cadmium sulfide（硫化镉）	8.0×10^{-27}
$AlPO_4$	Aluminum phosphate（磷酸铝）	9.84×10^{-21}	$CoCO_3$	Cobalt carbonate（碳酸钴）	1.4×10^{-13}
$BaCO_3$	Barium carbonate（碳酸钡）	2.58×10^{-9}	α-CoS	Cobalt sulfide（α- 硫化钴）	4.0×10^{-21}
$BaCrO_4$	Bariumchromate（铬酸钡）	1.17×10^{-10}	β-CoS	Cobalt sulfide（β- 硫化钴）	2.0×10^{-25}
BaF_2	Barium fluoride（氟化钡）	1.84×10^{-7}	$Cr(OH)_3$	Chromium(III) hydroxide（氢氧化铬(III)）	6.3×10^{-31}
$Ba(OH)_2 \cdot 8H_2O$	Barium hydroxide octahydrate（八水合氢氧化钡）	2.55×10^{-4}	CuI	Copper(I) iodide（碘化亚铜）	1.27×10^{-12}
$BaSO_4$	Barium Sulfate（硫酸钡）	1.08×10^{-10}	CuS	Copper(II) sulfide（硫化铜）	6.3×10^{-36}
$BeCO_3 \cdot 4H_2O$	Beryllium carbonate tetrahydrate（四水合碳酸铍）	1×10^{-3}	$Cu(OH)_2$	Copper(II) hydroxide（氢氧化铜）	2.2×10^{-20}
$Be(OH)_2$	Beryllium hydroxide（氢氧化铍）	6.92×10^{-22}	Cu_2S	Copper(I) sulfide（硫化亚铜）	2.5×10^{-48}

Formula (化学式)	Name (名称)	K_{sp}	Formula (化学式)	Name (名称)	K_{sp}
$Fe(OH)_2$	Iron(II) hydroxide (氢氧化亚铁)	4.87×10^{-17}	$NiCO_3$	Nickel(II) carbonate (碳酸镍)	1.42×10^{-7}
$Fe(OH)_3$	Iron(III) hydroxide (氢氧化铁 (III))	2.79×10^{-39}	$Ni(OH)_2$	Nickel(II) hydroxide (氢氧化镍)	5.48×10^{-16}
$FePO_4 \cdot 2H_2O$	Iron(III) phosphate dihydrate (二水合磷酸铁)	9.91×10^{-16}	$PbCO_3$	Lead(II) carbonate (碳酸铅)	7.4×10^{-14}
FeS	Iron(II) sulfide (硫化亚铁)	6.3×10^{-18}	$PbCl_2$	Lead(II) chloride (氯化铅 (II))	1.70×10^{-5}
Hg_2Cl_2	Mercury(I) chloride (氯化亚汞)	1.43×10^{-18}	$PbCrO_4$	Lead(II) chromate (铬酸铅 (II))	2.8×10^{-13}
$\alpha\text{-}HgS$	Mercury(II) sulfide (red) (α- 硫化汞 - 红)	4×10^{-53}	PbI_2	Lead(II) iodide (碘化铅)	9.8×10^{-9}
$\beta\text{-}HgS$	Mercury(II) sulfide (black) (β- 硫化汞 - 黑)	1.6×10^{-52}	$Pb(OH)_2$	Lead(II) hydroxide (氢氧化铅)	1.43×10^{-15}
Li_2CO_3	Lithium carbonate (碳酸锂)	2.5×10^{-2}	PbS	Lead(II) sulfide (硫化铅)	8.0×10^{-28}
Li_3PO_4	Lithium phosphate (磷酸锂)	2.37×10^{-11}	$PbSO_4$	Lead(II) sulfate (硫酸铅)	2.53×10^{-8}
$MgCO_3$	Magnesium carbonate (碳酸镁)	6.82×10^{-6}	$SrCO_3$	Strontium carbonate (碳酸锶)	5.60×10^{-10}
MgF_2	Magnesium fluoride (氟化镁)	5.16×10^{-11}	$SrSO_4$	Strontium sulfate (硫酸锶)	3.44×10^{-7}
$Mg(OH)_2$	Magnesium hydroxide (氢氧化镁)	5.61×10^{-12}	$Sn(OH)_2$	Tin(II) hydroxide (氢氧化锡)	5.45×10^{-28}
$MnCO_3$	Manganese(II) carbonate (碳酸锰)	2.34×10^{-11}	$ZnCO_3$	Zinc carbonate (碳酸锌)	1.46×10^{-10}
MnS	Manganese(II) sulfide (硫化锰)	2.5×10^{-13}	ZnS	Zinc sulfide (硫化锌)	1.6×10^{-24}

D. Complex-Ion Formation/Stability Constants (配离子生成 / 稳定常数)

Formula (化学式)	K_f	Formula (化学式)	K_f	Formula (化学式)	K_f
$[AuCl_4]^-$	1.5×10^{25}	$[Mn(EDTA)]^{2-}$	5×10^{13}	$[Co(NH_3)_6]^{2+}$	1.3×10^5
$[HgCl_4]^{2-}$	1.2×10^{15}	$[Ni(EDTA)]^{2-}$	3.6×10^{18}	$[Co(NH_3)_6]^{3+}$	2×10^{35}
$[PdCl_4]^{2-}$	5.0×10^{15}	$[Zn(EDTA)]^{2-}$	2.5×10^{16}	$[Cu(NH_3)_4]^{2+}$	2.1×10^{13}
$[PtCl_4]^{2-}$	1×10^{16}	$[Co(en)_3]^{2+}$	8.7×10^{13}	$[Ni(NH_3)_6]^{2+}$	5.5×10^8
$[Au(CN)_2]^-$	2.0×10^{38}	$[Co(en)_3]^{3+}$	4.9×10^{48}	$[Pt(NH_3)_6]^{2+}$	2×10^{35}
$[Cu(CN)_2]^-$	1×10^{24}	$[Cu(en)_2]^{2+}$	1.0×10^{20}	$[Zn(NH_3)_4]^{2+}$	2.9×10^9
$[Cu(CN)_4]^{3-}$	2.0×10^{30}	$[Ni(en)_3]^{2+}$	2.1×10^{18}	$[Bi(OH)_4]^-$	2×10^{35}
$[Fe(CN)_6]^{4-}$	1×10^{35}	$[Zn(en)_3]^{2+}$	1.3×10^{14}	$[Cu(OH)_4]^{2-}$	3.2×10^{18}
$[Fe(CN)_6]^{3-}$	1×10^{42}	$[AlF_6]^{3-}$	6.9×10^{19}	$[Al(OH)_4]^-$	1.1×10^{33}
$[Ni(CN)_4]^{2-}$	2.0×10^{31}	FeF_3	1.1×10^{12}	$[Be(OH)_3]^-$	1.6×10^{15}
$[Zn(CN)_4]^{2-}$	5.0×10^{16}	$[CuI_2]^-$	7.1×10^8	$[Fe(ox)_3]^{4-}$	1.7×10^5
$[Ca(EDTA)]^{2-}$	1×10^{11}	$[HgI_4]^{2-}$	6.8×10^{29}	$[Fe(ox)_3]^{3-}$	2×10^{20}
$[Cu(EDTA)]^{2-}$	5×10^{18}	$[Ag(NH_3)_2]^+$	1.1×10^7	$[Fe(SCN)_2]^+$	2.3×10^3

Table C.3　Standard Electrode (Reduction) Potentials at 298.15 K
表 C.3　298.15 K 时的标准电极（还原）电势

Reduction Half-Reaction（还原半反应）	E°/V
Standard acidic solution (pH = 0)	
$F_2(g) + 2H^+(aq) + 2e^- \rightarrow 2HF(aq)$	+3.053
$O_3(g) + 2H^+(aq) + 2e^- \rightarrow O_2(g) + H_2O(l)$	+2.076
$HFeO_4^-(aq) + 7H^+(aq) + 3e^- \rightarrow Fe^{3+}(aq) + 4H_2O(l)$	+2.07
$S_2O_8^{2-}(aq) + 2e^- \rightarrow 2SO_4^{2-}(aq)$	+2.010
$Co^{3+}(aq) + e^- \rightarrow Co^{2+}(aq)$	+1.92
$H_2O_2(aq) + 2H^+(aq) + 2e^- \rightarrow 2H_2O(l)$	+1.776
$N_2O(g) + 2H^+(aq) + 2e^- \rightarrow N_2(g) + H_2O(l)$	+1.766
$Au^+(aq) + e^- \rightarrow Au(s)$	+1.692
$HClO_2(aq) + 2H^+(aq) + 2e^- \rightarrow HClO(aq) + H_2O(l)$	+1.645
$2HClO(aq) + 2H^+(aq) + 2e^- \rightarrow Cl_2(g) + 2H_2O(l)$	+1.611
$Mn^{3+}(aq) + e^- \rightarrow Mn^{2+}(aq)$	+1.5415
$MnO_4^-(aq) + 8H^+(aq) + 5e^- \rightarrow Mn^{2+}(aq) + 4H_2O(l)$	+1.507
$Au^{3+}(aq) + 3e^- \rightarrow Au(s)$	+1.498
$PbO_2(s) + 4H^+(aq) + 2e^- \rightarrow Pb^{2+}(aq) + 2H_2O(l)$	+1.455
$ClO_3^-(aq) + 6H^+(aq) + 6e^- \rightarrow Cl^-(aq) + 3H_2O(l)$	+1.451
$2NH_3OH^+(aq) + H^+(aq) + 2e^- \rightarrow N_2H_5^+(aq) + 2H_2O(l)$	+1.42
$ClO_4^-(aq) + 8H^+(aq) + 8e^- \rightarrow Cl^-(aq) + 4H_2O(l)$	+1.389
$Cr_2O_7^{2-}(aq) + 14H^+(aq) + 6e^- \rightarrow 2Cr^{3+}(aq) + 7H_2O(l)$	+1.36
$Cl_2(g) + 2e^- \rightarrow 2Cl^-(aq)$	+1.35827
$O_2(g) + 4H^+(aq) + 4e^- \rightarrow 2H_2O(l)$	+1.229
$MnO_2(s) + 4H^+(aq) + 2e^- \rightarrow Mn^{2+}(aq) + 2H_2O(l)$	+1.224
$ClO_3^-(aq) + 3H^+(aq) + 2e^- \rightarrow HClO_2(aq) + H_2O(l)$	+1.214
$2IO_3^-(aq) + 12H^+(aq) + 10e^- \rightarrow I_2(s) + 6H_2O(l)$	+1.195
$ClO_4^-(aq) + 2H^+(aq) + 2e^- \rightarrow ClO_3^-(aq) + H_2O(l)$	+1.189
$Pt^{2+}(aq) + 2e^- \rightarrow Pt(s)$	+1.18
$Cu^{2+}(aq) + 2CN^-(aq) + e^- \rightarrow [Cu(CN)_2]^-(aq)$	+1.103
$IO_3^-(aq) + 6H^+(aq) + 6e^- \rightarrow I^-(aq) + 3H_2O(l)$	+1.085
$Br_2(l) + 2e^- \rightarrow 2Br^-(aq)$	+1.066
$N_2O_4(g) + 2H^+(aq) + 2e^- \rightarrow 2HNO_2(aq)$	+1.065
$[AuCl_4]^-(aq) + 3e^- \rightarrow Au(s) + 4Cl^-(aq)$	+1.002
$HNO_2(aq) + H^+(aq) + e^- \rightarrow NO(g) + H_2O(l)$	+0.983
$NO_3^-(aq) + 4H^+(aq) + 3e^- \rightarrow NO(g) + 2H_2O(l)$	+0.957
$NO_3^-(aq) + 3H^+(aq) + 2e^- \rightarrow HNO_2(aq) + H_2O(l)$	+0.934
$2Hg^{2+}(aq) + 2e^- \rightarrow Hg_2^{2+}(aq)$	+0.920
$Hg^{2+}(aq) + 2e^- \rightarrow Hg(l)$	+0.851
$2NO_3^-(aq) + 4H^+(aq) + 2e^- \rightarrow N_2O_4(g) + 2H_2O(l)$	+0.803
$Ag^+(aq) + e^- \rightarrow Ag(s)$	+0.7996
$Fe^{3+}(aq) + e^- \rightarrow Fe^{2+}(aq)$	+0.771

续表

Reduction Half-Reaction（还原半反应）	$E°/V$
$[PtCl_4]^{2-}(aq) + 2e^- \rightarrow Pt(s) + 4Cl^-(aq)$	+0.755
$O_2(g) + 2H^+(aq) + 2e^- \rightarrow H_2O_2(aq)$	+0.695
$I_3^-(aq) + 2e^- \rightarrow 3I^-(aq)$	+0.536
$I_2(s) + 2e^- \rightarrow 2I^-(aq)$	+0.5355
$H_2SO_3(aq) + 4H^+(aq) + 4e^- \rightarrow S(s) + 3H_2O(l)$	+0.449
$[Fe(CN)_6]^{3-}(aq) + e^- \rightarrow [Fe(CN)_6]^{4-}(aq)$	+0.358
$Cu^{2+}(aq) + 2e^- \rightarrow Cu(s)$	+0.3419
$Hg_2Cl_2(s) + 2e^- \rightarrow 2Hg(l) + 2Cl^-(aq)$	+0.26808
$AgCl(s) + e^- \rightarrow Ag(s) + Cl^-(aq)$	+0.22233
$SO_4^{2-}(aq) + 4H^+(aq) + 2e^- \rightarrow SO_2(g) + 2H_2O(l)$	+0.172
$Cu^{2+}(aq) + e^- \rightarrow Cu^+(aq)$	+0.153
$Sn^{4+}(aq) + 2e^- \rightarrow Sn^{2+}(aq)$	+0.151
$Sn(OH)_3^+(aq) + 3H^+(aq) + 2e^- \rightarrow Sn^{2+}(aq) + 3H_2O(l)$	+0.142
$S(s) + 2H^+(aq) + 2e^- \rightarrow H_2S(g)$	+0.142
$N_2(g) + 2H_2O(l) + 6H^+(aq) + 6e^- \rightarrow 2NH_4OH(aq)$	+0.092
$S_4O_6^{2-}(aq) + 2e^- \rightarrow 2S_2O_3^{2-}(aq)$	+0.08
$2H^+(aq) + 2e^- \rightarrow H_2(g)$	0
$TiOH^{3+}(aq) + H^+(aq) + e^- \rightarrow Ti^{3+}(aq) + H_2O(l)$	−0.055
$SnO_2(s) + 4H^+(aq) + 2e^- \rightarrow Sn^{2+}(aq) + 2H_2O(l)$	−0.094
$Pb^{2+}(aq) + 2e^- \rightarrow Pb(s)$	−0.1262
$Sn^{2+}(aq) + 2e^- \rightarrow Sn(s)$	−0.1375
$CO_2(g) + 2H^+(aq) + 2e^- \rightarrow HCOOH(aq)$	−0.199
$Ni^{2+}(aq) + 2e^- \rightarrow Ni(s)$	−0.257
$H_3PO_4(aq) + 2H^+(aq) + 2e^- \rightarrow H_3PO_3(aq) + H_2O(l)$	−0.276
$Co^{2+}(aq) + 2e^- \rightarrow Co(s)$	−0.28
$Ti^{3+}(aq) + e^- \rightarrow Ti^{2+}(aq)$	−0.369
$Cr^{3+}(aq) + e^- \rightarrow Cr^{2+}(aq)$	−0.407
$Fe^{2+}(aq) + 2e^- \rightarrow Fe(s)$	−0.447
$H_3PO_3(aq) + 2H^+(aq) + 2e^- \rightarrow H_3PO_2(aq) + H_2O(l)$	−0.499
$TiO_2(s) + 4H^+(aq) + 2e^- \rightarrow Ti^{2+}(aq) + 2H_2O(l)$	−0.502
$H_3PO_2(aq) + H^+(aq) + e^- \rightarrow P(s) + 2H_2O(l)$	−0.508
$Cr^{3+}(aq) + 3e^- \rightarrow Cr(s)$	−0.744
$Zn^{2+}(aq) + 2e^- \rightarrow Zn(s)$	−0.7618
$Mn^{2+}(aq) + 2e^- \rightarrow Mn(s)$	−1.185
$Ti^{2+}(aq) + 2e^- \rightarrow Ti(s)$	−1.628
$Al^{3+}(aq) + 3e^- \rightarrow Al(s)$	−1.676
$Mg^{2+}(aq) + 2e^- \rightarrow Mg(s)$	−2.372
$Na^+(aq) + e^- \rightarrow Na(s)$	−2.71
$Ca^{2+}(aq) + 2e^- \rightarrow Ca(s)$	−2.868
$Sr^{2+}(aq) + 2e^- \rightarrow Sr(s)$	−2.899
$K^+(aq) + e^- \rightarrow K(s)$	−2.931
$Cs^+(aq) + e^- \rightarrow Cs(s)$	−3.026
$Li^+(aq) + e^- \rightarrow Li(s)$	−3.0401
$3N_2(g) + 2H^+(aq) + 2e^- \rightarrow 2HN_3(aq)$	−3.09

续表

Reduction Half-Reaction（还原半反应）	$E°/V$
Standard basic solution (pH = 14)	
$F_2(g) + 2e^- \rightarrow 2F^-(aq)$	+2.866
$O_3(g) + H_2O(l) + 2e^- \rightarrow O_2(g) + 2OH^-(aq)$	+1.24
$HO_2^-(aq) + H_2O(l) + 2e^- \rightarrow 3OH^-(aq)$	+0.878
$N_2O_4(g) + 2e^- \rightarrow 2NO_2^-(aq)$	+0.867
$OCl^-(aq) + H_2O(l) + 2e^- \rightarrow Cl^-(aq) + 2OH^-(aq)$	+0.81
$ClO_2^-(aq) + H_2O(l) + 2e^- \rightarrow OCl^-(aq) + 2OH^-(aq)$	+0.66
$MnO_4^{2-}(aq) + 2H_2O(l) + 2e^- \rightarrow MnO_2(s) + 4OH^-(aq)$	+0.60
$MnO_4^-(aq) + e^- \rightarrow MnO_4^{2-}(aq)$	+0.558
$O_2(g) + 2H_2O(l) + 4e^- \rightarrow 4OH^-(aq)$	+0.401
$ClO_4^-(aq) + H_2O(l) + 2e^- \rightarrow ClO_3^-(aq) + 2OH^-(aq)$	+0.36
$Ag_2O(s) + H_2O(l) + 2e^- \rightarrow 2Ag(s) + 2OH^-(aq)$	+0.342
$ClO_3^-(aq) + H_2O(l) + 2e^- \rightarrow ClO_2^-(aq) + 2OH^-(aq)$	+0.33
$Co(OH)_3(s) + e^- \rightarrow Co(OH)_2(s) + OH^-(aq)$	+0.17
$Mn(OH)_3(s) + e^- \rightarrow Mn(OH)_2(s) + OH^-(aq)$	+0.15
$HgO(s) + H_2O(l) + 2e^- \rightarrow Hg(l) + 2OH^-(aq)$	+0.0977
$2Cu(OH)_2(s) + 2e^- \rightarrow Cu_2O(s) + 2OH^-(aq) + H_2O(l)$	−0.080
$CrO_4^{2-}(aq) + 4H_2O(l) + 3e^- \rightarrow Cr(OH)_3(s) + 5OH^-(aq)$	−0.13
$Cu_2O(s) + H_2O(l) + 2e^- \rightarrow 2Cu(s) + 2OH^-(aq)$	−0.360
$S(g) + 2e^- \rightarrow S^{2-}(aq)$	−0.47627
$B(OH)_3(s) + 7H^+(aq) + 8e^- \rightarrow BH_4^-(aq) + 3H_2O(l)$	−0.481
$NiO_2(s) + 2H_2O(l) + 2e^- \rightarrow Ni(OH)_2(s) + 2OH^-(aq)$	−0.490
$Fe(OH)_3(s) + e^- \rightarrow Fe(OH)_2(s) + OH^-(aq)$	−0.56
$2SO_3^{2-}(aq) + 3H_2O(l) + 4e^- \rightarrow S_2O_3^{2-}(aq) + 6OH^-(aq)$	−0.571
$Ag_2S(s) + 2e^- \rightarrow 2Ag(s) + S^{2-}(aq)$	−0.691
$Ni(OH)_2(s) + 2e^- \rightarrow Ni(s) + 2OH^-(aq)$	−0.72
$Co(OH)_2(s) + 2e^- \rightarrow Co(s) + 2OH^-(aq)$	−0.73
$2H_2O(l) + 2e^- \rightarrow H_2(g) + 2OH^-(aq)$	−0.8277
$2NO_3^-(aq) + 2H_2O(l) + 2e^- \rightarrow N_2O_4(g) + 4OH^-(aq)$	−0.85
$P(s) + 3H_2O(l) + 3e^- \rightarrow PH_3(g) + 3OH^-(aq)$	−0.87
$HSnO_2^-(aq) + H_2O(l) + 2e^- \rightarrow Sn(s) + 3OH^-(aq)$	−0.909
$Sn(OH)_6^{2-}(aq) + 2e^- \rightarrow HSnO_2^-(aq) + 3OH^-(aq) + H_2O(l)$	−0.93
$SO_4^{2-}(aq) + H_2O(l) + 2e^- \rightarrow SO_3^{2-}(aq) + 2OH^-(aq)$	−0.93
$PO_4^{3-}(aq) + 2H_2O(l) + 2e^- \rightarrow HPO_3^{2-}(aq) + 3OH^-(aq)$	−1.05
$ZnO_2^{2-}(aq) + 2H_2O(l) + 2e^- \rightarrow Zn(s) + 4OH^-(aq)$	−1.215
$SiF_6^{2-}(aq) + 4e^- \rightarrow Si(s) + 6F^-(aq)$	−1.24
$Zn(OH)_2(s) + 2e^- \rightarrow Zn(s) + 2OH^-(aq)$	−1.249
$ZnO(s) + H_2O(l) + 2e^- \rightarrow Zn(s) + 2OH^-(aq)$	−1.260
$Cr(OH)_3(s) + 3e^- \rightarrow Cr(s) + 3OH^-(aq)$	−1.48
$Mn(OH)_2(s) + 2e^- \rightarrow Mn(s) + 2OH^-(aq)$	−1.56
$HPO_3^{2-}(aq) + 2H_2O(l) + 2e^- \rightarrow H_2PO_2^-(aq) + 3OH^-(aq)$	−1.65
$SiO_3^{2-}(aq) + 3H_2O(l) + 4e^- \rightarrow Si(s) + 6OH^-(aq)$	−1.697
$HPO_3^{2-}(aq) + 2H_2O(l) + 3e^- \rightarrow P(s) + 5OH^-(aq)$	−1.71
$AlF_6^{3-}(aq) + 3e^- \rightarrow Al(s) + 6F^-(aq)$	−2.069
$Al(OH)_4^-(aq) + 3e^- \rightarrow Al(s) + 4OH^-(aq)$	−2.310
$Mg(OH)_2(s) + 2e^- \rightarrow Mg(s) + 2OH^-(aq)$	−2.690
$Sr(OH)_2(s) + 2e^- \rightarrow Sr(s) + 2OH^-(aq)$	−2.88